Where:

a	= Speed of sound	k	= Coefficient of thermal conductivity
c_p	= Specific heat	l	= Characteristic dimension
dp/dz	= Pressure gradient	q_w	= Surface heat flux
D	= Hydraulic diameter	$T_w - T_f$	= Characteristic temperature difference
g	= Gravitational acceleration	U	= Characteristic forced velocity
h_{fg}	= Latent heat		

α	= Thermal diffusivity	ρ	= Density
β	= Bulk coefficient	τ_w	= Surface shear stress
μ	= Coefficient of viscosity	σ	= Surface tension coefficient
ν	= Kinematic viscosity		

Forced Convection:

$$Nu = \text{function}[Re, Pr]$$

Natural Convection:

$$Nu = \text{function}[Gr, Pr]$$

An Introduction to Convective Heat Transfer Analysis

Patrick H. Oosthuizen
Heat Transfer Laboratory
Department of Mechanical Engineering
Queen's University

and

David Naylor
Department of Mechanical Engineering
Ryerson Polytechnic University

WCB McGraw-Hill

New York   St. Louis   San Francisco   Auckland   Bogotá   Caracas
Lisbon   London   Madrid   Mexico City   Milan   Montreal   New Delhi
San Juan   Singapore   Sydney   Tokyo   Toronto

WCB/McGraw-Hill
*A Division of The **McGraw·Hill** Companies*

AN INTRODUCTION TO CONVECTIVE HEAT TRANSFER ANALYSIS

Copyright © 1999 by The McGraw-Hill Companies, Inc. All rights reserved. Printed in the United States of America. Except as permitted under the United States Copyright Act of 1976, no part of this publication may be reproduced or distributed in any form or by any means, or stored in a data base or retrieval system, without the prior written permission of the publisher.

This book is printed on acid-free paper.

1 2 3 4 5 6 7 8 9 0 BKM/BKM 9 3 2 1 0 9 8

ISBN 0-07-048201-2

Vice president and editorial director: *Kevin T. Kane*
Publisher: *Tom Casson*
Senior sponsoring editor: *Debra Riegert*
Marketing manager: *John T. Wannemacher*
Project manager: *Jim Labeots*
Senior production supervisor: *Elizabeth LaManna*
Designer: *Kiera Cunningham*
Supplement coordinator: *Linda Huenecke*
Compositor: *Publication Services, Inc.*
Typeface: *10.5/12 Times Roman*
Printer: *Book-mart Press*

Library of Congress Cataloging-in-Publication Data
Oosthuizen, P. H.
    An introduction to convective heat transfer analysis / Patrick H. Oosthuizen and David Naylor.
    p.   cm.
    Includes bibliographical references and index.
    ISBN 0-07-048201-2
    1. Heat—Convection.   I. Naylor, David, Ph.D.   II. Title.
TJ260.O57   1999
621.402′2—dc21   98-5484

http://www.mhhe.com

McGraw-Hill Series in Mechanical Engineering

CONSULTING EDITORS

Jack P. Holman, Southern Methodist University

John R. Lloyd, Michigan State University

Anderson: *Computational Fluid Dynamics: The Basics with Applications*
Anderson: *Modern Compressible Flow: With Historical Perspective*
Arora: *Introduction to Optimum Design*
Borman and Ragland: *Combustion Engineering*
Burton: *Introduction to Dynamic Systems Analysis*
Culp: *Principles of Energy Conversion*
Dieter: *Engineering Design: A Materials & Processing Approach*
Doebelin: *Engineering Experimentation: Planning, Execution, Reporting*
Driels: *Linear Control Systems Engineering*
Edwards and McKee: *Fundamentals of Mechanical Component Design*
Gebhart: *Heat Conduction and Mass Diffusion*
Gibson: *Principles of Composite Material Mechanics*
Hamrock: *Fundamentals of Fluid Film Lubrication*
Heywood: *Internal Combustion Engine Fundamentals*
Hinze: *Turbulence*
Histand and Alciatore: *Introduction to Mechatronics and Measurement Systems*
Holman: *Experimental Methods for Engineers*
Howell and Buckius: *Fundamentals of Engineering Thermodynamics*
Jaluria: *Design and Optimization of Thermal Systems*
Juvinall: *Engineering Considerations of Stress, Strain, and Strength*
Kays and Crawford: *Convective Heat and Mass Transfer*
Kelly: *Fundamentals of Mechanical Vibrations*
Kimbrell: *Kinematics Analysis and Synthesis*
Kreider and Rabl: *Heating and Cooling of Buildings*
Martin: *Kinematics and Dynamics of Machines*
Mattingly: *Elements of Gas Turbine Propulsion*
Modest: *Radiative Heat Transfer*
Norton: *Design of Machinery*
Oosthuizen and Carscallen: *Compressible Fluid Flow*
Oosthuizen and Naylor: *Introduction to Convective Heat Transfer Analysis*
Phelan: *Fundamentals of Mechanical Design*
Reddy: *An Introduction to Finite Element Method*
Rosenberg and Karnopp: *Introduction to Physical Systems Dynamics*
Schlichting: *Boundary-Layer Theory*
Shames: *Mechanics of Fluids*
Shigley: *Kinematic Analysis of Mechanisms*
Shigley and Mischke: *Mechanical Engineering Design*
Shigley and Uicker: *Theory of Machines and Mechanisms*
Stiffler: *Design with Microprocessors for Mechanical Engineers*
Stoecker and Jones: *Refrigeration and Air Conditioning*
Turns: *An Introduction to Combustion: Concepts and Applications*
Ullman: *The Mechanical Design Process*
Wark: *Advanced Thermodynamics for Engineers*
Wark and Richards: *Thermodynamics*
White: *Viscous Fluid Flow*
Zeid: *CAD/CAM Theory and Practice*

To my loving wife Jane, for her support and encouragement.

P. H. O.

To my wife Kathryn, for all her love and support.

D. N.

CONTENTS

Preface xiii

Nomenclature xvii

1 Introduction 1
  1.1 Convective Heat Transfer 1
  1.2 Forced, Free, and Combined Convection 4
  1.3 External and Internal Flows 5
  1.4 The Convective Heat Transfer Coefficient 5
  1.5 Application of Dimensional Analysis to Convection 11
  1.6 Physical Interpretation of the Dimensionless Numbers 23
  1.7 Fluid Properties 26
  1.8 Concluding Remarks 26
      Problems 27
      References 29

2 The Equations of Convective Heat Transfer 31
  2.1 Introduction 31
  2.2 Continuity and Navier-Stokes Equations 32
  2.3 The Energy Equation for Steady Flow 35
  2.4 Similarity in Convective Heat Transfer 41
  2.5 Vorticity and Temperature Fields 46
  2.6 The Equations for Turbulent Convective Heat Transfer 49
  2.7 Simplifying Assumptions Used in the Analysis of Convection 59
  2.8 The Boundary Layer Equations for Laminar Flow 61
  2.9 The Boundary Layer Equations for Turbulent Flow 69
  2.10 The Boundary Layer Integral Equations 71
  2.11 Concluding Remarks 80
      Problems 80
      References 82

3 Some Solutions for External Laminar Forced Convection — 83

- **3.1** Introduction — 83
- **3.2** Similarity Solution for Flow over an Isothermal Plate — 83
- **3.3** Similarity Solutions for Flow over Flat Plates with Other Thermal Boundary Conditions — 98
- **3.4** Other Similarity Solutions — 106
- **3.5** Integral Equation Solutions — 114
- **3.6** Numerical Solution of the Laminar Boundary Layer Equations — 123
- **3.7** Viscous Dissipation Effects on Laminar Boundary Layer Flow over a Flat Plate — 140
- **3.8** Effect of Fluid Property Variations with Viscous Dissipation Effects on Laminar Boundary Layer Flow over a Flat Plate — 149
- **3.9** Solutions to the Full Governing Equations — 150
- **3.10** Concluding Remarks — 152
- Problems — 152
- References — 155

4 Internal Laminar Flows — 157

- **4.1** Introduction — 157
- **4.2** Fully Developed Laminar Pipe Flow — 158
- **4.3** Fully Developed Laminar Flow in a Plane Duct — 169
- **4.4** Fully Developed Laminar Flow in Ducts with Other Cross-Sectional Shapes — 179
- **4.5** Pipe Flow with a Developing Temperature Field — 189
- **4.6** Plane Duct Flow with a Developing Temperature Field — 197
- **4.7** Laminar Pipe Flow with Developing Velocity and Temperature Fields — 201
- **4.8** Laminar Flow in a Plane Duct with Developing Velocity and Temperature Fields — 212
- **4.9** Solutions to the Full Navier-Stokes Equations — 219
- **4.10** Concluding Remarks — 220
- Problems — 220
- References — 225

5 Introduction to Turbulent Flows 227

- 5.1 Introduction 227
- 5.2 Governing Equations 228
- 5.3 Mixing Length Turbulence Models 234
- 5.4 More Advanced Turbulence Models 239
- 5.5 Analogy Solutions for Heat Transfer in Turbulent Flow 244
- 5.6 Near-Wall Region 245
- 5.7 Transition from Laminar to Turbulent Flow 247
- 5.8 Concluding Remarks 250
- Problems 250
- References 252

6 External Turbulent Flows 254

- 6.1 Introduction 254
- 6.2 Analogy Solutions for Boundary Layer Flows 254
- 6.3 Integral Equation Solutions 272
- 6.4 Numerical Solution of the Turbulent Boundary Layer Equations 281
- 6.5 Effects of Dissipation on Turbulent Boundary Layer Flow Over a Flat Plate 296
- 6.6 Solutions to the Full Turbulent Flow Equations 299
- 6.7 Concluding Remarks 299
- Problems 300
- References 302

7 Internal Turbulent Flows 304

- 7.1 Introduction 304
- 7.2 Analogy Solutions for Fully Developed Pipe Flow 304
- 7.3 Thermally Developing Pipe Flow 322
- 7.4 Developing Flow in a Plane Duct 329
- 7.5 Solutions to the Full Governing Equations 336
- 7.6 Concluding Remarks 337
- Problems 337
- References 339

8 Natural Convection 342

- 8.1 Introduction 342
- 8.2 Boussinesq Approximation 342

8.3	Governing Equations		344
8.4	Similarity in Free Convective Flows		345
8.5	Boundary Layer Equations for Natural Convective Flows		349
8.6	Similarity Solutions for Free Convective Laminar Boundary Layer Flows		354
8.7	Numerical Solution of the Natural Convective Boundary Layer Equations		365
8.8	Free Convective Flow through a Vertical Channel		366
8.9	Natural Convective Heat Transfer across a Rectangular Enclosure		385
8.10	Horizontal Enclosures Heated from Below		403
8.11	Turbulent Natural Convective Flows		407
8.12	Concluding Remarks		416
	Problems		416
	References		420

9 Combined Convection — 426

9.1	Introduction	426
9.2	Governing Parameters	426
9.3	Governing Equations	430
9.4	Laminar Boundary Layer Flow Over an Isothermal Vertical Flat Plate	431
9.5	Numerical Solution of Boundary Layer Equations	442
9.6	Combined Convection Over a Horizontal Plate	446
9.7	Solutions to the Full Governing Equations	446
9.8	Correlation of Heat Transfer Results for Mixed Convection	449
9.9	Effect of Buoyancy Forces on Turbulent Flows	455
9.10	Internal Mixed Convective Flows	464
9.11	Fully Developed Mixed Convective Flow in a Vertical Plane Channel	466
9.12	Mixed Convective Flow in a Horizontal Duct	474
9.13	Concluding Remarks	477
	Problems	477
	References	480

10 Convective Heat Transfer Through Porous Media — 487

10.1	Introduction	487
10.2	Area Averaged Velocity	488

	10.3	Darcy Flow Model	490
	10.4	Energy Equation	495
	10.5	Boundary Layer Solutions for Two-Dimensional Forced Convective Heat Transfer	498
	10.6	Fully Developed Duct Flow	521
	10.7	Natural Convective Boundary Layer Flows	526
	10.8	Natural Convection in Porous Media-Filled Enclosures	531
	10.9	Stability of Horizontal Porous Layers Heated from Below	540
	10.10	Non-Darcy and Other Effects	545
	10.11	Concluding Remarks	547
		Problems	547
		References	550

11 Condensation 555

	11.1	Introduction	555
	11.2	Laminar Film Condensation on a Vertical Plate	558
	11.3	Wavy and Turbulent Film Condensation on a Vertical Surface	570
	11.4	Film Condensation on Horizontal Tubes	574
	11.5	Effect of Surface Shear Stress on Film Condensation on a Vertical Plate	579
	11.6	Effect of Noncondensible Gases on Film Condensation	585
	11.7	Improved Analyses of Laminar Film Condensation	586
	11.8	Nongravitational Condensation	597
	11.9	Concluding Remarks	600
		Problems	600
		References	602

Appendices

A	Properties of Saturated Water	606
B	Properties of Air at Standard Atmospheric Pressure	607

Index 609

PREFACE

Convective heat transfer occurs in almost all branches of engineering and a knowledge of the methods used to model convective heat transfer is therefore required by many practicing engineers. Most conventional introductory courses on heat transfer deal with some aspects of convective heat transfer but the treatment is usually relatively superficial. For this reason, many engineering schools offer a course dealing with convective heat transfer at the senior undergraduate or graduate level. The purpose of such courses is to expand and extend the coverage given in the basic course on heat transfer. The present book is intended to provide a clear presentation of the material covered in such courses. Completion of undergraduate courses in basic heat transfer, thermodynamics, and fluid mechanics is assumed.

Coverage

The book provides a comprehensive coverage of the subject giving a full discussion of forced, natural, and mixed convection including some discussion of turbulent natural and mixed convection. A comprehensive discussion of convective heat transfer in porous media flows and of condensation heat transfer is also provided. The book contains a large number of worked examples that illustrate the use of the derived results. All chapters in the book also contain an extensive set of problems.

Convective heat transfer has become a subject of very wide extent and the selection of material for inclusion in a book of the present type requires careful consideration. In this book, heat transfer during boiling and solidification have not been considered. This is not in any way meant to suggest that these topics are of lesser importance than those included in the book. Rather, it is felt that the student needs a good grounding in the topics covered in the present book before approaching the analysis of heat transfer with boiling and solidification. The book thus lays the foundation for more advanced courses on specialized aspects of convective heat transfer.

Approach

The basic aim of the book is to present a discussion of some currently available methods for predicting convective heat transfer rates. The main emphasis is, therefore, on the prediction of heat transfer rates rather than on the presentation of large amounts of experimental data. Attention is given to both analytical and numerical methods of analysis. Another aim of the book is to present a thorough discussion of the foundations of the subject in a clear, easy to follow, student-oriented style.

Because new methods of analyzing convection are constantly being introduced, the emphasis in the book is on developing a thorough understanding of the basic equations and of the processes involved in the analysis of convective heat transfer, together with developing an understanding of the assumptions on which existing methods of analysis are based and on developing a clear understanding of the limitations of these methods. Even at the graduate level, some students have difficulty dealing with the process of taking a complex real situation and, by introducing a series of carefully reasoned and documented assumptions, deriving a model of the real situation that is amenable to solution. In this book, emphasis has, therefore, been placed wherever possible on this modeling process.

The widespread use of computer software for the analysis of engineering problems has, in many ways, increased the need to understand the assumptions and the theory on which such analyses are based. Such an understanding is required to interpret the computer results and to judge whether a particular piece of software will give results that are of adequate accuracy for the application being considered. Therefore, while there is an extensive discussion of numerical methods in this book, the major emphasis is on developing an understanding of the material and of the assumptions conventionally used in analyzing convective heat transfer.

Compared to available textbooks on the subject, then, the present book is, it is hoped, distinguished by its attempt to develop a thorough understanding of the theory and of the assumptions on which this theory is based and by the breadth of its coverage.

Solutions Manual

A manual, written by the authors of this textbook, that contains detailed solutions to all of the problems in the book, is available. Each solution in this manual starts on a separate page thus making it easier to post solutions to individual problems.

Software

A number of computer programs are discussed in this book. These are all based on relatively simple finite-difference procedures that are developed in the book. While the numerical methods used are relatively simple, it is believed that if the students gain a good understanding of these methods and are exposed to the power of even simple numerical solution procedures, they will have little difficulty in understanding and using more advanced numerical methods. Examples of the use of the computer programs are included in the text.

The computer programs described in this book can all be downloaded from the web site of one of the authors (P. H. Oosthuizen) at Queen's University. For more information on these programs and on how they can be obtained, please contact this author via e-mail at oosthuiz@me.queensu.ca.

ACKNOWLEDGMENTS

This book has been developed from notes prepared over many years for use in courses on convective heat transfer taught to advanced undergraduate and graduate students at Queen's University. The input of students into a professor's understanding of a subject cannot be overemphasized and the material in this book and the way in which it is presented have been greatly influenced by the undergraduate and graduate students that have taken these courses. The authors wish to express their deep and sincere gratitude to all of these students.

Jane Paul undertook the tedious job of preparing and checking much of the text and her help and encouragement is gratefully acknowledged. The authors would like to thank John Lloyd and John D. Anderson for their advice and encouragement during the early stages of the development of this book. They would also like to express their gratitude to the following reviewers for their contributions to the development of this text:

A. M. Kanury
Oregon State University

Afshin J. Ghajar
Oklahoma State University

John Lloyd
Michigan State University

Jamal Seyed-Yagoobi
Texas A&M University

David G. Briggs
Rutgers University

Ramendra P. Roy
Arizona State University

Patrick H. Oosthuizen

David Naylor

NOMENCLATURE

The following is a list of some of the main symbols used in this book:

A	area
a	speed of sound
C_f	shear stress coefficient
c_p	constant pressure specific heat
D	diameter or drag force
D_h	hydraulic diameter
E	Eckert number
e	wall roughness
Fr	Froude number
G	buoyancy force parameter
Gr	Grashof number
Gr^*	heat flux Grashof number
g	gravitational acceleration
H	rate of enthalpy transport
h	heat transfer coefficient
$\overline{h}$	average heat transfer coefficient
h_{fg}	latent heat
Ja	Jakob number
K	turbulence kinetic energy or permeability
k	thermal conductivity
L	length
ℓ	mixing length
M	mass or Mach number or rate of momentum transport
$\dot{m}$	mass flow rate
Nu	Nusselt number
Nu_x	local Nusselt number
P	dimensionless pressure or perimeter
p	pressure
$\overline{p}$	time-averaged pressure
p'	fluctuating component of pressure
Pe	Peclet number
Pr	Prandtl number
Pr_T	turbulent Prandtl number
Q	heat transfer rate
q	heat transfer rate per unit area
Ra	Rayleigh number
Re	Reynolds number
r	radius
T	temperature
$\overline{T}$	time-averaged temperature
T'	fluctuating component of temperature

T_b	bulk temperature
T_f	fluid temperature
T_s	saturation temperature (also T_{sat})
T_w	surface temperature
T_{wad}	adiabatic wall temperature
t	time
U	forced velocity or dimensionless velocity
u	velocity component
$\bar{u}$	average velocity component
u'	fluctuating component of velocity
u_1	free-stream velocity (also u_∞)
u_c	center-line velocity
u^*	friction velocity
V	volume or dimensionless velocity
v	velocity component
$\bar{v}$	average velocity component
v'	fluctuating component of velocity
w	velocity component
$\bar{w}$	average velocity component
w'	fluctuating component of velocity
x	coordinate direction
y	coordinate direction
z	coordinate direction
α	thermal diffusivity
β	volumetric expansion coefficient
γ	specific heat ratio
Γ	mass flow rate per unit width
δ	velocity boundary layer thickness or condensate film thickness
δ_T	thermal boundary layer thickness
ϵ	eddy viscosity
ϵ_H	eddy diffusivity
θ	dimensionless temperature or angle
μ	dynamic viscosity
ν	kinematic viscosity
π	dimensionless parameter
ρ	density
η	similarity parameter or dimensionless coordinate
σ	normal stress
τ	shear stress
τ_m	molecular shear stress
τ_T	turbulent shear stress
Φ	dissipation function
ϕ	angular coordinate or dimensionless temperature or porosity
Ψ	dimensionless stream function
ψ	stream function
Ω	dimensionless vorticity
ω	vorticity

CHAPTER 1

Introduction

1.1
CONVECTIVE HEAT TRANSFER

Heat convection is the term applied to the process involved when energy is transferred from a surface to a fluid flowing over it as a result of a difference between the temperatures of the surface and the fluid. In convection therefore as indicated in Fig. 1.1, there is always a surface, a fluid flowing relative to this surface, and a temperature difference between the surface and the fluid and the concern is with the rate of heat transfer between the surface and the fluid.

Convective heat transfer occurs extensively in practice. The cooling of the cutting tool during a machining operation, the cooling of the electronic components in a computer, the generation and condensation of steam in a thermal power plant, the heating and cooling of buildings, cooking, and the thermal control of a re-entering spacecraft all, for example, involve convective heat transfer. Some examples of situations in which convective heat transfer is important are shown in Fig. 1.2.

Convection is one of the three so-called modes of heat transfer, the other two are conduction and radiation [1],[2],[3],[4]. In most real situations, the overall heat transfer is accomplished by a combination of at least two of these modes of heat transfer. However, it is possible, in many such cases, to consider the modes separately and then combine the solutions for each of the modes in order to obtain the overall heat transfer rate. For example, heat transfer from one fluid to another fluid through the walls of a pipe occurs in many practical devices. In this case, heat is transferred by convection from the hotter fluid to the one surface of the pipe. Heat is then transferred by conduction through the walls of the pipe. Finally, heat is transferred by convection from the other surface to the colder fluid. These heat transfer processes are shown in Fig. 1.3. The overall heat transfer rate can be calculated by considering the three processes separately and then combining the results.

2 Introduction to Convective Heat Transfer Analysis

FIGURE 1.1
Convective heat transfer.

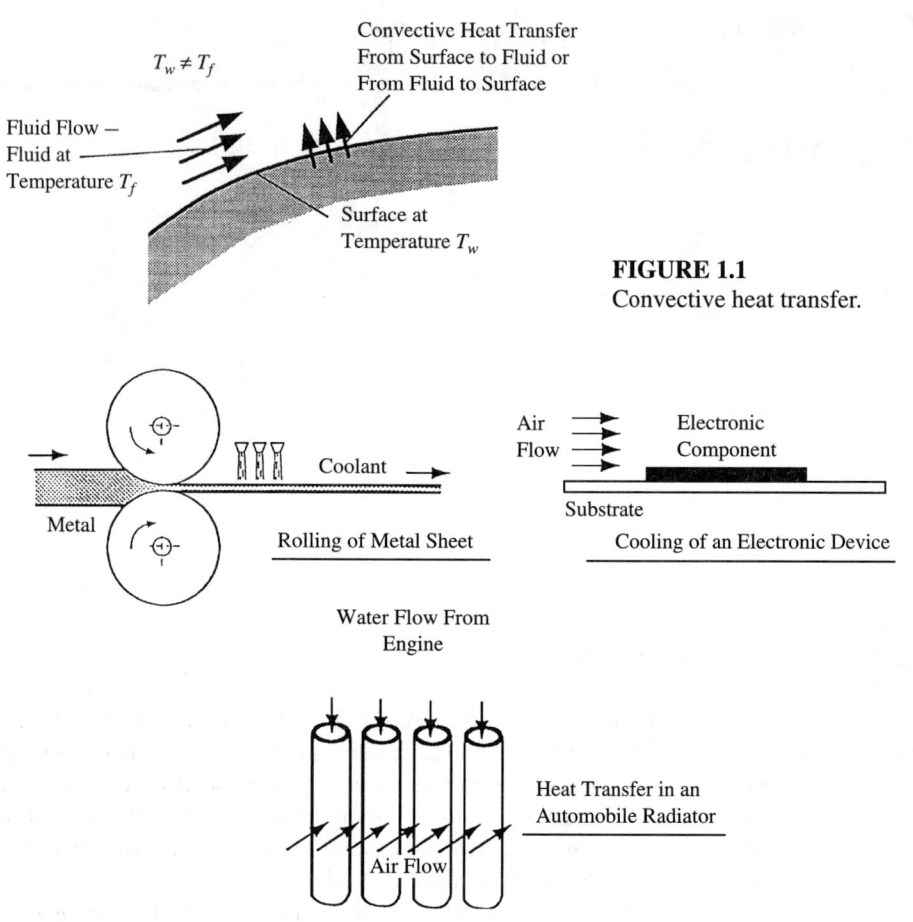

FIGURE 1.2
Some situations that involve convective heat transfer.

This book is concerned with a description of some methods of determining convective heat transfer rates in various flow situations, realizing that in many cases these methods will need to be combined with calculations for the other modes of heat transfer in order to predict the overall heat transfer rate.

In some cases it is not possible to consider the modes separately. For example, if a gas, such as water vapor or carbon dioxide, which absorbs and generates thermal radiation, flows over a surface at a higher temperature, heat is transferred from the surface to the gas by both convection and radiation. In this case, the radiant heat exchange influences the temperature distribution in the fluid. Therefore, because the convective heat transfer rate depends on this temperature distribution in the fluid, the radiant and convective modes interact with each other and cannot be considered separately. However, even in cases such as this, the calculation procedures developed for convection by itself form the basis of the calculation of the convective part of the overall heat transfer rate.

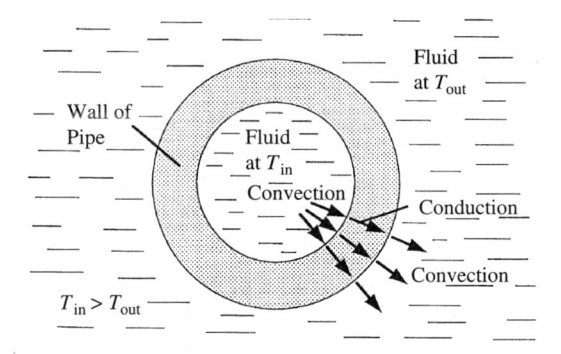

FIGURE 1.3
Combined conduction and convection.

Convective heat transfer rates depend on the details of the flow field about the surface involved as well as on the properties of the fluid. The determination of convective heat transfer rates is therefore, in general, an extremely difficult task since it involves the determination of both the velocity and temperature fields. It is only in comparatively recent times that any widespread success has been achieved in the development of methods of calculating convective heat transfer rates.

The transfer of heat by convection involves the transfer of energy from the surface to the fluid on a molecular scale and then the diffusion of this heat through the fluid by bulk mixing due to the fluid motion. The basic heat transport mechanism in convection is still conduction which is, of course, governed by Fourier's law. This law states that the heat transfer rate, q, in any direction, n, per unit area measured normal to n is given by:

$$q = -k\frac{\partial T}{\partial n} \qquad (1.1)$$

k being the coefficient of conductivity which is a property of the fluid involved.

Conduction of heat is always important close to the surface over which the fluid is flowing. However, when the flow is turbulent, the rate at which heat is "convected" by the turbulent eddies is usually much greater than the rate of heat conduction, and, therefore, in such flows, conduction can often be neglected except in a region lying close to the surface. At a solid surface, where the velocity is effectively zero, Fourier's law always, of course, applies as indicated in Fig. 1.4.

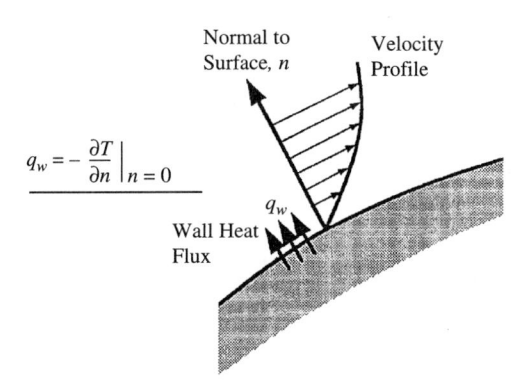

FIGURE 1.4
Heat transfer at wall.

1.2
FORCED, FREE, AND COMBINED CONVECTION

Since convective heat transfer rates depend on the details of the flow field they will be strongly dependent on how the flow is generated. It is important, therefore, to distinguish between forced and free (or natural) convection [5],[6],[7],[8],[9],[10]. In the case of forced convection, the fluid motion is caused by some external means such as a fan or pump. In the case of free convection, the flow is generated by the body forces that occur as a result of the density changes arising from the temperature changes in the flow field. These body forces are actually generated by pressure gradients imposed on the whole fluid. The most common source of this imposed pressure field is gravity, the pressure gradient then being the normal hydrostatic pressure gradient existing in any fluid bulk. The body forces in this case are usually termed buoyancy forces. Another source of imposed pressure gradients which can cause free-convective flows are the centrifugal forces which arise when there is an overall rotary motion such as exists in a rotating machine.

In all flows involving heat transfer and, therefore, temperature changes, the buoyancy forces arising from the gravitational field will, of course, exist. The term forced convection is only applied to flows in which the effects of these buoyancy forces are negligible. In some flows in which a forced velocity exists, the effects of these buoyancy forces will, however, not be negligible and such flows are termed combined- or mixed free and forced convective flows. The various types of convective heat transfer are illustrated in Fig. 1.5.

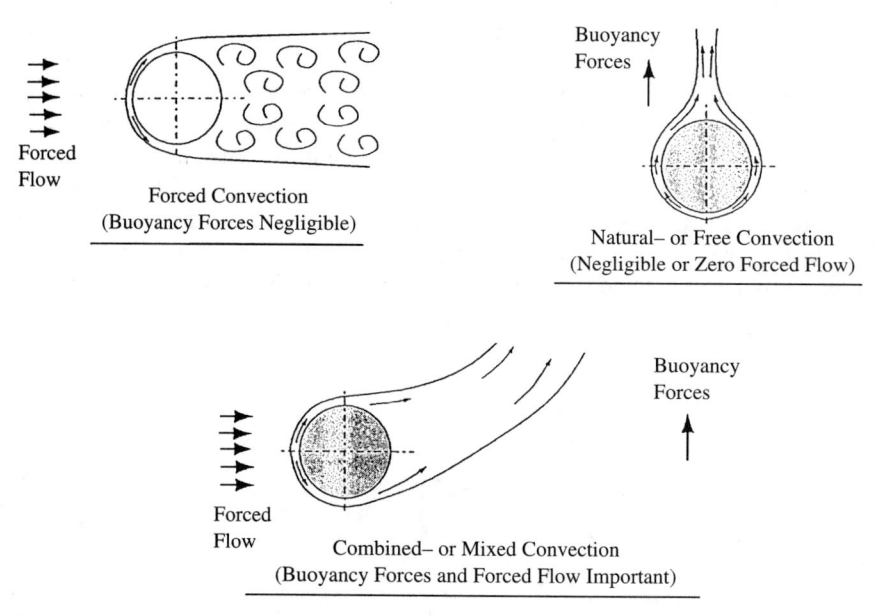

FIGURE 1.5
Forced, free, and combined convection.

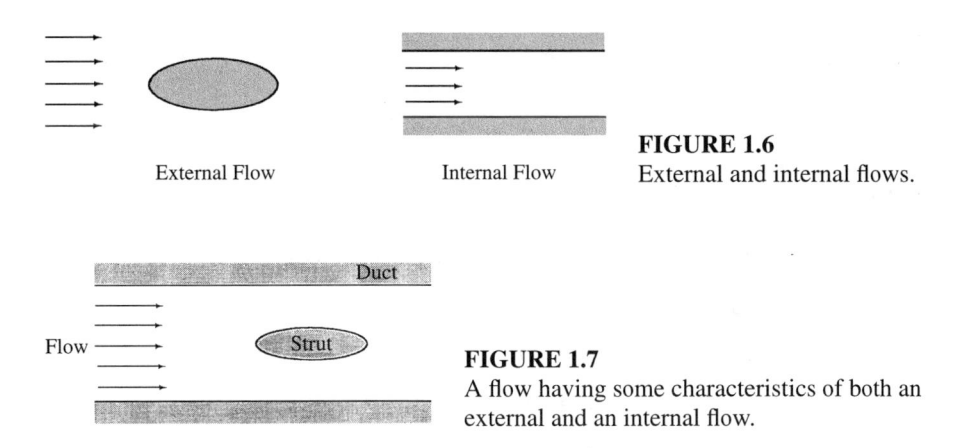

FIGURE 1.6
External and internal flows.

External Flow   Internal Flow

FIGURE 1.7
A flow having some characteristics of both an external and an internal flow.

Heat transfer involving a change of phase is classified as convective heat transfer even though when the solid phase is involved, the overall process involves combined and interrelated convection and conduction. Heat transfer during boiling, condensation, and solidification (freezing) all, thus, involve convective heat transfer.

1.3
EXTERNAL AND INTERNAL FLOWS

Convective flows are conventionally broken down into those that involve external flow and those that involve internal flow. These two types of flow are illustrated in Fig. 1.6.

External flows involve a flow, that is essentially infinite in extent, over the outer surface of a body. Internal flows involve the flow through a duct or channel. It is not always possible to clearly place a convective heat transfer problem into one of these categories since some problems have several of the characteristics of both an internal and an external flow. An example of such a case is shown in Fig. 1.7.

In predicting heat transfer rates one of the first steps in creating a model of the flow is to decide whether it involves internal or external flow because there are different assumptions that can be conveniently adopted in the two types of flow.

1.4
THE CONVECTIVE HEAT TRANSFER COEFFICIENT

Consider a fluid flowing over a surface which is maintained at a temperature that is different from that of the bulk of the fluid as shown in Fig. 1.8.

If the temperature at some point on the surface is T_w and if the rate at which heat is being transferred locally at this point from the surface to the fluid per unit surface

6 Introduction to Convective Heat Transfer Analysis

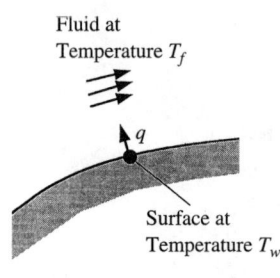

FIGURE 1.8 Convective heat transfer from a surface.

area is q, then it is usual to define a quantity, h, such that:

$$q = h(T_w - T_f) \tag{1.2}$$

In this equation T_f is some convenient fluid temperature which will be more precisely defined at a later stage [11],[12],[13],[14],[15].

The heat transfer rate, q, is taken as positive in the direction wall-to-fluid so that it will have the same sign as $(T_w - T_f)$ and h will always, therefore, be positive. A number of names have been applied to h including "convective heat transfer coefficient", "heat transfer coefficient", "film coefficient", "film conductance", and "unit thermal convective conductance". The heat transfer coefficient, h, has the units W/m^2-K or, since its definition only involves temperature differences, W/m^2°C, in the SI system of units. In the imperial system of units, h has the units Btu/ft^2-hr-°F.

It must be clearly understood that h is not only dependent on the fluid involved in the process but is also strongly dependent on, among other things, the shape of the surface over which the fluid is flowing and on the velocity of the fluid.

A large part of the present book will be concerned with a discussion of some of the analytical and numerical methods that can be used to try to predict h for various flow situations. Basically, these methods involve the simultaneous application of the principles governing viscous fluid flow, i.e., the principles of conservation of mass and momentum and the principle of conservation of energy. Although considerable success has been achieved with these methods, there still remain many cases in which experimental results have to be used to arrive at working relations for the prediction of h.

Consideration will now be given to how the fluid temperature, T_f, used in Eq. (1.2), is defined, the definition used depending of course, on the flow situation. Consider first, incompressible flow over the outside of a body, as shown in Fig. 1.9.

In this case, T_f is most conveniently taken as the temperature, T_1, of the fluid in the freestream ahead of the body. This is, of course, a very obvious choice for

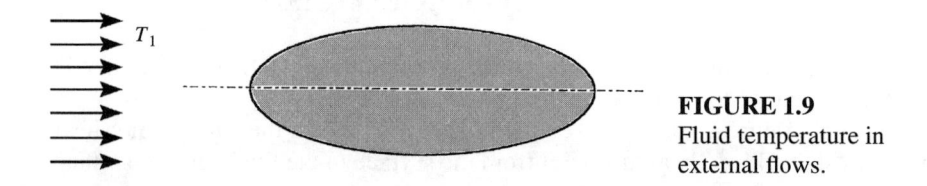

FIGURE 1.9 Fluid temperature in external flows.

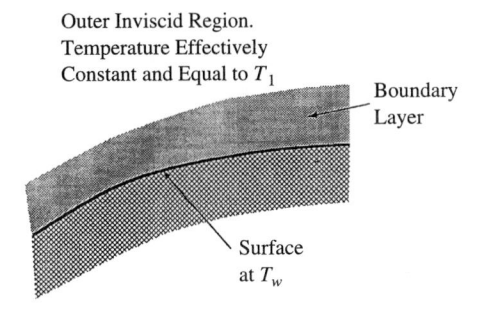

FIGURE 1.10
Boundary layer on a surface.

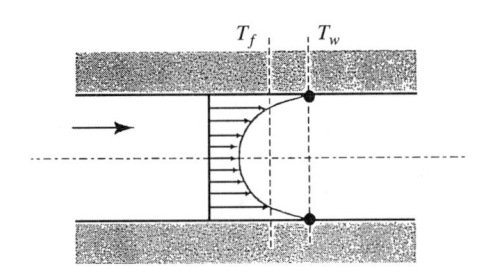

FIGURE 1.11
Fluid temperature in internal flows.

T_f when, as is the case of the majority of problems of engineering importance, a boundary layer-type flow exists, i.e., there exists adjacent to the surface of the body a comparatively thin region in which very rapid changes in velocity and temperature occur in the direction normal to the surface and where outside this region the fluid temperature is effectively equal to that existing in the freestream, as shown in Fig. 1.10. In the case of internal flows, such as flow through a pipe, the fluid temperature will, in general, vary continuously across the whole flow area, as indicated in Fig. 1.11.

In such cases, T_f is taken as some mean temperature. The most commonly used mean temperature is the so-called bulk temperature, T_b. Basically, this is defined such that:

$$\text{Mass flow rate} \times \text{specific enthalpy} = \frac{\text{Rate at which enthalpy crosses}}{\text{the selected section of duct}}$$

i.e.,

$$\dot{m} c_p T_b = H \tag{1.3}$$

where c_p is the specific heat at the constant pressure of the fluid, the specific enthalpy having thus been set equal to $c_p T$. The specific heat is, in general, temperature dependent. It is assumed to be constant here having been evaluated at some suitable mean temperature.

Now, if an elemental portion of the cross-sectional area, dA, of the duct as shown in Fig. 1.12 is considered and if the velocity and temperature at the place where dA

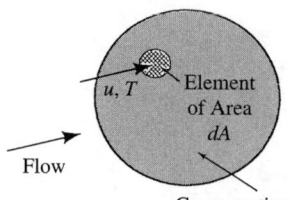

8 Introduction to Convective Heat Transfer Analysis

FIGURE 1.12 Element of duct cross-section.

is taken are u and T respectively then the rate at which enthalpy passes through dA will be given by:

$$dH = \rho u \, dA \, c_p T \tag{1.4}$$

Therefore, integrating this equation to get the total rate at which enthalpy passes through the cross-section of the duct gives:

$$H = \int_A \rho u c_p T \, dA \tag{1.5}$$

Substituting this into Eq. (1.3) then gives the following, because, as discussed above, c_p is being treated as a constant:

$$T_b = \int_A \rho u T \, dA / \dot{m} \tag{1.6}$$

By the same reasoning as that used to deduce Eq. (1.5), it can be shown that:

$$\dot{m} = \int_A \rho u \, dA \tag{1.7}$$

Therefore, if the density ρ is also assumed to be constant, Eq. (1.6) can be written as:

$$T_b = \frac{\int_A u T \, dA}{\int_A u \, dA} \tag{1.8}$$

The bulk temperature is, of course, equal to the temperature that would be attained if the fluid at the particular section of the duct being considered was discharged into a container and, without any heat transfer occurring, was mixed until a uniform temperature was obtained. For this reason it is sometimes referred to as the "mixing cup temperature".

EXAMPLE 1.1. Air flows through a pipe with a diameter D. The velocity distribution in the pipe is approximately given by $u = 20[(R-r)/R]^{1/7}$ m/s, r being the radial distance to the point at which u is the velocity and R is the radius of the pipe. The temperature distribution in the flow is approximately given by $70 - 40[(R-r)/R]^{1/7}$°C. Find the mean temperature in the flow.

CHAPTER 1: Introduction 9

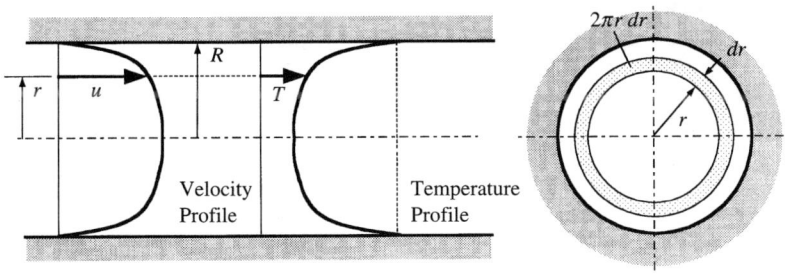

FIGURE E1.1

Solution. Using:

$$T_b = \frac{\int_A uT\,dA}{\int_A u\,dA}$$

and noting as indicated in Fig. E1.1 that:

$$dA = 2\pi r\,dr$$

it follows that:

$$T_b = \frac{\int_0^R uT 2\pi r\,dr}{\int_0^R u 2\pi r\,dr}$$

i.e.,

$$T_b = \frac{\int_0^1 uT(r/R)\,d(r/R)}{\int_0^1 u(r/R)\,d(r/R)}$$

Using the given expressions for the velocity and temperature distributions then gives:

$$T_b = 70 - 20 \times 40 \frac{\int_0^1 [1-(r/R)]^{2/7}(r/R)\,d(r/R)}{20\int_0^1 [1-(r/R)]^{1/7}(r/R)\,d(r/R)}$$

i.e., writing:

$$X = 1 - (r/R)$$

TABLE 1.1
Typical values for the mean heat transfer coefficient

Flow situation and fluid	Mean heat transfer coefficient (W/m²-K)
Forced convection in air	10–200
Forced convection in water	40–10,000
Forced convection in liquid metals	10,000–100,000
Free convection in air	3–20
Free convection in water	20–200
Boiling in water	2,000–100,000
Condensing in steam	5,000–50,000

the above equation becomes:

$$T_b = 70 - 20 \times 40 \frac{\int_1^0 X^{2/7}(1-X)\,dX}{20\int_1^0 X^{1/7}(1-X)\,dX}$$

$$T_b = 70 - \frac{20 \times 40[7/9 - 7/16]}{20[7/8 - 7/15]}$$

i.e.,

$$T_b = 36.7°C$$

Eq. (1.2) defined the local heat transfer coefficient, h, in terms of the local rate of heat transfer rate per unit area, the local wall temperature and, in those cases where it is changing, the local fluid temperature. In general, all of these quantities, i.e., h, q, T_w, and T_f, vary with position on the surface. For the majority of applications it is convenient to define, therefore, a mean or average heat transfer coefficient, $\bar{h}$, such that if Q is the total heat transfer rate from the entire surface of area A then:

$$Q = \bar{h}A(\bar{T}_w - \bar{T}_f) \tag{1.9}$$

where $\bar{T}_w$ is the mean surface temperature and $\bar{T}_f$ is the mean fluid temperature, the latter being defined in some suitable manner. If there is no possibility of any ambiguity, the bars on $\bar{h}$, $\bar{T}_w$, and $\bar{T}_f$ to indicate mean quantities are usually omitted.

Typical values of the mean convective heat transfer coefficient for various flow situations are listed in Table 1.1.

EXAMPLE 1.2. A pipe with a diameter of 2 cm is kept at a surface temperature of 40°C. Find the heat transfer rate per m length of this pipe if it is (a) placed in an air flow in which the temperature is 50°C and (b) placed in a tank of water kept at a temperature of 30°C. The heat transfer coefficients in these two situations, which involve (a) forced convection in air and (b) free convection in water, are estimated to be 20 W/m²K and 70 W/m²K, respectively.

Solution. The definition of the mean heat transfer coefficient gives:

$$Q = hA(T_w - T_f)$$

In the situation being considered, $T_w = 40°C$ and, since a 1-m length of the pipe is being considered, $A = \pi DL = \pi \times 0.02$. Hence:

$$Q = h\pi \times 0.02 \times (40 - T_f)$$

For case (a), this gives:

$$Q = 20 \times \pi \times 0.02 \times (40 - 50) = -12.57 \text{ W}$$

The negative sign indicates that the heat transfer is from the air to the cylinder.
For case (b), this gives:

$$Q = 70 \times \pi \times 0.02 \times (40 - 30) = 43.98 \text{ W}$$

This result is positive because in this case, the heat transfer is from the cylinder to the water.

1.5 APPLICATION OF DIMENSIONAL ANALYSIS TO CONVECTION

As previously mentioned, the major aim of the present book is to describe some typical analytical and numerical methods for determining h. However, there exist many cases of great practical importance for which it is not possible to obtain even approximate theoretical solutions. In such cases, the prediction of heat transfer rates has to be based on the results of previous experimental studies. In order that these experimental results be of the most usefulness they must be expressed in terms of general variables that then allow them to be applied to a much wider range of conditions than those under which the actual measurements were obtained. For example, if measurements of the forced convective heat transfer rates from cylinders to air and water are made, the results must be expressed if possible in terms of variables that will allow them to be used to predict the heat transfer rate from a cylinder to another fluid, e.g., fuel oil. Therefore, in order to be able to make the most use of experimental results, they must be expressed in terms of the "general" variables governing the problem. One way of determining these "general" variables is by the application of dimensional analysis [16],[17],[18],[19].

The heat transfer coefficient, h, is the variable whose value is being sought. Consider a series of bodies of the same geometrical shape, e.g., a series of elliptical cylinders (see Fig. 1.13), but of different size placed in various fluids. It can be deduced, either by physical argument or by considering available experimental results, that if, in the case of gas flows, the velocity is low enough for compressibility effects to be ignored, h will depend on:

- The conductivity, k, of the fluid with which the body is exchanging heat.
- The viscosity, μ, of the fluid with which the body is exchanging heat.

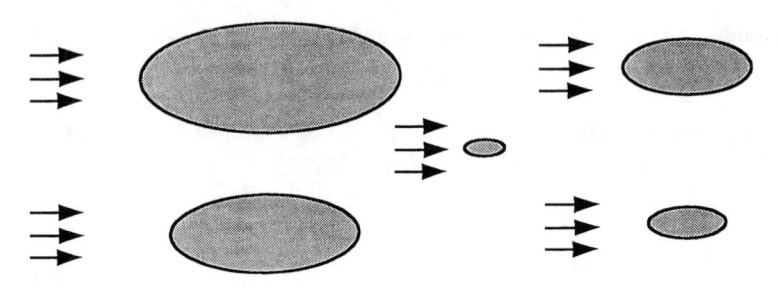

FIGURE 1.13
Series of bodies with same geometrical shape.

- The specific heat, c_p, of the fluid with which the body is exchanging heat. The reason for using the specific heat at constant pressure may be queried. For liquids, which may be assumed to be incompressible under the conditions here being considered, all specific heats are the same. For gases, c_p is used basically because it is the changes of enthalpy of the fluid that are of importance and the enthalpy of the fluid per unit mass at any point is $c_p T$.
- The density, ρ, of the fluid with which the body is exchanging heat.
- The size of the body as specified by some characteristic dimension, ℓ. Since a series of bodies having the same geometrical shape is being considered, any convenient dimension can be used for ℓ.
- The magnitude of the forced fluid velocity, U, relative to the body. The manner in which this is defined will depend on the nature of the problem as indicated in Fig. 1.14. In the case of external flows, the undisturbed freestream velocity is usually the most convenient to use for U, whereas in the case of internal flows it is usually more convenient to take U as the mean or mass average velocity in the duct.
- Some quantity which will act as a measure of the magnitude of the body force acting on the flow. Consideration here will only be given to buoyancy forces, i.e., to forces arising due to gravity. However, the method used here to derive a measure of this force can easily be applied when other force fields exist.

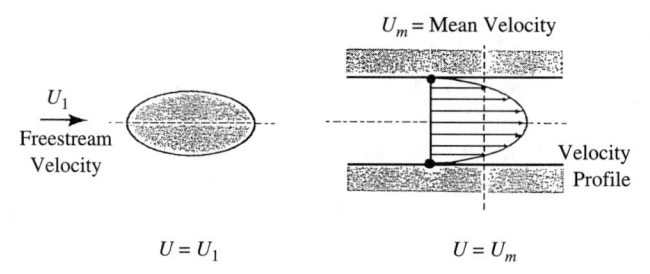

FIGURE 1.14
Definition of forced velocity.

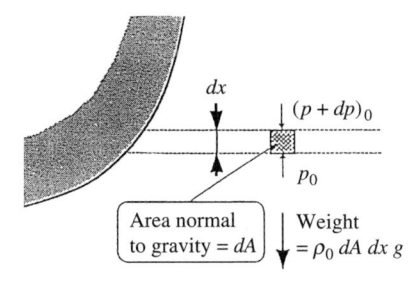

FIGURE 1.15
Control volume considered.

To derive an expression for a measure of the magnitude of the buoyancy force, consider an elemental volume of the fluid as shown in Fig. 1.15. First, consider the forces acting on this control volume when the fluid is unheated and at rest.

Since the fluid is at rest, the hydrostatic pressure forces must just balance the weight of the fluid. Hence, if dA is the horizontal cross-sectional area of the control volume and if ρ_0 is the density of the fluid in this unheated state, then the force balance requires that:

$$[p_0 - (p + dp)_0]\, dA = \rho_0\, dA\, dx\, g \tag{1.10}$$

Now consider the same control volume when the surface is at a different temperature from the fluid far away from it and the fluid is in motion. The fluid near the surface will also be at a different temperature from the fluid far away from the surface and its density will as a result also be different. Let the density of the fluid in the control volume in this case be ρ. The situation is shown in Fig. 1.16.

Applying the conservation of momentum principle for the x-direction to the control volume gives:

$$[p' - (p + dp)']\, dA + \begin{array}{c}\text{Net shearing force on surface of}\\ \text{control volume in x-direction}\end{array}$$
$$= \begin{array}{c}\text{Rate of momentum change in}\\ \text{x-direction through control volume}\end{array} + \rho\, dA\, dx\, g \tag{1.11}$$

Now it is convenient to measure the pressure relative to the hydrostatic pressure that would exist if there was no heating and no fluid motion, i.e., to define:

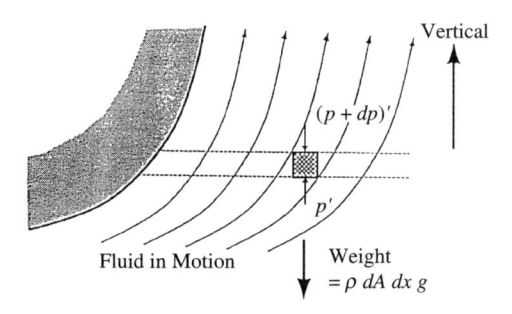

$$p = p' - p_0 \tag{1.12}$$

FIGURE 1.16
Generation of buoyancy forces.

14 Introduction to Convective Heat Transfer Analysis

from which it follows that:

$$(p + dp) = (p + dp)' - (p + dp)_0 \tag{1.13}$$

Using these two equations and Eq. (1.10) it then follows that:

$$[p' - (p + dp)] dA = [p_0 - (p + dp)_0] dA + [p - (p + dp)] dA \tag{1.14}$$
$$= \rho_0 \, dA \, dx \, g + [p - (p + dp)] dA$$

Substituting this result into Eq. (1.11) then gives:

$$[p - (p + dp)] dA + \begin{matrix}\text{Net shearing}\\\text{force on}\\\text{surface of}\\\text{control volume}\\\text{in x-direction}\end{matrix} + (\rho_0 - \rho) g \, dA \, dx = \begin{matrix}\text{Rate of momentum}\\\text{change in}\\\text{x-direction}\\\text{through control}\\\text{volume}\end{matrix}$$

$$\tag{1.15}$$

The third term on the left-hand side of this equation is called the buoyancy force. It arises, of course, as a result of the density change with temperature and is equal to the difference between the weight of the fluid in the control volume with no heating and the weight of the fluid in the control volume when heated. Its value could have been more simply derived using a less rigorous approach. However, difficulty can sometimes be encountered when such approaches are attempted for body forces arising due to effects other than gravity.

The above discussion shows, therefore, that the buoyancy force per unit volume acting on the fluid in the vertically upward direction is given by:

$$(\rho_0 - \rho)g \tag{1.16}$$

Now attention in the present section is being restricted to incompressible flows in which only density changes with temperature are significant, i.e., the density changes produced by the pressure changes in the flow are assumed to be negligible. It is convenient, therefore, to express this buoyancy force in terms of the temperature difference by introducing the coefficient of bulk expansion (sometimes called the coefficient of cubical expansion), β, of the fluid. This is defined such that if a mass M of the fluid occupies a volume V_i at a temperature T_i then, if its temperature is changed to T, its volume V at this temperature is given by:

$$\frac{(V - V_i)}{V_i} = \beta(T - T_i) \tag{1.17}$$

If the densities of the fluid at temperatures T_i and T are ρ_i and ρ respectively then, since:

$$\rho_i = \frac{M}{V_i}, \quad \rho = \frac{M}{V} \tag{1.18}$$

Eq. (1.17) can be written as:

$$\frac{\left[\left(\frac{M}{\rho}\right) - \left(\frac{M}{\rho_i}\right)\right]}{\left(\frac{M}{\rho_i}\right)} = \beta(T - T_i)$$

i.e.,

$$\rho_i - \rho = \beta \rho_i (T - T_i) \tag{1.19}$$

Applying this result between the temperature of the unheated fluid T_f and T then shows using Eq. (1.16) that the buoyancy force per volume is given by:

$$\beta g \rho (T - T_f) \tag{1.20}$$

Since the temperature difference $T - T_f$ at any point in the flow will depend on the overall temperature difference $(T_w - T_f)$ it is seen that a convenient measure of the buoyancy force acting on the fluid is $\beta g \rho (T_w - T_f)$. However, ρ has already been independently selected as a variable on which h depends so the variable that will be used as a measure of the buoyancy force will be taken as:

$$\beta g (T_w - T_f) \tag{1.21}$$

It is assumed, therefore, that:

$$h = \text{function} \, [k, \mu, c_p, \rho, \ell, U, \beta g (T_w - T_f)] \tag{1.22}$$

which can be written as:

$$f[h, k, \mu, c_p, \rho, \ell, U, \beta g (T_w - T_f)] = 0 \tag{1.23}$$

where f is some function.

There are thus, eight dimensional variables involved in the convection problem. The theory of dimensional analysis shows, however, that fewer generalized dimensionless variables are in fact required to describe the problem, the number required being equal to the number of dimensionless variables, i.e., eight in the present case, less the number of basic dimensions required to describe the problem. In the present situation, four basic dimensions, i.e., mass, M, length, L, time, t, and temperature, T are required and, therefore, four (i.e., 8 − 4) dimensionless variables are sufficient to describe the problem. If it had been assumed that the changes in the kinetic energy of the flowing fluid are associated with significant enthalpy changes, then temperature could not have been taken as an independent dimension and only three dimensions would have been required to define the problem. In this case, an extra dimensionless number will arise, this being the Mach number which is equal to the ratio of the characteristic velocity to the speed of sound. The fact that incompressible flow is being assumed indicates that the Mach number is low and this effect therefore negligible. Four basic dimensions will therefore be assumed in the present work.

16 Introduction to Convective Heat Transfer Analysis

The theory of dimensional analysis indicates that the dimensionless variables are formed by the products of powers of certain of the original dimensional variables. For this reason, the generalized dimensionless variables are often termed dimensionless products and are denoted by the symbol π. The four dimensionless variables required in the present problem will, therefore, be denoted by π_1, π_2, π_3, and π_4. Eq. (1.23) can, therefore, be reduced to the following form:

$$\text{function } (\pi_1, \pi_2, \pi_3, \pi_4) = 0 \tag{1.24}$$

The theory of dimensional analysis shows that each of these π's must contain one prime dimensional variable which does not occur in any of the other π's, together with other dimensional variables which occur in all the π's. In the present problem there are, of course, eight dimensional variables and four π's so four prime variables must be selected and there will also be four other dimensional variables which occur in all the π's.

The prime variables should be selected so that each dimensionless product characterizes some distinct feature of the flow. For example, the forced velocity, U, should be used as one prime variable since it determines whether or not the problem involves forced convection. Similarly, the buoyancy variable $\beta g(T_w - T_f)$ will determine the importance of free convective effects and should also be used as a prime variable. Also, since h is the variable whose value is required, it should be used as a prime variable. The fourth prime variable will be taken as c_p. This choice is not as obvious as the others but stems from the fact that c_p will determine the thermal capacity of the fluid and will, therefore, influence the relation between the velocity and temperature fields.

With these four prime variables, the π's will have the form:

$$\pi_1 = [hk^{a_1} \mu^{b_1} \rho^{c_1} \ell^{d_1}] \tag{1.25}$$

$$\pi_2 = [Uk^{a_2} \mu^{b_2} \rho^{c_2} \ell^{d_2}] \tag{1.26}$$

$$\pi_3 = [\beta g(T_w - T_f)k^{a_3} \mu^{b_3} \rho^{c_3} \ell^{d_3}] \tag{1.27}$$

$$\pi_4 = [c_p k^{a_4} \mu^{b_4} \rho^{c_4} \ell^{d_4}] \tag{1.28}$$

The dimensional variables k, μ, ρ, and ℓ which are not prime variables occur, of course, in all four π's. The possibility that some of the indices of these variables $a_1, b_1, c_1, \ldots, c_4, d_4$ are zero is not excluded.

It should be noted that no additional assumption is inherent in setting the index of the prime variables equal to unity since in place of π_1, for example, the following could have been used:

$$\pi_1' = [h^e k^f \mu^g \rho^h \ell^j]$$
$$= [h^e k^{ea} \mu^{eb} \rho^{ec} \ell^{ed}]$$
$$= \pi_1^e$$

so that if π_1 is dimensionless then so is π_1'.

CHAPTER 1: Introduction   17

The values of the indices $a_1, b_1, c_1, \ldots, c_4, d_4$ are determined by using the fact that the π's are dimensionless. In order to utilize this fact, the dimensions of all the dimensional variables must be known in terms of the basic dimensions. These are as follows, the way in which they are found from the definition of the variable being indicated in some cases. For example, since:

$$h = \frac{Q}{A(T_w - T_f)} \qquad (1.29)$$

it follows that:

$$\text{Dimensions of } h = \text{Dimensions of } \left[\frac{\text{Energy}}{\text{Area} \times \text{Time} \times \text{Temperature}}\right]$$

$$= \left(\frac{ML^2}{t^2}\right)/L^2 t T$$

$$= M t^{-3} T^{-1}$$

Similarly, it follows that since:

$$k = -\frac{Q}{A} \bigg/ \frac{\partial T}{\partial n}$$

$$\text{Dimensions of } k = \text{Dimensions of } \left[\frac{\text{Energy}}{\text{Area} \times \text{Time} \times (\text{Temperature/Length})}\right]$$

$$= \left(\frac{ML^2}{t^2}\right)\bigg/\left(L^2 t \frac{T}{L}\right)$$

$$= MLt^{-3}T^{-1}$$

The following can be derived using the same approach:

$$\text{Dimensions of } \mu = ML^{-1}t^{-1}$$
$$\text{Dimensions of } \rho = ML^{-3}$$
$$\text{Dimensions of } \ell = L$$
$$\text{Dimensions of } U = Lt^{-1}$$

Further, since

$$\beta = \frac{V - V_i}{V_i(T - T_i)}$$

it follows that β has dimensions $1/T$ so:

$$\text{Dimensions of } \beta g(T_w - T_0) = \left(\frac{1}{T}\right)\left(\frac{L}{t^2}\right)T = Lt^{-2}$$

18 Introduction to Convective Heat Transfer Analysis

Lastly, since

$$c_p = \frac{Q}{m(T_2 - T_1)}$$

it follows that:

$$\text{Dimensions of } c_p = \text{Dimensions of } \left[\frac{\text{Energy}}{\text{Mass} \times \text{Temperature}}\right]$$

$$= \left(\frac{ML^2}{t^2}\right)/(MT)$$

$$= L^2 t^{-2} T - 1$$

Substituting the relevant of the above results into Eq. (1.25) then shows that

$$\text{Dimensions of } \pi_1 = (Mt^{-3}T^{-1})(MLt^{-3}T^{-1})^{a_1}(ML^{-1}t^{-1})^{b_1}(ML^{-3})^{c_1}(L)^{d_1}$$
$$= M^{1+a_1+b_1+c_1} L^{a_1-b_1-3c_1+d_1} t^{-3-3a_1-b_1} T^{-1-a_1} \quad (1.30)$$

But the π's are dimensionless so it is known that:

$$\text{Dimensions of } \pi_1 = M^0 L^0 t^0 T^0 \quad (1.31)$$

Equating the indices of the various dimensions on the right-hand sides of Eqs. (1.30) and (1.31) then gives:

$$\begin{aligned}
\text{For } M: & \quad 1 + a_1 + b_1 + c_1 = 0 \\
\text{For } L: & \quad a_1 - b_1 - 3c_1 + d_1 = 0 \\
\text{For } t: & \quad -3 - 3a_1 - b_1 = 0 \\
\text{For } T: & \quad -1 - a_1 = 0
\end{aligned} \quad (1.32)$$

Solving between these equations then gives:

$$a_1 = -1, b_1 = 0, c_1 = 0, d_1 = 1 \quad (1.33)$$

Therefore, the first dimensionless product, π_1, must be given by:

$$\pi_1 = (h\ell/k) = Nu, \text{ the Nusselt number} \quad (1.34)$$

If a similar procedure is applied to the other three π's the following are obtained:

$$\pi_2 = (U\ell\rho/\mu) = (U\ell/\nu) = Re, \text{ the Reynolds number} \quad (1.35)$$

$$\pi_3 = \beta g(T_w - T_0)\rho^2 \ell^3/\mu^2 = \beta g(T_w - T_f)\ell^3/\nu^2$$
$$= Gr, \text{ the Grashof number} \quad (1.36)$$

$$\pi_4 = (c_p\mu/k) = Pr, \text{ the Prandtl number} \quad (1.37)$$

where in these equations $\nu = \mu/\rho$ is the kinematic viscosity.

Therefore, in the convection problem, in general:

$$\text{function } (Nu, Re, Gr, Pr) = 0 \qquad (1.38)$$

which can be rewritten as:

$$Nu = \text{function } (Re, Gr, Pr) \qquad (1.39)$$

Thus, in order to correlate experimental heat transfer results for a particular geometrical situation, only the four dimensionless variables Nu, Re, Gr, and Pr need to be used. This conclusion can be reached in other ways, it being demonstrated in the next chapter by using the governing differential equations and considering the conditions under which dynamic similarity will exist.

If the buoyancy forces have a negligible effect on the flow, i.e., if the flow is purely forced convective, then Eq. (1.39) reduces to:

$$Nu = \text{function } (Re, Pr) \qquad (1.40)$$

Similarly, if there is no forced velocity or if it is negligibly small, the flow is purely free convective and:

$$Nu = \text{function } (Gr, Pr) \qquad (1.41)$$

Available experimental data for various flow situations have been correlated in terms of the above dimensionless variables and the results fitted by empirical equations or simply presented in graphical form. Despite the reduction in the number of significant variables achieved by the introduction of the dimensionless products, a considerable amount of experimental work has to be carried out in order to arrive at a useable correlation for most geometries. For this reason it is usually worthwhile to attempt to carry out some form of analytical or numerical solution of the problem, even if the solution is a very simplified one, because this solution may indicate the general form of the correlation equation or, at least, indicate where the major emphasis should be placed in the experimental program.

In the above analysis, the dependence of h on the fluid properties μ, ρ, k, c_p, and β was considered. Now, in general, these properties are all temperature dependent. (They are also, in general, pressure dependent, but this is not for the moment being considered, attention being restricted to incompressible flow.) Therefore, since in any problem involving heat transfer, the fluid temperature will vary through the flow field, the question arises as to how this variation can be accounted for. It has been found both experimentally and numerically that, provided the overall temperature change is not very great, this can be done by evaluating these fluid properties at some suitable mean temperature. In the case of external flows, this mean temperature is usually taken as the average of the surface and fluid temperatures, i.e., as $(T_w + T_f)/2$, while in the case of internal flows it is usually adequate to evaluate the fluid properties at the bulk temperature.

The above analysis was based on the assumption that only one length scale, ℓ, was needed to define the problem. This is, however, not always true. Consider for example, heat transfer from the walls of short pipes, as shown in Fig. 1.17, to a fluid flowing through them.

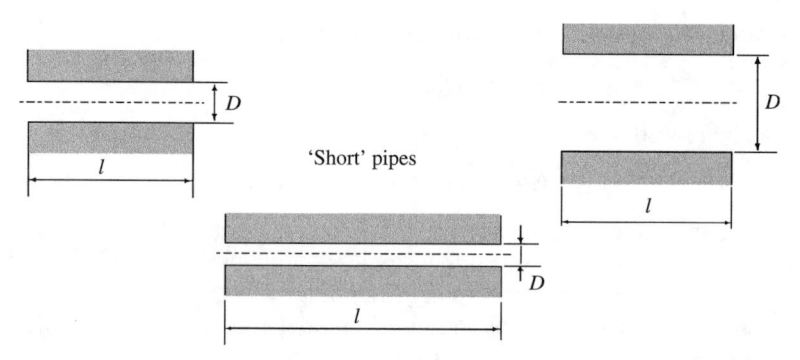

FIGURE 1.17
Series of short pipes.

In this case, the average heat transfer coefficient for the pipe will depend on both ℓ and D. If dimensionless analysis is applied to this problem, assuming for simplicity that the buoyancy force effects are negligible, then one possible result would be:

$$Nu_D = \text{function } [Re_D, Pr, (\ell/D)] \qquad (1.42)$$

where Nu_D and Re_D are the Nusselt and Reynolds numbers respectively based on the diameter D. Thus, in cases such as this, another dimensionless product of the form (ℓ/D) is required.

Other factors beside those considered above can influence the value of h. For example, as already mentioned, in high speed gas flows, the compressibilty of the gas, i.e., the changes in density caused by the changes in the pressure in the flow, can become important. In this case, h will also be dependent on the speed of sound in the gas, a. This will mean that an extra dimensionless variable will be involved, this being the Mach number, M, defined by:

$$U/a = M, \text{ the Mach number} \qquad (1.43)$$

Buoyancy force effects are seldom important when compressibilty effects are important so in high speed gas flows:

$$Nu = \text{function } [Re, Pr, M] \qquad (1.44)$$

In free surface flows such as that shown in Fig. 1.18 the gravitational acceleration g has an effect on h and this will again mean that an extra dimensionless variable will be involved, this being the Froude number, Fr, defined by:

$$g\ell/U^2 = Fr, \text{ the Froude number} \qquad (1.45)$$

In this case, if buoyancy forces are negligible,

$$Nu = \text{function } [Re, Pr, Fr] \qquad (1.46)$$

The surface tension can also sometimes influence the heat transfer rate and in this case another new dimensionless number will be involved in defining the Nusselt number.

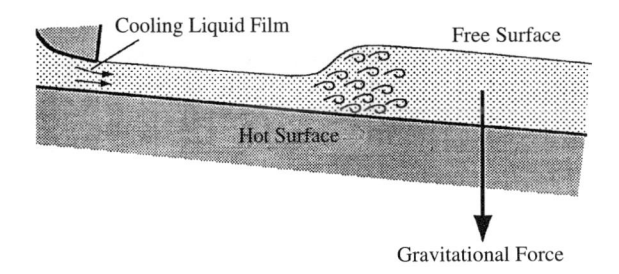

FIGURE 1.18
Flow with a free surface.

EXAMPLE 1.3. In the jet cooling of a surface, a cooling fluid is discharged at a mean velocity of U from a nozzle with a diameter, D, onto the surface that is to be cooled. If the distance of the nozzle from the surface is H, and if the characteristic size of the surface is ℓ, and if the jet is discharged at an angle, ϕ, to the surface, write down the dimensionless variables that will be involved in describing the heat transfer rate in this situation. Buoyancy effects (i.e., the effects of the density changes due to the temperature changes) may be ignored but the overall effect of the gravity on the flow should be included.

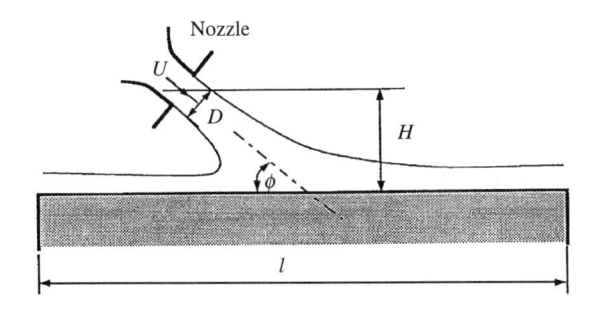

FIGURE E1.3

Solution. In this case the mean heat transfer coefficient for the surface on which the jet is impinging, h, will depend on:

- The conductivity, k, of the fluid
- The viscosity, μ, of the fluid
- The specific heat, c_p, of the fluid
- The density, ρ, of the fluid
- The size of the surface, ℓ
- The magnitude of the initial jet velocity, U
- The nozzle diameter, D
- The nozzle to surface distance, H
- The angle at which the jet is discharged, ϕ
- The gravitational acceleration, g

Hence:

$$h = \text{function}\,[k, \mu, c_p, \rho, \ell, U, D, H, \phi, g]$$

which can be written as:

$$f[h, k, \mu, c_p, \rho, \ell, U, D, H, \phi, g] = 0$$

22 Introduction to Convective Heat Transfer Analysis

There are, thus, 11 dimensional variables involved in the problem. The number of dimensionless variables required is therefore (11 − the number of basic dimensions), i.e., 11 − 4 = 7. Each of these dimensionless numbers must contain one prime dimensional variable which does not occur in any of the other dimensionless variables together with other dimensional variables which occur in all of the dimensionless variables. In the present problem then, seven variables must be selected as prime variables. The following will be used here as prime variables: h, U, c_p, ℓ, H, ϕ, and g. The dimensionless parameters will then have the form:

$$\pi_1 = [h k^{a_1} \mu^{b_1} \rho^{c_1} D^{d_1}]$$
$$\pi_2 = [U k^{a_2} \mu^{b_2} \rho^{c_2} D^{d_2}]$$
$$\pi_3 = [c_p k^{a_3} \mu^{b_3} \rho^{c_3} D^{d_3}]$$
$$\pi_4 = [\ell k^{a_4} \mu^{b_4} \rho^{c_4} D^{d_4}]$$
$$\pi_5 = [H k^{a_5} \mu^{b_5} \rho^{c_5} D^{d_5}]$$
$$\pi_6 = [\phi k^{a_6} \mu^{b_6} \rho^{c_6} D^{d_6}]$$
$$\pi_7 = [g k^{a_7} \mu^{b_7} \rho^{c_7} D^{d_7}]$$

It will be noted that the nozzle diameter has been used as the characteristic length scale that occurs in all the π's.

From the discussion given above it will be obvious that:

$$\pi_1 = (hD/k) = Nu, \text{ the Nusselt number}$$
$$\pi_2 = (UD\rho/\mu) = Re, \text{ the Reynolds number}$$
$$\pi_3 = (c_p \mu/k) = Pr, \text{ the Prandtl number}$$

Because both π_4 and π_5 have a length as the prime variable, it should be clear that:

$$\pi_4 = (\ell/D)$$
$$\pi_5 = (H/D)$$

Next consider π_6. Because ϕ is an angle it is dimensionless. Hence:

$$\pi_6 = \phi$$

Lastly consider π_7. Because the dimensions of g are L/t^2, it will be clear that $a_7 = 0$ because no other term will involve energy. Substituting for the dimensions of the other terms as discussed above then gives:

$$b_7 = -2, \quad c_7 = 2, \quad d_7 = 3$$

Hence:

$$\pi_7 = (g\rho^2 D^3/\mu^2) = (gD/U^2)(\rho UD/\mu)^2 = (gD/U^2)Re^2$$

Because the Reynolds number Re has already been taken as a governing dimensionless number, π_7 will be taken as U^2/gD i.e., as the Froude number, Fr. In some ways the use of U in this dimensionless number contradicts the use of U as a prime variable. However, this is not really a problem because the gravitational acceleration can only affect the heat transfer if there is a flow, i.e., if U is nonzero. Hence:

$$\text{function } [Nu, Re, Pr, (\ell/D), (H/D), Fr] = 0 \tag{1.38}$$

which can be rewritten as

$$Nu = \text{function } [Re, Pr, (\ell/D), (H/D), Fr] \tag{1.39}$$

CHAPTER 1: Introduction   23

TABLE 1.2
Dimensionless numbers

Name	Symbol	Definition	Interpretation
Brinkman number	Br	$V^2\mu/k(T_w - T_f)$	Dissipation/heat transfer rate
Eckert number	Ec	$V^2/c_p(T_w - T_f)$	Kinetic energy/enthalpy change
Froude number	Fr	V^2/gL	Inertial force/gravitational force
Grashof number	Gr	$\beta g(T_w - T_f)L^3/\nu^2$	Buoyancy force/viscous force
Mach number	M	V/a	Velocity/velocity of propagation of weak disturbances
Nusselt number	Nu	hL/k	Convective heat transfer rate/conduction heat transfer rate
Peclet number	Pe	VL/α	Re Pr
Prandtl number	Pr	ν/α	Rate of diffusion of viscous effects/rate of diffusion of heat
Rayleigh number	Ra	$\beta g(T_w - T_f)L^3/\nu\alpha$	Gr Pr
Reynolds number	Re	VL/ν	Inertia force/viscous force
Stanton number	St	$h/\rho c_p V$	Nu/Re Pr
Weber number	We	$\rho V^2 L/\sigma$	Inertia force/surface tension force

A list of dimensionless numbers that can arise in the analysis of convective heat transfer is given in Table 1.2.

1.6
PHYSICAL INTERPRETATION OF THE DIMENSIONLESS NUMBERS

The dimensionless parameters, such as the Nusselt and Reynolds numbers, can be thought of as measures of the relative importance of of certain aspects of the flow. For example, if the flow through an area dA is considered, as shown in Fig. 1.19, the rate momentum passes through this area is equal to the mass flow rate times the velocity, i.e., equal to $\dot{m}V\,dA$, i.e., equal to $\rho VV\,dA$, i.e., equal to $\rho V^2\,dA$. If, therefore, U is a "measure" of the velocity, the quantity ρU^2 is a "measure" of the magnitude of the momentum flux in the flow. This quantity is often termed the "inertia force". Further, since the Newtonian viscosity law indicates that the viscous shear stresses

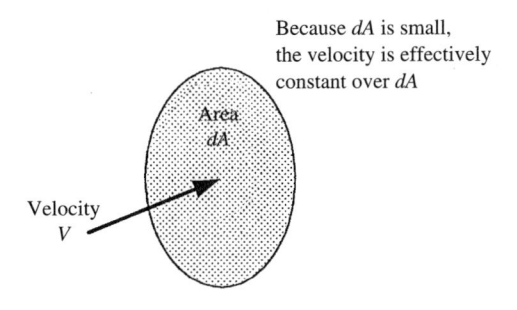

FIGURE 1.19
Flow area considered.

in the flow are given by an equation of the form $\tau = \mu(\partial U/\partial n)$, n being any chosen direction, it follows that the quantity $\mu U/\ell$ is a measure of the magnitude of the viscous stresses in the flow.

Now, the Reynolds number can be written as:

$$Re = \frac{\rho U \ell}{\mu} = \frac{\rho U^2 \ell}{\mu U/\ell}$$

i.e., the Reynolds number is a measure of the ratio of the magnitude of the inertia forces in the flow to the magnitude of the viscous forces in the flow. This means, therefore, that if the Reynolds number is relatively low, the viscous forces are high compared to the inertia forces and any disturbances that arise in the flow will tend to be damped out by the action of viscosity and laminar flow will tend to exist. If the Reynolds number is relatively high, however, the viscous forces are low compared to the inertia forces and any disturbances that arise in the flow will tend to grow, i.e., turbulent flow will tend to develop.

Next, consider the Nusselt number. The convective heat transfer from a surface will depend on the magnitude of $h(T_w - T_f)$. Also, if there was no flow, i.e., if the heat transfer was purely by conduction, Fourier's law indicates that the quantity $k(T_w - T_f)/\ell$ would be a measure of the heat transfer rate. Now, the Nusselt number can be written as:

$$Nu = \frac{h\ell}{k} = \frac{h(T_w - T_f)}{k(T_w - T_f)/\ell}$$

i.e., the Nusselt number is a measure of the ratio of the magnitude of the convective heat transfer rate to the magnitude of the heat rate that would exist in the same situation with pure conduction.

To obtain a similar interpretation of the Grashof number, it is necessary to obtain a measure of the magnitude of the flow velocity induced by the buoyancy forces. Now, as discussed above, a measure of the magnitude of the buoyancy forces in the flow per unit volume is the quantity $\beta g \rho(T_w - T_f)$. Therefore a measure of the work done by the buoyancy forces on the fluid as it flows over the body, i.e., a measure of the product of the buoyancy forces and the distance over which they act, will be $\beta g \rho(T_w - T_f)\ell$. But the work done by the buoyancy forces results in the fluid gaining kinetic energy. If U_b is a measure of velocity induced by the buoyancy forces then a measure of the kinetic energy per unit volume will be $\rho U_b^2/2$. Therefore, a measure of the magnitude of U_b is $\sqrt{\beta g(T_w - T_f)\ell}$. For the reasons discussed when considering the Reynolds number, a measure of the ratio of the inertia forces to the viscous forces in natural convection is the quantity:

$$\frac{\rho U_b^2}{\mu U_b/\ell} = \frac{\rho U_b \ell}{\mu}$$

i.e., using the expression for U_b derived above, a measure of of the ratio of the inertia forces to the viscous forces is the quantity:

$$\frac{\sqrt{\beta g(T_w - T_f)\ell}\, \rho \ell}{\mu} = \sqrt{\beta g(T_w - T_f)\rho^2 \ell^3/\mu^2} = Gr^{0.5}$$

CHAPTER 1: Introduction   25

Hence, the Grashof number is a measure of the magnitude of the ratio of the inertia forces induced in the flow by the buoyancy forces to the viscous stresses, i.e., it is effectively a measure of the magnitude of the ratio of the buoyancy forces to the viscous forces in the flow.

Attention will, lastly, be given to the Prandtl number. Consider a steady flow over a surface which is at a different temperature from the fluid flowing over the surface, the situation considered being shown in Fig. 1.20.

The viscous stress acting on the surface causes a reduction in the momentum of the fluid flow near the surface. If, as shown in Fig. 1.20, Δ_U is a measure of the thickness of the layer in the flow over which the effects of this momentum decrease occur, then a measure of the momentum decrease will be the product of the mass flux through this layer and the initial fluid velocity, U, i.e., will be $\rho U \Delta_U W \times U$, W being a measure of the width of the body. This momentum decrease is caused by the viscous force acting on the surface. A measure of this force will be the product of a measure of the shear stress acting on the surface and a measure of the area of the surface. A measure of the shear stress is $\mu U / \Delta_U$ and a measure of the surface area is ℓW. Hence, equating this measure of the shear force to the measure of the momentum decrease gives:

$$\mu \frac{U}{\Delta_U} \ell W = \rho U^2 \Delta_U W$$

This can be rearranged to give:

$$\frac{\Delta_U}{\ell} = \sqrt{\frac{\mu}{U \rho \ell}} = \frac{1}{\sqrt{Re}} \tag{1.47}$$

Next, consider the enthaply changes in the flow that result from the heat transfer from the surface. If, as shown in Fig. 1.18, Δ_T is a measure of the thickness of the layer in the flow over which the effects of this enthalpy change occur then a measure of the enthalpy change will be the product of the mass flux through this layer and a measure of the enthalpy change per unit mass. The fluid has a temperature T_f ahead of the body and a measure of the surface temperature is T_w. Therefore, a measure of the enthalpy change per unit mass will be $c_p(T_W - T_f)$. Hence, a measure of the enthalpy change will be $\rho U \Delta_T W \times c_p(T_W - T_f)$. This enthalpy change is caused by the heat transfer at the surface. A measure of the heat transfer rate at the surface

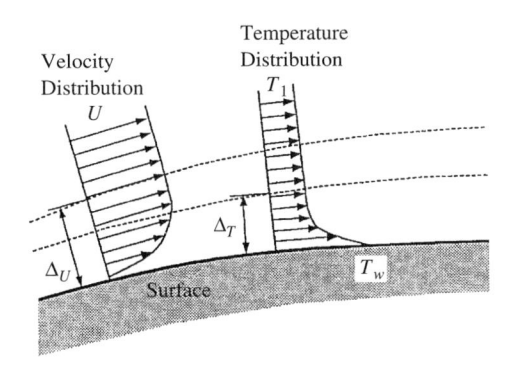

FIGURE 1.20
Momentum and temperature changes induced by presence of surface.

will be the product of a measure of the heat transfer rate per unit area at the surface and a measure of the area of the surface. A measure of the heat transfer rate per unit area from the surface is, by virtue of Fourier's law, $k(T_W - T_f)/\Delta_T$ and a measure of the surface area is ℓW. Hence, equating this measure of the surface heat transfer rate to the measure of the enthalpy change gives:

$$k \frac{(T_W - T_f)}{\Delta_T} \ell W = \rho U \Delta_T W \times c_p (T_W - T_f)$$

This can be rearranged to give:

$$\frac{\Delta_T}{\ell} = \sqrt{\frac{k}{c_p U \rho \ell}} = \sqrt{\frac{k}{\mu c_p}} \frac{1}{\sqrt{Re}} = \frac{1}{\sqrt{Pr\, Re}}$$

Dividing Eq. (1.47) by this equation gives:

$$\frac{\Delta_U}{\Delta_T} = \sqrt{Pr}$$

Hence, the Prandtl number is a measure of the ratio of the rate of spread of the effects of momentum changes in the flow to the rate of spread of the effects of temperature differences, i.e., of the effects of heat transfer in the flow.

Similar interpretations of the other commonly used dimensionless numbers can be derived. Some of these interpretations are given in Table 1.2.

1.7 FLUID PROPERTIES

The calculation of heat transfer rates will almost invariably require the determination of some or all of the fluid properties ρ, μ (or ν), k, β, c_p (or Pr), $\nu = \mu/\rho$ being the kinematic viscosity. These properties are all, in general, temperature and pressure dependent although the effect of pressure changes on μ, k, β, and c_p is often negligible. These properties are usually listed for various fluids in tabular form [1],[2],[20],[21], tables of air and water properties being given in the Appendix. The variation of some of the properties of air and water with temperature are shown in Figs. 1.21 and 1.22.

In many circumstances, it is more convenient to have equations describing the variations of these properties such equations being available for many common fluids [21].

1.8 CONCLUDING REMARKS

This chapter has been concerned with the meaning of "convective heat transfer" and the "heat transfer coefficient". The dimensionless variables on which convective heat transfer rates depend have also been introduced using dimensional analysis.

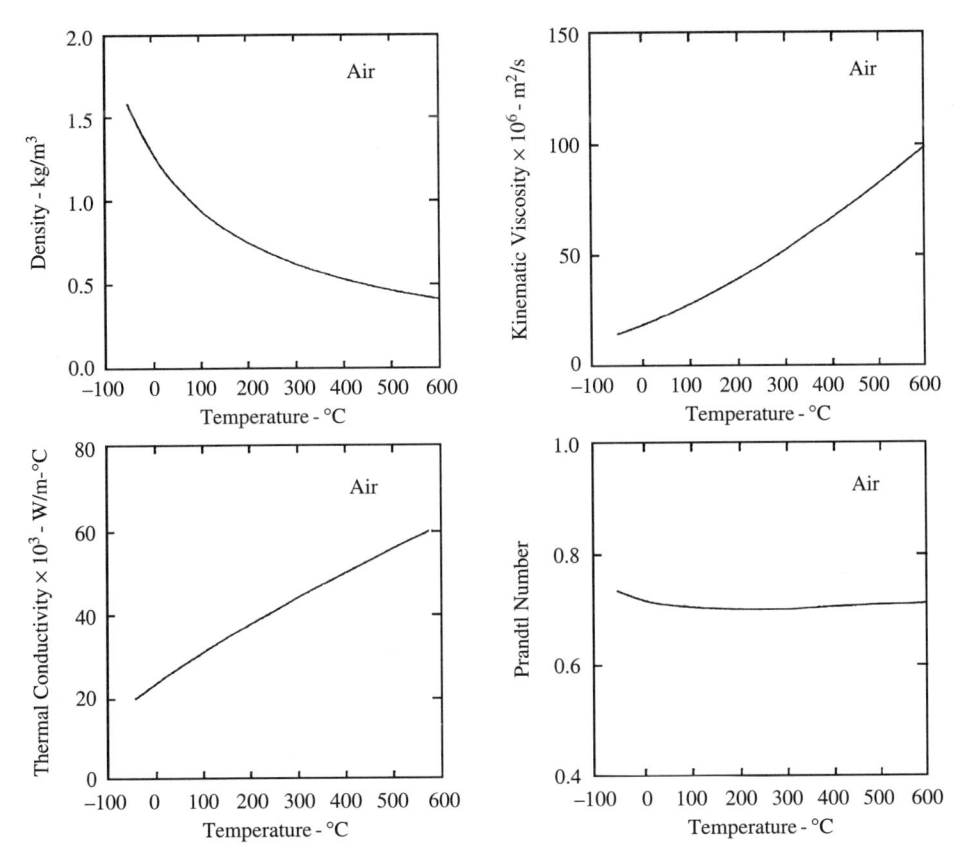

FIGURE 1.21
Variation of some properties of air at standard pressure with temperature.

PROBLEMS

1.1. Discuss briefly SIX practical situations in which convective heat transfer occurs.

1.2. Air flows through a plane duct, i.e., effectively between two large plates, with a width of $2w$. The flow can be assumed to be the same at all sections across the duct. The velocity distribution in the duct is approximately given by $u = 12[(w-y)/w]^{1/5}$ m/s, y being the distance from the center-plane to the point at which u is the velocity and the temperature distribution in the flow is approximately given by $60 - 40[(w-y)/w]^{1/5}$ °C. Find the mean temperature in the flow.

1.3. In an electrical heater, energy is generated in a 1.5-mm diameter wire at a rate of 1200 W. Air at a temperature of 30°C is blown over the wire at a velocity that gives a mean heat transfer coefficient of 50 W/m²°C. If the surface temperature of the wire is not to exceed 200°C, find the length of the wire.

1.4. Typical values of the mean heat transfer coefficient for a variety of situations were listed in Table 1.1. Discuss some of the physical reasons why these values vary so greatly from one situation to another.

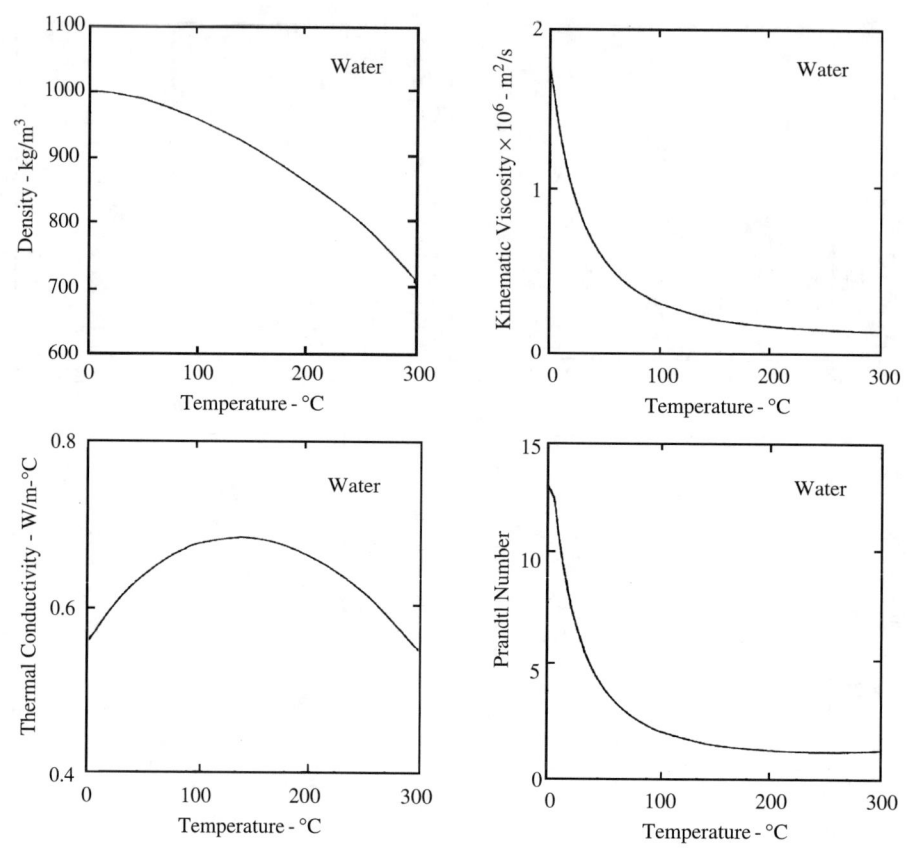

FIGURE 1.22
Variation of some properties of water with temperature.

1.5. In some forced convective flows it has been found that the Nusselt number is approximately proportional to the square root of the Reynolds number. If, in such a flow, it is found that h has a value of 15 W/m²K when the forced velocity has a magnitude of 5 m/s, find the heat transfer coefficient if the forced velocity is increased to 40 m/s.

1.6. The surface tension σ has the dimensions force per unit length. Consider a convective heat transfer situation in which surface tension is important. Show that the additional quantity $\rho U^2 \ell / \sigma$, termed the Weber number, We, is required to determine the Nusselt number.

1.7. If, in the case of gas flows, the heat transfer coefficient is assumed to depend on the speed of sound, a, in the gas in addition to the variables considered in this chapter, find the additional dimensionless number on which the Nusselt number will depend.

1.8. Discuss the physical interpretation of (i) the Weber number and (ii) the Stanton number.

1.9. Discuss why dimensionless variables are used in presenting convective heat transfer results.

1.10. Using the results given in the Appendix, draw a graph showing how the density at ambient pressure, ρ, the coefficient of viscosity, μ, the thermal conductivity, k, and the Prandtl number, Pr, of air vary with temperature for temperatures between 0 and 500°C.

1.11. Using the definition of β in conjunction with the perfect gas law, show that for a perfect gas, $\beta = 1/T$.

1.12. It is often assumed that for gases:

$$\frac{\mu}{\mu_{\text{ref}}} = \left(\frac{T}{T_{\text{ref}}}\right)^n$$

where T is the absolute temperature and μ_{ref} is the viscosity of the gas at some chosen temperature T_{ref}. Using the properties of air given in the Appendix and taking T_{ref} as 273 K, plot the variation of the index, n, for air with temperature for temperatures between -50 and $+200$°C.

1.13. Consider heat transfer from a circular cylinder whose axis is normal to a forced flow and which is rotating at an angular velocity, ω. If the surface of the cylinder is maintained at a uniform temperature, find the dimensionless parameters on which the Nusselt number depends.

1.14. One way of varying the the Grashof number in experimental studies involving air is to keep the model size fixed and to vary the air pressure. Consider a situation in which the model height is 5 cm, the model surface is at a temperature of 60°C, and the air at a temperature of 20°C. Plot the variation of Grashof number with air pressure for air pressures between 0.1 and 10 times standard atmospheric pressure.

1.15. Discuss the physical meaning of the Eckert number.

REFERENCES

1. Incropera, F.P. and DeWitt, D.P., *Fundamentals of Heat and Mass Transfer*, 3rd ed., Wiley, New York, 1990.
2. Mills, A.F., *Heat Transfer*, Irwin, Homewood, IL, 1992.
3. Holman, J.P., *Heat Transfer*, 5th ed., McGraw-Hill, New York, 1981.
4. Gebhart, B., *Heat Transfer*, 2nd ed., McGraw-Hill, New York, 1971.
5. Burmeister, L.C., *Convective Heat Transfer*, 2nd ed., Wiley-Interscience, New York, 1993.
6. Bejan, A., *Convection Heat Transfer*, 2nd ed., Wiley, New York, 1995.
7. Kakac, S. and Yener, Y., *Convective Heat Transfer*, 2nd ed., CRC Press, Boca Raton, FL, 1995.
8. Kays, W.M. and Crawford, M.E., *Convective Heat and Mass Transfer*, 3rd ed., McGraw-Hill, New York, 1993.
9. Kaviany, M., *Principles of Convective Heat Transfer*, Springer-Verlag, New York, 1994.
10. Arpaci, V.S. and Larsen, P.S., *Convective Heat Transfer*, Prentice-Hall, Englewood Cliffs, NJ, 1984.
11. King, W.J., "The Basic Laws and Data of Heat Transmission", *Mech. Eng.*, 54, 410-415, 1932.

12. Layton, E.T., Jr. and Lienhard, J.H., Editors, *History of Heat Transfer,* American Society of Mechanical Engineers, New York, 1988.
13. McAdams, W.H., *Heat Transmission,* 3rd ed., McGraw-Hill, New York, 1954.
14. Jakob, M., *Heat Transfer,* Vols. 1 and 2, Wiley, New York, 1949, 1957.
15. Jakob, M. and Hawkins, G.A., *Elements of Heat Transfer,* 3rd ed., Wiley, New York, 1957.
16. Bridgeman, P.W., *Dimensional Analysis,* Yale University Press, New Haven, CT, 1931.
17. Kreith, F. and Black, W.Z.M, *Basic Heat Transfer,* Harper & Row, New York, 1980.
18. Langhaar, H.L., *Dimensional Analysis and Theory of Models,* Wiley, New York, 1951.
19. Van Driest, E.R., "On Dimensional Analysis and the Presentation of Data in Fluid Flow Problems", *J. Appl. Mech.,* 13, A-34, 1940.
20. Chapman, A.J., *Heat Transfer,* 3rd ed., Macmillan, New York, 1974.
21. Ried, R.C., Prausnitz, J.M., and Poling, B. E., *The Properties of Gases and Liquids,* McGraw-Hill, New York, 1987.

CHAPTER 2

The Equations of Convective Heat Transfer

2.1 INTRODUCTION

In order to be able to predict convective heat transfer rates there are, in general, three variables whose distributions through the flow field must be determined. These variables (see Fig. 2.1) are:

- The pressure, p
- The velocity vector, $\vec{V}$
- The temperature, T

Once the distributions of these quantities are determined, the variation of any other required quantity, such as the heat transfer rate, can be obtained.

In order to determine the distributions of pressure, velocity, and temperature the principles of conservation of mass, conservation of momentum (Newton's Law) and conservation of energy (first law of Thermodynamics) are applied. These conservation principles represent empirical models of the behavior of the physical world. They do not, of course, always apply, e.g., there can be a conversion of mass into energy in some circumstances, but they are adequate for the analysis of the vast majority of engineering problems. These conservation principles lead to the so-called Continuity, Navier-Stokes and Energy equations respectively. These equations involve, beside the basic variables mentioned above, certain fluid properties, e.g., density, ρ; viscosity, μ; conductivity, k; and specific heat, c_p. Therefore, to obtain the solution to the equations, the relations between these properties and the pressure and temperature have to be known. (Non-Newtonian fluids in which μ depends on the velocity field are not considered here.) As discussed in the previous chapter, there are, however, many practical problems in which the variation of these properties across the flow field can be ignored, i.e., in which the fluid properties can be assumed to be constant in obtaining the solution. Such solutions are termed "constant

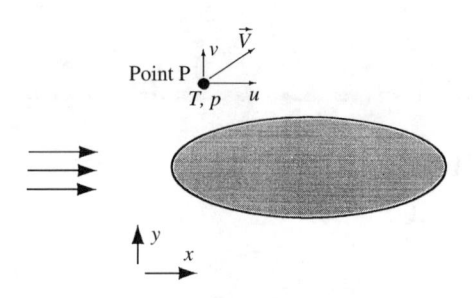

FIGURE 2.1
Variables required to determine convective heat transfer rates.

fluid property" solutions. They are found to adequately describe many flows in which the overall temperature changes are quite large provided that the fluid properties are evaluated at a suitable mean temperature. For this reason, it will be assumed in most of the present chapter and in much of the remainder of this book that the fluid properties are constant. Of course, many of the methods of analysis described here can quite easily be modified to account for variations in fluid properties and some discussion of this will be given in a later chapter.

In the present chapter, it will also be assumed that all body forces are negligible. A discussion of some flows in which buoyancy forces are important will be given in Chapters 8 and 9.

2.2
CONTINUITY AND NAVIER-STOKES EQUATIONS

The equations will here be presented in terms of a cartesian coordinate system, the velocity vector $\vec{V}$ having components u, v, and w in the three coordinate directions x, y, and z, respectively. For the moment, the discussion will also be restricted to steady laminar flow in which the flow variables u, v, w, p, T do not vary with time.

As mentioned above, the continuity equation is obtained by the application of the principle of conservation of mass to the fluid flow. A differentially small rectangular control volume with sides parallel to the three coordinate directions is introduced. This is shown in Fig. 2.2.

Since steady flow is being considered, the rate at which mass leaves this control volume can be set equal to the rate at which it enters the control volume. Taking the

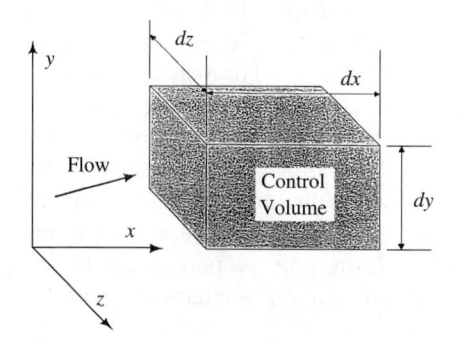

FIGURE 2.2
Control volume considered.

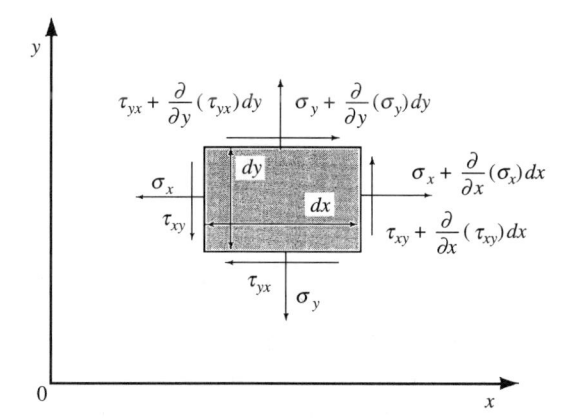

CHAPTER 2: The Equations of Convective Heat Transfer   33

FIGURE 2.3
Viscous stresses acting on control volume.

limiting form of this result as the size of the control volume tends to zero then gives, because density is being assumed constant [1],[2],[3]:

$$\frac{\partial u}{\partial x} + \frac{\partial v}{\partial y} + \frac{\partial w}{\partial z} = 0 \qquad (2.1)$$

This is the continuity equation for constant density flow.

As was mentioned above, the Navier-Stokes equations are obtained by the application of the conservation of momentum principle to the fluid flow. The same control volume that was introduced above in the discussion of the continuity equation is considered and the conservation of momentum in each of the three coordinate directions is separately considered. The net force acting on the control volume in any of these directions is then set equal to the difference between the rate at which momentum leaves the control volume in this direction and the rate at which it enters in this direction. The net force arises from the pressure forces and the shearing forces acting on the faces of the control volume. The viscous shearing forces for two-dimensional flow (see later) are shown in Fig. 2.3. They are expressed in terms of the velocity field by assuming the fluid to be Newtonian and are then given by [4],[5]:

$$\sigma_x = 2\mu \frac{\partial u}{\partial x} \qquad (2.2)$$

$$\sigma_y = 2\mu \frac{\partial v}{\partial y} \qquad (2.3)$$

$$\tau_{xy} = \mu \left(\frac{\partial u}{\partial y} + \frac{\partial v}{\partial x} \right) = \tau_{yx} \qquad (2.4)$$

(These relations are also required in the derivation of the energy equation as discussed later in this chapter.)

Using the expressions for the shearing stresses in the momentum balance for the control volume and taking the limit of the resulting equations as the size of the control volume tends to zero then gives the following set of equations [6],[7],[8],[9]:

34 Introduction to Convective Heat Transfer Analysis

$$u\frac{\partial u}{\partial x} + v\frac{\partial u}{\partial y} + w\frac{\partial u}{\partial z} = -\frac{1}{\rho}\frac{\partial p}{\partial x} + \nu\left(\frac{\partial^2 u}{\partial x^2} + \frac{\partial^2 u}{\partial y^2} + \frac{\partial^2 u}{\partial z^2}\right) \quad (2.5)$$

$$u\frac{\partial v}{\partial x} + v\frac{\partial v}{\partial y} + w\frac{\partial v}{\partial z} = -\frac{1}{\rho}\frac{\partial p}{\partial y} + \nu\left(\frac{\partial^2 v}{\partial x^2} + \frac{\partial^2 v}{\partial y^2} + \frac{\partial^2 v}{\partial z^2}\right) \quad (2.6)$$

$$u\frac{\partial w}{\partial x} + v\frac{\partial w}{\partial y} + w\frac{\partial w}{\partial z} = -\frac{1}{\rho}\frac{\partial p}{\partial z} + \nu\left(\frac{\partial^2 w}{\partial x^2} + \frac{\partial^2 w}{\partial y^2} + \frac{\partial^2 w}{\partial z^2}\right) \quad (2.7)$$

These are the Navier-Stokes equations for steady constant fluid property flow. They are sometimes termed the x-, y-, and z-momentum equations, respectively. Basically these equations state that the net rate of change of momentum per unit mass in any direction (the left-hand side) is the sum of the net pressure force and the net viscous force in that direction.

It should be clearly noted that since the fluid properties are being assumed constant, the set of eqs. (2.5) to (2.7) is independent of the temperature field and can, therefore, be solved independently of the energy equation to give the velocity and pressure distribution.

Many practical flows can be represented with sufficient accuracy by assuming the flow to be two-dimensional, i.e., by assuming that the axes can be so positioned that one of the velocity components, here always taken as w, is effectively zero. For this reason and because the study of two-dimensional flows forms the basis of the study of more complex three-dimensional flows, most of the analyses discussed in later chapters will be for two-dimensional flows. Setting w equal to zero gives the following as the equations governing the velocity and pressure fields in steady two-dimensional flow:

$$\frac{\partial u}{\partial x} + \frac{\partial v}{\partial y} = 0 \quad (2.8)$$

$$u\frac{\partial u}{\partial x} + v\frac{\partial u}{\partial y} = -\frac{1}{\rho}\frac{\partial p}{\partial x} + \nu\left(\frac{\partial^2 u}{\partial x^2} + \frac{\partial^2 u}{\partial y^2}\right) \quad (2.9)$$

$$u\frac{\partial v}{\partial x} + v\frac{\partial v}{\partial y} = -\frac{1}{\rho}\frac{\partial p}{\partial y} + \nu\left(\frac{\partial^2 v}{\partial x^2} + \frac{\partial^2 v}{\partial y^2}\right) \quad (2.10)$$

In some problems it is more convenient to have the flow equations expressed in terms of other coordinate schemes. One of the most important of these, and the only other one that will be used in the present book, is the cylindrical coordinate system which is defined in Fig. 2.4. Here the position of any point being considered is defined in terms of the coordinates z, r, ϕ and the velocity components in these three directions are here designated $u, v,$ and w, respectively.

In this coordinate system, the continuity equation becomes:

$$\frac{\partial u}{\partial z} + \frac{1}{r}\frac{\partial}{\partial r}(vr) + \frac{1}{r}\frac{\partial w}{\partial \phi} = 0 \quad (2.11)$$

CHAPTER 2: The Equations of Convective Heat Transfer   35

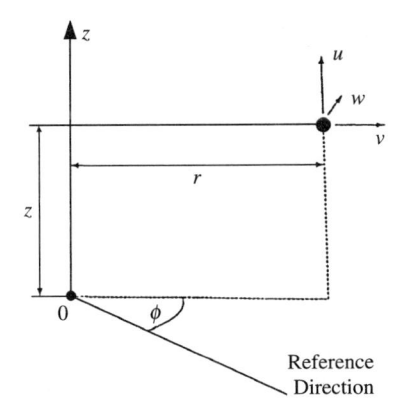

FIGURE 2.4
Cylindrical coordinate system.

In expressing the Navier-Stokes equations in these coordinates it is convenient to define:

$$\frac{D}{Dt} = u\frac{\partial}{\partial z} + v\frac{\partial}{\partial r} + \frac{w}{r}\frac{\partial}{\partial \phi} \qquad (2.12)$$

and

$$\nabla^2 = \frac{\partial^2}{\partial r^2} + \frac{1}{r}\frac{\partial}{\partial r} + \frac{1}{r^2}\frac{\partial^2}{\partial \phi^2} + \frac{\partial^2}{\partial z^2} \qquad (2.13)$$

Eq. (2.12) is used because steady flow is being considered.

Using this notation, the Navier-Stokes equations become in cylindrical coordinates:

$$\frac{Du}{Dt} = -\frac{1}{\rho}\frac{\partial p}{\partial z} + \nu\nabla^2 u \qquad (2.14)$$

$$\frac{Dv}{Dt} - \frac{w^2}{r} = -\frac{1}{\rho}\frac{\partial p}{\partial r} + \nu\left[\nabla^2 v - \frac{v}{r^2} - \frac{2}{r^2}\frac{\partial w}{\partial \phi}\right] \qquad (2.15)$$

$$\frac{Dw}{Dt} - \frac{vw}{r} = -\frac{1}{\rho r}\frac{\partial p}{\partial \phi} + \nu\left[\nabla^2 w + \frac{2}{r^2}\frac{\partial v}{\partial \phi} - \frac{w}{r^2}\right] \qquad (2.16)$$

Expressions for the flow equations in terms of general orthogonal coordinates will be found in books on fluid mechanics.

2.3
THE ENERGY EQUATION FOR STEADY FLOW

The present book is concerned with methods of predicting heat transfer rates. These methods basically utilize the continuity and momentum equations to obtain the velocity field which is then used with the energy equation to obtain the temperature field from which the heat transfer rate can then be deduced. If the variation of fluid properties with temperature is significant, the continuity and momentum equations must be solved simultaneously with the energy equation. As explained above, there are, however, many cases in which this fluid property variation is negligible. In such cases the velocity field can be solved for independently of the energy equation and many such solutions are given in books on viscous flow theory. These solutions can then be used with the energy equation to derive heat transfer rates.

The prediction of convective heat transfer rates will, however, always involve the solution of the energy equation. Therefore, because of its fundamental importance in the present work, a discussion of the way in which the energy equation is derived will be given here [2],[3],[5],[7]. For this purpose, attention will be restricted to two-dimensional, incompressible flow.

Consider flow through a rectangular control volume having sides of length dx and dy parallel to the x- and y-axes, respectively, and having unit depth, this control volume being shown in Fig. 2.5. (The equation can also be derived by using a closed system, i.e., using an element of fluid containing a fixed identifiable mass of fluid.)

For this control volume, the steady flow energy equation requires that:

$$\begin{matrix} \text{Rate at which} \\ \text{enthalpy leaves} \\ \text{control volume} \end{matrix} + \begin{matrix} \text{Rate at which} \\ \text{kinetic energy} \\ \text{leaves control volume} \end{matrix} - \begin{matrix} \text{Rate at which} \\ \text{enthalpy enters} \\ \text{control volume} \end{matrix} - \begin{matrix} \text{Rate at which} \\ \text{kinetic energy} \\ \text{enters control} \\ \text{volume} \end{matrix} \quad (2.17)$$

$$= \begin{matrix} \text{Net rate at which} \\ \text{heat is transferred} \\ \text{into control volume} \end{matrix} + \begin{matrix} \text{Net rate at which} \\ \text{work is done} \\ \text{on control volume} \end{matrix}$$

The left-hand side of this equation will be considered first. It is convenient to define:

$$I = \text{specific enthalpy} + \text{specific kinetic energy} = c_p T + (u^2 + v^2)/2 \quad (2.18)$$

I is sometimes called the total enthalpy.

Consider the x-direction. The rates at which the total enthalpy enters and leaves the control volume in the x-direction are as shown in Fig. 2.6.

If $\dot{m}$ is the mass flow rate through ab, the difference between the rate at which the sum of enthalpy and kinetic energy leave and enter the control volume in the

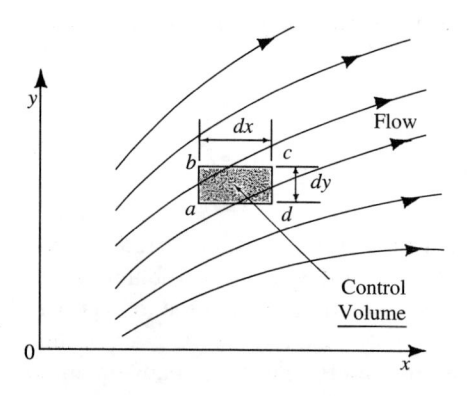

FIGURE 2.5
Control volume considered.

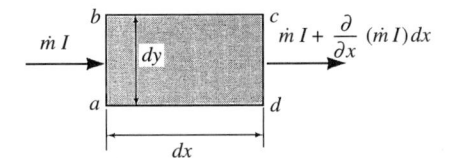

CHAPTER 2: The Equations of Convective Heat Transfer   37

FIGURE 2.6
Enthalpy flows into and out of control volume.

x-direction is, since dx is assumed to be small, given by:

$$\dot{m}I + \frac{\partial}{\partial x}(\dot{m}I)\,dx - \dot{m}I = \frac{\partial}{\partial x}(\dot{m}I)\,dx \tag{2.19}$$

Since the control volume has unit width

$$\dot{m} = \rho u\,dy \tag{2.20}$$

Eq. (2.19) can, therefore, be written as:

$$\left[\frac{\partial}{\partial x}(\rho u I)\right] dx\,dy \tag{2.21}$$

Exactly the same arguments can be applied in the y-direction and it follows from this that the difference between the rate at which the sum of enthalpy and kinetic energy leaves and enters the control volume in the y-direction is given by:

$$\left[\frac{\partial}{\partial y}(\rho v I)\right] dx\,dy \tag{2.22}$$

Adding Eqs. (2.21) and (2.22) and using Eq. (2.18) then shows that the left-hand side of Eq. (2.17) is equal to:

$$\frac{\partial}{\partial x}\left[\rho u\left\{c_p T + \left(\frac{u^2 + v^2}{2}\right)\right\}\right] dx\,dy + \frac{\partial}{\partial y}\left[\rho v\left\{c_p T + \left(\frac{u^2 + v^2}{2}\right)\right\}\right] dx\,dy \tag{2.23}$$

Next, consider the heat transfer term on the right-hand side of Eq. (2.17). Heat is transferred into the control volume by conduction through its surface. The x- and y-directions are again separately considered. If Q_x is the rate at which heat is transferred into the volume through the face, ab (see Fig. 2.7), then the difference between the rate at which heat is conducted into the control volume in the x-direction and the rate at which it is conducted out in this direction is given by:

$$Q_x - \left[Q_x + \frac{\partial}{\partial x}(Q_x)\,dx\right] = -\frac{\partial}{\partial x}(Q_x)\,dx \tag{2.24}$$

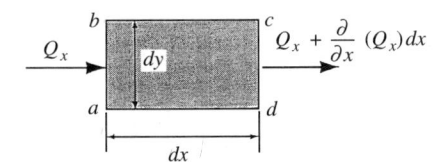

FIGURE 2.7
Heat flows into and out of control volume.

38 Introduction to Convective Heat Transfer Analysis

But again noting that the control volume has unit thickness and using Fourier's law gives:

$$Q_x = -k\,dy\,\frac{\partial T}{\partial x} \tag{2.25}$$

Substituting this into Eq. (2.24) gives the net rate at which heat is transferred into the control volume in the x-direction as:

$$\left[\frac{\partial}{\partial x}\left(k\frac{\partial T}{\partial x}\right)\right]dx\,dy \tag{2.26}$$

If the same procedure is applied in the y-direction it can be shown that the net rate at which heat is transferred into the control volume in this direction is:

$$\left[\frac{\partial}{\partial y}\left(k\frac{\partial T}{\partial y}\right)\right]dx\,dy \tag{2.27}$$

Combining these and noting that k, a fluid property, is being assumed constant then gives the net rate at which heat is transferred into the control volume in both the x- and y-directions as:

$$k\left(\frac{\partial^2 T}{\partial x^2} + \frac{\partial^2 T}{\partial y^2}\right)dx\,dy \tag{2.28}$$

The last term in Eq. (2.17) that has to be considered is the work term. This work is done by the fluid on the control volume by virtue of the forces exerted by the flowing fluid on the surface of the control volume. These forces arise due to the normal pressure and due to the previously discussed shearing stresses on its surface. The way in which this work is done and the way in which the work term is evaluated are perhaps most easily seen by considering the control volume to be moving through the fluid which is at rest.

Considering the stresses as defined in Fig. 2.3 then shows that the rate at which work is done on the control volume is given by:

$$\begin{aligned}
&\left\{(-u\sigma_x\,dy) + \left[(u\sigma_x\,dy) + \frac{\partial}{\partial x}(u\sigma_x)\,dx\,dy\right]\right\} \\
&+ \left\{(up\,dy) - \left[(up\,dy) + \frac{\partial}{\partial x}(up)\,dx\,dy\right]\right\} \\
&+ \left\{(-u\tau_{yx}\,dx) + \left[(u\tau_{yx}\,dx) + \frac{\partial}{\partial y}(u\tau_{yx})\,dx\,dy\right]\right\} \\
&+ \left\{(-v\sigma_y\,dx) + \left[(v\sigma_y\,dx) + \frac{\partial}{\partial y}(v\sigma_y)\,dx\,dy\right]\right\} \\
&+ \left\{(vp\,dx) - \left[(vp\,dx) + \frac{\partial}{\partial y}(vp)\,dx\,dy\right]\right\} \\
&+ \left\{(-v\tau_{xy}\,dy) + \left[(v\tau_{xy}\,dy) + \frac{\partial}{\partial x}(v\tau_{xy})\,dx\,dy\right]\right\}
\end{aligned} \tag{2.29}$$

CHAPTER 2: The Equations of Convective Heat Transfer 39

In setting up this equation it has been noted that the pressure acts in the opposite direction to that taken as positive for the normal stresses σ_x and σ_y.

Rearranging Eq. (2.29) then gives the rate at which work is done on the control volume as:

$$\left[\frac{\partial}{\partial x}(u\sigma_x) - \frac{\partial}{\partial x}(up) + \frac{\partial}{\partial y}(u\tau_{yx}) + \frac{\partial}{\partial y}(v\sigma_y) - \frac{\partial}{\partial y}(vp) + \frac{\partial}{\partial x}(v\tau_{xy})\right]dx\,dy$$

i.e.,

$$\left(u\frac{\partial \sigma_x}{\partial x} - u\frac{\partial p}{\partial x} + u\frac{\partial \tau_{yx}}{\partial y} + v\frac{\partial \sigma_y}{\partial y} - v\frac{\partial p}{\partial y} + v\frac{\partial \sigma_{xy}}{\partial x}\right)dx\,dy$$

$$- p\left(\frac{\partial u}{\partial x} + \frac{\partial v}{\partial y}\right)dx\,dy + \left(\sigma_x\frac{\partial u}{\partial x} + \tau_{yx}\frac{\partial u}{\partial y} + \sigma_y\frac{\partial v}{\partial y} + \tau_{xy}\frac{\partial v}{\partial x}\right)dx\,dy$$

(2.30)

Now, consider separately the various bracketed terms in this equation. Expanding the first of these using the expressions for the stresses given in Eqs. (2.8) to (2.10) gives:

$$2u\mu\frac{\partial^2 u}{\partial x^2} - u\frac{\partial p}{\partial x} + u\mu\left(\frac{\partial^2 u}{\partial y^2} + \frac{\partial^2 v}{\partial y\,\partial x}\right) + 2v\mu\frac{\partial^2 v}{\partial y^2} - v\frac{\partial p}{\partial y} + v\mu\left(\frac{\partial u}{\partial y} + \frac{\partial v}{\partial x}\right)$$

(2.31)

But from the continuity equation it follows that:

$$\frac{\partial^2 u}{\partial x^2} = -\frac{\partial^2 v}{\partial y\,\partial x}$$

and

$$\frac{\partial^2 v}{\partial y^2} = -\frac{\partial^2 u}{\partial x\,\partial y}$$

so that Eq. (2.31) can be written as:

$$u\left[-\frac{\partial p}{\partial x} + \mu\left(\frac{\partial^2 u}{\partial y^2} + \frac{\partial^2 u}{\partial x^2}\right)\right] + v\left[-\frac{\partial p}{\partial y} + \mu\left(\frac{\partial^2 v}{\partial y^2} + \frac{\partial^2 v}{\partial x^2}\right)\right] \quad (2.32)$$

If this equation is compared with the Navier-Stokes equations (2.6) and (2.7) it will be seen that it can be written as:

$$\rho u^2\frac{\partial u}{\partial x} + \rho uv\frac{\partial u}{\partial y} + \rho uv\frac{\partial v}{\partial x} + \rho v^2\frac{\partial v}{\partial y} \quad (2.33)$$

The second bracketed term in Eq. (2.30) is zero by virtue of the continuity equation and it follows, therefore, that the rate at which work is done on the control volume is given by:

$$\left(\rho u^2\frac{\partial u}{\partial x} + \rho uv\frac{\partial u}{\partial y} + \rho uv\frac{\partial v}{\partial x} + \rho v^2\frac{\partial v}{\partial y}\right)dx\,dy$$

$$+ \left(\sigma_x\frac{\partial u}{\partial x} + \tau_{yx}\frac{\partial u}{\partial y} + \sigma_y\frac{\partial v}{\partial y} + \tau_{xy}\frac{\partial v}{\partial x}\right)dx\,dy$$

(2.34)

40 Introduction to Convective Heat Transfer Analysis

Substituting Eqs. (2.23), (2.28), and (2.34) into Eq. (2.17) and dividing by $dx\,dy$ gives:

$$\frac{\partial}{\partial x}\left[\rho u\left\{c_p T+\left(\frac{u^2+v^2}{2}\right)\right\}\right]+\frac{\partial}{\partial y}\left[\rho v\left\{c_p T+\left(\frac{u^2+v^2}{2}\right)\right\}\right]$$

$$=k\left(\frac{\partial^2 T}{\partial x^2}+\frac{\partial^2 T}{\partial y^2}\right)+\left(\rho u^2\frac{\partial u}{\partial x}+\rho uv\frac{\partial u}{\partial y}+\rho uv\frac{\partial v}{\partial x}+\rho v^2\frac{\partial v}{\partial y}\right) \quad (2.35)$$

$$+\left(\sigma_x\frac{\partial u}{\partial x}+\tau_{yx}\frac{\partial u}{\partial y}+\sigma_y\frac{\partial v}{\partial y}+\tau_{xy}\frac{\partial v}{\partial x}\right)$$

Since ρ and c_p are fluid properties which are being assumed constant, this equation can be written as:

$$\left(u\frac{\partial T}{\partial x}+v\frac{\partial T}{\partial y}\right)+\frac{1}{c_p}u^2\left(\frac{\partial u}{\partial x}+uv\frac{\partial v}{\partial x}+vu\frac{\partial u}{\partial y}+v^2\frac{\partial v}{\partial y}\right)$$

$$+\left[T+\left(\frac{u^2+v^2}{2c_p}\right)\right]\left(\frac{\partial u}{\partial x}+\frac{\partial v}{\partial y}\right)$$

$$=\left(\frac{k}{\rho c_p}\right)\left(\frac{\partial^2 T}{\partial x^2}+\frac{\partial^2 T}{\partial y^2}\right)+\frac{1}{c_p}u^2\left(\frac{\partial u}{\partial x}+uv\frac{\partial v}{\partial x}+vu\frac{\partial u}{\partial y}+v^2\frac{\partial v}{\partial y}\right) \quad (2.36)$$

$$+\left(\frac{1}{\rho c_p}\right)\left(\sigma_x\frac{\partial u}{\partial x}+\tau_{yx}\frac{\partial u}{\partial y}+\sigma_y\frac{\partial v}{\partial y}+\tau_{xy}\frac{\partial v}{\partial y}\right)$$

The third term on the left-hand side of this equation is zero by virtue of the continuity equation and the second terms on the two sides are identical and therefore can be cancelled. The two-dimensional constant fluid property energy equation, therefore, has the form:

$$u\frac{\partial T}{\partial x}+v\frac{\partial T}{\partial y}=\left(\frac{k}{\rho c_p}\right)\left(\frac{\partial^2 T}{\partial x^2}+\frac{\partial^2 T}{\partial y^2}\right)+\left(\frac{\mu}{\rho c_p}\right)\Phi \quad (2.37)$$

where Φ is termed the dissipation function and is given by:

$$\Phi=\frac{1}{\mu}\left(\sigma_x\frac{\partial u}{\partial x}+\tau_{yx}\frac{\partial u}{\partial y}+\sigma_y\frac{\partial v}{\partial x}+\tau_{xy}\frac{\partial v}{\partial y}\right) \quad (2.38)$$

If the expressions for the relation between the shearing stresses and the velocity field given in Eqs. (2.2) to (2.4) are used, this dissipation function becomes:

$$\Phi=2\left(\frac{\partial u}{\partial x}\right)^2+2\left(\frac{\partial v}{\partial y}\right)^2+\left(\frac{\partial v}{\partial x}\right)\left(\frac{\partial u}{\partial y}\right)+\left(\frac{\partial v}{\partial x}\right)^2+\left(\frac{\partial u}{\partial y}\right)^2+\left(\frac{\partial u}{\partial y}\right)\left(\frac{\partial v}{\partial x}\right)$$

$$=2\left[\left(\frac{\partial u}{\partial x}\right)^2+\left(\frac{\partial v}{\partial y}\right)^2\right]+\left[\frac{\partial u}{\partial y}+\frac{\partial v}{\partial x}\right]^2 \quad (2.39)$$

This last term in Eq. (2.37), the dissipation term, arises, it will be noted, because of the work done by the viscous stresses. In low-speed flows this dissipation term will usually be negligible. A discussion of this point will be given later in this chapter.

The above derivation was presented for two-dimensional flow. Exactly the same approach can be used for three-dimensional flow and the resultant energy equation is as follows:

$$u\frac{\partial T}{\partial x} + v\frac{\partial T}{\partial y} + w\frac{\partial T}{\partial z} = \left(\frac{k}{\rho c_p}\right)\left(\frac{\partial^2 T}{\partial x^2} + \frac{\partial^2 T}{\partial y^2} + \frac{\partial^2 T}{\partial z^2}\right) + \left(\frac{\mu}{\rho c_p}\right)\Phi \quad (2.40)$$

where, for this three-dimensional flow case, the dissipation function is given by:

$$\Phi = 2\left[\left(\frac{\partial u}{\partial x}\right)^2 + \left(\frac{\partial v}{\partial y}\right)^2 + \left(\frac{\partial w}{\partial z}\right)^2\right] + \left[\left(\frac{\partial u}{\partial y}\right) + \left(\frac{\partial v}{\partial x}\right)\right]^2$$
$$+ \left[\left(\frac{\partial w}{\partial y}\right) + \left(\frac{\partial v}{\partial z}\right)\right]^2 + \left[\left(\frac{\partial u}{\partial z}\right) + \left(\frac{\partial w}{\partial x}\right)\right]^2 \quad (2.41)$$

In terms of the cylindrical coordinate system defined earlier in this chapter, the energy equation for steady constant fluid property flow is:

$$u\frac{\partial T}{\partial z} + v\frac{\partial T}{\partial r} + \frac{w}{r}\frac{\partial T}{\partial \phi} = \left(\frac{k}{\rho c_p}\right)\left[\frac{\partial^2 T}{\partial z^2} + \frac{1}{r}\frac{\partial}{\partial r}\left(r\frac{\partial T}{\partial r}\right) + \frac{1}{r^2}\frac{\partial^2 T}{\partial \phi^2}\right] + \left(\frac{\mu}{\rho c_p}\right)\Phi \quad (2.42)$$

where in this coordinate system the dissipation function is given by:

$$\Phi = 2\left[\left(\frac{\partial u}{\partial z}\right)^2 + \left(\frac{\partial v}{\partial r}\right)^2 + \left(\frac{1}{r}\frac{\partial w}{\partial \phi} + \frac{v}{r}\right)^2\right]$$
$$+ \left[\frac{1}{r}\frac{\partial u}{\partial \phi} + \frac{\partial w}{\partial z}\right]^2 + \left[\frac{\partial v}{\partial z} + \frac{\partial u}{\partial r}\right]^2 + \left[\frac{1}{r}\frac{\partial v}{\partial \phi} + \frac{\partial w}{\partial r} - \frac{w}{r}\right]^2 \quad (2.43)$$

It is perhaps worth pointing out that if the velocity components are zero the energy equation reduces, of course, to the heat conduction equation, i.e., in the case of Cartesian coordinates, to

$$\frac{\partial^2 T}{\partial x^2} + \frac{\partial^2 T}{\partial y^2} + \frac{\partial^2 T}{\partial z^2} = 0 \quad (2.44)$$

2.4 SIMILARITY IN CONVECTIVE HEAT TRANSFER

Consideration will now be given to what is basically the same question that was earlier dealt with by using dimensionless analysis. The problem is, essentially, that of determining under what conditions the flow and temperature distributions about

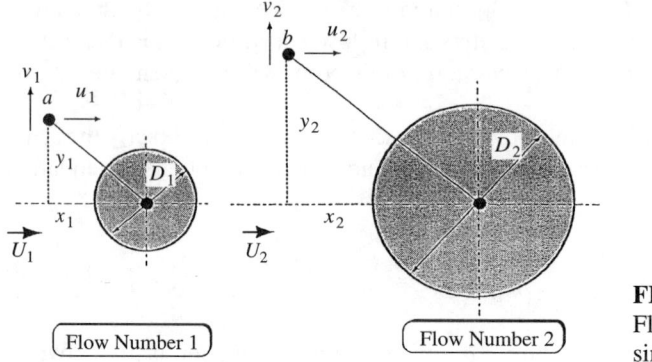

FIGURE 2.8
Flow over geometrically similar bodies.

geometrically similar bodies will be "similar", i.e., under what conditions the velocity and temperature fields about the bodies will be "similar" [7].

To understand what similarity means, consider the flow over two geometrically similar bodies such as those shown in Fig. 2.8. Assume photographs are taken of the flow and temperature fields about the first body and these photographs are then enlarged until the size of the body in the photograph is the same as that of the second body. If the enlarged photographs so obtained show the flow and temperature fields about the second body, the two flows are similar. Hence, by similar velocity and temperature fields, it is here meant that the velocity components and temperature when expressed in dimensionless form are related to the dimensionless coordinate scheme by the same functions in all of the flows considered. For example, consider two-dimensional flow about two long cylinders. Consider the points a and b in flows 1 and 2. These points are at geometrically similar positions, i.e., are such that:

$$x_1/D_1 = x_2/D_2 \quad \text{and} \quad y_1/D_1 = y_2/D_2$$

If the velocity fields in the two flows are similar then the velocity components at these two points will be such that:

$$u_1/U_1 = u_2/U_2, \quad v_1/U_1 = v_2/U_2 \quad (2.45)$$

Attention will first be given to the determination of conditions under which the velocity fields are similar, the temperature field being dealt with later. The following dimensionless coordinate system is introduced:

$$X = x/L, \quad Y = y/L, \quad Z = z/L \quad (2.46)$$

where L is some convenient reference dimension. Since the series of bodies being considered are of the same geometrical shape, any convenient dimension can be used for this purpose. In the case of flow over a sphere, for example, the diameter could conveniently be used for L.

The velocity components are also expressed in dimensionless form using the following:

$$U = u/U_1, \qquad V = v/U_1, \qquad W = w/U_1 \tag{2.47}$$

where U_1 is any convenient characteristic velocity. In the case of external flows, U_1 is usually most conveniently taken as the freestream velocity while in internal flows it is usually convenient to take it as the mean velocity.

The pressure is nondimensionalized using the dynamic pressure based on U_1 as follows:

$$P = p/\rho U_1^2 \tag{2.48}$$

ρU_1^2 being, of course, twice the dynamic pressure.

In terms of these dimensionless variables, the equations governing the velocity field, i.e., the continuity and Navier-Stokes equations, become:

$$\frac{\partial U}{\partial X} + \frac{\partial V}{\partial Y} + \frac{\partial W}{\partial Z} = 0 \tag{2.49}$$

$$U\frac{\partial U}{\partial X} + V\frac{\partial U}{\partial Y} + W\frac{\partial U}{\partial Z} = -\frac{\partial P}{\partial X} + \left(\frac{\mu}{\rho L U_1}\right)\left(\frac{\partial^2 U}{\partial X^2} + \frac{\partial^2 U}{\partial Y^2} + \frac{\partial^2 U}{\partial Z^2}\right) \tag{2.50}$$

$$U\frac{\partial V}{\partial X} + V\frac{\partial V}{\partial Y} + W\frac{\partial V}{\partial Z} = -\frac{\partial P}{\partial Y} + \left(\frac{\mu}{\rho L U_1}\right)\left(\frac{\partial^2 V}{\partial X^2} + \frac{\partial^2 V}{\partial Y^2} + \frac{\partial^2 V}{\partial Z^2}\right) \tag{2.51}$$

$$U\frac{\partial W}{\partial X} + V\frac{\partial W}{\partial Y} + W\frac{\partial W}{\partial Z} = -\frac{\partial P}{\partial Z} + \left(\frac{\mu}{\rho L U_1}\right)\left(\frac{\partial^2 W}{\partial X^2} + \frac{\partial^2 W}{\partial Y^2} + \frac{\partial^2 W}{\partial Z^2}\right) \tag{2.52}$$

It will be seen from this set of equations that for each value of the parameter $(\rho L U_1/\mu)$ there is a unique solution, i.e., a unique relation between the dimensionless velocity components and pressure and the position in terms of the dimensionless coordinate system. This is equivalent to saying that the flow fields about a series of geometrically similar bodies will be dynamically similar if the Reynolds numbers $(\rho L U_1/\mu)$ are the same in all the flows.

Consideration will next be given to the determination of the conditions under which the temperature fields are similar. In order to define a dimensionless temperature variable, a convenient reference fluid temperature, T_1, and some convenient measure of the wall temperature, T_{wr}, are introduced. In external flows, T_1 is usually most conveniently taken as the freestream temperature while in internal flows it is usually taken as a convenient mean temperature. Using these the following dimensionless temperature is defined:

$$\theta = (T - T_1)/(T_{wr} - T_1) \tag{2.53}$$

If the temperature fields in the fluid flowing about the series of geometrically similar bodies are to be similar, one obvious requirement is that the dimensionless surface temperature distributions must be the same, i.e., at geometrically similar positions on the surfaces, the values of $(T_w - T_1)/(T_{wr} - T_1)$ must be the same, T_w being the local wall temperature at the point being considered.

Introducing θ, U, V, W, X, and Z into the energy equation gives:

$$U\frac{\partial \theta}{\partial X} + V\frac{\partial \theta}{\partial Y} + W\frac{\partial \theta}{\partial Z} = \left(\frac{k}{\rho c_p U_1 L}\right)\left(\frac{\partial^2 \theta}{\partial X^2} + \frac{\partial^2 \theta}{\partial Y^2} + \frac{\partial^2 \theta}{\partial Z^2}\right)$$
$$+ \left[\frac{\mu U_1}{\rho c_p L(T_{wr} - T_1)}\right]\left\{2\left[\left(\frac{\partial U}{\partial X}\right)^2 + \left(\frac{\partial V}{\partial Y}\right)^2 + \left(\frac{\partial W}{\partial Z}\right)^2\right]\right.$$
$$\left.+ \left(\frac{\partial V}{\partial X} + \frac{\partial U}{\partial Y}\right)^2 + \left(\frac{\partial W}{\partial Y} + \frac{\partial V}{\partial Z}\right)^2 + \left(\frac{\partial U}{\partial Z} + \frac{\partial W}{\partial X}\right)^2\right\}$$
(2.54)

From this it follows that if the temperature fields are to be similar it is necessary that the following two quantities be the same in the flows

$$\left(\frac{\rho c_p U_1 L}{k}\right), \quad \left[\frac{\rho c_p L(T_{wr} - T_1)}{\mu U_1}\right] \quad (2.55)$$

It also, of course, follows from Eq. (2.54) that if the temperature fields are to be similar the velocity fields must be similar. The Reynolds number ($\rho U_1 L/\mu$) has, therefore, already been recognized as a parameter on which the similarity of the temperature fields depends. The two parameters given in Eq. (2.55) are, therefore, written as

$$\left(\frac{\mu c_p}{k}\right)\left(\frac{\rho U_1 L}{\mu}\right), \quad \left(\frac{\rho U_1 L}{\mu}\right) \bigg/ \left[\frac{U_1^2}{c_p(T_{wr} - T_1)}\right] \quad (2.56)$$

From this it is seen that for the temperature fields to be similar, i.e., for θ to be a unique function of X, Y, and Z, it is necessary that, besides the Reynolds number, the following two parameters be the same in the flows

$$\left(\frac{\mu c_p}{k}\right), \quad \left[\frac{U_1^2}{c_p(T_{wr} - T_1)}\right] \quad (2.57)$$

The first of these is, of course, the Prandtl number which was previously introduced. The second is termed the Eckert number, i.e.:

$$\text{Eckert number, } Ec = \frac{U_1^2}{c_p(T_{wr} - T_1)} \quad (2.58)$$

It should be noted that the Eckert number arises from the dissipation term and that it can be written as:

$$Ec = \left(\frac{U_1^2}{c_p T_1}\right) \bigg/ \left(\frac{T_{wr}}{T_1} - 1\right) \quad (2.59)$$

For the case of gas flows, if the gas can be assumed to be perfect:

$$c_p T_1 = \frac{\gamma R T_1}{(\gamma - 1)} = \frac{a_1^2}{(\gamma - 1)} \quad (2.60)$$

CHAPTER 2: The Equations of Convective Heat Transfer   45

where a_1 is the speed of sound at the temperature, T_1. Substituting Eq. (2.60) into Eq. (2.59) then gives:

$$Ec = \frac{(\gamma - 1)M_1^2}{\left(\dfrac{T_{wr}}{T_1} - 1\right)} \tag{2.61}$$

where $M_1 = U_1/a_1$ is the Mach number based on U_1 and T_1.

It follows from Eq. (2.61) that the Eckert number and, therefore, the dissipation term will only be significant if the Mach number is large or if (T_{wr}/T_1) is close to 1, i.e., T_{wr} very nearly equal to T_1. It is significant in the former case because the velocities are high and in the latter case it is significant because the heat conduction term is very small. In the limiting case where the surface temperature is everywhere equal to T_1, the temperature differences in the fluid are the result of dissipation alone. When T_{wr} is approximately equal to T_1 the heat transfer rate from the surface will, generally, be small and this case has, therefore, no practical significance in the present work. Thus, for the present purposes, it is concluded that the Eckert number and, therefore, the dissipation term will only be significant if the Mach number is large. However, constant fluid properties solutions will only be applicable if the Mach number is low so it is usually consistent with the assumption of constant fluid properties to ignore the dissipation term.

From the above discussion, it can be concluded that the dimensionless temperature field $\theta(X, Y)$ depends only on the dimensionless numbers Re, Pr, and Ec. In order to determine the implications this has for the local heat transfer rate at any point on the surface it is noted that this heat transfer rate is given by Fourier's law as:

$$q_w = -k \left.\frac{\partial T}{\partial n}\right|_w \tag{2.62}$$

where $\partial T/\partial n|_w$ is the local temperature gradient at the surface measured normal to the surface. Now this derivative can be written as:

$$\left.\frac{\partial T}{\partial n}\right|_w = \left.\frac{\partial T}{\partial y}\right|_w \left.\frac{\partial y}{\partial n}\right|_w + \left.\frac{\partial T}{\partial x}\right|_w \left.\frac{\partial x}{\partial n}\right|_w \tag{2.63}$$

It should be noted that the derivatives $\partial y/\partial n|_w$ and $\partial x/\partial n|_w$ will depend only on the shape of the surface at the point considered.

Now the right-hand side of Eq. (2.63) is conveniently rewritten in dimensionless form to give:

$$\left.\frac{\partial T}{\partial n}\right|_w = \left\{\left.\frac{\partial \theta}{\partial Y}\right|_w \left.\frac{\partial Y}{\partial N}\right|_w + \left.\frac{\partial \theta}{\partial X}\right|_w \left.\frac{\partial X}{\partial N}\right|_w\right\} \left(\frac{T_{wr} - T_1}{L}\right) \tag{2.64}$$

where

$$N = \frac{n}{L} \tag{2.65}$$

Since $\partial X/\partial N|_w$ and $\partial Y/\partial N|_w$ will depend only on the dimensionless shape of the surface and since the dimensionless temperature gradients $\partial \theta/\partial X|_w$ and $\partial \theta/\partial Y|_w$

must depend on the same variables as the dimensionless temperature field, i.e., on Re, Pr, and Ec, it follows that at any position on the surface

$$\left.\frac{\partial \theta}{\partial N}\right|_w = k\left(\frac{T_{wr} - T_1}{L}\right) \text{function}(Re, Pr, Ec) \tag{2.66}$$

Substituting this result into Eq. (2.62) then gives

$$\frac{q_w L}{(T_w - T_1)k} = \left(\frac{T_{wr} - T_1}{T_w - T_1}\right) \text{function}(Re, Pr, Ec) \tag{2.67}$$

Now, the left-hand side of this equation is the local Nusselt number, Nu. It follows, therefore, that for a given surface temperature distribution, i.e., for a given distribution of $(T_w - T_1)/(T_{wr} - T_1)$:

$$Nu = \text{function}(Re, Pr, Ec) \tag{2.68}$$

For constant fluid property flows, the Eckert number will, for the previously discussed reasons, have a negligible effect and in this case

$$Nu = \text{function}(Re, Pr) \tag{2.69}$$

This result was, of course, previously derived by dimensional analysis.

2.5
VORTICITY AND TEMPERATURE FIELDS

When a viscous fluid flows over a surface, vorticity is generated by the action of the viscous shearing stresses, the vorticity essentially being proportional to the angular momentum of the fluid particles. This vorticity is, in general, highest at the surface and decreases with increasing distance from the surface as indicated in Fig. 2.9. The vorticity, ω, which as noted above, is a measure of the angular momentum or rotational motion induced by the action of the viscous stresses acting tangentially around the surface of the fluid particles, is given in terms of the velocity components by:

$$\omega = \left(\frac{\partial v}{\partial x} - \frac{\partial u}{\partial y}\right) \tag{2.70}$$

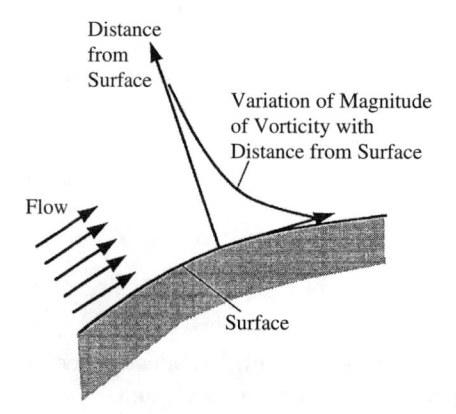

FIGURE 2.9
Vorticity variation in flow.

If the surface over which the fluid is flowing is held at a different temperature from the bulk of the fluid, there will also be temperature gradients in the fluid near the surface. The difference between the temperature of the fluid locally and that of the fluid far from the surface is, in general, highest at the surface and decreases with increasing distance from the surface. The purpose of the present section is to show that there is a similarity between these vorticity and temperature fields.

For the present purposes, attention will be restricted to steady, two-dimensional constant fluid property flow and, consistent with this assumption, the dissipation term in the energy equation will be neglected. The result will be derived using the governing equations expressed in Cartesian coordinates although a similar result can, of course, be obtained in terms of other coordinate systems.

Consider the velocity field first. It is governed by the continuity and Navier-Stokes equations which, subject to the assumptions introduced above, are, as previously presented:

$$\frac{\partial u}{\partial x} + \frac{\partial v}{\partial y} = 0 \tag{2.71}$$

$$u\frac{\partial u}{\partial x} + v\frac{\partial u}{\partial y} = -\frac{1}{\rho}\frac{\partial p}{\partial x} + \left(\frac{\mu}{\rho}\right)\left(\frac{\partial^2 u}{\partial x^2} + \frac{\partial^2 u}{\partial y^2}\right) \tag{2.72}$$

$$u\frac{\partial v}{\partial x} + v\frac{\partial v}{\partial y} = -\frac{1}{\rho}\frac{\partial p}{\partial y} + \left(\frac{\mu}{\rho}\right)\left(\frac{\partial^2 v}{\partial x^2} + \frac{\partial^2 v}{\partial y^2}\right) \tag{2.73}$$

The pressure is eliminated between Eqs. (2.72) and (2.73) by differentiating Eq. (2.72) with respect to y and Eq. (2.73) with respect to x and subtracting the results. This gives:

$$-\frac{\partial u}{\partial y}\frac{\partial u}{\partial x} - u\frac{\partial^2 u}{\partial y \partial x} - \frac{\partial v}{\partial y}\frac{\partial u}{\partial y} - v\frac{\partial^2 u}{\partial y^2} + \frac{\partial u}{\partial x}\frac{\partial v}{\partial x} + u\frac{\partial^2 v}{\partial x^2} + \frac{\partial v}{\partial x}\frac{\partial v}{\partial y} + v\frac{\partial^2 v}{\partial x \partial y}$$

$$= \left(\frac{\mu}{\rho}\right)\left[-\frac{\partial^3 u}{\partial y \partial x^2} - \frac{\partial^3 u}{\partial y^3} + \frac{\partial^3 v}{\partial x^3} + \frac{\partial^3 v}{\partial x \partial y^2}\right] \tag{2.74}$$

This equation can be arranged as follows:

$$u\left(\frac{\partial^2 v}{\partial x^2} - \frac{\partial^2 u}{\partial x \partial y}\right) + v\left(\frac{\partial^2 v}{\partial y \partial x} - \frac{\partial^2 u}{\partial y^2}\right) + \frac{\partial v}{\partial x}\left(\frac{\partial u}{\partial x} + \frac{\partial v}{\partial y}\right) - \frac{\partial u}{\partial y}\left(\frac{\partial u}{\partial x} + \frac{\partial v}{\partial y}\right)$$

$$= \left(\frac{\mu}{\rho}\right)\left[\left(\frac{\partial^3 v}{\partial x^3} - \frac{\partial^3 u}{\partial x^2 \partial y}\right) + \left(\frac{\partial^3 v}{\partial y^2 \partial x} - \frac{\partial^3 u}{\partial y^3}\right)\right] \tag{2.75}$$

But, as noted above, the vorticity, ω, is given by:

$$\omega = \left(\frac{\partial v}{\partial x} - \frac{\partial u}{\partial y}\right)$$

so using the continuity equation that Eq. (2.75) can be written as

$$u\frac{\partial \omega}{\partial x} + v\frac{\partial \omega}{\partial y} = \left(\frac{\mu}{\rho}\right)\left(\frac{\partial^2 \omega}{\partial x^2} + \frac{\partial^2 \omega}{\partial y^2}\right) \tag{2.76}$$

Now when the dissipation term is neglected, the two-dimensional energy equation becomes

$$u\frac{\partial T}{\partial x} + v\frac{\partial T}{\partial y} = \frac{1}{Pr}\left(\frac{\mu}{\rho}\right)\left(\frac{\partial^2 T}{\partial x^2} + \frac{\partial^2 T}{\partial y^2}\right) \tag{2.77}$$

Thus, the vorticity and temperature fields are governed by equations having the same basic form. When Pr is equal to 1, the equations have exactly the same form. Even in this case, however, the vorticity and temperature fields will not be identical because the boundary conditions on the two fields at the surface will not in general be identical. However, there will obviously be similarities between the two fields. Vorticity is generated in the flow by the action of viscosity due to the presence of the surface. The temperature differences arise in the flow because the surface is at a temperature which is different from the flowing fluid. Thus, Eq. (2.76) essentially describes the rate at which viscous effects spread into the fluid while Eq. (2.77) describes the rate at which the effects of the temperature changes at the surface spread into the fluid. It will be seen that the relative rates of spread depend on the value of the Prandtl number.

Because they do not contain the pressure as a variable, Eqs. (2.76) and (2.77) have been used quite extensively in solving problems for which the boundary layer equations (see later) cannot be used. For this purpose, instead of solving the Navier-Stokes and energy, simultaneously with the continuity equation, it is convenient to introduce the stream function, ψ, which is defined such that

$$u = \frac{\partial \psi}{\partial y}, \qquad v = -\frac{\partial \psi}{\partial x} \tag{2.78}$$

This function, of course, satisfies the continuity equation, because the left-hand side of the continuity equation is:

$$\frac{\partial u}{\partial x} + \frac{\partial v}{\partial y} = \frac{\partial^2 \psi}{\partial x \partial y} - \frac{\partial^2 \psi}{\partial x \partial y} = 0$$

In terms of the stream function, Eqs. (2.76) and (2.77) become

$$\frac{\partial \psi}{\partial y}\frac{\partial \omega}{\partial x} - \frac{\partial \psi}{\partial x}\frac{\partial \omega}{\partial y} = \left(\frac{\mu}{\rho}\right)\left(\frac{\partial^2 \omega}{\partial x^2} + \frac{\partial^2 \omega}{\partial y^2}\right) \tag{2.79}$$

$$\frac{\partial \psi}{\partial y}\frac{\partial T}{\partial x} - \frac{\partial \psi}{\partial x}\frac{\partial T}{\partial y} = \left(\frac{1}{Pr}\right)\left(\frac{\mu}{\rho}\right)\left(\frac{\partial^2 T}{\partial x^2} + \frac{\partial^2 T}{\partial y^2}\right) \tag{2.80}$$

while in terms of the stream function, the vorticity, as defined by Eq. (2.75), becomes

$$\omega = -\left(\frac{\partial^2 \psi}{\partial x^2} + \frac{\partial^2 \psi}{\partial y^2}\right) \tag{2.81}$$

These three equations in the three variables ψ, ω, and T are the set of equations that must be simultaneously solved subject to the correct boundary conditions to obtain the heat transfer rate. The original set of equations from which they were deduced, i.e., Eqs. (2.70), (2.71), (2.72), and (2.77), contained four unknowns u, v, p, and T and this reduction in the number of variables and, therefore, the number of governing equations in itself constitutes a considerable simplification. The governing equations in terms of u, v, p, and T are often said to be expressed in "primitive variable" form.

Eqs. (2.79) to (2.81) or their equivalent in other coordinate systems have been used in many numerical solutions of the two-dimensional Navier-Stokes and energy equations.

2.6
THE EQUATIONS FOR TURBULENT CONVECTIVE HEAT TRANSFER

In all of the above discussion it was assumed that the flow was laminar. The majority of flows encountered in practice are, however, turbulent. Consideration must, therefore, be given to the question of how these equations must be modified in order to apply them to turbulent flow.

In turbulent flows, even when the mean flow is steady (see below), the flow variables, i.e., velocity, pressure, and temperature, all fluctuate randomly with time due to the superposition of the turbulent "eddies" on the mean flow. For example, if the temperature at some point in the flow is measured by means of some device which has very good time response characteristics and the output of this device is displayed on a suitable instrument then a signal resembling that shown in Fig. 2.10 will be obtained in turbulent flow.

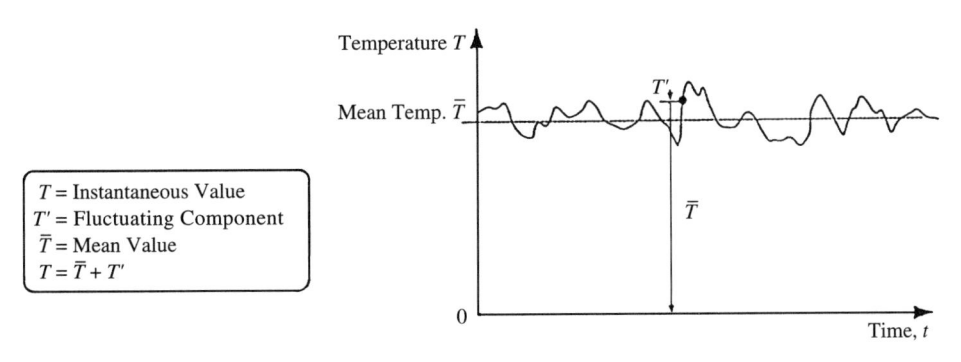

T = Instantaneous Value
T' = Fluctuating Component
$\bar{T}$ = Mean Value
$T = \bar{T} + T'$

FIGURE 2.10
Temperature variation with time in turbulent flow.

50   Introduction to Convective Heat Transfer Analysis

Because the flow variables are fluctuating with time in this way it is necessary to clearly define what is meant by "steady turbulent" flow. To do this, mean values of the variables are defined in the following way [3],[7],[10],[11],[12]:

$$\bar{u} = \frac{1}{t}\int_0^t u\,dt; \quad \bar{v} = \frac{1}{t}\int_0^t v\,dt; \quad \bar{w} = \frac{1}{t}\int_0^t w\,dt$$
$$\bar{p} = \frac{1}{t}\int_0^t p\,dt; \quad \bar{T} = \frac{1}{t}\int_0^t T\,dt \quad (2.82)$$

Here, t is some interval of time. The term "steady" turbulent flow implies that, provided a sufficiently long interval of time, t, is taken for this averaging process, constant values of the mean quantities $\bar{u}$, $\bar{v}$, $\bar{w}$, $\bar{p}$, and $\bar{T}$ will be obtained at all points in the flow. These mean values are thus often termed time-averaged values.

For many purposes, it is convenient to express the instantaneous values of the flow variables in terms of these constant mean values and the fluctuations from these values, i.e., to write:

$$u = \bar{u} + u'; \quad v = \bar{v} + v'; \quad w = \bar{w} + w';$$
$$p = \bar{p} + p'; \quad T = \bar{T} + T' \quad (2.83)$$

where the primes denote the fluctuating components.

It will be obvious that the mean (i.e., time averaged) values of these fluctuating components will, by virtue of the definition of the mean values of the variables, i.e., $\bar{u}$, $\bar{v}$, $\bar{w}$, $\bar{p}$, and $\bar{T}$, be equal to zero, i.e., that

$$\overline{u'} = \frac{1}{t}\int_0^t u'\,dt = 0; \quad \overline{v'} = \frac{1}{t}\int_0^t v'\,dt = 0; \quad \overline{w'} = \frac{1}{t}\int_0^t w'\,dt = 0$$
$$\overline{p'} = \frac{1}{t}\int_0^t p'\,dt = 0; \quad \overline{T'} = \frac{1}{t}\int_0^t T'\,dt = 0 \quad (2.84)$$

These results can be formally proved in the following way, the temperature, T, being selected as an example:

$$\bar{T} = \frac{1}{t}\int_0^t T\,dt = \frac{1}{t}\int_0^t \left(\bar{T} + T'\right)dt = \frac{1}{t}\int_0^t \bar{T}\,dt + \frac{1}{t}\int_0^t T'\,dt$$

But $\bar{T}$ is a constant so

$$\bar{T} = \frac{\bar{T}}{t}\int_0^t dt + \frac{1}{t}\int_0^t T'\,dt = \bar{T} + \frac{1}{t}\int_0^t T'\,dt = \bar{T} + \overline{T'}$$

therefore, $\overline{T'} = 0$.

The mean value of the product of two or more fluctuating components will, however, not in general be zero; e.g.,

$$\overline{u'u'} \neq 0; \quad \overline{u'T'} \neq 0$$
$$\overline{u'^2 p'} \neq 0; \quad \overline{u'v'T'} \neq 0 \quad (2.85)$$

In the derivation of the equations for turbulent flow given below, the following relations, and obvious extensions of them, concerning the mean and fluctuating values are required. All of these relations are really obvious but they can be formally proved using the same basic procedure that was used above to prove that $\overline{T'} = 0$. If q and r are any two of the flow variables, i.e., u, v, w, p, T and if n is any of the coordinate directions $x, y,$ and z, then

$$\overline{\frac{\partial q'}{\partial n}} = \frac{\partial}{\partial n}(\overline{q'}) = 0; \qquad \overline{\frac{\partial q}{\partial n}} = \frac{\partial \overline{q}}{\partial n} \qquad (2.86)$$

$$\overline{\overline{q}\,r} = \overline{q}\,\overline{r}; \qquad \overline{\overline{q}q'} = \overline{q}\,\overline{q'} = 0; \qquad \overline{q + r} = \overline{q} + \overline{r}$$

The continuity, Navier-Stokes, and energy equations for steady flow were given earlier in this chapter. These equations can be derived for unsteady flow using the same basic procedure as that discussed above, the resultant set of equations being as follows:

$$\frac{\partial u}{\partial x} + \frac{\partial v}{\partial y} + \frac{\partial w}{\partial z} = 0 \qquad (2.87)$$

$$\frac{\partial u}{\partial t} + u\frac{\partial u}{\partial x} + v\frac{\partial u}{\partial y} + w\frac{\partial u}{\partial z} = -\frac{1}{\rho}\frac{\partial p}{\partial x} + \nu\left(\frac{\partial^2 u}{\partial x^2} + \frac{\partial^2 u}{\partial y^2} + \frac{\partial^2 u}{\partial z^2}\right) \qquad (2.88)$$

$$\frac{\partial v}{\partial t} + u\frac{\partial v}{\partial x} + v\frac{\partial v}{\partial y} + w\frac{\partial v}{\partial z} = -\frac{1}{\rho}\frac{\partial p}{\partial y} + \nu\left(\frac{\partial^2 v}{\partial x^2} + \frac{\partial^2 v}{\partial y^2} + \frac{\partial^2 v}{\partial z^2}\right) \qquad (2.89)$$

$$\frac{\partial w}{\partial t} + u\frac{\partial w}{\partial x} + v\frac{\partial w}{\partial y} + w\frac{\partial w}{\partial z} = -\frac{1}{\rho}\frac{\partial p}{\partial z} + \nu\left(\frac{\partial^2 w}{\partial x^2} + \frac{\partial^2 w}{\partial y^2} + \frac{\partial^2 w}{\partial z^2}\right) \qquad (2.90)$$

$$\frac{\partial T}{\partial t} + u\frac{\partial T}{\partial x} + v\frac{\partial T}{\partial y} + w\frac{\partial T}{\partial z} = \frac{\nu}{Pr}\left(\frac{\partial^2 T}{\partial x^2} + \frac{\partial^2 T}{\partial y^2} + \frac{\partial^2 T}{\partial z^2}\right) \qquad (2.91)$$

The dissipation term has, for reasons previously given, been neglected in the energy equation.

In the derivation of these unsteady flow equations, no assumptions regarding the nature of the unsteadiness are made. In turbulent flow, therefore, the instantaneous values of the variables will satisfy these equations. Numerical solutions to the above equations are very difficult to obtain and for most purposes it is only the mean values of these variables and the mean heat transfer rate that are required. An attempt is, therefore, usually made to express them in terms of the mean values of the variables.

Attention here will be restricted to the case of two-dimensional mean flow, i.e., to flow in which $\overline{w} = 0$, it being noted that this does not necessarily mean that $w' = 0$. The following values of the variables are therefore substituted into the above equations:

$$u = \overline{u} + u'; \qquad v = \overline{v} + v'; \qquad w = w';$$
$$p = \overline{p} + p'; \qquad T = \overline{T} + T' \qquad (2.92)$$

52 Introduction to Convective Heat Transfer Analysis

Consider first the continuity equation (2.85). This becomes

$$\frac{\partial}{\partial x}(\bar{u} + u') + \frac{\partial}{\partial y}(\bar{v} + v') + \frac{\partial}{\partial z}(w') = 0 \qquad (2.93)$$

Taking the time average of this equation then gives

$$\frac{1}{t}\int_0^t \left[\frac{\partial}{\partial x}(\bar{u} + u')\right]dt + \frac{1}{t}\int_0^t \left[\frac{\partial}{\partial y}(\bar{v} + v')\right]dt$$

$$+ \frac{1}{t}\int_0^t \left[\frac{\partial}{\partial z}(w')\right]dt = 0$$

i.e.,

$$\overline{\frac{\partial}{\partial x}(\bar{u} + u')} + \overline{\frac{\partial}{\partial y}(\bar{v} + v')} + \overline{\frac{\partial w'}{\partial z}} = 0 \qquad (2.94)$$

Using the results given in Eq. (2.86) then leads to

$$\frac{\partial \bar{u}}{\partial x} + \frac{\partial \bar{v}}{\partial y} = 0 \qquad (2.95)$$

This is the continuity equation for turbulent flow when the mean motion is two-dimensional. It will be noted that this equation has exactly the same form as the continuity equation for two-dimensional steady laminar flow with the mean values of the velocity components substituted in place of the steady values that apply in laminar flow. This result can, in fact, be deduced by intuitive reasoning and simply states that if an elemental control volume through which the fluid flows is considered, then over a sufficiently long period of time, the fluctuating components contribute nothing to the mass transfer through this control volume.

Consider, next, the x-wise Navier-Stokes equation, i.e., Eq. (2.88). Substituting the values of the variables as given by Eq. (2.92) into this equation and taking the time average of each of the terms gives, after multiplying the terms out:

$$\overline{\frac{\partial \bar{u}}{\partial t}} + \overline{\frac{\partial u'}{\partial t}} + \overline{\bar{u}\frac{\partial \bar{u}}{\partial x}} + \overline{\bar{u}\frac{\partial u'}{\partial x}} + \overline{u'\frac{\partial \bar{u}}{\partial x}} + \overline{u'\frac{\partial u'}{\partial x}} + \overline{\bar{v}\frac{\partial \bar{u}}{\partial y}}$$

$$+ \overline{\bar{v}\frac{\partial u'}{\partial y}} + \overline{v'\frac{\partial \bar{u}}{\partial y}} + \overline{v'\frac{\partial u'}{\partial y}} + \overline{w'\frac{\partial \bar{u}}{\partial z}} + \overline{w'\frac{\partial u'}{\partial z}} \qquad (2.96)$$

$$= -\frac{1}{\rho}\overline{\frac{\partial \bar{p}}{\partial x}} - \frac{1}{\rho}\overline{\frac{\partial p'}{\partial x}} + \nu\left(\overline{\frac{\partial^2 \bar{u}}{\partial x^2}} + \overline{\frac{\partial^2 u'}{\partial x^2}} + \overline{\frac{\partial^2 \bar{u}}{\partial y^2}} + \overline{\frac{\partial^2 u'}{\partial y^2}} + \overline{\frac{\partial^2 \bar{u}}{\partial z^2}} + \overline{\frac{\partial^2 u'}{\partial z^2}}\right)$$

Applying rules of the type given in Eq. (2.86) to this equation then gives on rearrangement since the mean flow is steady, i.e., since

$$\frac{\partial \bar{u}}{\partial t} = \frac{\partial \bar{v}}{\partial t} = \frac{\partial \bar{w}}{\partial t} = 0 \qquad (2.97)$$

CHAPTER 2: The Equations of Convective Heat Transfer 53

the following as the x-wise momentum equation for turbulent flow:

$$\bar{u}\frac{\partial \bar{u}}{\partial x} + \bar{v}\frac{\partial \bar{u}}{\partial y} = -\frac{1}{\rho}\frac{\partial \bar{p}}{\partial x} + \nu\left(\frac{\partial^2 \bar{u}}{\partial x^2} + \frac{\partial^2 \bar{u}}{\partial y^2}\right) - \left(\overline{u'\frac{\partial u'}{\partial x}} + \overline{v'\frac{\partial u'}{\partial y}} + \overline{w'\frac{\partial u'}{\partial z}}\right) \quad (2.98)$$

The fluctuating components only occur in the last term of this equation. This term can be written as follows

$$\overline{u'\frac{\partial u'}{\partial x}} + \overline{v'\frac{\partial u'}{\partial y}} + \overline{w'\frac{\partial u'}{\partial z}} = \frac{\partial \overline{(u'^2)}}{\partial x} - \overline{u'\frac{\partial u'}{\partial x}}$$

$$+ \frac{\partial}{\partial y}\overline{(v'u')} - \overline{u'\frac{\partial v'}{\partial y}} + \frac{\partial}{\partial z}\overline{(w'u')} - \overline{u'\frac{\partial w'}{\partial z}} \quad (2.99)$$

Now if the continuity equation is multiplied by u and then applied to turbulent flow the following is obtained

$$(\bar{u} + u')\frac{\partial}{\partial x}(\bar{u} + u') + (\bar{u} + u')\frac{\partial}{\partial y}(\bar{v} + v') + (\bar{u} + u')\frac{\partial w'}{\partial z} = 0 \quad (2.100)$$

Time averaging the terms in this equation then gives

$$\bar{u}\frac{\partial \bar{u}}{\partial x} + \bar{u}\frac{\partial \bar{v}}{\partial y} + \overline{u'\frac{\partial u'}{\partial x}} + \overline{u'\frac{\partial v'}{\partial y}} + \overline{u'\frac{\partial w'}{\partial z}} = 0 \quad (2.101)$$

But by virtue of the continuity equation for turbulent flow, i.e., Eq. (2.95), the sum of the first two terms in this equation is zero so that Eq. (2.101) reduces to:

$$\overline{u'\frac{\partial u'}{\partial x}} + \overline{u'\frac{\partial v'}{\partial y}} + \overline{u'\frac{\partial w'}{\partial z}} = 0 \quad (2.102)$$

Substituting this result into Eq. (2.99) and the result so obtained back into Eq. (2.98) then gives:

$$\bar{u}\frac{\partial \bar{u}}{\partial x} + \bar{v}\frac{\partial \bar{v}}{\partial y} = -\frac{1}{\rho}\frac{\partial \bar{p}}{\partial x} + \nu\left(\frac{\partial^2 \bar{u}}{\partial x^2} + \frac{\partial^2 \bar{u}}{\partial y^2}\right) - \left[\frac{\partial \overline{(u'^2)}}{\partial x} + \frac{\partial}{\partial y}\overline{(v'u')} + \frac{\partial}{\partial z}\overline{(w'u')}\right]$$

$$(2.103)$$

This is the x-wise momentum equation for constant fluid property turbulent flow when the mean flow is steady and two-dimensional. Using exactly the same approach will give the following for the y-direction:

$$\bar{u}\frac{\partial \bar{v}}{\partial x} + \bar{v}\frac{\partial \bar{v}}{\partial y} = -\frac{1}{\rho}\frac{\partial \bar{p}}{\partial y} + \nu\left(\frac{\partial^2 \bar{v}}{\partial x^2} + \frac{\partial^2 \bar{v}}{\partial y^2}\right)$$

$$- \left[\frac{\partial}{\partial x}\overline{(u'v')} + \frac{\partial}{\partial y}\overline{(v'^2)} + \frac{\partial}{\partial z}\overline{(w'v')}\right] \quad (2.104)$$

and similarly for the z-direction because the mean flow is two-dimensional:

$$0 = \frac{\partial}{\partial x}\overline{(u'w')} + \frac{\partial}{\partial y}\overline{(v'w')} + \frac{\partial}{\partial z}\overline{(w'^2)} \quad (2.105)$$

The terms involving the derivatives with respect to z in these equations will in most circumstances be effectively zero since the mean flow is two-dimensional. In this case, the x- and y- momentum equations become:

$$\bar{u}\frac{\partial \bar{u}}{\partial x} + \bar{v}\frac{\partial \bar{u}}{\partial y} = -\frac{1}{\rho}\frac{\partial \bar{p}}{\partial x} + \nu\left(\frac{\partial^2 \bar{u}}{\partial x^2} + \frac{\partial^2 \bar{u}}{\partial y^2}\right) - \left[\frac{\partial}{\partial x}\overline{(u'^2)} + \frac{\partial}{\partial y}\overline{(v'u')}\right]$$

$$\bar{u}\frac{\partial \bar{v}}{\partial x} + \bar{v}\frac{\partial \bar{u}}{\partial y} = -\frac{1}{\rho}\frac{\partial \bar{p}}{\partial y} + \nu\left(\frac{\partial^2 \bar{v}}{\partial x^2} + \frac{\partial^2 \bar{v}}{\partial y^2}\right) - \left[\frac{\partial}{\partial x}\overline{(u'v')} + \frac{\partial}{\partial y}\overline{(v'^2)}\right]$$

(2.106)

Comparing these equations with the x- and y- Navier-Stokes equations for two-dimensional laminar flow shows that in turbulent flow extra terms arise due to the presence of the fluctuating velocity components. These extra terms, which arise because the Navier-Stokes equations contain nonlinear terms, are the result of the momentum transfer caused by the velocity fluctuating components and are often termed the turbulent or Reynolds stress terms because of their similarity to the viscous stress terms which arise due to momentum transfer on a molecular scale. This similarity can be clearly seen by noting that the x-wise momentum equation, for example, for laminar flow can be written as:

$$u\frac{\partial u}{\partial x} + v\frac{\partial u}{\partial y} = -\frac{1}{\rho}\frac{\partial p}{\partial x} + \frac{1}{\rho}\left(\frac{\partial \sigma_x}{\partial x} + \frac{\partial \tau_{xy}}{\partial y}\right) \quad (2.107)$$

where the stress terms are related to the velocity field by

$$\sigma_x = 2\mu\frac{\partial u}{\partial y}$$

$$\tau_{xy} = \mu\left(\frac{\partial u}{\partial y} + \frac{\partial v}{\partial x}\right)$$

The x-wise momentum equation for turbulent flow can be written in a similar way as follows:

$$\bar{u}\frac{\partial \bar{u}}{\partial x} + \bar{v}\frac{\partial \bar{u}}{\partial y} = -\frac{1}{\rho}\frac{\partial \bar{p}}{\partial x} + \frac{1}{\rho}\left[\frac{\partial}{\partial x}(\bar{\sigma}_x - \rho\overline{u'^2}) + \frac{\partial}{\partial y}(\bar{\tau}_{xy} - \rho\overline{u'^2})\right] \quad (2.108)$$

where $\bar{\sigma}_x$ and $\bar{\tau}_{xy}$ are the time averaged molecular stress terms which are given by:

$$\bar{\sigma}_x = 2\mu\frac{\partial \bar{u}}{\partial y}$$

$$\bar{\tau}_{xy} = \mu\left(\frac{\partial \bar{u}}{\partial y} + \frac{\partial \bar{v}}{\partial x}\right)$$

(2.109)

Thus, the extra terms arising from the fluctuating velocity components are such that their effects are the same as an increase in the viscous shear stress and it is for this reason that they are termed the turbulent or Reynolds stress terms.

The presence of these turbulent stress terms can be derived using a more physical approach. To do this, consider, as before, a control volume of the type shown in Fig. 2.11 through which fluid is flowing, the flow being turbulent.

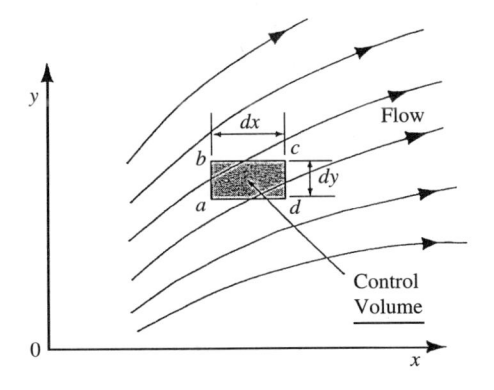

CHAPTER 2: The Equations of Convective Heat Transfer   55

FIGURE 2.11
Control volume considered.

Consider the flow through face ab of this control volume and let the area of this face be dA_x. The rate which x-wise momentum passes into the control volume through this face is given by:

$$\text{Mass flow rate through face} \times u = \rho u^2 \, dA_x$$

Setting $\bar{u}$ equal to $u + u'$ and time averaging then gives the mean rate at which x-wise momentum crosses the face as:

$$\rho(\bar{u}^2 + \overline{u'^2}) \, dA_x$$

Thus, the presence of the fluctuating turbulent velocity components causes the momentum transfer rate to be different from $\rho \times$ (mean velocity)$^2 \times dA$. But in applying the momentum conservation principle to the control volume the presence of additional momentum transfer is the equivalent of an additional force on the face of the control volume in the opposite direction to the momentum transfer. Thus, the additional momentum transfer due to the fluctuating velocity leads to an equivalent stress, i.e., force per unit area, of value $\overline{\rho u'^2}$ and this is what is termed the turbulent "stress" on the face.

If the lower face, ad, of the control volume is considered and if it has area dA_y, the mass transfer rate through it will be $\rho v \, dA_y$ and the rate of x-wise momentum transfer through it will, therefore, be $\rho v u \, dA_y$. Time averaging this then shows that on this face there will be an equivalent turbulent "shearing stress" of value $\overline{\rho u' v'}$. Using this type of argument, the presence of the turbulent "stresses" in the other directions can be deduced.

Consider, next, the energy equation. Substituting the instantaneous values of the variables into the unsteady form of this equation gives:

$$\frac{\partial}{\partial t}(\bar{T} + T') + (\bar{u} + u')\frac{\partial}{\partial x}(\bar{T} + T') + (\bar{v} + v')\frac{\partial}{\partial y}(\bar{T} + T') + (\bar{w} + w')\frac{\partial}{\partial z}(\bar{T} + T')$$

$$= \left(\frac{k}{\rho c_p}\right)\left[\frac{\partial^2}{\partial x^2}(\bar{T} + T') + \frac{\partial^2}{\partial y^2}(\bar{T} + T') + \frac{\partial^2}{\partial z^2}(\bar{T} + T')\right] \quad (2.110)$$

56 Introduction to Convective Heat Transfer Analysis

Time averaging this equation using the previously discussed rules then gives since the mean temperature distribution is steady:

$$\bar{u}\frac{\partial \bar{T}}{\partial x} + \bar{v}\frac{\partial \bar{T}}{\partial y} = \left(\frac{k}{\rho c_p}\right)\left(\frac{\partial^2 \bar{T}}{\partial x^2} + \frac{\partial^2 \bar{T}}{\partial y^2}\right) - \left(\overline{u'\frac{\partial T'}{\partial x}} + \overline{v'\frac{\partial T'}{\partial y}} + \overline{w'\frac{\partial T'}{\partial z}}\right) \quad (2.111)$$

Consider the last term in this equation. It can be rewritten as follows:

$$\overline{u'\frac{\partial T'}{\partial x}} + \overline{v'\frac{\partial T'}{\partial y}} + \overline{w'\frac{\partial T'}{\partial z}} = \frac{\partial}{\partial x}\overline{(u'T')} + \frac{\partial}{\partial y}\overline{(v'T')}$$
$$+ \frac{\partial}{\partial z}\overline{(w'T')} - \overline{T'\frac{\partial u'}{\partial x}} + \overline{T'\frac{\partial v'}{\partial y}} + \overline{T'\frac{\partial w'}{\partial z}} \quad (2.112)$$

But if the continuity equation is multiplied by T ($= \bar{T} + T'$) and the terms then time averaged, the following is obtained:

$$\bar{T}\frac{\partial \bar{u}}{\partial x} + \bar{T}\frac{\partial \bar{v}}{\partial y} + \overline{T'\frac{\partial u'}{\partial x}} + \overline{T'\frac{\partial v'}{\partial y}} + \overline{T'\frac{\partial w'}{\partial z}} = 0 \quad (2.113)$$

The sum of the first two terms in this equation, which can be written as $\bar{T}(\partial \bar{u}/\partial x + \partial \bar{v}/\partial y)$, is zero by virtue of the continuity equation so that:

$$\overline{T'\frac{\partial u'}{\partial x}} + \overline{T'\frac{\partial v'}{\partial y}} + \overline{T'\frac{\partial w'}{\partial z}} = 0 \quad (2.114)$$

Substituting this result into Eq. (2.112) and neglecting the term involving the z-derivative then gives:

$$\bar{u}\frac{\partial \bar{T}}{\partial x} + \bar{v}\frac{\partial \bar{T}}{\partial y} = \left(\frac{k}{\rho c_p}\right)\left(\frac{\partial^2 \bar{T}}{\partial x^2} + \frac{\partial^2 \bar{T}}{\partial y^2}\right) - \left[\frac{\partial}{\partial x}\overline{(u'T')} + \frac{\partial}{\partial y}\overline{(v'T')}\right] \quad (2.115)$$

This is the energy equation for two-dimensional turbulent flow. Comparing it with the equation for laminar flow shows that in turbulent flow extra terms arise because of the fluctuating velocity and temperature components. These terms arise because of the enthalpy transport caused by these fluctuating terms.

Now Eq. (2.115) can be written as

$$\bar{u}\frac{\partial \bar{T}}{\partial x} + \bar{v}\frac{\partial \bar{T}}{\partial y} = \left(\frac{1}{\rho c_p}\right)\left\{\frac{\partial}{\partial x}\left[\left(k\frac{\partial \bar{T}}{\partial x}\right) - (\rho c_p \overline{u'T'})\right]\right.$$
$$\left. + \frac{\partial}{\partial y}\left[\left(k\frac{\partial \bar{T}}{\partial y}\right) - (\rho c_p \overline{v'T'})\right]\right\} \quad (2.116)$$

Therefore, since $(-k\partial \bar{T}/\partial x)$ and $(-k\partial \bar{T}/\partial y)$ are the time-averaged heat conduction rates per unit area in the x- and y-directions respectively, it will be seen that the effects of the additional turbulence terms are the same as an increase in the heat transfer rate. For this reason, these extra terms $\rho c_p \overline{u'T'}$ and $\rho c_p \overline{v'T'}$ are often termed the turbulent heat transfer terms. Their presence can be demonstrated in a more physical manner using the same line of reasoning as was adopted in the discussion of the turbulent stresses. For example, considering the element shown in

CHAPTER 2: The Equations of Convective Heat Transfer 57

Fig. 2.11, the rate at which enthalpy enters the face ab parallel to the y-axis is equal to $\rho u c_p T \, dA_x$. If the instantaneous values of u and T are substituted into this equation and the time-average taken then the mean rate at which enthalpy enters this face per unit area will be found to be $(\rho c_p \bar{u}\bar{T} + \rho c_p \overline{u'T'})$. Thus, because of the fluctuating components, this rate is greater than ($\rho \times c_p \times$ mean velocity $\times$ mean temperature). This additional enthalpy transport is equivalent to additional heat transfer into this face at a rate of $\rho c_p \overline{u'T'}$ per unit area. Similarly, if the face of the element ad parallel to the x-axis is considered, the mean rate at which enthalpy enters it per unit area will be found to be $(\rho c_p \bar{v}\bar{T} + \rho c_p \overline{v'T'})$. There is, therefore, again an additional enthalpy transport due to the turbulent fluctuations, this addition being equivalent to an increase in the heat transfer rate per unit area of $\rho c_p \overline{v'T'}$.

The four equations governing two-dimensional turbulent flow are, therefore, Eqs. (2.95), (2.106), and (2.116). These contain, beside the four mean flow variables $\bar{u}$, $\bar{v}$, $\bar{p}$, and $\bar{T}$, additional terms which depend on the turbulence field. In order to solve this set of equations, therefore, information concerning these turbulence terms must be available and the difficulties associated with the solution of turbulent flow problems arise basically from the difficulty of analytically predicting the values of these terms. The set of equations effectively contains more variables than the number of equations. This is termed the turbulence "closure problem". In order to bring about closure of this set of equations, extra equations must be generated, these extra equations constituting a "turbulence model".

In the many traditional methods of calculating turbulent flows, these turbulence terms are empirically defined, i.e., turbulence models that are almost entirely empirical are used. Some success has, however, been achieved by using additional differential equations to help in the description of these terms. Empiricism is not entirely eliminated, at present, by the use of these extra equations but the empiricism can be introduced in a more systematic and logical manner than is possible if the turbulence terms in the momentum equation are completely empirically described. One of the most widely used additional equations for this purpose is the turbulence kinetic energy equation and its general derivation will now be discussed.

The x-wise momentum equation for two-dimensional mean flow is first multiplied by u' to give:

$$u'\frac{\partial}{\partial t}(\bar{u} + u') + u'(\bar{u} + u')\frac{\partial}{\partial x}(\bar{u} + u') + u'(\bar{v} + v')\frac{\partial}{\partial y}(\bar{u} + u') + u'w'\frac{\partial}{\partial z}(\bar{u} + u')$$

$$= -\frac{u'}{\rho}\frac{\partial}{\partial x}(\bar{p} + p') + (\frac{\mu}{\rho})u'\left[\frac{\partial^2}{\partial x^2}(\bar{u} + u') + \frac{\partial^2}{\partial y^2}(\bar{u} + u') + \frac{\partial^2}{\partial z^2}(\bar{u} + u')\right]$$

(2.117)

Time averaging the terms in this equation then gives for steady mean flow

$$\overline{\bar{u}u'\frac{\partial u'}{\partial x}} + \overline{u'^2\frac{\partial \bar{u}}{\partial x}} + \overline{u'^2\frac{\partial u'}{\partial x}} + \overline{\bar{v}u'\frac{\partial u'}{\partial y}} + \overline{u'v'\frac{\partial \bar{u}}{\partial y}} + \overline{u'v'\frac{\partial u'}{\partial y}} + \overline{u'w'\frac{\partial u'}{\partial z}}$$

$$= -\frac{1}{\rho}\overline{u'\frac{\partial p'}{\partial x}} + (\frac{\mu}{\rho})\overline{u'\left[\frac{\partial^2 u'}{\partial x^2} + \frac{\partial^2 u'}{\partial y^2} + \frac{\partial^2 u'}{\partial z^2}\right]}$$

which can be written as:

$$\bar{u}\frac{1}{2}\frac{\partial}{\partial x}(\overline{u'^2}) + \bar{v}\frac{1}{2}\frac{\partial}{\partial y}(\overline{u'^2}) + \overline{(u'^2)}\frac{\partial \bar{u}}{\partial x} + \overline{(u'v')}\frac{\partial \bar{u}}{\partial y} + \overline{\left(u'^2\frac{\partial u'}{\partial x} + u'v'\frac{\partial u'}{\partial y} + u'w'\frac{\partial u'}{\partial z}\right)}$$

$$= -\frac{1}{\rho}\overline{u'\frac{\partial p'}{\partial x}} + (\frac{\mu}{\rho})\overline{u'\nabla^2 u'}$$

(2.118)

where:

$$\nabla^2 = \frac{\partial^2}{\partial x^2} + \frac{\partial^2}{\partial y^2} + \frac{\partial^2}{\partial z^2} \qquad (2.119)$$

Now the bracketed term on the left-hand side of this equation can be written as:

$$\frac{1}{2}\overline{\left(\frac{\partial u'^3}{\partial x} + \frac{\partial u'^2 v'}{\partial y} + \frac{\partial u'^2 w'}{\partial z}\right)} - \frac{\overline{u'^2}}{2}\overline{\left(\frac{\partial u'}{\partial x} + \frac{\partial v'}{\partial y} + \frac{\partial w'}{\partial z}\right)} \qquad (2.120)$$

But multiplying the continuity equation by u'^2 and time averaging shows that:

$$\overline{u'^2\left(\frac{\partial u'}{\partial x} + \frac{\partial v'}{\partial y} + \frac{\partial w'}{\partial z}\right)} = 0 \qquad (2.121)$$

Using this result in Eq. (2.120) and then substituting back into Eq. (2.118) gives:

$$\bar{u}\frac{1}{2}\frac{\partial}{\partial x}(\overline{u'^2}) + \bar{v}\frac{1}{2}\frac{\partial}{\partial y}(\overline{u'^2}) = -\overline{(u'^2)}\frac{\partial \bar{u}}{\partial x} - \overline{(u'v')}\frac{\partial \bar{u}}{\partial y} - \frac{1}{\rho}\overline{u'\frac{\partial p'}{\partial x}}$$

$$- \frac{1}{2}\left[\frac{\partial \overline{(u'^3)}}{\partial x} + \frac{\partial}{\partial y}\overline{(u'^2 v'^2)} + \frac{\partial}{\partial z}\overline{(u'^2 w'^2)}\right] + \left(\frac{\mu}{\rho}\right)\overline{(u'\nabla^2 u')}$$

(2.122)

Similarly if the y-momentum equation is multiplied by v' and the result time averaged it can be shown that:

$$\bar{u}\frac{1}{2}\frac{\partial}{\partial x}\overline{(v'^2)} + \bar{v}\frac{1}{2}\frac{\partial}{\partial y}\overline{(v'^2)} = -\overline{(u'v')}\frac{\partial \bar{v}}{\partial x} - \overline{(v'^2)}\frac{\partial \bar{v}}{\partial y} - \frac{1}{\rho}\overline{v'\frac{\partial p'}{\partial y}}$$

$$- \frac{1}{2}\left[\frac{\partial}{\partial x}\overline{(v'^2 u')} + \frac{\partial}{\partial y}\overline{(v'^3)} + \frac{\partial}{\partial z}\overline{(v'^2 w')}\right] + \left(\frac{\mu}{\rho}\right)\overline{v'\nabla^2 v'}$$

(2.123)

Lastly, multiplying the z-momentum equation by w', time averaging, and rearranging gives:

$$\bar{u}\frac{1}{2}\frac{\partial}{\partial x}\overline{(w'^2)} + \bar{v}\frac{1}{2}\frac{\partial}{\partial y}\overline{(w'^2)} = -\frac{1}{\rho}\overline{w'\frac{\partial p'}{\partial z}}$$

$$- \frac{1}{2}\left[\frac{\partial}{\partial x}\overline{(w'^2 u')} + \frac{\partial}{\partial y}\overline{(w'^2 v')} + \frac{\partial}{\partial z}\overline{(w'^3)}\right] \quad (2.124)$$

$$+ \left(\frac{\mu}{\rho}\right)\overline{w'\nabla^2 w'}$$

CHAPTER 2: The Equations of Convective Heat Transfer 59

Adding these three equations, i.e., (2.122), (2.123), and (2.124), together and defining

$$K = (\overline{u'^2} + \overline{v'^2} + \overline{w'^2})/2 \tag{2.125}$$

K thus being the mean kinetic energy associated with the velocity fluctuations, gives:

$$\begin{aligned}\bar{u}\frac{\partial K}{\partial x} + \bar{v}\frac{\partial K}{\partial y} = &-\left[\overline{(u'^2)}\frac{\partial \bar{u}}{\partial x} + \overline{(u'v')}\frac{\partial \bar{u}}{\partial y} + \overline{(u'v')}\frac{\partial \bar{v}}{\partial x} + \overline{(v'^2)}\frac{\partial \bar{v}}{\partial y}\right] \\ &-\frac{1}{\rho}\left[\overline{u'\frac{\partial p'}{\partial x}} + \overline{v'\frac{\partial p'}{\partial y}} + \overline{w'\frac{\partial p'}{\partial z}}\right] \\ &-\frac{1}{2}\left[\frac{\partial}{\partial x}\overline{(u'^3)} + \frac{\partial}{\partial y}\overline{(u'^2 v')} + \frac{\partial}{\partial z}\overline{(u'^2 w')} + \frac{\partial}{\partial x}\overline{(v'^2 u')} \right. \\ &\left. + \frac{\partial}{\partial y}\overline{(v'^3)} + \frac{\partial}{\partial z}\overline{(v'^2 w')} + \frac{\partial}{\partial x}\overline{(w'^2 u')} + \frac{\partial}{\partial y}\overline{(w'^2 v')} + \frac{\partial}{\partial z}\overline{(w'^3)}\right] \\ &+ \left(\frac{\mu}{\rho}\right)\left[\overline{u'\nabla^2 u'} + \overline{v'\nabla^2 v'} + \overline{w'\nabla^2 w'}\right]\end{aligned} \tag{2.126}$$

It is, of course, not obvious, at this stage, how this equation can be used in the determination of the turbulence terms in the momentum equation. In order to utilize it for this purpose, some of the terms in the equation must be related to other flow quantities by approximate relations. A discussion of this will be given in Chapter 5.

2.7
SIMPLIFYING ASSUMPTIONS USED IN THE ANALYSIS OF CONVECTION

The general equations governing convective heat transfer, i.e., the continuity, Navier-Stokes, and energy equations, together form a very complex set of simultaneous partial differential equations. Analytical solutions to this set of equations have only been found for a few, relatively very simple cases. Numerical solutions to these equations can be obtained using relatively moderately sized computers although the computer times involved with three-dimensional flow solutions may be relatively large. There is still, therefore, considerable interest in obtaining approximate analytical or simplified numerical solutions to these equations. Now, in many flows of practical importance, some of the terms in the equations are negligibly small compared to the remaining terms. By dropping these terms, i.e., by assuming that the flow has certain simplifying characteristics, the complexity of the governing set of equations is often considerably reduced and its solution becomes much more feasible. The following two types of flow are, from a practical viewpoint, probably the most important in which such simplifying assumptions can be adopted.

- In the case of flow inside a duct, it can be assumed that the flow is fully developed. Basically this means that it can be assumed that certain properties of the flow do not change with distance along the duct. Real duct flows well away from the entrance or other fittings are very nearly fully developed in many cases.

For fully developed duct flows in which it can be assumed that the fluid properties are constant, the form of the velocity and temperature profiles do not change with distance along the duct, i.e., considering the variables as defined in Fig. 2.12, if the velocity and temperature profiles are expressed in the form

$$\left(\frac{u}{u_c}\right) = f\left(\frac{y}{R}\right), \quad \left(\frac{T - T_c}{T_w - T_c}\right) = g\left(\frac{y}{R}\right) \tag{2.127}$$

then in fully developed flow, the velocity and temperature profile functions f and g are independent of the distance along the duct. This will mean that the velocity at any distance y from the center line of the duct will remain constant with distance along the duct and that the temperature at this position will vary in such a way relative to the center line and wall temperatures that $(T - T_c)/(T_w - T_c)$ remains constant. In fully developed flow because u is not changing, it follows that the velocity components in the radial and tangential coordinate directions will be zero.

From this discussion it should be clear that many of the terms in the general governing equations will be zero when they are applied to fully developed flow. For this reason, analytical solutions for fully developed flows can relatively easily be derived.

- In the case of external flows at high Reynolds numbers, it can be assumed that the effects of viscosity (and, therefore, turbulence) and heat transfer are restricted to a thin region immediately adjacent to the surface over which the fluid is flowing and that outside this region the flow is effectively inviscid and at constant temperature. In this region, termed the boundary layer, in which the effects of viscosity and heat transfer are important, the velocity and temperature change from the values they have at the wall to the values applicable in the outer inviscid flow. Since the boundary layer is thin in a very large number of practically important flows, the gradients of velocity and temperature in the direction normal to the surface are high. As a result it is possible, as will be shown in the next section, to drop a number of terms from the governing equations when applying them to the flow in the boundary layer. Further, because the boundary layer is thin, it is possible to calculate the outer effectively inviscid flow by neglecting the presence of the

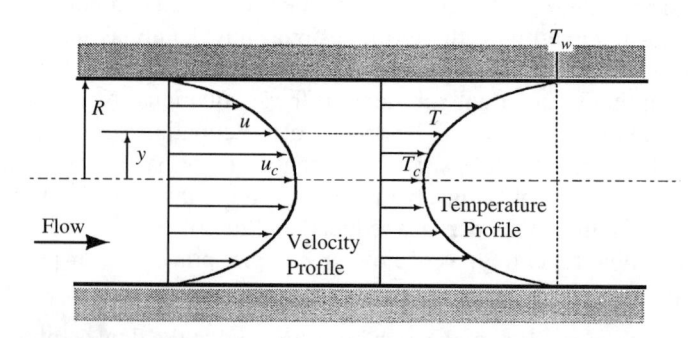

FIGURE 2.12
Velocity and temperature profiles in pipe flow.

boundary layer, i.e., by simply calculating the potential flow over the surface concerned. The velocity at the surface given by this potential flow solution is then used as the boundary condition on velocity at the outer edge of the boundary layer in the boundary layer solution. Because of the simplifications resulting from the introduction of the boundary layer concept, it is possible to obtain analytical solutions to such flows in a number of cases. Numerical solutions to the boundary layer equations are also readily obtained with extremely modest computer resources compared to those needed for the solution of the full equations.

2.8
THE BOUNDARY LAYER EQUATIONS FOR LAMINAR FLOW

As explained in the preceding section, it is possible in many external flows in which the Reynolds numbers are large (the reason for this stipulation is discussed later) to consider the flow to be split into two regions. It is not, of course, possible to define precisely where one region ends and the other begins, the two regions actually merging into each other with no sharp division between them. The two regions are:

- A thin layer adjacent to the surface in which the effects of viscosity and heat transfer are important. This region is termed the boundary layer.
- An outer region in which the flow is effectively inviscid and at uniform temperature. This region is often termed the freestream.

The two regions are shown schematically in Fig. 2.13.

The flow in the boundary layer can be either laminar or turbulent. In the present section, attention is being restricted to those cases in which the flow in the boundary layer is laminar.

Now, in general, the effects of viscosity and heat transfer do not extend to the same distance from the surface. For this reason, it is convenient to define both a velocity boundary layer thickness and a thermal or temperature boundary layer thickness as shown in Fig. 2.14. The velocity boundary layer thickness is a measure of the distance from the surface at which viscous effects cease to be important while the thermal boundary layer thickness is a measure of the distance from the wall at which heat transfer effects cease to be important.

For the flow of gases, the two boundary layer thicknesses are generally, approximately equal. For the flow of liquids, however, the two thicknesses are generally

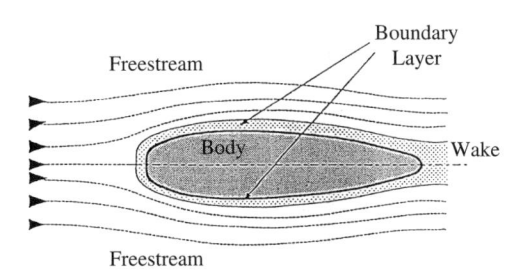

FIGURE 2.13
Boundary layer and freestream flows.

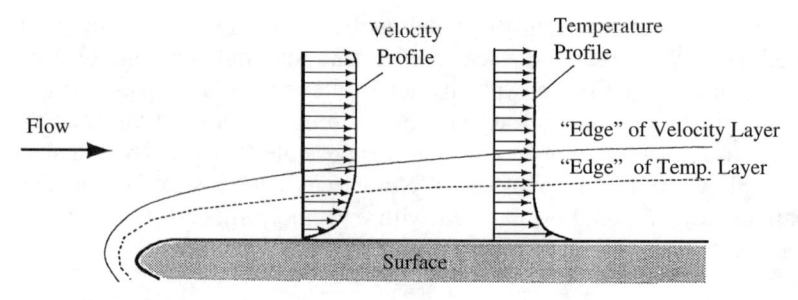

FIGURE 2.14
Velocity and temperature boundary layers.

very different although, except in the case of liquid metals, both are of the same order of magnitude.

The discussion of the present section is concerned with the simplifications to the general equations that can be introduced as a result of the fact that the boundary layer is thin. Due to the action of viscosity, the velocity of the fluid in contact with the surface over which it is flowing has the same velocity as this surface. In most cases, this surface velocity is zero in a coordinate system attached to the body. Similarly, the temperature of the fluid in contact with the surface is the same as that of the surface. At the outer "edge" of the boundary layer, the fluid velocity and temperature become equal to those in the freestream and so are very different from the values at the wall. Therefore, if the boundary layer is thin, the gradients of velocity and temperature across the boundary layer, i.e., normal to the surface, will be much greater than those in the direction parallel to the surface. It is as a result of this that certain of the terms in the governing equations are negligibly small compared to the remaining terms and can be ignored. The equations that result when these terms are dropped are called the boundary layer equations.

In order to illustrate how the boundary layer equations are derived, [7],[9],[13], [14],[15], consider two-dimensional constant fluid property flow over a plane surface which is set parallel to the x-axis. The following are then defined:

u_1 = Some measure of the order of magnitude of the velocity in the freestream outside the boundary layer.

L = Some characteristic dimension of the body, e.g., its length.

δ = Some measure of the order of magnitude of the boundary layer thickness.

No distinction is made at this stage between the velocity and temperature boundary layer thicknesses since both are assumed to be of the same order of magnitude. δ is, therefore, a measure of the order of magnitude of both boundary layer thicknesses.

Now, in the boundary layer, the velocity component u varies from zero at the surface, i.e., at $y = 0$, to approximately u_1 at the outer edge of the layer, where y is of the same order as δ. In the boundary layer, therefore, (u/u_1) has the order of magnitude of 1, i.e., $o(1)$. Similarly, since y varies from 0 to δ across the boundary layer, (y/δ) will have $o(1)$ in the boundary layer. Similarly, (x/L) has $o(1)$.

In light of these conclusions regarding the orders of magnitude of (u/u_1), (y/δ), and (x/L), all being $o(1)$, consideration can now be given to the derivation of the

CHAPTER 2: The Equations of Convective Heat Transfer 63

boundary layer equations by an order of magnitude analysis of the terms in the general equations. Consider, first, the continuity equation which for the case under consideration is:

$$\frac{\partial u}{\partial x} + \frac{\partial v}{\partial y} = 0$$

This can be rewritten as follows

$$\frac{\partial(u/u_1)}{\partial(x/L)} + \frac{\partial(v/u_1)}{\partial(y/\delta)} \times \frac{1}{(\delta/L)} = 0 \qquad (2.128)$$

From this it follows that

$$o(v/u_1) = \frac{o(u/u_1)}{o(x/L)} \times o(y/\delta) \times o(\delta/L) \qquad (2.129)$$

Therefore, since, as discussed above, $o(u/u_1)$, $o(y/\delta)$, and $o(x/L)$ are all unity:

$$o\left(\frac{v}{u_1}\right) = o\left(\frac{\delta}{L}\right) \qquad (2.130)$$

Since the basic assumption of boundary layer theory is that (δ/L) is small, it follows from this that (v/u_1) is also small. Thus, in the boundary layer the lateral velocity component, v, is very much smaller than the longitudinal component, u, which, of course, is what is physically to be expected.

Next consider the x-wise momentum equation. For the type of flow being considered, this has the form

$$u\frac{\partial u}{\partial x} + v\frac{\partial u}{\partial y} = -\frac{1}{\rho}\frac{\partial p}{\partial x} + \left(\frac{\mu}{\rho}\right)\left[\frac{\partial^2 u}{\partial x^2} + \frac{\partial^2 u}{\partial y^2}\right]$$

This can be rewritten as

$$\left(\frac{u}{u_1}\right)\frac{\partial(u/u_1)}{\partial(x/L)} + \left(\frac{v}{u_1}\right)\frac{\partial(u/u_1)}{\partial(y/\delta)} \times \frac{1}{(\delta/L)}$$

$$= -\left(\frac{L}{\rho u_1^2}\right)\frac{\partial p}{\partial x} + \left(\frac{\mu}{\rho u_1 L}\right)\left[\frac{\partial^2(u/u_1)}{\partial(x/L)^2} + \frac{\partial^2(u/u_1)}{\partial(y/\delta)^2}\frac{1}{(\delta/L)^2}\right] \qquad (2.131)$$

Now the pressure changes in the flow can at most be of the same order as the dynamic pressure ρu_1^2 and because x is of the same order as L, it follows that the first term on the right-hand side of Eq. (2.131) has, at most, the order of magnitude of unity, i.e., $o(1)$.

Therefore, defining

$$Re_L = \rho u_1 L/\mu \qquad (2.132)$$

Re_L thus being the Reynolds' number based on u_1 and L, the orders of magnitude of the various terms in Eq. (2.131) are as follows:

$$o(1)\frac{o(1)}{o(1)} + o(\delta/L)\frac{o(1)}{o(1)}\frac{1}{(\delta/L)} = o(1) + \frac{1}{Re_L}\left[\frac{o(1)}{o(1)} + \frac{o(1)}{o(1)}\right]\frac{1}{(\delta/L)^2}$$

64 Introduction to Convective Heat Transfer Analysis

i.e.,

$$o(1) + o(1) = o(1) + \left[o\left(\frac{1}{Re_L}\right) + \frac{o(1/Re_L)}{(\delta/L)^2} \right] \quad (2.133)$$

The last term in this equation shows that if (δ/L) is to be small, as it must be if the boundary layer approximations are to be applicable, then Re_L must be large, its order of magnitude being such that:

$$\frac{o(1/Re_L)}{o[(\delta/L)^2]} = o(1)$$

i.e.,

$$o(Re_L) = o\left[\frac{1}{(\delta/L)^2}\right] \quad (2.134)$$

Basically, in deriving the dimensionless equation (2.134), the magnitude of the viscous term has been compared with the magnitude of the inertia "force" of which ρu_1^2 is a measure. From this comparison, it then follows that if these two have the same orders of magnitude and δ/L is small, Re_L must be large. Note also from Eq. (2.135) it follows that $\delta/L = o[1/\sqrt{Re_L}]$.

As a consequence of the above result, it is seen that the first term within the square bracket on the right-hand side of Eq. (2.134), i.e., the term originating from $(\mu \partial^2 u/\partial x^2)$ will have $o(\delta/L)^2$ whereas the other terms in the equation have $o(1)$. For this reason, when dealing with boundary layer flows, this term in the x-wise Navier-Stokes equation is negligible. For such flows, therefore, the x-momentum equation is:

$$u\frac{\partial u}{\partial x} + v\frac{\partial u}{\partial y} = -\frac{1}{\rho}\frac{\partial p}{\partial x} + \left(\frac{\mu}{\rho}\right)\frac{\partial^2 u}{\partial y^2} \quad (2.135)$$

Next, consider the y-direction Navier-Stokes equation which for the flow situation being considered is:

$$u\frac{\partial v}{\partial x} + v\frac{\partial v}{\partial y} = -\frac{1}{\rho}\frac{\partial p}{\partial y} + \left(\frac{\mu}{\rho}\right)\left[\frac{\partial^2 v}{\partial x^2} + \frac{\partial^2 v}{\partial y^2}\right] \quad (2.136)$$

Expressing this in terms of the same dimensionless variables as were used with the x-direction equation and then multiplying through by (δ/L) gives:

$$(u/u_1)\frac{\partial(v/u_1)}{\partial(x/L)}(\delta/L) + (v/u_1)\frac{\partial(v/u_1)}{\partial(y/\delta)}$$

$$= -\frac{\partial(p/\rho u_1^2)}{\partial(y/\delta)} + \frac{1}{Re_L}\left[\frac{\partial^2(v/u_1)}{\partial(x/L)^2}(\delta/L) + \frac{\partial^2(v/u_1)}{\partial(y/\delta)^2(\delta/L)}\right] \quad (2.137)$$

Now it was shown above that (v/u_1) was of the order of (δ/L) and that for the boundary layer assumptions to be applicable Re_L was of the order of $1/(\delta/L)^2$. The orders of magnitude of the various terms in Eq. (2.137) are then as follows:

$$o[(\delta/L)^2] + o[(\delta/L)^2] = -\frac{\partial(p/\rho u_1^2)}{\partial(y/\delta)} + o[(\delta/L)^4] + o[(\delta/L)^2] \quad (2.138)$$

CHAPTER 2: The Equations of Convective Heat Transfer 65

Since terms of the order of magnitude of $(\delta/L)^2$ and smaller are being neglected, it follows from this equation that the y-momentum equation for boundary layer flow is:

$$\frac{\partial p}{\partial y} = 0 \qquad (2.139)$$

Eq. (2.139) indicates that the changes of pressure in the y-direction can be ignored in boundary layer flows. It follows that the x-direction momentum equation for two-dimensional boundary layer flows, i.e., Eq. (2.135), can be written as:

$$u\frac{\partial u}{\partial x} + v\frac{\partial u}{\partial y} = -\frac{1}{\rho}\frac{dp}{dx} + \left(\frac{\mu}{\rho}\right)\frac{\partial^2 u}{\partial y^2} \qquad (2.140)$$

because the pressure depends only on x. The pressure in the boundary layer will, therefore, be essentially the same as that acting along the outer "edge" of the boundary layer at the same value of x. As previously mentioned, however, because the boundary layer is thin, the potential flow over the surface can be calculated ignoring the presence of the boundary layer and the values of the variables that this solution gives at the surface can be used as the boundary conditions at the outer edge of the boundary layer. As a result of this, (dp/dx) becomes a known quantity in the boundary layer solution.

Next consider the energy equation which for the type of flow being considered is:

$$u\frac{\partial T}{\partial x} + v\frac{\partial T}{\partial y} = \left(\frac{k}{\rho c_p}\right)\left[\frac{\partial^2 T}{\partial x^2} + \frac{\partial^2 T}{\partial y^2}\right]$$
$$+ \left(\frac{\mu}{\rho c_p}\right)\left\{2\left[\frac{\partial u}{\partial x}\right]^2 + 2\left[\frac{\partial v}{\partial y}\right]^2 + \left[\frac{\partial u}{\partial y} + \frac{\partial v}{\partial x}\right]^2\right\} \qquad (2.141)$$

The following dimensionless temperature variable is introduced:

$$\theta = \frac{(T - T_1)}{(T_{wr} - T_1)} \qquad (2.142)$$

where T_{wr} is some measure of the order of magnitude of the surface temperature and T_1 is some measure of the order of magnitude of the freestream temperature. In the boundary layer, θ will be of the order of magnitude of unity. In terms of the dimensionless variables, Eq. (2.141) becomes:

$$(u/u_1)\frac{\partial \theta}{\partial (x/L)} + (v/u_1)\frac{\partial \theta}{\partial (y/\delta)}\frac{1}{(\delta/L)} = \left(\frac{k}{\rho c_p L u_1}\right)\left[\frac{\partial^2 \theta}{\partial (x/L)^2} + \frac{\partial^2 \theta}{\partial (y/\delta)^2}\frac{1}{(\delta/L)^2}\right]$$
$$+ \left[\frac{\mu u_1}{\rho c_p L (T_{wr} - T_1)}\right]\left\{2\left[\frac{\partial (u/u_1)}{\partial (x/L)}\right]^2 \qquad (2.143)\right.$$
$$\left.+2\left[\frac{\partial (v/u_1)}{\partial (y/\delta)}\right]^2\frac{1}{(\delta/L)^2} + \left[\frac{\partial (u/u_1)}{\partial (y/\delta)}\frac{1}{(\delta/L)} + \frac{\partial (v/u_1)}{\partial (x/L)}\right]^2\right\}$$

It will be noted that

$$\left(\frac{k}{\rho c_p L u_1}\right) = \frac{1}{Pr\, Re_L}$$

and

$$\frac{\mu u_1}{\rho c_p L (T_{wr} - T_1)} = \frac{Ec}{Re_L}$$

where Pr is the Prandtl number and Ec the Eckert number. Therefore, if these terms are at most of the order of 1 (Ec can of course be very small in which case the dissipation term is negligible) the various terms in Eq. (2.143) are as follows:

$$o(1) + \frac{o(\delta/L)}{o(\delta/L)} = o[(\delta/L)^2]\left\{o(1) + \frac{1}{o[(\delta/L)^2]}\right\} + o[(\delta/L)^2]$$

$$\left\{o(1) + \frac{o[(\delta/L)^2]}{o[(\delta/L)^2]} + \frac{1}{o[(\delta/L)^2]} + o[(\delta/L)^2] + o(1) + o[(\delta/L)^2]\right\}$$

(2.144)

the last term in the square bracket in Eq. (2.143) having been multiplied out.

If terms of the order $(\delta/L)^2$ and less are again neglected, it will be seen that the energy equation for laminar two-dimensional boundary layer flow becomes:

$$u\frac{\partial T}{\partial x} + v\frac{\partial T}{\partial y} = \left(\frac{k}{\rho c_p}\right)\frac{\partial^2 T}{\partial y^2} + \left(\frac{\mu}{\rho c_p}\right)\left(\frac{\partial u}{\partial y}\right)^2 \quad (2.145)$$

To summarize, therefore, the equations for two-dimensional, constant fluid property boundary layer flow over a plane surface are:

$$\frac{\partial u}{\partial x} + \frac{\partial v}{\partial y} = 0$$

$$u\frac{\partial u}{\partial x} + v\frac{\partial u}{\partial y} = -\frac{1}{\rho}\frac{dp}{dx} + \left(\frac{\mu}{\rho}\right)\frac{\partial^2 u}{\partial y^2}$$

$$u\frac{\partial T}{\partial x} + v\frac{\partial T}{\partial y} = \left(\frac{k}{\rho c_p}\right)\frac{\partial^2 T}{\partial y^2} + \left(\frac{\mu}{\rho c_p}\right)\left(\frac{\partial u}{\partial y}\right)^2$$

These equations were derived for flow over a plane surface. They may be applied to flow over a curved surface provided that the boundary layer thickness remains small compared to the radius of curvature of the surface. When applied to flow over a curved surface, x is measured along the surface and y is measured normal to it at all points as shown in Fig. 2.15, i.e., body-fitted coordinates are used.

In order to solve the above set of boundary layer equations, the boundary conditions on the variables involved, i.e., u, v, and T, must be known. At the wall, the

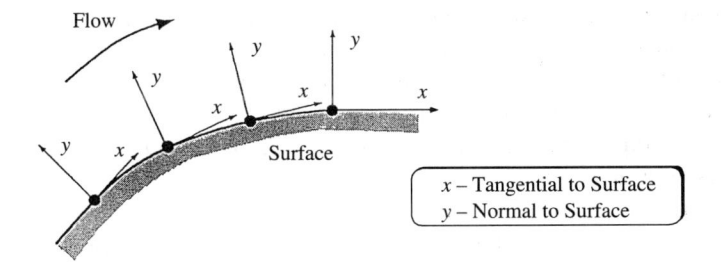

FIGURE 2.15
Coordinate system used for boundary layer flow over a curved surface.

no-slip condition requires that the velocity be the same as that of the wall, i.e., usually zero, so that one such boundary condition is

$$\text{At } y = 0: \quad u = v = 0 \tag{2.146}$$

If there is blowing or suction at the surface, the rate of blowing or sucking will determine the value of v at the wall.

The boundary conditions on temperature at the wall depend on the thermal conditions specified at the wall. If the distribution of the temperature of the wall is specified then, since the fluid in contact with the wall must be at the same temperature as the wall, the boundary condition on temperature is:

$$\text{At } y = 0: \quad T = T_w \tag{2.147}$$

where T_w is the temperature of the wall at the particular value of x being considered.

If, instead of the wall temperature, the heat transfer rate at the wall is specified, the boundary condition becomes, using Fourier's law:

$$\text{At } y = 0: \quad \frac{\partial T}{\partial y} = -\frac{q_w}{k} \tag{2.148}$$

where q_w is the specified local heat transfer rate from the wall per unit area at the particular value of x being considered.

The conditions on velocity at the outer "edge" of the boundary layer are a little more difficult to define because there is really an interaction between the boundary layer flow and the outer inviscid flow, i.e., because of the reduction in velocity near the surface, the outer flow is somewhat different from that in truly inviscid flow over the surface. However, as discussed above, in many cases, this effect can be ignored because the boundary layer remains thin and in such cases the inviscid flow over the surface considered is calculated and the value that this solution gives for the velocity on the surface at any position is used as the boundary condition at the outer "edge" of the boundary layer, i.e., if $u_1(x)$ is the surface velocity distribution given by the solution for inviscid flow over the surface then the boundary condition on the boundary layer solution is

$$\text{For large } y: \quad u \to u_1(x) \tag{2.149}$$

For example, the inviscid solution for flow over a flat plate is simply that the velocity is constant everywhere and equal to the velocity in the undisturbed flow ahead of the plate, say u_1. In calculating the boundary layer on a flat plate, therefore, the outer boundary condition is that u must tend to u_1 at large y. The term "large y" is meant to imply "outside the boundary layer", the boundary layer thickness, δ, being by assumption small.

There are some cases where this approach fails. One such case is that in which significant regions of separated flow exist. In this case, although the boundary layer equations are adequate to describe the flow upstream of the separation point, the presence of the separated region alters the "effective" body shape for the outer inviscid flow and the velocity outside the boundary layer will be different from that given by the inviscid flow solution over the solid surface involved. For example, consider flow over a circular cylinder as shown in Fig. 2.16. Potential theory gives the velocity, u_1, on the surface of the cylinder as:

$$(u_1/U) = 2\sin(x/R) \tag{2.150}$$

However, because of the large wake, the actual velocity distribution outside of the boundary layer is very different from this except in the vicinity of the stagnation point.

Another case where the boundary layer and the freestream interact occurs in the entrance region to ducts as shown in Fig. 2.17.

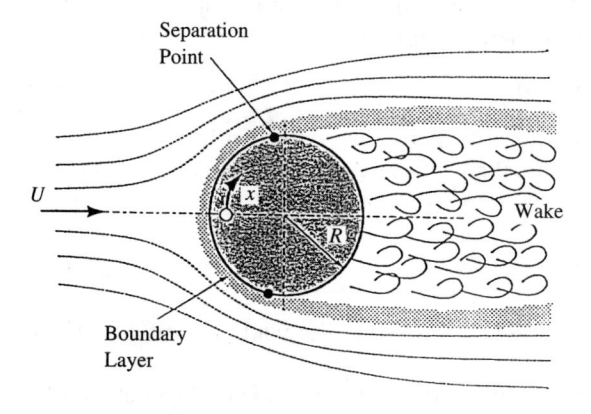

FIGURE 2.16
Two-dimensional flow over a circular cylinder.

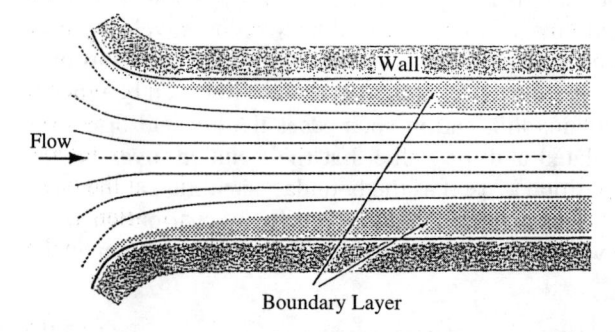

FIGURE 2.17
Flow in the entrance region to a duct.

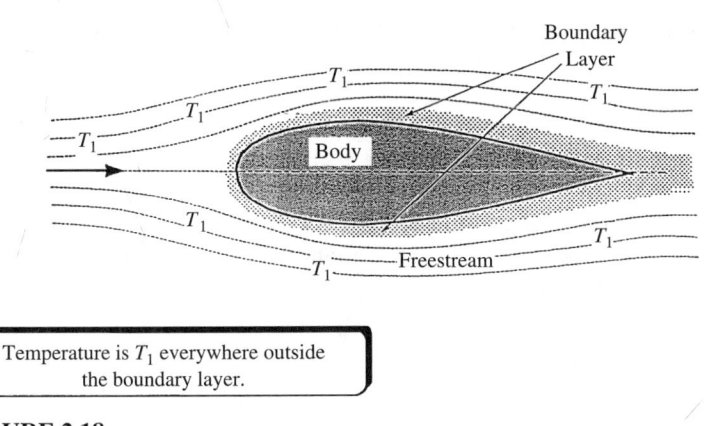

CHAPTER 2: The Equations of Convective Heat Transfer   69

FIGURE 2.18
Temperature distribution outside boundary layer.

Such flow can be treated adequately using the boundary layer assumptions but the freestream velocity gradients exist purely because of the boundary layer growth and the boundary layer and inviscid core flows must be simultaneously considered. There are, nevertheless, many important practical problems in which such interactions can be ignored.

In passing, it should also be noted that since the flow outside the boundary is assumed to be inviscid and since v_1 is of a lower order of magnitude than u_1 at the outer edge of the boundary layer, the Bernoulli equation gives:

$$\frac{\partial p}{\partial x} = -\rho u_1 \frac{\partial u_1}{\partial x} \qquad (2.151)$$

This equation then relates the pressure gradient term in the boundary layer equations to the freestream velocity distribution.

The last boundary condition that has to be considered is that on temperature at the outer "edge" of the boundary layer. Since, by assumption, the effects of heat transfer are negligible outside the boundary layer, the fluid temperature outside the boundary layer must everywhere be the same and, therefore, equal to the temperature in the undisturbed freestream ahead of the surface as shown in Fig. 2.18.

The outer boundary condition on temperature is, therefore:

$$\text{For large } y: \qquad T \to T_1 \qquad (2.152)$$

2.9
THE BOUNDARY LAYER EQUATIONS FOR TURBULENT FLOW

Compared to the x-wise Navier-Stokes equation for two-dimensional laminar flow, the equivalent equation for turbulent flow has, when the time averaged values of the variables are substituted in place of the steady values in laminar flow, the additional term:

$$-\left[\frac{\partial \overline{u'^2}}{\partial x} + \frac{\partial}{\partial y}(\overline{u'v'})\right] \quad (2.153)$$

If the turbulent momentum equation is expressed in nondimensional form in the same way as was done in deriving the laminar boundary layer equations then the additional term becomes:

$$-\left[\frac{\partial}{\partial(x/L)}\left(\frac{\overline{u'^2}}{u_1^2}\right) + \frac{\partial}{\partial(y/\delta)}\left(\frac{\overline{u'v'}}{u_1^2}\right)\frac{1}{(\delta/L)}\right] \quad (2.154)$$

Now, the rest of the terms retained in the boundary layer equations have the order of magnitude of unity and, therefore, for the boundary layer equations to apply, the dimensionless turbulence terms $(\overline{u'^2}/u_1^2)$ and $(\overline{u'v'}/u_1^2)$, which are assumed to have the same order of magnitude, will have the order of magnitude of (δ/L) at most. The first term in Eq. (2.154) is, therefore, negligible compared to the rest of the terms in the boundary layer equations. Therefore, the x-wise momentum equation for turbulent boundary layer flow is:

$$\bar{u}\frac{\partial \bar{u}}{\partial x} + \bar{v}\frac{\partial \bar{u}}{\partial y} = -\frac{1}{\rho}\frac{\partial \bar{p}}{\partial x} + \nu\frac{\partial^2 \bar{u}}{\partial y^2} - \frac{\partial}{\partial y}(\overline{u'v'}) \quad (2.155)$$

Consider next the y-momentum equation. Expressing this in dimensionless form in the same way as was done with the laminar flow equation and neglecting the appropriate mean flow terms, this equation then gives:

$$-\frac{\partial(\bar{p}/u_1^2)}{\partial(y/\delta)} - \frac{\partial}{\partial(x/L)}\left(\frac{\overline{u'v'}}{u_1^2}\right)(\delta/L) - \frac{\partial}{\partial(y/\delta)}\left(\frac{\overline{v'^2}}{u_1^2}\right) = 0 \quad (2.156)$$

But $\overline{v'^2}/u_1^2$ must have the same order of magnitude as the turbulence terms in the x-momentum equation so its order of magnitude is (δ/L). Hence, it follows that in turbulent flow, as in laminar flow:

$$\frac{\partial \bar{p}}{\partial y} = 0 \quad (2.157)$$

The same line of reasoning can be applied to the energy equation and if it is assumed that the turbulence terms $(\overline{u'T'})/u_1(T_{wr} - T_1)$ and $(\overline{v'T'})/u_1(T_{wr} - T_1)$ in the resultant equation have the same order of magnitude as the turbulence terms in the momentum equation, i.e., (δ/L), then the energy equation for turbulent boundary layer flow becomes

$$\bar{u}\frac{\partial \bar{T}}{\partial x} + \bar{v}\frac{\partial \bar{T}}{\partial y} = \left(\frac{k}{\rho c_p}\right)\frac{\partial^2 \bar{T}}{\partial y^2} - \frac{\partial}{\partial y}(\overline{v'T'}) \quad (2.158)$$

To summarize, the equations for two-dimensional constant fluid property, turbulent boundary layer flow are

$$\frac{\partial \bar{u}}{\partial x} + \frac{\partial \bar{v}}{\partial y} = 0$$

$$\bar{u}\frac{\partial \bar{u}}{\partial x} + \bar{v}\frac{\partial \bar{u}}{\partial y} = -\frac{1}{\rho}\frac{\partial \bar{p}}{\partial x} + \nu\frac{\partial^2 \bar{u}}{\partial y^2} - \frac{\partial}{\partial y}\overline{(u'v')}$$

$$\bar{u}\frac{\partial \bar{T}}{\partial x} + \bar{v}\frac{\partial \bar{T}}{\partial y} = \left(\frac{k}{\rho c_p}\right)\frac{\partial^2 \bar{T}}{\partial y^2} - \frac{\partial}{\partial y}\overline{(v'T')}$$

As was the case with the full equations, these contain beside the three mean flow variables $\bar{u}$, $\bar{v}$, and $\bar{T}$ (the pressure is, of course, by virtue of Eq. (2.157) again determined by the external inviscid flow) additional terms arising as a result of the turbulence. Therefore, as previously discussed, in order to solve this set of equations, there must be an additional input of information, i.e., a turbulence model must be used. Many turbulence models are based on the turbulence kinetic energy equation that was previously derived. When the boundary layer assumptions are applied to this equation, it becomes:

$$\bar{u}\frac{\partial}{\partial x}K + \bar{v}\frac{\partial}{\partial y}K = -\overline{(u'v')}\frac{\partial \bar{u}}{\partial y} - \frac{1}{\rho}\overline{\left(v'\frac{\partial p'}{\partial y}\right)}$$
$$- \frac{1}{2}\left[\frac{\partial}{\partial y}\overline{(u'^2 v')} + \frac{\partial}{\partial y}\overline{(v'^3)} + \frac{\partial}{\partial y}\overline{(w'^3 v')}\right] \quad (2.159)$$
$$+ \nu[\overline{(u'\nabla^2 u')} + \overline{(v'\nabla^2 v')} + \overline{(w'\nabla^2 w')}]$$

2.10
THE BOUNDARY LAYER INTEGRAL EQUATIONS

While the boundary layer equations that were derived in the preceding sections are much simpler than the general equations from which they were derived, they still form a complex set of simultaneous partial differential equations. Analytical solutions to this set of equations have been obtained in a few important cases. For the majority of flows, however, a numerical solution procedure must be adopted. Such solutions are readily obtained today using modest modern computing facilities. This was, however, not always so. For this reason, approximate solutions to the boundary layer equations have in the past been quite widely used. While such methods of solution are less important today, they are still used to some extent. One such approach will, therefore, be considered in the present text.

When only approximate values of the overall features of the flow, such as the surface heat transfer rate and surface shearing stress, are required, it is possible to apply the boundary layer assumptions in a different way to obtain, relatively easily, approximate solutions for these quantities. The derivation of the equations required in this approximate solution procedure will be discussed in the present section [2],[7],[13].

Now the Navier-Stokes and energy equations are derived by applying the conservation of momentum and conservation of energy principles to a differentially small control volume through which the fluid is flowing and then taking the limiting form of these equations as the dimensions of this control volume tend to zero. The resultant set of equations governs the conditions at any point in the flow. In the approximate method being discussed here these same conservation principles are applied to a control volume which spans the whole boundary layer as shown in Fig. 2.19. The height, ℓ, of the control volume is chosen to be greater than the boundary layer thickness.

The limiting forms of the equations that result from the application of these conservation principles to this control volume as $dx \to 0$ give a set of equations governing the "average" conditions across the boundary layer. These resultant equations are termed the boundary layer momentum integral and energy integral equations.

In order to illustrate how these integral equations are derived, attention will be given to two-dimensional, constant fluid property flow. First, consider conservation of momentum. It is assumed that the flow consists of a boundary layer and an outer inviscid flow and that, because the boundary layer is thin, the pressure is constant across the boundary layer. The boundary layer is assumed to have a distinct edge in the present analysis. This is shown in Fig. 2.20.

The width of the control volume is arbitrary and is taken as unity for convenience. The wall is assumed to be solid.

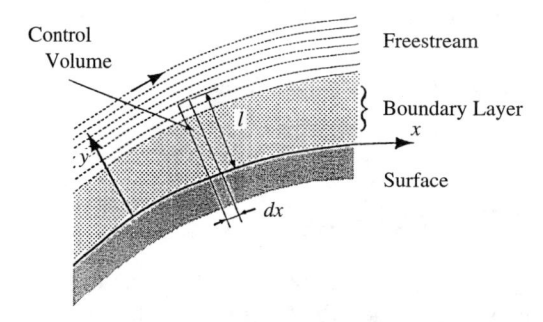

FIGURE 2.19
Control volume used in deriving the boundary layer integral equation.

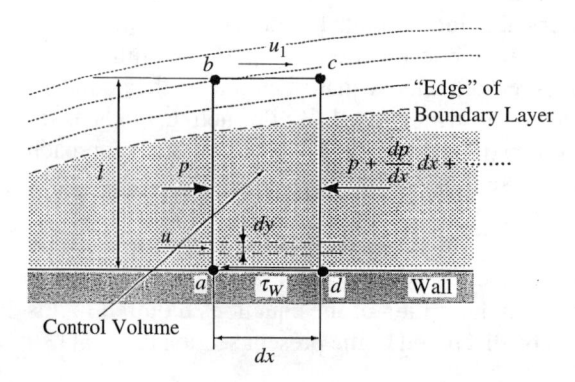

FIGURE 2.20
Assumed flow through control volume.

CHAPTER 2: The Equations of Convective Heat Transfer   73

Applying the conservation of momentum principle to the control volume indicated by *abcd* in Fig. 2.20 then gives:

$$\begin{array}{l}\text{Net force}\\\text{on control}\\\text{volume in}\\\text{positive}\\\text{x-direction}\end{array} = \begin{array}{l}\text{Rate at which}\\\text{x-momentum}\\\text{leaves through}\\\text{face } dc\end{array} - \begin{array}{l}\text{Rate at which}\\\text{x-momentum}\\\text{enters through}\\\text{face } ab\end{array} - \begin{array}{l}\text{Rate at which}\\\text{x-momentum}\\\text{enters through}\\\text{face } bc\end{array} \quad (2.160)$$

The various terms in this equation will now be separately considered starting with the net force term. Because *bc* is parallel to the wall and lies in the freestream where, by the basic assumptions, viscous forces are negligible, there can be no force acting on *bc*. Therefore:

$$\begin{array}{l}\text{Net force on control}\\\text{in x-direction, } F_x\end{array} = \begin{array}{l}\text{Pressure}\\\text{force on } ab\end{array} - \begin{array}{l}\text{Pressure}\\\text{force on } dc\end{array} - \begin{array}{l}\text{Shearing}\\\text{force on } ad\end{array}$$

$$= p\ell - \left(p + \frac{dp}{dx}dx\right)\ell - \tau_w\,dx = -\frac{dp}{dx}dx\,\ell - \tau_w dx \quad (2.161)$$

where τ_w is the shearing stress acting on the wall.

Next consider the momentum terms in Eq. (2.160). To find the rate at which momentum enters through *ab* it is noted that the rate at which mass crosses the strip on *ab* indicated in Fig. 2.20 is equal to $(\rho u\,dy)$, the control surface having a unit width. The rate at which momentum enters through this strip is, therefore, equal to $(\rho u^2\,dy)$. The rate at which momentum enters through the face *ab*, which will here be termed M_{ab}, is, therefore, given by

$$M_{ab} = \rho \int_o^\ell u^2\,dy \quad (2.162)$$

It should be noted, and the point will be discussed further at a later stage, that the equations derived in the present section are applicable to both laminar flow and turbulent flow. In the case of turbulent flow, the mean values of the variables are implied although strictly for turbulent flow, M_{ab} will be given by

$$\rho \int_o^\ell \bar{u}^2\,dy + \rho \int_o^\ell \overline{u'^2}\,dy$$

For reasons discussed in the derivation of the partial differential equations for turbulent boundary layer flow, the turbulence term is, however, neglected.

The rate at which momentum leaves the control volume through the face *dc*, which will here be termed M_{dc}, is, because dx is by assumption small, related to M_{ab} by:

$$M_{dc} = M_{ab} + \frac{d}{dx}(M_{ab})\,dx \quad (2.163)$$

Using Eq. (2.162) then gives

$$M_{dc} = \rho \int_o^\ell u^2\,dy + \rho \frac{d}{dx}\left[\int_o^\ell u^2\,dy\right]dx \quad (2.164)$$

74 Introduction to Convective Heat Transfer Analysis

Lastly, the rate at which momentum enters through bc, i.e., M_{bc}, must be obtained. This is done by noting that since bc lies in the freestream, any fluid that enters the control volume through it has a velocity u_1 in the x-direction. It follows, therefore, that:

$$M_{bc} = \text{Mass flow rate through } bc \times u_1$$

The mass flow rate through bc is obtained, in turn, by using the conservation of mass principle to give, since the flow is steady:

$$\begin{array}{c}\text{Mass flow rate}\\ \text{through } bc\end{array} = \begin{array}{c}\text{Mass flow rate}\\ \text{through } dc\end{array} - \begin{array}{c}\text{Mass flow rate}\\ \text{through } ab\end{array}$$

It was previously noted that the rate at which mass crosses the strip shown on ab is $u\,dy$. Therefore,

$$\text{Mass flow rate through } ab = \rho \int_o^\ell u\,dy$$

Using this then gives, because dx is small:

$$\begin{array}{c}\text{Mass flow rate}\\ \text{through } dc\end{array} = \rho \int_o^\ell u\,dy + \rho \frac{d}{dx}\left[\int_o^\ell u\,dy\right] dx$$

Combining these equations then gives:

$$M_{bc} = \left\{\frac{d}{dx}\left[\int_o^\ell u\,dy\right]\right\} \rho u_1\, dx \tag{2.165}$$

Substituting Eqs. (2.162), (2.164), (2.165), and (2.161) into Eq. (2.160) and dividing by dx then gives:

$$-\frac{dp}{dx}\ell - \tau_w = \rho \frac{d}{dx}\left[\int_o^\ell u^2\,dy\right] - \rho u_1 \frac{d}{dx}\left[\int_o^\ell u\,dy\right] \tag{2.166}$$

It is convenient to rearrange this equation by noting that

$$u_1 \frac{d}{dx}\left[\int_o^\ell u\,dy\right] = \frac{d}{dx}\left\{u_1\left[\int_o^\ell u\,dy\right]\right\} - \frac{du_1}{dx}\left[\int_o^\ell u\,dy\right] \tag{2.167}$$

Using this result, Eq. (2.166) can be rewritten as

$$\frac{d}{dx}\left[\int_o^\ell (u_1 - u)u\,dy\right] - \frac{du_1}{dx}\left[\int_o^\ell u\,dy\right] = \frac{1}{\rho}\frac{dp}{dx}\ell + \frac{\tau_w}{\rho} \tag{2.168}$$

Because the freestream flow is by assumption inviscid and, as previously discussed, the velocity component normal to the surface, v, is small compared to that parallel to the surface, u, the Bernoulli equation applied in the freestream gives as previously noted:

$$\frac{dp}{dx} = -\rho u_1 \frac{du_1}{dx} \tag{2.169}$$

CHAPTER 2: The Equations of Convective Heat Transfer   75

Substituting this into Eq. (2.168) and noting that:

$$\frac{1}{\rho}\frac{dp}{dx}\ell = -u_1\ell\frac{du_1}{dx} = -\frac{du_1}{dx}\int_o^\ell u_1\,dy \qquad (2.170)$$

then gives

$$\frac{d}{dx}\left[\int_o^\ell (u_1 - u)u\,dy\right] + \frac{du_1}{dx}\left[\int_o^\ell (u_1 - u)\,dy\right] = \frac{\tau_w}{\rho} \qquad (2.171)$$

Because u is equal to u_1 outside the boundary layer, i.e., because $(u - u_1) = 0$ for $y > \delta$, the integrals in the above equation need only be evaluated up to δ and the equation can, as a result, be written as:

$$\frac{d}{dx}\left[\int_o^\delta (u_1 - u)u\,dy\right] + \frac{du_1}{dx}\left[\int_o^\delta (u_1 - u)\,dy\right] = \frac{\tau_w}{\rho} \qquad (2.172)$$

This is termed the boundary layer momentum integral equation. As previously mentioned, it is equally applicable to laminar and turbulent flow. In laminar flow, u is the actual steady velocity while in turbulent flow it is the time averaged value.

The way in which the momentum integral equation is applied will be discussed in detail in the next chapter. Basically, it involves assuming the form of the velocity profile, i.e., of the variation of u with y in the boundary layer. For example, in laminar flow a polynomial variation is often assumed. The unknown coefficients in this assumed form are obtained by applying the known condition on velocity at the inner and outer edges of the boundary layer. For example, the velocity must be zero at the wall while at the outer edge of the boundary layer it must become equal to the freestream velocity, u_1. Thus, two conditions that the assumed velocity profile must satisfy are:

$$y = 0: \qquad u = 0$$
$$y = \delta: \qquad u = u_1$$

Other boundary conditions for laminar flow are discussed in the next chapter. In this way, the velocity profile is expressed in terms of u_1 and δ. If the wall shearing stress τ_w is then also related to these quantities, the momentum integral equation (2.173) will allow the variation of δ with x to be found for any specified variation of the freestream velocity, u_1.

Consider next the application of the conservation of energy principle to the control volume that was used above in the derivation of the momentum integral equation. The height, ℓ, of this control volume is taken to be greater than both the velocity and temperature boundary layer thicknesses as shown in Fig. 2.21.

As with the velocity boundary layer, the thermal boundary layer is assumed to have a definite thickness, δ_T, and outside this boundary layer the temperature is assumed to be constant.

Since attention is being restricted to low-speed, two-dimensional constant fluid property flow, dissipation effects on the energy balance are neglected. The conservation of energy principle therefore gives for the flow through the control surface:

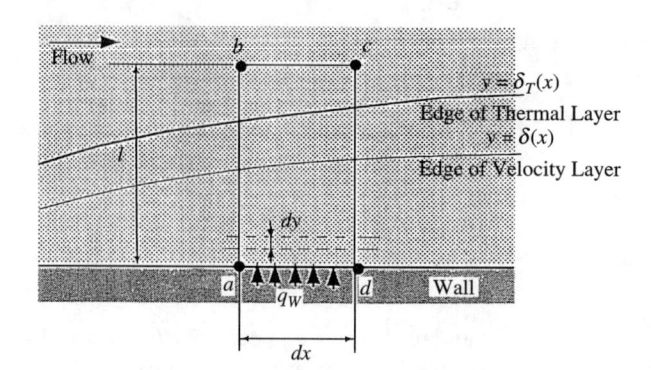

76 Introduction to Convective Heat Transfer Analysis

FIGURE 2.21
"Height" of control volume considered.

| Rate at which enthalpy leaves through dc | $-$ | Rate at which enthalpy enters through ab | $-$ | Rate at which enthalpy enters through bc | $=$ | Rate at which heat is transferred from the wall into the control volume through ad | (2.173) |

In writing this equation, it has been noted that since bc lies in the freestream where the temperature is constant, there can be no heat transfer into the control volume through it. Longitudinal conduction effects have also been ignored because the boundary layer is assumed to be thin. This is consistent with the neglect of the effects of longitudinal viscous forces in the derivation of the momentum integral equation.

If q_w is the heat transfer rate per unit area through the wall at the value of x being considered, i.e., is the local heat transfer rate, then the rate at which heat is transferred into the control volume, which has unit width, through ad is given by

$$q_w \times dx \tag{2.174}$$

If it is again noted that the rate at which mass crosses the strip on ab shown in Fig. 2.20 will be $\rho u\, dy$, it follows that the rate at which enthalpy enters the control volume through this strip is given by $\rho u c_p T\, dy$ if c_p is assumed to be constant. The rate at which enthalpy enters the control volume through the face ab, which will here be denoted by H_{ab}, is therefore given by

$$H_{ab} = \rho c_p \int_0^\ell uT\, dy \tag{2.175}$$

The rate at which enthalpy leaves the face dc, termed H_{dc} is, since dx is small, given by:

$$H_{dc} = H_{ab} + \frac{d}{dx}(H_{ab})\, dx \tag{2.176}$$

Using Eq. (2.175) this gives:

$$H_{dc} = \rho c_p \left\{ \int_0^\ell uT\, dy + \frac{d}{dx}\left[\int_0^\ell uT\, dy\right] dx \right\} \tag{2.177}$$

In order to determine the rate at which enthalpy crosses bc, i.e., H_{bc}, it is noted that since this face of the control volume lies in the freestream, any fluid that crosses it will have a temperature of T_1. Hence:

$$H_{bc} = \text{Mass flow rate through } bc \times c_p T_1 \qquad (2.178)$$

Therefore, using the expression for the mass flow rate through bc that was derived when dealing with the momentum integral equation gives:

$$H_{bc} = \left\{ \frac{d}{dx} \left[\int_0^\ell u\, dy \right] \right\} \rho c_p T_1\, dx \qquad (2.179)$$

Substituting Eqs. (2.174), (2.175), (2.177), and (2.179) into Eq. (2.173) and dividing the result by dx gives:

$$\rho c_p \frac{d}{dx}\left[\int_0^\ell uT\, dy\right] - \rho c_p T_1 \frac{d}{dx}\left[\int_0^\ell u\, dy\right] = q_w \qquad (2.180)$$

Since the freestream temperature T_1 is independent of x, this equation can be rewritten as

$$\frac{d}{dx}\left[\int_0^\ell u(T - T_1)\, dy\right] = \frac{q_w}{\rho c_p} \qquad (2.181)$$

Now outside the thermal boundary layer, i.e., for $y > \delta_T$, $T = T_1$. Eq. (2.181) can, therefore, be written as

$$\frac{d}{dx}\left[\int_0^{\delta_T} u(T - T_1)\, dy\right] = \frac{q_w}{\rho c_p} \qquad (2.182)$$

This is termed the boundary layer energy integral equation.

The energy integral equation is applied in basically the same way as the momentum integral equation. The form of the boundary layer temperature profile, i.e., of the variation of $(T - T_1)$ with y, is assumed. In the case of laminar flow, for example, a polynomial form is again often used. The unknown coefficients in this assumed temperature profile are then determined by applying known boundary conditions on temperature at the inner and outer edges of the boundary layer. For example, the variation of the wall temperature T_w with x may be specified. Therefore, because at the outer edge of the boundary layer the temperature must become equal to the freestream temperature T_1, two boundary conditions on the assumed temperature profile in this case are:

$$y = 0: \qquad T = T_w$$
$$y = T: \qquad T = T_1$$

Using these and other boundary conditions, some of which for laminar flow will be discussed in the next chapter, the temperature distribution is expressed as a function of δ_T. If q_w is then related to the wall thermal conditions and δ_T, the energy integral can be solved, using the solution to the momentum integral equation, to give

78   Introduction to Convective Heat Transfer Analysis

the variation of δ_T with x. Once this is found, the variation of the heat transfer rate q_w with x can be found.

The boundary layer integral equations have been derived above without recourse to the partial differential equations for boundary layer flow. They can, however, be determined directly from these equations. Consider, for example, the laminar momentum equation (2.140). Integrating this equation across the boundary layer to some distance ℓ from the wall, ℓ being greater than the boundary layer thickness, gives because $\partial u/\partial y$ is zero outside the boundary layer and because dp/dx is independent of y:

$$\int_0^\ell u \frac{\partial u}{\partial x} dy + \int_0^\ell v \frac{\partial u}{\partial y} dy = -\frac{1}{\rho} \frac{dp}{dx} \ell - \mu \left. \frac{\partial u}{\partial y} \right|_{y=0} \tag{2.183}$$

But $\mu \partial u/\partial y|_{y=0}$ is equal to the wall shearing stress, τ_w.

Also:

$$\int_0^\ell u \frac{\partial u}{\partial x} dy = \frac{1}{2} \int_0^\ell \frac{\partial u^2}{\partial x} dy = \frac{1}{2} \frac{d}{dx} \left[\int_0^\ell u^2 dy \right] \tag{2.184}$$

and:

$$\int_0^\ell v \frac{\partial u}{\partial y} dy = \int_0^\ell \frac{\partial}{\partial y}(vu) dy - \int_0^\ell u \frac{\partial v}{\partial y} dy$$

$$= v_1 u_1 - \int_0^\ell u \frac{\partial v}{\partial y} dy \tag{2.185}$$

where v_1 and u_1 are the values of the velocity components outside the boundary layer.

Now the continuity equation gives:

$$\frac{\partial v}{\partial y} = -\frac{\partial u}{\partial x} \tag{2.186}$$

which can be integrated to give

$$v_1 = -\int_0^\ell \frac{\partial u}{\partial x} dy = -\frac{d}{dx} \left[\int_0^\ell u \, dy \right] \tag{2.187}$$

Substituting Eqs. (2.186) and (2.187) into Eq. (2.185) then gives:

$$\int_0^\ell v \frac{\partial u}{\partial x} dy = -u_1 \frac{d}{dx} \left[\int_0^\ell u \, dy \right] + \int_0^\ell u \frac{\partial u}{\partial x} dy$$

$$= -u_1 \frac{d}{dx} \left[\int_0^\ell u \, dy \right] + \frac{1}{2} \frac{d}{dx} \left[\int_0^\ell u^2 dy \right] \tag{2.188}$$

Now substituting Eqs. (2.184) and (2.188) into Eq. (2.183) and remembering that:

$$-\frac{1}{\rho} \frac{dp}{dx} \ell = u_1 \frac{du_1}{dx} \ell$$

then gives:

$$\frac{d}{dx}\left[\int_0^\ell u^2\,dy\right] - u_1\frac{d}{dx}\left[\int_0^\ell u\,dy\right] = u_1\frac{du_1}{dx}\ell - \frac{\tau_w}{\rho} \qquad (2.189)$$

This equation can be rearranged, as before, to give:

$$\frac{d}{dx}\left[\int_0^\delta (u_1-u)u\,dy\right] + \frac{du_1}{dx}\left[\int_0^\delta (u_1-u)\,dy\right] = \frac{\tau_w}{\rho}$$

where it has again been noted that the upper limit of the integration can be taken as δ since the integrands are zero for $y > \delta$.

The energy integral equation can be derived in a similar way from the energy equation (2.145). When this equation is integrated across the boundary layer it gives if the dissipation term is ignored:

$$\int_0^\ell u\frac{\partial T}{\partial x}\,dy + \int_0^\ell v\frac{\partial T}{\partial y}\,dy = -\left(\frac{k}{\rho c_p}\right)\frac{\partial T}{\partial y}\bigg|_{y=0} \qquad (2.190)$$

But $-k\partial T/\partial y|_{y=0}$ is the heat transfer rate at the wall q_w. Also

$$\int_0^\ell u\frac{\partial T}{\partial x}\,dy = \frac{d}{dx}\left[\int_0^\ell uT\,dy\right] - \int_0^\ell T\frac{\partial u}{\partial x}\,dy \qquad (2.191)$$

and:

$$\int_0^\ell v\frac{\partial T}{\partial y}\,dy = \int_0^\ell \frac{\partial}{\partial y}(vT)\,dy - \int_0^\ell T\frac{\partial v}{\partial y}\,dy$$
$$= v_1 T_1 - \int_0^\ell T\frac{\partial v}{\partial y}\,dy \qquad (2.192)$$

Therefore, substituting Eqs. (2.191) and (2.192) into Eq. (2.190) and using Eq. (2.187) gives:

$$\frac{d}{dx}\left[\int_0^\ell uT\,dy\right] - T_1\frac{d}{dx}\left[\int_0^\ell u\,dy\right] - \int_0^\ell T\left(\frac{\partial u}{\partial x}+\frac{\partial v}{\partial y}\right)dy = \frac{q_w}{\rho c_p} \qquad (2.193)$$

Because T_1 is a constant, it follows by using the continuity equation that this equation can be written as:

$$\frac{d}{dx}\left[\int_0^{\delta_T} u(T-T_1)\,dy\right] = \frac{q_w}{\rho c_p} \qquad (2.194)$$

The upper limit of the integration has been written as δ_T since the integrand is zero for $y > \delta_T$.

The quite extensive attention given above to the boundary layer integral equations must not be taken as indicating that these equations are, today, widely used. A study of the derivations does help in gaining an understanding of the meaning of the equations governing convective heat transfer and are, to some extent, the basis

80 Introduction to Convective Heat Transfer Analysis

of some methods of numerically solving the general equations governing convective heat transfer. It is for these reasons that they have been presented in detail here.

2.11
CONCLUDING REMARKS

A flow is completely defined if the values of the velocity vector, the pressure, and the temperature are known at every point in the flow. The distributions of these variables can be described by applying the principles of conservation of mass, momentum, and energy, these conservation principles leading to the *continuity*, the *Navier-Stokes*, and the *energy* equations, respectively. If the fluid properties can be assumed constant, which is very frequently an adequate assumption, the first two of these equations can be simultaneously solved to give the velocity vector and pressure distributions. The energy equation can then be solved to give the temperature distribution. Fourier's law can then be applied at the surface to get the heat transfer rates.

The flows over a series of bodies of the same geometrical shape will be similar, i.e., will differ from each other only in scale, if the Reynolds and Prandtl numbers are the same in all the flows. From this it follows that the Nusselt number in forced convection will depend only on the Reynolds and Prandtl numbers.

The unsteady governing equations that apply in laminar flow also apply in turbulent flow but essentially cannot be solved in their full form at present. In turbulent flow, therefore, the variables are conventionally split into a time averaged value plus a fluctuating component and an attempt is then made to express the governing equations in terms of the time averaged values alone. However, as a result of the nonlinear terms in the governing equations, the resultant equations contain "extra variables" which depend on the nature of the turbulence, i.e., there are, effectively, more variables than equations. There is, thus, a closure problem and to bring about closure, extra equations which constitute a "turbulence model" must be introduced.

The amount of effort required to solve the full governing equations can be considerably reduced if certain assumptions can be introduced that simplify these equations. The most commonly used assumptions are that the flow in a duct is fully developed or that the flow has a boundary layer character.

PROBLEMS

2.1. Using the control volume discussed in this chapter, derive the continuity equation for unsteady flow allowing for density variations in the flow.

2.2. Derive the two-dimensional energy equation in cylindrical coordinates using the control volume shown in Fig. P2.1.

2.3. Starting with the two-dimensional energy equation, i.e.:

$$\frac{\partial T}{\partial t} + u\frac{\partial T}{\partial x} + v\frac{\partial T}{\partial y} = \left(\frac{k}{\rho c_p}\right)\left(\frac{\partial^2 T}{\partial x^2} + \frac{\partial^2 T}{\partial y^2}\right)$$

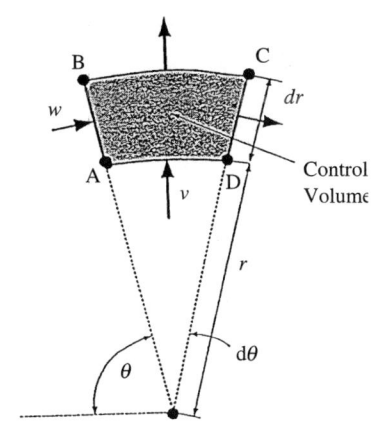

CHAPTER 2: The Equations of Convective Heat Transfer   81

FIGURE P2.1

and using:

$$u = \bar{u} + u'$$
$$v = \bar{v} + v'$$
$$T = \bar{T} + T'$$

derive the time-averaged energy equation for turbulent flow.

2.4. Write out the continuity, Navier-Stokes, and energy equations in cylindrical coordinates for steady, laminar flow with constant fluid properties. The dissipation term in the energy equation can be ignored. Using this set of equations, investigate the parameters that determine the conditions under which "similar" velocity and temperature fields will exist when the flow over a series of axisymmetric bodies of the same geometrical shape but with different physical sizes is considered.

2.5. Discuss the meaning of the following terms: (i) The turbulence closure problem. (ii) A turbulence model. (iii) A boundary layer. (iv) Similar flow fields.

2.6. Discuss the assumptions used in deriving the boundary layer equations.

2.7. Consider two-dimensional flow over an axisymmetric body. Write the governing equations in terms of a suitably defined stream function and vorticity.

2.8. Two-dimensional flow over a series of geometrically similar bodies having a specified surface temperature was discussed in this chapter. If the surface heat flux rather than the temperature is specified, a dimensional temperature of the form $(T - T_1)/(q_{wr}L/k)$ should be used. Derive the parameters on which the mean surface temperature will depend in this situation. Viscous dissipation effects can be ignored.

2.9. Write the governing equations in Cartesian coordinates given in this chapter in tensor notation, i.e., using $u_i = u$, $u_j = v$, $u_l = w$, $x_i = x$, $x_j = y$, and $x_l = z$. Use the Einstein summation convention.

2.10. The turbulence kinetic energy equation for forced convection was derived in this chapter. Rederive this turbulence kinetic energy equation by starting with the momentum equations that include the buoyancy term $\beta g(T - T_{\text{ref}})$ in the x-momentum equation, T_{ref} being a suitable reference temperature.

2.11. Consider the derivation, presented in this chapter, of the energy integral equation from the governing boundary layer energy equation. Repeat this derivation but include the viscous dissipation term in the energy equation.

2.12. In some flows there is a volumetric heat generation due to chemical or other effects. In such situations, the heat generation rate per unit volume is usually known throughout the flow, its value generally varying with position in the flow. Explain what changes in the steady state energy equation are required to account for such heat generation.

2.13. Consider two-dimensional laminar boundary layer flow over a flat isothermal surface. Very close to the surface, the velocity components are very small. If the pressure changes are assumed to be negligible in the flow being considered, derive an expression for the temperature distribution near the wall. Viscous dissipation effects should be included in the analysis.

REFERENCES

1. Bird, R.B., Stewart, W.E., and Lightfoot, E.N., *Transport Phenomena,* Wiley, New York, 1960.
2. Burmeister, L.C., *Convective Heat Transfer,* 2nd ed., Wiley-Interscience, New York, 1993.
3. Eckert, E.R.G. and Drake, R.M., Jr., *Analysis of Heat and Mass Transfer,* McGraw-Hill, New York, 1973.
4. Fox, R.W. and McDonald, A.T., *Introduction to Fluid Mechanics,* 3rd ed., Wiley, New York, 1985.
5. Kakac, S. and Yener, Y., *Convective Heat Transfer,* 2nd ed., CRC Press, Boca Raton, FL, 1995.
6. Kaviany, M., *Principles of Convective Heat Transfer,* Springer-Verlag, New York, 1994.
7. Schlichting, H., *Boundary Layer Theory,* 7th ed., McGraw-Hill, New York, 1979.
8. Sherman, F.S., *Viscous Flow,* McGraw-Hill, New York, 1990.
9. White, F.M., *Viscous Fluid Flow,* 2nd ed., McGraw-Hill, New York, 1991.
10. Hinze, J.O., *Turbulence,* 2nd ed., McGraw-Hill, New York, 1975.
11. Patankar, S.V., *Numerical Heat Transfer and Fluid Flow,* Hemisphere, Washington, D.C., 1980.
12. Tennekes, H. and Lumley, J.L., *A First Course in Turbulence,* The MIT Press, Cambridge, MA, 1972.
13. Goldstein, S., Ed., *Modern Developments in Fluid Dynamics,* Oxford University Press, London, 1938.
14. Kays, W.M. and Crawford, M.E., *Convective Heat and Mass Transfer,* 3rd ed., McGraw-Hill, New York, 1993.
15. Rosenhead, L., *Laminar Boundary Layers,* Oxford University Press, London, 1963.

CHAPTER 3

Some Solutions for External Laminar Forced Convection

3.1
INTRODUCTION

The purpose of this chapter is to illustrate some of the ways in which the equations derived in the previous chapter can be used to obtain heat transfer rates for situations involving external laminar flows. External flows involve a flow, which is essentially infinite in extent, over the outer surface of a body as shown in Fig. 3.1.

The problems chosen to illustrate the methods of solution are not directly, in most cases, of great practical significance but they serve to illustrate the basic ideas involved in the solution procedures. Some of the flows considered, although apparently highly idealized, are, however, good models of situations that are of great practical importance. For example, flow over a flat plate aligned with the flow will be extensively considered. This flow is a good model of many situations involving flow over fins that are relatively widely spaced.

In all the solutions given in the present chapter, the fluid properties will be assumed to be constant and the flow will be assumed to be two-dimensional. In addition, dissipation effects in the energy equation will be neglected in most of this chapter, these effects being briefly considered in a last section of this chapter. Also, solutions to the full Navier-Stokes and energy equations will be dealt with only relatively briefly, the majority of the solutions considered being based on the use of the boundary layer equations.

3.2
SIMILARITY SOLUTION FOR FLOW OVER AN ISOTHERMAL PLATE

Consider the flow of a fluid at a velocity of u_1 over a flat plate whose entire surface is held at a uniform temperature of T_w which is different from that of the fluid ahead of the body which is T_1. The flow situation is thus as shown in Fig. 3.2.

84 Introduction to Convective Heat Transfer Analysis

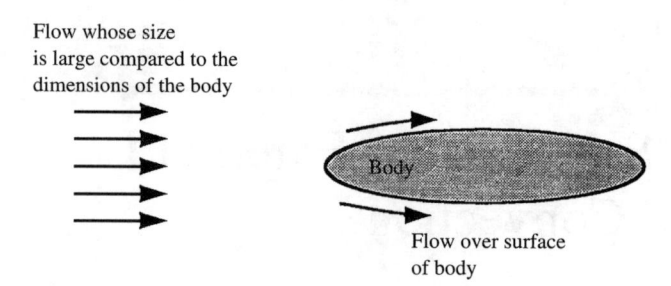

FIGURE 3.1
External flow.

It will be assumed that the Reynolds number is large enough for the boundary layer assumptions to be applicable. It will further be assumed that the flow is two-dimensional which means that the plate is assumed to be wide compared to its longitudinal dimension. As a result of these assumptions, the equations governing the problems are:

$$\frac{\partial u}{\partial x} + \frac{\partial v}{\partial y} = 0 \tag{3.1}$$

$$u\frac{\partial u}{\partial x} + v\frac{\partial u}{\partial y} = \nu\frac{\partial^2 u}{\partial y^2} \tag{3.2}$$

$$u\frac{\partial T}{\partial x} + v\frac{\partial T}{\partial y} = \left(\frac{\nu}{Pr}\right)\frac{\partial^2 T}{\partial y^2} \tag{3.3}$$

In writing these equations it has been noted that the solution for inviscid flow over a flat plate of zero thickness is that the velocity is everywhere the same and equal to the undisturbed freestream velocity, u_1, i.e., that, if the effects of viscosity are ignored, a flat plate aligned with the flow will have no effect on the flow. As a result, the pressure gradient, dp/dx, is everywhere zero. But the conditions outside the boundary layer are assumed to be those that exist in inviscid flow over the body considered and the pressure gradient in the boundary layer is assumed to be equal to that existing in this outer inviscid flow. Therefore, in obtaining the boundary layer solution for flow over a flat plate, it will be assumed that the velocity outside the boundary is equal to u_1 and that the pressure gradient is everywhere equal to zero. In the case of a real plate of finite thickness, a pressure gradient will exist but this should only be significant in the immediate vicinity of the leading edge. Since the boundary layer equations themselves are not applicable very near the leading edge, the longitudinal gradients of velocity and temperature being comparable to the lateral ones in this region, this effect will not be considered here.

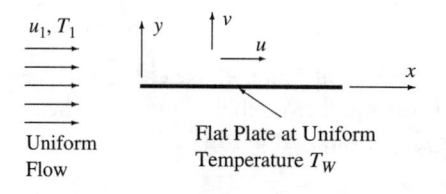

FIGURE 3.2
Flow over a flat plate.

CHAPTER 3: Some Solutions for External Laminar Forced Convection 85

In writing Eq. (3.2), the kinematic viscosity, ν, defined by:

$$\nu = \frac{\mu}{\rho} \qquad (3.4)$$

has been introduced as before. It has also been noted in writing Eq. (3.3) that:

$$\frac{k}{\rho c_p} = \frac{\nu}{Pr} \qquad (3.5)$$

Equations (3.1) to (3.3) must be solved subject to the following boundary conditions:

$$\begin{aligned} y = 0: \ u = 0, v = 0, T = T_w \\ y \text{ large: } u \to u_1, T \to T_1 \end{aligned} \qquad (3.6)$$

A discussion of how these conditions are derived was given in Chapter 2.

It is noted that Eqs. (3.1) and (3.2), which together allow the velocity distribution to be determined, are independent of the temperature because the fluid properties are being assumed constant. Therefore, as previously discussed, these two equations can be solved independently of Eq. (3.3) to give the velocity distribution. Once this velocity distribution has been obtained, Eq. (3.3) can be solved for the temperature distribution. The heat transfer rate can then be determined from this distribution.

The similarity solution uses as a starting point the assumption that the boundary layer profiles are similar at all values of x, i.e., that the basic form of the velocity profiles at different values of x, as shown in Fig. 3.3, are all the same [1],[2],[3],[4],[5].

The velocity profiles at all points in the boundary are therefore assumed to be such that:

$$\frac{u}{u_1} = \text{function}\left(\frac{y}{\delta}\right) \qquad (3.7)$$

The velocity profiles are thus assumed to be similar in the sense that although δ varies with x and although the velocity at a fixed distance y from the wall varies with x, the velocity at fixed values of y/δ, where δ is the local boundary layer thickness, remains constant. The proof that similar profiles do exist actually follows from the analysis given below.

It was shown in the derivation of the boundary layer assumptions that for the boundary layer assumptions to apply it is necessary that:

$$\frac{\delta}{x} = o\left(\frac{1}{\sqrt{Re_x}}\right) \qquad (3.8)$$

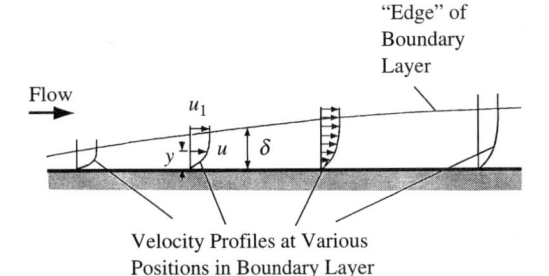

FIGURE 3.3
Velocity profiles at various positions in the boundary layer on a flat plate.

86 Introduction to Convective Heat Transfer Analysis

the characteristic length having here been taken as x. Re_x is, of course, the Reynolds number based on x, i.e. $(u_1 x/\nu)$.

Therefore, the similar velocity profile assumption given in Eq. (3.7) can be written as:

$$\frac{u}{u_1} = F\left(\frac{y}{x}\sqrt{Re_x}\right) = F\left(y\sqrt{\frac{u_1}{\nu x}}\right) \tag{3.9}$$

It is convenient to define:

$$\eta = \frac{y}{x}\sqrt{Re_x} = y\sqrt{\frac{u_1}{\nu x}} \tag{3.10}$$

this variable, η, being termed the "similarity variable".

Eq. (3.9) can, therefore, be written as:

$$\frac{u}{u_1} = F(\eta) \tag{3.11}$$

Consider the continuity equation (3.1). Because the boundary conditions given in Eq. (3.6) give v as zero at the wall, this continuity equation can be integrated with respect to y leading to the following expression for v at any point in the boundary layer.

$$v = -\int_0^y \frac{\partial u}{\partial x} dy = -\sqrt{\frac{x\nu}{u_1}} \int_0^\eta u_1 \frac{dF}{d\eta} \frac{\partial \eta}{\partial x} d\eta \tag{3.12}$$

But

$$\frac{\partial \eta}{\partial x} = -\frac{1}{2}\frac{y}{x}\sqrt{\frac{u_1}{x\nu}} = -\frac{\eta}{2x} \tag{3.13}$$

so Eq. (3.12) can be written as:

$$v = \frac{1}{2}\sqrt{\frac{\nu}{xu_1}} \int_0^\eta \frac{dF}{d\eta} \eta \, d\eta \tag{3.14}$$

Because of the form of this integral, it is convenient in the following work to define a new function, f of η, which is related to F by:

$$\frac{df}{d\eta} = f' = F \tag{3.15}$$

Rewriting Eq. (3.14) in terms of f and carrying out the integration then gives:

$$\frac{v}{u_1} = \frac{1}{2}\sqrt{\frac{\nu}{xu_1}}(\eta f' - f) \tag{3.16}$$

Next consider the momentum equation (3.2) which can be written as:

$$\left(\frac{u}{u_1}\right)\frac{\partial(u/u_1)}{\partial \eta}\frac{\partial \eta}{\partial x} + \left(\frac{v}{u_1}\right)\frac{\partial(u/u_1)}{\partial \eta}\frac{\partial \eta}{\partial y} = \left(\frac{\nu}{u_1}\right)\frac{\partial^2(u/u_1)}{\partial \eta^2}\left(\frac{\partial \eta}{\partial y}\right)^2 \tag{3.17}$$

CHAPTER 3: Some Solutions for External Laminar Forced Convection   87

but by virtue of Eq. (3.15):

$$\frac{u}{u_1} = f' \tag{3.18}$$

so it follows using these and Eq. (3.16) that the momentum equation reduces to:

$$f'f''\left(-\frac{1}{2}\frac{y}{x}\sqrt{\frac{u_1}{x\nu}}\right) + \left[\frac{1}{2}\sqrt{\frac{\nu}{xu_1}}(\eta f' - f)\right]f''\sqrt{\frac{u_1}{x\eta}} = \frac{\nu}{u_1}f'''\left(\frac{u_1}{x\nu}\right)$$

i.e., to:

$$2f''' + ff'' = 0 \tag{3.19}$$

The primes, of course, denote differentiation with respect to η, i.e., Eq. (3.19) could have been written as:

$$2\frac{d^3 f}{d\eta^3} + f\frac{d^2 f}{d\eta^2} = 0$$

The problem of determining the velocity profile in the boundary layer has, therefore, been reduced to that of solving the ordinary differential equation given in Eq. (3.19). In terms of the similarity function, the boundary conditions on the solution that are given in Eq. (3.6) become:

When $y = 0$, $u = 0$ becomes, using Eq. (3.18), when $\eta = 0$, $f' = 0$
When $y = 0$, $v = 0$ becomes, using Eq. (3.16), when $\eta = 0$, $f = 0$
When y is large, $u \to u_1$ becomes, using Eq. (3.18), when η is large, $f' \to 1$
(3.20)

There are, therefore, two boundary conditions at $\eta = 0$ and one boundary condition that applies at large values of η [1],[6],[7],[8]. The solution to Eq. (3.19) subject to these boundary conditions is relatively easily found. One way of finding the solution involves, basically, guessing the value of f'' at the wall and then, using the first two of the boundary conditions given in Eq. (3.20), numerically integrating Eq. (3.19), thus giving the variation of f with η. This solution will of course not, in general, satisfy the outer boundary condition, i.e., the boundary condition for large η. The solution is, therefore, repeated for another guessed value of f'' at the wall. The solution that satisfies the outer boundary condition can then be obtained by combining these two solutions or by using an iterative procedure. Because the iterative ("shooting") method will be used elsewhere in this book, it will be adopted here. The iterative method is based on Newton's method, i.e., if f''_w is the value of f'' at $\eta = 0$ and if f'_∞ is the value of f' for large η obtained by using this value, then a better estimate of f''_w is:

$$f''_w + \frac{f_\infty - 1}{d(f'_\infty)/d(f''_w)}$$

In obtaining the solution, Eq. (3.19) is conveniently written as three simultaneous first-order differential equations; i.e., by defining $g = f'$ and $h = g' = f''$

Eq. (3.19) can be used to give the following set of simultaneous equations:

$$2h' = -0.5fh$$
$$g' = h$$
$$f' = g$$

The boundary conditions on the functions involved in these equations are:

$$\eta = 0, \ g = 0; \ \eta = 0, \ f = 0; \ \eta \text{ large}, \ g \to 1$$

The first step of the solution thus involves guessing the value of h at $\eta = 0$ and then integrating the above set of equations. The Newton equation is then used to progressively work towards the value of h at $\eta = 0$ that gives $g \to 1$ for η large. In applying this method, the value of $d(f'_\infty)/d(f''_w)$ is obtained by obtaining f'_∞ using a particular value of f''_w and then incrementing f''_w by a small amount, then getting a new value of f'_∞ and then using a finite difference approximation for the derivative.

In this way the variation of f and hence f' with η can be found. The variation resembles that shown in Fig. 3.4. A simple program to obtain the solution is available as discussed in the Preface.

The predicted velocity profile shown in this figure is in very good agreement with experimentally measured profiles.

The fact that the original two partial differential equations governing the velocity components, u and v, were reduced to an ordinary differential equation justifies the assumption of similar velocity profiles.

While there is no distinct "edge" to the boundary layer, it is convenient to have some measure of the distance from the wall over which significant effects of viscosity exist. For this reason, it is convenient to arbitrarily define the boundary layer thickness, δ_u, as the distance from the wall at which u reaches to within 1% of the freestream velocity, i.e., to define δ as the value of y at which $u = 0.99u_1$. Using the result given in Fig. 3.4 then shows that $u = 0.99u_1$, i.e., $f' = 0.99$, when $\eta = 5$ which indicates, in view of the definition of η, that δ_u is approximately given by:

$$\delta_u \sqrt{\frac{u_1}{\nu x}} = 5$$

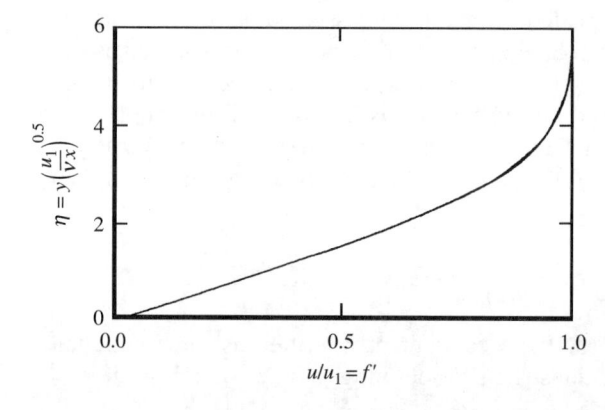

FIGURE 3.4
Variation of the velocity profile function f' with the similarity variable η for the boundary layer on a flat plate.

CHAPTER 3: Some Solutions for External Laminar Forced Convection 89

i.e.,

$$\frac{\delta_u}{x} = \frac{5}{\sqrt{Re_x}} \tag{3.21}$$

Having established the form of the velocity profile, attention can now be turned to the determination of the temperature profile, i.e., to the solution of Eq. (3.3). Because the wall temperature is uniform in the situation being considered, it is logical to assume that the temperature profiles are similar in the same sense as the velocity profiles. For this reason, the following dimensionless temperature function is introduced

$$\theta = \frac{T_w - T}{T_w - T_1} \tag{3.22}$$

The assumption that the temperature profiles are similar is equivalent to assuming that θ depends only on the similarity variable, η, because the thermal boundary layer thickness is also of order $x/\sqrt{Re_x}$.

Since both T_w and T_1 are constant, the energy equation (3.3) can be written in terms of θ as

$$u\frac{\partial \theta}{\partial x} + v\frac{\partial \theta}{\partial y} = \frac{\nu}{Pr}\frac{\partial^2 \theta}{\partial y^2} \tag{3.23}$$

while in terms of this variable, the boundary conditions become:

$$\begin{aligned} y = 0: \theta = 0 \\ y \text{ large}: \theta \to 1 \end{aligned} \tag{3.24}$$

Because the temperature profiles are being assumed to be similar, i.e., it is being assumed that θ is a function of η alone, it follows using the relations for the velocity components previously derived, that Eq. (3.23) gives:

$$f'\frac{d\theta}{d\eta}\frac{\partial \eta}{\partial x} + \left[\frac{1}{2}\sqrt{\frac{\nu}{u_1 x}}(\eta f' - f)\right]\frac{d\theta}{d\eta}\frac{\partial \eta}{\partial y} = \frac{1}{Pr}\frac{\nu}{u_1}\frac{d^2\theta}{d\eta^2}\left[\frac{\partial \eta}{\partial y}\right]^2 \tag{3.25}$$

On rearrangement this equation becomes:

$$\theta'' + \frac{Pr}{2}\theta' f = 0 \tag{3.26}$$

while the boundary conditions on the solution given in Eq. (3.24) can be written as:

$$\begin{aligned} \eta = 0: \theta = 0 \\ \eta \text{ large}: \theta \to 1 \end{aligned} \tag{3.27}$$

Thus, as was the case with the velocity distribution, the partial differential equation governing the temperature distribution has been reduced to an ordinary differential equation. This confirms the assumption that the temperature profiles are similar.

Now Eq. (3.26) could be solved, subject to the boundary conditions given in Eq. (3.27), using a similar procedure to that used to solve Eq. (3.19) which was briefly outlined above. The solution is, however, more easily obtained in the following way. It is noted that Eq. (3.26) can be written as:

$$\frac{1}{\theta'}\frac{d}{d\eta}\theta' = -\frac{Pr}{2}f \tag{3.28}$$

This can be integrated to give:

$$\theta' = c_1 \exp\left[-\frac{Pr}{2}\int_0^\eta f\,d\eta\right] \tag{3.29}$$

where c_1 is a constant of integration which has still to be determined.

Integrating Eq. (3.29) then gives:

$$\theta = c_1 \int_0^\eta \left[\exp\left(\frac{-Pr}{2}\int_0^\eta f\,d\eta\right)\right]d\eta + c_2 \tag{3.30}$$

where c_2 is another constant of integration. It will, of course, be equal to the value of θ when η is zero and so by virtue of the first of the boundary conditions given in Eq. (3.27) c_2 is zero. c_1 is determined from the second of the boundary conditions given in Eq. (3.27) and is, therefore, given by

$$c_1 = 1\Big/\int_0^\infty \left[\exp\left(-Pr\int_0^\eta f\,d\eta\right)\right]d\eta \tag{3.31}$$

The boundary condition for large η has been applied by taking the integration to ∞. The final equation obtained is, however, such that once the upper limit of the integration is taken above a certain value, the result becomes independent of the actual value used for this upper limit. This point is discussed further below.

Substituting Eq. (3.31) into Eq. (3.30) then gives since c_2 is zero

$$\theta = \frac{\int_0^\eta \left[\exp\left(\frac{-Pr}{2}\int_0^\eta f\,d\eta\right)\right]d\eta}{\int_0^\infty \left[\exp\left(\frac{-Pr}{2}\int_0^\eta f\,d\eta\right)\right]d\eta} \tag{3.32}$$

Since f is a known function of η being given by the solution for the velocity field, Eq. (3.32) can be integrated to give the variation of θ with η. This solution is more easily obtained by noting that Eq. (3.19) gives:

$$\frac{f'''}{f''} = -\frac{f}{2} \tag{3.33}$$

which can be integrated to give

$$\log_e f'' = -\frac{1}{2}\int_0^\eta f\,d\eta + \text{constant}$$

i.e.,

$$\exp\left[-\frac{1}{2}\int_0^\eta f\,d\eta\right] = \frac{f''}{c_3} \qquad (3.34)$$

c_3 being, of course, a constant of integration. Integrating this equation again then gives

$$\int_0^\eta \exp\left[-Pr\int_0^\eta f\,d\eta\right]d\eta = \int_0^\eta \left(\frac{f''}{c_3}\right)^{Pr} d\eta = \frac{1}{c_3^{Pr}}\int_0^\eta (f'')^{Pr}\,d\eta \qquad (3.35)$$

Substituting this result into Eq. (3.32) then gives

$$\theta = \frac{\left[\int_0^\eta f''^{Pr}\,d\eta\right]}{\left[\int_0^\infty f''^{Pr}\,d\eta\right]} = \frac{\left[\int_0^\infty f''^{Pr}\,d\eta - \int_\eta^\infty f''^{Pr}\,d\eta\right]}{\left[\int_0^\infty f''^{Pr}\,d\eta\right]} = 1 - \frac{\left[\int_\eta^\infty f''^{Pr}\,d\eta\right]}{\left[\int_0^\infty f''^{Pr}\,d\eta\right]} \qquad (3.36)$$

This allows the variation of θ with η to be derived for any chosen value of Pr.

Since the variation of f'' with η is given by the solution for the velocity profile, this equation is easily integrated to give the variation of θ with η. It will be noted that f'', which is equal to $d(u/u_1)/d\eta$, tends to zero outside the boundary layer, i.e., at large values of η. The actual value of the upper limit of the integrals in Eq. (3.36) will not, therefore, affect the result provided it is sufficiently large.

A simple computer program, SIMPLATE, written in FORTRAN that implements the above procedures for obtaining the similarity profiles for velocity and temperature is available as discussed in the Preface. In this program, the solution to the set of equations defining the velocity profile function is obtained using the basic Runge-Kutta procedure.

Some typical variations of θ with η for various values of Pr obtained using this program are shown in Fig. 3.5.

The curve for a Prandtl number of 1 is identical to that giving the variation of f' with η which was given in Fig. 3.4. This is discussed below.

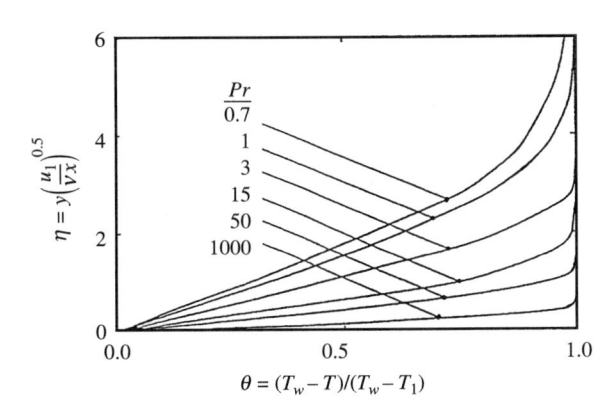

FIGURE 3.5
Variation of the temperature profile function, θ, with the similarity variable, η, for various values of Pr for the boundary layer on a flat plate.

92 Introduction to Convective Heat Transfer Analysis

It will be seen from the results given in Fig. 3.5 that, if the thermal boundary layer "thickness," δ_T, is defined in a similar way to the velocity boundary layer "thickness" as the distance from the wall at which θ becomes equal to 0.99, i.e., reaches to within 1% of its freestream value, then:

$$\frac{\delta_T}{x} = \frac{\Delta(Pr)}{\sqrt{Re_x}}$$

but:

$$\frac{\delta_u}{x} = \frac{5}{\sqrt{Re_x}}$$

It follows from these that:

$$\frac{\delta_T}{\delta_u} = \Delta(Pr)$$

i.e., the ratio of the two boundary layer thicknesses depends only on the value of the Prandtl number. It will be noted from the results given in Fig. 3.5 that the thermal boundary layer is thicker than the velocity boundary layer when Pr is less than one and thinner than the velocity boundary layer when Pr is greater than one.

The heat transfer rate at the wall is given by

$$q_w = -k \left.\frac{\partial T}{\partial y}\right|_{y=0} \tag{3.37}$$

Hence, using Eq. (3.22)

$$\frac{q_w}{k(T_w - T_1)} = \left.\frac{\partial \theta}{\partial y}\right|_{y=0} = \left.\frac{d\theta}{d\eta}\right|_{\eta=0} \frac{\partial \eta}{\partial y} \tag{3.38}$$

i.e.,

$$\frac{q_w x}{k(T_w - T_1)} = \left.\theta'\right|_{\eta=0} \sqrt{Re_x} \tag{3.39}$$

i.e.,

$$Nu_x = \left.\theta'\right|_{\eta=0} \sqrt{Re_x} \tag{3.40}$$

Nu_x and Re_x being, of course, the local Nusselt and Reynolds numbers based on x.

Because θ depends only on η for a given Pr, $\left.\theta'\right|_{\eta=0}$ depends only on Pr and its value can be obtained from the solution for the variation θ with η for any value of Pr. It is convenient to define:

$$A(Pr) = \left.\theta'\right|_{\eta=0} \tag{3.41}$$

In terms of this function A, Eq. (3.40) can be written as

$$Nu_x = A\sqrt{Re_x} \tag{3.42}$$

Values of A for various values of Pr are shown in Table 3.1.

CHAPTER 3: Some Solutions for External Laminar Forced Convection   93

TABLE 3.1
Values of A for various values of *Pr*

Pr	$A = \theta'|_{\eta=0}$	$0.332Pr^{1/3}$
0.6	0.276	0.280
0.7	0.293	0.295
0.8	0.307	0.308
0.9	0.320	0.321
1.0	0.332	0.332
1.1	0.344	0.343
7.0	0.645	0.635
10.0	0.730	0.715
15.0	0.835	0.819

Over the range of Prandtl numbers covered in the table, it has been found that A varies very nearly as $Pr^{1/3}$ and, as will be seen from the results given in the table, is quite closely represented by the approximate relation:

$$A = 0.332 Pr^{1/3} \tag{3.43}$$

For values of *Pr* very different from those in the table, i.e., very different from one, the errors involved in the use of this approximate equation may be unacceptably large.

Now, in practical situations, the concern is more likely to be with the total heat transfer rate from the entire surface than with the local heat transfer rate. Consideration is, therefore, now given to the total heat transfer rate from a plate of length, L. Unit width of the plate is considered because the flow is, by assumption, two-dimensional. The total heat transfer rate per unit width Q_w will, of course, be related to the local heat transfer rate, q_w by:

$$Q_w = \int_0^L q_w \, dx \tag{3.44}$$

But Eq. (3.39) gives the local heat transfer rate as:

$$q_w = Ak(T_w - T_1)\sqrt{\frac{u_1}{x\nu}} \tag{3.45}$$

Substituting this result into Eq. (3.44) then gives on carrying the integration:

$$Q_w = 2Ak(T_w - T_1)\sqrt{\frac{u_1 L}{\nu}} \tag{3.46}$$

If a mean heat transfer coefficient for the whole plate, $\bar{h}$, is defined such that:

$$\bar{h} = \frac{Q_w}{L(T_w - T_1)} \tag{3.47}$$

then, since unit width of the plate is being considered, Eq. (3.46) gives:

$$\bar{h} = \frac{2Ak}{L}\sqrt{\frac{u_1 L}{\nu}} \tag{3.48}$$

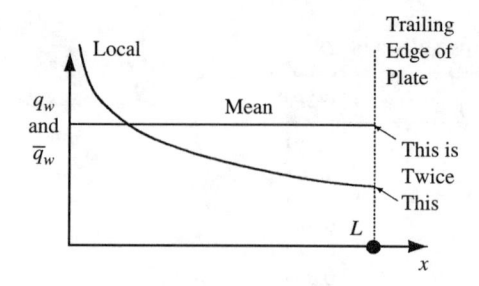

FIGURE 3.6
Relation between local and mean heat transfer rates.

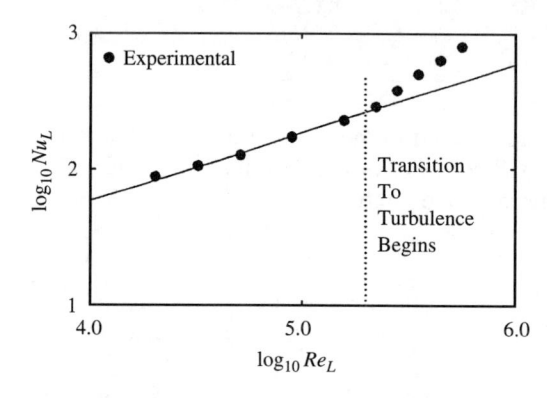

FIGURE 3.7
Comparison between predicted and experimental mean Nusselt numbers.

The mean Nusselt number for the whole plate $Nu_L (= \bar{h}L/k)$ is, therefore, given by:

$$Nu_L = 2A Re_L^{1/2} \tag{3.49}$$

where Re_L is the Reynolds number based on the plate length, L.

It will be seen from the above that the average heat transfer rate for the entire plate surface is twice the local heat transfer rate existing at the end of the plate, i.e., at $x = L$. The relation between the local and average heat transfer coefficients is shown diagrammatically in Fig. 3.6.

It should be noted that if the approximate expression for A given in Eq. (3.43) is utilized, the local and mean Nusselt numbers are given by:

$$Nu_x = 0.332 Pr^{1/3} Re_x^{1/2} \tag{3.50}$$

$$Nu_L = 0.664 Pr^{1/3} Re_L^{1/2} \tag{3.51}$$

These expressions give results that are in reasonably good agreement with experimental results, a comparison of some measurements of mean Nusselt number with the values predicted by Eq. (3.51) being shown in Fig. 3.7.

EXAMPLE 3.1. Air flows at a velocity of 5 m/s over a wide flat plate that has a length of 20 cm in the flow direction. The air ahead of the plate has a temperature of 20°C while the surface of the plate is kept at 80°C. Using the similarity solution to the laminar boundary layer equations, plot the variation of local heat transfer rate in W/m² along the plate. Show the mean heat transfer rate from the plate on this plot. Also plot the velocity and temperature profiles in the boundary layer at the end of the plate. Use the equations and graphical results given above to answer this question—it is not necessary to solve the governing equations.

CHAPTER 3: Some Solutions for External Laminar Forced Convection   95

Solution. The mean temperature of the air is:

$$T_{mean} = \frac{T_w + T_1}{2} = \frac{80 + 20}{2} = 50°C$$

At this temperature for air:

$$k = 0.0278 \text{W/m-°C}, \quad \nu = 0.0000179 \text{ m}^2/\text{s}$$

Hence, since here:

$$u_1 = 5 \text{ m/s}$$

it follows that the Reynolds number based on the length of the plate, i.e., 0.2 m, is:

$$Re_L = \frac{u_1 L}{\nu} = \frac{5 \times 0.2}{0.0000179} = 55{,}866$$

The boundary layer on the plate will therefore remain laminar (see discussion of transition Reynolds number given later).

Air has a Prandtl number of approximately 0.7 and for this value of Pr, Table 3.1 shows that:

$$A = 0.293$$

Hence, since:

$$q_w = Ak(T_w - T_1)\sqrt{\frac{u_1}{x\nu}}$$

it follows that:

$$q_w = 0.293k(T_w - T_1)\sqrt{\frac{u_1}{x\nu}}$$

Using the value of k given above, it then follows that:

$$q_w = 0.293 \times 0.0278 \times (80 - 20) \times \sqrt{\frac{5}{x \cdot 0.0000179}} = \frac{258.2}{\sqrt{x}} \text{ W/m}^2$$

In this equation, x is in m.

The variation of q_w with x given by this equation is shown in Fig. E3.1a.

Because the length of the plate is 0.2 m and because the mean heat transfer rate is twice the local value at the end of the plate, it follows that:

$$\overline{q_w} = 2 \times \frac{258.2}{\sqrt{0.2}} = 1154.7 \text{ W/m}^2$$

This value is also shown in Fig. E3.1a.

Figures 3.4 and 3.5 show the variations of f' with η and θ with η for $Pr = 0.7$. But:

$$\eta = y\sqrt{\frac{u_1}{\nu x}}$$

so:

$$y = \eta\sqrt{\frac{\nu x}{u_1}}$$

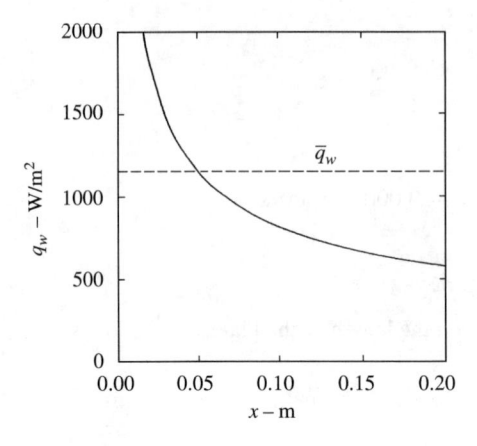

96 Introduction to Convective Heat Transfer Analysis

[Figure E3.1a: plot of $\bar{q}_w$ – W/m² vs. x – m, showing a decreasing curve with a horizontal dashed line labeled $\bar{q}_w$.]

FIGURE E3.1a

In the present case, the velocity and temperature profiles at the end of the plate, i.e., at $x = 0.2$ m, are required, so, using the value of ν given previously, it follows that at the trailing edge of the plate:

$$y = \eta \sqrt{\frac{0.0000179 \times 0.2}{5}} = 0.0008462\eta \text{ m}$$

Also, recalling that:

$$\frac{u}{u_1} = f'$$

it follows that:

$$u = 5f' \text{ m/s}$$

Also:

$$\theta = \frac{T_w - T}{T_w - T_1} = \frac{80 - T}{80 - 20}$$

from which it follows that:

$$T = 80 - 60\theta$$

Using the variations of f' with η and θ with η for $Pr = 0.7$ given in the above figures or by using the program discussed above and then using the equations for y, u, and T derived above, a table of the following form can be constructed:

η	f'	θ	y (cm)	u (m/s)	T (°C)
0	0	0	0	0	80.0
0.5	0.166	0.146	0.0423	0.829	73.0
1.0	0.330	0.291	0.0846	1.649	62.5
2.0	0.630	0.564	0.169	3.149	46.2
3.0	0.846	0.780	0.254	4.230	33.2
4.0	0.956	0.914	0.339	4.778	25.2
5.0	0.992	0.975	0.423	4.958	21.5
6.0	0.999	0.995	0.508	4.995	20.7
8.0	1.000	1.000	0.677	5.000	20.0

FIGURE E3.1b

FIGURE E3.1c

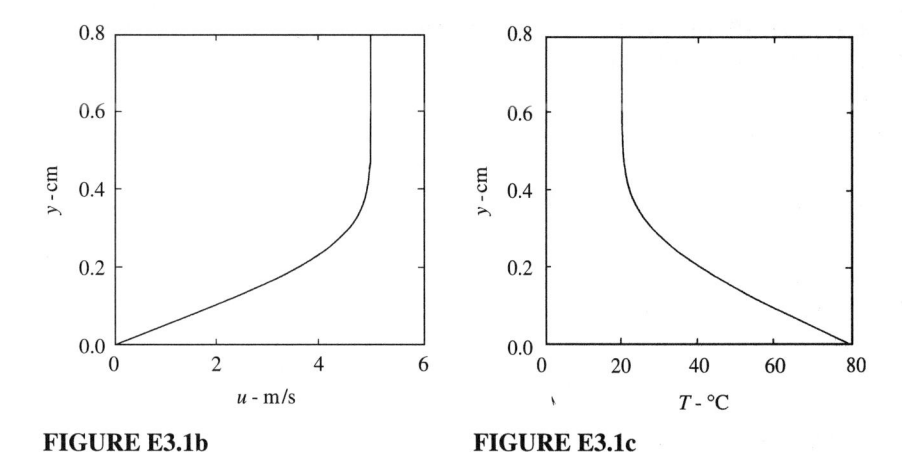

The variations of velocity and temperature given in this table are plotted in Figs. E3.1b and E3.1c. More values than are given in the above table were actually used in constructing these figures.

It should be noted that for the particular case of a fluid having a Prandtl number equal to 1, Eq. (3.23) reduces to

$$\text{For } Pr = 1: u\frac{\partial \theta}{\partial x} + v\frac{\partial \theta}{\partial y} = \nu\frac{\partial^2 \theta}{\partial y^2} \tag{3.52}$$

The boundary conditions on the solution to this equation are still those given in Eq. (3.24).

Now the momentum equation (3.2) can be written as

$$u\frac{\partial (u/u_1)}{\partial x} + v\frac{\partial (u/u_1)}{\partial y} = \nu\frac{\partial^2 (u/u_1)}{\partial y^2} \tag{3.53}$$

While the boundary conditions listed in Eq. (3.6) are:

$$\begin{aligned} y &= 0: u/u_1 = 0 \\ y \text{ large}: u/u_1 &\to 1 \end{aligned} \tag{3.54}$$

A comparison of Eqs. (3.52) and (3.53) and also of their boundary conditions as given in Eqs. (3.24) and (3.54) respectively, shows that these equations are identical in all respects. Therefore, for the particular case of Pr equal to one, the distribution of θ through the boundary layer is identical to the distribution of (u/u_1). In this particular case, therefore, Fig. 3.4 also gives the temperature distribution and the two boundary layer thicknesses are identical in this case. Now many gases have Prandtl numbers which are not very different from 1 and this relation between the velocity and temperature fields and the results deduced from it will be approximately correct for them.

As discussed above, the heat transfer at the wall is given by:

$$\frac{q_w}{k(T_w - T_1)} = \theta'|_{\eta=0}\sqrt{Re_x} \tag{3.55}$$

But since the distributions of θ and u/u_1 have been shown to be identical for $Pr = 1$ it follows that for this case:

$$\theta'|_{\eta=0} = \left.\frac{d(u/u_1)}{d\eta}\right|_{\eta=0} = f''|_{\eta=0} \tag{3.56}$$

From the results given in Fig. 3.4 it follows that

$$f''|_{\eta=0} = 0.332 \tag{3.57}$$

Using this result then gives:

$$\text{For } Pr = 1: \ Nu_x = \frac{q_w x}{k(T_w - T_1)} = 0.332\sqrt{Re_x} \tag{3.58}$$

Nu_x being as before the Nusselt number based on x.

It is also worth noting that the shearing stress on the wall, τ_w, is given by

$$\tau_w = \mu \left.\frac{\partial u}{\partial y}\right|_{y=0} \tag{3.59}$$

This can be rearranged to give:

$$\frac{\tau_w}{\rho u_1^2} = \left(\frac{\mu}{\rho u_1}\right) \left.\frac{d(u/u_1)}{d\eta}\right|_{\eta=0} \frac{\partial \eta}{\partial y} = \frac{1}{\sqrt{Re_x}} f''\bigg|_{\eta=0} \tag{3.60}$$

Combining this result with Eqs. (3.56) and (3.55) then shows that:

$$\text{For } Pr = 1: \ Nu_x = (\tau_w/\rho u_1^2) Re_x \tag{3.61}$$

This is, basically, the Reynolds analogy for laminar flow. It relates the dimensionless heat transfer rate to the dimensionless shear stress. It will not be pursued further at this stage.

Equation (3.58) which, it will be recalled, was deduced without solving the energy equation, gives results which agree reasonably well with the exact results for gases with Prandtl numbers near one. If the Prandtl number of air is assumed to be 0.7, then the value of Nu_x given by Eq. (3.58) is about 12% greater than the true value.

3.3
SIMILARITY SOLUTIONS FOR FLOW OVER FLAT PLATES WITH OTHER THERMAL BOUNDARY CONDITIONS

The above similarity solution was for the case where the plate has a uniform surface temperature, i.e., for which:

$$T_w - T_1 = \text{constant}$$

Similarity solutions for a few cases of flow over a flat plate where the plate temperature varies with x in a prescribed manner can also be obtained. In all such cases the solution for the velocity profile is, of course, not affected by the boundary condition

CHAPTER 3: Some Solutions for External Laminar Forced Convection 99

on temperature and is the same as for the uniform temperature case. For example, consider the case where:

$$T_w = T_1 + Cx^n$$

i.e.:

$$T_w - T_1 = Cx^n \tag{3.62}$$

C and n being constants.

The following dimensionless temperature is again introduced:

$$\theta = \frac{T_w - T}{T_w - T_1} = 1 - \frac{T - T_1}{T_w - T_1} \tag{3.63}$$

and it is again assumed that θ depends only on the similarity variable, η.

In the present case, $T_w - T_1$ is a function of x because:

$$T - T_1 = (1 - \theta)(T_w - T_1)$$

the energy equation (3.3) can be written in terms of θ as

$$-u\frac{\partial}{\partial x}[(1-\theta)(T_w - T_1)] + v\frac{\partial \theta}{\partial y}(T_w - T_1) = \left[\frac{\nu}{Pr}\right]\frac{\partial^2 \theta}{\partial y^2}(T_w - T_1) \tag{3.64}$$

In terms of θ, the boundary conditions again are:

$$y = 0: \theta = 0$$
$$y \text{ large}: \theta \to 1 \tag{3.65}$$

Using the relations for the velocity components previously derived, Eq. (3.64) gives:

$$f'\frac{d\theta}{d\eta}\frac{\partial \eta}{\partial x} - \frac{nf'(1-\theta)}{x} + \left[\frac{1}{2}\sqrt{\frac{\nu}{u_1 x}}(\eta f' - f)\right]\frac{d\theta}{d\eta}\frac{\partial \eta}{\partial y} = \frac{1}{Pr}\frac{\nu}{u_1}\frac{d^2\theta}{d\eta^2}\left[\frac{\partial \eta}{\partial y}\right]^2 \tag{3.66}$$

On rearrangement, this equation becomes:

$$\theta'' + nPrf'(1-\theta) + \frac{Pr}{2}\theta' f = 0 \tag{3.67}$$

while the boundary conditions on the solution given in Eq. (3.65) can be written as:

$$\eta = 0: \theta = 0$$
$$\eta \text{ large}: \theta \to 1 \tag{3.68}$$

Thus, as was the case with uniform plate temperature, the partial differential equation governing the temperature distribution has been reduced to an ordinary differential equation. This confirms the assumption that the temperature profiles are similar. For any prescribed values of Pr and n, the variation of θ with η can be obtained by solving Eq. (3.67). A computer program, SIMVART, which is an extension of that for the uniform temperature surface case, obtains this solution. The program

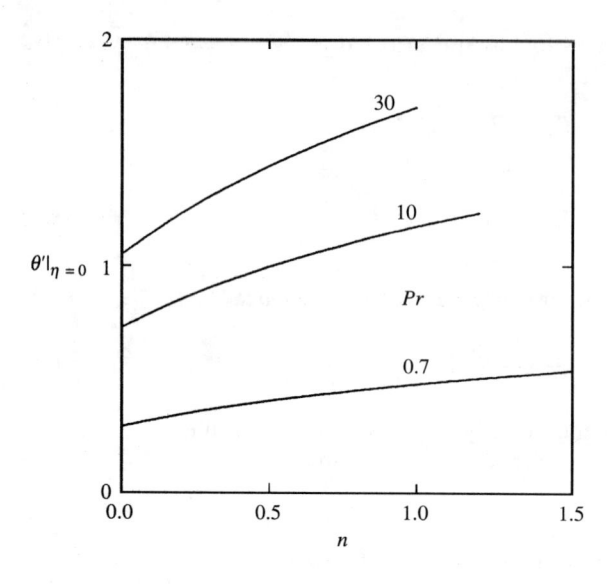

FIGURE 3.8 Variation of $\theta'|_{\eta=0}$ with n for various values of Pr.

first obtains the velocity profile solution and then uses the same procedure to solve Eq. (3.67). The program is available in the way discussed in the Preface.

The heat transfer rate at the wall is as before given by

$$q_w = -k \left.\frac{\partial T}{\partial y}\right|_{y=0} \qquad (3.69)$$

Hence, using Eq. (3.63), it again follows that:

$$\frac{q_w x}{k(T_w - T_1)} = \theta'|_{\eta=0} \sqrt{Re_x} \qquad (3.70)$$

For any prescribed values of Pr and n, $\theta'|_{\eta=0}$ will have a specific value. It therefore follows that q_w will be proportional to $(T_w - T_1)/x^{0.5}$. Hence, the case where the heat flux at the surface of the plate is uniform corresponds to the case where $n = 0.5$, i.e., a similarity solution exists for flow over a flat plate with a uniform surface heat flux.

Some typical variations of $\theta'|_{\eta=0}$ with n for various values of Pr obtained using the above program are shown in Fig. 3.8.

EXAMPLE 3.2. Air flows at a velocity of 3 m/s over a wide flat plate that has a length of 30 cm in the flow direction. The air ahead of the plate has a temperature of 20°C while the temperature of the surface of the plate is given by $[20 + 40(x/30)]°C$, x being the distance measured along the plate in cm. Using the similarity solution results, plot the variation of local heat transfer rate in W/m² along the plate.

Solution. The plate temperature varies linearly from 20 to 60°C. Its mean temperature is therefore 40°C. The mean temperature of the air in the boundary layer is therefore,

$$T_{mean} = \frac{T_{w_{mean}} + T_1}{2} = \frac{40 + 20}{2} = 30°C$$

At this temperature for air:

$$k = 0.0264 \text{ W/m-°C}, \quad \nu = 0.0000160 \text{ m}^2/\text{s}$$

CHAPTER 3: Some Solutions for External Laminar Forced Convection 101

Hence, since here:

$$u_1 = 3 \text{ m/s}$$

it follows that the Reynolds number based on the length of the plate, i.e., 0.3 m, is:

$$Re_L = \frac{u_1 L}{\nu} = \frac{3 \times 0.3}{0.0000160} = 56{,}250$$

The boundary layer on the plate will therefore remain laminar (see discussion of the transition Reynolds number given later).

Because the plate temperature variation can be written:

$$T_w - 20 = \frac{40}{0.3} x \text{ °C}$$

x being in m, the plate temperature variation is of the form:

$$T_w - T_1 = Cx$$

Hence, in this case $n = 1$. Now, air has a Prandtl number of approximately 0.7 and for this value of Pr, for $n = 1$, Fig. 3.8 gives:

$$\theta'|_{\eta=0} = 0.480$$

Hence since:

$$\frac{q_w x}{k(T_w - T_1)} = \theta'|_{\eta=0} \sqrt{Re_x}$$

it follows that:

$$q_w = 0.480 k(T_w - T_1) \sqrt{\frac{u_1}{x\nu}}$$

Using the value of k given above, it follows that:

$$q_w = 0.480 \times 0.0264 \times (40/0.3)x \sqrt{\frac{3}{x \cdot 0.0000160}} = 731.6 \sqrt{x} \text{ W/m}^2$$

In this equation, x is in m.

The variation of q_w with x given by this equation is shown in Fig. E3.2.

In this case, because the plate temperature is equal to the air temperature at $x = 0$, the heat transfer rate is zero at the leading edge of the plate and increases with x.

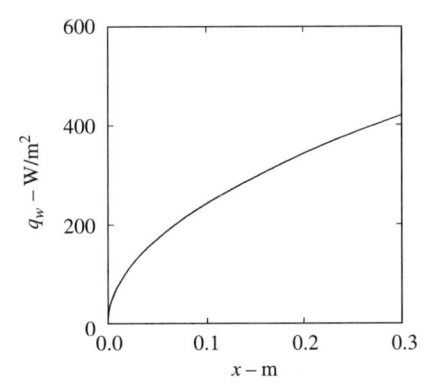

FIGURE E3.2

EXAMPLE 3.3. Air flows at a velocity of 7 m/s over a wide flat plate that has a length of 10 cm in the flow direction. The air ahead of the plate has a temperature of 20°C. There is a uniform heat flux of 2 kW/m² at the surface of the plate. Using the similarity solution results, plot the variation of local temperature along the plate.

Solution. Here, the plate temperature is not known so the mean film temperature at which the air properties are determined is not initially known. The air properties will therefore first be evaluated at the freestream temperature 20°C and the plate surface temperature variation will be evaluated. This will allow an estimate of the mean temperature to be made and the air properties at this mean temperature can be found and the calculation can then be repeated using these air properties. If necessary, a new mean air temperature can be determined using this new temperature and the calculation again repeated. This third step is, however, seldom required.

At 20°C, air has the following properties:

$$k = 0.0256 \text{ W/m-°C}, \quad \nu = 0.0000151 \text{ m}^2\text{/s}$$

Hence, since here:

$$u_1 = 7 \text{ m/s}$$

it follows that the Reynolds number based on the length of the plate, i.e., 0.1 m, is:

$$Re_L = \frac{u_1 L}{\nu} = \frac{7 \times 0.1}{0.0000151} = 46{,}358$$

The boundary layer on the plate will therefore remain laminar (see discussion of the transition Reynolds number given later).

Now, it was shown that if there is a uniform heat flux at the surface:

$$T_w - T_1 = C x^{0.5}$$

i.e., when there is a uniform surface heat flux, $n = 0.5$. Air has a Prandtl number of approximately 0.7 and for this value of Pr, for $n = 0.5$, Fig. 3.8 gives:

$$\theta'|_{\eta=0} = 0.406$$

Hence, since:

$$\frac{q_w x}{k(T_w - T_1)} = \theta'|_{\eta=0} \sqrt{Re_x}$$

it follows that in the present case where $q_w = 2000$ W/m²:

$$\frac{2000 x}{0.0256(T_w - T_1)} = 0.406 \sqrt{\frac{7x}{0.0000151}}$$

i.e., since $T_1 = 20$°C:

$$T_w = 20 + 282.6 x^{0.5}$$

The variation of T_w with x given by this equation is shown in Fig. E3.3.
Now the mean plate temperature is given by:

$$\overline{T_w} = \frac{1}{L} \int_0^L T_w \, dx$$

which using the expression for T_w derived above gives:

$$T_{w_{\text{mean}}} = 20 + 59.6 = 79.6\text{°C}$$

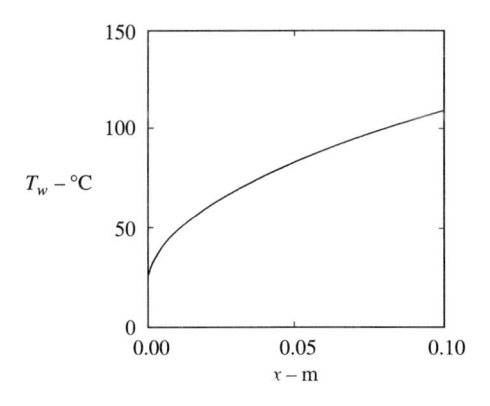

CHAPTER 3: Some Solutions for External Laminar Forced Convection   103

FIGURE E3.3

The mean air temperature is therefore

$$T_{\text{mean}} = \frac{T_{w\text{mean}} + T_1}{2} = \frac{79.6 + 20}{2} = 49.8°C$$

At this temperature for air:

$$k = 0.0277 \text{ W/m-°C}, \quad \nu = 0.0000178 \text{ m}^2/\text{s}$$

Using these values for the air properties, the equation for the surface temperature becomes:

$$T_w = 20 + 283.7 x^{0.5}$$

This gives values that differ from those derived before by less than 0.5%. The results given in Fig. E3.3, therefore, adequately describe the surface temperature variation and there is no need to repeat the calculation with an improved mean air temperature.

Because of the linearity of the energy equation, solutions for different wall temperature variations can be combined to get solutions for other temperature variations, i.e., if T_A and T_B are similarity solutions to the energy equations for two different wall temperature variations then $T_A + T_B$ is also a solution to the energy-energy equation and applies to the case where the wall temperature is the sum of those giving the solutions T_A and T_B, i.e., since:

$$(T - T_1)_A = (1 - \theta_A)(T_w - T_1)_A$$

and:

$$(T - T_1)_B = (1 - \theta_B)(T_w - T_1)_B$$

where θ_A and θ_B are the similarity solutions for the two wall temperature variations then the solution for the wall temperature variation:

$$(T_w - T_1)_A + (T_w - T_1)_B$$

is:

$$(T - T_1) = (1 - \theta_A)(T_w - T_1)_A + (1 - \theta_B)(T_w - T_1)_B$$

For example, if θ_A is the solution for a plate with a uniform surface temperature, T_{wo}, and if θ_B is the solution for $T_w = T_1 + Cx^n$, then:

$$(T - T_1) = (1 - \theta_A)(T_{wo} - T_1) + (1 - \theta_B)Cx^n \quad (3.71)$$

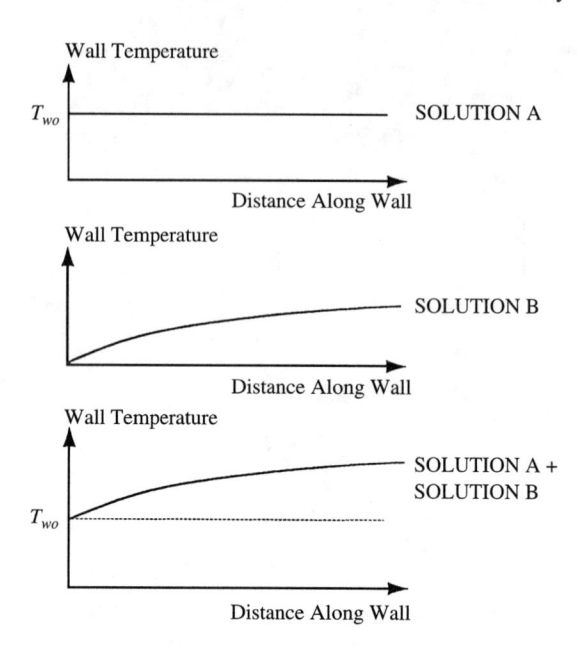

104 Introduction to Convective Heat Transfer Analysis

FIGURE 3.9
Combination of temperature field solutions for different boundary conditions.

will be the solution for the case where:

$$(T_w - T_1) = (T_{wo} - T_1) + Cx^n \tag{3.72}$$

This is illustrated in Fig. 3.9.

The heat transfer rate at the wall is given by:

$$\frac{q_w x}{k(T_{wo} - T_1)} = \left[\theta_A'|_{\eta=0} + \theta_B'|_{\eta=0} \frac{Cx^n}{T_{wo} - T_1}\right] \sqrt{Re_x} \tag{3.73}$$

EXAMPLE 3.4. Air flows at a velocity of 3 m/s over a wide flat plate that has a length of 30 cm in the flow direction. The air ahead of the plate has a temperature of 20°C while the surface of the plate is equal to $[40 + 40(x/30)]$°C, x being the distance measured along the plate in cm.

Using the similarity solution results, plot the variation of local heat transfer rate in W/m² along the plate.

Solution. The plate temperature varies linearly from 40 to 80°C. Its mean temperature is therefore 60°C. The mean temperature of the air in the boundary layer is:

$$T_{mean} = \frac{T_{w\,mean} + T_1}{2} = \frac{60 + 20}{2} = 40°C$$

At this temperature for air:

$$k = 0.0271 \text{ W/m-°C}, \quad \nu = 0.0000170 \text{ m}^2/\text{s}$$

Hence, since here:

$$u_1 = 3 \text{ m/s}$$

it follows that the Reynolds number based on the length of the plate, i.e., 0.3 m, is:

$$Re_L = \frac{u_1 L}{\nu} = \frac{3 \times 0.3}{0.0000170} = 52{,}941$$

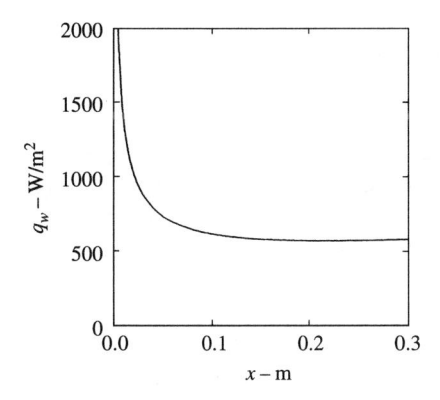

FIGURE E3.4

The boundary layer on the plate will therefore remain laminar (see discussion of the transition Reynolds number given later).

The surface temperature variation is made up of a step jump in temperature at the leading edge of the plate plus a linearly increasing temperature along the plate. The surface temperature can be written as:

$$(T_w - T_1) = (60 - 20) + 133.33x \text{ °C}$$

where, in this equation, x is in m. In the present case therefore $(T_{wo} - T_1) = 40°C$ and $n = 1$.

As discussed above, the solution is obtained by combining the similarity solutions for a plate with a uniform temperature and for a plate with a linearly increasing temperature. These two solutions give for air ($Pr = 0.7$) (see Fig. 3.8):

$$\theta'|_{\eta=0} = 0.293$$

and

$$\theta'|_{\eta=0} = 0.406$$

respectively.

Now it was shown that:

$$\frac{q_w x}{k(T_{wo} - T_1)} = \left[\theta_A'|_{\eta=0} + \theta_B'|_{\eta=0} \frac{Cx^n}{T_{wo} - T_1}\right] \sqrt{Re_x}$$

Hence, in the present case where $C = 133.33$ and $n = 1$:

$$\frac{q_w x}{0.0271 \times 40} = \left[0.293 + \frac{0.406 \times 133.33x}{40}\right]\sqrt{\frac{3x}{0.0000170}}$$

i.e.,

$$q_w = \frac{133.4}{x^{0.5}} + 616.1x^{0.5}$$

The variation of q_w with x given by this equation is shown in Fig. E3.4.

It will be seen from Fig. E3.4 that near the leading edge of the plate the effect of the temperature jump at the leading edge predominates and q_w is approximately proportional to $1/\sqrt{x}$ whereas near the trailing edge of the plate the effect of the linearly increasing plate temperature predominates and q_w is only very weakly dependent on x.

3.4
OTHER SIMILARITY SOLUTIONS

In the preceding sections, the solution for boundary layer flow over a flat plate was obtained by reducing the governing set of partial differential equations to a pair of ordinary differential equations. This was possible because the velocity and temperature profiles were similar in the sense that at all values of x, (u/u_1) and $(T_w - T)/(T_w - T_1)$ were functions of a single variable, η, alone. Now, for flow over a flat plate, the freestream velocity, u_1, is independent of x. The present section is concerned with a discussion of whether there are any flow situations in which the freestream velocity, u_1, varies with x and for which similarity solutions can still be found [1],[10].

Consideration is again first given to the velocity profile solution. The problem is to determine for what distributions of freestream velocity, u_1, these profiles will be similar, i.e., for what distributions of u_1 is it possible to assume that:

$$\frac{u}{u_1} = \text{function}(\eta^*) = f'(\eta^*) \tag{3.74}$$

where η^* is some similarity variable whose form has also still to be determined. Notice that the function relating (u/u_1) to η^* has again been denoted by f', the prime now denoting differentiation with respect to η^*.

Now the form of the similarity variable, η, for flow over a flat plate was arrived at by noting that if the velocity profiles were similar then:

$$\frac{u}{u_1} = \text{function}(y/\delta) \tag{3.75}$$

where δ was some measure of the local boundary layer thickness. It was then noted that, from the order of magnitude analysis used in the derivation of the boundary layer equations, the order of magnitude of δ has to be $x/\sqrt{Re_x}$ and this was substituted for δ in Eq. (3.75). This then gave η as $y\sqrt{Re_x}/x$. In the case being studied in the present section where u_1 varies, the above result will be generalized by assuming that δ will depend on $Le(\bar{x})/\sqrt{Re_L}$ where $\bar{x}$ is equal to x/L, e is some function of $\bar{x}$, L is some convenient reference length, and Re_L is the Reynolds number based on L and some convenient reference velocity U, i.e.,

$$Re_L = UL/\nu \tag{3.76}$$

Substituting this as the measure of the boundary layer thickness into Eq. (3.75) then gives the following as the form of the generalized similarity variable:

$$\eta^* = \frac{y\sqrt{Re_L}}{Le(\bar{x})} \tag{3.77}$$

The function $e(\bar{x})$ remains to be determined.

The basic procedure for arriving at the similarity solution is, of course, the same as that for flat plate flow. The continuity equation is first used to express v in terms of the similarity function f'. For this purpose it is written as:

$$\frac{\partial v}{\partial y} = -\frac{\partial u}{\partial x}$$

CHAPTER 3: Some Solutions for External Laminar Forced Convection   107

i.e.,

$$\frac{\partial v}{\partial \eta^*}\frac{\partial \eta^*}{\partial y} = -\frac{\partial}{\partial \bar{x}}(u_1 f') = -f'\frac{du_1}{d\bar{x}} - u_1 f''\frac{\partial \eta^*}{\partial \bar{x}} \qquad (3.78)$$

where, as before

$$\bar{x} = \frac{x}{L} \qquad (3.79)$$

Using Eq. (3.77) in Eq. (3.78) then gives:

$$\sqrt{Re_L}\frac{\partial v}{\partial \eta^*} = -f'e\frac{du_1}{d\bar{x}} + u_1\frac{de}{d\bar{x}}\eta^* f'' \qquad (3.80)$$

This is integrated with respect to η^* noting that $v = 0$ when $\eta^* = 0$, to give:

$$\sqrt{Re_L}\, v = -fe\frac{du_1}{d\bar{x}} + u_1\frac{de}{d\bar{x}}(\eta^* f' - f)$$

i.e.,

$$-\sqrt{Re_L}\, v = f\frac{d}{d\bar{x}}(eu_1) - \eta^* u_1 f'\frac{de}{d\bar{x}} \qquad (3.81)$$

Now, since flows in which u_1, the freestream velocity, in general, varies with x are being considered, the momentum equation to be solved has the form

$$u\frac{\partial u}{\partial x} + v\frac{\partial u}{\partial y} = u_1\frac{du_1}{dx} + v\frac{\partial^2 u}{\partial y^2} \qquad (3.82)$$

Using the expressions for the variables given above, this equation becomes:

$$u_1 f'\frac{\partial}{\partial \bar{x}}(f' u_1) - \frac{L}{\sqrt{Re_L}} f\frac{d}{d\bar{x}}(eu_1) - \eta^* u_1 f'\frac{de}{d\bar{x}}\frac{\partial}{\partial y}(f' u_1)$$

$$= u_1\frac{du_1}{d\bar{x}} + v u_1 L f'''\left(\frac{\partial \eta^*}{\partial y}\right)^2 \qquad (3.83)$$

With η^* as defined in Eq. (3.77), this equation can be rewritten as:

$$f'^2\frac{du_1}{d\bar{x}} - ff''\frac{1}{e}\frac{d}{d\bar{x}}(eu_1) - \frac{du_1}{d\bar{x}} - \frac{U}{e^2}f''' = 0 \qquad (3.84)$$

If the following are defined, for convenience:

$$\alpha = \frac{e}{U}\frac{d}{d\bar{x}}(eu_1) \qquad (3.85)$$

$$\beta = \frac{e^2}{U}\frac{du_1}{d\bar{x}} \qquad (3.86)$$

Eq. (3.84) can be written as:

$$f''' + \alpha ff'' + \beta(1 - f'^2) = 0 \qquad (3.87)$$

Now, if similar velocity profiles do exist, Eq. (3.87) must allow f to be determined as a function of the similarity variable η^* alone. This will only be possible if α and β are independent of $\bar{x}$. Since they don't depend on η^*, this means that it is only for flows in which:

$$\alpha = \text{constant}$$
$$\beta = \text{constant} \quad (3.88)$$

that similarity solutions can be found. This requirement on α and β then allows the freestream velocity distributions $u_1(\bar{x})$ occurring in the expression for the similarity variable η^* to be found. For this purpose it is noted that Eqs. (3.85) and (3.86) give:

$$2\alpha - \beta = \frac{1}{U}\frac{d}{d\bar{x}}(e^2 u_1) \quad (3.89)$$

Thus, the velocity distribution and the function, e, for similar solutions must be such that $d(e^2 u_1)/d\bar{x}$ is a constant equal to $(2\alpha - \beta)U$. Now one possibility is that $(2\alpha - \beta)$ be equal to zero. However, the variation of u_1 with x that provides this situation seems to have little practical significance and it will not be considered here. Therefore, $(2\alpha - \beta)$ will be assumed to be nonzero.

Integrating Eq. (3.89) gives:

$$e^2 \frac{u_1}{U} = (2\alpha - \beta)\bar{x} \quad (3.90)$$

Now it will also be noted from Eqs. (3.85) and (3.86) that:

$$\alpha - \beta = e\frac{u_1}{U}\frac{de}{d\bar{x}} \quad (3.91)$$

which can be rearranged using Eq. (3.86) to give:

$$(\alpha - \beta)\frac{1}{u_1}\frac{du_1}{d\bar{x}} = \frac{e}{U}\frac{du_1}{d\bar{x}}\frac{de}{d\bar{x}} = \frac{\beta}{e}\frac{de}{d\bar{x}} \quad (3.92)$$

This, in turn, can be integrated to give:

$$u_1^{\alpha-\beta} = Ke^\beta \quad (3.93)$$

where K is a constant of integration.

The function e is now eliminated between Eqs.(3.90) and (3.93) to give:

$$\sqrt{(2\alpha - \beta)\bar{x}U/u_1} = u_1^{(\alpha-\beta)/\beta} K^{1/\beta}$$

i.e.,

$$u_1 = K^{(2/\alpha-\beta)}[(2\alpha - \beta)U]^{\beta/(2\alpha-\beta)} \bar{x}^{\beta/(2\alpha-\beta)} \quad (3.94)$$

Since α and β are both constants and both contain the function e, it is possible, without any loss of generality, to set α equal to 1, any common factor between α and β then being incorporated into e. The case $\alpha = 0$ is, of course, being excluded from the discussion. With α set equal to 1 for this reason, Eq. (3.94) becomes:

$$u_1 = K^{2/(1-\beta)}[(2 - \beta)U/L]^{[\beta/(2-\beta)]} x^{\beta/(2-\beta)} \quad (3.95)$$

CHAPTER 3: Some Solutions for External Laminar Forced Convection   109

It is convenient to express β in terms of a new constant, m, which is such that:

$$m = \beta/(2 - \beta) \tag{3.96}$$

This can, of course, be rearranged to give

$$\beta = 2m/(m + 1) \tag{3.97}$$

In terms of m, the freestream velocity distributions for which similarity solutions can be found, these being given in Eq. (3.95), can be written as:

$$u_1 = K^{1+m}\left[\left(\frac{2}{m+1}\right)\frac{U}{L}\right]^m x^m \tag{3.98}$$

Since K, U, and L are all constants this equation has the form:

$$u_1 = cx^m \tag{3.99}$$

where c is a constant. Therefore, similarity solutions to the boundary layer equations can be found for flows over bodies whose freestream velocity varies as x^m.

Having established the form of the freestream velocity distribution that gives similarity solutions, consideration can be given to the similarity variable function $e(\bar{x})$. This is found by using Eq. (3.90). Noting that α has already been set equal to one, this equation gives:

$$e = \sqrt{(2-\beta)\frac{U}{L}}\sqrt{\frac{x}{u_1}} = \sqrt{\left(\frac{2}{m+1}\right)\frac{U}{L}}\sqrt{\frac{x}{u_1}} \tag{3.100}$$

Substituting this into the expression for the similarity variable η^* then gives:

$$\eta^* = \sqrt{\left(\frac{m+1}{2}\right)\left(\frac{u_1}{x\nu}\right)} = \frac{y}{x}\sqrt{\left(\frac{m+1}{2}\right)Re_x} \tag{3.101}$$

where Re_x is the Reynolds number based on x and the local freestream velocity is u_1.

With α set equal to one and β defined in terms of m by Eq. (3.97), the reduced momentum Eq. (3.87) becomes

$$f''' + ff'' + \left(\frac{2m}{m+1}\right)(1 - f'^2) = 0 \tag{3.102}$$

The boundary conditions on the solution to this equation being, of course

$$\begin{array}{l}\eta^* = 0: \ f = 0, \ f' = 0 \\ \eta^* \text{ large: } f' \to 1\end{array} \tag{3.103}$$

Before discussing any solutions to this equation, it is useful to consider what body shapes will give a velocity distribution of the type for which similarity solutions can be found, i.e., of the type given in Eq. (3.99). A velocity distribution of this type will exist with flow over wedge-shaped bodies having an included angle, ϕ, which is equal to $\pi\beta$ as shown in Fig. 3.10.

Because m is related to β by Eq. (3.96) it is related to the included angle ϕ by:

$$m = \phi/(2\pi - \phi) \tag{3.104}$$

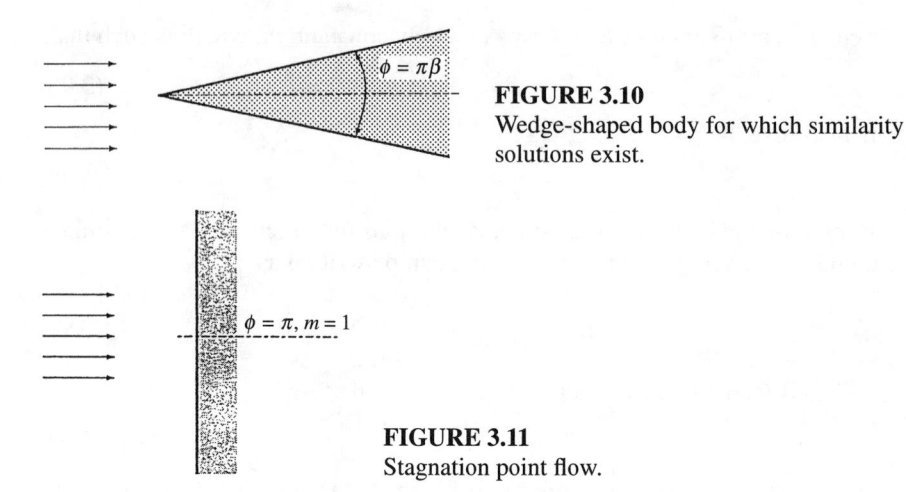

FIGURE 3.10
Wedge-shaped body for which similarity solutions exist.

FIGURE 3.11
Stagnation point flow.

It may be noted that the case ϕ equal to zero corresponds to flow over a flat plate while the case of ϕ equal to π, i.e., m equal to one, corresponds to flow over a plate set normal to the direction of the undisturbed freestream as shown in Fig. 3.11. The latter flow closely represents flow near the stagnation point on a bluff body.

For example, in the case of flow over a circular cylinder the freestream velocity is given by

$$u_1/U = 2\sin(x/R) \tag{3.105}$$

the variables being defined in Fig. 3.12.

For small values of x/R, i.e., near the stagnation point, this equation reduces approximately to:

$$u_1/U = 2x/R \tag{3.106}$$

Therefore, near the stagnation point, u_1 is proportional to x so that the similarity solution with m equal to 1 will apply in this region.

Of course freestream velocity distributions of the form $u_1 = cx^m$ can also be obtained in other situations, e.g., in a duct with varying cross-sectional area.

Solutions to Eq. (3.102) subject to the boundary conditions given in Eq. (3.103) have been obtained for a number of different values of the velocity distribution index, m. Some typical results are shown in Fig. 3.13.

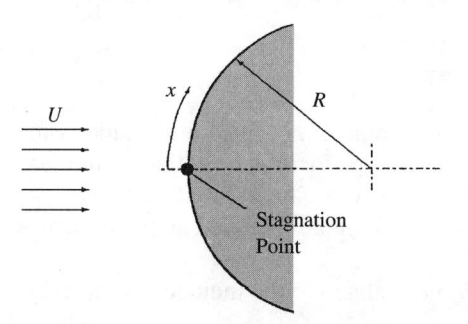

FIGURE 3.12
Flow over a circular cylinder.

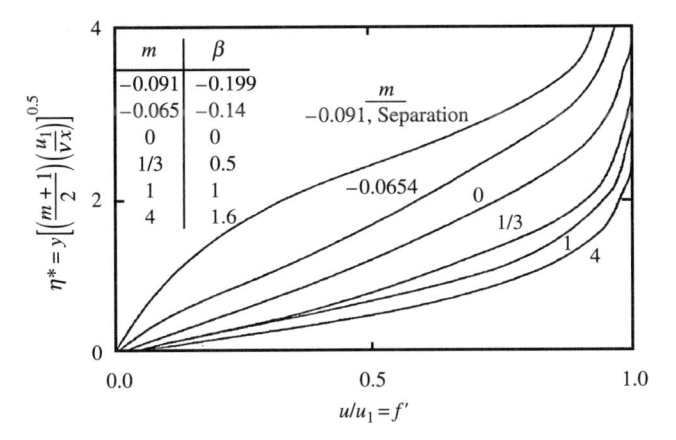

CHAPTER 3: Some Solutions for External Laminar Forced Convection   111

FIGURE 3.13
Results for various values of the index, m.

The profile for m equal to zero corresponds to flow over a flat plate and is identical to the solution given in the previous section for this type of flow. In making a comparison with this solution it should be noted that for $m = 0$, η^* is equal to $\sqrt{u_1/2x_1\nu}$, i.e., $\eta^* = \eta/\sqrt{2}$, η being the similarity variable used in deriving the flat plate solution.

It should also be noted that when m is less than zero, du_1/dx is negative i.e., there is an adverse pressure gradient (the pressure is rising). These solutions for $m \leq 0$ do not, of course, correspond to flow over real wedges, but such velocity distributions can be generated in other ways. The velocity profiles for values of m less than zero have an inflection point as will be noted from the results shown in Fig. 3.13. When m is equal to -0.091, the slope of the velocity profile at the wall is zero, i.e., $d(u/u_1)/d\eta^* = 0$ and so $du/dy = 0$, and the boundary layer is, therefore, on the point of separating. Solutions for more negative values of m have no meaning because they exhibit reversed flow in the region adjacent to the wall and the boundary layer equations are not applicable to such flows.

Having established that similarity solutions for the velocity profile can be found for certain flows involving a varying freestream velocity, attention must now be turned to the solutions of the energy equation corresponding to these velocity solutions. The temperature is expressed in terms of the same nondimensional variable that was used in obtaining the flat plate solution, i.e., in terms of $\theta = (T_w - T)/(T_w - T_1)$ and it is assumed that θ is also a function of η^* alone. Attention is restricted to flow over isothermal surfaces, i.e., with T_w a constant, and T_1, of course, is also constant.

In terms of the variables introduced above, the energy equation is:

$$u_1 f'\theta' \frac{\partial \eta^*}{\partial x} + v\theta \frac{\partial \eta^*}{\partial y} = \left(\frac{\nu}{\text{Pr}}\right)\theta''\left(\frac{\partial \eta^*}{\partial y}\right)^2 \qquad (3.107)$$

112 Introduction to Convective Heat Transfer Analysis

But substituting Eq. (3.100) and (3.101) into the expression derived for v gives:

$$v \frac{\partial \eta^*}{\partial y} = \left[\eta^* u_1 f' \frac{d}{dx}\left(\sqrt{\frac{x}{u_1}}\right) - f \frac{d}{dx}\left(\sqrt{u_1 x}\right) \right] \sqrt{\frac{u_1}{x}} \quad (3.108)$$

Substituting this into Eq. (3.107) and using Eq. (3.99) then gives on rearrangement

$$\theta'' + Pr f \theta' = 0 \quad (3.109)$$

The boundary conditions on θ are the same as those for flow over a flat plate, i.e.,

$$\begin{array}{l} \eta^* = 0: \ \theta = 0 \\ \eta^* \text{ large}: \ \theta \to 1 \end{array} \quad (3.110)$$

For any selected values of m and Pr, the variation of f with η^* derived using the procedure previously outlined, can be used to obtain the solution to Eq. (3.109) subject to the boundary conditions given in Eq. (3.110). To obtain this solution, the same procedure that was used to find the flat plate solution and which was discussed above is utilized. This solution will give the variation of θ with η^* corresponding to the selected values of m and Pr.

With the temperature distribution determined in this way, the heat transfer rate can be determined by noting that

$$q_w = -k \left.\frac{\partial T}{\partial y}\right|_w = k(T_w - T_1)\left[\frac{d\theta}{d\eta^*}\frac{\partial \eta^*}{\partial y}\right]_{y=0} \quad (3.111)$$

This can be rearranged using Eq. (3.101) as

$$Nu_x = \sqrt{\frac{m+1}{2}} \sqrt{Re_x} \, \theta' \bigg|_{\eta^*=0} \quad (3.112)$$

where Nu_x is again the local Nusselt number and Re_x is the local Reynolds number. It is convenient to write

$$\alpha^* = \sqrt{\frac{m+1}{2}} \, \theta' \bigg|_{\eta^*=0} \quad (3.113)$$

and Eq. (3.112) can then be written as

$$Nu_x = \alpha^* \sqrt{Re_x} \quad (3.114)$$

Values of α^* for various values of m and Pr have been derived from the calculated variations of θ with η^*. Typical values are shown in Table 3.2.

The values for $m = 0$ are, of course, the same as the values of α given in the previous section for flow over a flat plate.

For $m = 1$, which corresponds, as previously indicated, to stagnation point flow, the values of α^* given in the table can be closely represented by

$$\alpha^*_{m=1} = 0.57 Pr^{0.4} \quad (3.115)$$

CHAPTER 3: Some Solutions for External Laminar Forced Convection 113

TABLE 3.2
Values of α^* for various values of m and Pr

m	0.7	0.8	1.0	5.0	10.0
−0.0753	0.242	0.253	0.272	0.457	0.570
0.0	0.293	0.307	0.332	0.585	0.730
0.111	0.331	0.348	0.378	0.669	0.851
0.333	0.384	0.403	0.440	0.792	1.013
1.0	0.496	0.523	0.570	1.043	1.344
4.0	0.813	0.858	0.938	1.736	2.236

Substituting this into Eq. (3.114) then gives the local heat transfer rate in the region of the stagnation point as

$$Nu_x = 0.57 Pr^{0.4} Re_x^{0.5} \tag{3.116}$$

In calculating stagnation point heat transfer rates, it is more convenient to use a Reynolds number based on the undisturbed freestream velocity ahead of the body, U, rather than one that is based on the local freestream velocity, u_1. The actual relation between u_1 and U will depend on the body shape. For the case of flow over a cylinder it is, as previously given in Eq. (3.106):

$$u_1 = 4Ux/D \tag{3.117}$$

This relationship only applies for small values of x. D is the diameter of the cylinder. While this relation is strictly only applicable for flow over a cylinder, it closely describes the freestream velocity distribution in the region of the stagnation point for flow over any rounded body, D, then being twice the radius of curvature of the leading edge. Substituting this value of u_1 into Eq. (3.116) then gives:

$$Nu_x = 1.14 Pr^{0.4} Re_D^{0.5} (x/D) \tag{3.118}$$

where Re_D is the Reynolds number based on U and D, i.e.,

$$Re_D = UD/\nu \tag{3.119}$$

If the Nusselt number is also based on D, i.e., if the following is defined

$$Nu_D = hD/k \tag{3.120}$$

h being the local heat transfer coefficient which, by virtue of Eq. (3.118), is a constant in the stagnation point region, then Eq. (3.118) gives

$$Nu_D = 1.14 Pr^{0.4} Re_D^{0.5} \tag{3.121}$$

The values of Nu_D predicted by this equation are in good agreement with measured heat transfer rates in the region of the stagnation point on rounded bodies.

EXAMPLE 3.5. Consider two-dimensional air flow over a circular cylinder with a diameter of 4 cm and a surface temperature of 50°C. If the temperature in the air stream ahead of the cylinder is 10°C and if the air velocity in this stream is 3 m/s, determine the heat transfer rate in the vicinity of the stagnation point.

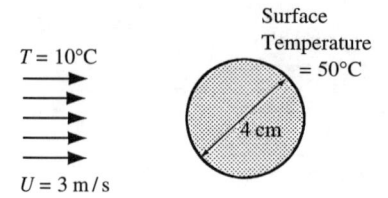

FIGURE E3.5

Solution. The flow situation being considered is shown in Fig. E3.5. The mean temperature of the air is:

$$T_{mean} = \frac{T_w + T_1}{2} = \frac{50 + 10}{2} = 30°C$$

At this temperature for air:

$$k = 0.0264 \text{W/m-°C}, \quad \nu = 0.0000160 \text{ m}^2/\text{s}$$

It was shown above that near the stagnation point:

$$Nu_D = 1.14 Pr^{0.4} Re_D^{0.5}$$

i.e.,

$$\frac{q_w D}{k} = 1.14 Pr^{0.4} \left[\frac{UD}{\nu}\right]^{0.5}$$

Here, $D = 0.04$ m and $U = 3$ m/s so this equation gives:

$$\frac{q_w \times 0.04}{0.0264} = 1.14 \times 0.7^{0.4} \left[\frac{3 \times 0.04}{0.0000160}\right]^{0.5}$$

which gives:

$$q_w = 56.5 \text{ W/m}^2$$

Hence, in the stagnation point region, the heat flux is 56.5 W/m². The heat flux is constant in this region, i.e., in this region q_w does not vary with x.

3.5
INTEGRAL EQUATION SOLUTIONS

Similarity solutions of the type discussed above cannot, of course, be obtained for arbitrary distributions of freestream velocity and surface temperature. Therefore, while the similarity solutions do give results which are of considerable practical significance and while the exact results they give are very useful for checking the accuracy of approximate methods of solving the boundary layer equations, other methods of solving the boundary layer equations for arbitrary distributions of freestream velocity and surface temperature must be sought. Approximate methods, based on use of the integral equations developed in the previous chapter, can be applied to flows with such arbitrary boundary conditions and they will be discussed in the present section [1]. As mentioned in Chapter 2, these integral equation methods have largely been superceded by purely numerical methods. However, they are still sometimes

CHAPTER 3: Some Solutions for External Laminar Forced Convection 115

used and are therefore briefly discussed here. Attention will be restricted to two-dimensional constant fluid property flow.

In order to illustrate the main ideas involved in the procedure, attention will first be given to flow over an isothermal flat plate.

The variation of velocity boundary layer thickness, δ, with the distance along the plate from the leading edge, x, is first determined by solving the momentum integral equation. In order to obtain this solution, it is assumed that the velocity profile can be represented by a third-order polynomial, i.e., by

$$u = a + by + cy^2 + dy^3 \qquad (3.122)$$

There is, of course, no special reason for adopting a third-order polynomial. Experience has, however, shown that it is the lowest-order polynomial that leads to a solution of acceptable accuracy.

The values of the coefficients a, b, c, and d in the assumed velocity profile are obtained, as mentioned in Chapter 2, by applying the boundary conditions on velocity at the inner and outer edges of the boundary layer. Three such boundary conditions are:

$$\text{at:} \quad \begin{aligned} y &= 0: & u &= 0 \\ y &= \delta: & u &= u_1 \\ y &= \delta: & \frac{\partial u}{\partial y} &= 0 \end{aligned} \qquad (3.123)$$

The first of these conditions follows from the "no-slip" at the wall requirement while the second simply follows from the fact that the velocity must be continuous at the outer edge of the boundary layer. The third condition follows from the requirement that the boundary layer profile blend smoothly into the freestream velocity distribution in which the viscous stresses are zero.

Equation (3.123) provides three boundary conditions which can be used to find three of the coefficients in Eq. (3.122). A fourth boundary condition is obtained by applying the momentum equation (2.136) for the boundary layer to conditions at the wall. Since $u = v = 0$ at the wall, this equation gives:

$$y = 0: \quad \frac{\partial^2 u}{\partial y^2} = \frac{1}{\mu}\frac{dp}{dx} \qquad (3.124)$$

This relation can also be derived by applying the conservation of momentum principle to a control volume that is in contact with the wall and is of small size Δy measured normal to the wall. Since the velocities will be negligibly small, the momentum flux through this control volume will be negligible and the forces acting on it must balance. Applying this force balance requirement and then taking the limiting form of the resultant equation as the size of the control volume goes to zero leads to Eq. (3.124). Since, for the moment, only the flow over a flat plate for which $dp/dx = 0$ is being considered, Eq. (3.124) reduces in this case:

$$\text{At } y = 0: \quad \frac{\partial^2 u}{\partial y^2} = 0 \qquad (3.125)$$

116 Introduction to Convective Heat Transfer Analysis

Applying the boundary conditions given in Eqs. (3.123) and (3.125) to Eq. (3.122) then gives the following three equations:

$$0 = a$$
$$u_1 = a + b\delta + c\delta^2 + d\delta^3$$
$$0 = b + 2c\delta + 3d\delta^2 \tag{3.126}$$
$$0 = 2c$$

This is a set of four algebraic equations in the four unknowns a, b, c, and d. Solving for these gives:

$$a = 0, \quad b = 3u_1/2\delta, \quad c = 0, \quad d = -u_1/2\delta^3 \tag{3.127}$$

Substituting these values into Eq. (3.122) and rearranging gives the velocity profile as:

$$\left(\frac{u}{u_1}\right) = \frac{3}{2}\left(\frac{y}{\delta}\right) - \frac{1}{2}\left(\frac{y}{\delta}\right)^3 \tag{3.128}$$

It should be realized that there is no real purpose in comparing this velocity profile with that given by the exact similarity solution since the integral equation method does not seek to accurately predict the details of the velocity and temperature profiles. The method seeks rather, by satisfying conservation of mean momentum and energy, to predict with reasonable accuracy the overall features of the flow.

Now for the case of flow over a flat plate for which u_1 is constant, the momentum integral equation (2.173) can be written as

$$\frac{d}{dx}\left[\delta \int_0^1 \left(1 - \frac{u}{u_1}\right)\left(\frac{u}{u_1}\right) d\left(\frac{y}{\delta}\right)\right] = \frac{\tau_w}{\rho u_1^2} \tag{3.129}$$

Using the velocity distribution given in Eq. (3.128) gives the following:

$$\int_0^1 \left(1 - \frac{u}{u_1}\right)\left(\frac{u}{u_1}\right) d\left(\frac{y}{\delta}\right) = \frac{39}{280} \tag{3.130}$$

Also:

$$\tau_w = \mu \left.\frac{\partial u}{\partial y}\right|_{y=0} = \left(\frac{\mu u_1}{\delta}\right) \left.\frac{\partial (u/u_1)}{\partial (y/\delta)}\right|_{y=0} \tag{3.131}$$

so that again using Eq. (3.128) gives:

$$\tau_w = \frac{3}{2}\left(\frac{\mu u_1}{\delta}\right) \tag{3.132}$$

Substituting Eqs. (3.130) and (3.132) into Eq. (3.129) then gives:

$$\frac{13}{140}\delta\frac{d\delta}{dx} = \frac{\mu}{\rho u_1} \tag{3.133}$$

This equation can be directly integrated using the following initial condition:

$$\text{When } x = 0: \; \delta = 0 \tag{3.134}$$

CHAPTER 3: Some Solutions for External Laminar Forced Convection   117

The integration of Eq. (3.133) then results in:

$$\frac{13}{280}\delta^2 = \frac{\mu}{\rho u_1} x \tag{3.135}$$

This is conveniently rearranged as

$$\frac{\delta}{x} = \frac{4.64}{\sqrt{Re_x}} \tag{3.136}$$

where Re_x is again the local Reynolds number $(u_1 x/\nu)$.

It may be recalled that it was deduced from the similarity solution for flow over a flat plate that $(\delta/x) = 5/\sqrt{Re_x}$. The difference between the value of the coefficient in this equation, i.e., 5, and the value in Eq. (3.136), i.e., 4.64, has no real significance since, in deriving the similarity solution result, it was arbitrarily assumed that the boundary layer thickness was the distance from the wall at which u became equal to $0.99\, u_1$.

Having obtained the solution to the momentum integral equation, attention must now be turned to the energy integral equation. To solve this, the form of the temperature profile must be assumed. As with the velocity, a third-order polynomial will be used for this purpose, i.e., it will be assumed that:

$$T = e + fy + gy^2 + hy^3 \tag{3.137}$$

The coefficients in this equation, i.e., e, f, g, and h, are determined by applying the boundary conditions on temperature at the inner and outer edges of the thermal boundary layer. Three such boundary conditions, which are analogous to those given for the velocity in Eq. (3.123), are:

$$\begin{array}{lll} y = 0: & T = T_w & \\ y = \delta_T: & T = T_1 & (3.138) \\ y = \delta_T: & \partial T/\partial y = 0 & \end{array}$$

δ_T is, of course, the thickness of the thermal boundary layer. The first of these conditions follows from the requirement that the fluid in contact with the wall must attain the same temperature as the wall. The other two conditions follow from the requirement that the boundary layer temperature profile must blend smoothly into the freestream temperature distribution at the outer edge of the boundary layer.

A fourth boundary condition is obtained by applying the boundary layer energy equation (2.146) to conditions at the wall. Since dissipation is being neglected this gives:

$$\text{At } y = 0: \quad \partial^2 T/\partial y^2 = 0 \tag{3.139}$$

Applying the four boundary conditions given in Eqs. (3.138) and (3.139) to Eq. (3.137) then gives:

$$\begin{aligned} T_w &= e \\ T_1 &= e + f\delta_T + g\delta_T^2 + h\delta_T^3 \\ 0 &= f2g\delta_T + 3h\delta_T^2 \\ 0 &= 2g \end{aligned} \tag{3.140}$$

118   Introduction to Convective Heat Transfer Analysis

This is, of course, again a set of four equations in the four unknown coefficients. Solving between them then gives

$$e = T_w, \quad f = -3(T_w - T_1)/2\delta_T$$
$$g = 0, \quad h = (T_w - T_1)/2\delta_T^3 \tag{3.141}$$

Substituting these values into Eq. (3.137) and rearranging gives the temperature profile as:

$$(T - T_1) = \left[1 - \frac{3}{2}\frac{y}{\delta_T} + \frac{1}{2}\left(\frac{y}{\delta_T}\right)^3\right](T_w - T_1) \tag{3.142}$$

The heat transfer rate at the wall is given by Fourier's law as:

$$q_w = -k\left.\frac{\partial T}{\partial y}\right|_{y=0} = \frac{-k(T_w - T_1)}{\delta_T}\frac{\partial}{\partial(y/\delta_T)}\left[\frac{T - T_1}{T_w - T_1}\right]_{y=0} \tag{3.143}$$

Using Eq. (3.142) to obtain the derivative then gives:

$$q_w = \frac{3k(T_w - T_1)}{2\delta_T} \tag{3.144}$$

Now, since u_1 and T_w are constants in the problem being considered, the energy integral equation (2.183) can be written as:

$$u_1(T_w - T_1)\frac{d}{dx}\left[\int_0^{\delta_T}\left(\frac{u}{u_1}\right)\left(\frac{T - T_1}{T_w - T_1}\right)dy\right] = \frac{q_w}{\rho c_p} \tag{3.145}$$

Using Eq. (3.144) and rearranging then gives:

$$\frac{d}{dx}\left[\int_0^{\delta_T}\left(\frac{u}{u_1}\right)\left(\frac{T - T_1}{T_w - T_1}\right)dy\right] = \frac{3}{2Pr\delta_T}\frac{\mu}{\rho u_1} \tag{3.146}$$

In evaluating the integral, care has to be exercised because the velocity and temperature profiles are really discontinuously described, one relation being used inside the boundary layer and another outside the boundary layer. The velocity profile, for example, is actually described by:

$$y \leq \delta: \quad \frac{u}{u_1} = \frac{3}{2}\frac{y}{\delta} - \frac{1}{2}\left(\frac{y}{\delta}\right)^3$$
$$y > \delta: \quad \frac{u}{u_1} = 1 \tag{3.147}$$

This offers no difficulty when $\delta_T < \delta$ because $(T - T_1)/(T_w - T_1)$ is zero for $y > \delta_T$. However when $\delta_T > \delta$, the integral in Eq. (3.146) has to be evaluated in the following way:

$$\int_0^{\delta_T}\left(\frac{u}{u_1}\right)\left(\frac{T - T_1}{T_w - T_1}\right)dy = \int_0^{\delta}\left(\frac{u}{u_1}\right)\left(\frac{T - T_1}{T_w - T_1}\right)dy + \int_{\delta}^{\delta_T}\left(\frac{u}{u_1}\right)\left(\frac{T - T_1}{T_w - T_1}\right)dy$$

$$\tag{3.148}$$

The solutions for $\delta_T < \delta$ and $\delta_T > \delta$ will, therefore, be separately obtained.

CHAPTER 3: Some Solutions for External Laminar Forced Convection   119

Consider, first, the case $\delta_T < \delta$. Here

$$\int_0^{\delta_T} \left(\frac{u}{u_1}\right)\left(\frac{T-T_1}{T_w-T_1}\right) dy = \int_0^{\delta_T} \left[\frac{3}{2}\frac{y}{\delta} - \frac{1}{2}\left(\frac{y}{\delta}\right)^3\right]\left[1 - \frac{3}{2}\frac{y}{\delta_T} + \frac{1}{2}\left(\frac{y}{\delta_T}\right)^3\right] dy$$

$$= \delta_T \int_0^1 \left[\frac{3}{2}\frac{y}{\delta_T}\frac{\delta_T}{\delta} - \frac{1}{2}\left(\frac{y}{\delta_T}\right)^3\left(\frac{\delta_T}{\delta}\right)^3\right]$$

$$\times \left[1 - \frac{3}{2}\frac{y}{\delta_T} + \frac{1}{2}\left(\frac{y}{\delta_T}\right)^3\right] d\left(\frac{y}{\delta_T}\right)$$

$$= \delta\left(\frac{\delta_T}{\delta}\right)\left[\frac{3}{20}\frac{\delta_T}{\delta} - \frac{3}{280}\left(\frac{\delta_T}{\delta}\right)^3\right] \quad (3.149)$$

If the following is defined for convenience

$$\Delta = \delta_T/\delta \quad (3.150)$$

substituting the value of the integral given in Eq. (3.149) back into Eq. (3.146) leads to

$$\frac{d}{dx}\left[\delta\left(\frac{3}{20}\Delta^2 - \frac{3}{280}\Delta^4\right)\right] = \frac{3}{2Pr\Delta\delta}\left(\frac{\mu}{\rho u_1}\right) \quad (3.151)$$

But the momentum integral equation result given in Eq. (3.133) can be written as:

$$\frac{39}{280}\frac{d\delta}{dx} = \left(\frac{3}{2\delta}\right)\left(\frac{\mu}{\rho u_1}\right) \quad (3.152)$$

Comparing Eq. (3.151) and (3.152) then shows that since δ and δ_T are both equal to zero at the leading edge, i.e., when $x = 0$, Δ is a constant and is given by:

For $\delta_T < \delta$, i.e., $\Delta < 1$: $\quad \frac{3}{20}\Delta^3 - \frac{3}{280}\Delta^5 = \frac{39}{280Pr} \quad (3.153)$

This equation allows the variation of Δ with Pr to be found for values of Δ that are less than 1.

Next, consider the solution of the energy integral equation for the case where $\delta_T > \delta$. In this case since (u/u_1) is equal to 1 for values of y greater than δ, the integral is, as previously discussed, evaluated in two parts as follows:

$$\int_0^{\delta_T}\left(\frac{u}{u_1}\right)\left(\frac{T-T_1}{T_w-T_1}\right)dy = \int_0^{\delta}\left[\frac{3}{2}\frac{y}{\delta} - \frac{1}{2}\left(\frac{y}{\delta}\right)^3\right]\left[1 - \frac{3}{2}\frac{y}{\delta_T} + \frac{1}{2}\left(\frac{y}{\delta_T}\right)^3\right]dy$$

$$+ \int_\delta^{\delta_T}\left[1 - \frac{3}{2}\frac{y}{\delta_T} + \frac{1}{2}\left(\frac{y}{\delta_T}\right)^3\right]dy$$

$$= \delta\int_0^1\left[\frac{3}{2}\frac{y}{\delta} - \frac{1}{2}\left(\frac{y}{\delta}\right)^3\right]\left[1 - \frac{3}{2}\frac{y}{\delta}\frac{1}{\Delta} + \frac{1}{2}\left(\frac{y}{\delta}\right)^3\frac{1}{\Delta^3}\right]d\left(\frac{y}{\delta}\right)$$

$$+ \delta_T\int_{1/\Delta}^1\left[1 - \frac{3}{2}\frac{y}{\delta_T} + \frac{1}{2}\left(\frac{y}{\delta_T}\right)^3\right]d\left(\frac{y}{\delta_T}\right)$$

$$= \delta\left[\frac{3}{8}\Delta - \frac{3}{8} + \frac{3}{20\Delta} - \frac{3}{280\Delta^3}\right] \quad (3.154)$$

Substituting this result back into Eq. (3.146) then gives:

$$\frac{d}{dx}\left[\delta\left(\frac{3}{8}\Delta - \frac{3}{8} + \frac{3}{20\Delta} - \frac{3}{280\Delta^3}\right)\right] = \left(\frac{3}{2Pr\Delta\delta}\right)\left(\frac{\mu}{\rho u_1}\right) \qquad (3.155)$$

Comparing this with the momentum integral equation result given in Eq. (3.152) shows that Δ is again a constant and is given by

$$\text{For } \delta_T > \delta, \text{ i.e., } \Delta > 1: \quad \frac{3}{8}\Delta^2 - \frac{3}{8}\Delta + \frac{3}{20} - \frac{3}{280}\Delta^2 = \frac{39}{280Pr} \qquad (3.156)$$

This equation then allows the variation of Δ with Pr to be found for values of Δ that are greater than 1.

Since the variation of δ with x has been derived by solving the momentum integral equation, Eqs. (3.153) and (3.156) together constitute the solution of the energy equation. The variation of Δ with Pr that they together give is shown in Fig. 3.14.

When Pr is equal to 1, Δ is, of course, equal to 1 because the form of the assumed velocity and temperature profiles are identical and when Pr is equal to one the momentum and energy integral equations have the same form for flow over a flat plate.

It will also be noted from Eq. (3.153) that because this equation only applies for $\Delta < 1$, the second term on the left-hand side can, if an approximate solution will suffice, be neglected when compared to the first term. Therefore:

$$\text{For } \Delta < 1: \quad \frac{3}{20}\Delta^3 \approx \frac{39}{280Pr} \qquad (3.157)$$

$$\Delta^3 \approx \frac{13}{14Pr}$$

It is seen, therefore, that for values of Δ less than and equal to 1, Δ is approximately given by:

$$\Delta = \frac{1}{Pr^{1/3}} \qquad (3.158)$$

It has been found that this equation also gives results of sufficient accuracy for most purposes for values of Δ greater than 1 provided that Pr is not very small. This can be seen from the results in Fig. 3.13.

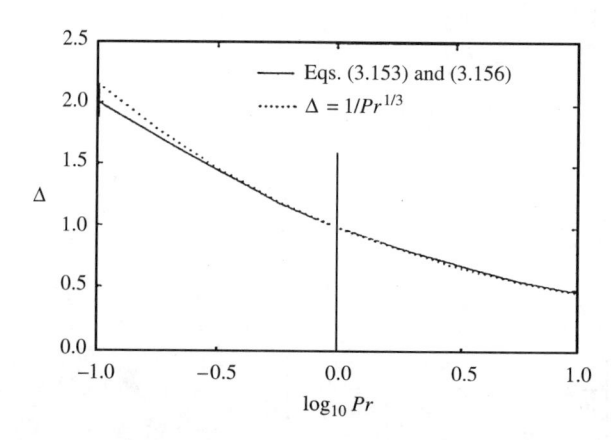

FIGURE 3.14
Variation of boundary layer thickness ratio Δ with Prandtl number.

CHAPTER 3: Some Solutions for External Laminar Forced Convection 121

Now the heat transfer rate as given by Eq. (3.144) can be written in terms of Δ as:

$$q_w = 3k(T_w - T_1)/2\Delta\delta \qquad (3.159)$$

Substituting the variation of δ with x given by Eq. (3.136) into this equation and rearranging then gives:

$$Nu_x = 0.343 \sqrt{Re_x}/\Delta \qquad (3.160)$$

If the approximate expression for Δ given in Eq. (3.158) is used, Eq. (3.160) becomes:

$$Nu_x = 0.343 Re_x^{1/2} Pr^{1/3} \qquad (3.161)$$

This result can be compared with that deduced from the similarity solution and given in Eq. (3.50). It will be seen that the integral equation method gives the correct form of the result, i.e., $Nu_x \propto Re_x^{1/2} Pr^{1/3}$ but that the coefficient is somewhat in error. This is typical of what can be expected of the integral equation method.

The problem to which the integral equation method was applied in the above discussion, i.e., flow over an isothermal plate, is, of course, one for which a similarity solution can be found. The usefulness of the integral equation method, however, arises mainly from the fact that it can be applied to problems for which similarity solutions cannot easily be found. In order to illustrate this ability, consider flow over a flat plate which has an unheated section adjacent to the leading edge as shown in Fig. 3.15.

The plate is, therefore, unheated up to a distance of x_0 downstream of the leading edge. Beyond this it is heated to a uniform temperature of T_w.

The solution of the momentum integral equation is, of course, unaffected by the surface conditions on temperature and is still, therefore, given by Eq. (3.136). Since the thermal boundary layer only starts growing downstream of the leading edge, attention will be restricted to the solution of the energy integral equation for $\delta_T < \delta$. It is, of course, possible, if Pr is less than one, for δ_T to eventually exceed δ at large values of x. This will not, however, be discussed here.

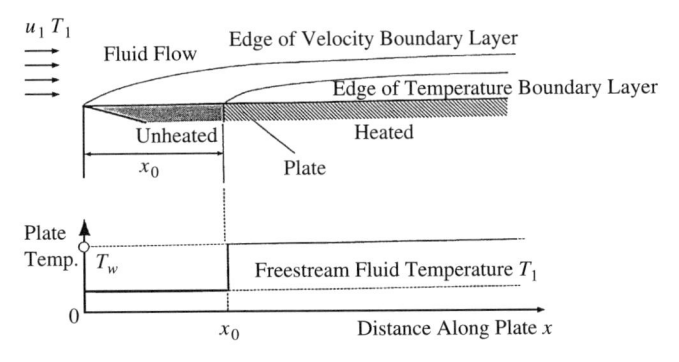

FIGURE 3.15
Flat plate with unheated leading edge section.

Now, for $\delta_T < \delta$, i.e., $\Delta < 1$, it was shown [see the derivation of Eq. (3.151)] that the energy integral equation becomes the following when third-order polynomials are assumed to describe the velocity and temperature profiles:

$$\frac{d}{dx}\left[\delta\left(\frac{3}{20}\Delta^2 - \frac{3}{280}\Delta^4\right)\right] = \left(\frac{3}{2Pr\Delta\delta}\right)\left(\frac{\mu}{\rho u_1}\right)$$

Because only the case of $\Delta < 1$ is being considered, the factor $(3\Delta^4/280)$ will be neglected compared to $(3\Delta^2/20)$. This equation then reduces to:

$$\frac{d}{dx}\left(\delta\Delta^2\right) = \left[\frac{10}{Pr\Delta\delta}\right]\left[\frac{\mu}{\rho u_1}\right] \quad (3.162)$$

But the momentum integral equation gives

$$\delta = \sqrt{\frac{280}{13}\frac{\mu}{\rho u_1}x} \quad (3.163)$$

Substituting this into Eq. (3.162) then gives:

$$2\Delta^2\frac{280}{13}\frac{\mu}{\rho u_1}x\frac{d\Delta}{dx} + \frac{\Delta^3}{2}\frac{280}{13}\frac{\mu}{\rho u_1} = \frac{10}{Pr}\frac{\mu}{\rho u_1}$$

i.e.:

$$4\Delta^2 x\frac{d\Delta}{dx} + \Delta^3 = \frac{13}{14}\frac{1}{Pr} \quad (3.164)$$

Now, since:

$$\Delta^2\frac{d\Delta}{dx} = \frac{1}{3}\frac{d\Delta^3}{dx}$$

Eq. (3.164) can be written as:

$$\frac{4}{3}x\frac{d\Delta^3}{dx} = \frac{13}{14}\frac{1}{Pr} - \Delta^3 \quad (3.165)$$

This equation can then be directly integrated to give:

$$\frac{13}{14Pr} - \Delta^3 = \frac{K}{x^{0.75}} \quad (3.166)$$

where K is a constant of integration. Now since:

$$\text{when } x = x_0, \; \Delta = 0 \quad (3.167)$$

this constant, K, is given by:

$$K = \frac{13}{14Pr}x_0^{0.75} \quad (3.168)$$

Eq. (3.166) therefore gives:

$$\Delta^3 = \frac{13}{14Pr}\left[1 - \left(\frac{x_0}{x}\right)^{0.75}\right] \quad (3.169)$$

This shows that for large x, Δ tends to a constant, this constant being of course, the same as that previously derived for flow over an isothermal plate.

Substituting Eq. (3.169) into Eq. (3.160) then gives:

$$Nu_x = 0.352 Re_x^{1/2} Pr^{1/3} \left[1 - \left(\frac{x_0}{x}\right)^{0.75} \right]^{-1/3} \qquad (3.170)$$

The integral equation method can also easily be extended to situations that involve a varying freestream velocity.

3.6
NUMERICAL SOLUTION OF THE LAMINAR BOUNDARY LAYER EQUATIONS

Similarity and integral equation methods for solving the boundary layer equations have been discussed in the previous sections. In the similarity method, it will be recalled, the governing partial differential equations are reduced to a set of ordinary differential equations by means of a suitable transformation. Such solutions can only be obtained for a very limited range of problems. The integral equation method can, basically, be applied to any flow situation. However, the approximations inherent in the method give rise to errors of uncertain magnitude. Many attempts have been made to reduce these errors but this can only be done at the expense of a considerable increase in complexity, and, therefore, in the computational effort required to obtain the solution.

Another way of solving the boundary layer equations involves approximating the governing partial differential equations by algebraic finite-difference equations [11]. The main advantages of this type of solution procedure are:

- The method can be applied to problems involving arbitrary surface thermal conditions and arbitrary freestream velocity and is easily extended to cover the effects of variable fluid properties and dissipation effects.
- The errors involved in the procedure are purely numerical and their magnitude can be estimated and can, in general, always be reduced to an acceptable level by reducing the numerical step size.

The disadvantages of finite-difference and other numerical methods are that quite a considerable amount of computational effort is usually required in order to obtain the solution and that they do not, in general, reveal certain unifying features of the solutions, such as the fact that the profiles are similar under certain conditions. The widespread availability of modern computer facilities has, however, made these disadvantages relatively unimportant.

There are a number of schemes for numerically approximating the boundary layer equations and many different solution procedures based on these various schemes have been developed. In the present section, one of the simpler finite-difference schemes will be described. The solution procedure based on this scheme should give quite acceptable results for most problems. The scheme is easily extended to deal with turbulent flows as will be discussed in Chapter 6. Before outlining this solution procedure, some general points concerning the numerical solution of the boundary layer equations will be discussed.

In order to solve the boundary layer equations by means of finite difference approximations, a series of nodal points is introduced. The values of the variables are then only determined at these nodal points and not continuously across the whole flow field as is the case with an analytical method. In order to describe the position of the nodal points, a series of grid lines running parallel to the two coordinate directions as shown in Fig. 3.16 is introduced, the nodal points lying at the intersection of these grid lines.

By introducing finite difference approximations to the derivative terms in the boundary layer equations, a set of algebraic equations can be obtained. Because of the so-called parabolic nature of the governing equations, this set of equations is of such a form that if the values of the variables are known at the nodal points on one y-grid line, then the variables at the nodal points on the next y-grid line can be found. Having determined the conditions on this grid line, the same procedure can be used to determine the conditions at the points on the next y-line and so on, the solution advancing from grid line to grid line in the x-direction as shown in Fig. 3.17.

The finite difference approximations can either be applied to the derivatives on the line *from* which the solution is advancing or on the line *to* which it is advancing, the former giving an explicit finite difference scheme and the latter an implicit scheme. The type of solution procedure obtained with the two schemes is illustrated in Fig. 3.18.

In the explicit scheme, the values of the variables at the point under consideration are directly determined from conditions at the points on the preceding line. In the implicit scheme, the values of the variables at adjacent points on the line on

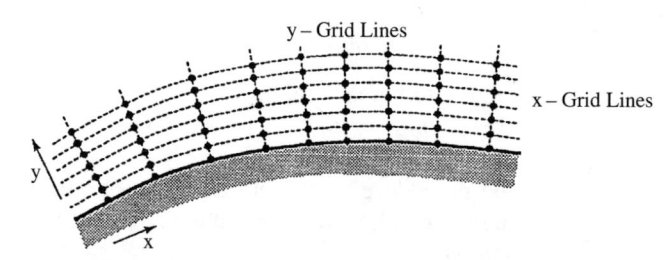

FIGURE 3.16
Grid lines and nodal points.

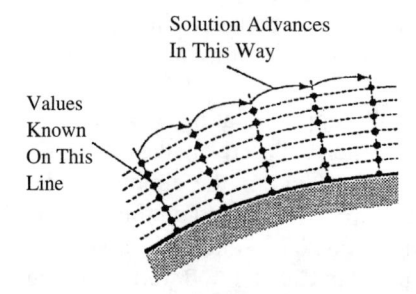

FIGURE 3.17
Forward "marching" solution procedure used.

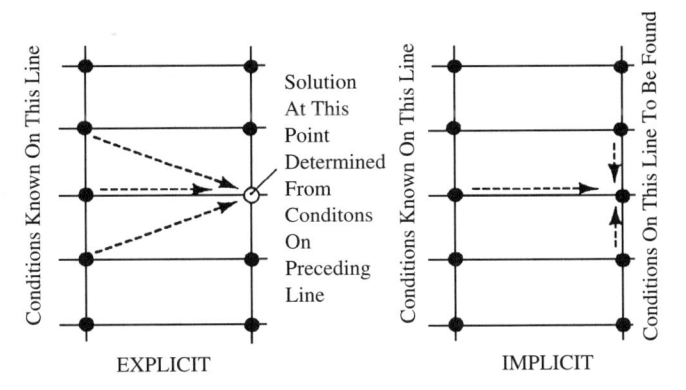

CHAPTER 3: Some Solutions for External Laminar Forced Convection

FIGURE 3.18
Implicit and explicit solution schemes.

which the solution is being sought are related to each other and to the values of the variables on the preceding line. By considering each nodal point in turn on the line to which the solution is advancing, a set of equations is obtained which must be simultaneously solved to give the values at all the nodal points. It is possible to obtain such a solution because the values of the variables at the point lying on the wall and at the outermost grid point, which is always selected to lie outside the boundary layer, are given by the boundary conditions.

Explicit finite difference schemes lead to comparatively simple solution procedures which are usually very easy to implement. However, with an explicit scheme, it is possible for the solution to become unstable, i.e., for small errors such as those resulting from numerical round-off to become magnified as the solution progresses leading to a useless solution. In order to avoid instability, there is a maximum numerical step size that can be used in the x-direction for any given minimum y-step size. Since, particularly when dealing with turbulent flows, it is necessary to use a small y-step size near the wall in order to obtain an accurate solution, small step sizes will have to be used in the x-direction in order to avoid instability which will mean that long computer times will be required. Solution procedures based on implicit finite difference schemes are usually somewhat more complex than those based on explicit schemes. However, they suffer from no inherent stability problems and consequently much larger x-wise steps can be used and the computing time required is usually much less than with an explicit scheme. In the present section, therefore, a simple implicit finite difference scheme will be described.

For the purpose of illustrating the ideas involved, attention will initially be restricted to two-dimensional constant fluid property flow over a flat surface and the dissipation term in the energy equation will be neglected. The equations to be solved are, therefore, as before:

$$\frac{\partial u}{\partial x} + \frac{\partial v}{\partial y} = 0 \tag{3.171}$$

$$u\frac{\partial T}{\partial x} + v\frac{\partial T}{\partial y} = -\frac{1}{\rho}\frac{dp}{dx} + \nu\frac{\partial^2 u}{\partial y^2} \tag{3.172}$$

$$u\frac{\partial T}{\partial x} + v\frac{\partial T}{\partial y} = \frac{\nu}{Pr}\frac{\partial^2 T}{\partial y^2} \tag{3.173}$$

The boundary conditions on these equations are, of course,

$$y = 0: u = v = 0, T = T_w(x)$$
$$y \text{ large}: u \to u_1(x), T \to T_1 \tag{3.174}$$

Blowing or sucking at the wall which gives a nonzero value of v at y equal to zero is easily incorporated into the solution scheme, but will not be considered here. Also, in some cases, the heat flux distribution at the wall rather than the wall temperature distribution is known and this is also easily incorporated into the solution scheme but will, for the moment, not be considered.

Although not necessary, it is convenient, in most cases, to write these equations in dimensionless form before deriving the finite difference approximations to them. For the case of laminar flow here being considered, the following dimensionless variables are convenient to use:

$$U = u/u_0 \qquad V = v\sqrt{Re_L}/u_0$$
$$X = x/L \qquad Y = y\sqrt{Re_L}/L \tag{3.175}$$
$$P = (p - p_1)/\rho u_0^2 \qquad \theta = (T - T_1)/(T_{wR} - T_1)$$

where u_0 is some convenient reference velocity, L is some convenient reference length, and T_{wR} some convenient reference wall temperature. Re_L is the Reynolds number based on u_0 and L, i.e., $(u_0 L/\nu)$ and T_1 is, as before, the freestream temperature.

In terms of these variables, Eqs. (3.171) to (3.173) become:

$$\frac{\partial U}{\partial X} + \frac{\partial V}{\partial Y} = 0 \tag{3.176}$$

$$U\frac{\partial U}{\partial X} + V\frac{\partial U}{\partial Y} = -\frac{dP}{dX} + \frac{\partial^2 U}{\partial Y^2} \tag{3.177}$$

$$U\frac{\partial \theta}{\partial X} + V\frac{\partial \theta}{\partial Y} = \frac{1}{Pr}\frac{\partial^2 \theta}{\partial Y^2} \tag{3.178}$$

and the boundary conditions given in Eq. (3.174) become:

$$Y = 0: U = V = 0, \theta = \theta_w(X)$$
$$Y \text{ large}: U \to U_1(X), \theta \to 0 \tag{3.179}$$

If the wall temperature is uniform, it will usually be convenient to set T_{wR} equal to this uniform wall temperature and θ_w will then be equal to 1.

In order to express the above set of equations in finite difference form, the grid lines are labeled as shown in Fig. 3.19, i lines running in the Y-direction normal to the surface and j lines running in the X-direction parallel to the surface. Although not necessary, a uniform grid spacing, ΔY in the Y-direction, will be used here for simplicity.

The conditions at nodal point lying at the intersection of the i and j grid lines are denoted by the subscript i, j. In order to derive the finite difference equations, attention is given to conditions at the four grid points shown in Fig. 3.20.

Consider first the finite difference approximation for $\partial U/\partial Y$ at the point i, j. In order to derive this finite-difference approximation it is noted that the values of U at

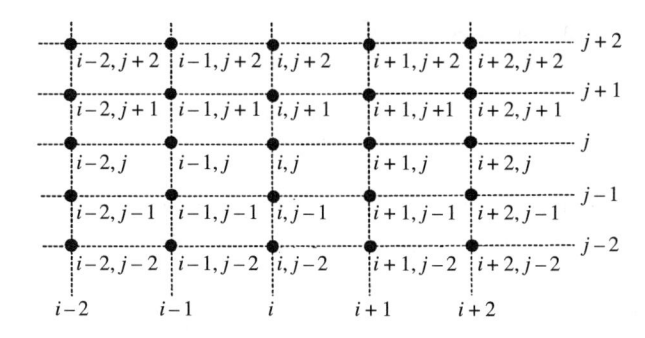

CHAPTER 3: Some Solutions for External Laminar Forced Convection 127

FIGURE 3.19
i- and j-grid lines.

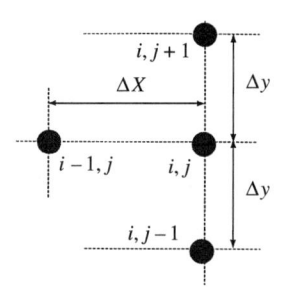

FIGURE 3.20
Grid points used in deriving finite-difference approximations.

points $i, j+1$ and $i, j-1$ can be related to the value at point i, j by Taylor expansions as follows, terms of order ΔY^3 and higher being ignored:

$$U_{i,j+1} = U_{i,j} + \left.\frac{\partial U}{\partial Y}\right|_{i,j} \Delta Y + \left.\frac{\partial^2 U}{\partial Y^2}\right|_{i,j} \frac{\Delta Y^2}{2!} \qquad (3.180)$$

$$U_{i,j-1} = U_{i,j} - \left.\frac{\partial U}{\partial Y}\right|_{i,j} \Delta Y + \left.\frac{\partial^2 U}{\partial Y^2}\right|_{i,j} \frac{\Delta Y^2}{2!} \qquad (3.181)$$

Subtracting these two equations and dividing the result by $2\Delta Y$ gives:

$$\left.\frac{\partial U}{\partial Y}\right|_{i,j} = \frac{U_{i,j+1} - U_{i,j-1}}{2\Delta Y} \qquad (3.182)$$

This is the required finite difference approximation for $\partial U/\partial Y$ at point i, j.

To obtain the difference approximation for $\partial^2 U/\partial Y^2$ at point i, j, Eq. (3.181) is added to Eq. (3.180) and the result is divided by ΔY^2. This gives:

$$\left.\frac{\partial^2 U}{\partial Y^2}\right|_{i,j} = \frac{U_{i,j+1} + U_{i,j-1} - 2U_{i,j}}{\Delta Y^2} \qquad (3.183)$$

In the solution procedure here being used, the X-derivatives are approximated to a lower order in ΔX than the Y-derivatives are in ΔY, i.e., the following backward difference approximation in the X-direction being used:

$$\left.\frac{\partial U}{\partial X}\right|_{i,j} = \frac{U_{i,j} - U_{i-1,j}}{\Delta X} \qquad (3.184)$$

Now the derivative terms on the left-hand side of the momentum equation have coefficients U and V. To illustrate how these are dealt with consider $U\partial U/\partial X$. Because:

$$U_{i,j} = U_{i-1,j} + \frac{\partial U}{\partial X}\Delta X + \frac{\partial^2 U}{\partial X^2}\frac{\Delta X^2}{2!} \qquad (3.185)$$

and since, in deriving the approximation given in Eq. (3.184), terms involving ΔX and higher have been neglected, it is seen that it is consistent to adopt the following approximation:

$$\left[U\frac{\partial U}{\partial X}\right]_{i,j} = U_{i-1,j}\left[\frac{U_{i,j} - U_{i-1,j}}{\Delta X}\right] \qquad (3.186)$$

Similarly, using Eq. (3.182), the following is obtained:

$$\left[V\frac{\partial U}{\partial Y}\right]_{i,j} = V_{i-1,j}\left[\frac{U_{i,j+1} - U_{i,j-1}}{2\Delta Y}\right] \qquad (3.187)$$

If the finite difference approximations given in Eqs. (3.183), (3.186), and (3.187) are substituted into the momentum equation, an equation that has the following form is obtained on rearrangement:

$$A_j U_{i,j} + B_j U_{i,j+1} + C_j U_{i,j-1} = D_j \qquad (3.188)$$

where the coefficients are given by:

$$A_j = \left(\frac{U_{i-1,j}}{\Delta X}\right) + \left(\frac{2}{\Delta Y^2}\right) \qquad (3.189)$$

$$B_j = \left(\frac{V_{i-1,j}}{\Delta Y}\right) - \left(\frac{1}{\Delta Y^2}\right) \qquad (3.190)$$

$$C_j = -\left(\frac{V_{i-1,j}}{\Delta Y}\right) - \left(\frac{1}{\Delta Y^2}\right) \qquad (3.191)$$

$$D_n = \left(\frac{U_{i-1,j}^2}{\Delta X}\right) - \left.\frac{dP}{dX}\right|_i \qquad (3.192)$$

Now the boundary conditions give $U_{i,1}$, the value of U at the wall, and $U_{i,N}$ the value of U at the outermost grid point which, as previously mentioned, is always chosen to lie outside the boundary layer. Therefore, since $dP/dX|_i$ is a known quantity, the application of Eq. (3.188) to each of the points $j = 1, 2, 3, \ldots, N-1, N$ gives a set of N equations in the N unknown values of U, i.e., $U_1, U_2, U_3, U_4, \ldots, U_{N-1}, U_N$. This set of equations has the following form since $U_{i,1}$ is zero

$$\begin{aligned} U_{i,1} &= 0 \\ A_2 U_{i,2} + B_2 U_{i,3} + C_2 U_{i,1} &= D_2 \\ A_3 U_{i,3} + B_3 U_{i,4} + C_3 U_{i,2} &= D_3 \\ &\vdots \\ A_{N-1} U_{i,N-1} + B_{N-1} U_{i,N} + C_{N-1} U_{i,N-2} &= D_{N-1} \\ U_{i,N} &= U_1 \end{aligned} \qquad (3.193)$$

CHAPTER 3: Some Solutions for External Laminar Forced Convection 129

The set of equations thus has the form:

$$\begin{bmatrix} 1 & 0 & 0 & 0 & 0 & . & 0 & 0 & 0 \\ C_2 & A_2 & B_2 & 0 & 0 & . & 0 & 0 & 0 \\ 0 & C_3 & A_3 & B_3 & 0 & . & 0 & 0 & 0 \\ 0 & 0 & C_4 & A_4 & B_4 & . & 0 & 0 & 0 \\ . & . & . & . & . & . & . & . & . \\ . & . & . & . & . & . & . & . & . \\ . & . & . & . & . & . & . & . & . \\ 0 & 0 & 0 & 0 & 0 & . & C_{N-1} & A_{N-1} & B_{N-1} \\ 0 & 0 & 0 & 0 & 0 & . & 0 & 0 & 1 \end{bmatrix} \begin{bmatrix} U_{i,1} \\ U_{i,2} \\ U_{i,3} \\ U_{i,4} \\ . \\ . \\ . \\ U_{i,N-1} \\ U_{i,N} \end{bmatrix} = \begin{bmatrix} 0 \\ D_2 \\ D_3 \\ D_4 \\ . \\ . \\ . \\ D_{N-1} \\ U_1 \end{bmatrix}$$

i.e., has the form:

$$Q_U U_{i,j} = R_U \tag{3.194}$$

where Q_U is a tridiagonal matrix. This equation can be solved using the standard tridiagonal matrix solver algorithm which is often termed the Thomas algorithm.

Consider next the energy equation. The terms in this equation have the same form as those in the momentum equation and they are, therefore, approximated in the same way. The following finite difference approximations are therefore used:

$$\left(U \frac{\partial \theta}{\partial X} \right)\bigg|_{i,j} = U_{i-1,j} \frac{(\theta_{i,j} - \theta_{i-1,j})}{\Delta X} \tag{3.195}$$

$$\left(V \frac{\partial \theta}{\partial Y} \right)\bigg|_{i,j} = V_{i-1,j} \frac{(\theta_{i,j+1} - \theta_{i,j-1})}{2\Delta Y} \tag{3.196}$$

$$\frac{\partial^2 \theta}{\partial Y^2}\bigg|_{i,j} = \frac{(\theta_{i,j+1} + \theta_{i,j-1} - 2\theta_{i,j})}{\Delta Y^2} \tag{3.197}$$

Substituting these into the energy equation then gives, on rearrangement an equation that has the form:

$$E_j \theta_{i,j} + F_j \theta_{i,j+1} + G_j \theta_{i,j-1} = H_j \tag{3.198}$$

where the coefficients in this equation are given by:

$$E_j = \left(\frac{U_{i-1,j}}{\Delta X} \right) + \left(\frac{2}{Pr \Delta Y^2} \right) \tag{3.199}$$

$$F_j = -\left(\frac{V_{i-1,j}}{\Delta Y} \right) - \left(\frac{1}{Pr \Delta Y^2} \right) \tag{3.200}$$

$$G_j = -\left(\frac{V_{i-1,j}}{\Delta Y} \right) - \left(\frac{1}{Pr \Delta Y^2} \right) \tag{3.201}$$

$$H_j = \left(\frac{U_{i-1,j} \theta_{i-1,j}}{\Delta X} \right) \tag{3.202}$$

Because the boundary conditions give $\theta_{i,1}$ and $\theta_{i,N}$, the outermost grid point being chosen to lie outside both the velocity and temperature boundary layers, the application of Eq. (3.188) to each of the internal points on the i-line, i.e., $j = 1, 2, 3, 4, \ldots, N-2, N-1, N$, again gives a set of N equations in the N unknown values of θ. Because $\theta_{i,N}$ is zero, this set of equations has the following form:

$$\theta_{i,1} = \theta_w$$
$$E_2 \theta_{i,2} + F_2 \theta_{i,3} + G_2 \theta_{i,1} = H_2$$
$$E_3 \theta_{i,3} + F_3 \theta_{i,4} + G_3 \theta_{i,2} = H_3$$
$$\vdots \qquad (3.203)$$
$$E_{N-1} \theta_{i,N-1} + F_{N-1} \theta_{i,N} + G_{N-1} \theta_{i,N-2} = H_{N-1}$$
$$\theta_{i,N} = 0$$

This set of equations has the same form as that derived from the momentum equation, i.e., has the form:

$$\begin{bmatrix} 1 & 0 & 0 & 0 & 0 & \cdot & 0 & 0 & 0 \\ G_2 & E_2 & F_2 & 0 & 0 & \cdot & 0 & 0 & 0 \\ 0 & G_3 & E_3 & F_3 & 0 & \cdot & 0 & 0 & 0 \\ 0 & 0 & G_4 & E_4 & F_4 & \cdot & 0 & 0 & 0 \\ \cdot & \cdot & \cdot & \cdot & \cdot & \cdot & \cdot & \cdot & \cdot \\ 0 & 0 & 0 & 0 & 0 & \cdot & G_{N-1} & E_{N-1} & F_{N-1} \\ 0 & 0 & 0 & 0 & 0 & \cdot & 0 & 0 & 1 \end{bmatrix} \begin{bmatrix} \theta_{i,1} \\ \theta_{i,2} \\ \theta_{i,3} \\ \theta_{i,4} \\ \cdot \\ \theta_{i,N-1} \\ \theta_{i,N} \end{bmatrix} = \begin{bmatrix} \theta_w \\ H_2 \\ H_3 \\ H_4 \\ \cdot \\ H_{N-1} \\ 0 \end{bmatrix}$$

i.e., which has the form:

$$Q_T \theta_{i,j} = R_T \qquad (3.204)$$

where Q_T is again a tridiagonal matrix. Thus, the same form of equation is obtained as that obtained from the momentum equation. This equation can also be solved using the standard tridiagonal matrix solver algorithm.

To deal with the continuity equation, the grid points shown in Fig. 3.21 are used. The continuity equation is applied to the point (not a nodal point) denoted by $(i, j - 1/2)$ which lies on the i-line, halfway between the $(j - 1)$ and the j-points.

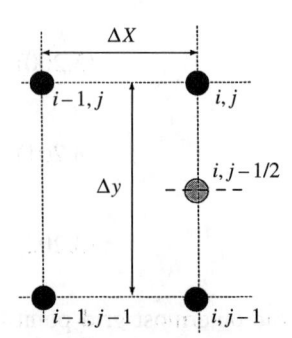

FIGURE 3.21
Nodal points used in finite-difference solution of the continuity equation.

CHAPTER 3: Some Solutions for External Laminar Forced Convection   **131**

Now, to the same order of accuracy as previously used:

$$V_{i,j} = V_{i,j-1/2} + \left.\frac{\partial V}{\partial Y}\right|_{i,j-1/2} \frac{\Delta Y}{2} + \left.\frac{\partial^2 V}{\partial Y^2}\right|_{i,j-1/2} \frac{\Delta Y^2}{4} \qquad (3.205)$$

$$V_{i,j-1} = V_{i,j-1/2} - \left.\frac{\partial V}{\partial Y}\right|_{i,j-1/2} \frac{\Delta Y}{2} + \left.\frac{\partial^2 V}{\partial Y^2}\right|_{i,j-1/2} \frac{\Delta Y^2}{4}$$

Subtracting these two equations gives:

$$\left.\frac{\partial V}{\partial Y}\right|_{i,j-1/2} = \frac{V_{i,j} - V_{i,j-1}}{\Delta Y} \qquad (3.206)$$

It is also assumed that the X-derivative at the point $(i, j - 1/2)$ is equal to the average of the X-derivatives at the points (i, j) and $(i, j - 1)$, i.e., it is assumed that:

$$\left.\frac{\partial U}{\partial X}\right|_{i,j-1/2} = \frac{1}{2}\left[\left.\frac{\partial U}{\partial X}\right|_{i,j} + \left.\frac{\partial U}{\partial X}\right|_{i,j-1}\right] \qquad (3.207)$$

The finite-difference approximation for $\partial U/\partial X$ given in Eq. (3.184) is used to determine the right-hand side of this equation. Therefore, since the continuity equation can be written as:

$$\frac{\partial V}{\partial Y} = -\frac{\partial U}{\partial X} \qquad (3.208)$$

the following finite difference approximation to this continuity equation is obtained:

$$\frac{V_{i,j} - V_{i,j-1}}{\Delta Y} = -\frac{1}{2}\left[\left(\frac{U_{i,j} - U_{i-1,j}}{\Delta X}\right) + \left(\frac{U_{i,j-1} - U_{i-1,j-1}}{\Delta X}\right)\right] \qquad (3.209)$$

This equation can be rearranged to give

$$V_{i,j} = V_{i,j-1} - \left(\frac{\Delta Y}{2\Delta X}\right)(U_{i,j} - U_{i-1,j} + U_{i,j-1} - U_{i-1,j-1}) \qquad (3.210)$$

Therefore, if the distribution of U across the i-line is first determined, this equation can be used to give the distribution of V across this line. This is possible because $V_{i,1}$ is given by the boundary conditions and the equation can, therefore, be applied progressively outward across the i-line starting at point $j = 2$ and then going to point $j = 3$ and so on across the line.

Thus, to summarize, the computational procedure based on the above finite difference equations involves the following steps.

1. The conditions along some initial $i = 1$ line must be specified.
2. Eq. (3.194) can then be used to find U at all the nodal points on the $i = 2$ line.
3. Eq. (3.204) can then be used to find θ at all the nodal points on the $i = 2$ line.

4. Starting at the first point away from the wall, i.e., $j = 2$ and working progressively outward, Eq. (3.210) can be used to determine the distribution of V on the $i = 2$ line.
5. Having in this way determined the values of all the variables on this $i = 2$ line, the same procedure can then be used to find the values on the $i = 3$ line and so on.

It should be noted that at any stage of the procedure, it is only necessary to deal with the values of the variables on two adjacent grid lines. Therefore, it is only necessary to allocate storage spaces for the values on the line on which conditions are known, i.e., the $(i - 1)$-line, and the line on which conditions are being calculated, i.e., the i-line. As soon as the values on the i-line have been calculated, they can be transferred into the storage spaces previously holding the values on the $(i - 1)$-line. The storage spaces that held the values on the i-line are then used to store the values on the $(i + 1)$-line as they are calculated and so on.

In the above procedure, it was assumed that the outermost grid point, i.e., the N-point, was always outside both the velocity and thermal boundary layers. One way of ensuring this is the case is, of course, to simply estimate the maximum boundary layer thickness expected and then select the number of grid points and their positioning such that the outermost grid line is at a greater distance from the surface than this maximum boundary layer thickness. This is illustrated in Fig. 3.22.

There are two disadvantages to this approach. Firstly, it may not be possible to estimate the boundary layer thickness with any degree of certainty and it may then be found during the calculation that the boundary layer grows out to the N-line and the whole calculation has then to be started again. Secondly, if the boundary layer grows at all appreciably during the calculation, a considerable number of grid points lie outside the boundary layer during the initial part of the calculation and computer time is wasted in carrying out the calculation at these points since they will lie in the freestream and the values of the variables are actually known at these points.

One way of overcoming these disadvantages is to start with a small number of j-grid-lines and monitor the boundary layer growth. Then, when the boundary layer has reached to within a few nodal points of the outermost point, the number of j-lines is increased. Since the additional points so generated initially lie outside the boundary layer the values of the variables on these points are initially known. The procedure is illustrated in Fig. 3.23.

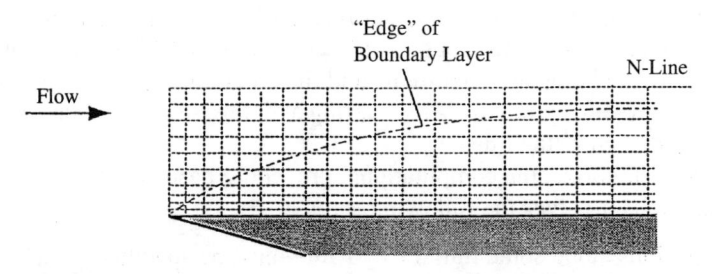

FIGURE 3.22
Choosing outermost grid line to be always outside boundary layer.

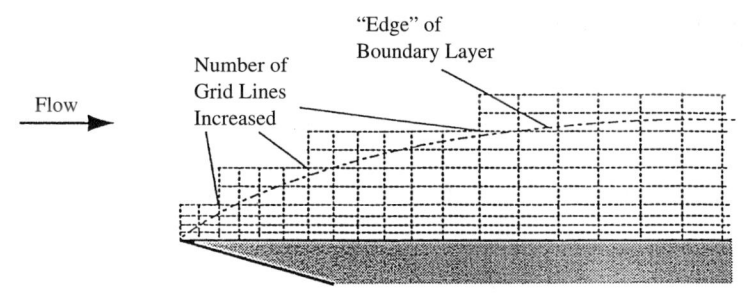

CHAPTER 3: Some Solutions for External Laminar Forced Convection   **133**

FIGURE 3.23
Adding points to keep outermost grid point outside the boundary layer.

Additional savings in computer time can be achieved if, instead of increasing the number of grid lines, their spacing is increased and in this way the outermost grid point is moved further from the surface. In the computer program discussed later in this section this has, however, not been done because of the added complication involved, the number of grid points simply being increased.

Once the distributions of U, V, and θ have been determined using the procedure outlined above, any other property of the flow can be determined. In the present discussion, the heat transfer rate at the wall, q_w, is the most important such property. This is, of course, given by Fourier's law as:

$$q_w = -k \left.\frac{\partial T}{\partial y}\right|_{y=0} \tag{3.211}$$

In terms of the dimensionless variables being used in the present work this gives:

$$\frac{q_w L}{(T_{wR} - T_1)k} = -\left.\frac{\partial \theta}{\partial Y}\right|_{Y=0} \sqrt{Re_L} \tag{3.212}$$

This can be rearranged as:

$$\frac{Nu_x}{\sqrt{Re_x}} = -\left.\frac{\partial \theta}{\partial Y}\right|_{Y=0} \left(\frac{\sqrt{X}}{\theta_w}\right) \tag{3.213}$$

where Nu_x is the local Nusselt number and θ_w is the local dimensionless wall temperature. Now since point $j = 1$ lies on the wall:

$$\left.\frac{\partial \theta}{\partial Y}\right|_{Y=0} = \left(\frac{\partial \theta}{\partial Y}\right)_{i,1} \tag{3.214}$$

In order to determine this from the values of θ calculated at the nodal points, it is noted that to the same approximation as previously used:

$$\theta_{i,2} = \theta_{i,1} + \left(\frac{\partial \theta}{\partial Y}\right)_{i,1} \Delta Y + \left(\frac{\partial^2 \theta}{\partial Y^2}\right)_{i,1} \frac{\Delta Y^2}{2!} \tag{3.215}$$

But the application of the boundary layer energy equation to conditions at the wall gives

$$\left.\frac{\partial^2 \theta}{\partial Y^2}\right|_{Y=0} = \left(\frac{\partial^2 \theta}{\partial Y^2}\right)_{i,1} = 0 \tag{3.216}$$

Therefore Eq. (3.215) can be rearranged to give

$$\left.\frac{\partial \theta}{\partial Y}\right|_{i,1} = \left(\frac{\theta_{i,2} - \theta_{i,1}}{\Delta Y}\right) \tag{3.217}$$

and Eq. (3.213) then gives

$$\frac{Nu_x}{\sqrt{Re_x}} = \left(\frac{\theta_{i,1} - \theta_{i,2}}{\Delta Y}\right)\left(\frac{\sqrt{X}}{\theta_w}\right) \tag{3.218}$$

θ_w being, of course, the same as $\theta_{i,1}$.

Some consideration must be given to the conditions existing along the initial $i = 1$ line which were assumed to be known in the above discussion. The actual conditions will depend on the nature of the problem. For flow over a flat plate, because the boundary layer equations are parabolic in form, the use of these equations requires that the plate have no effect on the flow upstream of the plate. Hence, in this case, the variables will have their freestream values at all the nodal points (except at the point which lies on the surface) on the initial line which is coincident with the leading edge. At the nodal point on the surface, the known conditions at the surface must apply. This is illustrated in Fig. 3.24.

The reference velocity is taken as the freestream velocity u_1, which is, of course, constant for flow over a flat plate. Hence, the conditions on this initial line are as follows:

$$Y = 0, \text{ i.e., } j = 1: U_{1,1} = 0, \ \theta_{1,1} = \theta_w(0)$$
$$Y > 0, \text{ i.e., } j > 1: U_{1,j} = 1, \ \theta_{1,j} = 0 \tag{3.219}$$

The values of V at the points on this initial line must also be specified. It is usually adequate to put

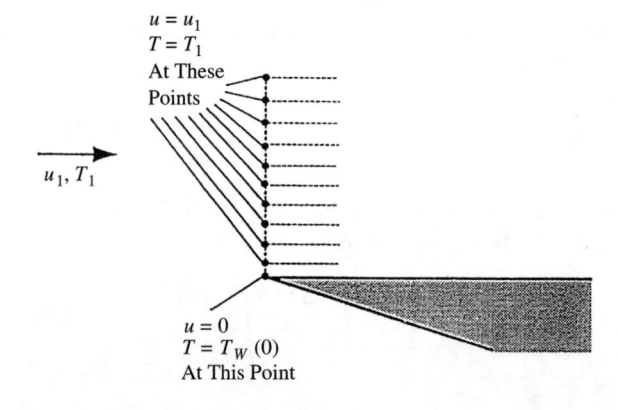

$$\text{For all } Y, \text{ i.e., all } j: V_{1,j} = 0 \tag{3.220}$$

FIGURE 3.24
Conditions on first grid line.

This is actually considerably in error because V, in fact, has its maximum value at the leading edge. However, the effects of this erroneous assumption quickly die out and if the initial spacing of the *i*-lines is chosen to be small, it has a negligible overall effect on the solution. These initial conditions are incorporated into the computer program discussed below.

A computer program, LAMBOUN, written in FORTRAN that is based on the above procedure is available as discussed in the Preface.

The program, as available, will calculate flow over a plate with a varying freestream velocity and varying surface temperature. These variations are both assumed to be described by a third-order polynomial, i.e., by:

$$U_1 = A_V + B_V X + C_V X^2 + D_V X^3$$

and:

$$\theta_w = A_T + B_T X + C_T X^2 + D_T X^3$$

It follows from the assumed form of the freestream velocity distribution that since:

$$\frac{dp}{dx} = -\rho u_1 \frac{du_1}{dx}$$

that:

$$\frac{dP}{dx} = -U_1 \frac{dU_1}{dX} = U_1(B_U + 2C_U X + 3D_U X^2)$$

EXAMPLE 3.6. Numerically determine the heat transfer rate variation with two-dimensional laminar boundary layer air flow over an isothermal flat plate. Compare the numerical results with those given by the similarity solution.

Solution. Because flow over a flat plate with an isothermal surface is being considered, the freestream velocity and surface temperature are constant. Therefore, the inputs to the program are:

$$X_{max} = 1, \; Pr = 0.7, \; A_T, B_T, C_T, D_T = 1, 0, 0, 0$$

$$A_V, B_V, C_V, D_V = 1, 0, 0, 0$$

The program, when run with these inputs, gives the values of $Nu_x/Re_x^{0.5}$ at various X values along the plate. These are shown in Fig. E3.6.

For a Prandtl number of 0.7, the similarity solution for flow over an isothermal flat plate gives:

$$\frac{Nu_x}{\sqrt{Re_x}} = \text{constant} = 0.293$$

This is also shown in Fig. E3.6. It will be seen that while the numerical results are in excellent agreement with the similarity solution at larger X values, there are considerable differences between the two solutions at small X. These differences could have been reduced by using smaller ΔX and ΔY values.

EXAMPLE 3.7. Air flows at a velocity of 4 m/s over a wide flat plate that has a length of 20 cm in the flow direction. The air ahead of the plate has a temperature of 20°C while

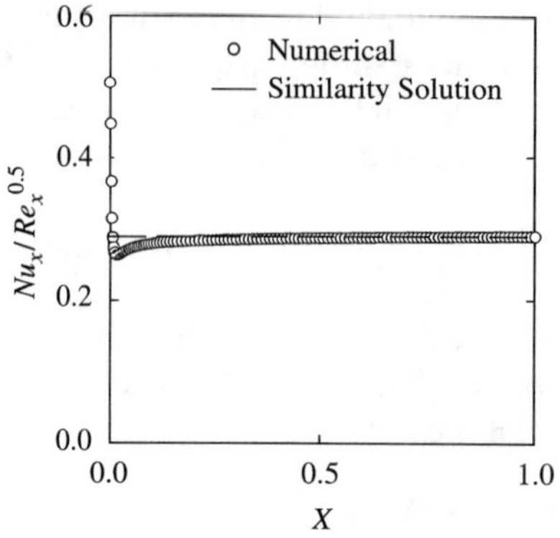

[Figure E3.6: Plot of $Nu_x/Re_x^{0.5}$ versus X comparing Numerical (circles) and Similarity Solution (line), values around 0.3 across X from 0.0 to 1.0]

FIGURE E3.6

the surface temperature of the plate is given by:

$$x < 10 \text{ cm}: T_w = 40°C$$

$$10 \text{ cm} < x < 20 \text{ cm}: T_w = 80°C$$

By numerically solving the two-dimensional laminar boundary layer equations, determine how the local heat transfer rate in W/m² varies along the plate.

Solution. The mean plate temperature is $(40 + 80)/2 = 60°C$. The mean temperature of the air in the boundary layer is therefore:

$$T_{mean} = \frac{T_{w\,mean} + T_1}{2} = \frac{60 + 20}{2} = 40°C$$

At this temperature for air:

$$k = 0.0271 \text{ W/m-°C}, \quad \nu = 0.0000170 \text{ m}^2/\text{s}$$

Hence, since here:

$$u_1 = 4 \text{ m/s}$$

it follows that the Reynolds number based on the length of the plate, i.e., 0.2 m, is:

$$Re_L = \frac{u_1 L}{\nu} = \frac{4 \times 0.2}{0.0000170} = 47{,}059$$

The boundary layer on the plate will, therefore, remain laminar (see the discussion of the transition Reynolds number given later).

The plate surface temperature variation is such that:

$$x < 10 \text{ cm}: T_w - T_1 = 20°C$$

$$10 \text{ cm} < x < 20 \text{ cm}: T_w - T_1 = 60°C$$

Hence, using the plate length, 0.2 m, as the reference length, L, and the value of the temperature difference $T_w - T_1$ on the first 10 cm of the plate as the reference temperature difference $T_{wR} - T_1$, and recalling that the dimensionless temperature used in the computer program is:

$$\theta = (T - T_1)/(T_{wR} - T_1)$$

CHAPTER 3: Some Solutions for External Laminar Forced Convection   137

it follows that the surface temperature difference is such that:

$$X < 0.5: \ \theta_w = 1$$
$$0.5 < X < 1.0: \ \theta_w = 3$$

A simple although not very elegant way of modifying the available computer program LAMBOUN to deal with the situation being considered in this example is to delete the lines:

```
WRITE(6,6002)
READ(5,*) AT,BT,CT,DT
```

and to change the subroutine TEMP to the following:

```
SUBROUTINE TEMP(X,TW)
COMMON AT,BT,CT,DT,AV,BV,CV,DV
*
*********** THIS DETERMINES THE WALL TEMPERATURE *******************
*                                  ----
*
    IF (X.LT.0.5) THEN
    TW = 1.0
    ELSE
    TW = 3.0
    ENDIF
    RETURN
    END
```

Running the program with these modifications gives the variation of $Nu_x/Re_x^{0.5}$ along the plate. The local heat transfer rate can then be calculated from this variation by recalling that:

$$\frac{Nu_x}{\sqrt{Re_x}} = \frac{q_w x}{(T_w - T_1)k}\left(\frac{\nu}{u_1 x}\right)^{0.5}$$

where T_w is the local wall temperature. This can be rearranged to give:

$$q_w = \frac{Nu_x}{\sqrt{Re_x}}(T_w - T_1)k\left(\frac{u_1}{\nu x}\right)^{0.5}$$

i.e.:

$$q_w = \frac{Nu_x}{\sqrt{Re_x}}(T_w - T_1) \times 0.0271 \times \left(\frac{4}{0.0000170 x}\right)^{0.5}$$

i.e.:

$$q_w = \frac{Nu_x}{\sqrt{Re_x}}(T_w - T_1)\frac{29.39}{X^{0.5}}$$

But:

$$(T_w - T_1) = \theta_w(T_{wR} - T_1)$$

Therefore, because $T_{wR} - T_1$ was chosen to be $40 - 20 = 20°C$, it follows that:

$$q_w = \frac{Nu_x}{\sqrt{Re_x}}\theta_w \frac{587.8}{X^{0.5}}$$

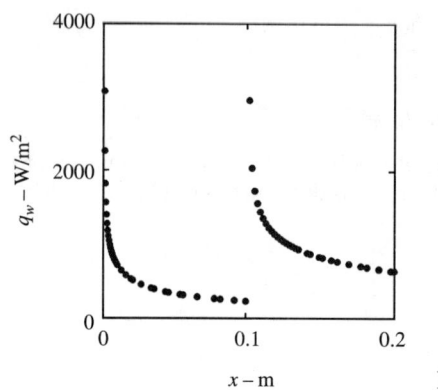

FIGURE E3.7

The program with these modifications when run with the following inputs:

$$X_{\max} = 1, \ Pr = 0.7, \ A_V, B_V, C_V, D_V = 1, 0, 0, 0$$

gives the values of $Nu_x/Re_x^{0.5}$ at various X values along the plate. Since the dimensionless plate temperature, θ_w, is known at these points, the above equation then allows q_w to be calculated at these X values. Results are shown in Fig. E3.7. Only some of the calculated points are shown in this figure.

In the above discussion, it was assumed that the surface temperature variation of the plate was specified. The procedure is easily extended to deal with other thermal boundary conditions at the surface. For example, if the heat flux distribution at the surface is specified, it is convenient to define the following dimensionless temperature:

$$\theta^* = \frac{T - T_1}{(q_{wR}L/k \sqrt{Re_L})} \tag{3.221}$$

where q_{wR} is some convenient reference wall heat flux. The energy equation in terms of this dimensionless temperature has the same form as that obtained with the dimensionless temperature used when the surface temperature is specified, i.e., the dimensionless energy equation in this case is:

$$U \frac{\partial \theta^*}{\partial X} + V \frac{\partial \theta^*}{\partial Y} = \frac{1}{Pr} \frac{\partial^2 \theta^*}{\partial Y^2} \tag{3.222}$$

Using Fourier's law, the boundary condition on temperature at the wall in the specified heat flux case is:

$$y = 0: \ -k \frac{\partial T}{\partial y} = q_w(x) \tag{3.223}$$

i.e.,

$$Y = 0: \ -\frac{\partial \theta^*}{\partial Y} = \frac{q_w(x)}{q_{wR}} = Q_R(X) \tag{3.224}$$

where $Q_R = q_w/q_{wR}$.

CHAPTER 3: Some Solutions for External Laminar Forced Convection 139

In order to express this in finite difference form it is recalled that if the wall point and the point adjacent to the wall on a given i-line are considered, then a Taylor expansion gives to order of accuracy ΔY^2:

$$\theta^*_{i,2} = \theta^*_{i,1} + \left.\frac{\partial \theta^*}{\partial Y}\right|_{i,1} \Delta Y + \left.\frac{\partial^2 \theta^*}{\partial Y^2}\right|_{i,1} \frac{\Delta Y^2}{2!} \tag{3.225}$$

But since U and V are zero at the wall, Eq. (3.222) gives when applied at the wall:

$$\left.\frac{\partial^2 \theta^*}{\partial Y^2}\right|_{i,1} = 0 \tag{3.226}$$

Substituting this into Eq. (3.225) then gives the following finite-difference approximation:

$$\left.\frac{\partial \theta^*}{\partial Y}\right|_{i,1} = \frac{\theta^*_{i,2} - \theta^*_{i,1}}{\Delta Y} \tag{3.227}$$

Substituting this result into Eq. (3.224) then gives:

$$-\frac{\theta^*_{i,2} - \theta^*_{i,1}}{\Delta Y} = Q_R \tag{3.228}$$

i.e.:

$$\theta^*_{i,1} = \theta^*_{i,2} + Q_R \Delta Y \tag{3.229}$$

This boundary condition is easily incorporated into the solution procedure that was outlined above, the set of equations governing the dimensionless temperature in this case having the form:

$$\begin{aligned}
\theta^*_{i,1} &= \theta^*_{i,2} + Q_R \Delta Y \\
E_2 \theta_{i,2} + F_2 \theta_{i,3} + G_2 \theta_{i,1} &= H_2 \\
E_3 \theta_{i,3} + B_3 \theta_{i,4} + C_3 \theta_{i,2} &= D_3 \\
&\vdots \\
E_{N-1} \theta_{i,N-1} + F_{N-1} \theta_{i,N} + G_{N-1} \theta_{i,N-2} &= H_{N-1} \\
\theta_{i,N} &= 0
\end{aligned} \tag{3.230}$$

the coefficients having the same values as previously defined. The matrix equation that gives the dimensionless temperature therefore has the form.

$$\begin{bmatrix}
1 & 1 & 0 & 0 & 0 & . & 0 & 0 & 0 \\
G_2 & E_2 & F_2 & 0 & 0 & . & 0 & 0 & 0 \\
0 & G_3 & E_3 & F_3 & 0 & . & 0 & 0 & 0 \\
0 & 0 & G_4 & E_4 & F_4 & . & 0 & 0 & 0 \\
. & . & . & . & . & . & . & . & . \\
. & . & . & . & . & . & . & . & . \\
0 & 0 & 0 & 0 & 0 & . & G_{N-1} & E_{N-1} & F_{N-1} \\
0 & 0 & 0 & 0 & 0 & . & 0 & 0 & 1
\end{bmatrix}
\begin{bmatrix} \theta^*_{i,1} \\ \theta^*_{i,2} \\ \theta^*_{i,3} \\ \theta^*_{i,4} \\ . \\ . \\ \theta^*_{i,N-1} \\ \theta^*_{i,N} \end{bmatrix}
=
\begin{bmatrix} Q_r \Delta Y \\ H_2 \\ H_3 \\ H_4 \\ . \\ . \\ H_{N-1} \\ 0 \end{bmatrix}$$

Thus, as before, a tridiagonal matrix is obtained.

The program discussed above is therefore easily modified to deal with the specified wall heat flux case and a program, LAMBOUQ, for this situation can be obtained in the way indicated in the Preface.

3.7 VISCOUS DISSIPATION EFFECTS ON LAMINAR BOUNDARY LAYER FLOW OVER A FLAT PLATE

The effects of viscous dissipation on the temperature field have been ignored in the discussion up to this point in the chapter. However, viscous dissipation effects can be important particularly if the viscosity of the fluid is high or if the velocities are high. An example of the first case would be the flow of the lubricant through a journal bearing. In this case the lubricant temperature rise caused by viscous dissipation can be quite high. The second case usually involves the flow of gases at relatively high Mach numbers and it is to this case that attention will be given in the present section. In the present section, then, an analysis of the effects of dissipation on the laminar boundary layer flow over a flat plate which is aligned with the flow will be presented [9],[12],[13],[14],[15],[16]. The flow situation considered is, therefore, as shown in Fig. 3.25.

The effects of fluid property variations will again be ignored. It should, however, be noted that if viscous dissipation effects are important, fluid property changes are usually quite large. These changes can usually be adequately accounted for by evaluating the fluid properties at some suitable mean temperature and then treating them as constant in the analysis of the flow as is being done here. This will be discussed in the next section.

The equations governing the flow were discussed in Chapter 2. Because two-dimensional flow over a flat plate is being considered, these equations are:

$$\frac{\partial u}{\partial x} + \frac{\partial v}{\partial y} = 0 \tag{3.231}$$

$$u\frac{\partial u}{\partial x} + v\frac{\partial u}{\partial y} = -\frac{1}{\rho}\frac{dp}{dx} + \left(\frac{\mu}{\rho}\right)\frac{\partial^2 u}{\partial y^2} \tag{3.232}$$

$$u\frac{\partial T}{\partial x} + v\frac{\partial T}{\partial y} = \left(\frac{k}{\rho c_p}\right)\frac{\partial^2 T}{\partial y^2} + \left(\frac{\mu}{\rho c_p}\right)\left(\frac{\partial u}{\partial y}\right)^2 \tag{3.233}$$

The last term in the energy equation, i.e., in Eq. (3.233), is the viscous dissipation term.

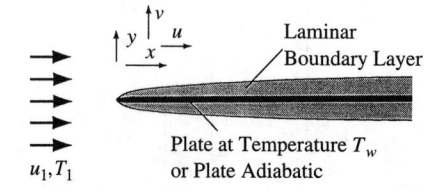

FIGURE 3.25 Laminar boundary layer flow with viscous dissipation.

CHAPTER 3: Some Solutions for External Laminar Forced Convection   141

The boundary conditions on the velocity distribution are:

$$\text{At } y = 0: \quad u = v = 0$$
$$\text{For large } y: \quad u \to u_1(x) \tag{3.234}$$

Two possible boundary conditions on temperature at the wall will be considered in the present section, i.e., it will be assumed that either:

$$\text{At } y = 0: \quad \frac{\partial T}{\partial y} = 0 \tag{3.235}$$

or that:

$$\text{At } y = 0: \quad T = T_w \tag{3.236}$$

where T_w is the temperature of the wall which is assumed constant. The first of these boundary conditions, i.e., Eq. (3.235), applies when the wall is adiabatic because Fourier's law gives the heat transfer rate at the wall, in general, as:

$$q_w = -k \left.\frac{\partial T}{\partial y}\right|_{y=0} \tag{3.237}$$

Therefore, since $q_w = 0$ if the wall is adiabatic, the gradient of temperature at the wall must be zero if the wall is adiabatic.

Outside the boundary layer, the boundary condition on temperature is:

$$\text{For large } y: \quad T \to T_1 \tag{3.238}$$

Dissipation has no effect on the continuity and momentum equations because the fluid properties are being assumed constant. As a result, Eqs. (3.231) and (3.232) are the same as those considered earlier in this chapter in Section 3.2. As discussed in that section, a similarity solution to these equations can be obtained. To do this, the following "similarity" variable was introduced:

$$\eta = \frac{y}{x}\sqrt{Re_x} = y\sqrt{\frac{u_1}{\nu x}} \tag{3.239}$$

In terms of this variable, the distribution of u in the boundary layer is assumed to be:

$$\frac{u}{u_1} = f'(\eta) \tag{3.240}$$

Using the continuity equation, it was then shown that:

$$\frac{v}{u_1} = \frac{1}{2}\sqrt{\frac{\nu}{xu_1}}(\eta f' - f) \tag{3.241}$$

The momentum equation could then be written in terms of η as:

$$2f''' + ff'' = 0 \tag{3.242}$$

The primes, of course, denote differentiation with respect to η.

In terms of the similarity function, the boundary conditions on the solution become:

$$\eta = 0, \ f' = 0 \quad \eta = 0, \ f = 0$$
$$\text{large}, \ f' \to 1$$

142 Introduction to Convective Heat Transfer Analysis

The solution of Eq. (3.242) subject to these boundary conditions was discussed in Section 3.2 and given in graphical form in Figs. 3.4 and 3.5. This solution, it must be stressed, also applies when viscous dissipation is included.

It was also shown in Section 3.2 that a similarity solution could be obtained for the temperature distribution for the case where the wall temperature is constant. To do this, a dimensionless temperature defined as follows was introduced:

$$\phi = \frac{T_w - T}{T_w - T_1} \tag{3.243}$$

and it was assumed that ϕ depended only on η. In terms of this variable, the energy equation without viscous dissipation became:

$$\phi'' + \frac{Pr}{2}\phi' f = 0 \tag{3.244}$$

While the boundary conditions on the solution can be written as:

$$\begin{aligned}\eta = 0: \quad &\phi = 0 \\ \eta \text{ large}: \quad &\phi \to 1\end{aligned} \tag{3.245}$$

Thus, as was the case with the velocity distribution, the partial differential equation governing the temperature distribution was shown to reduce to an ordinary differential equation.

It seems reasonable to assume that similar temperature profiles will also exist when viscous dissipation is important. Attention will first be given to the adiabatic wall case. If the wall is adiabatic and viscous dissipation is neglected, then the solution to the energy equation will be $T = T_1$ everywhere in the flow. However, when viscous dissipation effects are important, the work done by the viscous forces leads to a rise in fluid temperature in the fluid. This temperature will be related to the kinetic energy of the fluid in the freestream flow, i.e., will be related to $u_1^2/2c_p$. For this reason, the similarity profiles in the adiabatic wall case when viscous dissipation is important are assumed to have the form:

$$\frac{T - T_1}{u_1^2/2c_p} = \theta_a(\eta) \tag{3.246}$$

Substituting this and the velocity profile results into the energy equation, i.e., Eq. (3.233), then gives:

$$u_1 f' \theta'_a \frac{\partial \eta}{\partial x} + \frac{u_1}{2}\sqrt{\frac{\nu}{xu_1}}(\eta f' - f)\theta'_a \frac{\partial \eta}{\partial y} = \frac{k}{\rho c_p}\theta''_a \frac{\partial \eta^2}{\partial y} + \frac{\mu}{\rho c_p}(f'')^2 \frac{u_1^2}{u_1^2/2c_p}\left(\frac{\partial \eta}{\partial y}\right)^2$$

i.e.:

$$-\frac{f'\theta'_a}{2x}y\sqrt{\frac{u_1}{x\nu}} + \frac{1}{2}\sqrt{\frac{\nu}{xu_1}}(\eta f' - f)\theta'_a\sqrt{\frac{u_1}{x\nu}} = \frac{k}{\rho c_p}\theta''_a \frac{u_1}{x\nu}\frac{1}{u_1} + \frac{\mu}{\rho c_p}(f'')^2\frac{2c_p}{u_1}\left[\frac{u_1}{x\nu}\right]$$

i.e.:

$$-\frac{f'\theta'_a \eta}{2} + \frac{\eta f'\theta'_a}{2} - \frac{f\theta'_a}{2} = \frac{\theta''_a}{Pr} + 2(f'')^2$$

CHAPTER 3: Some Solutions for External Laminar Forced Convection   143

i.e.:

$$\theta_a'' + \frac{Pr}{2} f \theta_a' + 2\, Pr(f'')^2 = 0 \tag{3.247}$$

This is an ordinary differential equation which shows that a similarity solution does, indeed, exist. This equation contains the Prandtl number as a parameter, i.e., the variation of θ_a with η depends on Pr.

The boundary conditions on θ_a are:

$$\eta = 0: \quad \theta_a' = 0$$
$$\eta \text{ large}: \quad \theta_a \to 0$$

The solution of Eq. (3.247) subject to these boundary conditions can be obtained by numerical integration or an analytical solution can be obtained by introducing an integrating factor. The latter procedure leads to the following equation for the value of θ_a at the wall:

$$\theta_a(0) = \frac{\int_0^\eta \exp\left(\frac{Pr}{2} \int_0^\eta f\, d\eta\right) 2\, Pr(f'')^2\, d\eta}{\int_0^\infty \exp\left(\frac{Pr}{2} \int_0^\eta f\, d\eta\right) d\eta} \tag{3.248}$$

This equation can be integrated numerically to give the value of $\theta_a(0)$ for any chosen value of the Prandtl number. Some typical results are shown in Fig. 3.26.

It will be seen from the results given in Fig. 3.26 that for Prandtl numbers between approximately 0.5 and 10, the variation of $\theta_a(0)$ with Pr is approximately described by:

$$\theta_a(0) = Pr^{1/2} \tag{3.249}$$

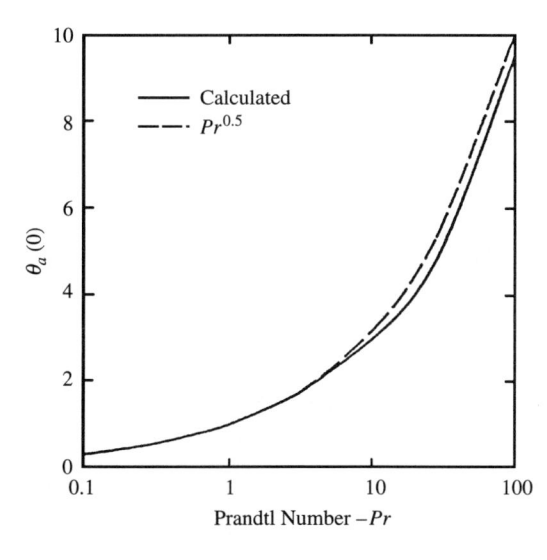

FIGURE 3.26
Variation of dimensionless adiabatic wall temperature with Prandtl number.

Considering the definition of θ_a as given in Eq. (3.246), it will be seen that:

$$\theta_a(0) = \frac{T_{w_{ad}} - T_1}{u_1^2/2c_p} \tag{3.250}$$

where $T_{w_{ad}}$ is the adiabatic wall temperature. This equation gives:

$$T_{w_{ad}} = T_1 + \theta_a(0) u_1^2/2c_p \tag{3.251}$$

Now for a perfect gas:

$$R = c_p - c_v = c_p(1 - 1/\gamma)$$

where $\gamma = c_p/c_v$ is the specific heat ratio of the gas involved and R is the gas constant. From this it follows that:

$$c_p = \frac{\gamma R}{\gamma - 1}$$

Hence:

$$\frac{u_1^2}{2c_p} = \frac{\gamma - 1}{2}\left(\frac{T_1 u_1^2}{\gamma R T_1}\right) \tag{3.252}$$

But:

$$a = \sqrt{\gamma R T}$$

where a is the speed of sound in the gas. Therefore Eq. (3.252) gives:

$$\frac{u_1^2}{2c_p} = \frac{\gamma - 1}{2} T_1 M_1^2 \tag{3.253}$$

where the Mach number, M, is defined by:

$$M = \frac{u}{a} = \frac{u}{\sqrt{\gamma R T}} \tag{3.254}$$

Therefore, Eq. (3.251) gives:

$$\frac{T_{w_{ad}}}{T_1} = 1 + \theta_a(0)\frac{\gamma - 1}{2} M_1^2 \tag{3.255}$$

It is conventional to write this equation for the adiabatic wall temperature as:

$$\frac{T_{w_{ad}}}{T_1} = 1 + r\left(\frac{\gamma - 1}{2}\right) M_1^2 \tag{3.256}$$

where:

$$r = \theta_a(0) \tag{3.257}$$

r being termed the recovery factor. The results given above therefore show that for Prandtl numbers between approximately 0.5 and 10, r is approximately given by:

$$r = Pr^{1/2} \tag{3.258}$$

This result agrees very well with experimental results for air as shown in Fig. 3.27.

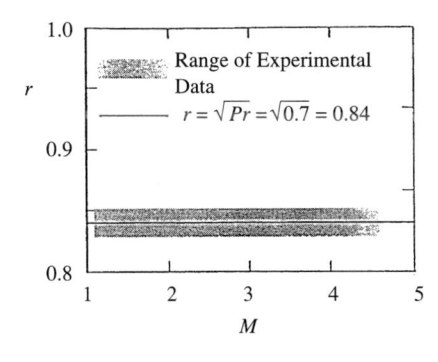

FIGURE 3.27
Typical effect of Mach number on the recovery factor for a laminar boundary layer.

EXAMPLE 3.8. Consider the flow of air which has a freestream temperature of 0°C over an adiabatic flat plate as shown in Fig. E3.8. If the flow in the boundary layer can be assumed to be laminar, determine how the temperature of the plate surface varies with Mach number.

Solution. Since the boundary layer flow is laminar and since it can be assumed that for air $Pr = 0.7$, it follows that here:

$$r = 0.7^{1/2} = 0.837$$

In this case then:

$$\frac{T_{w_{ad}}}{T_1} = 1 + 0.167 M_1^2$$

But here, $T_1 = 0°C = 273$ K so:

$$T_{w_{ad}} = 273(1 + 0.167 M_1^2)$$

Using this equation, the following are obtained:

M	$T_{w_{ad}}/T_1$	$T_{w_{ad}}$ K	$T_{w_{ad}}$ C
0	1.000	273	0
1	1.167	319	46
2	1.669	456	183
3	2.506	684	441
4	3.677	1004	731

It will be seen, therefore, that high surface temperatures can exist at Mach numbers above about 2. These surface temperatures may, in fact, be so high that special high-temperature materials such as titanium alloys rather than conventional aluminum alloys must be used for structural components in aircraft designed to fly at high Mach numbers.

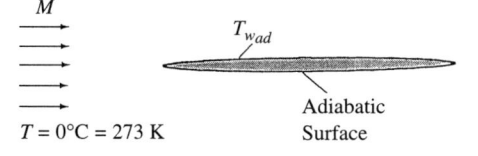

FIGURE E3.8

146 Introduction to Convective Heat Transfer Analysis

The above analysis dealt with the effects of dissipation on the flow over an adiabatic flat plate. Attention is now turned to flow over a plate that is kept at a uniform temperature of T_w.

Now, as discussed earlier in this section, when viscous dissipation effects are neglected, if a dimensionless temperature defined as follows is introduced:

$$\phi = \frac{T_w - T}{T_w - T_1}$$

the energy equation without viscous dissipation becomes:

$$\phi'' + \frac{Pr}{2}\phi' f = 0$$

While the boundary conditions on the solution can be written as:

$$\eta = 0: \quad \phi = 0$$
$$\eta \text{ large:} \quad \phi \to 1$$

The variation of ϕ obtained by solving this equation was discussed earlier in this chapter.

The temperature distribution in flow over a plate kept at a uniform temperature when there are no dissipation effects and the temperature distribution in flow over an adiabatic plate accounting for dissipation effects have thus now been obtained. These two solutions were denoted by ϕ and θ_a, respectively. Because the energy equation is linear, it is to be expected that the solution for flow over a plate kept at a uniform temperature when dissipation is accounted for will be some form of linear combination of these two solutions [9]. To investigate this possibility, the following dimensionless temperature will be introduced:

$$\theta(\eta) = \frac{T - T_1}{u_1^2/2c_p} \tag{3.259}$$

Now it will be noted that:

$$\phi = \frac{T_w - T_1 + T_1 - T}{T_w - T_1} = 1 - \frac{T - T_1}{T_w - T_1} = 1 - \frac{\theta}{\theta(0)}$$

i.e.:

$$\theta = (1 - \phi)\theta(0) \tag{3.260}$$

Hence, it seems reasonable to assume that in the case of flow with viscous dissipation over a plate kept at a uniform temperature the solution will have the form:

$$\theta_T = \theta_a + C_1(1 - \phi)\theta_T(0) + C_2 \tag{3.261}$$

where θ_T is the actual solution for flow with viscous dissipation over a plate kept at a uniform temperature and where C_1 and C_2 are constants.

If the derivation of Eq. (3.247) is considered, it will be seen that the boundary conditions do not influence the form of the equation obtained. Hence, the temperature profile for flow with viscous dissipation over a plate kept at a uniform temperature must satisfy the following equation:

$$\theta_T'' + \frac{Pr}{2}f\theta_T' + 2Pr(f'')^2 = 0 \tag{3.262}$$

CHAPTER 3: Some Solutions for External Laminar Forced Convection 147

Therefore, to check that Eq. (3.261) does represent a solution, it is substituted into the left-hand side of Eq. (3.262) to give:

$$\theta_a'' + C_1\theta_T(0)\phi'' + \frac{Pr}{2}f(\theta_a' + C_1\theta_T(0)\phi') + 2\,Pr(f'')^2$$

i.e.,

$$\left[\theta_a'' + \frac{Pr}{2}f(\theta_a' + 2\,Pr(f'')^2\right] - C_1\theta_T(0)\left[\phi'' + \frac{Pr}{2}f\phi'\right] \qquad (3.263)$$

Now the first term in this equation is zero by virtue of Eq. (3.247) while the second term is zero by virtue of Eq. (3.244). Hence, Eq. (3.261) is a solution of the energy equation. The constants C_1 and C_2 are found by applying the boundary conditions. These give:

$$\eta = 0: \theta_T = 1, \theta_a' = 0, \phi = 0$$
$$\eta \text{ large}: \theta_T \to 0, \theta_a \to 0, \phi = 1 \qquad (3.264)$$

Hence, applying these boundary conditions in Eq. (3.261) gives:

$$\theta_T(0) = \theta_a(0) + C_1\theta_T(0) + C_2 \qquad (3.265)$$

and:

$$0 = 0 + C_2 \qquad (3.266)$$

These equations give:

$$C_2 = 0, \quad C_1 = 1 - \frac{\theta_a(0)}{\theta_T(0)} \qquad (3.267)$$

Substituting these into Eq. (3.261) then gives:

$$\theta_T = \theta_a + [\theta_T(0) - \theta_a(0)](1 - \phi) \qquad (3.268)$$

Now consider the heat transfer rate from the surface. It is given by Fourier's law as:

$$q_w = -k\left.\frac{\partial T}{\partial y}\right|_w$$

In terms of the variables introduced above this gives:

$$q_w = -k\frac{u_1^2}{2c_p}\left[\frac{d\theta_T}{d\eta}\frac{\partial \eta}{\partial y}\right]_{y=0}$$

i.e.,

$$q_w = -k\frac{u_1^2}{2c_p}\frac{1}{x}\sqrt{\frac{xu_1}{\nu}}\left.\theta_T'\right|_{\eta=0}$$

which can be rearranged to give:

$$\frac{q_w x/k}{\sqrt{Re_x}} = -\frac{u_1^2}{2c_p}\left.\theta_T'\right|_{\eta=0}$$

148 Introduction to Convective Heat Transfer Analysis

Using Eq. (3.268), this gives:

$$\frac{q_w x/k}{\sqrt{Re_x}} = \frac{u_1^2}{2c_p}[-\theta_a'|_{\eta=0} + (\theta_T(0) - \theta_a(0))\phi'|_{\eta=0}] \qquad (3.269)$$

But:

$$\theta_a'|_{\eta=0} = 0$$

and:

$$\frac{u_1^2}{2c_p}(\theta_T(0) - \theta_a(0)) = (T_w - T_1) - (T_{w_{ad}} - T_1) = T_w - T_{w_{ad}}$$

Hence, Eq. (3.269) becomes:

$$\frac{q_w x/k}{\sqrt{Re_x}} = (T_w - T_{w_{ad}})\phi'|_{\eta=0} \qquad (3.270)$$

But it was shown in Section 3.1 that $\phi'|_{\eta=0}$, which is only dependent on the Prandtl number, is approximately given by:

$$\phi'|_{\eta=0} = 0.332 Pr^{1/3}$$

so, Eq. (3.270) gives, using this approximation:

$$\frac{q_w x/k}{\sqrt{Re_x}} = 0.332 Pr^{1/3}(T_w - T_{w_{ad}})$$

i.e.:

$$\frac{q_w x}{(T_w - T_{w_{ad}})k} = 0.332 Re_x^{1/2} Pr^{1/3} \qquad (3.271)$$

If, therefore, a local Nusselt number based on $T_w - T_{w_{ad}}$ is introduced, i.e., if the following is defined:

$$Nu_x = \frac{q_w x}{(T_w - T_{w_{ad}})k}$$

then Eq. (3.271) can be written as:

$$Nu_x = 0.332 Re_x^{1/2} Pr^{1/3} \qquad (3.272)$$

This is identical to the equation that applies when dissipation effects are negligible except that the heat transfer coefficient is now based on the difference between the wall and the adiabatic wall temperatures, i.e., in gas flows in which viscous dissipation effects are important, the same results as obtained by neglecting dissipation can be used to find the heat transfer coefficient provided that the heat transfer coefficient is defined by:

$$q_w = h(T_w - T_{w_{ad}})$$

where h is the heat transfer coefficient. This equation is really the same as that used when dissipation effects are neglected, i.e.:

$$q_w = h(T_w - T_1)$$

because if dissipation effects are negligible, $T_{w_{ad}} = T_1$.

CHAPTER 3: Some Solutions for External Laminar Forced Convection   149

Equation (3.272) can be integrated as shown in Chapter 2 to give the following expression for the average Nusselt number for a plate of length L as:

$$Nu_L = 0.664 Re_L^{1/2} Pr^{1/3} \qquad (3.273)$$

where Nu_L and Re_L are the mean Nusselt number and the Reynolds number based on L, respectively.

EXAMPLE 3.9. Consider the flow of air which has a freestream temperature of 0°C over a flat plate that is kept at a temperature of 30°C. If the Reynolds number based on the length of the plate is low enough for the flow in the boundary layer to be assumed to be laminar and if the freestream Mach number is 0.9, find whether heat is being transferred to or from the plate.

Solution. The flow situation being considered is shown in Fig. E3.9. It will be assumed that the Prandtl number for the air is 0.7. Hence, using Eq. (3.258):

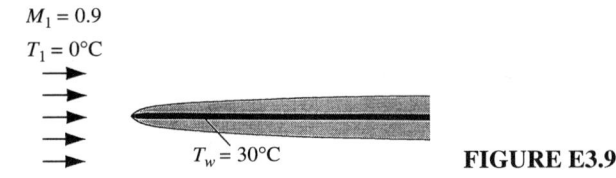

$M_1 = 0.9$
$T_1 = 0°C$

$T_w = 30°C$ **FIGURE E3.9**

$$r = (0.7)^{1/2} = 0.837$$

Therefore, since Eq. (3.256) gives:

$$\frac{T_{w_{ad}}}{T_1} = 1 + r\left(\frac{\gamma - 1}{2}\right)M_1^2$$

it follows that, since for air $\gamma = 1.4$:

$$\frac{T_{w_{ad}}}{T_1} = 1 + 0.837 \times 0.2 M_\infty^2 = 1 + 0.167 \times 0.9^2 = 1.135$$

Hence, since the freestream temperature is 0°C, it follows that:

$$T_{w_{ad}} = 1.135 \times 273 = 309.9 \text{ K}$$

Therefore, the adiabatic wall temperature is 36.9°C. Hence, since the wall is kept at a temperature of 30°C, i.e., since $T_w - T_{w_{ad}}$ is negative, heat is being transferred from the air to the plate.

3.8
EFFECT OF FLUID PROPERTY VARIATIONS WITH VISCOUS DISSIPATION EFFECTS ON LAMINAR BOUNDARY LAYER FLOW OVER A FLAT PLATE

As mentioned above, because of the large temperature variations that often exist in flows in which viscous dissipation is important, there can be large variations in the fluid properties across such flows. In dealing with external flows in which the effects of viscous dissipation are not important, fluid property variations can usually be adequately accounted for by evaluating the properties at the mean film temperature,

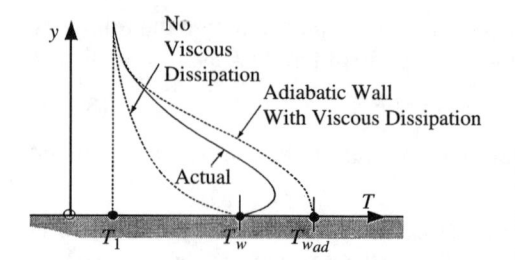

FIGURE 3.28
Temperature distributions in a laminar boundary layer with and without viscous dissipation.

i.e., at the average of the surface and freestream temperatures, i.e., at $(T_w + T_1)/2$, i.e., at:

$$T_{prop} = T_1 + 0.5(T_w - T_1) \quad (3.274)$$

When viscous dissipation effects are important, however, there are three temperatures that have an influence on the temperature distribution and therefore on the fluid properties, these three temperatures being the wall temperature, the freestream temperature, and the adiabatic wall temperature, i.e., T_w, T_1, and $T_{w_{ad}}$. This is illustrated in Fig. 3.28.

When viscous dissipation effects are important it is to be expected therefore, that fluid property variations could be accounted for by evaluating these properties at a mean fluid temperature that is given by an equation of the form [9],[12],[15]:

$$T_{prop} = T_1 + C_1(T_w - T_1) + C_2(T_{w_{ad}} - T_1) \quad (3.275)$$

where C_1 and C_2 are constants. Because in flows in which viscous dissipation effects are negligible, $T_{w_{ad}} = T_1$, a comparison of Eqs. (3.274) and (3.275) shows that $C_1 = 0.5$. Therefore, in flows in which viscous dissipation is important, the fluid properties should be evaluated at a temperature that is given by:

$$T_{prop} = T_1 + 0.5(T_w - T_1) + C_2(T_{w_{ad}} - T_1) \quad (3.276)$$

In order to find the "best" value to use for the constant C_2 in this equation, numerical solutions in which the effects of property variations are accounted for can be obtained for some simple situations such as for flow over a flat plate and these numerical results can then be used to deduce the value of C_2 that leads to the best agreement between the results obtained accounting for property variations and results obtained assuming constant fluid properties evaluated at T_{prop}. This procedure indicates that C_2 should be taken as 0.22 [17],[18],[19],[20],[21],[22], i.e., that:

$$T_{prop} = T_1 + 0.5(T_w - T_1) + 0.22(T_{w_{ad}} - T_1) \quad (3.277)$$

3.9
SOLUTIONS TO THE FULL GOVERNING EQUATIONS

Solutions based on the use of the boundary layer approximations to the full governing equations have been discussed in the above sections. These boundary layer approximations are, however, often not applicable. For example, the boundary layer equations do not apply if there are significant areas of reversed flow or if the Reynolds number is low. Even with two-dimensional flow over a circular cylinder, the boundary layer equations can only be used to predict the heat transfer rate near the stagnation point. They do not apply after flow separation occurs and even well ahead of this separation point there is significant interaction between the outer inviscid flow and the wake with the result that the pressure distribution on the cylinder cannot be obtained by ignoring the boundary layer and calculating the inviscid flow over the cylinder as discussed before.

When the boundary layer and other related approximate solutions cannot be applied, it is necessary to solve the full governing equations. In almost all cases it is necessary to do this using numerical methods [23],[24],[25]. However, because the full equations are a set of nonlinear, simultaneous partial differential equations, this usually requires a significant computational effort. Commercial computer packages based on finite element methods or on various types of finite-difference methods are available for obtaining such solutions. These packages usually contain pre-processing and post-processing modules that make it easy to set up the required nodal system for complex geometrical situations and easy to display the calculated results in a convenient form.

As a simple example of a flow situation in which the boundary layer equations do not apply, consider two-dimensional steady flow over a square cylinder placed in a uniform flow as shown in Fig. 3.29. This flow situation is not of great practical significance. It is used purely to illustrate the type of solution that can be obtained.

The flow in the wake of the cylinder will, in general, be unsteady due to the shedding of vortices from the rear portion of the cylinder as shown schematically in Fig. 3.30. This shedding causes the wake flow to be asymmetric about an axis that is parallel to the upstream flow and drawn through the center of the cylinder as shown in Fig. 3.30.

However, for many purposes it is adequate to assume that the flow is steady and symmetrical. The solution obtained by making these assumptions will in many circumstances provide a good description of the time-averaged flow about the cylinder.

Some results obtained using the program EXTSQCYL that obtains a solution for this flow and that is available in the way discussed in the Preface are shown in Fig. 3.31.

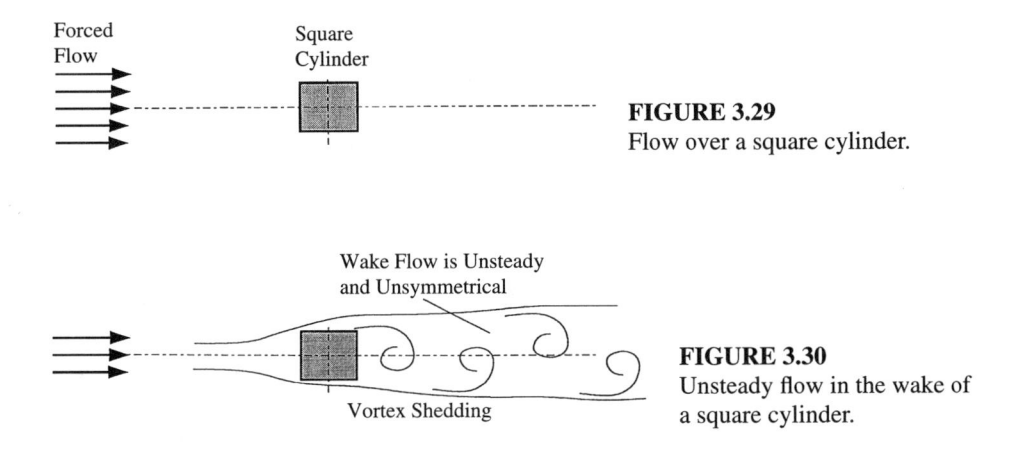

FIGURE 3.29
Flow over a square cylinder.

FIGURE 3.30
Unsteady flow in the wake of a square cylinder.

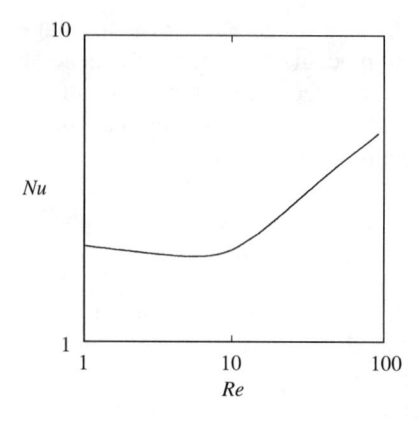

FIGURE 3.31
Predicted variation of mean Nusselt number for a square cylinder with Reynolds number for $Pr = 0.7$.

3.10 CONCLUDING REMARKS

Some of the commonly used methods for obtaining solutions to problems involving laminar external flows have been discussed in this chapter. Many such problems can be treated with adequate accuracy using the boundary layer equations and similarity; integral and numerical methods of solving these equations have been discussed. A brief discussion of the solution of the full governing equations has also been presented.

PROBLEMS

3.1. Air flows at a velocity of 9 m/s over a wide flat plate that has a length of 6 cm in the flow direction. The air ahead of the plate has a temperature of 10°C while the surface of the plate is kept at 70°C. Using the similarity solution results given in this chapter, plot the variation of local heat transfer rate in W/m^2 along the plate and the velocity and temperature profiles in the boundary layer on the plate at a distance of 4 cm from the leading edge of the plate. Also calculate the mean heat transfer rate from the plate.

3.2. Air at a temperature of 20°C flows at a velocity of 1 m/s over a surface which can be modeled as a wide 30-mm long flat plate. The entire surface of this plate is kept at a temperature of 60°C. Plot a graph showing how the local heat transfer rate in W/m^2 and the boundary layer thickness in mm varies along the plate. Also plot the temperature profile at the trailing edge of the plate. Assume two-dimensional flow.

3.3. Air at a temperature of 40°C flows at a velocity of 5 m/s over a surface which can be modeled as a wide 100-mm long flat plate. The entire surface of this plate is kept at a temperature of 0°C. Plot a graph showing how the local heat transfer rate varies along the plate. Also plot the temperature profile in the boundary layer on the plate at a distance of 60 mm from the leading edge of the plate.

3.4. Air at 300 K and 1 atm flows at a velocity of 2 m/s along a flat plate which has a length of 0.2 m. The plate is kept at a temperature of 330 K. Plot the variations of the velocity and thermal boundary layer thicknesses along the plate.

CHAPTER 3: Some Solutions for External Laminar Forced Convection 153

3.5. Glycerin at a temperature of 30°C flows over a 30-cm long flat plate at a velocity of 1 m/s. The surface of the plate is kept at a temperature of 20°C. Find the mean heat transfer rate per unit area to the plate.

3.6. Air flows at a velocity of 6 m/s over a wide flat plate that has a length of 2 cm in the flow direction. There is a uniform heat flux of 3 kW/m^2 at the surface of the plate. Using the similarity solution results, plot the variation of local temperature along the plate. The air ahead of the plate has a temperature of 10°C.

3.7. Air flows at a velocity of 8 m/s over a wide flat plate that has a length of 10 cm in the flow direction. The air ahead of the plate has a temperature of 10°C while the surface of the plate is equal to $[10 + 50(x/10)]$°C, x being the distance measured along the plate in cm. Using the similarity solution results, plot the variation of local heat transfer rate in W/m^2 along the plate.

3.8. Air flows at a velocity of 4 m/s over a wide flat plate that has a length of 20 cm in the flow direction. The temperature of the surface of the plate is given by $[30 + 30(x/20)^{0.7}]$°C, x being the distance measured along the plate in cm. The air ahead of the plate has a temperature of 20°C. Using the similarity solution results, plot the variation of local heat transfer rate in W/m^2 along the plate.

3.9. Air flows at a velocity of 2 m/s normal to the axis of a circular cylinder with a diameter of 2.5 cm. The surface of the cylinder is kept at a uniform surface temperature of 50°C and the temperature in the air stream ahead of the cylinder is 10°C. Assuming that the flow is two-dimensional, find the heat transfer rate in the vicinity of the stagnation point.

3.10. Consider two-dimensional air flow normal to a plane surface. If the initial air temperature is 20°C, the surface temperature 80°C, and the air velocity in the freestream ahead of the plate is 1 m/s, plot the variation of heat transfer rate in the vicinity of the stagnation point.

3.11. Consider laminar forced convective flow over a flat plate at whose surface the heat transfer rate per unit area, q_w is constant. Assuming a Prandtl number of 1, use the integral equation method to derive an expression for the variation of surface temperature. Assume two-dimensional flow.

3.12. Air with a temperature of -10°C flows steadily at a velocity of 8 m/s parallel to a flat plate that is 6 cm long in the flow direction. The first 2 cm of the surface of the plate is adiabatic and the remainder of the plate surface is kept at a temperature of 50°C. Assuming two-dimensional flow and using the integral equation result for a plate with an unheated leading edge section, plot the variation of the heat flux along the heated portion of the plate.

3.13. Air at 5°C and 70 kPa flows over a flat plate at 6 m/s. A heater strip 2.5 cm long is placed on the plate at a distance of 5 cm from the leading edge. Calculate the heat lost from the strip per unit depth of plate for a heater surface temperature of 65°C. Use the appropriate integral equation result.

3.14. Air flows parallel to the surface of a flat plate which is unheated (adiabatic) up to a distance of x_0 from the leading edge. Downstream of this point, there is a uniform heat flux at the surface of the plate. Assuming steady, two-dimensional, constant property flow, use the integral method to find the surface temperature along the heated portion of the plate. Show that if $x_0 = 0$, the local wall temperature is proportional to $x^{0.5}$.

3.15. Consider the two-dimensional laminar boundary flow of air over a wide 15-cm long flat plate whose surface temperature varies linearly from 20°C at the leading edge to 40°C at the trailing edge. This plate is placed in an airstream with a velocity of 2 m/s and a temperature of 10°C. Numerically determine how the surface heat flux varies along this plate.

3.16. Air flows at a velocity of 3 m/s over a wide flat plate that has a length of 30 cm in the flow direction. The air ahead of the plate has a temperature of 10°C while the surface temperature of the plate is given by:

$$x < 5 \text{ cm}: \quad T_w = 40°C$$
$$5 \text{ cm} < x < 25 \text{ cm}: \quad T_w = 60°C$$
$$25 \text{ cm} < x < 30 \text{ cm}: \quad T_w = 40°C$$

By numerically solving the two-dimensional laminar boundary layer equations, determine how the local heat transfer rate in W/m² varies along the plate.

3.17. Numerically determine the heat transfer rate variation with two-dimensional laminar boundary layer air flow over a flat plate with a uniform heat flux at the surface. Compare the numerical results with those given by the similarity solution.

3.18. Liquid films are used for cooling in a number of industrial situations. Consider the following simple case:

Assuming the flow remains laminar and has a boundary layer-like characteristic, write down the governing equations together with the boundary and initial conditions. If the y coordinate is replaced by the stream function derived by:

$$y = \int_0^y \frac{\partial \psi}{u}$$

show that the x-momentum equation becomes:

$$u \frac{\partial u}{\partial x} + vu \frac{\partial u}{\partial \psi} = \nu \left(\frac{\partial^2 u}{\partial \psi^2} \right) u^2$$

and write down the boundary conditions in this case. Discuss how you could numerically solve these equations using the finite difference method.

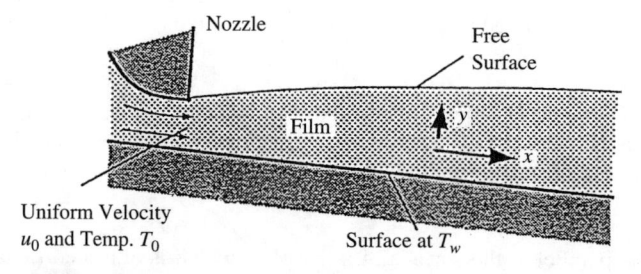

FIGURE P3.17

CHAPTER 3: Some Solutions for External Laminar Forced Convection   155

3.19. An implicit finite-difference procedure for solving the laminar boundary layer equations was discussed in this chapter. Discuss how these boundary layer equations could be solved using an explicit procedure. In such a procedure, the terms $\partial u/\partial y$ and $\partial^2 u/\partial y^2$ are evaluated on the "$(i - 1)$" line. The continuity equation is treated in the same way in both procedures. Write a computer program based on this procedure and show by numerical experimentation with this program that instability develops if:

$$\Delta x > K \Delta y^2$$

where K is a constant. Estimate the value of K.

3.20. Show how the numerical method for solving the laminar boundary layer equations discussed in this chapter can be modified to allow for viscous dissipation. Use a computer program based on this modified procedure to estimate the importance of this dissipation on the heat transfer rate along an isothermal flat plate in low speed flow.

3.21. Air at a Mach number of 3 and a temperature of $-30°C$ flows over a flat plate that is aligned with the flow. The plate is kept at a temperature of 25°C. The flow in the boundary layer is laminar. Is heat transferred to or from the plate surface?

3.22. Air at a pressure of 5 kPa and a temperature of $-30°C$ flows at a Mach number of 2.5 over a flat plate that is aligned with the flow. The plate is kept at a uniform temperature of 5°C. Find the heat transfer from the plate surface to the air.

3.23. A flat plate of length L is heated to a uniform surface temperature and dragged through water which is at a temperature of 10°C at a velocity, V. Plot the variation of power required to pull the plate through the water and of the total heat transfer rate from the plate with plate surface temperature for surface temperatures between 10 and 95°C. Comment on the results obtained. The boundary layer on the plate can be assumed to remain laminar.

3.24. In order to measure the velocity of a stream of air, a flat plate of length 2 cm in the flow direction is placed in the flow. This plate is electrically heated, the heat dissipation rate being uniform over the plate surface. The plate is wide so a two-dimensional laminar boundary layer flow can be assumed to exist. The velocity is to be deduced by measuring the temperature of the plate at its trailing edge. If this temperature is to be at least 40°C when the air temperature is 20°C and the air velocity is 3 m/s, find the required rate of heat dissipation in the plate per unit surface area.

REFERENCES

1. Schlichting, H., *Boundary Layer Theory*, 7th ed., McGraw-Hill, New York, 1979.
2. Howarth, L., "On the Solution to the Laminar Boundary Layer Equations", *Proc. R. Soc. London Ser. A*, Vol. 164, 1938, p. 547.
3. Howarth, L., Editor, *Modern Developments in Fluid Dynamics High Speed Flow*, Vols. 1 and 2, Oxford University Press, London, 1953.
4. White, F.M., *Viscous Fluid Flow*, 2nd. ed., McGraw-Hill, New York, 1991.
5. Sherman, F.S., *Viscous Flow*, McGraw-Hill, New York, 1990.
6. Bejan, A., *Convection Heat Transfer*, 2nd ed., Wiley, New York, 1995.

7. Fox, L., *The Numerical Solution of Two-Point Boundary Problems*, Oxford University Press, London, 1957.
8. Van Dyke, M., *Perturbation Methods in Fluid Mechanics*, Parabolic Press, Stanford, CA, 1975.
9. Kays, W.M. and Crawford, M.E., *Convective Heat and Mass Transfer*, 3rd ed., McGraw-Hill, New York, 1993.
10. Falkner, V.M. and Skan, S.W., "Some Approximate Solutions of the Boundary Layer Equations", *Philos. Mag.*, Vol. 12, 1931, p. 865.
11. Patankar, S.V. and Spalding, D.B., *Heat and Mass Transfer in Boundary Layers*, 2d. ed., International Textbook Co., London, 1970.
12. DeJarnette, F.R., Hamilton, H.H., Weilmuenster, K.L., and Cheatwook, F.M., "A Review of Some Approximate Methods Used in Aerodynamic Heating Analyses", *J. Thermophys. Heat Transf.*, Vol. 1, No. 1, pp. 5–12, 1987.
13. Dorrance, W.H., *Viscous Hypersonic Flow*, McGraw-Hill, New York, 1962.
14. Kaye, J., "Survey of Friction Coefficients, Recovery Factors, and Heat Transfer Coefficients for Supersonic Flow", *J. Aeronaut. Sci.*, Vol. 21, No. 2, pp. 117–229, 1954.
15. Truitt, R.W., *Fundamentals of Aerodynamic Heating*, Ronald Press, New York, 1960.
16. van Driest, E.R., "The Problem of Aerodynamic Heating", *Aeronaut. Eng. Rev.*, Vol. 15, pp. 26–41, 1956.
17. Beckwith, I.E. and Gallagher J.J., "Local Heat Transfer and Recovery Temperatures on a Yawed Cylinder at a Mach Number of 4.15 and High Reynolds Numbers", NASA TR-R-104, 1962.
18. Cohen, C.B. and Reshotko, E., "Similar Solutions for the Compressible Laminar Boundary Layer with Heat Transfer and Pressure Gradient", NASA Report 1293, Houston, TX, 1956.
19. Eckert, E.R.G., "Engineering Relations for Heat Transfer and Friction in High-Velocity Laminar and Turbulent Boundary Layer Flow over Surfaces with Constant Pressure and Temperature", *Trans. ASME*, Vol. 78, pp. 1273–1284, 1956.
20. Fischer, W.W. and Norris, R., "Supersonic Convective Heat Transfer Correlation from Skin Temperature Measurement on V-2 Rocket in Flight", *Trans. ASME*, Vol. 71, pp. 457–469, 1949.
21. Levy, S., "Effect of Large Temperature Changes (Including Viscous Heating) upon Laminar Boundary Layers with Variable Free-Stream Velocity", *J. Aeronaut. Sci.*, Vol. 21, No. 7, pp. 459–474, 1954.
22. Rubesin, M.W. and Johnson, H.A., "Aerodynamic Heating and Convective Heat Transfer—Summary of Literature Survey", *Trans. ASME*, Vol. 71, pp. 383–388, 1949.
23. Patankar, S.V., *Numerical Heat Transfer and Fluid Flow*, Hemisphere Publ., Washington, D. C., 1980.
24. Gosman, A.D., Pun, W.M., Runchal, A.K., Spalding, D.B., and Wolfshtein, M., *Heat and Mass Transfer in Recirculating Flows*, Academic Press, New York, 1969.
25. Chow, L.C. and Tien, C.L. "An Examination of Four Differencing Schemes for Some Elliptic-Type Convection Equations", *Numerical Heat Transfer*, Vol. 1, 1978, pp. 87–100.

CHAPTER 4

Internal Laminar Flows

4.1 INTRODUCTION

Attention will be given in this chapter to heat transfer from the walls of a duct to a fluid flowing through the duct. Attention will be restricted to situations in which the flow is laminar. Ducts of various cross-sectional shape such as those illustrated in Fig. 4.1 will be considered.

Internal flows of the type here being considered occur in heat exchangers, for example, where the fluid may flow through pipes or between closely spaced plates that effectively form a duct. Although laminar duct flows do not occur as extensively as turbulent duct flows, they do occur in a number of important situations in which the size of the duct involved is small or in which the fluid involved has a relatively high viscosity. For example, in an oil cooler the flow is usually laminar. Conventionally, it is usual to assume that a higher heat transfer rate is achieved with turbulent flow than with laminar flow. However, when the restraints on possible solutions to a particular problem are carefully considered, it often turns out that a design that involves laminar flow is the most efficient from a heat transfer viewpoint.

Near the inlet to a duct or following the disturbance produced by a fitting such as a bend, there will be a region in which the characteristics of the flow and the heat transfer rate are changing relatively rapidly with distance along the duct. However, following this "developing flow" region, the flow reaches a "fully developed" state in which the basic characteristics of the flow are not changing with distance along the duct. This is illustrated in Fig. 4.2. In this chapter, attention will first be given to heat transfer in fully developed duct flows. Heat transfer in the developing region will then be considered.

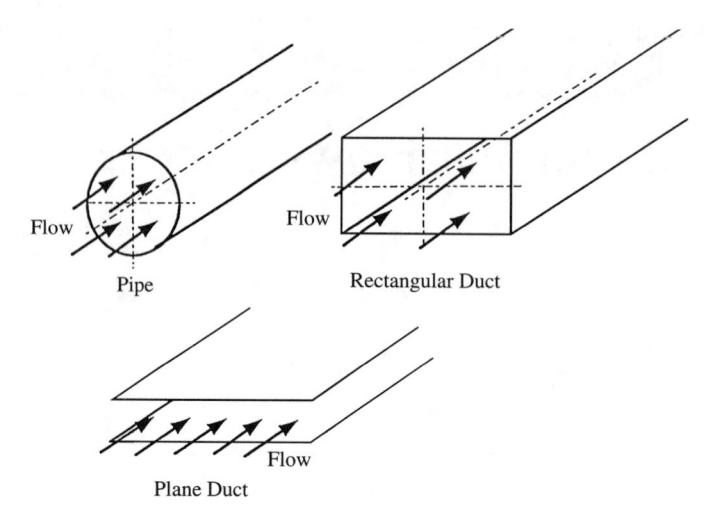

FIGURE 4.1
Internal laminar flows.

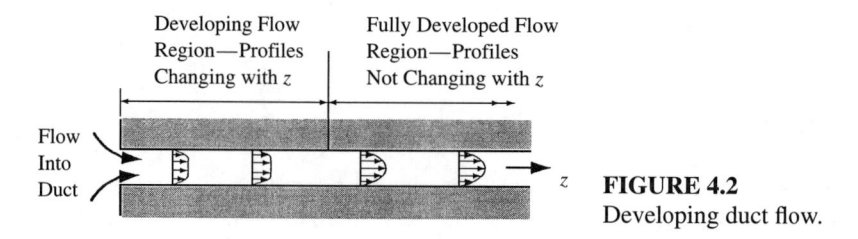

FIGURE 4.2
Developing duct flow.

4.2
FULLY DEVELOPED LAMINAR PIPE FLOW

Fully developed flow in a pipe, i.e., a duct with a circular cross-sectional shape, will first be considered [1],[2],[3]. The analysis is, of course, carried out using the governing equations written in cylindrical coordinates. The z-axis is chosen to lie along the center line of the pipe and the velocity components are defined in the same way that they were in Chapter 2, i.e., as shown in Fig. 4.3.

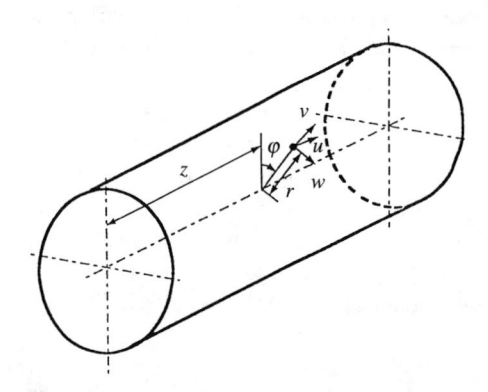

FIGURE 4.3
Coordinate system used in analysis of pipe flow.

Assuming the fluid properties can be treated as constant [4],[5],[6] and that the swirl velocity component w is 0 and that the flow is, therefore, symmetrical about the center line, the equations governing the flow are

$$\frac{1}{r}\frac{\partial}{\partial r}(vr) + \frac{\partial u}{\partial z} = 0 \tag{4.1}$$

$$u\frac{\partial u}{\partial z} + v\frac{\partial u}{\partial r} = -\frac{1}{\rho}\frac{\partial p}{\partial z} + \nu\left(\frac{\partial^2 u}{\partial z^2} + \frac{\partial^2 u}{\partial r^2} + \frac{1}{r}\frac{\partial u}{\partial r}\right) \tag{4.2}$$

$$u\frac{\partial v}{\partial z} + v\frac{\partial v}{\partial r} = -\frac{1}{\rho}\frac{\partial p}{\partial r} + \nu\left(\frac{\partial^2 v}{\partial z^2} + \frac{\partial^2 v}{\partial r^2} + \frac{1}{r}\frac{\partial v}{\partial r} - \frac{v}{r^2}\right) \tag{4.3}$$

$$u\frac{\partial T}{\partial z} + v\frac{\partial T}{\partial r} = \left(\frac{\nu}{Pr}\right)\left(\frac{\partial^2 T}{\partial z^2} + \frac{\partial^2 T}{\partial r^2} + \frac{1}{r}\frac{\partial T}{\partial r}\right) \tag{4.4}$$

Attention is being restricted to fully developed flow in the present section which means that the forms of the velocity and temperature profiles are not changing with distance along the pipe, i.e., that

$$\frac{u}{u_c} = F\left(\frac{r}{r_o}\right), \qquad \frac{T_w - T}{T_w - T_c} = G\left(\frac{r}{r_o}\right) \tag{4.5}$$

the functions F and G being independent of the distance along the pipe, z. In these expressions, r_o is the radius of the pipe, u_c is the velocity on the center line, T_c is the temperature on the center line, and T_w is the wall temperature.

Now the mass flow rate through the pipe is given by

$$\dot{m} = \rho \int_0^{r_o} u 2\pi r \, dr \tag{4.6}$$

i.e., using Eq. (4.5)

$$\dot{m} = u_c 2\pi \rho r_o^2 \int_0^1 F \, d\left(\frac{r}{r_o}\right) \tag{4.7}$$

Because the mass flow rate remains constant along the pipe as does, by assumption, the velocity profile function F, this shows that:

$$u_c = \text{constant, i.e.,} \quad \frac{du_c}{dz} = 0 \tag{4.8}$$

From Eq. (4.5) it then follows that:

$$\frac{\partial u}{\partial z} = 0 \tag{4.9}$$

Substituting this into Eq. (4.1) then gives on integrating:

$$vr = \text{constant} \tag{4.10}$$

But the boundary conditions give $v = 0$ when $r = r_o$ and this equation shows therefore that:

$$v = 0 \tag{4.11}$$

Using this result in Eq. (4.3) then gives:

$$\frac{\partial p}{\partial r} = 0 \tag{4.12}$$

and Eq. (4.2) then finally reduces to:

$$\nu\left(\frac{d^2 u}{dr^2} + \frac{1}{r}\frac{du}{dr}\right) = \frac{1}{\rho}\frac{dp}{dz} \tag{4.13}$$

This is conveniently rewritten as:

$$\frac{1}{r}\frac{d}{dr}\left(r\frac{du}{dr}\right) = \frac{1}{\mu}\frac{dp}{dz} \tag{4.14}$$

Integrating this equation subject to the boundary condition $du/dr = 0$ when $r = 0$ then gives on rearrangement:

$$\frac{du}{dr} = \frac{r}{2\mu}\frac{dp}{dz} \tag{4.15}$$

Integrating this equation subject to the boundary condition $u = 0$ when $r = r_o$ then gives on rearrangement:

$$u = -\frac{1}{4\mu}\frac{dp}{dz}\left(r_o^2 - r^2\right) \tag{4.16}$$

This gives the center line velocity as:

$$u_c = -\frac{1}{4\mu}\frac{dp}{dz}r_o^2 \tag{4.17}$$

Dividing Eq. (4.16) by Eq. (4.17) then gives:

$$\frac{u}{u_c} = 1 - \left(\frac{r}{r_o}\right)^2 \tag{4.18}$$

This is, of course, the well-known parabolic velocity profile for fully developed laminar pipe flow.

Having established the form of the velocity profile, attention must now be turned to the solution of the energy equation (4.4). Since v was shown to be equal to 0 in fully developed flow, this equation reduces in this case to

$$u\frac{\partial T}{\partial z} = \left(\frac{\nu}{Pr}\right)\left[\frac{1}{r}\frac{\partial}{\partial r}\left(r\frac{\partial T}{\partial r}\right) + \frac{\partial^2 T}{\partial z^2}\right] \tag{4.19}$$

For the flow of most fluids under practically significant conditions, the rate at which heat is conducted down the pipe will be negligible compared to the rate at which it is conducted in a radial direction, this being similar to the situation existing in a boundary layer flow. For this reason, the second term on the right-hand side of Eq. (4.19) is ignored compared to the first and the energy equation for fully developed pipe flow is assumed to be:

$$u \frac{\partial T}{\partial z} = \left(\frac{\nu}{Pr}\right) \frac{1}{r} \frac{\partial}{\partial r}\left[r \frac{\partial T}{\partial r}\right] \quad (4.20)$$

This is the equation that must be used to give the form of the temperature profile function G given in Eq. (4.5). Now, because G does not depend on z, it follows that:

$$\frac{\partial}{\partial z}\left[\frac{T_w - T}{T_w - T_c}\right] = 0$$

i.e., that:

$$\frac{\partial T}{\partial z} = \frac{\partial T_w}{\partial z} - \left[\frac{T_w - T}{T_w - T_c}\right]\left[\frac{dT_w}{dz} - \frac{dT_c}{dz}\right] \quad (4.21)$$

To proceed further with the solution, the wall boundary conditions on temperature must be specified. Consideration will first be given to the case where the heat flux at the wall, q_w, is uniform and specified. Some discussion of the solution for the case where the wall temperature is kept uniform will be given later.

Now using Fourier's law, the heat transfer rate at the wall is given by:

$$q_w = +k \left.\frac{\partial T}{\partial r}\right|_{r=r_o} \quad (4.22)$$

The positive sign arises because r is measured from the center toward the wall whereas the heat flux, q_w, is taken as positive in the inward direction, i.e., in the wall-to-fluid direction.

Eq. (4.22) can be written using the definition of the function G given in Eq. (4.5) as:

$$q_w = -\frac{k(T_w - T_c)}{r_o} \left.\frac{dG}{d(r/r_o)}\right|_{r=r_o} \quad (4.23)$$

Because the case where q_w is uniform is being considered, this equation shows that $T_w - T_c =$ constant. From this it follows that:

$$\frac{dT_w}{dz} = \frac{dT_c}{dz} \quad (4.24)$$

Substituting this result into Eq. (4.21) then shows that when the wall heat flux is uniform:

$$\frac{\partial T}{\partial z} = \frac{dT_w}{dz} \quad (4.25)$$

Having established this result from the boundary conditions, attention can now be returned to the governing equation (4.20). Substituting Eq. (4.25) into this equation gives:

$$u \frac{dT_w}{dz} = \left(\frac{\nu}{Pr}\right) \frac{1}{r} \frac{\partial}{\partial r}\left(r \frac{\partial T}{\partial r}\right) \quad (4.26)$$

If the velocity profile as given in Eq. (4.18) is substituted into this equation the following is obtained on rearrangement:

$$\frac{\partial}{\partial(r/r_o)}\left[\left(\frac{r}{r_o}\right)\frac{\partial T}{\partial(r/r_o)}\right] = \left(\frac{Pr}{\nu}\right)u_c r_o^2 \frac{dT_w}{dz}\left[\left(\frac{r}{r_o}\right) - \left(\frac{r}{r_o}\right)^3\right] \quad (4.27)$$

This equation can be integrated, subject to the boundary conditions, to give the variation of T with r. The boundary conditions on the solution are:

$$\begin{aligned} r = 0: &\quad \partial T/\partial r = 0 \\ r = r_o: &\quad T = T_w \end{aligned} \quad (4.28)$$

The first of these conditions follows, of course, from the requirement that the profile be symmetrical about the center line.

Integrating Eq. (4.27) once and using the first boundary condition gives:

$$\frac{\partial T}{\partial(r/r_o)} = \left(\frac{Pr}{\nu}\right) u_c r_o^2 \frac{dT_w}{dz}\left[\frac{1}{2}\left(\frac{r}{r_o}\right) - \frac{1}{4}\left(\frac{r}{r_o}\right)^3\right] \quad (4.29)$$

Integrating this equation and applying the second boundary condition then gives

$$T = T_w - \left(\frac{Pr}{\nu}\right) u_c r_o^2 \frac{dT_w}{dz}\left[\frac{3}{16} - \frac{1}{4}\left(\frac{r}{r_o}\right)^2 + \frac{1}{16}\left(\frac{r}{r_o}\right)^4\right] \quad (4.30)$$

This is, then, the temperature distribution for fully developed laminar pipe flow when the heat flux at the wall is uniform. It can be written in terms of the specified uniform wall heat flux, q_w, by noting that when Eq. (4.30) is used to give the value of $\partial T/\partial r|_{r=r_o}$ in Eq. (4.22), the following is obtained:

$$\frac{q_w r_o}{k} = \left(\frac{u_c r_o^2 Pr}{4}\right) \frac{dT_w}{dz} \quad (4.31)$$

Substituting this result back into Eq. (4.30) then allows the temperature distribution to be written as:

$$T_w - T = \left(\frac{q_w r_o}{k}\right)\left[\frac{3}{4} - \left(\frac{r}{r_o}\right)^2 + \frac{1}{4}\left(\frac{r}{r_o}\right)^4\right] \quad (4.32)$$

From this equation it follows that the center line temperature, T_c, is given by:

$$T_w - T_c = \frac{3}{4}\left(\frac{q_w r_o}{k}\right) \quad (4.33)$$

Therefore, the temperature distribution can be written in the form initially introduced for fully developed flow as follows:

$$G = \frac{T_w - T}{T_w - T_c} = 1 - \frac{4}{3}\left(\frac{r}{r_o}\right)^2 + \frac{1}{3}\left(\frac{r}{r_o}\right)^4 \tag{4.34}$$

The forms of the velocity and temperature distributions for fully developed flow through a pipe with a uniform heat flux at the wall as given by Eqs. (4.18) and (4.34) are shown in Fig. 4.4.

Now, in fully developed flow it is usually convenient to utilize the mean fluid temperature, T_m, rather than the center line temperature in defining the Nusselt number. This mean or bulk temperature is given, as explained in Chapter 1, by:

$$T_m = \frac{1}{\text{Mass Flow Rate}} \int_A \rho u T \, dA \tag{4.35}$$

where A is the cross-sectional area of the duct. Now:

$$\text{Mass Flow Rate} = \int_A \rho u \, dA$$

Hence, for flow in a pipe where dA is set equal to $2\pi r \, dr$:

$$\text{Mass Flow Rate} = \rho u_c r_o^2 \int_0^1 \left(\frac{u}{u_c}\right) 2\pi \left(\frac{r}{r_o}\right) d\left(\frac{r}{r_o}\right) \tag{4.36}$$

Using the velocity distribution given in Eq. (4.18) and integrating then gives:

$$\text{Mass Flow Rate} = \frac{\pi}{2} \rho u_c r_o^2 \tag{4.37}$$

This shows, incidentally, that the mean velocity in fully developed laminar pipe flow is half of the center line velocity, a well-known result.

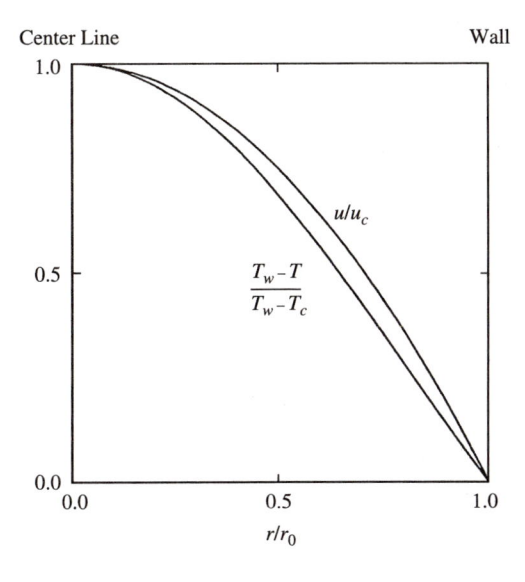

FIGURE 4.4
Velocity and temperature profiles in fully developed flow with a uniform wall heat flux.

164 Introduction to Convective Heat Transfer Analysis

Similarly, since:

$$\int_A \rho u T \, dA = \rho u_c r_o^2 \int_0^1 \left(\frac{u}{u_c}\right)[T_w - (T_w - T)] 2\pi \left(\frac{r}{r_o}\right) d\left(\frac{r}{r_o}\right) \quad (4.38)$$

it follows using the velocity and temperature profiles given in Eqs. (4.18) and (4.32) respectively that:

$$\int_A \rho u T \, dA = \frac{\pi}{2} \rho u_c r_o^2 \left[T_w - \frac{11}{24}\left(\frac{q_w r_o}{k}\right)\right] \quad (4.39)$$

Substituting Eqs. (4.37) and (4.39) into Eq. (4.35) then gives the mean temperature as:

$$T_m = T_w - \frac{11}{24}\left(\frac{q_w R}{k}\right) \quad (4.40)$$

This can be rearranged to give:

$$Nu_D = \frac{q_w D}{(T_w - T_m)k} = \frac{48}{11} = 4.363 \quad (4.41)$$

where Nu_D is the Nusselt number based on the difference between the wall temperature and the mean temperature, i.e., $(T_w - T_m)$ and on the pipe diameter, D. This Nusselt number is seen, therefore, to be constant in fully laminar flow through a circular pipe with a constant heat transfer rate at the wall.

EXAMPLE 4.1. In some situations it is possible to find the heat transfer rate with adequate accuracy by assuming that the velocity is constant across the duct, i.e., to assume that so-called "slug flow" exists. Find the temperature distribution and the Nusselt number in slug flow in a pipe when the thermal field is fully developed and when there is a uniform wall heat flux.

Solution. The temperature distribution will still be described by Eq. (4.26). In the flow here being considered, $u = u_m$ everywhere so this equation gives:

$$u_m \frac{dT_w}{dz} = \left(\frac{\nu}{Pr}\right)\frac{1}{r}\frac{\partial}{\partial r}\left[r\frac{\partial T}{\partial r}\right] \quad (i)$$

The boundary conditions on the solution are as above:

When $r = 0$: $\quad \partial T/\partial r = 0$
When $r = r_o$: $\quad T = T_w$

Integrating Eq. (*i*) once and using the first boundary condition gives:

$$\frac{\partial T}{\partial r} = \left(\frac{Pr}{\nu}\right) u_m \frac{dT_w}{dz} \frac{r}{2} \quad (ii)$$

Integrating this equation and applying the second boundary condition then gives:

$$T = T_w - \left(\frac{Pr}{\nu}\right) u_m \frac{r_o^2}{4} \frac{dT_w}{dz}\left[1 - \left(\frac{r}{r_o}\right)^2\right] \quad (iii)$$

This can be written in terms of the specified uniform wall heat flux, q_w, by again noting that because $q_w/k = \partial T/\partial r|_{r=r_o}$, Eq. (ii) gives:

$$\left(\frac{q_w r_o}{k}\right) = \left(\frac{u_m r_o^2 \, Pr}{2\nu}\right)\frac{dT_w}{dz}$$

Substituting this result into Eq. (iii) then gives:

$$T_w - T = \left(\frac{q_w r_o}{2k}\right)\left[1 - \left(\frac{r}{r_o}\right)^2\right]$$

From this equation it follows that the center line temperature, T_c, is given by:

$$T_w - T_c = \left(\frac{q_w r_o}{2k}\right)$$

Therefore, the temperature distribution can be written as:

$$\frac{T_w - T}{T_w - T_c} = 1 - \left(\frac{r}{r_o}\right)^2$$

The mean or bulk temperature is given:

$$T_m = \left(\frac{1}{\text{Mass Flow Rate}}\right)\int_A \rho u T \, dA$$

i.e., since, in the case here being considered, u is constant and equal to u_m:

$$T_m = \frac{1}{A}\int_A T \, dA$$

i.e.:

$$T_m = \int_0^1 2T\left(\frac{r}{r_o}\right) d\left(\frac{r}{r_o}\right)$$

i.e.:

$$T_m = T_w - \left(\frac{q_w r_o}{2k}\right)\int_0^1 2\left[\frac{r}{r_o} - \left(\frac{r}{r_o}\right)^3\right] d\left(\frac{r}{r_o}\right)$$

Integrating then gives:

$$T_m = T_w - \left(\frac{q_w r_o}{4k}\right)$$

This can be rearranged to give:

$$Nu_D = \frac{q_w D}{(T_w - T_m)k} = 8$$

where Nu_D is the Nusselt number based on the difference between the wall temperature and the mean temperature, i.e., $(T_w - T_m)$ and on the pipe diameter, D. This is higher than the mean Nusselt number that exists with a parabolic velocity profile because of the higher velocity near the wall with slug flow.

166 Introduction to Convective Heat Transfer Analysis

Next, consider fully developed laminar flow through a pipe whose wall temperature is kept constant. For this boundary condition, since $dT_w/dz = 0$, Eq. (4.21) gives:

$$\frac{\partial T}{\partial z} = \left(\frac{T_w - T}{T_w - T_c}\right)\frac{dT_c}{dz} \tag{4.42}$$

Substituting this result into Eq. (4.20) which has to be solved to give the temperature distribution and using the velocity distribution given in Eq. (4.18) which is, of course, independent of the temperature distribution and, hence, of the boundary conditions on temperature, then gives:

$$u_1\left[1 - \left(\frac{r}{r_o}\right)^2\right]\left(\frac{T_w - T}{T_w - T_c}\right)\frac{dT_c}{dz} = \left(\frac{\nu}{Pr}\right)\frac{1}{r}\frac{\partial}{\partial r}\left(r\frac{\partial T}{\partial r}\right) \tag{4.43}$$

Since T_w is constant, this equation can be written in terms of the temperature function, G, defined in Eq. (4.34) as:

$$\left[\frac{Pr\, r_o^2 u_1}{\nu(T_w - T_1)}\frac{dT_1}{dz}\right]G\left[1 - \left(\frac{r}{r_o}\right)^2\right]\left(\frac{r}{R}\right) = -\frac{d}{d(r/r_o)}\left[\left(\frac{r}{r_o}\right)\frac{dG}{d(r/r_o)}\right] \tag{4.44}$$

This equation must be solved to give the variation of G with r/r_o subject to the following boundary conditions:

$$\begin{aligned}\frac{r}{r_o} &= 0: \quad \frac{dG}{d(r/r_o)} = 0 \\ \frac{r}{r_o} &= 0: \quad G = 1, \qquad \frac{r}{r_o} = 1: \quad G = 0\end{aligned} \tag{4.45}$$

No simple closed form solution can be obtained. However, the variation of G with (r/r_o) can be quite easily obtained to any required degree of accuracy by using an iterative procedure. This starts with the temperature profile for the constant heat flux case. This profile is used to give a first approximation for the left-hand side of Eq. (4.44) and this equation can then be integrated to give a second approximation for G and so on until acceptable convergence is obtained. In this procedure, Eq. (4.44) is integrated once to give $dG/d(r/r_o)$ using the boundary condition

$$\left(\frac{r}{r_o}\right) = 0: \quad \frac{dG}{d(r/r_o)} = 0$$

This result is then integrated to give the variation of G with (r/r_o). The remaining boundary conditions can be used to determine the constant integration and to eliminate the unknown quantity:

$$\frac{Pr\, r_o^2 u_c}{\nu(T_w - T_c)}\frac{dT_c}{dz}$$

in Eq. (4.44).

Once the temperature profile function G is obtained, the Nusselt number can be determined because, using the results derived in dealing with the constant wall

CHAPTER 4: Internal Laminar Flows   167

$$(T_w - T_m) = 4(T_w - T_c) \int_0^1 G\left(\frac{u}{u_c}\right)\left(\frac{r}{r_o}\right) d\left(\frac{r}{r_o}\right) \quad (4.46)$$

The Nusselt number is therefore given using Eq. (4.23) by:

$$Nu_D = -\left.\frac{dG}{d(r/r_o)}\right|_{(r=r_o)} \bigg/ 4\int_0^1 G\left[1 - \left(\frac{r}{r_o}\right)^2\right]\left(\frac{r}{r_o}\right) d\left(\frac{r}{r_o}\right) \quad (4.47)$$

The above procedure can be undertaken using a series-type solution or it can be carried out numerically. A simple computer program PIPETEM that implements the procedure is available as discussed in the Preface. Alternatively, the procedure can be carried out using a spreadsheet.

If the solution procedure is carried through as outlined above, the following is obtained for fully developed laminar flow through a pipe with constant wall temperature:

$$Nu_D = 3.658 \quad (4.48)$$

This is some 16% lower than the Nusselt number for flow with a constant wall heat transfer rate.

EXAMPLE 4.2. A device is cooled by clamping an aluminum block to it and passing cooling air through small diameter holes in this block. The length of the block in the flow direction is fixed by practical consideration and the block can be assumed to be at a uniform temperature of T_w. The air enters the holes in the block at a known temperature of T_o and the total mass flow rate of air through the holes in the block, $\dot{M}$, is fixed. Investigate the effect of the number of holes on the heat transfer rate from the block to the air passing through it. Assume that the flow in the cooling holes is laminar and fully developed.

Solution. The flow situation being considered is shown in Fig. E4.2.

If n is the number of holes and D is the diameter of the holes then the mass flow rate through each channel is given by:

$$\dot{m} = \frac{\dot{M}}{n} \quad (i)$$

If T_m is the mean bulk temperature at any point in a flow channel then the rate of change of T_m with respect to distance z along the duct is given by applying an energy

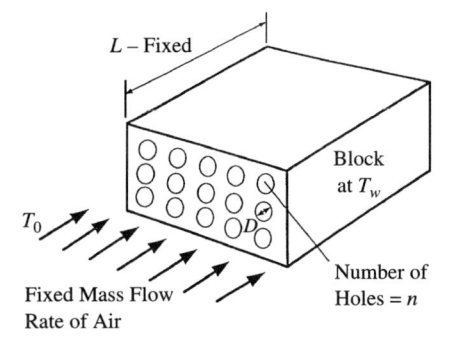

FIGURE E4.2

168 Introduction to Convective Heat Transfer Analysis

balance by:

$$\dot{m}c_p \frac{dT_m}{dz} = h\pi D(T_w - T_m) \qquad (ii)$$

c_p being the specific heat of the air and h the heat transfer coefficient.

Because the flow is being assumed to be laminar and fully developed with a uniform wall temperature, Eq. (4.48) gives:

$$Nu_D = \frac{hD}{k} = 3.658$$

i.e.:

$$h = 3.658 \frac{k}{D}$$

Substituting this into Eq. (*ii*) and using Eq. (*i*) gives on rearrangement:

$$\frac{1}{(T_w - T_m)} \frac{dT_m}{dz} = \frac{3.658 k\pi n}{\dot{M}c_p}$$

which can be written because T_w is constant as:

$$\frac{1}{(T_w - T_m)} \frac{d}{dz}(T_w - T_m) = -11.81 \frac{kn}{\dot{M}c_p}$$

At the inlet to the block $T_m = T_o$, so integrating the above equation gives, if T_{me} is the mean temperature at exit:

$$\frac{T_w - T_{me}}{T_w - T_o} = e^{-11.81(kn/\dot{M}c_p)L} \qquad (iii)$$

where L is the length of the block.

If the following is defined:

$$L^* = \frac{knL}{\dot{M}c_p}$$

then Eq. (*iii*) can be written as:

$$\frac{T_w - T_{me}}{T_w - T_o} = e^{-11.81 L^*} \qquad (iii)$$

Now the heat transfer rate to the air as it flows through the block is given by:

$$Q = \dot{M}c_p(T_{me} - T_o) = \dot{M}c_p[(T_w - T_o) - (T_w - T_{me})]$$

i.e., using Eq. (*iii*):

$$Q = \dot{M}c_p(T_w - T_o)[1 - e^{-11.81 L^*}] \qquad (iv)$$

If the block was infinitely long the air would be heated to the wall temperature, i.e., T_{me} would equal T_w. Therefore, the maximum possible heat transfer rate is:

$$Q_{max} = \dot{M}c_p(T_w - T_o)$$

Hence:

$$\frac{Q}{Q_{max}} = 1 - e^{-11.81 L^*}$$

This equation shows that as L^* tends to infinity, Q/Q_{max} tends to 1, and that the larger L^* the closer Q is to Q_{max}. Consider the definition of L^*, i.e.:

$$L^* = n \frac{kL}{\dot{M} c_p}$$

This shows that L^* is proportional to the number of holes, n. Therefore, a design that involves a large number of holes will produce the highest heat transfer rate.

Consider a design that will give:

$$Q = 0.95 Q_{max}$$

It will be seen from Eq. (*iv*) that this requires:

$$1 - e^{-11.81 L^*} = 0.95$$

i.e.:

$$L^* = 0.2537$$

Hence, using the definition of L^*:

$$n \frac{kL}{\dot{M} c_p} = 0.2537$$

i.e.:

$$n = 0.2537 \frac{\dot{M} c_p}{kL}$$

4.3
FULLY DEVELOPED LAMINAR FLOW IN A PLANE DUCT

Fully developed flow in a very wide duct is considered in this section, the flow situation considered being shown in Fig. 4.5. Because the duct is wide, changes in the flow properties in the x-direction are negligible, i.e., flow between two large parallel plates is effectively being considered. The analysis is, of course, similar to that adopted in dealing with pipe flow.

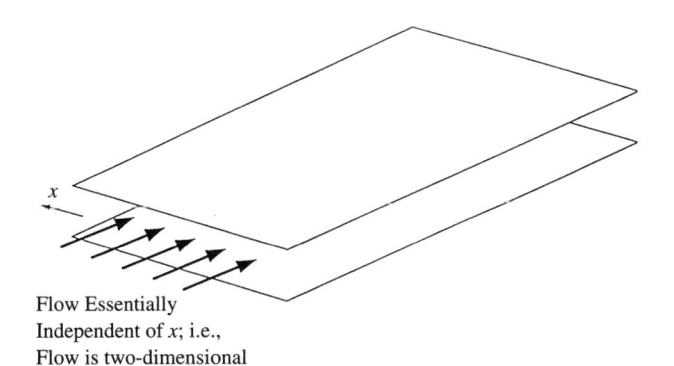

Flow Essentially Independent of x; i.e., Flow is two-dimensional

FIGURE 4.5
Flow in a plane duct.

170 Introduction to Convective Heat Transfer Analysis

The equations governing the flow are, since there is no velocity component in the x-direction:

$$\frac{\partial u}{\partial z} + \frac{\partial v}{\partial y} = 0 \tag{4.49}$$

$$u\frac{\partial u}{\partial z} + v\frac{\partial u}{\partial y} = -\frac{1}{\rho}\frac{\partial p}{\partial z} + \nu\left(\frac{\partial^2 u}{\partial z^2} + \frac{\partial^2 u}{\partial y^2}\right) \tag{4.50}$$

$$u\frac{\partial v}{\partial z} + v\frac{\partial v}{\partial y} = -\frac{1}{\rho}\frac{\partial p}{\partial y} + \nu\left(\frac{\partial^2 v}{\partial z^2} + \frac{\partial^2 v}{\partial y^2}\right) \tag{4.51}$$

$$u\frac{\partial T}{\partial z} + v\frac{\partial T}{\partial y} = \left(\frac{\nu}{Pr}\right)\left(\frac{\partial^2 T}{\partial z^2} + \frac{\partial^2 T}{\partial y^2}\right) \tag{4.52}$$

The flow is assumed to be symmetrical about the center plane so the boundary conditions on the solution are:

$$y = 0: \quad \frac{\partial u}{\partial y} = 0, \ v = 0, \ \frac{\partial T}{\partial y} = 0 \tag{4.53}$$

$$y = w: \quad u = 0, \ v = 0, \ T = T_w$$

Here, $w = W/2$ is the half width of the duct.

Because attention is being restricted to fully developed flow in this section, the forms of the velocity and temperature profiles do not change with distance along the duct, i.e.:

$$\frac{u}{u_c} = F\left(\frac{y}{w}\right), \quad \frac{T_w - T}{T_w - T_c} = G\left(\frac{y}{w}\right) \tag{4.54}$$

the functions F and G being independent of the distance along the duct, z. In Eq. (4.54), u_c is the velocity on the center line, T_c is the temperature on the center line, and T_w is, as before, the wall temperature.

Now the mass flow rate through the duct per unit width of the duct is given by:

$$\dot{m} = 2\rho\int_0^w u\,dy \tag{4.55}$$

i.e., using Eq. (4.54)

$$\dot{m} = u_c W \int_0^1 F\,d\left(\frac{y}{w}\right) \tag{4.56}$$

Because the mass flow rate remains constant along the duct as does, by assumption, the profile function F, Eq. (4.56) shows that:

$$u_c = \text{constant, i.e., } \frac{du_c}{dz} = 0 \tag{4.57}$$

From Eq. (4.54) it then follows that:

$$\frac{\partial u}{\partial z} = 0 \tag{4.58}$$

Substituting this into Eq. (4.49) then gives after integrating:

$$v = \text{constant} \tag{4.59}$$

But the boundary conditions give $v = 0$ when $y = 0$ so this equation therefore shows that:

$$v = 0 \tag{4.60}$$

Using this result in Eq. (4.51) then gives:

$$\frac{\partial p}{\partial y} = 0 \tag{4.61}$$

and Eq. (4.50) then finally reduces to:

$$\nu \frac{d^2 u}{dy^2} = \frac{1}{\rho}\frac{dp}{dz} \tag{4.62}$$

Integrating this equation using the boundary condition $du/dy = 0$ when $y = 0$ gives on rearrangement:

$$\nu \frac{du}{dy} = \frac{1}{\rho}\frac{dp}{dz} y \tag{4.63}$$

Integrating this equation using the boundary condition $u = 0$ when $y = w$ then gives:

$$\nu u = -\frac{1}{\rho}\frac{dp}{dz}\left[\frac{w^2}{2} - \frac{y^2}{2}\right] \tag{4.64}$$

This gives:

$$\nu u_c = -\frac{1}{\rho}\frac{dp}{dz}\frac{w^2}{2} \tag{4.65}$$

Dividing the above two equations then gives:

$$\frac{u}{u_c} = 1 - \left(\frac{y}{w}\right)^2 \tag{4.66}$$

Therefore, as was the case with fully developed pipe flow, the velocity profile in fully developed plane duct flow is parabolic.

Having established the form of the velocity profile, attention can now be turned to the solution of the energy equation, i.e., Eq. (4.52). Because v was shown to be equal to zero in fully developed flow, this equation reduces in this case to:

$$u\frac{\partial T}{\partial z} = \left(\frac{\nu}{Pr}\right)\left(\frac{\partial^2 T}{\partial z^2} + \frac{\partial^2 T}{\partial y^2}\right) \tag{4.67}$$

As was the case with pipe flow, for the flow of most fluids under practically significant conditions:

$$\frac{\partial^2 T}{\partial z^2} \ll \frac{\partial^2 T}{\partial y^2}$$

172 Introduction to Convective Heat Transfer Analysis

i.e., the first term on the right-hand side of Eq. (4.67) can be ignored compared to the second and the energy equation for fully developed duct flow can be assumed to be:

$$u \frac{\partial T}{\partial z} = \left(\frac{\nu}{Pr}\right) \frac{\partial^2 T}{\partial y^2} \quad (4.68)$$

This is the equation that must be solved to give the form of the temperature profile function G given in Eq. (4.54). Now, because G does not depend on z, it follows that, just as with fully developed pipe flow:

$$\frac{\partial}{\partial z}\left[\frac{T_w - T}{T_w - T_c}\right] = 0$$

i.e., that:

$$\frac{\partial T}{\partial z} = \frac{\partial T_w}{\partial z} - \left[\frac{T_w - T}{T_w - T_c}\right]\left[\frac{dT_w}{dz} - \frac{dT_c}{dz}\right] \quad (4.69)$$

To proceed further with the solution, the wall boundary condition on temperature must, as with pipe flow, be specified. Consideration will again first be given to the case where the heat transfer rate at the wall is uniform. Now the heat transfer rate at the wall is given using Fourier's Law by:

$$q_w = +k \left.\frac{\partial T}{\partial y}\right|_{y=w} \quad (4.70)$$

The positive sign arises because y is measured from the center line of the duct towards the wall.

Combining this equation with Eq. (4.54) then gives:

$$q_w = -k(T_w - T_c) \left.\frac{dG}{dy}\right|_{y=w} \quad (4.71)$$

Because q_w is a constant, this equation shows that $T_w - T_c = $ constant, i.e., that:

$$\frac{dT_w}{dz} = \frac{dT_c}{dz} \quad (4.72)$$

Substituting this result into Eq. (4.69) then shows that when the wall heat flux is constant:

$$\frac{\partial T}{\partial z} = \frac{dT_w}{dz} \quad (4.73)$$

Having established this result from the boundary condition, attention can now be returned to the governing equation, i.e., Eq. (4.68). Substituting Eq. (4.73) into this equation gives:

$$u\frac{dT_w}{dz} = \left(\frac{\nu}{Pr}\right)\frac{\partial^2 T}{\partial y^2} \quad (4.74)$$

CHAPTER 4: Internal Laminar Flows 173

If the velocity profile given in Eq. (4.66) is substituted into this equation the following is obtained on rearrangement:

$$\frac{\partial^2 T}{\partial y^2} = \left(\frac{Pr}{\nu}\right) u_1 \frac{dT_w}{dz}\left[1 - \left(\frac{y}{w}\right)^2\right] \quad (4.75)$$

This equation can be integrated, subject to the boundary condition $y = 0$, $\partial T/\partial y = 0$ to give:

$$\frac{\partial T}{\partial y} = \left(\frac{Pr}{\nu}\right) u_c \frac{dT_w}{dz}\left[\left(\frac{y}{w}\right) - \frac{1}{3}\left(\frac{y}{w}\right)^3\right] \quad (4.76)$$

Integrating this in turn subject to the boundary condition $y = w$, $T = T_w$ then gives:

$$T = T_w - \left[\left(\frac{Pr}{\nu}\right) u_c \frac{dT_w}{dz}\right]\left[\frac{5}{12} - \frac{1}{2}\left(\frac{y}{w}\right)^2 + \frac{1}{12}\left(\frac{y}{w}\right)^4\right] \quad (4.77)$$

This equation describes the temperature distribution in fully developed laminar plane duct flow when the wall heat flux is a constant. It can be written in terms of the specified wall heat flux, q_w, by noting that when Eq. (4.77) is used to give the value of $\partial T/\partial y|_{y=w}$ in Eq. (4.71), the following is obtained:

$$\left(\frac{q_w w}{k}\right) = \left(\frac{2u_c Pr}{3\nu}\right)\frac{dT_w}{dz} \quad (4.78)$$

Substituting this result back into Eq. (4.77) then allows the temperature distribution to be written as:

$$(T_w - T) = \left(\frac{q_w w}{k}\right)\left[\frac{5}{8} - \frac{3}{4}\left(\frac{y}{w}\right)^2 + \frac{1}{8}\left(\frac{y}{w}\right)^4\right] \quad (4.79)$$

From this equation it follows that the center line temperature, T_c, is given by:

$$T_w - T_c = \frac{5}{8}\left(\frac{q_w w}{k}\right) \quad (4.80)$$

The temperature distribution can therefore be written in the form initially introduced as follows:

$$G = \frac{T_w - T}{T_w - T_c} = 1 - \frac{6}{5}\left(\frac{y}{w}\right)^2 + \frac{1}{5}\left(\frac{y}{w}\right)^4 \quad (4.81)$$

Now, in duct flows, as previously discussed, it is usually convenient to utilize the mean fluid temperature, T_m, in defining the Nusselt number. This mean or bulk temperature is given as explained in Chapter 1 by:

$$T_m = \left(\frac{1}{\text{Mass Flow Rate}}\right)\int_A \rho u T \, dA \quad (4.82)$$

174 Introduction to Convective Heat Transfer Analysis

where A is the cross-sectional area of the duct. Now:

$$\text{Mass Flow Rate} = \int_A \rho u \, dA = 2w\rho u_c \int_0^1 \left(\frac{u}{u_c}\right) d\left(\frac{y}{w}\right) \tag{4.83}$$

Hence, using the velocity distribution given in Eq. (4.66) and integrating gives:

$$\text{Mass Flow Rate} = \frac{4w\rho u_c}{3} \tag{4.84}$$

Because the area of the duct per unit width is $2w$ and because, by definition, the mass flow rate is equal to the (density × area × mean velocity), this shows that the mean velocity in fully developed laminar plane duct flow is 6/4 (= 1.5) times the center line velocity.

Similarly because:

$$\int_A \rho u T \, dA = 2\rho u_c w \int_0^1 \left(\frac{u}{u_c}\right)[T_w - (T_w - T)]d\left(\frac{y}{w}\right) \tag{4.85}$$

it follows using the velocity and temperature profiles given in Eqs. (4.66) and (4.77) respectively and integrating that:

$$\int_A \rho u T \, dA = 2\rho u_c w \left[\frac{2}{3}T_w - \frac{1}{7}\left(\frac{q_w w}{k}\right)\right] \tag{4.86}$$

Substituting Eqs. (4.84) and (4.86) into Eq. (4.82) then gives the mean temperature as:

$$T_m = T_w - \frac{17}{35}\left(\frac{q_w w}{k}\right) \tag{4.87}$$

This can be rearranged to give:

$$Nu_W = \frac{q_w W}{(T_w - T_m)k} = \frac{70}{17} = 4.118 \tag{4.88}$$

where Nu_W is the Nusselt number based on the difference between the wall temperature and the mean temperature, i.e., on $(T_w - T_m)$ and on the duct width, W. This Nusselt number is seen, therefore, to be constant for fully developed laminar flow through a plane duct with a constant heat transfer rate at the wall.

EXAMPLE 4.3. Consider fully developed flow in a plane duct in which a uniform heat flux, q_w, is applied at one wall and where the other wall is unheated and heavily insulated. Derive expressions for the temperature distribution in the duct and the Nusselt number.

Solution. The flow situation being considered is shown in Fig. E4.3.
The heat flux at the unheated wall will be 0, i.e., at $y = +w$, $q = 0$. Here, $w = W/2$ is the duct half-width and y is measured from the center line of the duct.

FIGURE E4.3

CHAPTER 4: Internal Laminar Flows   175

The full duct has to be considered because the temperature distribution will not be symmetrical about the center line. The velocity distribution is not dependent on the temperature boundary conditions and is given by Eq. (4.66) as:

$$\frac{u}{u_c} = 1 - \left(\frac{y}{w}\right)^2$$

u_c being the center line velocity which is equal to 1.5 times the mean velocity, u_m.

It will be assumed that in the fully developed flow the temperature distribution has the form:

$$\frac{T_{w1} - T}{T_{w1} - T_{w2}} = G\left(\frac{y}{w}\right)$$

where T_{w1} and T_{w2} are the temperatures of the heated and unheated walls respectively which are at $y = -w$ and at $y = +w$, respectively. From this equation it follows that:

$$\frac{\partial}{\partial z}\left[\frac{T_{w1} - T}{T_{w1} - T_{w2}}\right] = 0$$

i.e., that:

$$\frac{\partial T}{\partial z} = \frac{\partial T_{w1}}{\partial z} - \left[\frac{T_{w1} - T}{T_{w1} - T_{w2}}\right]\left[\frac{dT_{w1}}{dz} - \frac{dT_{w2}}{dz}\right]$$

Now the heat transfer rate at the wall is given using Fourier's law by:

$$q_{w1} = -k \left.\frac{\partial T}{\partial y}\right|_{y=-w}$$

i.e., using the assumed form of temperature distribution:

$$q_w = k(T_{w1} - T_{w2}) \left.\frac{dG}{dy}\right|_{y=-w}$$

Because q_w is a constant, this equation shows that $T_{w1} - T_{w2}$ = constant. Consideration of the assumed form of the temperature distribution then shows that $T_{w1} - T$ is a function of y alone. From these results it follows that:

$$\frac{dT}{dz} = \frac{dT_{w1}}{dz} = \frac{dT_{w2}}{dz}$$

The equation governing the temperature distribution is:

$$u\frac{\partial T}{\partial z} = \left(\frac{\nu}{Pr}\right)\frac{\partial^2 T}{\partial y^2}$$

which can therefore be written as:

$$u\frac{dT_{w1}}{dz} = \left[\frac{\nu}{Pr}\right]\frac{\partial^2 T}{\partial y^2}$$

If the velocity profile is substituted into this equation the following is obtained on rearrangement:

$$\frac{\partial^2 T}{\partial y^2} = \left(\frac{Pr}{\nu}\right)u_c\frac{dT_{w1}}{dz}\left[1 - \left(\frac{y}{w}\right)^2\right]$$

176 Introduction to Convective Heat Transfer Analysis

Integrating this equation once gives:

$$\frac{\partial T}{\partial y} = \left(\frac{Pr}{\nu}\right) u_c \frac{dT_{w1}}{dz} \left[\left(\frac{y}{w}\right) - \frac{1}{3}\left(\frac{y}{w}\right)^3\right] + C_1$$

where C_1 is a constant of integration. Now, the solution must satisfy the boundary condition $y = w$, $\partial T/\partial y = 0$ which gives:

$$C_1 = -\frac{2}{3}\left(\frac{Pr}{\nu}\right) u_1 \frac{dT_{w1}}{dz}$$

hence:

$$\frac{\partial T}{\partial y} = \left(\frac{Pr}{\nu}\right) u_c \frac{dT_{w1}}{dz} \left[\left(\frac{y}{w}\right) - \frac{1}{3}\left(\frac{y}{w}\right)^3 - \frac{2}{3}\right]$$

Integrating this in turn, subject to the boundary condition $y = -w$, $T = T_{w1}$, then gives:

$$T_{w1} - T = \left[\left(\frac{Pr}{w\nu}\right) u_c \frac{dT_{w1}}{dz}\right]\left[\frac{13}{12} - \frac{1}{2}\left(\frac{y}{w}\right)^2 + \frac{1}{12}\left(\frac{y}{w}\right)^4 + \frac{2}{3}\left(\frac{y}{w}\right)\right]$$

Because at $y = +w$, $T = T_{w2}$ the above equation gives:

$$T_{w1} - T_{w2} = \left[\left(\frac{Pr}{w\nu}\right) u_c \frac{dT_w}{dz}\right] \frac{4}{3}$$

Dividing the above two results then gives:

$$\frac{T_{w1} - T}{T_{w1} - T_{w2}} = \left[\frac{13}{16} - \frac{3}{8}\left(\frac{y}{w}\right)^2 + \frac{1}{16}\left(\frac{y}{w}\right)^4 + \frac{1}{2}\left(\frac{y}{w}\right)\right]$$

This equation gives the temperature variation in terms of the wall temperatures. Now $q_w = -k\partial T/\partial y|_{y=-w}$ so using the equation for $\partial T/\partial y$ derived above gives:

$$\left(\frac{q_w w}{k}\right) = -\frac{4}{3}\left(\frac{Pr}{\nu}\right) u_c \frac{dT_{w1}}{dz}$$

Using this then allows the temperature distribution to be written as:

$$T_{w1} - T = \frac{3}{4}\left(\frac{q_w w}{k}\right)\left[\frac{13}{12} - \frac{1}{2}\left(\frac{y}{w}\right)^2 + \frac{1}{12}\left(\frac{y}{w}\right)^4 + \frac{2}{3}\left(\frac{y}{w}\right)\right]$$

This mean or bulk temperature is given by:

$$T_m = \left(\frac{1}{\text{Mass Flow Rate}}\right) \int_A \rho u T \, dA$$

But as shown above:

$$\text{Mass Flow Rate} = \frac{4w\rho u_1}{3}$$

Further:

$$\int_A \rho u T \, dA = \rho u_c w \int_{-1}^{+1} \left(\frac{u}{u_c}\right)[T_{w1} - (T_{w1} - T)] \, d\left(\frac{y}{w}\right)$$

so it follows that:

$$\int_A \rho u T \, dA = \rho u_c W \left[\frac{4}{3} T_{w1} - 0.9905 \left(\frac{q_w W}{k} \right) \right]$$

Hence:

$$T_m = T_{w1} - 1.346 \left(\frac{q_w W}{k} \right)$$

This can be rearranged to give:

$$N_W = \frac{q_w W}{(T_w - T_m) k} = 2.692$$

where N_W is the Nusselt number based on the difference between the wall temperature and the mean temperature, i.e., on $(T_w - T_m)$ and on the duct width, W. This Nusselt number is seen, therefore, also to be constant.

When dealing with noncircular ducts it is common to assume that the equations for the heat transfer rate for a circular pipe can be applied to the noncircular duct provided that the correct "equivalent diameter" is used for the noncircular duct. Now for a circular pipe:

$$\frac{\text{Area}}{\text{Perimeter}} = \frac{\pi D^2 / 4}{\pi D} = \frac{D}{4} \tag{4.89}$$

It is therefore usual to take as the "equivalent diameter" the following:

$$D_H = 4 \frac{\text{Area}}{\text{Wetted Perimeter}} \tag{4.90}$$

this being termed the "hydraulic diameter", the wetted perimeter being the perimeter in contact with the fluid. For a plane duct (see Fig. 4.6), the hydraulic diameter is given by:

$$D_H = 4 \frac{W \times L}{2 \times L} = 2W \tag{4.91}$$

where L is the arbitrary width of the duct considered as shown in Fig. 4.6.

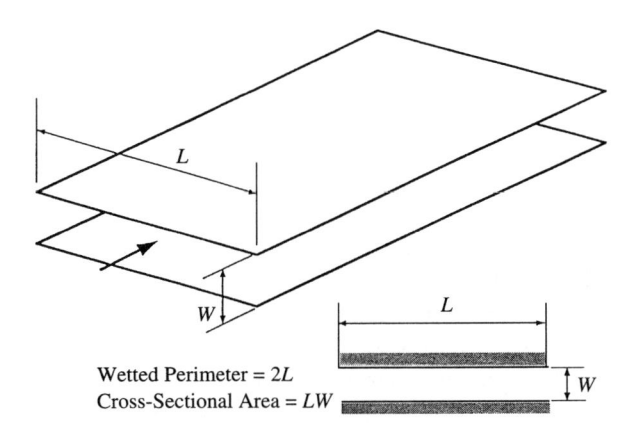

Wetted Perimeter = $2L$
Cross-Sectional Area = LW

FIGURE 4.6
Hydraulic diameter of a plane duct.

Now for a circular duct, i.e., a pipe with a uniform heat flux at the wall, the analysis discussed in the previous section gave $Nu_D = 4.364$. Therefore, for fully developed flow in a plane duct with a uniform heat flux at the wall, this would indicate using the hydraulic diameter concept, that:

$$Nu_W = Nu_D \frac{W}{D_H} = \frac{4.364}{2} = 2.182 \tag{4.92}$$

This is very different from the actual value of 4.118 derived above showing that the use of the hydraulic diameter concept can sometimes give very erroneous results for the heat transfer rate.

Next consider fully developed laminar flow through a plane duct whose wall temperature is kept constant. As with pipe flow, for this boundary condition, Eq. (4.69) gives:

$$\frac{\partial T}{\partial z} = \left(\frac{T_w - T}{T_w - T_c}\right) \frac{dT_c}{dz} \tag{4.93}$$

Substituting this result into Eq. (4.68), which must be solved to give the temperature distribution, and using the velocity distribution given in Eq. (4.66), which is, of course, independent of the temperature distribution, hence, of the boundary conditions on temperature, then gives:

$$u_c \left[1 - \left(\frac{y}{w}\right)^2\right] \left(\frac{T_w - T}{T_w - T_c}\right) \frac{dT_c}{dz} = \left(\frac{\nu}{Pr}\right) \frac{\partial^2 T}{\partial y^2} \tag{4.94}$$

Because T_w is here a constant, this equation can be written in terms of the temperature profile function, G, defined in Eq. (4.54) as:

$$\left[\left(\frac{Pr\, w^2 u_c}{\nu(T_w - T_c)}\right) \frac{dT_c}{dz}\right] G \left[1 - \left(\frac{y}{w}\right)^2\right] = -\frac{d^2 G}{d(y/w)^2} \tag{4.95}$$

This equation must be solved to give the variation of G with y/w subject to the following boundary conditions:

$$\begin{array}{ll} \left(\dfrac{y}{w}\right) = 0: & \dfrac{dG}{d(y/w)} = 0 \\[6pt] \left(\dfrac{y}{w}\right) = 0: \quad G = 1, & \left(\dfrac{y}{w}\right) = 1: \quad G = 0 \end{array} \tag{4.96}$$

No simple closed form solution to Eq. (4.95) subject to these boundary conditions can be obtained. However, just as with pipe flow, the variation of G with (y/w) can be quite easily obtained to any required degree of accuracy by using an iterative procedure. This procedure starts with the temperature profile for the uniform heat flux case. This profile is used to give a first approximation for the left-hand side of Eq. (4.95) and this equation can then be integrated to give a second approximation for G and so on until acceptable convergence is obtained. In this procedure, Eq. (4.95) is integrated once to give $dG/d(y/w)$ using the boundary condition $(y/w) = 0 : dG/d(y/w) = 0$. The resulting equation is then integrated to give the variation of G with (y/w) using the remaining boundary conditions to determine the constant of integration and to eliminate the unknown quantity:

$$\left[\left(\frac{Pr\,w^2 u_c}{\nu(T_w - T_c)}\right)\frac{dT_c}{dz}\right]$$

in Eq. (4.95).

Once the temperature profile function G is obtained, the Nusselt number can be determined since, using the results derived in dealing with the constant wall heat flux case, it can be easily shown that:

$$(T_w - T_m) = \frac{3}{2}(T_w - T_c)\int_0^1 G\left(\frac{u}{u_c}\right)d\left(\frac{y}{w}\right)$$

The Nusselt number is, therefore, given by:

$$Nu_W = \frac{-\frac{4}{3}\frac{dG}{d(y/w)}\bigg|_{(y/w)=1}}{\int_0^1 G\left[1 - \left(\frac{y}{w}\right)^2\right]d\left(\frac{y}{w}\right)} \quad (4.97)$$

The above procedure can be undertaken using a series type solution or it can be carried out numerically. A simple computer program that implements the procedure is available as discussed in the Preface. Alternatively, the procedure can be carried out using a spreadsheet.

If the solution procedure is carried through as outlined above, the following is obtained for fully developed laminar flow through a plane duct with uniform wall temperature:

$$Nu_D = 3.771 \quad (4.98)$$

This is some 8% lower than the Nusselt number for flow with a uniform wall heat transfer rate.

4.4
FULLY DEVELOPED LAMINAR FLOW IN DUCTS WITH OTHER CROSS-SECTIONAL SHAPES

In fully developed flow in a pipe and in a plane duct, as discussed above, the velocity and temperature profiles could be expressed in terms of a single cross stream coordinate, i.e., in terms of either r/R or y/W. In many other situations, however, the cross-sectional shape of the duct is such that the profiles will depend on two cross stream coordinates, e.g., consider fully developed flows in the ducts with the cross-sectional shapes shown in Fig. 4.7.

To illustrate how such situations can be analyzed, consider fully developed flow in a duct with a rectangular cross-sectional shape, i.e., with a duct that has the shape shown in Fig. 4.8 [2],[7],[8],[9].

Because the flow is fully developed, the velocity components in the x and y coordinate directions will be zero and the pressure will depend only on z. If the

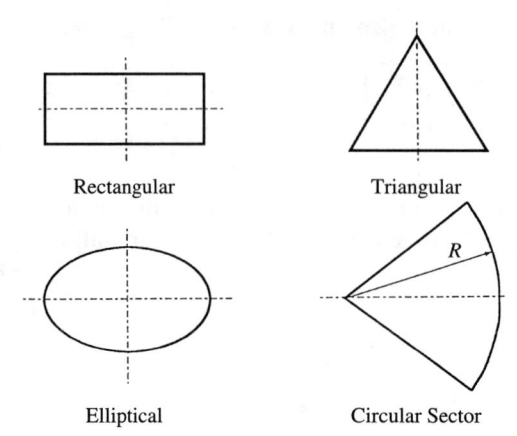

FIGURE 4.7
Ducts in which the fully developed flow is two-dimensional.

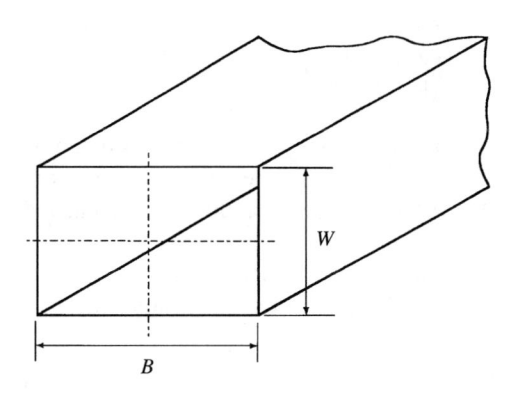

FIGURE 4.8
Duct with rectangular cross-sectional shape.

same assumptions as used in dealing with pipe flow and with plane duct flow are adopted, i.e., if the diffusion of heat in the z-direction is neglected compared to the rates in x and y directions, the governing equations are:

$$\left(\frac{\partial^2 u}{\partial x^2} + \frac{\partial^2 u}{\partial y^2}\right) = \frac{1}{\mu}\frac{dp}{dz} \tag{4.99}$$

$$u\frac{\partial T}{\partial z} = \left(\frac{\nu}{Pr}\right)\left(\frac{\partial^2 T}{\partial x^2} + \frac{\partial^2 T}{\partial y^2}\right) \tag{4.100}$$

Because the flow is fully developed:

$$\frac{u}{u_c} = F\left(\frac{x}{W}, \frac{y}{W}\right), \qquad \frac{T_w - T}{T_w - T_c} = G\left(\frac{x}{W}, \frac{y}{W}\right) \tag{4.101}$$

the functions F and G being independent of the distance along the duct, z. In these expressions u_c is the velocity on the center line, T_c is the temperature on the center line, T_w is, as before, the wall temperature, and W is the representative duct cross-sectional size. Because the flow is fully developed, the center line velocity, u_c, is not changing.

Attention will first be given to the velocity profile. In terms of the profile function, F, the momentum equation, i.e., Eq. (4.99), becomes:

$$\left(\frac{\partial^2 F}{\partial X^2} + \frac{\partial^2 F}{\partial Y^2}\right) = \left[\frac{W^2}{u_c \mu}\frac{dp}{dz}\right] \quad (4.102)$$

where:

$$X = \frac{x}{W}, \quad Y = \frac{y}{W} \quad (4.103)$$

Because the solution will be symmetrical about the vertical and horizontal center lines, the solution in only a quarter of the duct need be considered, this being shown in Fig. 4.9.

The boundary conditions on the solution are then:

$$X = 0: \quad \frac{\partial F}{\partial X} = 0, \quad Y = 0: \quad \frac{\partial F}{\partial Y} = 0 \quad (4.104)$$

$$X = 0.5A: \quad F = 0, \quad Y = 0.5: \quad F = 0$$

Here, $A = W/B$ is the aspect ratio of the duct cross-section.

In addition, because of the way in which F is defined, the following applies:

$$X = 0, \quad Y = 0: \quad F = 1 \quad (4.105)$$

There are a number of analytical and numerical approaches that can be used to solve Eq. (4.102) subject to these boundary conditions. A very simple numerical approach will be discussed here. Because it is a linear equation in F, the solution is linearly dependent on the value of the right-hand side of the equation, i.e., if F_1 is a solution with a particular value of the right-hand side, then if the value of the right-hand side is C times as large, the solution will be $F_2 = CF_1$. Hence, the solution to Eq. (4.102) can be obtained basically by setting the right-hand side equal to an arbitrary value, here taken to be -1 (a negative value being used because the pressure gradient is negative so the right-hand side of Eq. (4.102) will be negative) and obtaining the solution to the equation subject to the boundary conditions given in

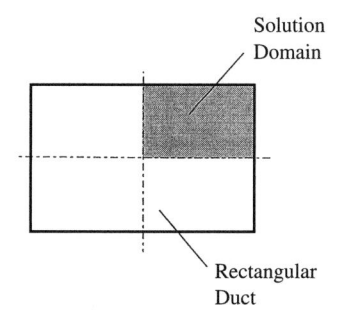

FIGURE 4.9
Solution domain being considered.

Eq. (4.104), i.e., by solving the following equation subject to the boundary conditions given in Eq. (4.104):

$$\left(\frac{\partial^2 F}{\partial X^2} + \frac{\partial^2 F}{\partial Y^2}\right) = -1 \qquad (4.106)$$

This solution will be denoted by F_1. This solution will not, in general, satisfy the boundary condition given in Eq. (4.105), giving instead a value of F on the center line, say F_c, that is not equal to 1. The correct solution is then given by F_1/F_c and the correct value of the right-hand side that gives a solution that satisfies Eq. (4.105) is given by:

$$-\left[\frac{W^2}{u_1 \mu}\frac{dp}{dz}\right] = \frac{1}{F_c} \qquad (4.107)$$

which can be written as:

$$-\left[\frac{W}{\rho u_1^2}\frac{dp}{dz}\right] = \frac{1/F_c}{Re_c} \qquad (4.108)$$

where $Re_c = \rho u_c W/\mu$.

The solution to Eq. (4.106) will here be obtained using a finite difference method. The grid system shown in Fig. 4.10 is introduced for this purpose.

Using the values of F at the five nodal points shown in Fig. 4.10, the following finite-difference form of Eq. (4.106) is obtained using the finite-difference approximations introduced in the previous chapter in the discussion of the numerical solution of external flows:

$$\left(\frac{F_{i+1,j} + F_{i-1,j} - 2F_{i,j}}{\Delta X^2}\right) + \left(\frac{F_{i,j+1} + F_{i,j-1} - 2F_{i,j}}{\Delta Y^2}\right) = -1 \qquad (4.109)$$

This equation applies to the internal points, i.e., for $i = 2$ to $NX - 1$ and for $j = 2$ to $NY - 1$. On the boundary points, the following are given by the boundary

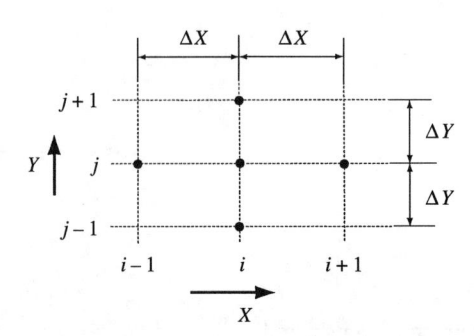

FIGURE 4.10
Nodal points used in obtaining finite-difference solution.

conditions:

$$\begin{aligned}
&\text{For } i = 1 \text{ to } NX: \quad F_{i,1} = F_{i,2}\\
&\text{For } j = 1 \text{ to } NY: \quad F_{1,j} = F_{2,j}\\
&\text{For } i = 1 \text{ to } NX: \quad F_{i,NY} = 0\\
&\text{For } j = 1 \text{ to } NY: \quad F_{NX,j} = 0
\end{aligned} \quad (4.110)$$

Eq. (4.109) subject to the boundary conditions given in Eq. (4.110) can be solved iteratively. The values of $F_{i,j}$ are first approximated. For example, using the plane duct result, the initial values could be taken as:

$$F_{i,j} = (1 - X^2)(1 - Y^2)$$

Eq. (4.109) can be then be applied sequentially to give updated values at the internal points and the boundary conditions can be used to update the boundary point values. In obtaining the solution, Eq. (4.109) can be written as:

$$F_{i,j} = \left(\frac{F_{i+1,j} + F_{i-1,j}}{\Delta X^2}\right) + \left(\frac{F_{i,j+1} + F_{i,j-1}}{\Delta Y^2} + 1\right) \Big/ \left(\frac{2}{\Delta X^2} + \frac{2}{\Delta Y^2}\right) \quad (4.111)$$

This is applied sequentially over and over to all points until convergence is obtained to any specified degree of accuracy. This gives the values of F_1. As discussed above, the actual solution is then given by:

$$F_{i,j} = \frac{F_{i,j}}{F_{1,1}}$$

Consideration will next be given to the solution for the temperature function, G. As with fully developed pipe and plane duct flows, the solution depends on the nature of the thermal boundary conditions at the wall. In the case of flow in a rectangular duct there are a variety of possible boundary conditions, some of these being shown in Fig. 4.11. Here, attention will be restricted to the case where the wall

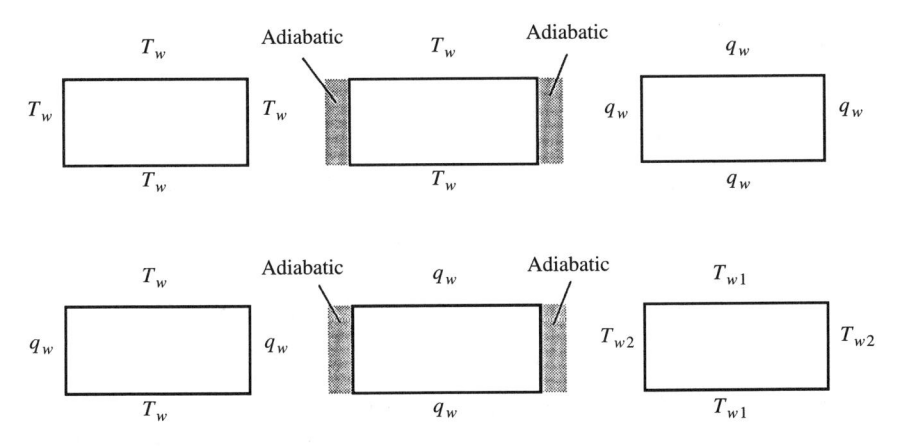

FIGURE 4.11
Some possible wall boundary conditions in a duct with rectangular cross-sectional shape.

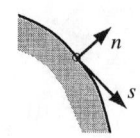

FIGURE 4.12
Definitions of n and s in Eq. (4.112).

temperature at any section of the duct is uniform but in which the total heat flux from the wall around the section of the duct is not changing with z. Now, with this boundary condition:

$$\int_{\text{Perimeter}} k \frac{\partial T}{\partial n} ds = q_w \times \text{length of perimeter} = \text{constant} \quad (4.112)$$

where n is the coordinate normal to the surface considered and s is the distance along the surface considered as shown in Fig. 4.12.

In terms of the temperature function G, and defining:

$$N = \frac{n}{W}, \quad S = \frac{s}{W} \quad (4.113)$$

and recalling that k is being assumed constant, Eq. (4.112) becomes:

$$(T_W - T_1) \int_P \frac{\partial G}{\partial N} dS = \text{constant} \quad (4.114)$$

P being the actual perimeter divided by W.

From Eq. (4.114) it follows that:

$$(T_W - T_1) = \text{constant} \quad (4.115)$$

Hence, using the definition of G, it follows that:

$$\frac{\partial T}{\partial z} = \frac{dT_w}{dz} = \frac{dT_c}{dz} \quad (4.116)$$

From this, it follows from Eq. (4.102), that the equation governing the temperature function is:

$$\left(\frac{\partial^2 G}{\partial X^2} + \frac{\partial^2 G}{\partial Y^2} \right) = -\left[\frac{W^2 u_c \, Pr}{\nu (T_w - T_c)} \frac{dT_w}{dz} \right] F \quad (4.117)$$

The boundary conditions on the solution are then:

$$X = 0: \quad \frac{\partial G}{\partial X} = 0, \quad Y = 0: \quad \frac{\partial G}{\partial Y} = 0$$
$$X = 0.5A: \quad G = 0, \quad Y = 0.5: \quad G = 0 \quad (4.118)$$

Here, again, $A = W/B$ is the aspect ratio of the duct cross-section.

In addition, because of the way in which G is defined, the following applies:

$$X = 0, \quad Y = 0: \quad G = 1 \quad (4.119)$$

The solution will, here, be obtained using the same approach as used in obtaining the solution for the velocity profile function, F. Because, when solving Eq. (4.117), F is a known quantity, Eq. (4.117) is a linear equation in G. Hence, the solution is linearly dependent on the right-hand side of the equation, i.e., if G_1 is a solution with a particular value of the right-hand side, then if the value of the right-hand side is C times as large, the solution will be $G_2 = CG_1$. Therefore, the solution to Eq. (4.117) can be obtained basically by setting the right-hand side equal to an arbitrary value, here taken to be F and obtaining the solution to the equation subject to the boundary conditions given in Eq. (4.118), i.e., by solving the following equation subject to the boundary conditions given in Eq. (4.118):

$$\left(\frac{\partial^2 G}{\partial X^2} + \frac{\partial^2 G}{\partial Y^2} \right) = -F \tag{4.120}$$

This solution will be denoted by G_1. This solution will not, in general, satisfy the boundary condition given in Eq. (4.119) giving instead a value of G on the center line, say G_c, that is not equal to 1. The correct solution is then given by G_1/G_c. The solution to Eq. (4.120) will here be obtained using a finite difference method.

Once the solution for G is obtained in this way, the heat transfer rate can be obtained by noting that, at any point on the wall, Fourier's law gives:

$$q_w = +k \left. \frac{\partial T}{\partial n} \right|_w \tag{4.121}$$

which can be written as:

$$\frac{q_w W}{(T_w - T_c)k} = + \left. \frac{\partial G}{\partial N} \right|_w \tag{4.122}$$

The mean heat transfer rate is then given by:

$$\frac{\overline{q_w} W}{(T_w - T_c)k} = \int_{\text{Perimeter}} \left. \frac{\partial G}{\partial N} \right|_w dS \tag{4.123}$$

This can be derived from the numerically determined variation of G.

The Nusselt number based on $(T_w - T_m)$ can then be obtained by recalling that:

$$T_m = \left(\frac{1}{\text{Mass Flow Rate}} \right) \int_A \rho u T \, dA \tag{4.124}$$

where A is the cross-sectional area of the duct. Now:

$$\text{Mass Flow Rate} = \int_A \rho u \, dA = \rho u_c \int_A F \, dA \tag{4.125}$$

Similarly, since:

$$\int_A \rho u T \, dA = \rho u_c \int_A F[T_w - (T_w - T)] \, dA \tag{4.126}$$

186 Introduction to Convective Heat Transfer Analysis

Hence:

$$T_w - T_m = (T_w - T_c)\frac{\int_A FG\,dA}{\int_A F\,dA} \qquad (4.127)$$

This equation can be used to determine $(T_w - T_m)/(T_w - T_c)$ by numerical integration of the numerical results. This result can then be combined with that given by Eq. (4.123) to give

$$Nu = \frac{\overline{q_w}W}{(T_w - T_m)k} \qquad (4.128)$$

The numerical solution to Eq. (4.120) is obtained using the same basic procedure as used in obtaining the solution for the velocity profile function, F. The grid system shown in Fig. 4.10 is again used. Using the values of G at the five nodal points indicated in Fig. 4.10, the following finite-difference form of Eq. (4.120) is obtained:

$$\left(\frac{G_{i+1,j} + G_{i-1,j} - 2G_{i,j}}{\Delta X^2}\right) + \left(\frac{G_{i,j+1} + G_{i,j-1} - 2G_{i,j}}{\Delta Y^2}\right) = -F_{i,j} \qquad (4.129)$$

This equation applies to the internal points, i.e., for $i = 2$ to $NX - 1$ and for $j = 2$ to $NY - 1$. On the boundary points, the following are given by the boundary conditions:

$$\begin{aligned}
&\text{For } i = 1 \text{ to } NX: && G_{i,1} = G_{i,2} \\
&\text{For } j = 1 \text{ to } NY: && G_{1,j} = G_{2,j} \\
&\text{For } i = 1 \text{ to } NX: && G_{i,NY} = 0 \\
&\text{For } j = 1 \text{ to } NY: && G_{NX,j} = 0
\end{aligned} \qquad (4.130)$$

Eq. (4.129), subject to the boundary conditions given in Eq. (4.130), can be solved iteratively. The values of $G_{i,j}$ are first approximated. For example, since the velocity profile function is determined prior to obtaining the solution for G, the following can be used to start the solution:

$$G_{i,j} = F_{i,j}$$

Eq. (4.129) can then be applied sequentially to give updated values at the internal points and the boundary conditions can be used to update the boundary point values. In doing this, Eq. (4.109) can be written as:

$$G_{i,j} = \left(\frac{G_{i+1,j} + G_{i-1,j}}{\Delta X^2}\right) + \left(\frac{G_{i,j+1} + G_{i,j-1}}{\Delta Y^2}\right) + F_{i,j}\right) \bigg/ \left(\frac{2}{\Delta X^2} + \frac{2}{\Delta Y^2}\right) \qquad (4.131)$$

This is applied sequentially over and over to all points until convergence is obtained to any specified degree of accuracy. This gives the values of G_1. As discussed above, the actual solution is then given by:

$$G_{i,j} = \frac{G_{i,j}}{G_{1,1}} \tag{4.132}$$

A simple computer program, RECDUCT, written in FORTRAN that implements the procedure outlined above is available as discussed in the Preface. Over-relaxation is used in this program, i.e., if $F_{i,j}^n$ is the new value of the velocity function given by Eq. (4.111), the actual new value is taken to be:

$$F_{i,j} = F_{i,j} + r(F_{i,j}^n - F_{i,j}) \tag{4.133}$$

where r is the relaxation factor, here taken to be greater than 1. Over-relaxation is also used in finding $G_{i,j}$, i.e., if $F_{i,j}^n$ is the new value of the temperature function given by Eq. (4.131), the actual new value is taken to be:

$$G_{i,j} = G_{i,j} + r(G_{i,j}^n - G_{i,j}) \tag{4.134}$$

The use of over-relaxation increases the rate of convergence.

The variation of the mean Nusselt number with aspect ratio as given by this program is shown in Fig. 4.13. This Nusselt number is based on the duct height, W.

Now, for a rectangular duct, the hydraulic diameter is given by:

$$D_H = 4\frac{\text{Area}}{\text{Wetted Perimeter}} = 4\frac{WB}{2B + 2W} = \frac{4W}{2 + 2/A} \tag{4.135}$$

Using this, the Nusselt number based on the hydraulic diameter can be found and the variation of this with aspect ratio, A, is also shown in Fig. 4.13. It will again be seen

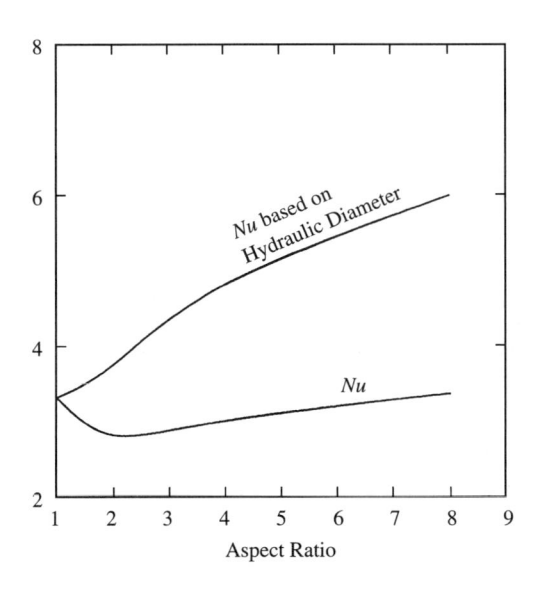

FIGURE 4.13
Variation of Nusselt number with aspect ratio for a rectangular duct.

TABLE 4.1
Effect of wall thermal boundary condition on the Nusselt number for fully developed flow in a rectangular duct (Nusselt number based on hydraulic diameter)

Aspect ratio	Nu–uniform peripheral temperature, uniform axial heat flux	Nu–uniform peripheral heat flux, uniform axial heat flux	Nu–uniform peripheral and axial temperature
1	3.608	3.091	2.976
2	4.123	3.017	3.391
4	5.331	2.940	4.439
6	6.049	2.930	5.137
8	6.490	2.904	5.597
Plane duct	8.235	8.235	7.541

that the use of the hydraulic diameter as the length scale does not give a constant Nusselt number, this again indicating the inadequecy of this concept for predicting the heat transfer rate.

The above results were for a rectangular duct with a wall temperature that is uniform around any section of the duct but in which the total heat flux from the wall around the section of the duct is not changing with distance, z, along the duct. A few results for ducts with other thermal boundary conditions [2],[10] are given in Table 4.1.

Solutions for fully developed flow in ducts with various other cross-sectional shapes have been obtained using techniques similar to those outlined above [2],[10]. The results given by some of these solutions are presented in Table 4.2.

TABLE 4.2
Nusselt numbers for fully developed flow in ducts with various cross-sectional shapes (Nusselt number based on hydraulic diameter)

Cross-sectional shape of duct	Nu–uniform peripheral temperature, uniform axial heat flux	Nu–uniform peripheral heat flux, uniform axial heat flux	Nu–uniform peripheral and axial temperature
Equilateral triangle	3.111	1.892	2.470
Hexagonal	4.002	3.862	3.340
Circular	4.364	4.364	3.657
Elliptical-eccentricity = 0.9	5.099	4.350	3.660
Plane duct	8.235	8.235	7.541

4.5
PIPE FLOW WITH A DEVELOPING TEMPERATURE FIELD

The preceding sections were concerned with fully developed flow in which the forms of both the velocity profile and the temperature profile were not changing with distance along the duct. Fully developed flow is, however, only attained in the flow, well downstream of the entrance to the duct or a bend or other fitting in the duct or of a region over which there is a change in the conditions at the duct wall. The region in which the flow is developing, i.e., moving towards the fully developed state, is termed the "entrance" region as discussed earlier. In general, both the velocity and temperature fields are simultaneously developing, i.e., both are changing with distance along the duct. However, in some cases, there is a long section of duct in which there is no heat transfer prior to the duct section in which heat transfer takes place. In many such cases, the velocity profile is then essentially fully developed before the heat transfer occurs and it is then only the temperature that is developing, i.e., there is only a "thermal entrance region" [1],[2],[11]. This is illustrated in Fig. 4.14.

The situation considered in this section is traditionally termed the "Graetz" problem.

Because there is no heat transfer in the initial portion of the duct flow, the fluid will have a uniform temperature, T_e at the point at which heat transfer starts, i.e.:

$$\text{At } z = 0: \quad T = T_e \tag{4.136}$$

Attention will in this section be given to thermally developing flow in a pipe. In this case, the velocity profile, which is not changing with z, is given, as discussed in Section 4.2, by:

$$u = u_c \left[1 - \left(\frac{r}{R} \right)^2 \right]$$

where u_c is, as before, the velocity on the center line of the pipe. This profile, it must be stressed, exists everywhere in the thermal entrance region.

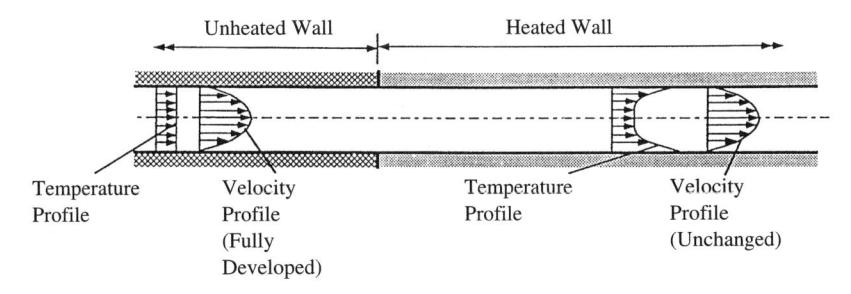

FIGURE 4.14
Thermal entrance region.

The temperature profile is changing with distance z along the duct, the temperature being assumed to be governed by:

$$u\frac{\partial T}{\partial z} = \left(\frac{\nu}{Pr}\right)\left(\frac{\partial^2 T}{\partial z^2} + \frac{\partial^2 T}{\partial r^2} + \frac{1}{r}\frac{\partial T}{\partial r}\right) \quad (4.137)$$

the velocity component in the radial direction, v, being zero because the velocity field is fully developed. The diffusion of heat in the axial direction will again be neglected compared to that in the radial direction, i.e., the equation governing the temperature is assumed to have the form:

$$u\frac{\partial T}{\partial z} = \left(\frac{\nu}{Pr}\right)\left(\frac{\partial^2 T}{\partial r^2} + \frac{1}{r}\frac{\partial T}{\partial r}\right) \quad (4.138)$$

i.e.:

$$u\frac{\partial T}{\partial z} = \left(\frac{\nu}{Pr}\right)\left[\frac{1}{r}\frac{\partial}{\partial r}\left(r\frac{\partial T}{\partial r}\right)\right] \quad (4.139)$$

In order to illustrate the type of solution obtained in the thermal entrance region, attention will be given first to the case where the wall temperature of the pipe is kept constant in this thermal entrance region, i.e., to the case where:

$$\text{For } z > 0: \quad T = T_w \text{ at } r = r_o \quad (4.140)$$

It is then convenient to introduce the following dimensionless variables:

$$U = \frac{u}{u_m}, \quad \theta = \frac{(T_w - T)}{(T_w - T_e)}$$
$$R = \frac{r}{D}, \quad Z = \frac{z/D}{Re\, Pr} \quad (4.141)$$

where u_m is the mean velocity in the pipe and Re is the Reynolds number based on this mean velocity and the diameter of the pipe, D, which is equal to $2r_o$. The dimensionless variable Z is termed the Graetz number.

In terms of these variables, the equation governing the developing temperature field is:

$$U\frac{\partial \theta}{\partial Z} = \frac{1}{R}\frac{\partial}{\partial R}\left(R\frac{\partial \theta}{\partial R}\right) \quad (4.142)$$

It should be noted that if the second term on the right-hand side of the energy equation, i.e., $\partial^2 T/\partial z^2$, had been retained, the right-hand side of the above equation would have been:

$$\frac{1}{R}\frac{\partial}{\partial R}\left(R\frac{\partial \theta}{\partial R}\right) + \frac{1}{(Re\, Pr)^2}\frac{\partial^2 \theta}{\partial Z^2} \quad (4.143)$$

This shows that the second term, i.e., the longitudinal diffusion term, is negligible compared to the first term, i.e., the radial diffusion term, if the quantity $Re\,Pr$ is large. The quantity $Re\,Pr$ is termed the Peclet number, Pe. Generally, if the Peclet number is greater than 10, the effects of longitudinal diffusion are, indeed, negligible.

Now the velocity profile for fully developed pipe flow discussed above can be written in terms of the dimensionless variables as:

$$U = 2\left[1 - \left(\frac{R}{0.5}\right)^2\right]$$

Therefore, Eq. (4.142) becomes:

$$2\left[1 - \left(\frac{R}{0.5}\right)^2\right]\frac{\partial \theta}{\partial Z} = \frac{1}{R}\frac{\partial}{\partial R}\left(R\frac{\partial \theta}{\partial R}\right) \tag{4.144}$$

This equation must be solved subject to the following boundary conditions:

$$R = 0: \quad \frac{\partial \theta}{\partial R} = 0, \qquad R = 0.5: \quad \theta = 0.0 \tag{4.145}$$

while the initial conditions become in terms of the dimensionless variables:

$$Z = 0: \quad \theta = 1 \tag{4.146}$$

The solution to Eq. (4.144) has traditionally been obtained by using the method of separation of variables. The solution is assumed to have the form:

$$\theta = F(R)G(Z) \tag{4.147}$$

where, as indicated, G is a function of Z alone and F is a function of R alone. Substituting this into Eq. (4.144) then gives:

$$2\left[1 - \left(\frac{R}{0.5}\right)^2\right]F\frac{dG}{dz} = \frac{G}{R}\frac{d}{dR}\left(R\frac{dF}{dR}\right)$$

i.e.:

$$\frac{1}{G}\frac{dG}{dz} = \frac{1}{2RF\left[1 - \left(\frac{R}{0.5}\right)^2\right]}\left[\frac{dF}{dR} + R\frac{d^2F}{dR^2}\right] \tag{4.148}$$

Because G is a function of Z alone and F is a function of R alone, this requires that:

$$\frac{1}{G}\frac{dG}{dz} = -\lambda^2 \tag{4.149}$$

and:

$$\frac{1}{2RF\left[1-\left(\frac{R}{0.5}\right)^2\right]}\left[\frac{dF}{dR} + R\frac{d^2F}{dR^2}\right] = -\lambda^2$$

i.e.:

$$R\frac{d^2F}{dR^2} + \frac{dF}{dR} + 2\lambda^2[R - 4R^3]F = 0 \quad (4.150)$$

The solution to the first of these equations, i.e., Eq. (4.149), is:

$$G = Ce^{-\lambda^2 Z} \quad (4.151)$$

where C is a constant of integration. When the form of the second equation is considered, it is found that there are an infinite number of C and λ values and that associated with each of these values is a solution for F. These are denoted by C_n, λ_n, and $F_n(R)$.

The values of F_n are given by the second equation, i.e., Eq. (4.150), which can be written as:

$$R\frac{d^2F_n}{dR^2} + \frac{dF_n}{dR} + 2\lambda_n^2[R - 4R^3]F_n = 0 \quad (4.152)$$

The solution of this equation must be such that when $R = 0$, F_n remains finite and when $R = 1$ $F_n = 0$.

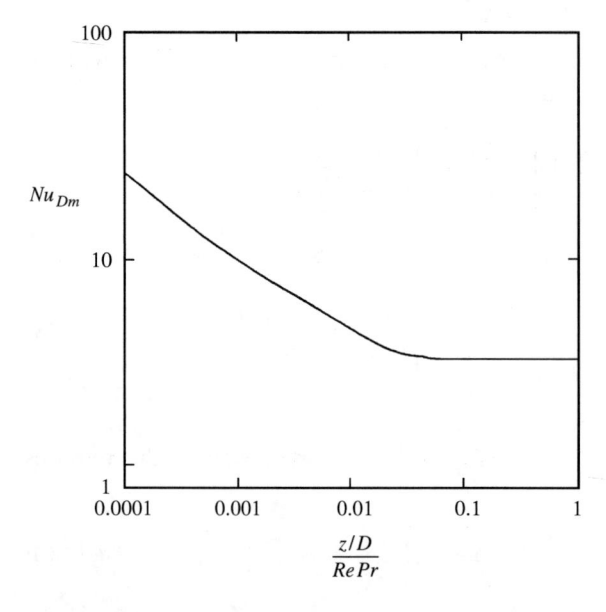

FIGURE 4.15
Nusselt number variation in thermally developing flow in a pipe.

The actual full solution is:

$$\theta = \sum_{n=0}^{\infty} C_n R_n e^{-\lambda_n^2 Z} \tag{4.153}$$

The solution for $n = 0$ is basically the fully developed flow solution discussed above which applies at large values of Z.

Once θ has been determined, the heat flux at the wall and the mean temperature can be found and the mean Nusselt number can then be found. Exact solutions for values of n up to 4 can be relatively easily obtained and approximate solutions for higher values of n can be obtained. The variation of the mean Nusselt number with Z given by these solutions is shown in Fig. 4.15.

Instead of using the method of separation of variables, a numerical solution to Eq. (4.144) can be easily obtained. Here, a numerical finite-difference solution will be discussed. A series of grid lines in the Z and R directions are introduced as shown in Fig. 4.16, a uniform grid spacing, ΔR, being used in the radial coordinate direction.

Consider the four adjacent nodal points shown in Fig. 4.17.

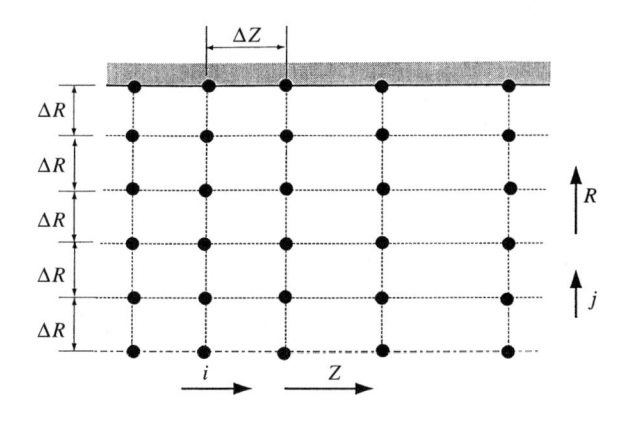

FIGURE 4.16
Nodal system used in obtaining numerical solution.

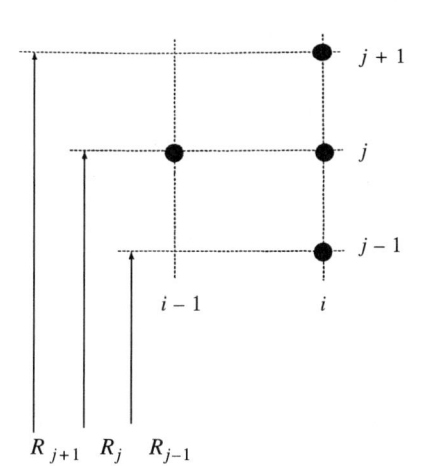

FIGURE 4.17
Nodal points used in obtaining finite-difference approximations.

194 Introduction to Convective Heat Transfer Analysis

Using the values at these points, the following finite-difference form of Eq. (4.144) applied at point, i, j, is obtained:

$$2\left[1-\left(\frac{R_j}{0.5}\right)^2\right]\left(\frac{\theta_{i,j}-\theta_{i-1,j}}{\Delta Z}\right)$$

$$= \frac{1}{R_j}\left[\frac{(R_{j+1}+R_j)(\theta_{i,j+1}-\theta_{i,j})}{2\Delta R} - \frac{(R_j+R_{j-1})(\theta_{i,j}-\theta_{i,j-1})}{2\Delta R}\right] \quad (4.154)$$

This can be rearranged to give:

$$A_j\theta_{i,j} + B_j\theta_{i,j+1} + C_j\theta_{i,j-1} = D_j \quad (4.155)$$

where:

$$A_j = \frac{2}{\Delta Z}\left[1-(R_j/0.5)^2\right] + \frac{R_{j+1}+R_j}{2R_j\Delta R^2} + \frac{R_j+R_{j-1}}{2R_j\Delta R^2} \quad (4.156)$$

$$B_j = -\frac{R_{j+1}+R_j}{2R_j\Delta R^2} \quad (4.157)$$

$$C_j = \frac{R_j+R_{j-1}}{2R_j\Delta R^2} \quad (4.158)$$

$$D_j = \frac{2}{\Delta Z}\left[1-(R_{i,j}/0.5)^2\right]\theta_{i-1,j} \quad (4.159)$$

Since the case of a constant wall temperature is being considered, i.e., since $\theta_{i,N}=0$ and since the symmetry condition on the center line gives, in finite-difference form, $\theta_{i,1}=\theta_{i,2}$, the following two results are obtained by considering the form of Eq. (4.155):

$$A_1 = 1, \quad B_1 = -1, \quad C_1 = 0, \quad D_1 = 0 \quad (4.160)$$
$$A_N = 1, \quad B_N = 0, \quad C_N = 0, \quad D_N = 0 \quad (4.161)$$

Therefore, the finite-difference approximation to the governing equation leads to a set of N simultaneous linear algebraic equations whose solution requires the inversion of a tri-diagonal matrix just as was the case in the finite-difference solution of the boundary layer equations discussed in Chapter 3. The solution is easily obtained using the same method as used in obtaining the boundary layer solution. This solution gives the values of $\theta_{i,j}$. The local heat transfer rate at the wall is then given by using Fourier's law as:

$$q_w = +k\left.\frac{\partial T}{\partial r}\right|_{r=r_o} \quad (4.162)$$

There is a "+" in this equation because r is measured radially outward in the opposite direction to that used in defining a positive value of q_w.

CHAPTER 4: Internal Laminar Flows   195

In terms of the dimensionless variables introduced above, Eq. (4.162) becomes:

$$\frac{q_w D}{k(T_W - T_e)} = \left.\frac{\partial \theta}{\partial R}\right|_{R=0.5} \tag{4.163}$$

The left-hand side of this equation is the local Nusselt number, Nu_D, i.e., Eq. (4.163) gives:

$$Nu_D = \left.\frac{\partial \theta}{\partial R}\right|_{R=0.5} \tag{4.164}$$

The finite difference approximation to this equation is:

$$Nu_D = \frac{\theta_{i,N} - \theta_{i,N-1}}{\Delta R} \tag{4.165}$$

It should be noted that the Nusselt number being used here is based on the difference between the wall and the inlet temperatures, i.e., on $T_W - T_e$, and not on the difference between the wall and the mean temperatures. Now the mean temperature, i.e., the bulk mean temperature, is given by:

$$T_m = \frac{\int_0^{r_o} uT 2\pi r\, dr}{\int_0^{r_o} u 2\pi r\, dr} \tag{4.166}$$

i.e., rewriting this equation in terms of the dimensionless variables:

$$\frac{T_W - T_m}{T_W - T_e} = 8 \int_0^{0.5} U\theta R\, dR \tag{4.167}$$

The Nusselt number based on the difference between the wall and the mean temperature is then given by:

$$Nu_{Dm} = Nu_D \frac{T_W - T_e}{T_W - T_m} = \frac{Nu_D}{8 \int_0^{0.5} U\theta R\, dR} \tag{4.168}$$

The value of the integral can be determined by numerical integration.

A computer program, DEVPIPE, written in FORTRAN based on the above procedure is available as discussed in the Preface.

The variations of Nu_D, Nu_{Dm}, and the dimensionless center line temperature, θ_c, with Z given by this program are shown in Fig. 4.18.

Next, consider thermally developing flow in the case where the wall heat flux, q_w, rather than the wall temperature is uniform. In this case, the following dimensionless temperature is used:

$$\theta = \frac{(T - T_e)}{(q_w D/k)} \tag{4.169}$$

The initial conditions at the beginning of the thermally developing region in this case is because of the way θ is defined:

$$Z = 0: \theta = 0 \tag{4.170}$$

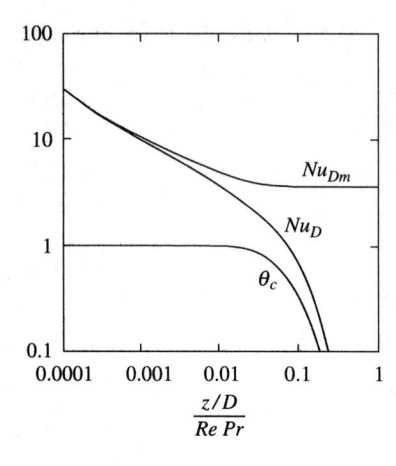

FIGURE 4.18
Nusselt number and center line temperature variation in developing flow in a pipe with a uniform wall temperature.

and the wall boundary condition is obtained by recalling that:

$$q_w = +k \left.\frac{\partial T}{\partial r}\right|_{r=r_o} \tag{4.171}$$

which can be written in terms of the dimensionless temperature as

$$\left.\frac{\partial \theta}{\partial R}\right|_{R=0.5} = 1 \tag{4.172}$$

The finite-difference approximation to this equation is:

$$\theta_{i,N} = \theta_{i,N-1} + \Delta R \tag{4.173}$$

This means that in this case:

$$A_N = 1, \; B_N = 0, \; C_N = -1, \; D_N = \Delta R \tag{4.174}$$

In the uniform wall heat flux case, the definition of the dimensionless temperature is such that:

$$\theta_w = \frac{T_w - T_e}{q_w D/k} = \frac{1}{Nu_D}$$

Further, since in this uniform heat flux case:

$$T - T_e = \theta(q_w D/k)$$

it follows that:

$$T_m - T_e = \frac{q_w D}{k} \frac{\int_0^{0.5} U\theta R \, dR}{\int_0^{0.5} U R \, dR} \tag{4.175}$$

i.e., using the known form of the variation of U with R:

$$\frac{T_m - T_e}{q_w D/k} = \theta_m = 8 \int_0^{0.5} U\theta R \, dR \tag{4.176}$$

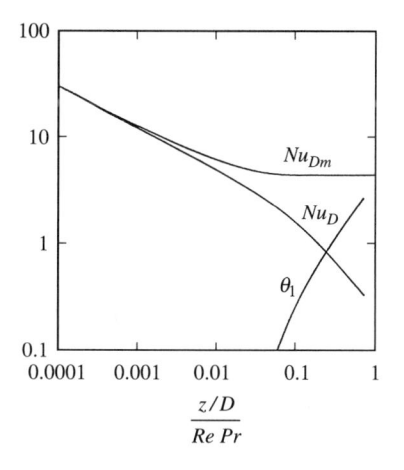

FIGURE 4.19
Nusselt number and center line temperature variation in developing flow in a pipe with a uniform wall heat flux.

This allows θ_m to be determined at any value of Z. The value of the integral can be determined by numerical integration. The Nusselt number based on the difference between the wall and the mean temperatures is then given by noting that:

$$\frac{T_w - T_m}{q_w D/k} = \theta_w - \theta_m$$

From which it follows that:

$$Nu_{Dm} = \frac{1}{\theta_w - \theta_m} \tag{4.177}$$

The computer program DEVPIPE discussed above allows either a uniform wall heat flux or a uniform wall temperature to be considered. The variations of Nu_D, Nu_{Dm}, and the dimensionless center line temperature, θ_1, with Z given by this program for the uniform wall heat flux case are shown in Fig. 4.19.

A solution for the uniform wall heat flux case can also be obtained using the separation of variables approach discussed above.

4.6
PLANE DUCT FLOW WITH A DEVELOPING TEMPERATURE FIELD

Flow with a developing temperature field in a plane duct is dealt with using the same procedure as used to deal with pipe flow. The flow situation is as shown in Fig. 4.20.

In this case, using the same assumptions as adopted in dealing with pipe flow, the temperature variation will be governed by:

$$u \frac{\partial T}{\partial z} = \left(\frac{\nu}{Pr}\right) \frac{\partial^2 T}{\partial y^2} \tag{4.178}$$

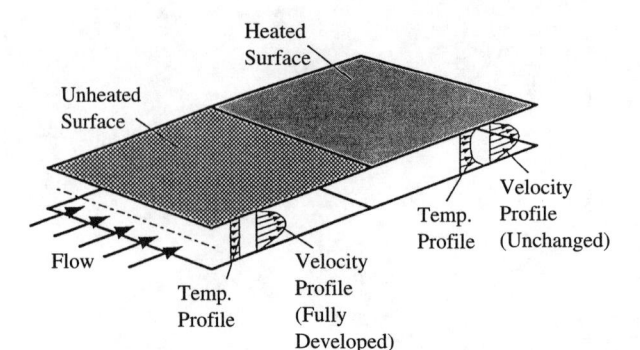

198 Introduction to Convective Heat Transfer Analysis

FIGURE 4.20
Thermally developing plane duct flow.

The following dimensionless variables are then introduced:

$$U = \frac{u}{u_m}, \quad \theta = \frac{(T_w - T)}{(T_w - T_e)}$$
$$Y = \frac{y}{W}, \quad Z = \frac{z/W}{Re\,Pr} \qquad (4.179)$$

where u_m is the mean velocity in the pipe, Re is the Reynolds number based on this mean velocity, and the width of the duct is W. In terms of these variables, the equation governing the developing temperature field is:

$$U \frac{\partial \theta}{\partial Z} = \frac{\partial^2 \theta}{\partial Y^2} \qquad (4.180)$$

Now the velocity profile for fully developed plane duct flow can be written in terms of the dimensionless variables as:

$$U = 1.5\left[1 - \left(\frac{Y}{0.5}\right)^2\right]$$

Using this, Eq. (4.180) becomes:

$$1.5\left[1 - \left(\frac{Y}{0.5}\right)^2\right] \frac{\partial \theta}{\partial Z} = \frac{\partial^2 \theta}{\partial Y^2} \qquad (4.181)$$

This equation must be solved subject to the following boundary conditions:

$$Y = 0: \frac{\partial \theta}{\partial Y} = 0, \quad Y = 0.5: \theta = 0.0 \qquad (4.182)$$

while the initial conditions are in terms of the dimensionless variables:

$$Z = 0: \theta = 1 \qquad (4.183)$$

The solution to Eq. (4.181) can either be obtained by using the method of separation of variables or numerically using the same type of approach as used in dealing with pipe flow. Here, a numerical finite-difference solution will be discussed. The procedure is identical to that used in dealing with pipe flow. A series of grid lines in the Z and Y directions are introduced, a uniform grid spacing ΔY being used in the Y-coordinate direction. Using the values of the variables at the same nodal points as used in dealing with pipe flow, the following finite-difference form of Eq. (4.181) applied at point i, j is obtained:

$$1.5\left[1 - \left(\frac{Y_j}{0.5}\right)^2\right]\left(\frac{\theta_{i,j} - \theta_{i-1,j}}{\Delta Z}\right) = \frac{\theta_{i,j+1} + \theta_{i,j-1} - 2\theta_{i,j}}{\Delta Y^2} \tag{4.184}$$

This can be rearranged to give, as with pipe flow:

$$A_j \theta_{i,j} + B_j \theta_{i,j+1} + C_j \theta_{i,j-1} = D_j \tag{4.185}$$

where in this case:

$$A_j = 1.5\left[1 - \left(\frac{y_j}{0.5}\right)^2\right]\frac{1}{\Delta Z} + \frac{2}{\Delta Y^2} \tag{4.186}$$

$$B_j = -\frac{1}{\Delta Y^2} \tag{4.187}$$

$$C_j = -\frac{1}{\Delta Y^2} \tag{4.188}$$

$$D_j = 1.5\left[1 - \left(\frac{Y_j}{0.5}\right)^2\right]\frac{\theta_{i-1,j}}{\Delta Z} \tag{4.189}$$

Because the case of a constant wall temperature is being considered, i.e., because $\theta_{i,N} = 0$, and since the symmetry condition on the center line gives, in finite-difference form, $\theta_{i,1} = \theta_{i,2}$, the following two results are obtained by considering the form of Eq. (4.185):

$$A_1 = 1, \quad B_1 = -1, \quad C_1 = 0, \quad D_1 = 0 \tag{4.190}$$

$$A_N = 1, \quad B_N = 0, \quad C_N = 0, \quad D_N = 0 \tag{4.191}$$

Therefore the finite-difference approximation to the governing equation again leads to a set of N simultaneous linear algebraic equations whose solution requires the inversion of a tri-diagonal matrix. The solution is easily obtained using the same method as used in dealing with pipe flow. This solution gives the values of $\theta_{i,j}$. The local heat transfer rate at the wall is then given by using Fourier's law as:

$$q_w = +k\left.\frac{\partial T}{\partial y}\right|_{y=w} \tag{4.192}$$

In terms of the dimensionless variables introduced above, Eq. (4.192) becomes:

$$\frac{q_w W}{k(T_W - T_e)} = \left.\frac{\partial \theta}{\partial Y}\right|_{Y=0.5} \tag{4.193}$$

The left-hand side of this equation is the local Nusselt number, Nu_W, i.e., Eq. (4.194) gives:

$$Nu_W = \left.\frac{\partial \theta}{\partial Y}\right|_{Y=0.5} \tag{4.194}$$

The finite-difference approximation to this equation is:

$$Nu_W = \frac{\theta_{i,N} - \theta_{i,N-1}}{\Delta Y} \quad (4.195)$$

The Nusselt number being used here is based on the difference between the wall and the inlet temperatures, i.e., on $(T_w - T_e)$ and not on the difference between the wall and the mean temperatures. Now the mean temperature, i.e., the bulk mean temperature, is given by:

$$T_m = \frac{\int_0^w uT\,dy}{\int_0^w u\,dy} \quad (4.196)$$

i.e., rewriting this equation in terms of the dimensionless variables and using the known form of the variation of U with R gives:

$$\frac{T_m - T_e}{T_w - T_e} = \frac{16}{11}\int_0^{0.5} U\theta\,dY$$

The Nusselt number based on the difference between the wall and the mean temperatures is then given by:

$$Nu_{Wm} = Nu_W \frac{T_w - T_e}{T_w - T_m} = \frac{Nu_W}{(16/11)\int_0^{0.5} U\theta\,dY} \quad (4.197)$$

The value of the integral can be determined by numerical integration thus allowing the variation of Nu_{Wm} with Z to be found. A computer program, DEVDUCT, based on this procedure written in FORTRAN is available as discussed in the Preface.

The variations of Nu_W, Nu_{Wm}, and the dimensionless center line temperature, θ_1, with Z given by the above program are shown in Fig. 4.21.

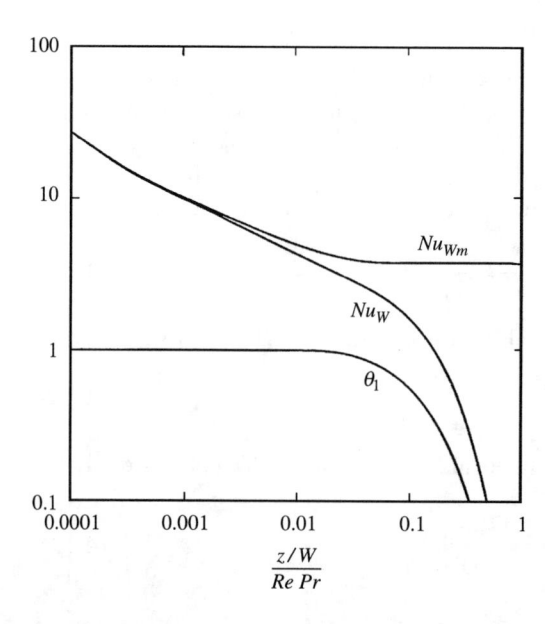

FIGURE 4.21
Nusselt number and center line temperature variation in developing flow in a plane duct with a uniform wall temperature.

4.7
LAMINAR PIPE FLOW WITH DEVELOPING VELOCITY AND TEMPERATURE FIELDS

The modifications to the above analysis to deal with the uniform wall heat flux case are basically the same as that required for pipe flow and will not be discussed here.

Flow in a pipe when both the velocity and temperature fields are developing is considered in this section, the flow situation thus being as shown in Fig. 4.22 [12],[13],[14],[15].

The z-axis is again chosen to lie along the center line of the pipe and the velocity components are defined in the same way as in Chapter 2, i.e., as shown in Fig. 4.23.

Assuming that the swirl velocity component, w, is zero and that the flow is therefore symmetrical about the center line, the equations governing the flow are:

$$\frac{1}{r}\frac{\partial}{\partial r}(vr) + \frac{\partial u}{\partial z} = 0 \tag{4.198}$$

$$u\frac{\partial u}{\partial z} + v\frac{\partial u}{\partial r} = -\frac{1}{\rho}\frac{\partial p}{\partial z} + \nu\left(\frac{\partial^2 u}{\partial z^2} + \frac{\partial^2 u}{\partial r^2} + \frac{1}{r}\frac{\partial u}{\partial r}\right) \tag{4.199}$$

$$u\frac{\partial v}{\partial z} + v\frac{\partial v}{\partial r} = -\frac{1}{\rho}\frac{\partial p}{\partial r} + \nu\left(\frac{\partial^2 v}{\partial z^2} + \frac{\partial^2 v}{\partial r^2} + \frac{1}{r}\frac{\partial v}{\partial r} - \frac{v}{r^2}\right) \tag{4.200}$$

$$u\frac{\partial T}{\partial z} + v\frac{\partial T}{\partial r} = \left(\frac{\nu}{Pr}\right)\left(\frac{\partial^2 T}{\partial z^2} + \frac{\partial^2 T}{\partial r^2} + \frac{1}{r}\frac{\partial T}{\partial r}\right) \tag{4.201}$$

It is assumed that the radius of the pipe is relatively small compared to the values of z being considered. In this case, the same type of order of magnitude analysis as used in deriving the boundary layer equations indicates that $v \ll u$ and that

FIGURE 4.22
Developing velocity and temperature fields.

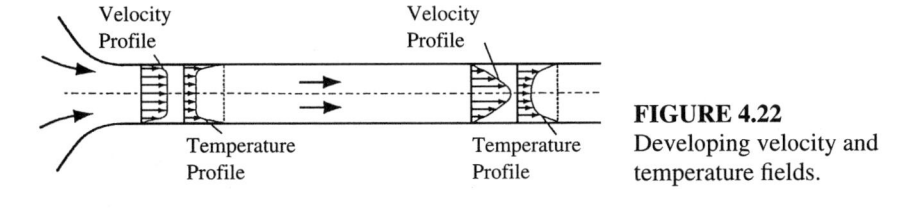

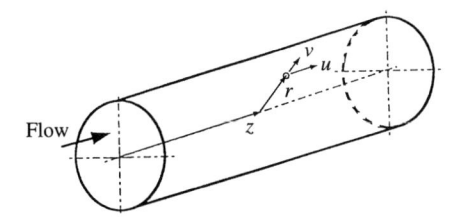

FIGURE 4.23
Coordinate system used.

$\partial/\partial z \ll \partial/\partial r$. As a result, the r-momentum equation gives:

$$\frac{\partial p}{\partial r} = 0 \tag{4.202}$$

The governing equations can therefore be written as:

$$\frac{1}{r}\frac{\partial}{\partial r}(vr) + \frac{\partial u}{\partial z} = 0 \tag{4.203}$$

$$u\frac{\partial u}{\partial z} + v\frac{\partial u}{\partial r} = -\frac{1}{\rho}\frac{dp}{dz} + \nu\left(\frac{\partial^2 u}{\partial r^2} + \frac{1}{r}\frac{\partial u}{\partial r}\right) \tag{4.204}$$

$$u\frac{\partial T}{\partial z} + v\frac{\partial T}{\partial r} = \left(\frac{\nu}{Pr}\right)\left(\frac{\partial^2 T}{\partial r^2} + \frac{1}{r}\frac{\partial T}{\partial r}\right) \tag{4.205}$$

These equations are parabolic in form whereas the full equations are elliptic in form. Eqs. (4.203) to (4.205) are therefore referred to as the parabolized form of the governing equations.

The numerical solution to these equations for the case where both the velocity and temperature profiles are developing and where the wall temperature is uniform will be considered in the present section.

The velocity and temperature profiles are assumed to be symmetrical about the center line of the pipe and the radial velocity component, v, is therefore zero on the center line. The boundary conditions on the solution on the center line are therefore:

$$\text{At } r = 0: \quad \frac{\partial u}{\partial r} = 0, \, v = 0, \, \frac{\partial T}{\partial r} = 0 \tag{4.206}$$

At the wall, the velocity components are zero and the wall temperature is constant and specified, i.e., the boundary conditions at the wall are:

$$\text{At } r = r_o: \quad u = 0, \quad v = 0, \quad T = T_W \tag{4.207}$$

r_o being the radius of the pipe, i.e., $r_o = D/2$ where D is the diameter of the pipe.

Although it is not a necessary step in obtaining the solution, the governing equations will be written in terms of the following dimensionless variables:

$$\begin{aligned} U = u/u_m, \quad V = vRePr/u_m, \quad P = (p - p_0)/\rho u_m^2 \\ Z = z/RePrD, \quad R = r/D, \quad \theta = (T - T_0)/(T_W - T_0) \end{aligned} \tag{4.208}$$

where u_m is the mean velocity across the pipe, Re is the Reynolds number based on u_m, and D, i.e., $Re = u_m D/\nu$, T_0 is the temperature on the inlet plane, and p_0 is some reference pressure. The value used for p_0 is irrelevant to the solution but, if actual values of the pressure are required, it is often convenient to take p_0 as the pressure upstream of inlet to the pipe as shown in Fig. 4.24.

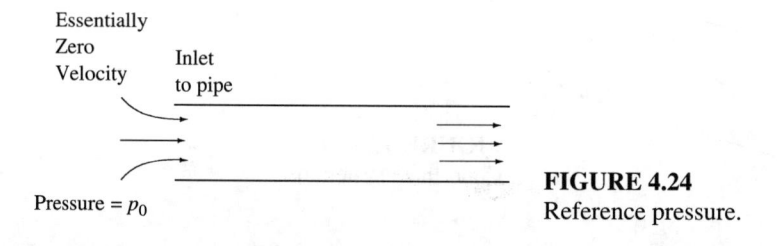

FIGURE 4.24
Reference pressure.

In terms of these dimensionless variables, the governing equations become:

$$\frac{1}{R}\frac{\partial}{\partial R}(VR) + \frac{\partial U}{\partial Z} = 0 \qquad (4.209)$$

$$\frac{1}{Pr}\left(U\frac{\partial U}{\partial Z} + V\frac{\partial U}{\partial R}\right) = -\frac{1}{Pr}\frac{dP}{dZ} + \left(\frac{\partial^2 U}{\partial R^2} + \frac{1}{R}\frac{\partial U}{\partial R}\right) \qquad (4.210)$$

$$U\frac{\partial \theta}{\partial Z} + V\frac{\partial \theta}{\partial R} = \left(\frac{\partial^2 \theta}{\partial R^2} + \frac{1}{R}\frac{\partial \theta}{\partial R}\right) \qquad (4.211)$$

In terms of the dimensionless variables, the boundary conditions are:

$$\text{At } R = 0: \quad \frac{\partial U}{\partial R} = 0, \quad V = 0, \quad \frac{\partial \theta}{\partial R} = 0 \qquad (4.212)$$

and:

$$\text{At } R = 0.5: \quad U = 0, \quad V = 0, \quad \theta = 1 \qquad (4.213)$$

It will be noted that the only parameter in the above set of equations is the Prandtl number.

In order to obtain a solution to these equations, the conditions on the inlet plane have to be known. It will be assumed here that the velocity and temperature distributions are uniform on the inlet plane to the pipe as illustrated in Fig. 4.25, i.e., that:

$$\text{At } z = 0: \quad u = u_m, \quad v = 0, \quad T = T_0 \qquad (4.214)$$

the velocity on the inlet plane being equal to u_m because, since the density is being assumed constant, the mean velocity does not change with z with the result that if the velocity across the pipe section is uniform, the velocity must be equal to u_m. Consistent with this is the assumption that v is zero on the inlet plane.

If the fluid flows into the pipe through a bell-shaped inlet section as shown in Figure 4.25, the losses in this inlet section will be small. In this case, if p_0 is taken as the pressure ahead of the inlet as shown in Fig. 4.26 and p_i is the pressure on the inlet plane then Bernoulli's equation applied across the inlet gives:

$$p_i + \rho u_m^2/2 = p_0 \qquad (4.215)$$

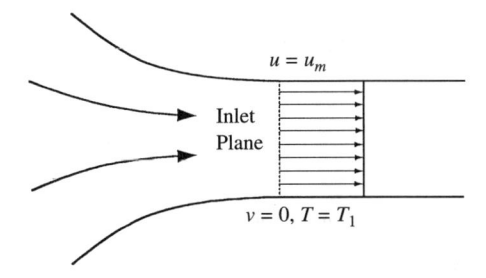

FIGURE 4.25
Inlet plane conditions.

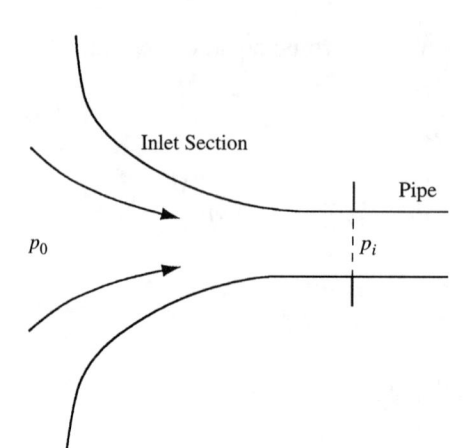

FIGURE 4.26
Low loss inlet section.

i.e.:

$$(p_i - p_0)/\rho u_m^2 = -1/2 \quad (4.216)$$

In terms of the dimensionless variables introduced above, these inlet conditions are:

$$\text{At } Z = 0, \quad U = 1, \quad V = 0, \quad \theta = 0, \quad P = -0.5 \quad (4.217)$$

A simple finite-difference solution to the above set of equations will be discussed here. A series of nodal lines in the Z- and R-directions are introduced.

Because of the parabolic nature of the assumed form of the governing equations, a forward-marching solution in the Z-direction can be used. The solution starts with known conditions on the first R-line. Finite-difference forms of the governing equations are then used to obtain the solution on the next R-line. Once this is obtained, the same procedure can be used to get the solution on the next R-line and so on. This is the same procedure as was used in Chapter 3 in obtaining a solution to the boundary layer equations.

An explicit finite-difference procedure will be used here in dealing with the momentum and energy equations. Consider the nodal points shown in Fig. 4.27. It will be seen that a uniform grid spacing is used in the R-direction.

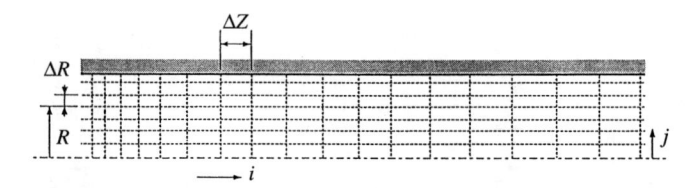

FIGURE 4.27
Nodal points used in dealing with momentum and energy equations.

CHAPTER 4: Internal Laminar Flows   205

First consider the momentum equation. The finite-difference form of this equation is:

$$\frac{1}{Pr} U_{i-1,j} \left(\frac{U_{i,j} - U_{i-1,j}}{\Delta Z} \right) + \frac{1}{Pr} V_{i-1,j} \left(\frac{U_{i-1,j+1} - U_{i-1,j-1}}{2 \Delta R} \right)$$

$$= -\frac{1}{Pr} \frac{(P_i - P_{i-1})}{\Delta Z} + \frac{(U_{i-1,j+1} + U_{i-1,j-1}) - 2U_{i-1,j}}{\Delta R^2}$$

$$+ \frac{1}{R_j} \frac{(U_{i-1,j+1} - U_{i-1,j-1})}{2 \Delta R} \quad (4.218)$$

which gives:

$$U_{i,j} = U_{i-1,j} - \frac{P_i}{U_{i-1,j}} + \frac{P_{i-1}}{U_{i-1,j}} - \frac{V_{i-1,j}}{U_{i-1,j}} \left(\frac{U_{i-1,j+1} - U_{i-1,j-1}}{2 \Delta R} \right) \Delta Z$$

$$+ Pr \left[\frac{(U_{i-1,j+1} + U_{i-1,j-1}) - 2U_{i-1,j}}{\Delta R^2} + \frac{1}{R_j} \frac{(U_{i-1,j+1} - U_{i-1,j-1})}{2 \Delta R} \right] \frac{\Delta Z}{U_{i-1,j}}$$

(4.219)

i.e.:

$$U_{i,j} = U_{i-1,j} - \frac{P_i}{U_{i-1,j}} + A_j \quad (4.220)$$

where:

$$A_j = \frac{P_{i-1}}{U_{i-1,j}} - \frac{V_{i-1,j}}{U_{i-1,j}} \left(\frac{U_{i-1,j+1} - U_{i-1,j-1}}{2 \Delta R} \right) \Delta Z$$

$$+ Pr \left[\frac{(U_{i-1,j+1} + U_{i-1,j-1}) - 2U_{i-1,j}}{\Delta R^2} + \frac{1}{R_j} \frac{(U_{i-1,j+1} - U_{i-1,j-1})}{2 \Delta R} \right] \frac{\Delta Z}{U_{i-1,j}}$$

(4.221)

A_j is therefore a known quantity at any stage of the calculation because the values of the variables on the $i - 1$ line are known.

Next consider the continuity equation written as:

$$\frac{\partial}{\partial R}(VR) = -R \frac{\partial U}{\partial Z} \quad (4.222)$$

This is applied to the nodal points shown in Fig. 4.28.

The equation is actually applied to the point p, shown in Fig. 4.28. This lies halfway between nodal points i, j and $i, j - 1$. The value of $\partial U/\partial Z$ at point p is taken as the average of the values of $\partial U/\partial Z$ at points i, j and $i, j - 1$. Hence, the continuity equation gives:

$$\frac{R_{i,j} V_{i,j} - R_{i,j-1} V_{i,j-1}}{\Delta R} = -\frac{1}{2} \left[\frac{U_{i,j} - U_{i-1,j}}{\Delta Z} + \frac{U_{i,j-1} - U_{i-1,j-1}}{\Delta Z} \right]$$

i.e.:

$$R_{i,j} V_{i,j} - R_{i,j-1} V_{i,j-1} = -\frac{\Delta R}{2 \Delta Z} [U_{i,j} - U_{i-1,j} + U_{i,j-1} - U_{i-1,j-1}] \quad (4.223)$$

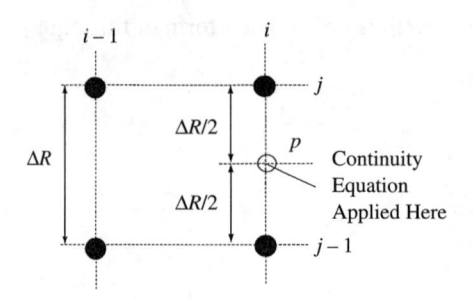

FIGURE 4.28
Nodal points used in dealing with continuity equation.

Substituting Eq. (4.220) into this equation then gives:

$$R_{i,j}V_{i,j} - R_{i,j-1}V_{i,j-1} = -\frac{\Delta R}{2\Delta Z}\left[-\frac{P_i}{U_{i-1,j}} + A_j - \frac{P_i}{U_{i-1,j-1}} + A_{j-1}\right] \quad (4.224)$$

which can be written as:

$$R_{i,j}V_{i,j} - R_{i,j-1}V_{i,j-1} = -P_i\left[\frac{\Delta R}{2\Delta Z}\left(\frac{1}{U_{i-1,j}} + \frac{1}{U_{i-1,j-1}}\right)\right] - \frac{\Delta R}{2\Delta Z}(A_j + A_{j-1}) \quad (4.225)$$

i.e.:

$$R_{i,j}V_{i,j} - R_{i,j-1}V_{i,j-1} = -B_j P_i - C_j \quad (4.226)$$

where:

$$B_j = \frac{\Delta R}{2\Delta Z}\left[\frac{1}{U_{i-1,j}} + \frac{1}{U_{i-1,j-1}}\right] \quad (4.227)$$

and:

$$C_j = \frac{\Delta R}{2\Delta Z}(A_j + A_{j-1}) \quad (4.228)$$

It will be noted that since at any stage of the solution the conditions on the $(i-1)$ line are known, the coefficients A_j, A_{j-1}, B_j, and C_j are known quantities.

Now, on the center line, i.e., at $j = 1$, the quantity $R_{i,j}V_{i,j}$ is 0. Hence, if Eq. (4.226) is applied sequentially outward from point $j = 2$ to the point $j = N$ which lies on the wall, the following is obtained:

$$R_{i,2}V_{i,2} - 0 = -B_2 P_i - C_2$$
$$R_{i,3}V_{i,3} - R_{i,2}V_{i,2} = -B_3 P_i - C_3$$
$$R_{i,4}V_{i,4} - R_{i,3}V_{i,3} = -B_4 P_i - C_4$$
$$R_{i,5}V_{i,5} - R_{i,4}V_{i,4} = -B_5 P_i - C_5$$
$$\vdots$$
$$R_{i,N}V_{i,N} - R_{i,N-1}V_{i,N-1} = -B_N P_i - C_N$$

Adding this set of equations together then gives:

$$R_{i,N}V_{i,N} = -P_i \sum_{k=2}^{N} B_k - \sum_{k=2}^{N} C_k \quad (4.229)$$

But the boundary conditions give $V_{i,N} = 0$ so Eq. (4.229) gives:

$$P_i = \frac{\sum_{k=2}^{N} C_k}{\sum_{k=2}^{N} B_k} \tag{4.230}$$

This allows the dimensionless pressure, P_i, to be determined. Once this has been found, Eq. (4.220) can be used to find the $U_{i,j}$ values and Eq. (4.229) applied sequentially outward from $j = 2$ as discussed above allows the $V_{i,j}$ values to be found. The boundary conditions give $U_{i,N} = 0$ and $U_{i,1} = U_{i,2}$.

Lastly, the energy equation, i.e., Eq. (4.211), is used to determine the distribution of $\theta_{i,j}$. The finite-difference form of Eq. (4.211) is obtained using the same nodal points as used in dealing with the momentum equation. The energy equation then gives:

$$U_{i-1,j}\left(\frac{\theta_{i,j} - \theta_{i-1,j}}{\Delta Z}\right) + V_{i-1,j}\left(\frac{\theta_{i-1,j+1} - \theta_{i-1,j-1}}{2\Delta R}\right) =$$

$$\frac{(\theta_{i-1,j+1} + \theta_{i-1,j-1}) - 2\theta_{i-1,j}}{\Delta R^2} + \frac{1}{R_j}\frac{(\theta_{i-1,j+1} - \theta_{i-1,j-1})}{2\Delta R} \tag{4.231}$$

which gives:

$$\theta_{i,j} = \theta_{i-1,j} - \frac{V_{i-1,j}}{U_{i-1,j}}\left(\frac{\theta_{i-1,j+1} - \theta_{i-1,j-1}}{2\Delta R}\right)\Delta Z$$

$$+ \frac{(\theta_{i-1,j+1} + \theta_{i-1,j-1}) - 2\theta_{i-1,j}}{\Delta R^2} + \frac{1}{R_j}\frac{(\theta_{i-1,j+1} - \theta_{i-1,j-1})}{2\Delta R} \tag{4.232}$$

This equation allows the $\theta_{i,j}$ values to be found. The boundary conditions give $\theta_{i,N} = 1$ and $\theta_{i,1} = \theta_{i,2}$. Once the θ values are determined, the heat transfer rate can be found. The local heat transfer rate at the wall at any value of Z is given by using Fourier's law as:

$$q_w = +k\left.\frac{\partial T}{\partial r}\right|_{r=r_o} \tag{4.233}$$

There is a plus (+) in this equation because r is measured radially outward, i.e., in the opposite direction to that used in defining a positive value of q_w.

In terms of the dimensionless variables introduced above, Eq. (4.233) becomes:

$$\frac{q_w D}{k(T_W - T_0)} = \left.\frac{\partial \theta}{\partial R}\right|_{R=0.5} \tag{4.234}$$

where as before $D = 2r_o$ is the pipe diameter. The left-hand side of this equation is the local Nusselt number, Nu_D, i.e., Eq. (4.234) gives:

$$Nu_D = \left.\frac{\partial \theta}{\partial R}\right|_{R=0.5} \tag{4.235}$$

The finite-difference approximation to this equation is:

$$Nu_D = \frac{\theta_{i,N} - \theta_{i,N-1}}{\Delta R} \tag{4.236}$$

The Nusselt number being used here is based on the difference between the wall and the inlet temperatures, i.e., on $T_W - T_0$, and not on the difference between the wall and the mean temperatures.

The mean heat transfer rate up to the value of Z being considered can be obtained by integration of the local wall heat transfer rate or by noting that the total heat transfer rate up to the point considered will be equal to the increase in the enthalpy flux between the inlet and the z value considered, i.e., that:

$$\overline{q_w} \pi D z = \int_0^{r_0} \rho u c_p T 2\pi r \, dr - \rho u_m c_p T_0 \pi r_0^2 \tag{4.237}$$

i.e., rewriting this equation in terms of the dimensionless variables:

$$\frac{\overline{q_w} Z Re Pr}{T_w - T_0} = \rho c_p u_m \int_0^{0.5} U\theta R \, dR$$

i.e.:

$$\frac{\overline{q_w} D}{k(T_w - T_0)} = \frac{1}{Z} \int_0^{0.5} U\theta R \, dR \tag{4.238}$$

The left-hand side of this equation is the mean Nusselt number up to dimensionless distance Z from the inlet considered. Hence, this equation can be written as:

$$\overline{Nu_D} = \frac{1}{Z} \int_0^{0.5} U\theta R \, dR \tag{4.239}$$

The value of the integral can be determined by numerical integration.

The Nusselt numbers introduced above were based on the temperature difference $T_W - T_0$. The Nusselt numbers based on the temperature difference $T_W - T_m$ where T_m is the mean temperature at any value of Z can be deduced by noting that the definition of T_m is such that:

$$\overline{q_w} \pi D z = \dot{m} c_p (T_m - T_0) \tag{4.240}$$

the left-hand side being the rate at which heat is transferred to the fluid between the inlet and a point distance z from the inlet. Hence, since $\dot{m}$ is given by:

$$\dot{m} = \rho u_m \pi D^2 / 4$$

it follows that:

$$Re Pr (T_m - T_0)/4 = \overline{q_w} z / k$$

i.e.:

$$\frac{T_m - T_0}{T_w - T_0} = 4 \overline{Nu_D} Z$$

i.e.:

$$\frac{T_w - T_m}{T_w - T_0} = 1 - 4 \overline{Nu_D} Z \tag{4.241}$$

This can be used to find the local Nusselt number based on $T_m - T_0$ which will be given by:

$$Nu_{Dm} = Nu_D \frac{T_w - T_0}{T_w - T_m} = \frac{Nu_D}{1 - 4\,\overline{Nu_D}\,Z/Pr} \qquad (4.242)$$

To summarize, the calculation procedure is as follows:

1. Define the values of the variables on the inlet plane. As discussed above, these will here be assumed to be:

$$U_{1,j} = 1.0, \qquad V_{1,j} = 0.0, \qquad \theta_{1,j} = 0.0, \qquad P_1 = -0.5$$

2. Use Eq. (4.230) to determine P_2.
3. Use Eq. (4.220) to find the $U_{2,j}$ values
4. Use Eq. (4.226) applied sequentially outward from $j = 2$ to give the $V_{2,j}$ values to be found.
5. Use Eq. (4.232) to find the $\theta_{2,j}$ values.
6. Use Eqs. (4.236), (4.239), and (4.242) to get the local and mean Nusselt numbers.
7. Having determined the values of the variables on the $i = 2$ line, the same procedure is used to advance the solution to the next i line and so on to the maximum Z value for which a solution is required. It will be noted that at any stage of the solution it is only necessary to know the values of the variables on two adjacent i lines.

A computer program, PIPEFLOW, written in FORTRAN, that is based on this procedure is available as discussed in the Preface.

Because an explicit finite-difference procedure is being used to solve the momentum and energy equations, the solution can become unstable, i.e., as the solution proceeds it can diverge increasingly from the actual solution as indicated in Fig. 4.29.

A simplified analysis of the conditions under which instability occurs is presented in the next section in which the numerical solution of developing flow in a plane duct is considered. The results obtained there indicate that stability exists if:

$$\Delta Z < \frac{0.5\Delta R^2 U_{i-1,N-1}}{Pr} \qquad (4.243)$$

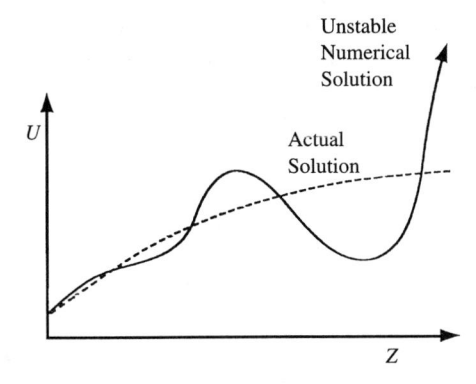

FIGURE 4.29
Development of the instability.

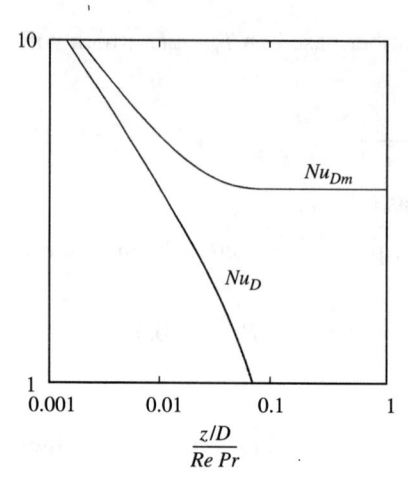

FIGURE 4.30
Nusselt number variations for developing pipe flow for $Pr = 0.7$.

will give a good indication of the maximum value of ΔZ that can be used without instability developing and this has been incorporated into the program listed above.

Typical variations of Nu_D and Nu_{Dm} with Z given by the above program are shown in Fig. 4.30.

EXAMPLE 4.4. Oil enters a 12-mm diameter pipe at a temperature of 10°C with a uniform velocity of 0.8 m/s. The walls of the pipe are kept at a uniform temperature of 30°C. If the pipe is 60 cm long, plot the variation of the temperature in the oil at the exit of the pipe. Assume that for the oil $\rho = 910$ kg/m^3, $\mu = 0.008$ kg/ms, $k = 0.14$ W/m-K, and $Pr = 1100$.

Solution. The Reynolds number is given by:

$$Re = \frac{\rho u_m D}{\mu} = \frac{910 \times 0.8 \times 0.012}{0.008} = 1092$$

Hence, if z is the axial coordinate measured from the inlet to the pipe then at the exit of the pipe if ℓ is the length of the pipe:

$$Z = \frac{\ell}{Re\,Pr\,D} = \frac{0.6}{1092 \times 1100 \times 0.012} = 0.00004163$$

The above results show that the flow is laminar and that the flow will not be fully developed at the exit of the pipe. The program discussed above gives the variation of dimensionless temperature with dimensionless radius at the pipe exit. The program has therefore been run up to a maximum Z value of 0.00004163. This gives the dimensionless temperature variation with dimensionless radius at the exit listed in Table E4.4. The actual radius and temperature are then obtained by recalling that:

$$R = r/D, \qquad \theta = (T - T_0)/(T_w - T_0)$$

from which it follows that:

$$r = DR = 12D \text{ mm}$$

and that:

$$T = \theta(T_w - T_0) + T_0 = \theta(30 - 10) + 10 = (20\theta + 10)°C$$

CHAPTER 4: Internal Laminar Flows   **211**

TABLE E4.4

R	θ	r (mm)	T (°C)
0.0000	0.0000	0.000	10.000
0.3370	0.0000	4.044	10.000
0.3587	0.0001	4.304	10.001
0.3804	0.0005	4.565	10.009
0.4022	0.0028	4.826	10.056
0.4239	0.0145	5.087	10.291
0.4457	0.0690	5.348	11.380
0.4674	0.2813	5.609	15.626
0.4783	0.4886	5.740	19.772
0.4891	0.7401	5.869	24.802
0.5000	1.0000	6.000	30.000

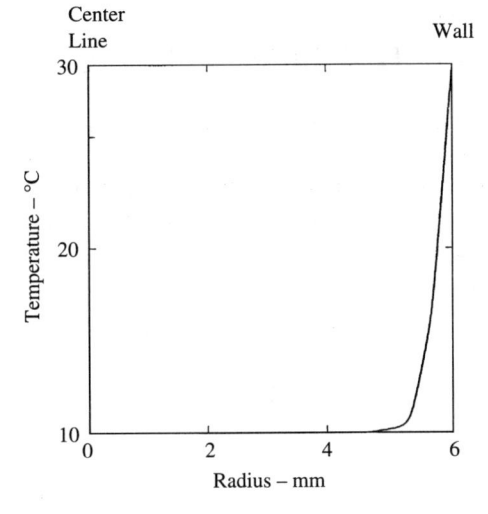

FIGURE E4.4

The values of the radius, r and T given by these equations are also shown in Table E4.4.

The variation of temperature with radius on the exit plane is plotted in Fig. E4.4.

In the above discussion, it was assumed that the temperature of the wall of the pipe was uniform and specified. If instead of this, the wall heat flux, q_w, was uniform and specified, the dimensionless temperature would be taken as:

$$\theta = \frac{(T - T_0)}{q_w D/k} \tag{4.244}$$

Since:

$$q_w = +k \left.\frac{\partial T}{\partial r}\right|_{r=r_o}$$

the boundary condition on the temperature at the wall in this case would be,

$$\left.\frac{\partial \theta}{\partial R}\right|_{R=0.5} = 1 \tag{4.245}$$

212 Introduction to Convective Heat Transfer Analysis

In finite-difference form this gives:

$$\frac{\theta_{i,N} - \theta_{i,N-1}}{\Delta R} = 1$$

i.e.:

$$\theta_{i,N} = \theta_{i,N-1} + \Delta R \tag{4.246}$$

Thus, the only difference required when the wall heat flux is specified is that instead of setting $\theta_{i,N} = 1$, Eq. (4.246) is used to determine $\theta_{i,N}$. The computer program discussed above allows for either a uniform wall temperature or a uniform wall heat flux.

4.8 LAMINAR FLOW IN A PLANE DUCT WITH DEVELOPING VELOCITY AND TEMPERATURE FIELDS

Here, attention will be given to developing flow in a wide duct as shown in Fig. 4.31. Basically, developing two-dimensional flow between two plates is therefore being considered. If the same "parabolic flow" assumptions as adopted in dealing with pipe flow in the previous section are used, the governing equations for this flow become, if the coordinate system shown in Fig. 4.32 is used:

$$\frac{\partial v}{\partial y} + \frac{\partial u}{\partial z} = 0 \tag{4.247}$$

$$u\frac{\partial u}{\partial z} + v\frac{\partial u}{\partial y} = -\frac{1}{\rho}\frac{dp}{dz} + \nu\left(\frac{\partial^2 u}{\partial y^2}\right) \tag{4.248}$$

$$u\frac{\partial T}{\partial z} + v\frac{\partial T}{\partial y} = \left(\frac{\nu}{Pr}\right)\left(\frac{\partial^2 T}{\partial y^2}\right) \tag{4.249}$$

Because the flow is assumed to be symmetrical about the center line, it follows that:

$$\text{At } y = 0: \quad \frac{\partial u}{\partial y} = 0, \quad v = 0, \quad \frac{\partial T}{\partial y} = 0 \tag{4.250}$$

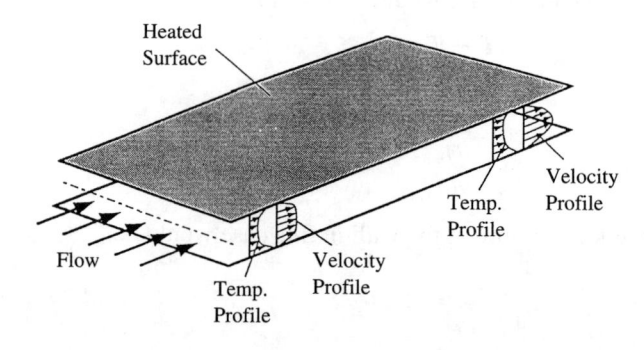

FIGURE 4.31
Flow in a plane duct.

CHAPTER 4: Internal Laminar Flows   213

FIGURE 4.32
Coordinate system used.

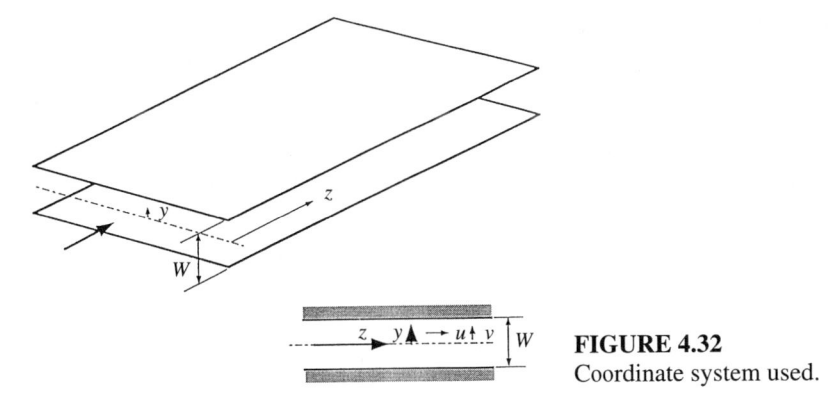

At the wall, the velocity components are zero and the wall temperature is uniform and its value is specified, i.e., the boundary conditions at the wall are:

$$\text{At } y = W/2: \quad u = 0, \quad v = 0, \quad T = T_W \qquad (4.251)$$

The governing equations are written in terms of the following dimensionless variables:

$$U = u/u_m, \quad V = vRePr/u_m, \quad P = (p - p_0)/\rho u_m^2$$
$$Z = z/RePrW, \quad Y = y/W, \quad \theta = (T - T_0)/(T_W - T_0) \qquad (4.252)$$

where u_m is the mean velocity across the duct, Re is the Reynolds number based on u_m and W, i.e., is equal to $u_m W/\nu$, T_0 is the temperature on the inlet plane, and p_0 is again some reference pressure.

In terms of these dimensionless variables, the governing equations become:

$$\frac{\partial V}{\partial Y} + \frac{\partial U}{\partial Z} = 0 \qquad (4.253)$$

$$\frac{1}{Pr}\left(U\frac{\partial U}{\partial Z} + V\frac{\partial U}{\partial Y}\right) = -\frac{1}{Pr}\frac{dP}{dZ} + \left(\frac{\partial^2 U}{\partial Y^2}\right) \qquad (4.254)$$

$$U\frac{\partial \theta}{\partial Z} + V\frac{\partial \theta}{\partial Y} = \frac{\partial^2 \theta}{\partial Y^2} + \frac{\partial^2 \theta}{\partial Y^2} \qquad (4.255)$$

In terms of these dimensionless variables, the boundary conditions are:

$$\text{At } Y = 0: \quad \frac{\partial U}{\partial Y} = 0, \quad V = 0, \quad \frac{\partial \theta}{\partial Y} = 0 \qquad (4.256)$$

and:

$$\text{At } Y = 0.5: \quad U = 0, \quad V = 0, \quad \theta = 1 \qquad (4.257)$$

In order to obtain the solutions to these equations, the conditions on the inlet plane have to be known. It will again be assumed that the velocity and temperature distributions are uniform on the inlet plane and that the losses in the inlet are negligible, i.e., that:

$$\text{At } z = 0: \quad u = u_m, \quad v = 0, \quad T = T_0, \quad p_i + \rho u_m^2/2 = p_0 \qquad (4.258)$$

214 Introduction to Convective Heat Transfer Analysis

the uniform velocity on the inlet plane being, as with pipe flow, equal to u_m. In terms of the dimensionless variables introduced above, these inlet conditions are:

$$\text{At } Z = 0: \quad U = 1, \quad V = 0, \quad \theta = 0, \quad P = -0.5 \quad (4.259)$$

First consider the momentum equation. Using the same nodal points as those used in dealing with flow in a pipe, the finite-difference form of this equation is:

$$\frac{1}{Pr} U_{i-1,j} \left(\frac{U_{i,j} - U_{i-1,j}}{\Delta Z} \right) + \frac{1}{Pr} V_{i-1,j} \left(\frac{U_{i-1,j+1} - U_{i-1,j-1}}{2\Delta Y} \right)$$

$$= -\frac{1}{Pr} \frac{(P_i - P_{i-1})}{\Delta Z} + \frac{(U_{i-1,j+1} + U_{i-1,j-1} - 2U_{i-1,j})}{\Delta Y^2} \quad (4.260)$$

i.e.:

$$U_{i,j} = U_{i-1,j} - \frac{P_i}{U_{i-1,j}} + A_j \quad (4.261)$$

where:

$$A_j = \frac{P_{i-1}}{U_{i-1,j}} - \frac{V_{i-1,j}}{U_{i-1,j}} \left(\frac{U_{i-1,j+1} - U_{i-1,j-1}}{2\Delta Y} \right) \Delta Z$$

$$+ Pr \left[\frac{(U_{i-1,j+1} + U_{i-1,j-1} - 2U_{i-1,j})}{\Delta Y^2} \right] \frac{\Delta Z}{U_{i-1,j}} \quad (4.262)$$

A_j is therefore a known quantity at any stage of the calculation because the values of the variables on the $i - 1$ line are known.

Next consider the continuity equation which can be written as:

$$\frac{\partial V}{\partial Y} = -\frac{\partial U}{\partial Z} \quad (4.263)$$

This equation, as with pipe flow, is applied to the nodal points shown in Fig. 4.28. The equation is actually applied to the point p, shown in Fig. 4.28 which lies halfway between nodal points i, j and $i, j - 1$. The value of $\partial U/\partial Z$ at point p is, as before, taken as the average of the values of $\partial U/\partial Z$ at these two points. Hence, the continuity equation gives:

$$\frac{V_{i,j} - V_{i,j-1}}{\Delta Y} = -\frac{1}{2} \left[\frac{U_{i,j} - U_{i-1,j}}{\Delta Z} + \frac{U_{i,j-1} - U_{i-1,j-1}}{\Delta Z} \right]$$

i.e.:

$$V_{i,j} - V_{i,j-1} = -\frac{\Delta Y}{2\Delta Z} [U_{i,j} - U_{i-1,j} + U_{i,j-1} - U_{i-1,j-1}] \quad (4.264)$$

Substituting Eq. (4.261) into this equation then gives:

$$V_{i,j} - V_{i,j-1} = -\frac{\Delta Y}{2\Delta Z} \left[-\frac{P_i}{U_{i-1,j}} + A_j - \frac{P_i}{U_{i-1,j-1}} + A_{j-1} \right] \quad (4.265)$$

which can be written as:

$$V_{i,j} - V_{i,j-1} = -P_i \left[\frac{\Delta Y}{2\Delta Z} \left(\frac{1}{U_{i-1,j}} + \frac{1}{U_{i-1,j-1}} \right) \right] - \frac{\Delta Y}{2\Delta Z}(A_j + A_{j-1}) \quad (4.266)$$

i.e.:

$$V_{i,j} - V_{i,j-1} = -B_j P_i - C_j \quad (4.267)$$

where:

$$B_j = \frac{\Delta Y}{2\Delta Z} \left[\frac{1}{U_{i-1,j}} + \frac{1}{U_{i-1,j-1}} \right] \quad (4.268)$$

and:

$$C_j = \frac{\Delta Y}{2\Delta Z}(A_j + A_{j-1}) \quad (4.269)$$

It will again be noted that since at any stage of the solution the conditions on the $(i-1)$ line are known, the coefficients A_j, A_{j-1}, B_j, and C_j are known quantities.

Now, on the center line, i.e., at $j = 1$, $V_{i,j}$ is 0. Hence, if Eq. (4.267) is applied sequentially outward from point $j = 2$ to the point $j = N$ which lies on the wall, the following is obtained:

$$V_{i,2} - 0 = -B_2 P_i - C_2$$
$$V_{i,3} - V_{i,2} = -B_3 P_i - C_3$$
$$V_{i,4} - V_{i,3} = -B_4 P_i - C_4$$
$$V_{i,5} - V_{i,4} = -B_5 P_i - C_5$$
$$\vdots$$
$$V_{i,N} - V_{i,N-1} = -B_N P_i - C_N$$

Adding this set of equations together then gives:

$$V_{i,N} = -P_i \sum_{k=2}^{N} B_k - \sum_{k=2}^{N} C_k \quad (4.270)$$

But the boundary conditions give $V_{i,N} = 0$ so Eq. (4.270) gives:

$$P_i = \frac{\sum_{k=2}^{N} C_k}{\sum_{k=2}^{N} B_k} \quad (4.271)$$

This allows the dimensionless pressure, P_i, to be determined. Once this has been found, Eq. (4.261) can be used to find the $U_{i,j}$ values and Eq. (4.267) applied sequentially outward from $j = 2$ as discussed above allows the $V_{i,j}$ values to be found. The boundary conditions give $U_{i,N} = 0$ and $U_{i,1} = U_{i,2}$.

Lastly, the energy equation, i.e., Eq. (4.255), is used to determine the distribution of $\theta_{i,j}$. The finite-difference form of Eq. (4.255) is obtained using the same nodal

216 Introduction to Convective Heat Transfer Analysis

points as used in dealing with the momentum equation and is:

$$U_{i-1,j}\left(\frac{\theta_{i,j} - \theta_{i-1,j}}{\Delta Z}\right) + V_{i-1,j}\left(\frac{\theta_{i-1,j+1} - \theta_{i-1,j-1}}{2\Delta Y}\right)$$

$$= \frac{(\theta_{i-1,j+1} + \theta_{i-1,j-1} - 2\theta_{i-1,j})}{\Delta Y^2} \quad (4.272)$$

which gives:

$$\theta_{i,j} = \theta_{i-1,j} - \frac{V_{i-1,j}}{U_{i-1,j}}\left(\frac{\theta_{i-1,j+1} - \theta_{i-1,j-1}}{2\Delta Y}\right)\Delta Z$$

$$+ \frac{(\theta_{i-1,j+1} + \theta_{i-1,j-1}) - 2\theta_{i-1,j}}{\Delta Y^2} \quad (4.273)$$

This equation allows the $\theta_{i,j}$ values to be found. The boundary conditions give $\theta_{i,1} = \theta_{i,2}$ and either $\theta_{i,N} = 1$ or $\theta_{i,N} = \theta_{i,N-1} + \Delta Y$ depending on whether the wall temperature or the wall heat flux is specified. Once these values are determined, the heat transfer rate can be found. The local heat transfer rate at the wall at any value of Z is given by using Fourier's law as:

$$q_w = +k\left.\frac{\partial T}{\partial y}\right|_{y=W/2} \quad (4.274)$$

In terms of the dimensionless variables introduced above, Eq. (4.210) becomes:

$$\frac{q_w W}{k(T_W - T_0)} = \left.\frac{\partial \theta}{\partial Y}\right|_{Y=0.5} \quad (4.275)$$

The left-hand side of this equation is the local Nusselt number, Nu_W; i.e., Eq. (4.275) gives:

$$Nu_W = \left.\frac{\partial \theta}{\partial Y}\right|_{Y=0.5} \quad (4.276)$$

The finite-difference approximation to this equation is:

$$Nu_W = \frac{\theta_{i,N} - \theta_{i,N-1}}{\Delta Y} \quad (4.277)$$

The mean heat transfer rate up to the value of Z being considered can be obtained by integration of the local wall heat transfer rate or by noting that the total heat transfer rate up to the point considered will be equal to the increase in the enthalpy flux between the inlet and the z value considered, i.e., that:

$$\overline{q_w}z = \int_0^{W/2} \rho u c_p T \, dy - \rho u_m c_p T_0 \frac{W}{2} \quad (4.278)$$

i.e., rewriting this equation in terms of the dimensionless variables, and recalling that:

$$u_m \frac{W}{2} = \int_0^{W/2} u \, dy$$

gives:

$$\frac{\overline{q_w}zRe\,Pr}{T_w - T_0} = \rho c_p u_m \int_0^{0.5} U\theta\,dY$$

i.e.:

$$\frac{\overline{q_w}W}{k(T_w - T_0)} = \frac{1}{Z}\int_0^{0.5} U\theta\,dY \qquad (4.279)$$

The left-hand side of this equation is the mean Nusselt number up to dimensionless distance Z from the inlet considered. Hence, this equation can be written as:

$$\overline{Nu_W} = \frac{1}{Z}\int_0^{0.5} U\theta\,dY \qquad (4.280)$$

The value of the integral can be determined by numerical integration.

The Nusselt numbers introduced above were based on the temperature difference $T_W - T_0$. The Nusselt numbers based on the temperature difference $T_W - T_m$ where T_m is the mean temperature at any value of Z can be deduced by noting that the definition of T_m is such that considering unit width of the channel:

$$2\overline{q_w}z = \dot{m}c_p(T_m - T_0)$$

the left-hand side being the rate at which heat is transferred to the fluid between the inlet and a point distance, z, from the inlet. Hence, since $\dot{m}$ is given by:

$$\dot{m} = \rho u_m W$$

it follows that:

$$RePr(T_m - T_0) = 2\overline{q_w}z/k$$

i.e.:

$$\frac{T_m - T_0}{T_w - T_0} = 2\overline{Nu_W}Z$$

i.e.:

$$\frac{T_w - T_m}{T_w - T_0} = 1 - 2\overline{Nu_W}Z$$

This can be used to find the local Nusselt number based on $T_m - T_0$ which will be given by:

$$Nu_{Wm} = Nu_W \frac{T_w - T_0}{T_w - T_m} = \frac{Nu_W}{1 - 2\overline{Nu_W}Z} \qquad (4.281)$$

The calculation procedure based on the above equations is the same as that used for pipe flow and a computer program, DUCTSYM, written in FORTRAN, based on this procedure is available as discussed in the Preface.

As mentioned in the discussion of pipe flow, because an explicit finite-difference procedure is being used to solve the momentum and energy equations, the solution can become unstable, i.e., as the solution proceeds it can diverge increasingly from the actual solution as indicated in Fig. 4.29. To determine the conditions under

218 Introduction to Convective Heat Transfer Analysis

which such instability will occur, consider the following simplified form of the momentum equation:

$$U\frac{\partial U}{\partial Z} = Pr\left(\frac{\partial^2 U}{\partial Y^2}\right) \quad (4.282)$$

The finite-difference form of this equation is:

$$U_{i-1,j}\left(\frac{U_{i,j} - U_{i-1,j}}{\Delta Z}\right) = Pr\frac{(U_{i-1,j+1} + U_{i-1,j-1} - 2U_{i-1,j})}{\Delta Y^2}$$

which gives:

$$U_{i,j} = U_{i-1,j} + \frac{(U_{i-1,j+1} + U_{i-1,j-1} - 2U_{i-1,j})}{\Delta Y^2}\left(\frac{Pr\Delta Z}{U_{i-1,j}}\right) \quad (4.283)$$

Consider the case where:

$$U_{i-1,j+1} = U_{i-1,j} + \Delta, \quad U_{i-1,j-1} = U_{i-1,j} + \Delta$$

In this case, Eq. (4.283) gives:

$$U_{i,j} = U_{i-1,j} + 2\Delta\left(\frac{Pr\Delta Z}{\Delta Y^2 U_{i-1,j}}\right) \quad (4.284)$$

From this equation it follows that if

$$\left(\frac{Pr\Delta Z}{\Delta Y^2 U_{i-1,j}}\right) > 0.5 \quad (4.285)$$

then:

$$U_{i,j} > U_{i-1,j} + \Delta \quad (4.286)$$

which is physically impossible because it implies that the presence of a velocity excess of Δ on each side of point i, j increases the velocity at point i, j by more than Δ. This is shown in Fig. 4.33.

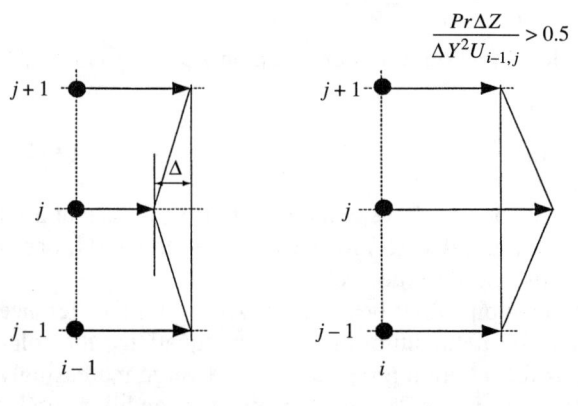

FIGURE 4.33
Induced change in velocity.

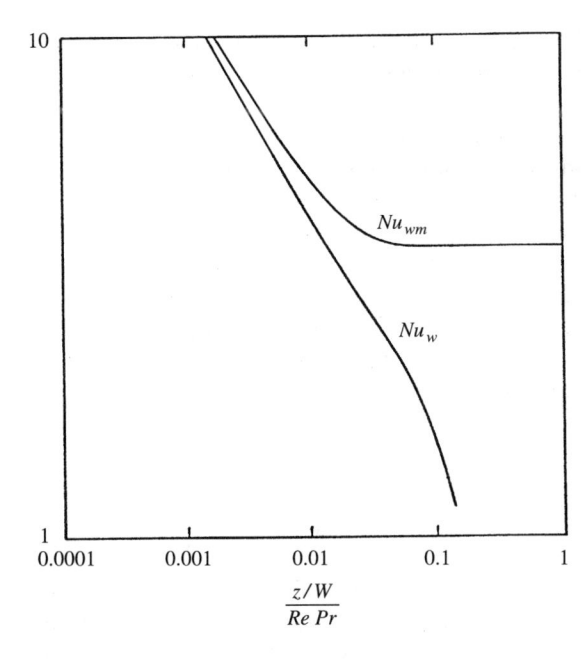

FIGURE 4.34
Nusselt number variations for developing plane duct flow for $Pr = 0.7$.

Hence, for a physically meaningful solution to be obtained, it is necessary that:

$$\Delta Z < \frac{0.5 \, \Delta Y^2 U_{i-1,j}}{Pr} \tag{4.287}$$

If ΔZ is greater than this, a physically meaningless, unstable solution will be obtained. Now, the lowest nodal point velocity will be at the nodal point adjacent to the wall, i.e., at $j = N-1$, since Eq. (4.287) does not apply at the wall itself. Hence, for stability:

$$\Delta Z < \frac{0.5 \, \Delta Y^2 U_{i-1,N-1}}{Pr} \tag{4.288}$$

Because this criterion was obtained using an approximate analysis, it is usual to assume that for stability:

$$\Delta Z < K \frac{\Delta Y^2 U_{i-1,N-1}}{Pr} \tag{4.289}$$

where K is less than 0.5. This stability criterion is incorporated into the above program.

Typical variations of Nu_W and Nu_{Wm} with Z given by the program are shown in Fig. 4.34.

4.9
SOLUTIONS TO THE FULL NAVIER-STOKES EQUATIONS

Approximate solutions to the full governing equations have been discussed in the above sections. The approximations on which these solutions are based are often not applicable. For example, the parabolic forms of the full governing equations which are the basis of the analyses given above do not apply if there are significant areas of reversed flow or if the Reynolds number is low.

When the parabolic flow assumptions cannot be applied, it is necessary to solve the full governing equations. In almost all cases it is necessary to do this using numerical methods [16],[17],[18],[19]. However, since the full equations are a set of nonlinear, simultaneous partial differential equations, this usually requires a significant computational effort. Commercial computer packages based on finite element methods or on various types of finite-difference methods are available for obtaining such solutions. These packages usually contain preprocessing and postprocessing modules that make it easy to set up the required nodal system for complex geometrical situations and to display the calculated results in a convenient form.

4.10
CONCLUDING REMARKS

This chapter has been concerned with the analysis of laminar flows in ducts with various cross-sectional shapes. If the flow is far from the inlet to the duct or from anything else causing a disturbance in the flow, a fully developed state is reached in many situations, the basic characteristics of the flow in this state not changing with distance along the duct. If the diffusion of heat down the duct can be neglected, which is true in most practical situations, it was shown that in such fully developed flows, the Nusselt number based on the difference between the local wall and bulk mean temperatures is constant. Values of the Nusselt number for fully developed flow in ducts of various cross-sectional shape were discussed.

In some practical situations, the velocity field becomes fully developed before changes to the temperature field occur. The analysis of such thermally developing flows was discussed, attention being given to flow in a circular pipe and between parallel plates, i.e., in a plane duct. In most real situations, the velocity and temperature fields develop simultaneously. The numerical analysis of such developing flows was discussed, attention again being restricted to flow in a pipe and a plane duct. The analysis was based on the parabolized form of the governing equations.

Although many internal flows that are of practical importance can be analyzed to an adequate degree of accuracy using the parabolized form of the governing equations, there are many situations in which the full governing equations must be solved.

PROBLEMS

4.1. Consider fully developed laminar flow fluid through a circular pipe with a uniform wall heat flux. If heat is generated uniformly in the fluid, perhaps as the result of a chemical reaction, at a rate of $\dot{q}$ per unit volume, determine the value of the Nusselt number based on the difference between the wall temperature and the mean fluid temperature in the pipe.

CHAPTER 4: Internal Laminar Flows   221

4.2. An organic fluid is to be heated from an initial temperature of 10°C to an exit temperature of 50°C by passing it through a heated pipe with a diameter of 12 mm and a length of 2 m. The pipe is heated by wrapping an electric resistance heating element uniformly about its surface. The mass flow rate of the fluid is 0.1 kg/s. The properties of the organic fluid can be treated as constant and the following values can be assumed: $Pr = 10$, $\rho = 800$ kg/m³, $k = 0.12$ W/m K, and $\mu = 0.008$ kg/ms. By assuming that the flow is fully developed, find and plot the variations of the pipe surface temperature and the mean fluid temperature with distance along the pipe from the inlet.

4.3. Consider fully developed flow in a circular tube with a uniform heat transfer rate at the wall. If the mean fluid velocity in the tube is 1 m/s, find the value of the heat transfer coefficient for the following fluids:
  (a) air at a mean temperature of 80°C and at standard atmospheric pressure flowing through a 2.5-cm diameter tube
  (b) air at a mean temperature of 80°C and at standard atmospheric pressure flowing through a 5.0-cm diameter tube
  (c) air at a mean temperature of 80°C and at a pressure of 1 MPa flowing through a 0.5-cm diameter tube
  (d) hydrogen at a mean temperature of 80°C and at a pressure of twice standard atmospheric pressure flowing through a 2.5-cm diameter tube
  (e) water at a mean temperature of 40°C flowing through a 2-mm diameter tube
  (f) liquid sodium at a mean temperature of 80°C flowing through a 1-mm diameter tube

4.4. Water at a mean bulk temperature of 35°C flows through a pipe with a diameter of 3 mm whose wall is maintained at a uniform temperature of 55°C. The tube length is 2 m. If the mean velocity of the water in the pipe is 0.05 m/s, find the heat-transfer rate to the water assuming fully developed flow.

4.5. In a laminar flow in a pipe with a diameter of 2 cm, the velocity and temperature profiles are given by:

$$u = 0.1[1 - (r/0.01)^2] \text{m/s}$$

and:

$$T = 100 - 3 \times 10^6(1.875 \times 10^{-5} - 0.25r^2 + 624r^4)\text{°C}$$

for radius, r, in meters. Determine the bulk temperature.

4.6. Oil is heated by passing it through a pipe with an inside diameter of 1 cm and a length of 1.5 m. An electrical heating element is wrapped around the surface of this pipe and provides a uniform heat transfer rate at the pipe wall. The oil is to be heated from a temperature of 10°C to a temperature of 40°C and the mean velocity of the oil in the pipe is 1.3 m/s. Find the heat transfer rate required at the wall and, assuming that the flow is laminar and fully developed, i.e., ignoring the entrance region, show how the wall and mean oil temperatures vary along the pipe. Assume that the oil has a density of 890 kg/m³, a specific heat of 1.9 kJ/kg°C, a coefficient of viscosity of 0.1 N-s/m², a thermal conductivity of 0.15 W/m K, and a Prandtl number of 500.

4.7. Oil is heated by passing it at a mean velocity of 1 m/s through a pipe. A uniform heat flux exists at the wall of this pipe. At the exit end of the pipe the wall has a temperature of 50°C and the oil a mean temperature of 30°C. Find the lowest temperature and highest velocity in the oil flow at this exit end of the pipe. Assume that at the exit, the flow is laminar and fully developed.

4.8. Derive an expression for the Nusselt number in fully developed laminar slug flow through an annulus when the inner and outer surfaces of the annulus have diameters of D_i and D_o respectively and when there is a uniform heat flux applied at the inner surface and when the outer surface is adiabatic.

4.9. In some situations it is possible to find the heat transfer rate with adequate accuracy by assuming that the velocity is constant across the duct, i.e., to assume that so-called "slug flow" exists. Find the temperature distribution and the Nusselt number in slug flow in a plane duct when the thermal field is fully developed and when there is a uniform wall heat flux.

4.10. Air flows at a mean velocity of 1 m/s between flat plates that can be assumed to be at a uniform temperature of 50°C. The plates have a length of 15 cm in the direction of air flow. The gap between the plates is 4 mm and the initial air temperature is 20°C. Assuming that the flow between the plates is laminar and fully developed, estimate the mean temperature of the air leaving the system.

4.11. Water is heated by passing it through an array of parallel, wide heated plates. The plates thus form a series of parallel plane channels. The distance between the plates is 1 mm and the mean velocity in the channels is 1 m/s. The plates are electrically heated, the two sides of each plate together transferring heat to the water flow at a rate of 1600 W/m². The water properties can be assumed to be constant and they can be evaluated at a temperature of 50°C. If the flow is assumed to be fully developed, find the rate of increase of mean water temperature with distance along the channels.

4.12. Consider fully developed flow in a plane duct in which uniform heat fluxes q_{w1} and q_{w2} are applied at the two walls. Derive expressions for the temperature distribution in the duct and the Nusselt number.

4.13. The cooling channels in a metal block which is held at a uniform temperature of 50°C have a rectangular cross-sectional shape with a width of 6 mm and a height of 2 mm. Air at an initial temperature of 10°C flows through these channels at a mean velocity of 1.5 m/s. The block has a length of 22 cm in the flow direction. Find the heat transfer rate to the air in one channel. Assume fully developed laminar flow.

4.14. Water flows in a rectangular 5 mm × 10 mm duct with a mean bulk temperature of 20°C. If the duct wall is kept at a uniform temperature of 40°C and if fully developed laminar flow is assumed to exist, find the heat transfer rate per unit length of the duct.

4.15. Consider a series of rectangular channels of width W, and height H, all with the same cross-sectional area whose walls are kept at a uniform temperature. The flow in these ducts can be assumed to be fully developed and laminar.
  (i) If it is assumed that the hydraulic diameter concept can be used, i.e., that the Nusselt number for fully developed flow in a pipe applies provided the hydraulic diameter is used as the length scale, find how the quantity $Q/2(H + W)(T_w - T_m)$ varies with W/H for values of W/H between 0.5 and 3. Q is the heat transfer rate per unit length of the duct.

(ii) Determine the variation of the quantity $Q/2(H + W)(T_w - T_m)$ with W/H for values of W/H between 0.5 and 3 using the actual Nusselt numbers for flow in a rectangular duct for various values of W/H.

4.16. Consider laminar fully developed flow through either a circular pipe or a rectangular duct whose cross-section has a length/width ratio of 8. Both of these ducts have the same cross-sectional area and the average velocity in them is the same. Determine the ratio of the (heat-transfer coefficient × surface area) product for these two flow situations.

4.17. Consider air flow in a plane channel when there is uniform heat flux at one wall and when the other wall is adiabatic. If the inlet air temperature is known, find how the temperature of the heated wall at the exit end of the duct varies with the distance, W, between the two walls. Assume that the mean air velocity is the same in all cases and that the flow is fully developed.

4.18. Find the variation of the local heat transfer rate with axial distance for the thermally developing flow of oil through a pipe with an inside diameter of 1 cm and a heated length of 10 cm. The wall of the pipe is kept at a temperature of 50°C and the oil enters the heated section of the pipe at a temperature of 10°C. The mean velocity of the oil in the pipe is 1.3 m/s. Assume that the oil has a density of 890 kg/m^3, a specific heat of 1.9 kJ/kg°C, a coefficient of viscosity of 0.1 kg/ms, and a thermal conductivity of 0.15 W/m K.

4.19. Discuss how the computer program for flow in the thermal entrance region to a plane duct must be modified in order to apply to the case where slug flow exists, i.e., for the situation in which the velocity can be assumed to be constant across the pipe and equal, of course, to the mean velocity in the pipe.

4.20. Discuss the modifications required to the computer program given for flow in the thermal entrance region of a plane duct to deal with the case where there is a uniform heat flux at one wall of the duct and where there is a uniform specified temperature at the other wall of the duct.

4.21. Air at 20 kPa and 5°C enters a 1.5-cm diameter tube at a uniform velocity of 1.5m/s. The tube walls are maintained at a uniform temperature of 45°C. Estimate the distance from the entrance at which the flow becomes fully developed.

4.22. Water at 15°C enters a plane duct (i.e., effectively flows between two large parallel plates) at a uniform velocity of 1.5 m/s. The walls of the duct are separated by a distance of 1 mm and are kept at a uniform temperature of 25°C. Estimate the distance from the entrance at which the flow can be assumed to be fully developed.

4.23. Engine oil enters a 5.0-mm diameter tube at a temperature of 120°C at a uniform velocity whose value is such that the Reynolds number is 1000. The tube wall is kept at a uniform temperature of 60°C. Calculate the mean exit oil temperatures for the tube lengths of 10, 20, and 50 cm.

4.24. Engine oil at a temperature of 20°C enters a 2-mm diameter tube at a uniform velocity of 1.2 m/s. The tube is 1.0 m long. The tube wall is maintained at a uniform temperature of 60°C. Calculate the mean exit oil temperature.

4.25. The program for developing flow in a plane channel discussed in this chapter assumed that both walls were at the same uniform temperature. Modify the program to allow for the possibility that the upper and lower walls of the duct are at different uniform temperatures. Use this modified program to find the variation of the heat transfer rate with distance along the upper and lower walls for the case of air flowing at a mean velocity of 0.6 m/s through a duct in which the upper and lower surfaces are 1 cm apart. The upper surface is kept at a temperature of 40°C and the lower surface is kept at a temperature of 30°C. The air enters the channel at a temperature of 15°C. Evaluate the properties of the air at a temperature of 20°C.

4.26. Discuss the modifications to the program for developing flow in a pipe that are necessary to allow it to calculate developing flow in an annulus when the inner and outer surfaces of the annulus have diameters of D_i and D_o, respectively, and when both the inner and outer surfaces are kept at the same uniform temperature.

4.27. Oil enters a 10-mm diameter pipe at a temperature of 20°C with a uniform velocity of 1 m/s. The wall of the pipe is kept at a uniform temperature of 40°C. If the pipe is 70 cm long, plot the variation of the temperature and velocity in the oil at the exit of the pipe. Assume that for the oil $\rho = 890$ kg/m^3, $\mu = 0.0075$ kg/ms, $k = 0.15$ W/m K, and $Pr = 900$.

4.28. Air flows at a mean velocity of 1.5 m/s between flat plates that can be assumed to be at a uniform temperature of 40°C. The plates have a length of 20 cm in the direction of air flow and are wide in the other direction and the gap between the plates is 4 mm. The initial air temperature is 20°C and the flow can be assumed to have a uniform velocity at the inlet to the system. Can the flow between the plates be assumed to be fully developed? Evaluate the air properties at a temperature of 30°C.

4.29. In a heat exchanger, air flows through a pipe of length, ℓ. The air enters the pipe at a temperature, T_i. The heat flux at the wall of this pipe increases linearly from zero at the inlet of the pipe to a value of q_w at the end of the pipe. The velocity in the pipe is such that $\ell/Re\,PrD = 0.07$ where D is the diameter of the pipe and Re is the Reynolds number. Determine how the Nusselt number based on the local wall heat transfer rate and on the difference between the local wall temperature and the inlet temperature varies with the dimensionless distance, Z, along the pipe.

4.30. Air at a temperature of 20°C enters a 25-mm diameter pipe with a uniform velocity of 0.7 m/s. The pipe has a length of 30 cm and has a uniform wall temperature of 50°C. Determine the air outlet temperature
(i) assuming fully developed flow throughout the pipe and
(ii) accounting for entrance region effects.

4.31. Water flows through a pipe of length, ℓ. The water enters the pipe with a uniform velocity and at a uniform temperature of T_i. The heat flux is uniform around the pipe but varies with distance z along the pipe, this variation being described by:

$$q_w = q_{wo}\left[1 + A\sin(\pi z/\ell)\right]$$

A being a constant. The velocity in the pipe is such that $\ell/RePrD = 0.06$ where D is the diameter of the pipe and Re is the Reynolds number. Numerically determine how the Nusselt number based on the local wall heat transfer rate and on the difference between the local wall temperature and the inlet temperature varies with the dimensionless distance, Z, along the pipe.

4.32. In a heat exchanger, air essentially flows through a plane channel of length, ℓ. The air enters the channel at a temperature, T_i. The temperature of the walls of this duct increases linearly from T_i to T_R over the first half of the duct and then remains constant at T_R. The velocity in the duct is such that $\ell/Re\,Pr\,W = 0.06$ where W is the width of the duct and Re is the Reynolds number. Determine how the Nusselt number based on the local wall heat transfer rate and on the difference between the local wall temperature and the inlet temperature varies with the dimensionless distance, Z, along the duct.

4.33. Consider constant-property, fully developed laminar flow between two large parallel plates, i.e., in a wide plane duct. One plate is adiabatic and the other is isothermal and the velocity is high enough for viscous dissipation effects to be significant. Determine the temperature distribution in the flow.

4.34. Consider laminar flow of a high-viscosity oil in a circular tube with a uniform wall temperature. If viscous dissipation effects are significant, determine the temperature profile far from the inlet, i.e., in fully developed flow. Assume constant fluid properties.

4.35. A gas flows between large parallel porous plates. The same gas is blown through one wall and exhausted through the other in such a way that the same normal velocity component, v_s, exists at both walls. The plates are kept at uniform but different temperatures. Assuming fully developed, constant fluid-property flow, find equations for the velocity and temperature profiles.

REFERENCES

1. Schlichting, H., *Boundary Layer Theory*, 7th ed., McGraw-Hill, New York, 1979.
2. Shah, R.K. and London, A.L., *Laminar Forced Convection in Ducts*, Academic Press, New York, 1978.
3. Siegel, R., Sparrow, E.M., and Hallman, T.M., "Steady Laminar Heat Transfer in a Circular Tube with Prescribed Wall Heat Flux", *Appl. Sci. Res., Sect. A*, Vol. 7, p. 386, 1958.
4. Worsoe-Schmidt, P.M., "Heat Transfer and Friction for Laminar Flow of Helium and Carbon Dioxide in a Circular Tube at High Heating Rate", *Int. J. Heat-Mass Transfer*, Vol. 9, pp. 1291–1295, 1966.
5. Yang, K.T., "Laminar Forced Convection of Liquids in Tubes with Variable Viscosity", *J. Heat Transfer*, Vol. 84, pp. 353–362, 1962.
6. Swearingen, T.W. and McEligot, D.M., "Internal Laminar Heat Transfer with Gas Property Variation", *Trans. ASME, Ser. C, J. Heat Transfer*, Vol. 93, pp. 432–440, 1971.
7. Clark, S.H. and Kays, W.M., "Laminar Flow Forced Convection in Rectangular Tubes", *Trans. ASME*, Vol. 75, p. 859, 1953.
8. Shah, R.K., "Laminar Flow Friction and Forced Convection Heat Transfer in Ducts of Arbitrary Geometry", *Int. J. Heat Mass Transfer*, Vol. 18, pp. 849–862, 1975.
9. Shah, R.K. and London, A.L., "Thermal Boundary Conditions and Some Solutions for Laminar Duct Flow Forced Convection", *J. Heat Transfer*, Vol. 96, pp. 159–165, 1974.
10. Burmeister, L.C., "Convective Heat Transfer", 2nd ed., Wiley-Interscience, New York, 1993.
11. Sellars, J.R., Tribus, M., and Klein, J.S., "Heat Transfer to Laminar Flow in a Round Tube or Flat Conduit—The Graetz Problem Extended", *Trans. ASME*, Vol. 78, p. 441, 1956.
12. Hornbeck, R.W., "An All-Numerical Method for Heat Transfer in the Inlet of a Tube", *Am. Soc. Mech. Eng.*, paper 65-WA/HT-36, 1965.

13. Kays, W.M., "Numerical Solution for Laminar Flow Heat Transfer in Circular Tubes", *Trans. ASME,* Vol. 77, pp. 1265–1274, 1955.
14. Shah, R.K., "A Correlation for Laminar Hydrodynamic Entry Length Solutions for Circular and Non-Circular Ducts", *J. Fluids Eng. Trans. ASME,* Vol. 100, p. 177, 1978.
15. Sparrow, E.M., "Analysis of Laminar Forced Convection Heat Transfer in the Entrance Region of Flat Rectangular Ducts", *NACA* TN 3331, 1955.
16. Gosman, A.D., Pun, W.M., Runchal, A.K., Spalding, D.B., and Wolfshtein, M., *Heat and Mass Transfer in Recirculating Flows,* Academic Press, New York, 1969.
17. Patankar, S.V., *Numerical Heat Transfer and Fluid Flow,* Hemisphere Publ., Washington, D.C., 1980.
18. Oosthuizen, P.H., "Laminar Forced Convective Heat Transfer from Rectangular Blocks Mounted on Opposite Walls of a Channel", Numerical Methods in Thermal Problems, Vol. VI, Part 1, Proc., 6th Int. Conf., Swansea, U.K., July 3–7, pp. 451–461, 1989.
19. Janssen, L.A.M. and Hoogendoorn, C.J., "Laminar Convective Heat Transfer in Helically Coiled Tubes", *Int. J. Heat Mass Transfer,* Vol. 21, pp. 1197–1206, 1978.

CHAPTER 5

Introduction to Turbulent Flows

5.1 INTRODUCTION

As discussed in Chapter 2, the values of the variables such as velocity and temperature fluctuate with time when the flow is turbulent. This is illustrated for the case of temperature in Fig. 5.1.

It is not, at the present time, possible on a routine basis to solve the governing equations to obtain the variation of the flow variables with time in turbulent flow. In most analyses of turbulent flow it is therefore usual to express the variables in terms of a time-averaged mean value plus a time-varying deviation from this mean value, i.e., as discussed in Chapter 2 [1],[2],[3],[4],[5], to express the variables in the following way:

$$u = \bar{u} + u', \quad v = \bar{v} + v', \quad w = \bar{w} + w',$$

$$p = \bar{p} + p', \quad T = \bar{T} + T' \tag{5.1}$$

where the primes denote the fluctuating components and where the mean values, indicated by the over-bar, are defined, for the case where the mean flow is steady, by:

$$\bar{u} = \frac{1}{t}\int_0^t u\,dt, \quad \bar{v} = \frac{1}{t}\int_0^t v\,dt, \quad \bar{w} = \frac{1}{t}\int_0^t w\,dt,$$

$$\bar{p} = \frac{1}{t}\int_0^t p\,dt, \quad \bar{T} = \frac{1}{t}\int_0^t T\,dt \tag{5.2}$$

Here t is a suitably long integration time. A solution for the mean values of the flow variables is then sought.

In order to illustrate the main features of the analysis of turbulent flow, attention will be restricted to two-dimensional boundary layer flows and to axially symmetric pipe flows. It will also be assumed that the fluid properties are constant and that the mean flow is steady.

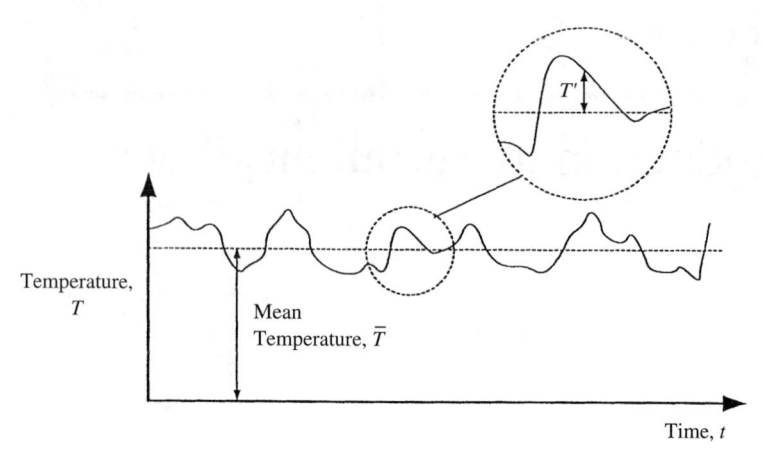

FIGURE 5.1
Variation of temperature with time in turbulent flow.

The equations governing the variation of the mean flow variables in such flows were discussed in Chapter 2. These equations contain, besides the mean flow variables, additional variables that depend on the fluctuating turbulence quantities (see below). There are thus, more variables in these governing sets of equations than there are equations. This is termed the "turbulence closure problem". In order to solve turbulent flow problems, it is necessary therefore to relate the extra turbulence terms to the mean flow variables, i.e., to introduce a "turbulence model". This, at the present, essentially always involves some degree of empiricism. The degree to which empiricism is introduced and the way in which it is introduced varies considerably between models. Some highly involved equations for these turbulence terms have been developed in which it is only necessary to specify the experimentally measured values of a relatively small number of fundamental constants. Attention will be restricted to relatively simple turbulence models in the present book. In discussing these turbulence models it will be necessary to introduce empiricism at certain points in the discussion. While every effort will be made to justify the way in which this empiricism is introduced, this will not always be possible without a more thorough discussion of the background experimental results than can be given here.

In the present chapter and in the following two chapters, which are concerned with turbulent boundary layer flows and with turbulent duct flows, respectively, consideration will be restricted to forced flows, i.e., the effect of buoyancy forces on the mean flow and on the turbulence structure will be assumed to be negligible. Some discussion of the effect of buoyancy forces on turbulent flows will be given in Chapter 9.

5.2
GOVERNING EQUATIONS

As mentioned above, attention will be restricted to constant fluid property boundary layer and pipe flows. The equations governing these types of flow were discussed briefly in Chapter 2. In the present section these equations will be reviewed and some points concerning their form will be discussed.

The equations governing two-dimensional, constant fluid property turbulent boundary layer flow are [2],[3]:

$$\frac{\partial \bar{u}}{\partial x} + \frac{\partial \bar{v}}{\partial y} = 0 \tag{5.3}$$

$$\bar{u}\frac{\partial \bar{u}}{\partial x} + \bar{v}\frac{\partial \bar{u}}{\partial y} = -\frac{1}{\rho}\frac{d\bar{p}}{dx} + \nu\frac{\partial^2 \bar{u}}{\partial y^2} - \frac{\partial}{\partial y}(\overline{v'u'}) \tag{5.4}$$

$$\bar{u}\frac{\partial \bar{T}}{\partial x} + \bar{v}\frac{\partial \bar{T}}{\partial y} = \left(\frac{k}{\rho c_p}\right)\frac{\partial^2 \bar{T}}{\partial y^2} - \frac{\partial}{\partial y}(\overline{v'T'}) \tag{5.5}$$

As discussed above, this set of three equations contains, in addition to the three mean flow variables $\bar{u}$, $\bar{v}$, and $\bar{T}$, the two turbulence quantities $\overline{v'u'}$ and $\overline{v'T'}$. Now since the molecular shearing stress, τ_m, and the molecular heat transfer, q_m, are given by:

$$\tau_m = \mu\frac{\partial \bar{u}}{\partial y} = \rho\nu\frac{\partial \bar{u}}{\partial y} \tag{5.6}$$

and:

$$q_m = -k\frac{\partial \bar{T}}{\partial y} = -\rho c_p\frac{\partial \bar{T}}{\partial y} \tag{5.7}$$

Eqs. (5.4) and (5.5) can be written as:

$$\rho\left(\bar{u}\frac{\partial \bar{u}}{\partial x} + \bar{v}\frac{\partial \bar{u}}{\partial y}\right) = -\frac{d\bar{p}}{dx} + \frac{\partial \tau_m}{\partial y} - \frac{\partial}{\partial y}(\overline{\rho v'u'}) \tag{5.8}$$

and:

$$\rho c_p\left(\bar{u}\frac{\partial \bar{T}}{\partial x} + \bar{v}\frac{\partial \bar{T}}{\partial y}\right) = -\frac{\partial q_m}{\partial y} - \frac{\partial}{\partial y}(\overline{\rho c_p v'T'}) \tag{5.9}$$

A consideration of the right-hand sides of these two equations indicates that the turbulence terms in these equations have, as discussed in Chapter 2, the form of additional shearing stress and heat transfer terms although they arise, of course, from the momentum transfer and enthalpy transfer produced by the mixing that arises from the turbulence. Because of their similarity to the molecular terms, the turbulence terms are usually called the turbulent shear stress and turbulent heat transfer rate respectively. Thus, the following are defined:

$$\tau_T = -\overline{\rho v'u'} \tag{5.10}$$

and:

$$q_T = \overline{\rho c_p v'T'} \tag{5.11}$$

τ_T being the turbulent shear stress and q_T being the turbulent heat transfer rate.

In terms of these quantities, the momentum and energy equations (5.8) and (5.9) can be written as:

$$\rho\left(\bar{u}\frac{\partial \bar{u}}{\partial x} + \bar{v}\frac{\partial \bar{u}}{\partial y}\right) = -\frac{d\bar{p}}{dx} + \frac{\partial}{\partial y}(\tau_m + \tau_T) \tag{5.12}$$

and:

$$\rho c_p\left(\bar{u}\frac{\partial \bar{T}}{\partial x} + \bar{v}\frac{\partial \bar{T}}{\partial y}\right) = -\frac{\partial}{\partial y}(q_m + q_T) \tag{5.13}$$

In most turbulent flows, τ_T is much greater than τ_m and q_T is much greater than q_m over most of the flow field although this will not be the case near the wall.

By analogy with the form of the molecular shearing stress and molecular heat transfer rate relations, as given in Eqs. (5.6) and (5.7), it is often convenient to express the turbulent shearing stress and turbulent heat transfer in terms of the velocity and temperature gradients in the following way:

$$\tau_T = -\overline{\rho v' u'} = \rho \epsilon \frac{\partial \bar{u}}{\partial y} \tag{5.14}$$

and:

$$q_T = \overline{\rho c_p v' T'} = -\rho c_p \epsilon_H \frac{\partial \bar{T}}{\partial y} \tag{5.15}$$

where ϵ is known as the eddy viscosity and ϵ_H is known as the eddy diffusivity (sometimes named eddy conductivity).

It must, of course, be clearly understood that ϵ and ϵ_H are not, like ν and α, properties of the fluid involved alone but depend primarily on the turbulence structure at the point under consideration and hence on the mean velocity and temperature at this point and the derivatives of these quantities as well as on the type of flow being considered. The use of ϵ and ϵ_H does not, in itself, constitute the use of an empirical turbulence model. It is only when attempts are made to describe the variation of ϵ and ϵ_H through the flow field on the basis of certain usually rather limited experimental measurements that the term eddy viscosity turbulence model is applicable. In fact, even when advanced turbulence models are used, it is often convenient to express the end results in terms of the eddy viscosity and eddy diffusivity.

If Eqs. (5.14) and (5.15) are used to describe τ_T and q_T, the momentum and energy equations (5.8) and (5.9) can be rewritten in the following way because the fluid properties are being assumed constant:

$$\bar{u}\frac{\partial \bar{u}}{\partial x} + \bar{v}\frac{\partial \bar{u}}{\partial y} = -\frac{1}{\rho}\frac{d\bar{p}}{dx} + \frac{\partial}{\partial y}\left[(\nu + \epsilon)\frac{\partial \bar{u}}{\partial y}\right] \tag{5.16}$$

and:

$$\bar{u}\frac{\partial \bar{T}}{\partial x} + \bar{v}\frac{\partial \bar{T}}{\partial y} = \frac{\partial}{\partial y}\left[(\alpha + \epsilon_H)\frac{\partial \bar{T}}{\partial y}\right] \tag{5.17}$$

CHAPTER 5: Introduction to Turbulent Flows   231

Now it will be noted that the Prandtl number is defined such that:

$$Pr = \frac{\mu c_p}{k} = \frac{\mu/\rho}{k/\rho c_p} = \frac{\nu}{\alpha} \qquad (5.18)$$

By analogy with this, it is often convenient to define the following:

$$Pr_T = \frac{\epsilon}{\epsilon_H} \qquad (5.19)$$

Pr_T then being termed the turbulent Prandtl number. In order to avoid confusion, Pr is then sometimes referred to as the molecular Prandtl number.

Using these expressions, the energy equation (5.17) can be written as:

$$\bar{u}\frac{\partial \bar{T}}{\partial x} + \bar{v}\frac{\partial \bar{T}}{\partial y} = \nu \frac{\partial}{\partial y}\left[\left(\frac{1}{Pr} + \frac{\epsilon}{\nu}\frac{1}{Pr_T}\right)\frac{\partial \bar{T}}{\partial y}\right] \qquad (5.20)$$

Eqs. (5.3), (5.16), and (5.20) are basically the form of the governing equations that will be used in the analysis of turbulent boundary layer flows. As mentioned before, attention will also be given to turbulent pipe flows. If the same coordinate system that was used in the discussion of laminar pipe flows is adopted, i.e., if the coordinate system shown in Fig. 5.2 is used, the equations governing turbulent pipe flow are, if assumptions similar to those used in dealing with boundary layer flows are adopted and if it is assumed that there is no swirl, as follows:

$$\frac{\partial \bar{u}}{\partial z} + \frac{1}{r}\frac{\partial}{\partial r}(\bar{v}r) = 0 \qquad (5.21)$$

$$\bar{u}\frac{\partial \bar{u}}{\partial z} + \bar{v}\frac{\partial \bar{u}}{\partial r} = -\frac{1}{\rho}\frac{d\bar{p}}{dz} + \frac{\nu}{r}\frac{\partial}{\partial r}\left(r\frac{\partial \bar{u}}{\partial r}\right) - \frac{1}{r}\frac{\partial}{\partial r}(r\overline{v'u'}) \qquad (5.22)$$

$$\bar{u}\frac{\partial \bar{T}}{\partial x} + \bar{v}\frac{\partial \bar{T}}{\partial r} = \left(\frac{k}{\rho c_p}\right)\frac{1}{r}\frac{\partial}{\partial r}\left(r\frac{\partial \bar{T}}{\partial r}\right) - \frac{1}{r}\frac{\partial}{\partial r}(r\overline{v'T'}) \qquad (5.23)$$

The above set of three equations also involves, beside the three mean flow variables $\bar{u}$, $\bar{v}$, and $\bar{T}$, the two turbulence quantities $\overline{v'u'}$ and $\overline{v'T'}$. These quantities also have the form of additional shear stress and heat transfer, respectively. For this reason the momentum and energy equations are often written in the following way just as

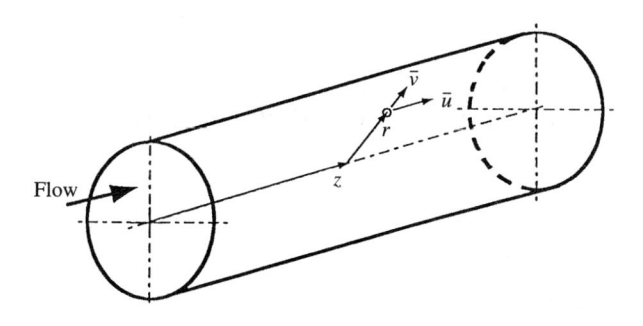

FIGURE 5.2
Coordinate system used in the analysis of turbulent pipe flow.

232 Introduction to Convective Heat Transfer Analysis

was done with the boundary layer equations:

$$\rho\left(\bar{u}\frac{\partial \bar{u}}{\partial z} + \bar{v}\frac{\partial \bar{u}}{\partial y}\right) = -\frac{d\bar{p}}{dz} + \frac{1}{r}\frac{\partial}{\partial r}\left[r(\tau_m + \tau_T)\right] \quad (5.24)$$

and:

$$\rho c_p\left(\bar{u}\frac{\partial \bar{T}}{\partial z} + \bar{v}\frac{\partial \bar{T}}{\partial y}\right) = \frac{1}{r}\frac{\partial}{\partial r}\left[r(q_m + q_T)\right] \quad (5.25)$$

where τ_m and τ_T are again the molecular shear stress and turbulent shear stress respectively, and q_m and q_T are the molecular heat transfer rate and turbulent heat transfer rate respectively.

In turbulent pipe flow it is again also often convenient to write the turbulence quantities in terms of the eddy viscosity and diffusivity and when this is done the momentum and energy equations become:

$$\bar{u}\frac{\partial \bar{u}}{\partial z} + \bar{v}\frac{\partial \bar{u}}{\partial y} = -\frac{1}{\rho}\frac{d\bar{p}}{dz} + \frac{1}{r}\frac{\partial}{\partial r}\left[r(\nu + \epsilon)\frac{\partial \bar{u}}{\partial r}\right] \quad (5.26)$$

and:

$$\bar{u}\frac{\partial \bar{T}}{\partial z} + \bar{v}\frac{\partial \bar{T}}{\partial y} = \frac{1}{r}\frac{\partial}{\partial r}\left[r(\alpha + \epsilon_H)\frac{\partial \bar{T}}{\partial r}\right] \quad (5.27)$$

The energy equation, i.e., Eq. (5.25), can alternatively be written as:

$$\bar{u}\frac{\partial \bar{T}}{\partial z} + \bar{v}\frac{\partial \bar{T}}{\partial y} = \frac{\nu}{r}\frac{\partial}{\partial r}\left[r\left(\frac{1}{Pr} + \frac{\epsilon}{\nu}\frac{1}{Pr_T}\right)\frac{\partial \bar{T}}{\partial r}\right] \quad (5.28)$$

where the turbulent Prandtl number as defined in Eq. (5.19) has been introduced in this equation.

EXAMPLE 5.1. Consider fully developed flow in a plane duct, i.e., essentially fully developed flow between parallel plates. If the velocity distribution in the flow near the center line can be approximately represented by:

$$\frac{\bar{u}}{\bar{u}_1} = \left(\frac{y}{w}\right)^{1/7}$$

where $\bar{u}_1$ is the mean center line velocity, y is the distance from the wall, and w is the half-width of the duct, find the variation of ϵ and ϵ_H in this portion of the flow.

Solution. The flow situation being considered is shown in Fig. E5.1a.

In fully developed flow, $\bar{u}$ will not depend on z and hence by virtue of the continuity equation $\bar{v}$ will be 0. Therefore, for the fully developed flow, the momentum equation gives:

$$0 = -\frac{d\bar{p}}{dz} + \frac{d}{dy}(\tau_m + \tau_T)$$

i.e.:

$$\frac{d}{dy}(\tau_m + \tau_T) = \frac{d\bar{p}}{dz}$$

CHAPTER 5: Introduction to Turbulent Flows   233

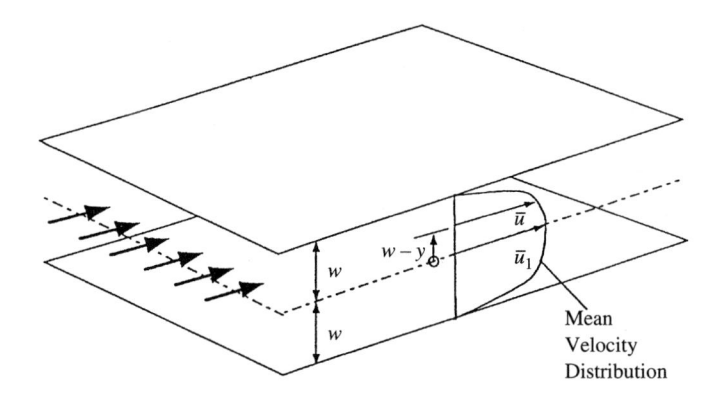

FIGURE E5.1a

But $d\bar{p}/dz$ is a constant in fully developed flow, i.e., is independent of z. Therefore, the above equation indicates that the total shear stress, i.e., $\tau_m + \tau_T$, varies linearly with y. But, on the center line, i.e., at $y = w$, $d\bar{u}/dy = 0$, so the total shear stress is zero on the center line. Therefore, if the wall shear stress is τ_w, it follows that:

$$\frac{(\tau_m + \tau_T)}{\tau_w} = 1 - \frac{y}{w}$$

Except in the region very close to the wall $\tau_m \ll \tau_T$. Hence, in the region here being considered:

$$\frac{\tau_T}{\tau_w} = 1 - \frac{y}{w}$$

It is next noted that in the case here being considered:

$$\tau_T = \rho \epsilon \frac{d\bar{u}}{dy}$$

From the above two equations it follows that:

$$\epsilon = \tau_w \frac{(1 - y/w)}{\rho(d\bar{u}/dy)}$$

But the assumed velocity profile gives:

$$\frac{d\bar{u}}{dy} = \frac{d}{dy}\left[\bar{u}_1 \left(\frac{y}{w}\right)^{1/7}\right] = \frac{\bar{u}_1}{7w}\left(\frac{y}{w}\right)^{-6/7}$$

From the above equations, it then follows that:

$$\epsilon = \left(\frac{7\tau_w w}{\rho \bar{u}_1}\right) \frac{1 - y/w}{(y/w)^{-6/7}}$$

i.e.:

$$\frac{\epsilon}{w\bar{u}_1} = 7\frac{\tau_w}{\rho \bar{u}_1^2}\left[1 - \left(\frac{y}{w}\right)\right]\left(\frac{y}{w}\right)^{6/7}$$

The variation of ϵ across the duct as given by this equation is shown in Fig. E5.1b. It should be noted that the equation does not really apply for values of y/w near 0, i.e., near the wall.

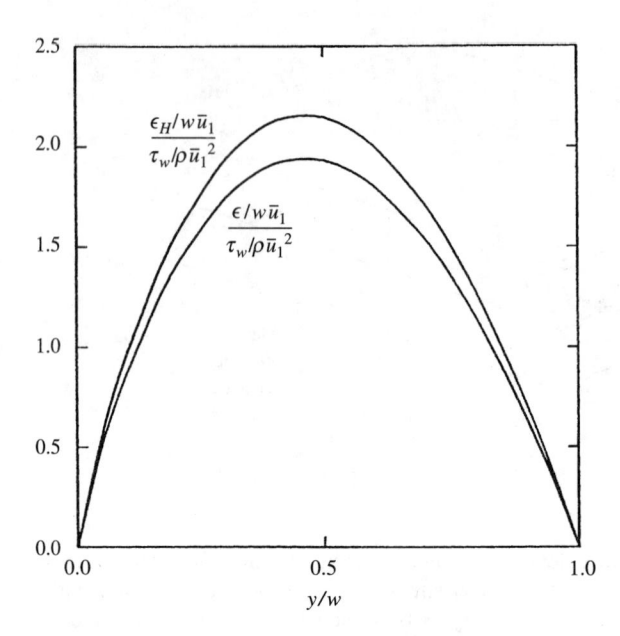

FIGURE E5.1b

[Figure showing $\epsilon_H/w\bar{u}_1 \over \tau_w/\rho\bar{u}_1^2$ and $\epsilon/w\bar{u}_1 \over \tau_w/\rho\bar{u}_1^2$ plotted against y/w]

If the turbulent Prandtl number is assumed constant then since:

$$\epsilon_H = \frac{\epsilon}{Pr_T}$$

it follows that:

$$\frac{\epsilon_H}{w\bar{u}_1} = \frac{7}{Pr_T} \frac{\tau_w}{\rho\bar{u}_1^2}\left[1 - \left(\frac{y}{w}\right)\right]\left(\frac{y}{w}\right)^{6/7}$$

The variation of ϵ_H as given by this equation assuming $Pr_T = 0.9$ (see later) is also shown in Fig. E5.16.

5.3
MIXING LENGTH TURBULENCE MODELS

The simplest turbulence model is based on the assumption that relations between ϵ and ϵ_H (or Pr_T) and the mean flow variables can be obtained from measurements in relatively simple flow situations and that these relations are then more universally applicable. Such an approach has, however, not met with a great deal of success except in the analysis of free boundary flows, e.g., in jets and wakes.

Another method of trying to describe the turbulence terms in the above equations is by means of Prandtl's mixing length theory. The mixing length concept will be introduced in this section and some simple turbulence models based on this concept will be discussed [1],[2],[3],[6],[7].

In the mixing length theory it is assumed basically that "lumps" of fluid are carried transversely across the fluid flow by the turbulent eddies and during this motion they preserve their initial momentum and enthalpy. The motion continues over a transverse distance, ℓ_m, after which the "lumps" interact with other fluid "lumps"

CHAPTER 5: Introduction to Turbulent Flows   235

giving rise to the observed velocity and temperature fluctuations in turbulent flow. In order to derive an expression for the turbulence terms using this idea, consider a two-dimensional flow and consider conditions on three planes 0, 1, and 2 shown in Figure 5.3, planes 1 and 2 lying at a distance ℓ_m from plane 0.

If the mixing length, ℓ_m, is assumed to be small compared to the overall extent of the flow then the velocity on plane 1, i.e., u_1, will be related to that on plane 0, i.e., u_0, by:

$$u_1 = u_0 + \left(\frac{\partial \bar{u}}{\partial y}\right)_0 \ell_m$$

Therefore, when the fluid "lump" originating on plane 1 arrives on plane 0 it has, since its x-momentum is assumed to have been conserved during this motion, a velocity excess that is given by:

$$\Delta u_1 = +\ell_m \left(\frac{\partial \bar{u}}{\partial y}\right)_0$$

Similarly a fluid "lump" originating on plane 2 arrives on plane 0 with a velocity excess which is given by:

$$\Delta u_2 = -\ell_m \left(\frac{\partial \bar{u}}{\partial y}\right)_0$$

In the mixing length model it is next assumed that the magnitude of the velocity fluctuation at 0 is proportional to the average of $|\Delta u_1|$ and $|\Delta u_2|$, i.e.:

$$|u'| \propto \frac{1}{2}[|\Delta u_1| + |\Delta u_2|]$$

i.e.:

$$|u'| \propto \ell_m \left|\frac{\partial \bar{u}}{\partial y}\right|$$

i.e.:

$$|u'| = K_u \ell_m \left|\frac{\partial \bar{u}}{\partial y}\right| \quad (5.29)$$

In the mixing length model it is next assumed that the transverse velocity fluctuation v' arises because of the "collision" of the fluid "lumps" arriving on plane 0 from planes 1 and 2 with different momentums. In order to satisfy continuity

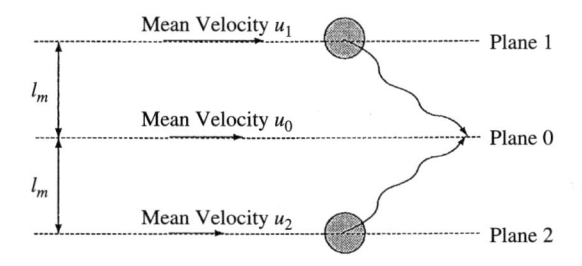

FIGURE 5.3
Planes in flow considered in mixing length analysis.

requirements, it follows that:

$$|v'| \propto |u'|$$

i.e., that:

$$|v'| = K_v \ell_m \left|\frac{\partial \bar{u}}{\partial y}\right| \tag{5.30}$$

Lastly, it is assumed that turbulent stress is proportional to $|v'||u'|$, i.e., that:

$$\tau_T = -\rho \overline{v'u'} \propto \rho |v'||u'|$$

i.e.:

$$\tau_T = \rho K_\tau K_u K_v \ell_m^2 \left|\frac{\partial \bar{u}}{\partial y}\right| \frac{\partial \bar{u}}{\partial y} \tag{5.31}$$

In writing these equations, account has been taken of the fact that the sign of τ_T depends on the sign of $\partial \bar{u}/\partial y$. Hence, $|\partial \bar{u}/\partial y|(\partial \bar{u}/\partial y)$ has been used instead of $(\partial \bar{u}/\partial y)^2$.

The above equation for the turbulent shear stress is conveniently written as:

$$\tau_T = \rho \ell^2 \left|\frac{\partial \bar{u}}{\partial y}\right| \frac{\partial \bar{u}}{\partial y} \tag{5.32}$$

where the constants of proportionality K_τ, K_u, and K_v have been combined with ℓ_m to give ℓ. ℓ is what is conventionally termed the "mixing length".

Of course, Eq. (5.32) and the analysis that led to it, gives no idea of the distribution of the mixing length in the flow field. However, it has been found that experimental measurements of the distribution of ℓ in certain simple flows can often be used as the basis for deducing the distribution in more complex flows. This point will be discussed later.

A comparison of Eqs. (5.14) and (5.32) shows that the eddy viscosity and mixing length are related by:

$$\epsilon = \ell^2 \left|\frac{\partial \bar{u}}{\partial y}\right| \tag{5.33}$$

Next consider the temperature fluctuations. In moving from plane 1 to plane 0 shown in Figure 5.3, the fluid "lump" is assumed to preserve its enthalpy and, hence, because the fluid properties are being assumed constant, its temperature. Therefore, the "lumps" arriving at 0 from 1 have a temperature excess of:

$$\Delta T_1 = \ell_m \left(\frac{\partial \bar{T}}{\partial y}\right)_0 \tag{5.34}$$

Similarly, the "lumps" arriving at 0 from 2 have a temperature deficit of:

$$\Delta T_2 = -\ell_m \left(\frac{\partial \bar{T}}{\partial y}\right)_0 \tag{5.35}$$

CHAPTER 5: Introduction to Turbulent Flows   237

As with the velocity fluctuation, it is assumed that the magnitude of the temperature fluctuation at ◊ is proportional to the average of $|\Delta T_1|$ and $|\Delta T_2|$, i.e.:

$$|T'| \propto \frac{1}{2}[|\Delta T_1| + |\Delta T_2|]$$

i.e.:

$$|T'| \propto \ell_m \left| \left(\frac{\partial \overline{T}}{\partial y} \right)_0 \right| \tag{5.36}$$

It is also assumed that the turbulent heat transfer rate is is proportional to $|v'||T'|$, i.e., that:

$$q_T = -\rho c_p \overline{v'T'} \propto \rho c_p |v'||T'| \tag{5.37}$$

i.e.:

$$q_T = -\rho c_p K_T \ell_m^2 \left| \frac{\partial \overline{u}}{\partial y} \right| \frac{\partial \overline{T}}{\partial y}$$

i.e.:

$$q_T = -\rho c_p C \ell^2 \left| \frac{\partial \overline{u}}{\partial y} \right| \frac{\partial \overline{T}}{\partial y} \tag{5.38}$$

where K_T and C are constants of proportionality.

In writing Eq. (5.38), account has been taken of the fact that the sign of q_T depends on the sign of $\partial \overline{T}/\partial y$.

A comparison of Eqs. (5.15) and (5.38) shows that the eddy diffusivity and mixing length are related by:

$$\epsilon_H = C\ell^2 \left| \frac{\partial \overline{u}}{\partial y} \right| \tag{5.39}$$

Using Eqs. (5.33) and (5.39) it follows that according to the mixing length theory the turbulent Prandtl number is given by:

$$Pr_T = \frac{\epsilon}{\epsilon_H} = \frac{1}{C} \tag{5.40}$$

i.e., according to the mixing length theory, the turbulent Prandtl number is a constant. In many flows this constant can be taken as 1. However, in some circumstances Pr_T can be very different from 1 and can vary significantly through the flow field which is, of course, in disagreement with the simple mixing length theory.

The mixing length analysis presented above does not directly provide a turbulence model. It is only when the mixing length distribution and the value of the turbulent Prandtl number in the flow are specified based on existing experimental studies that it yields a turbulence model. The advantage of the mixing length concept is that it should be far easier to deduce fairly general relations for the mixing length distribution from experimental results obtained in relatively simple flows than it would be to deduce, say, general relations for the eddy viscosity and diffusivity distributions from such measurements.

The mixing length relation can also be deduced by starting with the assumption that the turbulence at any point in the flow can be characterized by a single length scale, ℓ_T, and a single velocity scale, u_T. From this it follows that:

$$\epsilon = \text{function}\,(\ell_T, u_T) \qquad (5.41)$$

and:

$$\epsilon_H = \text{function}\,(\ell_T, u_T) \qquad (5.42)$$

Dimensional analysis then indicates from this that:

$$\frac{\epsilon}{\ell_T u_T} = \text{constant}, \quad \text{i.e.,}\ \epsilon = C_1 \ell_T u_T \qquad (5.43)$$

and:

$$\frac{\epsilon_H}{\ell_T u_T} = \text{constant}, \quad \text{i.e.,}\ \epsilon_H = C_2 \ell_T u_T \qquad (5.44)$$

If it is then assumed that because the length scale, ℓ_T, is small compared to the overall size of the flow:

$$u_T = \ell_T \left|\frac{\partial \bar{u}}{\partial y}\right| \qquad (5.45)$$

Combining the above results then gives:

$$\epsilon = C_1 \ell_T^2 \left|\frac{\partial \bar{u}}{\partial y}\right| \qquad (5.46)$$

and:

$$\epsilon_H = C_2 \ell_T^2 \left|\frac{\partial \bar{u}}{\partial y}\right| \qquad (5.47)$$

or setting:

$$\ell^2 = C_1 \ell_T^2 \qquad (5.48)$$

the above equations give:

$$\epsilon = \ell^2 \left|\frac{\partial \bar{u}}{\partial y}\right| \qquad (5.49)$$

$$\epsilon_H = \frac{C_2}{C_1} \ell^2 \left|\frac{\partial \bar{u}}{\partial y}\right| \qquad (5.50)$$

These are the same as the equations derived above if $C = C_2/C_1$.

EXAMPLE 5.2. Consider flow in the outer portion of the turbulent boundary layer on a flat plate. If the velocity distribution in this flow is approximately given by:

$$\frac{\bar{u}}{u_1} = \left(\frac{y}{\delta}\right)^{1/7}$$

where u_1 is the free-stream velocity, y is the distance from the wall, and δ is the local boundary layer thickness, find the variation of ϵ in this portion of the flow by assuming that the mixing length is a constant and equal to 0.09δ.

Solution. Using:

$$\epsilon = \ell^2 \left|\frac{\partial \bar{u}}{\partial y}\right|$$

gives:

$$\epsilon = 0.09^2 \delta^2 \frac{u_1}{7} \left(\frac{y}{\delta}\right)^{-6/7} \frac{1}{\delta}$$

i.e.:

$$\frac{\epsilon}{\delta u_1} = 0.00116 \left(\frac{y}{\delta}\right)^{-6/7}$$

5.4
MORE ADVANCED TURBULENCE MODELS

The mixing length model, with the mixing length distribution specified using empirical relations, has given satisfactory solutions to many important problems but has the disadvantage that the mixing length distribution varies considerably from one type of flow to another and also can vary quite considerably from one part of a flow to another. It is, therefore, only possible to extrapolate in a very limited way from situations in which the mixing length has been experimentally measured. This was not of great consequence before the widespread availability of digital computers because it was only possible then to solve the governing equations for certain simple flow situations, which were similar to those in which the mixing length distribution had been experimentally measured. However, since the equations governing turbulent flow can now be numerically solved for complex flows, an obvious need has arisen for a turbulence model that is applicable to all flow situations, the only empirical inputs into this model being the values of one or more universal constants whose values can be determined from measurements in simple flows. No such truly general model has yet been developed but considerable progress has been made in the development of models that are more general than the mixing length, see [8] to [21]. Almost all of these advanced turbulence models use additional differential equations to describe the turbulence terms. The most widely used such equation is that governing the turbulent kinetic energy which will here be denoted by K. This quantity is, as discussed in Chapter 2, given by:

$$K = \left(\overline{u'^2} + \overline{v'^2} + \overline{w'^2}\right)/2 \tag{5.51}$$

An equation governing this quantity was derived in Chapter 2. If the mean flow is two-dimensional, this equation is:

$$\bar{u}\frac{\partial}{\partial x}K + \bar{v}\frac{\partial}{\partial y}K = -\left[\overline{(u'^2)}\frac{\partial \bar{u}}{\partial x} + \overline{(u'v')}\frac{\partial \bar{u}}{\partial y} + \overline{(u'v')}\frac{\partial \bar{v}}{\partial x} + \overline{(v'^2)}\frac{\partial \bar{v}}{\partial y}\right]$$
$$-\frac{1}{\rho}\left[\overline{u'\frac{\partial p'}{\partial x}} + \overline{v'\frac{\partial p'}{\partial y}} + \overline{w'\frac{\partial p'}{\partial z}}\right]$$
$$-\frac{1}{2}\left[\frac{\partial}{\partial x}\overline{(u'^3)} + \frac{\partial}{\partial y}\overline{(u'^2 v')} + \frac{\partial}{\partial z}\overline{(u'^2 w')} + \frac{\partial}{\partial x}\overline{(v'^2 u')}\right.$$
$$\left. + \frac{\partial}{\partial y}\overline{(v'^3)} + \frac{\partial}{\partial z}\overline{(v'^2 w')} + \frac{\partial}{\partial x}\overline{(w'^2 u')} + \frac{\partial}{\partial y}\overline{(w'^2 v')} + \frac{\partial}{\partial z}\overline{(w'^3)}\right]$$
$$+ \left(\frac{\mu}{\rho}\right)\left[\overline{u'\Delta^2 u'} + \overline{v'\Delta^2 v'} + \overline{w'\Delta^2 w'}\right] \tag{5.52}$$

In order to utilize this equation it is necessary to use other equations to describe some of the terms in this equation and/or to model some of the terms in this equation. To illustrate how this is done, attention will be given to two-dimensional boundary layer flow. For two-dimensional boundary layer flows the turbulence kinetic energy equation, Eq. (5.52), has the following form, some further rearrangement having been undertaken:

$$\bar{u}\frac{\partial K}{\partial x} + \bar{v}\frac{\partial K}{\partial y} = -\overline{(u'v')}\frac{\partial \bar{u}}{\partial y} - \frac{1}{\rho}\overline{\frac{\partial p'v'}{\partial y}} - \frac{\partial}{\partial y}\overline{(q'^2 v')}$$
$$+ \nu\frac{\partial^2 K}{\partial y^2} - \nu\left[\overline{\frac{\partial u'}{\partial y}\frac{\partial u'}{\partial y}} + \overline{\frac{\partial v'}{\partial y}\frac{\partial v'}{\partial y}} + \overline{\frac{\partial w'}{\partial y}\frac{\partial w'}{\partial y}}\right] \tag{5.53}$$

Here:
$$q'^2 = u'^2 + v'^2 + w'^2 \tag{5.54}$$

Eq. (5.53) basically states that at any point in the flow:

Rate of convection of K = Rate of production of K
+ Rate of diffusion of K + Rate of dissipation of K
$$\tag{5.55}$$

The diffusion and dissipation terms contain unknown turbulence quantities and must be modeled. Here, a very basic type of model will be discussed. The diffusion term is assumed to have the same form as the diffusion terms in the other conservation equation, i.e., it is assumed that:

$$\text{Net rate of diffusion of } K = c_1 \frac{\partial}{\partial y}\left(\epsilon \frac{\partial K}{\partial y}\right) \tag{5.56}$$

c_1 being a constant.

The dissipation term can be modeled by considering the work done against the drag force on a fluid lump as it moves with the turbulent motion [22]. Now, if a body has a characteristic size, R, and is moving with velocity, V, relative to a fluid, it experiences a drag force, D, whose magnitude is given by an equation that has the form:

$$D = \frac{1}{2} C_D \rho V^2 R^2 \tag{5.57}$$

where C_D is the drag coefficient.

The rate at which work is done on this body will then be given by:

$$\dot{W} = DV = \frac{1}{2} C_D \rho V^3 R^2 \tag{5.58}$$

The rate at which work is done per unit mass is then given by:

$$\dot{w} = \frac{(1/2) C_D \rho V^3 R^2}{C_V \rho R^3} = C \frac{V^3}{R} \tag{5.59}$$

where the volume of the body has been expressed as:

$$\text{Volume} = C_V R^3$$

C_V being a constant whose value depends on the body shape.

Now, consider a fluid lump moving in a turbulent flow. Relative to the mean flow its mean velocity is:

$$\sqrt{\overline{u'^2} + \overline{v'^2} + \overline{w'^2}} = \sqrt{2K}$$

Therefore, if the characteristic size of the fluid "lumps" is L, the rate at which the mean flow does work against the turbulent motion is from Eq. (5.59) given by:

$$\frac{c_2 K^{3/2}}{L} \tag{5.60}$$

This result could have been obtained by simply assuming that the rate of dissipation depends on K and L alone. Eq. (5.60) could then have been derived by dimensional analysis.

Substituting Eqs. (5.57) and (5.60) into Eq. (5.53) gives the following modeled form of the turbulent kinetic energy equation for two-dimensional boundary layer flow:

$$\bar{u} \frac{\partial K}{\partial x} + \bar{v} \frac{\partial K}{\partial y} = -\overline{(u'v')} \frac{\partial \bar{u}}{\partial y} + c_1 \frac{\partial}{\partial y} \left(\epsilon \frac{\partial K}{\partial y} \right) - \frac{c_2 K^{3/2}}{L} \tag{5.61}$$

If the turbulent stress in the production term is written in terms of the eddy viscosity this equation becomes:

$$\bar{u} \frac{\partial K}{\partial x} + \bar{v} \frac{\partial K}{\partial y} = \epsilon \left(\frac{\partial \bar{u}}{\partial y} \right)^2 + c_1 \frac{\partial}{\partial y} \left(\epsilon \frac{\partial K}{\partial y} \right) - \frac{c_2 K^{3/2}}{L} \tag{5.62}$$

In order to utilize this equation to determine the turbulent shear stress it is necessary to obtain an additional equation relating, for example, the eddy viscosity to the quantities involved in the turbulent kinetic energy equation. If it is assumed that:

$$\epsilon = \text{function } (K, L) \tag{5.63}$$

then dimensional analysis gives:

$$\epsilon = c_T K^{1/2} L \tag{5.64}$$

where c_T is a constant.

Eqs. (5.62) and (5.64) together then allow the variation of ϵ to be obtained by solving these equations simultaneously with the momentum and continuity equations.

The turbulence model discussed above contains three constants c_1, c_2, and c_T together with the unknown length scale, L. It is assumed that the values of these constants and the distribution of L, are universal, i.e., once their values have been determined from measurements in simple flows it is assumed that these values will be applicable in all flows. This has not however, proved to be the case. The constants have been found to vary slightly with the type of flow and also to be Reynolds number-dependent in some circumstances, viscous effects having been neglected in the above derivation. The length scale distribution has also been found to vary quite considerably with the type of flow. Despite these failings the turbulent kinetic energy model has proved to be superior to simpler models in some circumstances.

It is of interest to note that if the convection and diffusion terms are negligible in the turbulence kinetic energy equation, i.e., if the rate of production of kinetic energy is just equal to the rate of dissipation of turbulence kinetic energy, Eq. (5.62) reduces to:

$$\epsilon \left(\frac{\partial \bar{u}}{\partial y}\right)^2 = \frac{c_2 K^{3/2}}{L} \tag{5.65}$$

If Eq. (5.64) is used to give

$$K = \left(\frac{\epsilon}{c_T L}\right)^2$$

Eq. (5.65) gives:

$$\epsilon \left(\frac{\partial \bar{u}}{\partial y}\right)^2 = \frac{c_2}{L} \left(\frac{\epsilon}{c_T L}\right)^3$$

i.e.:

$$\epsilon = \left(\frac{c_T^3}{c_2}\right)^{1/2} L^2 \left|\frac{d\bar{u}}{dy}\right| \tag{5.66}$$

This is the same form as the result given by the basic mixing length theory, i.e., the turbulence kinetic energy equation gives the same result as the mixing length model when the convection and diffusion terms are neglected in the turbulence kinetic energy equation. Except near a wall, however, all the terms in the turbulent kinetic equation are significant in most cases.

In applying the turbulence kinetic energy model it is common to assume that the turbulent Prandtl number, Pr_T, is constant.

In the case of axially symmetrical pipe flow, the turbulent kinetic energy equation has the following form when the terms are modeled in some way as was done with the equation for boundary layer flow:

$$\bar{u}\frac{\partial K}{\partial x} + \bar{v}\frac{\partial K}{\partial r} = \epsilon\left(\frac{\partial \bar{u}}{\partial r}\right)^2 + \frac{c_1}{r}\frac{\partial}{\partial y}\left(r\epsilon\frac{\partial K}{\partial r}\right) - \frac{c_2 K^{3/2}}{L} \tag{5.67}$$

Eq. (5.64) being applicable without modification in this coordinate system.

In an effort to overcome the deficiencies in the turbulent kinetic energy equation-based turbulence model discussed above, particularly with regard to the need to specify the distribution of the length scale, other differential equations have been developed to supplement the kinetic energy equation. The most commonly used such additional equation is obtained by making the dissipation of turbulent kinetic energy a variable; i.e., if attention is again given to boundary layer flow, to write the turbulent kinetic energy equation as:

$$\bar{u}\frac{\partial K}{\partial x} + \bar{v}\frac{\partial K}{\partial y} = \epsilon\left(\frac{\partial \bar{u}}{\partial y}\right)^2 + c_1\frac{\partial}{\partial y}\left(\epsilon\frac{\partial K}{\partial y}\right) - E \tag{5.68}$$

where E is the rate of dissipation of turbulent kinetic energy. An additional differential equation for E is then developed, this equation having a similar form to the turbulent kinetic energy equation. For example, for boundary layer flow, the dissipation equation has the form:

$$\bar{u}\frac{\partial E}{\partial x} + \bar{v}\frac{\partial E}{\partial y} = c_3\epsilon\left(\frac{\partial \bar{u}}{\partial y}\right)^2 \frac{E}{K} + c_4\frac{\partial}{\partial y}\left(\epsilon\frac{\partial E}{\partial y}\right) - c_5\frac{E^2}{K} \tag{5.69}$$

where c_3, c_4, and c_5 are additional empirical constants.

Now Eqs. (5.60) and (5.64) give:

$$E = \frac{c_2 K^{3/2}}{L}$$

and:

$$\epsilon = c_T K^{1/2} L$$

Eliminating the length scale, L, between these two equations then gives:

$$\epsilon = c_e \frac{K^2}{E} \tag{5.70}$$

where $c_e = c_2 c_T$.

Eq. (5.70) could have been derived by simply assuming that:

$$\epsilon = \text{function}(K, E)$$

Dimensional analysis then directly gives Eq. (5.70).

The K-E turbulence model discussed above, which is often termed the k-ϵ model because of the symbols originally used for K and E, contains a number of empirical constants. Typically assumed values for these constants are:

$$c_1 = 1, \; c_3 = 1.44, \; c_4 = 0.77$$
$$c_5 = 1.92, \; c_e = 0.09$$

244 Introduction to Convective Heat Transfer Analysis

In order to use this model in the prediction of heat transfer rates it is usual to also assume that:

$$Pr_T = 0.9$$

The K-E turbulence model discussed above only applies when $\epsilon \gg \nu$. This will not be true near the wall. The most common way of dealing with this problem is to assume that there is a "universal" velocity distribution adjacent to the wall and the K-E turbulence model is then only applied outside of the region in which this wall region velocity distribution applies. Alternatively, more refined versions of the K-E turbulence model have been developed that apply under all conditions, i.e., across the entire boundary layer.

5.5
ANALOGY SOLUTIONS FOR HEAT TRANSFER IN TURBULENT FLOW

Many early efforts at trying to theoretically predict heat transfer rates in turbulent flow concentrated on trying to relate the wall heat transfer rate to the wall shearing stress. The reason for this was that a considerable body of experimental and semi-theoretical knowledge concerning this shearing stress in various flow situations was available and that the mechanism of heat transfer in turbulent flow is obviously similar to the mechanism of momentum transfer. Such solutions, which give the heat transfer rate in turbulent flow in terms of the wall shearing stress, are termed analogy solutions [23],[24],[25].

In outlining the main steps in obtaining an analogy solution, attention will here be given to two-dimensional flow. The "total" shear stress and "total" heat transfer rate are made up of the molecular and turbulent contributions, i.e.:

$$\tau = \tau_m + \tau_T = \mu \frac{\partial \bar{u}}{\partial y} - \overline{\rho v' u'} \tag{5.71}$$

and:

$$q = q_m + q_T = -k \frac{\partial \bar{T}}{\partial y} - \overline{\rho c_p v' T'} \tag{5.72}$$

which can be written as:

$$\tau = \rho(\nu + \epsilon)\frac{\partial \bar{u}}{\partial y} \tag{5.73}$$

and:

$$q = -\rho c_p (\alpha + \epsilon_H)\frac{\partial \bar{T}}{\partial y} \tag{5.74}$$

In the usual analogy solution approach, Eqs. (5.73) and (5.74) are rearranged to give:

$$\frac{\partial \bar{u}}{\partial y} = \frac{\tau}{\rho(\nu + \epsilon)} \tag{5.75}$$

and:

$$\frac{\partial \bar{T}}{\partial y} = -\frac{q}{\rho c_p(\alpha + \epsilon_H)} \tag{5.76}$$

and these equations are then integrated outward from the wall, i.e., $y = 0$, to some point in the flow at distance y from the wall. This gives:

$$\bar{u} = \int_0^y \frac{\tau}{\rho(\nu + \epsilon)} dy \tag{5.77}$$

$$T_W - \bar{T} = \int_0^y \frac{q}{\rho c_p(\alpha + \epsilon_H)} dy \tag{5.78}$$

The relationship between ϵ and ϵ_H and between τ and q is then assumed. The assumed relation between τ and q will usually involve the values of these quantities at the wall. The integrals in Eqs. (5.77) and (5.78) can then be related and eliminated between the two equations leaving a relationship between the wall heat transfer rate and the wall shear stress and certain mean flow field quantities.

The application of the analogy approach to turbulent boundary flow and to turbulent duct flow will be discussed in Chapters 6 and 7, respectively.

5.6
NEAR-WALL REGION

The presence of the solid wall has a considerable influence on the turbulence structure near the wall. Because there can be no flow normal to the wall near the wall, v' decreases as the wall is approached and as a result the turbulent stress and turbulent heat transfer rate are negligible in the region very near the wall. This region in which the effects of the turbulent stress and turbulent heat transfer rate can be neglected is termed the "sublayer" or, sometimes, the "laminar sublayer" [1],[2], [26],[27],[28],[29]. In this sublayer:

$$\tau = \rho \nu \frac{\partial \bar{u}}{\partial y} \tag{5.79}$$

and:

$$q = -\rho c_p \alpha \frac{\partial \bar{T}}{\partial y} \tag{5.80}$$

y being measured from the wall into the flow.

Further, because the sublayer is normally very thin, the variations of the shear stress, τ and q, through this layer are usually negligible; i.e., in the sublayer it can be assumed that:

$$\tau = \tau_w, \quad q = q_w \tag{5.81}$$

the subscript w denoting conditions at the wall. Substituting these values into Eqs. (5.79) and (5.80) and integrating the resultant equations outward from the wall gives:

$$\bar{u} = \frac{\tau_w}{\rho \nu} y \tag{5.82}$$

and:

$$T_w - \bar{T} = \frac{q_w}{\rho c_p \alpha} y \tag{5.83}$$

The velocity and temperature distributions in the sublayer are thus linear.
If the following are defined:

$$u^* = \sqrt{\frac{\tau_w}{\rho}}, \quad u^+ = \frac{\bar{u}}{u^*}, \quad y^+ = \frac{y u^*}{\nu} \tag{5.84}$$

u^* being termed the "friction velocity", then the velocity distribution in the sublayer as given in Eq. (5.82) can be written as:

$$u^+ = y^+ \tag{5.85}$$

Eqs. (5.82) and (5.83) have been found to adequately describe the mean velocity and temperature distributions from the wall out to $y^+ = 5$; i.e., the sublayer extends from $y^+ = 0$ to $y^+ = 5$.

For $y^+ > 5$ the turbulence stress and heat transfer rate become important. However, near the wall the total shear stress and total heat transfer rate will remain effectively constant and equal to the wall shear stress and wall heat transfer rate, respectively.

The size of the turbulent "eddies" near the wall is determined by the distance from the wall; i.e., near the wall it is to be expected that their size will increase linearly with distance from the wall. Now, the mixing length is related to the scale of the turbulence, i.e., to the size of the "eddies," and it is to be expected therefore that near the wall:

$$\ell = K_v y \tag{5.86}$$

K_v being a constant termed the von Karman constant.

Using the assumptions discussed above then gives:

$$\tau_w = \rho \left(\nu + K_v^2 y^2 \frac{\partial \bar{u}}{\partial y} \right) \frac{\partial \bar{u}}{\partial y} \tag{5.87}$$

If it is further assumed that $\nu \ll \epsilon$, the above equation becomes:

$$\tau_w = \rho K_v^2 y^2 \left(\frac{\partial \bar{u}}{\partial y} \right)^2$$

i.e.,

$$K_v y \frac{\partial \bar{u}}{\partial y} = \sqrt{\frac{\tau_w}{\rho}} (= u^*)$$

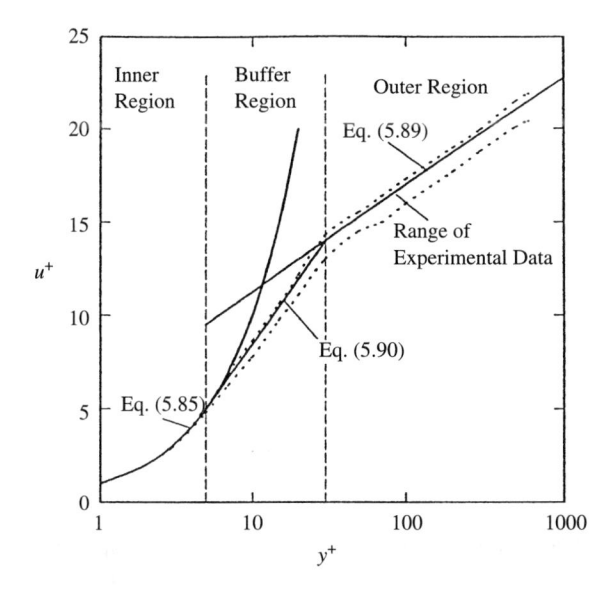

FIGURE 5.4
Mean velocity distribution near wall.

i.e.,

$$K_v y^+ \frac{\partial u^+}{\partial y^+} = 1 \tag{5.88}$$

Integrating this equation gives:

$$u^+ = \frac{1}{K_v} \ln y^+ + C \tag{5.89}$$

where C is a constant whose value cannot directly be determined because this expression, which is based on the assumption that $\nu \ll \epsilon$, does not apply very near the wall. With K_v set equal to 0.4 and C set equal to 5.5, Eq. (5.89) provides a good description of the mean velocity distribution for $y^+ > 30$ as shown in Figure 5.4.

Thus, Eq. (5.85) applies from the wall out to $y^+ = 5$ while Eq. (5.89) applies for $y^+ > 30$. Between $y^+ = 5$ and $y^+ = 30$, where both the molecular and the turbulent stresses are important, experiments indicate that the velocity distribution is given by:

$$u^+ = 5 \ln y^+ - 3.05 \tag{5.90}$$

This region between $y^+ = 5$ and $y^+ = 30$ is termed the "buffer" region.

The velocity variations given by Eqs. (5.85) and (5.90) are also shown in Figure 5.4.

5.7
TRANSITION FROM LAMINAR TO TURBULENT FLOW

In predicting heat transfer rates it is important to know if the flow remains laminar or whether transition to turbulence occurs. If transition occurs, it is usually also

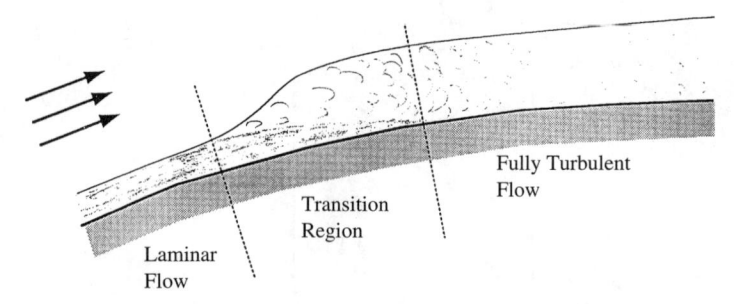

FIGURE 5.5
Laminar flow, transition, and fully turbulent regions.

important to know where the transition occurs. It is also important to realize that there is a transition region between the region of laminar flow and the region of fully turbulent flow, as illustrated in Figure 5.5.

The conditions under which transition occurs depend on the geometrical situation being considered, on the Reynolds number, and on the level of unsteadiness in the flow well away from the surface over which the flow is occurring [2], [30]. For example, in the case of flow over a flat plate as shown in Figure 5.6, if the level of unsteadiness in the freestream flow ahead of the plate is very low, transition from laminar to turbulent boundary layer flow occurs approximately when:

$$Re_x (= \bar{u}_1 x / \nu) = 2.8 \times 10^6 \tag{5.91}$$

and fully turbulent flow is achieved approximately when:

$$Re_x = 3.9 \times 10^6 \tag{5.92}$$

The level of unsteadiness in the freestream is usually specified using the following quantity:

$$J = \frac{1}{\bar{u}_1} \left[\frac{(\overline{u'^2} + \overline{v'^2} + \overline{w'^2})}{3} \right]^{1/2} \tag{5.93}$$

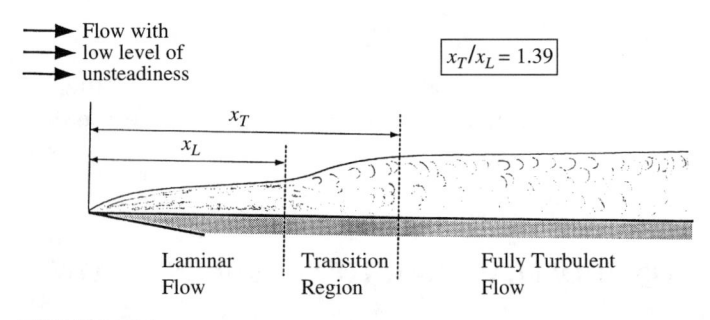

FIGURE 5.6
Transition in boundary layer flow over a flat plate.

where $\bar{u}_1$ is mean freestream velocity and u', v', and w' are the instantaneous deviations of the velocity from the mean value. J is often termed the freestream turbulence level and is commonly expressed as a percentage.

For the values of J normally encountered in practice, i.e., approximately 1%, the transition Reynolds number is usually assumed to be:

$$Re_x = 3.5 \times 10^5 \text{ to } 5 \times 10^5 \tag{5.94}$$

and fully turbulent flow is usually assumed to be achieved when:

$$Re_x = 9 \times 10^5 \text{ to } 1.3 \times 10^6 \tag{5.95}$$

For boundary layer flows on bodies of other shape, the transition Reynolds number based on the distance around the surface from the leading edge of the body is usually increased if the pressure is decreasing, i.e., if there is a "favorable" pressure gradient, and is usually decreased if the pressure is increasing, i.e., if there is an "unfavorable" pressure gradient.

EXAMPLE 5.3. Air at a temperature of 30°C flows at a velocity of 100 m/s over a wide flat plate that is aligned with the flow. The surface of the plate is maintained at a uniform wall temperature of 50°C. Estimate the distance from the leading edge of the plate at which transition to turbulence begins and when transition is complete.

Solution. The mean air temperature is $(30 + 50)/2 = 40°C$. At 40°C, for air at standard ambient pressure:

$$\nu = 16.96 \times 10^{-6} \text{ m}^2/\text{s}$$

The Reynolds number at distance, x, from the leading edge is then given by:

$$Re_x = u_1 x/\nu = 100x/16.96 \times 10^{-6} = 5.896 \times 10^6 x$$

It will be assumed that transition begins when:

$$Re_x = 4 \times 10^5$$

Therefore, the distance from the leading edge at which transition begins is given by:

$$5.896 \times 10^6 x = 4 \times 10^5, \text{ i.e., } x = 0.0678 \text{ m}$$

It will be assumed that transition is complete when:

$$Re_x = 1 \times 10^6$$

Therefore, the distance from the leading edge at which transition is complete is given by:

$$5.896 \times 10^6 x = 10^6, \text{ i.e., } x = 0.1696 \text{ m}$$

Therefore, transition begins at 0.0678 m from the leading edge and is complete at a distance of 0.1696 m from the leading edge of the plate.

Attention is next turned to transition in pipe flow. It is usual to assume that, with the level of unsteadiness in the flow that is usually encountered in practice, transition will occur if:

$$Re_D(= U_m D/\nu) = 2300 \tag{5.96}$$

Here, U_m is the mean velocity across the section of the pipe, i.e.:

$$U_m = \frac{\text{Volume flow rate through pipe}}{\pi D^2/4} \tag{5.97}$$

For ducts with other cross-sectional shapes, it is common to assume that the same criterion as that given in Eq. (5.96) applies provided the "hydraulic diameter" is used in place of the diameter. The hydraulic diameter, as previously discussed, is defined by:

$$D_h = \frac{4 \times \text{Flow area}}{\text{Wetted perimeter}} \tag{5.98}$$

This is only a relatively rough guide as very complex flows in the transitional region can arise in some cases involving noncircular ducts.

5.8 CONCLUDING REMARKS

This chapter has been concerned with certain fundamental aspects of turbulent flows. The time-averaged forms of the governing equations have been reviewed. The concept of a turbulent stress and a turbulent heat transfer rate have been introduced and the eddy viscosity and diffusivity have been defined together with the turbulent Prandtl number. A turbulence model has been defined and some basic turbulence models, including those based on the mixing-length concept, have been considered. The idea of a universal near-wall region has been introduced and the basic ideas involved in an analogy-type solution have been discussed. Some attention has also been given to the conditions under which transition from laminar to turbulent flow occurs under various circumstances.

PROBLEMS

5.1. Assuming that the velocity distribution in fully developed pipe flow near the center line can be approximately represented by:

$$\frac{\bar{u}}{\bar{u}_1} = \left(\frac{y}{R}\right)^{\frac{1}{n}}$$

where $\bar{u}_1$ is the mean center line velocity, y is the distance from the wall, R is the radius of the pipe, and n is a constant, find the variation of ϵ and ϵ_H in this portion of the flow. Assume a turbulent Prandtl number of 0.9.

5.2. The mean velocity distribution in the outer portion of the turbulent boundary layer on a flat plate is approximately given by:

$$\frac{\bar{u}}{u_1} = \left(\frac{y}{\delta}\right)^{\frac{1}{5}}$$

where u_1 is the freestream velocity, y is the distance from the wall, and δ is the local boundary layer thickness. Find the variation of ϵ, ϵ_H, the turbulence shear stress, and the turbulent heat transfer rate in this portion of the flow by assuming that the mixing length is a constant and equal to $0.09\,\delta$ and that the turbulent Prandtl number is 0.9.

5.3. Consider air flow at a velocity of 35 m/s over a wide flat plate that is aligned with the flow. The air is at a temperature of 20°C and the plate is kept at a uniform wall temperature of 40°C. Estimate the distance from the leading edge of the plate at which transition to turbulence begins and when transition is complete.

5.4. Air flows at a velocity of 25 m/s over a wide flat plate that is aligned with the flow. The mean air temperature in the boundary layer is 30°C. Plot the mean velocity distribution near the wall in the boundary layer assuming that flow is turbulent and that the wall shear stress is given by:

$$\frac{\tau_w}{\rho u_1^2} = \frac{0.029}{Re_x^{0.2}}$$

5.5. Consider transition in the boundary layer flow over a flat plate. Using the expression for the thickness of a laminar boundary layer on a flat plate given in Chapter 3, find the value of the Reynolds number based on the boundary layer thickness at which transition begins.

5.6. Air flows through a duct with a square cross-sectional shape, the side length of the square section being 3 cm. If the mean air temperature in the duct is 40°C, find the air velocity in the duct at which the flow becomes turbulent.

5.7. In the mixing length turbulence model as discussed in this chapter, it is assumed that the "lumps" of fluid maintain their temperature while they move through a distance ℓ_m. In fact, because, during this motion, the "lumps" are at a different temperature from the surrounding fluid, there is heat transfer from the "lumps" leading to a temperature change during the motion. Assuming that the heat transfer, and therefore the temperature change, is proportional to the mean temperature difference, examine the changes that are required to the mixing length model to account for this effect.

5.8. Consider fully developed flow between parallel plates. The velocity distribution in the flow near the center line can be approximately represented by:

$$\frac{\bar{u}}{\bar{u}_C} = 1 - \left(\frac{y}{w}\right)^{\frac{1}{n}}$$

where $\bar{u}_c$ is the mean center line velocity, y is the distance from the center line, and w is half the distance between the plates. Determine the variations of ϵ and ϵ_H in this portion of the flow. Assume a turbulent Prandtl number of 0.9

5.9. Air at a temperature of 20°C flows at a velocity of 100 m/s over a wide flat plate that is aligned with the flow. The surface of the plate is maintained at a uniform wall temperature. Plot the variation of the distances from the leading edge of the plate at which transition to turbulence begins and when transition is complete against surface temperature for surface temperatures between 20°C and 100°C.

5.10. Derive an expression for the time-averaged mass flow in the sublayer of a turbulent boundary layer on a plane surface. How does this mass flow rate vary with the distance along the surface?

REFERENCES

1. Hinze, J.O., *Turbulence,* 2nd ed., McGraw-Hill, New York, 1975.
2. Schlichting, H., *Boundary Layer Theory,* 7th ed., McGraw-Hill, New York, 1979.
3. Tennekes, H. and Lumley, J.L., *A First Course in Turbulence,* The MIT Press, Cambridge, MA, 1972.
4. Townsend, A.A., *The Structure of Turbulent Shear Flow,* 2nd ed., Cambridge University Press, London, 1976.
5. Bradshaw, P., Editor, *Turbulence,* Springer-Verlag, Berlin, 1978.
6. Adams, J.C., Jr and Hodge, B.K., "The Calculation of Compressible, Transitional, Turbulent, and Relaminarizational Boundary Layers over Smooth and Rough Surfaces Using an Extended Mixing Length Hypothesis", AIAA Paper 77-682, Albuquerque, NM, 1977.
7. Lin, C.C., Editor, "Turbulent Flows and Heat Transfer", in *High Speed Aerodynamics and Jet Propulsion,* Vol. V, Princeton University Press, Princeton, N.J., 1959.
8. Bradshaw, P.T., Cebeci, and Whitelaw, J., *Engineering Calculational Methods for Turbulent Flow,* Academic Press, New York, 1981.
9. Chyou, Sh.-W., and Sleicher, C.A.," Can One- and Two-Equation Turbulence Models be Modified to Calculate Turbulent Heat Transfer with Variable Properties?" *Industrial and Engineering Chemistry Research,* Vol. 31, pp. 756–759, 1992.
10. Kolmogorov, A.N., "Equations of Turbulent Motion of an Incompressible Turbulent Fluid", *Izv. Akad, Nauk SSSR, Ser. Fiz.,* VI, No. 1–2, pp. 56–58, 1942.
11. Launder, B.E. and Spalding, D.B., *Mathematical Models of Turbulence,* Academic Press, New York, 1972.
12. Launder, B.E. and Spalding, D.B., "The Numerical Computation of Turbulent Flows", *Comput. Meth. Appl. Mech. Eng.,* Vol. 3, pp. 269–289, 1974.
13. Launder, B.W., "On the Computation of Convective Heat Transfer in Complex Turbulent Flows", *ASME J. Heat Transfer,* 110, pp. 1112–1128, 1988.
14. Maciejewski, P.K. and Anderson, A.M., "Elements of a General Correlation for Turbulent Heat Transfer", *Journal of Heat Transfer,* Vol. 118, pp. 287–293, May 1996.
15. Patankar, S.V., *Numerical Heat Transfer and Fluid Flow,* Hemisphere Publ., Washington, D.C., 1980.
16. Pletcher, R.H., "Progress in Turbulent Forced Convection", *ASME J. Heat Transfer,* 110, 1129–1144, 1988.
17. Rodi, W., "Examples of Turbulence Models for Incompressible Flows", *AIAA J.,* Vol. 20, pp. 872–879, 1982.
18. Saffman, P.G. and Wilcox, D.C., "Turbulence Model Prediction for Turbulent Boundary Layers", *AIAA J.,* Vol. 12, pp. 541–546, 1974.
19. Wilcox, D.C. and Traci, R.M., "A Complete Model of Turbulence", AIAA Paper 76-351, San Diego, CA, 1976.
20. Youssef, M.S., Nagano, Y., and Tagawa, M., "A Two-Equation Heat Transfer Model for Predicting Turbulent Thermal Fields Under Arbitrary Wall Thermal Conditions", *Int. J. Heat Mass Transfer,* Vol. 35, pp. 3095–3104, 1992.
21. Cho, J.R. and Chung, M.K., "A k-ϵ Equation Turbulence Model", *J. Fluid Mech.,* Vol. 237, pp. 301–322, 1992.
22. Eckert, E.R.G. and Drake, R.M., Jr., *Analysis of Heat and Mass Transfer,* McGraw-Hill, New York, 1973.
23. Boelter, L.M.K., Martinelli, R.C., and Jonassen, F., "Remarks on the Analogy between Heat and Momentum Transfer", *Trans. ASME,* Vol. 63, pp. 447–455, 1941.
24. Colburn, A.P., A Method for Correlating Forced Convection Heat Transfer Data and a Comparison with Fluid Friction, *Trans. Am. Inst. Chem. Eng.,* Vol. 29, pp. 174–210, 1933; reprinted in *Int. J. Heat Mass Transfer,* Vol. 7, pp. 1359–1384, 1964.

25. von Karman, T., "The Analogy Between Fluid Friction and Heat Transfer", *Trans. ASME,* Vol. 61, pp. 705–710, 1939.
26. Spalding, D.B., "A Single Formula for the Law of the Wall", *J. Appl. Mech.,* Vol. 28, pp. 455–457, 1961.
27. Kozlu, H., Mikic, B.B., and Patera, A.T., "Turbulent Heat Transfer Augmentation Using Microscale Disturbances Inside the Viscous Sublayer", *J. Heat Transfer,* Vol. 114, pp. 348–353, 1992.
28. van Driest, E.R., "On Turbulent Flow Near a Wall", *J. Aero. Sci.,* Vol. 23, pp. 1007–1011, 1956.
29. Kasagi, N., Kuroda, A., and Hirata, M., "Numerical Investigation of Near-Wall Turbulent Heat Transfer Taking Into Account the Unsteady Heat Conduction in the Solid Wall", *J. of Heat Transfer,* Vol. 111, pp. 385–392, 1989.
30. Chandrasekhar, S., *Hydrodynamic and Hydromagnetic Stability,* Clarendon Press, Oxford, 1961.

CHAPTER 6

External Turbulent Flows

6.1
INTRODUCTION

This chapter is concerned with the prediction of the heat transfer rate from the surface of a body placed in a large fluid stream when the flow over the surface is partly or entirely turbulent. Some examples of the type of situation being considered in this chapter are shown in Fig. 6.1.

Attention will mainly be given to turbulent boundary layer flows in this chapter. A brief discussion of the analysis of flows in which it is not possible to use these boundary layer assumptions will be given at the end of the chapter.

6.2
ANALOGY SOLUTIONS FOR BOUNDARY LAYER FLOWS

As discussed in the previous chapter, most early efforts at trying to theoretically predict heat transfer rates in turbulent flow concentrated on trying to relate the wall heat transfer rate to the wall shear stress [1],[2],[3],[4]. The reason for this is that a considerable body of experimental and semi-theoretical knowledge concerning the shear stress in various flow situations is available and that the mechanism of heat transfer in turbulent flow is obviously similar to the mechanism of momentum transfer. In the present section an attempt will be made to outline some of the simpler such analogy solutions for boundary layer flows, attention mainly being restricted to flow over a flat plate.

The simplest analogy solution is that due to Reynolds. This analogy has many limitations but does have some practical usefulness and serves as the basis for more refined analogy solutions.

Consider turbulent boundary layer flow over a plate which has a local temperature of T_w. The flow situation is shown in Fig. 6.2.

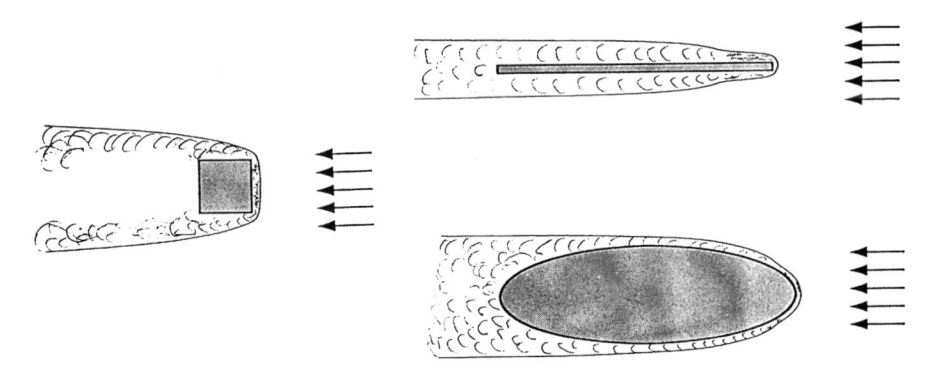

FIGURE 6.1
External turbulent flows.

u_1 and T_1 are the free-stream velocity and temperature respectively.

If τ and q are defined as the total shear stress and heat transfer rate respectively, at any point in the flow, then, using the definitions given in Chapter 5, it follows that:

$$\tau = \tau_m + \tau_T = \mu \frac{\partial \bar{u}}{\partial y} - \rho \overline{v'u'} \tag{6.1}$$

and:

$$q = q_m + q_T = -k \frac{\partial \bar{T}}{\partial y} - \rho c_p \overline{v'T'} \tag{6.2}$$

These equations can be written as:

$$\tau = \rho(\nu + \epsilon) \frac{\partial \bar{u}}{\partial y} \tag{6.3}$$

and:

$$q = -\rho c_p (\alpha + \epsilon_H) \frac{\partial \bar{T}}{\partial y} \tag{6.4}$$

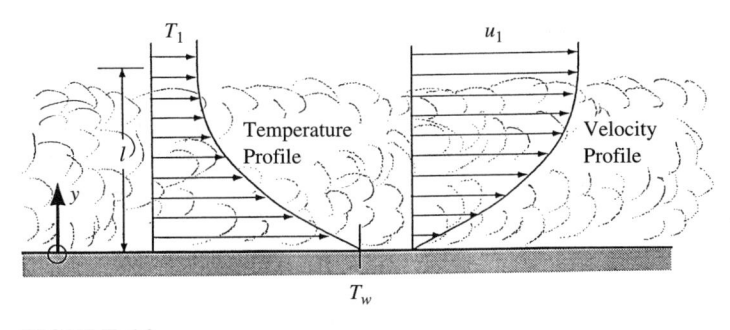

FIGURE 6.2
Flow situation being considered.

These equations can be rearranged to give:

$$\frac{\partial \bar{u}}{\partial y} = \frac{\tau}{\rho(\nu + \epsilon)} \tag{6.5}$$

and:

$$\frac{\partial \bar{T}}{\partial y} = -\frac{q}{\rho c_p(\alpha + \epsilon_H)} \tag{6.6}$$

Integrating these two equations outward from the wall to some point in the free-stream at a distance of ℓ from the wall then gives:

$$u_1 = \int_0^\ell \frac{\tau}{\rho(\nu + \epsilon)} \, dy \tag{6.7}$$

and:

$$T_w - T_1 = \int_0^\ell \frac{q}{\rho c_p(\alpha + \epsilon_H)} \, dy \tag{6.8}$$

Because both ϵ and ϵ_H depend on the structure of the turbulence it is to be expected that they will have similar values, a conclusion that is confirmed by experiment. In the present simple analysis it will be assumed, therefore, that they are equal, i.e., that:

$$\epsilon_H = \epsilon \tag{6.9}$$

This is, of course, equivalent to assuming that the turbulent Prandtl number, Pr_T, is equal to 1.

In the Reynolds analogy, attention is further restricted to the flow of fluids for which:

$$\alpha = \nu \tag{6.10}$$

i.e., to fluids that have a Prandtl number of 1. No such real fluids exist, but as mentioned earlier, most gases have Prandtl numbers which are close to unity.

Using the assumptions given in Eqs. (6.9) and (6.10), Eq. (6.8) can be rewritten as:

$$T_w - T_1 = \int_0^\ell \frac{q}{\rho c_p(\nu + \epsilon)} \, dy \tag{6.11}$$

Now both the shear stress, τ, and the heat transfer rate, q, have their highest values at the wall and both are zero in the free-stream. It will therefore be assumed that the distributions of total shearing stress and heat transfer rate are similar, i.e., that at any point in the boundary layer:

$$\frac{q}{q_w} = \frac{\tau}{\tau_w} \tag{6.12}$$

where τ_w and q_w are the values of the shear stress and the heat transfer rate at the wall. This assumption is supported by available measurements and intuitive reasoning.

This assumption, in conjunction with the previous assumptions, is equivalent to assuming that the temperature and velocity profiles are similar (see Example 6.1 below).

Substituting Eq. (6.12) into Eq. (6.11) then gives:

$$T_w - T_1 = \frac{q_w}{\tau_w} \int_0^\ell \frac{\tau}{\rho c_p(\nu + \epsilon)} dy \qquad (6.13)$$

Dividing this equation by Eq. (6.7) then gives:

$$\frac{T_w - T_1}{u_1} = \frac{q_w}{c_p \tau_w} \qquad (6.14)$$

Introducing x, the distance along the plate from the leading edge to the point being considered, then allows this equation to be rewritten as:

$$\frac{q_w}{(T_w - T_1)} \frac{x}{k} = \left(\frac{\tau_w}{\rho u_1^2}\right)\left(\frac{\rho u_1 x}{\mu}\right)\left(\frac{\mu c_p}{k}\right)$$

i.e.:

$$Nu_x = \left(\frac{C_f}{2}\right) Re_x Pr \qquad (6.15)$$

where Nu_x is the local Nusselt number, Re_x is the local Reynolds number, and C_f is the local shearing stress coefficient, i.e., $\tau_w/(1/2 \rho u_1^2)$.

Now it was assumed in the derivation of Eq. (6.15) that the Prandtl number, Pr, was equal to 1. Hence, the final form of the Reynolds analogy is:

$$Nu_x = \left(\frac{C_f}{2}\right) Re_x \qquad (6.16)$$

Therefore the distribution of Nu_x can be determined from a knowledge of the distribution of C_f. For example, experiment has indicated that for moderate values of the Reynolds number [1]:

$$C_f = \frac{0.058}{Re_x^{0.2}} \qquad (6.17)$$

Substituting this into Eq. (6.16) then gives:

$$Nu_x = 0.029 Re_x^{0.8} \qquad (6.18)$$

EXAMPLE 6.1. Show that the Reynolds analogy indicates that the velocity and temperature profiles in the boundary layer are similar.

Solution. If Eqs. (6.5) and (6.6) are integrated from the wall out to some arbitrary distance, y, from the wall where the mean velocity is $\bar{u}$ and the mean temperature is $\bar{T}$, the following are obtained:

$$\bar{u} = \int_0^y \frac{\tau}{\rho(\nu + \epsilon)} dy \qquad (i)$$

and:

$$T_w - \overline{T} = \int_0^y \frac{q}{\rho c_p(\alpha + \epsilon_H)} \, dy \qquad (ii)$$

If the assumptions discussed above, i.e.:

$$\epsilon_H = \epsilon, \alpha = \nu$$

$$\frac{q}{q_w} = \frac{\tau}{\tau_w}$$

are introduced, Eq. (*ii*) above becomes:

$$T_w - \overline{T} = \frac{q_w}{\tau_w} \int_0^y \frac{\tau}{\rho c_p(\nu + \epsilon)} \, dy \qquad (iii)$$

Dividing Eq. (*i*) above by Eq. (6.7) and Eq. (*iii*) above by Eq. (6.13) then gives:

$$\frac{\overline{u}}{u_1} = \frac{\int_0^y \frac{\tau}{\rho(\nu + \epsilon)} \, dy}{\int_0^\ell \frac{\tau}{\rho(\nu + \epsilon)} \, dy}$$

and:

$$\frac{T_w - \overline{T}}{T_w - T_1} = \frac{\int_0^y \frac{\tau}{\rho(\nu + \epsilon)} \, dy}{\int_0^\ell \frac{\tau}{\rho(\nu + \epsilon)} \, dy}$$

Defining:

$$F(y) = \frac{\int_0^y \frac{\tau}{\rho(\nu + \epsilon)} \, dy}{\int_0^\ell \frac{\tau}{\rho(\nu + \epsilon)} \, dy}$$

it will be seen that the above two equations give:

$$\frac{\overline{u}}{u_1} = F$$

and:

$$\frac{T_w - \overline{T}}{T_w - T_1} = F$$

Hence, the mean velocity profile and the mean temperature profile are described by the same function, F, i.e., the two profiles are "similar".

An expression for the local Nusselt number was derived above using the Reynolds analogy. In order to determine the average Nusselt number for a plate of length L, it must be remembered that the flow near the leading edge of the plate is laminar as shown in Fig. 6.3.

As discussed in the previous chapter, transition from laminar to turbulent flow in the boundary layer does not occur sharply at a point. Instead there is a region of "mixed" flow over which the transition occurs, the extent of this region being

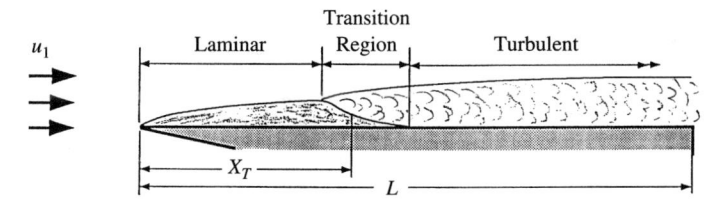

FIGURE 6.3
Transition in a turbulent boundary layer on a flat plate.

influenced by a number of factors. However, for many purposes, such as the present one of trying to determine the average heat transfer rate for the whole surface, it is adequate to assume that the flow is laminar up to some distance, x_T, from the leading edge and turbulent from there on, x_T being chosen roughly as the distance to the center of the transition region.

In view of the above assumption, if $q_{w\ell}$ is the local rate of heat transfer at any point in the laminar part of the flow and q_{wt} is the local rate of heat transfer at any point in the turbulent part of the flow, then the average rate of heat transfer for the entire surface is given by:

$$\bar{q}_w = \frac{1}{L}\left[\int_0^{x_T} q_{w\ell}\,dx + \int_{x_T}^L q_{wt}\,dx\right] \tag{6.19}$$

As discussed in Chapter 3, because a fluid with a Prandtl number of 1 is being considered, $q_{w\ell}$ is given by:

$$q_{w\ell} = 0.332\left(\frac{u_1}{x\nu}\right)^{0.5} k(T_w - T_1) \tag{6.20}$$

Therefore, using the value of q_{wt} given by Eq. (6.18), Eq. (6.19) gives, if the surface is isothermal:

$$\frac{\bar{q}_w L}{k(T_w - T_1)} = \left[\int_0^{x_T} 0.332\left(\frac{u_1}{x\nu}\right)^{0.5} dx + \int_{x_T}^L \frac{0.029}{x^{0.2}}\left(\frac{u_1}{\nu}\right)^{0.8} dx\right]$$

i.e.:

$$\overline{Nu}_L = 0.664 Re_T^{0.5} + 0.036(Re_L^{0.8} - Re_T^{0.8}) \tag{6.21}$$

where Re_T is the Reynolds number based on x_T, Re_L is the Reynolds number based on the plate length, L, and $\overline{Nu}_L$ is the mean Nusselt number for the entire plate.

If Re_T is taken as 10^6, an approximate experimental value as discussed in Chapter 5, then Eq. (6.21) gives:

$$\overline{Nu}_L = 0.036 Re_L^{0.8} - 1600 \tag{6.22}$$

If Re_L is large, the second term is negligible compared to the first and in this case the above equation reduces effectively to:

$$\overline{Nu}_L = 0.036 Re_L^{0.8} \tag{6.23}$$

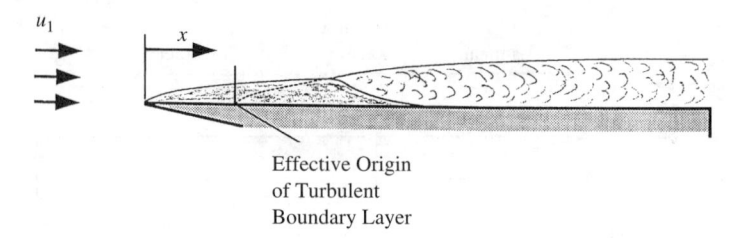

260 Introduction to Convective Heat Transfer Analysis

FIGURE 6.4
Effective origin of turbulent boundary layer.

which is, of course, the result that would have been obtained had the flow been assumed to be turbulent from the leading edge.

In deriving the above equations it was assumed that after transition:

$$\frac{q_w x}{k(T_w - T_1)} = 0.029 \left(\frac{u_1 x}{\nu}\right)^{0.8}$$

where x is the distance from the leading edge of the plate. However, the effective origin of the turbulent boundary layer after transition is not in general at the leading edge of the plate, i.e., as indicated in Fig. 6.4, in the portion of the flow with a turbulent boundary layer, the heat transfer rate should have been expressed as:

$$\frac{q_w(x - x_0)}{k(T_w - T_1)} = 0.029 \left[\frac{u_1(x - x_0)}{\nu}\right]^{0.8}$$

where x_0 is the distance of the effective origin of the turbulent boundary layer from the leading edge of the plate as shown in Fig. 6.4. The effect of this displaced origin is however usually small and will not be considered here.

EXAMPLE 6.2. Air flows over a flat plate which has a uniform surface temperature of 50°C, the temperature of the air ahead of the plate being 30°C. The air velocity is such that the Reynolds number based on the length of the plate is 5×10^6, the length of the plate being 30 cm. Using the Reynolds analogy, determine the variation of the local heat transfer rate from the wall, q_w, with x/L assuming that (i) the boundary layer flow remains laminar, (ii) the boundary layer flow is turbulent from the leading edge of the plate, and (iii) boundary layer transition occurs at Re_T of 10^6. x is the distance from the leading edge of the plate and L is the length of the plate.

Solution. The mean air temperature is $(30 + 50)/2 = 40°C$. At this temperature, air has a thermal conductivity of 0.0271 W/m-C.

At any distance, x, from the leading edge, the local Reynolds number will be given by:

$$Re_x = \frac{x}{L} Re_L = 5\frac{x}{L} \times 10^6$$

If the boundary layer flow remains laminar assuming $Pr = 1$:

$$Nu_x = 0.332 Re_x^{0.5}$$

i.e.:
$$\frac{q_w x}{k(T_w - T_1)} = 0.332\left(5\frac{x}{L} \times 10^6\right)^{0.5}$$

i.e.:
$$q_w = 0.332\left(5\frac{x}{L} \times 10^6\right)^{0.5} k(T_w - T_1)\frac{L}{x}\frac{1}{L}$$

i.e.:
$$q_w = 0.332\left(5 \times 10^6\right)^{0.5} \times 0.0271 \times (50 - 30)\left(\frac{x}{L}\right)^{-0.5}\frac{1}{0.3} = 1341\left(\frac{x}{L}\right)^{-0.5} \text{ W/m}^2 \quad (i)$$

If the boundary layer flow is turbulent from the leading edge:
$$Nu_x = 0.029 Re_x^{0.8}$$

i.e.:
$$\frac{q_w x}{k(T_w - T_1)} = 0.029\left(5\frac{x}{L} \times 10^6\right)^{0.8}$$

i.e.:
$$q_w = 0.029\left(5\frac{x}{L} \times 10^6\right)^{0.8} k(T_w - T_1)\frac{L}{x}\frac{1}{L}$$

i.e.:
$$q_w = 0.029\left(5 \times 10^6\right)^{0.8} \times 0.0271 \times (50 - 30)\left(\frac{x}{L}\right)^{-0.2}\frac{1}{0.3}$$
$$= 11980\left(\frac{x}{L}\right)^{-0.2} \text{ W/m}^2 \quad (ii)$$

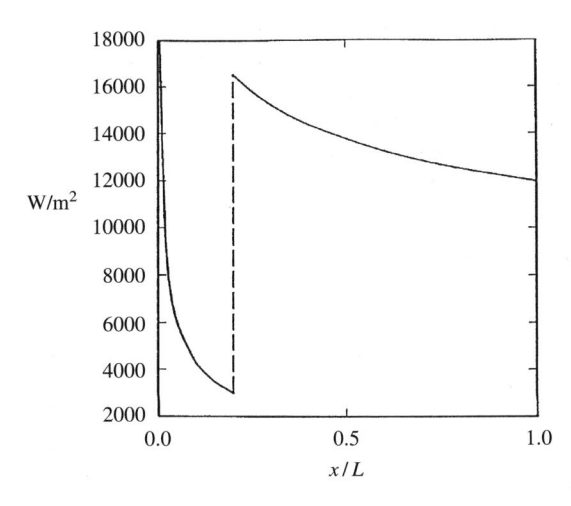

FIGURE E6.2

If boundary layer transition occurs at Re_T of 10^6 then transition occurs at:

$$\frac{x_T}{L} = \frac{10^6}{5 \times 10^6} = 0.2$$

In this case then it will be assumed that:

$$\frac{x}{L} < 0.2: \quad q_w = 1341 \left(\frac{x}{L}\right)^{-0.5} \text{W/m}^2 \qquad (iii)$$

$$\frac{x}{L} > 0.2: \quad q_w = 11980 \left(\frac{x}{L}\right)^{-0.2} \text{W/m}^2 \qquad (iv)$$

The variations of q_w with x/L as given by Eqs. (*iii*) and (*iv*) are shown in Fig. E6.2.

Experiments in air suggest that the form of the relation for $\overline{Nu}_L$ given by the Reynolds analogy is approximately correct, i.e., that it does vary approximately as $Re_L^{0.8}$. The coefficient, i.e., 0.036, has, however, not been found to be in good agreement with experiment. For fluids with Prandtl numbers very different from unity, the agreement is even less satisfactory. Thus, the Reynolds analogy gives results that are reasonably satisfactory for fluids with Prandtl numbers near unity but gives relatively poor results for other fluids. For this reason, other analyses have been developed which utilize the same basic ideas as those used in the derivation of the Reynolds analogy but which try to relax some of the assumptions made in its derivation. A discussion of the simplest of these extended analogy solutions will now be given.

It was previously mentioned that the turbulent shearing stress and heat transfer rate are very much greater than the molecular values over most of the turbulent boundary layer flow. It is only in a region close to the surface, in which the velocities are very low, that the molecular terms become more important than the turbulent terms. The extension of the Reynolds analogy presently being discussed is based on the assumption that the flow can, therefore, be split into two distinct regions [1]. In the inner region, which lies adjacent to the wall, it is assumed that the turbulent shearing stress and heat transfer rate are negligible compared to the molecular values, while in the outer region, which covers the rest of the boundary layer, it is assumed that the molecular shearing stress and heat transfer rate are negligible compared to the turbulent values. In reality there are, of course, not two such clear and distinct regions. The assumed flow pattern is shown in Fig. 6.5.

Because the molecular components are being neglected in the outer region, the total shearing stress and heat transfer rate in this region are given by:

$$\tau = \tau_T = \rho \epsilon \frac{\partial \overline{u}}{\partial y} \qquad (6.24)$$

and:

$$q = q_T = -\rho c_p \epsilon_H \frac{\partial \overline{T}}{\partial y} \qquad (6.25)$$

These two equations can now be integrated outward from the outer edge of the inner layer, i.e., from $y = \Delta_s$, to some point in the free-stream which lies at a

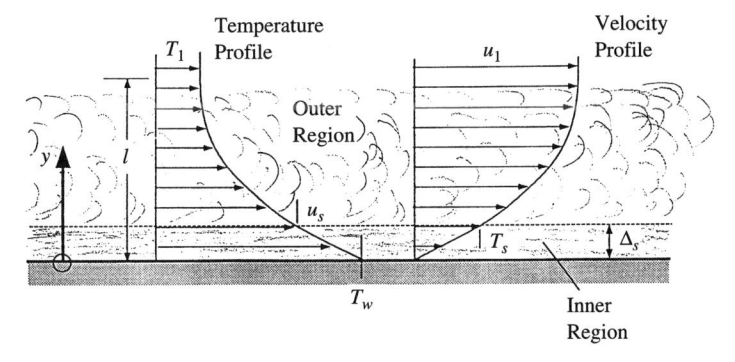

FIGURE 6.5
Taylor-Prandtl model for turbulent boundary layer flow.

distance, ℓ, from the wall. This integration gives:

$$u_1 - u_s = \int_{\Delta_s}^{\ell} \frac{\tau}{\rho \epsilon} dy \qquad (6.26)$$

and:

$$T_s - T_1 = \int_{\Delta_s}^{\ell} \frac{q}{\rho c_p \epsilon_H} dy \qquad (6.27)$$

where u_s and T_s are the mean velocity and temperature at the outer edge of the inner layer and u_1 and T_1 are, as before, the velocity and temperature in the free-stream.

Similar assumptions to those used in the Reynolds analogy are now introduced; i.e., it is assumed in the outer layer that:

$$\epsilon_H = \epsilon, \frac{q}{q_s} = \frac{\tau}{\tau_s}$$

where τ_s and q_s are the total shearing stress and heat transfer rate at the outer edge of the inner layer. Using these assumptions, Eq. (6.27) can be written:

$$T_s - T_1 = \frac{q_s}{\tau_s} \int_{\Delta_s}^{\ell} \frac{\tau}{\rho c_p \epsilon} dy \qquad (6.28)$$

Dividing this equation by Eq. (6.26) then gives:

$$\frac{T_s - T_1}{u_1 - u_s} = \frac{q_s}{c_p \tau_s} \qquad (6.29)$$

Next consider the inner layer. Because it is being assumed that turbulent shearing stress and heat transfer rate are negligible in this region, the total shearing stress and heat transfer rate in this layer are given by:

$$\tau = \rho \nu \frac{\partial \overline{u}}{\partial y} \qquad (6.30)$$

264 Introduction to Convective Heat Transfer Analysis

and:
$$q = -\rho c_p \alpha \frac{\partial \overline{T}}{\partial y} \tag{6.31}$$

It is next noted that because the velocities are very low in the inner, i.e., near-wall, region, the convective terms in the boundary layer equations can be neglected in this region, i.e., in this region the momentum and energy equations can be assumed to have the form:

$$0 = \rho \nu \frac{\partial^2 \overline{u}}{\partial y^2}, \quad \text{i.e., } 0 = \frac{\partial \tau}{\partial y} \tag{6.32}$$

and:

$$0 = \frac{k}{\rho c_p} \frac{\partial^2 \overline{u}}{\partial y^2}, \quad \text{i.e., } 0 = \frac{\partial q}{\partial y} \tag{6.33}$$

These equations show that in the inner region the shearing stress and heat transfer rate can be assumed to be constant, i.e., that in the inner region:

$$\tau = \tau_w = \tau_s \tag{6.34}$$

and:

$$q = q_w = q_s \tag{6.35}$$

Using these in Eqs. (6.30) and (6.31) and integrating across the inner layer then gives:

$$\frac{\tau_w}{\rho \nu} \Delta_s = u_s \tag{6.36}$$

and:

$$\frac{q_w}{\rho c_p \alpha} \Delta_s = T_w - T_s \tag{6.37}$$

Dividing these two equations then gives:

$$\frac{T_w - T_s}{u_s} = \frac{q_w}{\tau_w} \frac{\nu}{c_p \alpha} \tag{6.38}$$

Now, in view of Eqs. (6.34) and (6.35), the result for the outer region, given in Eq. (6.29), can be written as:

$$\frac{T_s - T_1}{u_1 - u_s} = \frac{q_w}{c_p \tau_w} \tag{6.39}$$

The temperature, T_s, can now be eliminated between Eqs. (6.38) and (6.39) to give:

$$\frac{q_w}{c_p \tau_w} = \frac{T_w - T_1}{u_1} \frac{1}{\left[1 + \frac{u_s}{u_1}(Pr - 1)\right]} \tag{6.40}$$

It should be noted that when $Pr = 1$, this equation reduces to the Reynolds analogy result given in Eq. (6.14).

If the distance, x, from the leading edge to the point being considered is again introduced, Eq. (6.40) can be written as:

$$Nu_x = \frac{\left(\frac{C_f}{2}\right)Re_x Pr}{\left[1 + \frac{u_s}{u_1}(Pr - 1)\right]} \quad (6.41)$$

where, as before, Nu_x is the local Nusselt number, Re_x is the local Reynolds number, and C_f is the local shearing stress coefficient, i.e., $\tau_w/(1/2\rho u_1^2)$.

This is known as the Taylor-Prandtl analogy. In order to apply it, the value of u_s/u_1 has to be known and the way in which this can be found will now be discussed.

As discussed in Chapter 5, the wall region extends out to a distance from the wall that is defined by the value of y_s^+ where:

$$y_s^+ = \left(\frac{\tau_w}{\rho}\right)^{0.5} \frac{\Delta_s}{\nu}$$

From this, it follows that:

$$\Delta_s = \frac{y_s^+ \nu}{(\tau_w/\rho)^{0.5}} \quad (6.42)$$

Hence, using Eq. (6.36), i.e.:

$$u_s = \frac{\tau_w}{\rho \nu}\Delta_s$$

it follows that:

$$u_s = \left(\frac{\tau_w}{\rho}\right)^{0.5} y_s^+$$

i.e.:

$$\frac{u_s}{u_1} = \left(\frac{C_f}{2}\right)^{0.5} y_s^+ \quad (6.43)$$

Substituting this into Eq. (6.40) then gives:

$$Nu_x = \frac{\left(\frac{C_f}{2}\right)Re_x Pr}{\left[1 + \left(\frac{C_f}{2}\right)^{0.5} y_s^+ (Pr - 1)\right]} \quad (6.44)$$

Using Eq. (6.17), i.e., assuming:

$$\frac{C_f}{2} = \frac{0.029}{Re_x^{0.2}}$$

allows Eq. (6.44) to be written as:

$$Nu_x = \frac{0.029 Re_x^{0.8} Pr}{\left[1 + \frac{0.17}{Re_x^{0.1}} y_s^+ (Pr - 1)\right]} \qquad (6.45)$$

If, as discussed in Chapter 5, it is assumed, based on available measurements, that:

$$y_s^+ = 12 \qquad (6.46)$$

then Eq. (6.45) gives:

$$Nu_x = \frac{0.029 Re_x^{0.8} Pr}{\left[1 + \frac{2}{Re_x^{0.1}} (Pr - 1)\right]} \qquad (6.47)$$

This equation can be written as:

$$Nu_x = 0.029 Re_x^{0.8} F \qquad (6.48)$$

Thus, compared to the Reynolds analogy there is the following additional factor which is dependent on both the Reynolds and Prandtl numbers:

$$F = \frac{Pr}{\left[1 + \frac{2}{Re_x^{0.1}} (Pr - 1)\right]} \qquad (6.49)$$

The variation of F with Pr for some typical values of Re_x is shown in Fig. 6.6. It will be seen from this figure that for Prandtl numbers near one, the factor F is approximately given by:

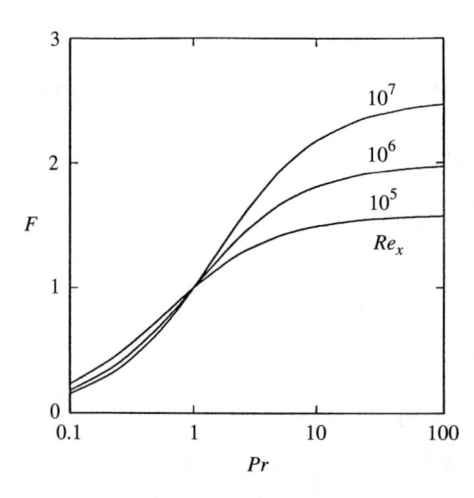

$$F = Pr^{0.4} \qquad (6.50)$$

FIGURE 6.6
Variation of function F with Prandtl number.

and in this case the Taylor-Prandtl analogy for turbulent boundary layer flow over a flat plate gives:

$$Nu_x = 0.029 Re_x^{0.8} Pr^{0.4} \tag{6.51}$$

Integrating Eq. (6.51) over the entire plate as was done with the Reynolds analogy equation and assuming that the flow is turbulent from the leading edge of the plate then gives the following expression for the mean Nusselt number:

$$\overline{Nu_L} = 0.036 Re_L^{0.8} Pr^{0.4} \tag{6.52}$$

The Taylor-Prandtl analogy indicates that the effect of the Prandtl number can be accounted for by splitting the boundary layer into layers in which certain effects predominate. In the Taylor-Prandtl analogy, two such layers were used. However, as indicated in Chapter 5, a better description of the flow is obtained by considering three layers, i.e., the inner or wall region, the buffer region, and the outer or fully turbulent region. An analysis based on the use of such a three-layer model will now be presented. In such an analysis, it is necessary to make assumptions about the nature of the flow in the buffer region. Here, it will be assumed that the mean velocity distribution in this buffer region is described by a universal and known function.

It is convenient to write the shear stress and heat transfer equations, i.e.:

$$\tau = \rho(\nu + \epsilon)\frac{\partial \overline{u}}{\partial y} \tag{6.3}$$

and:

$$q = -\rho c_p (\alpha + \epsilon_H)\frac{\partial \overline{T}}{\partial y} \tag{6.4}$$

in terms of the following variables which were introduced in Chapter 5:

$$u^* = \sqrt{\frac{\tau_w}{\rho}}, \quad u^+ = \frac{\overline{u}}{u^*}, \quad y^+ = \frac{y u^*}{\nu} \tag{6.53}$$

u^* being termed the "friction velocity". It is also convenient to define:

$$T^+ = \frac{T_w - \overline{T}}{q_w/\rho c_p} u^* \tag{6.54}$$

In terms of the variables defined in Eqs. (6.53) and (6.54), Eqs. (6.3) and (6.4) become:

$$\frac{\tau}{\tau_w} = \left(1 + \frac{\epsilon}{\nu}\right)\frac{\partial u^+}{\partial y^+} \tag{6.55}$$

and:

$$\frac{q}{q_w} = \left(\frac{1}{Pr} + \frac{\epsilon/\nu}{Pr_T}\right)\frac{\partial T^+}{\partial y^+} \tag{6.56}$$

First, consider the inner layer which, as discussed in Chapter 5, is assumed to cover the region $0 \leq y^+ \leq 5$. In this region it is again assumed that the effects of

268 Introduction to Convective Heat Transfer Analysis

the turbulence on the heat transfer rate are negligible and that the heat transfer rate is constant in this region and equal, therefore, to the heat transfer rate at the wall, q_w; i.e., it is assumed that in the inner layer:

$$\frac{q}{q_w} = 1, \epsilon = 0 \tag{6.57}$$

Therefore, in the inner region, Eq. (6.56) gives:

$$1 = \frac{1}{Pr}\frac{\partial T^+}{\partial y^+} \tag{6.58}$$

This can be integrated to give the temperature distribution in the inner layer as:

$$T^+ = Pr\, y^+ \tag{6.59}$$

Applying this at the outer edge of the inner layer where $y^+ = 5$ gives:

$$T_s^+ = 5Pr \tag{6.60}$$

where T_s^+ is the value of T^+ at the outer edge of the inner layer.

Next, consider the buffer region. As discussed in Chapter 5, this is assumed to cover the region $5 \leq y^+ \leq 30$. In this region it is again assumed that the shear stress and the heat transfer rate are constant and, in view of the assumptions made about the inner layer, they are therefore equal to the shear stress at the wall, τ_w, and to the heat transfer rate at the wall, q_w, respectively; i.e., it is assumed that in the buffer layer:

$$\frac{\tau}{\tau_w} = 1, \quad \frac{q}{q_w} = 1 \tag{6.61}$$

It will also be assumed that the mean velocity distribution in the buffer region is given by:

$$u^+ = 5 + 5\ln\left(\frac{y^+}{5}\right) \tag{6.62}$$

Differentiating Eq. (6.62) gives:

$$\frac{\partial u^+}{\partial y^+} = \frac{5}{y^+} \tag{6.63}$$

Substituting this result into Eq. (6.55) and using Eq. (6.61) then gives:

$$1 = \left(1 + \frac{\epsilon}{\nu}\right)\frac{5}{y^+}$$

i.e.:

$$\frac{\epsilon}{\nu} = \frac{y^+}{5} - 1 \tag{6.64}$$

Substituting Eq. (6.64) and Eq. (6.61) into Eq. (6.56) then gives for the buffer region:

$$1 = \left[\frac{1}{Pr} + \frac{(y^+/5 - 1)}{Pr_T}\right]\frac{\partial T^+}{\partial y^+} \tag{6.65}$$

This equation can be integrated across the buffer region to give:

$$T^+ - T_s^+ = \int_5^{y^+} \frac{dy^+}{\left[\frac{1}{Pr} + \frac{(y^+/5 - 1)}{Pr_T}\right]}$$

i.e.:

$$T^+ - T_s^+ = 5Pr_T \ln\left[1 + \frac{Pr}{Pr_T}\left(\frac{y^+}{5}\right) - 1\right] \tag{6.66}$$

Applying Eq. (6.66) across the entire buffer region, i.e., out to $y^+ = 30$, then gives:

$$T_b^+ - T_s^+ = 5Pr_T \ln\left[1 + \frac{Pr}{Pr_T}\left(\frac{30}{5}\right) - 1\right] = 5Pr_T \ln\left[1 + 5\frac{Pr}{Pr_T}\right] \tag{6.67}$$

where T_b^+ is the value of T^+ at the outer edge of the buffer region.

Lastly, consider the outer, fully turbulent region which is assumed to exist when $y^+ > 30$. In this region it is assumed that $\epsilon \gg \nu$ and $\epsilon_H \gg \nu$, i.e., that in this region:

$$\frac{\tau}{\tau_w} = \frac{\epsilon}{\nu}\frac{\partial u^+}{\partial y^+} \tag{6.68}$$

and:

$$\frac{q}{q_w} = \frac{\epsilon/\nu}{Pr_T}\frac{\partial T^+}{\partial y^+} \tag{6.69}$$

It is further assumed that the distributions of shear stress and heat transfer rate are similar in the outer region, i.e., that:

$$\frac{\tau}{\tau_b} = \frac{q}{q_b} \tag{6.70}$$

where τ_b and q_b are the values of the shear stress and heat transfer rate at the outer edge of the buffer layer. Because the shear stress and heat transfer rate were assumed to be constant across the inner two layers, it follows that it is being assumed that:

$$\tau_b = \tau_w, \quad q_b = q_w$$

Eq. (6.70) can therefore be written as:

$$\frac{\tau}{\tau_w} = \frac{q}{q_w} \tag{6.71}$$

Eqs. (6.68) and (6.69) together then give:

$$\frac{\epsilon}{\nu}\frac{\partial u^+}{\partial y^+} = \frac{\epsilon/\nu}{Pr_T}\frac{\partial T^+}{\partial y^+}$$

i.e.:

$$\frac{\partial T^+}{\partial y^+} = Pr_T \frac{\partial u^+}{\partial y^+} \tag{6.72}$$

270 Introduction to Convective Heat Transfer Analysis

Integrating this equation across the outer layer then gives:

$$\int_{T_b^+}^{T_1^+} dT^+ = Pr_T \int_{u_b^+}^{u_1^+} du^+$$

i.e.:

$$T_1^+ - T_b^+ = Pr_T(u_1^+ - u_b^+) \tag{6.73}$$

where T_1^+ and u_1^+ are the values of T^+ and u^+ in the free-stream. But $y_b^+ = 30$ so Eq. (6.62) gives:

$$u_b^+ = 5 + 5\ln\left(\frac{30}{5}\right) = 5(1 + \ln 6)$$

Substituting this into Eq. (6.73) then gives:

$$T_1^+ - T_b^+ = Pr_T\left[u_1^+ - 5(1 + \ln 6)\right] \tag{6.74}$$

Adding Eqs. (6.60), (6.67), and (6.74) then gives:

$$\begin{aligned}T_1^+ &= 5Pr + 5Pr_T \ln\left[1 + 5\frac{Pr}{Pr_T}\right] + Pr_T\left[u_1^+ - 5(1 + \ln 6)\right]\\&= 5Pr + 5Pr_T \ln\left[\frac{1 + Pr/Pr_T}{6}\right] + Pr_T(u_1^+ - 5)\end{aligned} \tag{6.75}$$

Now:

$$T^+ = \frac{T_w - \overline{T}}{q_w/\rho c_p}u^* = \frac{T_w - \overline{T}}{q_w/\rho c_p}\sqrt{\frac{\tau_w}{\rho}} = \frac{Pr}{Nu_x}Re_x\sqrt{\frac{C_f}{2}} \tag{6.76}$$

Eq. (6.75) can therefore be written as:

$$Nu_x = \frac{Re_x\sqrt{C_f/2}}{5 + 5\dfrac{Pr_T}{Pr}\ln\left[\dfrac{1 + 5Pr/Pr_T}{6}\right] + \dfrac{Pr_T}{Pr}(u_1^+ - 5)} \tag{6.77}$$

i.e., since:

$$u_1^+ = \frac{u_1}{u^*} = \frac{u_1}{\sqrt{\tau_w/\rho}}$$

Eq. (6.77) gives:

$$Nu_x = \frac{Re_x\sqrt{C_f/2}}{5 + 5\dfrac{Pr_T}{Pr}\ln\left[\dfrac{1 + 5Pr/Pr_T}{6}\right] + \dfrac{Pr_T}{Pr}\left(\dfrac{1}{\sqrt{\tau_w/\rho u_1^2}} - 5\right)}$$

i.e.:

$$\frac{Nu_x}{PrRe_x} = \frac{C_f/2}{\sqrt{\dfrac{C_f}{2}}\left[5Pr + 5Pr_T\ln\left(\dfrac{1 + 5Pr/Pr_T}{6}\right) - 5Pr_T\right] + Pr_T} \tag{6.78}$$

CHAPTER 6: External Turbulent Flows   271

It will be noted that if $Pr = 1$ and $Pr_T = 1$ Eq. (6.78) gives:

$$\frac{Nu_x}{Re_x} = \frac{C_f}{2}$$

which is the Reynolds analogy result.

If the turbulent Prandtl number, Pr_T, is taken as 1 and if flow over a flat plate is considered, C_f then being assumed to be given by Eq. (6.17), i.e., by:

$$C_f = \frac{0.058}{Re_x^{0.2}}$$

Eq. (6.78) gives:

$$Nu_x = \frac{0.029 Re_x^{0.8} Pr}{\sqrt{\frac{0.029}{Re_x^{0.2}}} \left[5Pr + 5\ln\left(\frac{1+5Pr}{6}\right) - 5 \right] + 1} \quad (6.79)$$

This equation can be written as:

$$Nu_x = 0.029 Re_x^{0.8} G \quad (6.80)$$

Thus, compared to the Reynolds analogy there is the following additional factor which is dependent on both the Reynolds and Prandtl numbers:

$$G = \frac{Pr}{\sqrt{\frac{0.029}{Re_x^{0.2}}} \left[5Pr + 5\ln\left(\frac{1+5Pr}{6}\right) - 5 \right] + 1} \quad (6.81)$$

The variation of G with Pr for some typical values of Re_x is shown in Fig. 6.7.

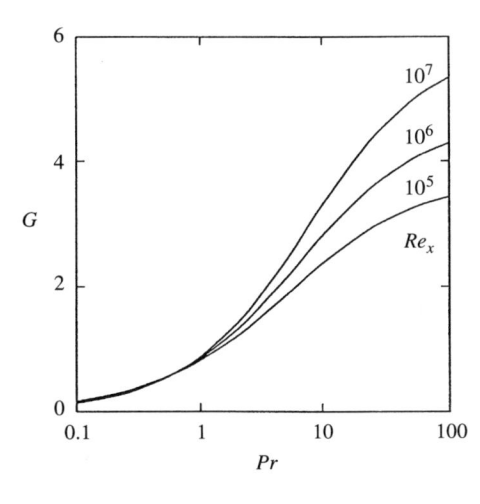

FIGURE 6.7
Variation of function G with Prandtl number.

6.3 INTEGRAL EQUATION SOLUTIONS

The analogy solutions discussed in the previous section use the value of the wall shear stress to predict the wall heat transfer rate. In the case of flow over a flat plate, this wall shear stress is given by a relatively simple expression. However, in general, the wall shear stress will depend on the pressure gradient and its variation has to be computed for each individual case. One approximate way of determining the shear stress distribution is based on the use of the momentum integral equation that was discussed in Chapter 2 [1],[2],[3],[5]. As shown in Chapter 2 (see Eq. 2.172), this equation has the form:

$$\frac{d}{dx}\left[\int_0^\delta (u_1 - \bar{u})\bar{u}\,dy\right] + \frac{du_1}{dx}\left[\int_0^\delta (u_1 - \bar{u})\,dy\right] = \frac{\tau_w}{\rho} \qquad (6.82)$$

where δ is the local boundary layer thickness, u_1 is the local freestream velocity, and τ_w is the local wall shear stress.

Eq. (6.82) is, for the present purposes, conveniently written as:

$$\frac{d}{dx}\left[u_1^2\int_0^\delta \left(1 - \frac{\bar{u}}{u_1}\right)\frac{\bar{u}}{u_1}\,dy\right] + u_1\frac{du_1}{dx}\left[\int_0^\delta \left(1 - \frac{\bar{u}}{u_1}\right)\right] = \frac{\tau_w}{\rho} \qquad (6.83)$$

The following are then defined:

$$\delta_1 = \int_0^\delta \left(1 - \frac{\bar{u}}{u_1}\right) dy \qquad (6.84)$$

and:

$$\delta_2 = \int_0^\delta \left(1 - \frac{\bar{u}}{u_1}\right)\frac{\bar{u}}{u_1}\,dy \qquad (6.85)$$

δ_1 is termed the boundary layer displacement thickness while δ_2 is termed the boundary layer momentum thickness. In terms of these quantities, Eq. (6.83) can be written as:

$$\frac{d}{dx}\left[u_1^2 \delta_2\right] + u_1 \delta_1 \frac{du_1}{dx} = \frac{\tau_w}{\rho}$$

i.e.:

$$\frac{d\delta_2}{dx} + (2\delta_2 + \delta_1)\frac{1}{u_1}\frac{du_1}{dx} = \frac{\tau_w}{\rho u_1^2}$$

i.e.:

$$\frac{d\delta_2}{dx} + \delta_2(2 + H)\frac{1}{u_1}\frac{du_1}{dx} = \frac{C_f}{2} \qquad (6.86)$$

where:

$$H = \frac{\delta_1}{\delta_2} \qquad (6.87)$$

CHAPTER 6: External Turbulent Flows 273

Empirical equations for the variations of the wall shear stress and of the "form factor", H, are now introduced. The following very approximate relations [1] will be used here in order to illustrate the method:

$$\frac{C_f}{2} = \frac{\tau_w}{\rho u_1^2} = \frac{0.123 \times 10^{-0.678H}}{(u_1 \delta_2/\nu)^{0.268}} \tag{6.88}$$

and:

$$\left(\frac{u_1 \delta_2}{\nu}\right)^{1/6} \delta_2 \frac{dH}{dx} = e^{5(H-1.4)} \left[-\left(\frac{u_1 \delta_2}{\nu}\right)^{1/6} \delta_2 \frac{1}{u_1} \frac{du_1}{dx} - 0.0135(H - 1.4) \right] \tag{6.89}$$

These two equations effectively constitute the turbulence model. They are solved simultaneously with Eq. (6.86) to give the variation of $\tau_w/\rho u_1^2$ with x. The selected analogy solution equation is then used to give the local heat transfer rate variation.

For flow over a flat plate:

$$\frac{du_1}{dx} = 0$$

and in this case Eq. (6.89) shows that H is a constant and given by:

$$H = 1.4 \tag{6.90}$$

Therefore, for flow over a flat plate, Eq. (6.88) gives:

$$\frac{\tau_w}{\rho u_1^2} = \frac{0.0138}{(u_1 \delta_2/\nu)^{0.268}} \tag{6.91}$$

and Eq. (6.86) becomes:

$$\frac{d\delta_2}{dx} = \frac{\tau_w}{\rho u_1^2} \tag{6.92}$$

Combining the above two equations then gives:

$$\frac{d\delta_2}{dx} = \frac{0.0138}{(u_1 \delta_2/\nu)^{0.268}} \tag{6.93}$$

This equation can be integrated using $\delta_2 = 0$ when $x = 0$ to give:

$$\frac{\delta_2^{1.268}}{1.268} = \frac{0.0138}{(u_1/\nu)^{0.268}} x$$

i.e.:

$$\frac{\delta_2}{x} = \frac{0.0412}{(u_1 x/\nu)^{0.211}} \tag{6.94}$$

Substituting this result back into Eq. (6.91) then gives:

$$\frac{\tau_w}{\rho u_1^2} = \frac{0.0324}{Re_x^{0.211}} \tag{6.95}$$

274 Introduction to Convective Heat Transfer Analysis

This equation gives values very close to those given by Eq. (6.17), which was used in the discussion of analogy solutions for flow over a flat plate presented in the previous section.

The present section is not really concerned with flow over a flat plate. It is instead concerned with flows in which the free-stream velocity is varying with x. The solution in such a case usually has to be obtained by numerically solving the governing equations. For this purpose, it is convenient to introduce the following dimensionless variables:

$$U = \frac{u_1}{u_\infty}, \quad X = \frac{x}{L}, \quad \Delta_2 = \frac{\delta_2}{L} \tag{6.96}$$

where u_∞ is some representative free-stream velocity, such as that in the flow upstream of the surface, and L is some characteristic size of the surface.

In terms of these variables, Eqs. (6.86), (6.88), and (6.89) become:

$$\frac{d\Delta_2}{dx} + \Delta_2(2 + H)\frac{1}{U}\frac{dU}{dX} = \frac{\tau_w}{\rho u_1^2} \tag{6.97}$$

$$\frac{\tau_w}{\rho u_1^2} = \frac{0.123 \times 10^{-0.678H}}{Re_L^{0.268}(U\Delta_2)^{0.268}} \tag{6.98}$$

$$Re_L^{1/6}(U\Delta_2)^{1/6}\Delta_2 \frac{dH}{dX} = e^{5(H-1.4)}\left[-Re_L^{1/6}(U\Delta_2)^{1/6}\frac{\Delta_2}{U}\frac{dU}{dX} - 0.0135(H-1.4)\right] \tag{6.99}$$

where:

$$Re_L = \frac{u_\infty L}{\nu} \tag{6.100}$$

These three equations must be simultaneously solved using a suitable numerical method to give the variations of H and Δ_2 with X. A simple program, TURBINRK, for obtaining such a solution based on the use of the Runge-Kutta method to numerically integrate the simultaneous differential equation is available as discussed in the Preface. This program is written in FORTRAN.

This program assumes that the surface is isothermal and that the dimensionless free-stream velocity variation can be represented by a third-order polynomial, i.e., by:

$$U = A + BX + CX^2 + DX^3 \tag{6.101}$$

where A, B, C, and D are known constants. The inputs to the program are then the values of Re_L, A, B, C, D, and Pr.

The program assumes the flow is turbulent from the leading edge and that $\delta_2 = 0$ when $x = 0$. The program can easily be modified to use a laminar boundary layer equation solution procedure to provide initial conditions for the turbulent boundary layer solution which would then be started at some assumed transition point.

The program as available uses the basic Reynolds analogy, i.e.:

$$\frac{T_w - T_1}{u_1} = \frac{q_w}{c_p T_w}$$

to obtain the heat transfer rate from the calculated wall shear stress. The local heat transfer rate is expressed in terms of a local Nusselt number based on x.

EXAMPLE 6.3. Air flows through a large plane duct with isothermal walls. The Reynolds number based on the length of the duct and the inlet air velocity is 10^7. Assuming that the boundary layer is turbulent and thin compared to the size of the duct, determine how the local Nusselt number varies with distance along the duct if the duct cross-sectional area is (i) constant, (ii) varies in such a way that the velocity increases linearly by 80% over the length of the duct, and (iii) decreases linearly by 80% over the length of the duct.

Solution. In all three cases:

$$Re_L = 10^7, \quad Pr = 0.7$$

Also because:

$$U = \frac{u_1}{u_{\text{inlet}}}, \quad X = \frac{x}{L}$$

where L is the length of the duct, it follows that in the three cases, the variation of U with X is given by:
Case (i):

$$U = 1$$

Case (ii):

$$U = 1 + 0.8X$$

Case (iii):

$$U = 1 - 0.8X$$

Therefore, because the assumed form of the variation of U with X is:

$$U = A + BX + CX^2 + DX^2$$

it follows that:
Case (i):

$$A = 1, B = 0, C = 0, D = 0$$

Case (ii):

$$A = 1, B = 0.8, C = 0, D = 0$$

Case (iii):

$$A = 1, B = -0.8, C = 0, D = 0$$

The available computer program TURBINRK has been run for each of the three sets of input values. The program gives the variation of the local Nusselt number, Nu_x, with X and the variations for each of the three cases considered is shown in Fig. E6.3.

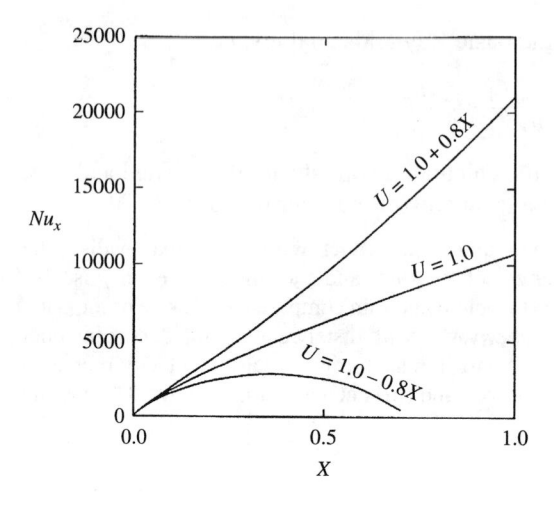

276 Introduction to Convective Heat Transfer Analysis

FIGURE E6.3

It will be seen that when the velocity is increasing along the duct the heat transfer rate is higher than when the velocity is constant. Simultaneously, when the velocity is decreasing along the duct the heat transfer rate is lower than when the velocity is constant. In the latter case, i.e., where the velocity is decreasing along the duct, boundary layer separation occurs at approximately $X = 0.73$ and the calculation is stopped just before separation occurs.

The solution procedure discussed above obtained the heat transfer rate by using the analogy solutions discussed in the previous section. These solutions assume that the velocity and thermal boundary layer thicknesses are of similar orders of magnitude. When this is not the case, the energy integral equation [3] has to be used to determine the variation of the thickness of the thermal boundary layer. To illustrate the procedure consider flow over a flat plate with an unheated leading edge section, often referred to as an unheated starting length, as illustrated in Fig. 6.8. The boundary layer is assumed to be turbulent from the leading edge of the plate. No heating of the flow occurs upstream of the heated section of the plate so the thermal boundary layer starts to grow at $x = x_0$, i.e., at the beginning of the heated section of the plate.

It is assumed that the temperature and velocity profiles in the boundary layer are given by:

$$\frac{T - T_1}{T_w - T_1} = 1 - \left(\frac{y}{\delta_T}\right)^{1/n} \tag{6.102}$$

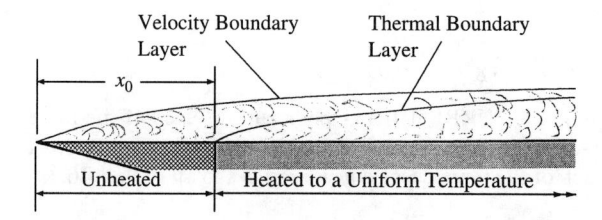

FIGURE 6.8
Turbulent boundary layer flow over a flat plate with an unheated leading edge section.

and:

$$\frac{\bar{u}}{u_1} = \left(\frac{y}{\delta}\right)^{1/n} \tag{6.103}$$

Here δ and δ_T are the thicknesses of the velocity and thermal boundary layers respectively and n is an integer constant. These assumed forms of the profiles will not apply very near the wall but this is not of consequence because of the way in which these profiles are applied in this analysis.

In order to decide what value of n to use in Eqs. (6.102) and (6.103), it is recalled that, as discussed above, experiment indicates that for flow over a flat plate:

$$H(=\delta_1/\delta_2) = 1.4 \tag{6.104}$$

Using the assumed velocity profile given in Eq. (6.103) and recalling that:

$$\delta_1 = \int_0^\delta \left[1 - \frac{\bar{u}}{u_1}\right] dy$$

it follows that:

$$\delta_1 = \delta \int_0^1 \left[1 - \left(\frac{y}{\delta}\right)^{1/n}\right] dy = \frac{\delta}{n+1} \tag{6.105}$$

Similarly, recalling that:

$$\delta_2 = \int_0^\delta \frac{\bar{u}}{u_1}\left[1 - \frac{\bar{u}}{u_1}\right] dy$$

it follows that:

$$\delta_2 = \delta \int_0^1 \left[\left(\frac{y}{\delta}\right)^{1/n} - \left(\frac{y}{\delta}\right)^{2/n}\right] dy = \frac{n}{(n+1)(n+2)} \delta \tag{6.106}$$

Using Eqs. (6.105) and (6.106) then gives:

$$H = \frac{\delta_1}{\delta_2} = \frac{n+2}{n} \tag{6.107}$$

Hence, for $H = 1.4$, it follows that:

$$n = \frac{2}{H-1} = 5 \tag{6.108}$$

This value will be used in the present analysis.
Using the assumed forms of the velocity and temperature profiles gives:

$$\tau = \rho\epsilon\frac{\partial \bar{u}}{\partial y} = \frac{\rho\epsilon u_1}{n\delta}\left(\frac{y}{\delta}\right)^{1/n-1} \tag{6.109}$$

and:

$$q = -\rho c_p \epsilon_H \frac{\partial \bar{T}}{\partial y} = \frac{\rho c_p \epsilon_H (T_w - T_1)}{n\delta_T}\left(\frac{y}{\delta_T}\right)^{1/n-1} \tag{6.110}$$

Dividing Eq. (6.110) by Eq. (6.109) then gives:

$$\frac{q}{\tau} = c_p \frac{\epsilon_H}{\epsilon} \frac{(T_w - T_1)}{u_1} \left(\frac{\delta}{\delta_T}\right)^{1/n}$$

Applying this equation at $y = 0$ (the equation is actually applied at the outer edge of the inner layer but because the shear stress and heat transfer rate can be assumed constant across the inner layer and equal to their values at the wall the same result is obtained) then gives:

$$\frac{q_w}{\tau_w} = c_p \frac{\epsilon_H}{\epsilon} \frac{(T_w - T_1)}{u_1} \left(\frac{\delta}{\delta_T}\right)^{1/n}$$

It will again be assumed that $\epsilon_H = \epsilon$, i.e., $Pr_T = 1$. The above equation then gives:

$$\frac{q_w}{\tau_w} = c_p \frac{(T_w - T_1)}{u_1} \left(\frac{\delta}{\delta_T}\right)^{1/n} \tag{6.111}$$

Now the analysis of the velocity boundary layer on a flat plate presented above gave:

$$\frac{\tau_w}{\rho u_1^2} = \frac{0.0324}{(u_1 x/\nu)^{0.211}}$$

and:

$$\frac{\delta_2}{x} = \frac{0.0412}{(u_1 x/\nu)^{0.211}}$$

But Eq. (6.106) gave:

$$\delta_2 = \frac{n}{(n+1)(n+2)} \delta$$

i.e., using $n = 5$ then gives:

$$\frac{\delta_2}{\delta} = \frac{5}{42}$$

Using this in Eq. (6.94) gives:

$$\frac{\delta}{x} = \frac{0.346}{(u_1 x/\nu)^{0.211}} \tag{6.112}$$

Now in the situation being considered here, $\delta_T < \delta$ so the energy integral equation, i.e., Eq. (2.194) gives:

$$\frac{d}{dx}\left[\int_0^{\delta_T} \bar{u}(\bar{T} - T_1)\,dy\right] = \frac{q_w}{\rho c_p}$$

This equation can be written as:

$$\frac{d}{dx}\left[u_1(T_w - T_1)\int_0^{\delta_T} \left(\frac{\bar{u}}{u_1}\right)\left(\frac{\bar{T} - T_1}{T_w - T_1}\right)dy\right] = \frac{q_w}{\rho c_p}$$

i.e., because, in the flow being considered, u_1 and T_w are constants, this equation gives:

$$\frac{d}{dx}\left[\int_0^{\delta_T}\left(\frac{\bar{u}}{u_1}\right)\left(\frac{\bar{T}-T_1}{T_w-T_1}\right)dy\right] = \frac{q_w}{\rho c_p u_1 (T_w - T_1)} \quad (6.113)$$

Using Eqs. (6.102), (6.103), and (6.108) then gives:

$$\frac{d}{dx}\left\{\delta_T \int_0^1 \left(\frac{y}{\delta}\right)^{1/5}\left[1 - \left(\frac{y}{\delta_T}\right)^{1/5}\right]d\left(\frac{y}{\delta_T}\right)\right\} = \frac{q_w}{\rho c_p u_1 (T_w - T_1)}$$

Defining:

$$\zeta = \frac{\delta_T}{\delta} \quad (6.114)$$

this equation can be written as:

$$\frac{d}{dx}\left\{\delta \zeta^{6/5} \int_0^1 \left(\frac{y}{\delta_T}\right)^{1/5}\left[1 - \left(\frac{y}{\delta_T}\right)^{1/5}\right]d\left(\frac{y}{\delta_T}\right)\right\} = \frac{q_w}{\rho c_p u_1 (T_w - T_1)} \quad (6.115)$$

Carrying out the integration and rearranging then gives:

$$\frac{\delta}{7}\zeta^{1/5}\frac{d\zeta}{dx} + \frac{5}{42}\zeta^{6/5}\frac{d\delta}{dx} = \frac{q_w}{\rho c_p u_1 (T_w - T_1)} \quad (6.116)$$

Now Eq. (6.111) can be written as:

$$\frac{q_w}{\rho c_p u_1 (T_w - T_1)} = \frac{\tau_w}{\rho u_1^2}\left(\frac{\delta}{\delta_T}\right)^{1/5} \quad (6.117)$$

Substituting this into Eq. (6.116) and rearranging then gives:

$$\frac{\delta}{7}\frac{d\zeta}{dx} + \frac{5}{42}\zeta\frac{d\delta}{dx} = \frac{\tau_w}{\rho u_1^2} \quad (6.118)$$

Substituting Eqs. (6.95) and (6.112) into this equation then gives:

$$\frac{0.0494}{(u_1 x/\nu)^{0.211}} \times \frac{d\zeta}{dx} + \frac{0.0325}{(u_1 x/\nu)^{0.211}}\zeta = \frac{0.0324}{(u_1 x/\nu)^{0.211}}$$

i.e.:

$$\frac{d\zeta}{dx} + 0.658\frac{\zeta}{x} = \frac{0.658}{x} \quad (6.119)$$

This equation must be integrated subject to the condition that $\zeta = 0$ when $x = x_0$ as discussed earlier. Eq. (6.119) therefore gives:

$$\int_0^\zeta \frac{d\zeta}{0.658 - 0.658\zeta} = \int_{x_0}^\zeta \frac{dx}{x} \quad (6.120)$$

280 Introduction to Convective Heat Transfer Analysis

Carrying out the integration and rearranging then gives:

$$\zeta = 1 - \left(\frac{x_0}{x}\right)^{0.658} \tag{6.121}$$

Substituting this result into Eq. (6.117) then gives:

$$\frac{q_w}{\rho c_p u_1 (T_w - T_1)} = \frac{\tau_w}{\rho u_1^2} \left[1 - \left(\frac{x_0}{x}\right)^{0.658}\right]^{-1/5}$$

which can be rearranged to give:

$$Nu_x = \frac{0.0323 Re_x^{0.789} Pr}{\left[1 - \left(\frac{x_0}{x}\right)^{0.658}\right]^{0.2}} \tag{6.122}$$

This equation shows, of course, that as x increases, with the result that x_0/x decreases, Nu_x tends to the value it would have if the plate had been heated from the leading edge.

EXAMPLE 6.4. Air at a temperature of 10°C flows at a velocity of 100 m/s over a 4-m long wide flat plate which is aligned with the flow. The first quarter of the plate is unheated and the remainder of the plate is maintained at a uniform wall temperature of 50°C. Plot the variation of the local heat transfer rate along the heated section of the plate. Evaluate the air properties at a temperature of 30°C.

Solution. The flow situation being considered is shown in Fig. E6.4a.

At 30°C, the mean temperature, air at standard ambient pressure has the following properties:

$$\nu = 16.01 \times 10^{-6} \text{ m}^2/\text{s}, \ k = 0.02638 \text{ W/m-°C}, \ Pr = 0.71$$

In the situation being considered:

$$x_0 = 1 \text{ m}$$

Using these values, Eq. (6.122) gives:

$$\frac{q_w x}{0.02638 \times (50 - 10)} = \frac{0.0323 \times (100 x/16.01 \times 10^{-6})^{0.789} \times 0.71}{\left[1 - \left(\frac{1}{x}\right)^{0.658}\right]^{0.2}}$$

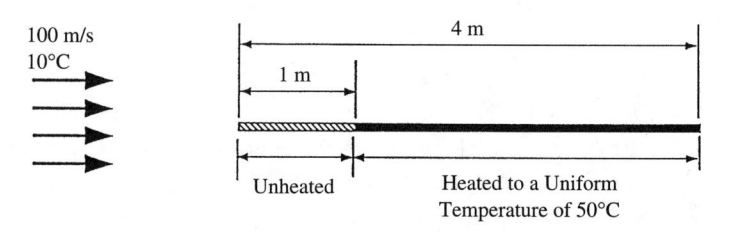

FIGURE E6.4a

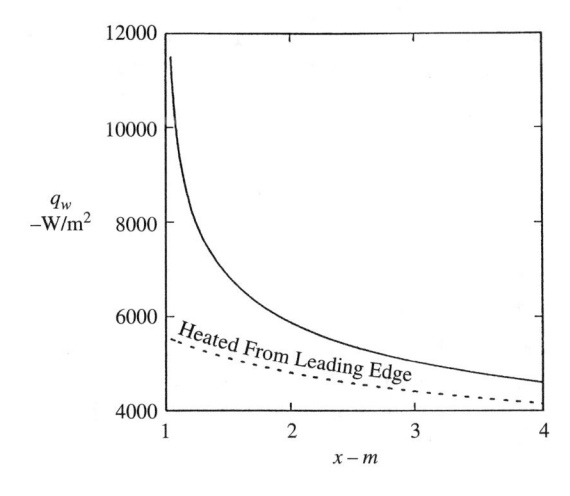

FIGURE E6.4b

i.e.:

$$q_w = \frac{5566/x^{0.211}}{\left[1 - \left(\frac{1}{x}\right)^{0.658}\right]^{0.2}} \text{ W/m}^2$$

This equation applies for $x > 1$. The variation of q_w with x given by this equation is shown in Fig. E6.4b. This figure also shows the variation of q_w that would exist if the plate were heated from its leading edge, i.e., if $x_0 = 0$.

6.4
NUMERICAL SOLUTION OF THE TURBULENT BOUNDARY LAYER EQUATIONS

Solutions to the boundary layer equations are, today, generally obtained numerically [6],[7],[8],[9],[10],[11],[12]. In order to illustrate how this can be done, a discussion of how the simple numerical solution procedure for solving laminar boundary layer problems that was outlined in Chapter 5 can be modified to apply to turbulent boundary layer flows. For turbulent boundary layer flows, the equations given earlier in the present chapter can, because the fluid properties are assumed constant, be written as:

$$\frac{\partial \bar{u}}{\partial x} + \frac{\partial \bar{v}}{\partial y} = 0 \tag{6.123}$$

$$\bar{u}\frac{\partial \bar{u}}{\partial x} + \bar{v}\frac{\partial \bar{u}}{\partial y} = -\frac{1}{\rho}\frac{d\bar{p}}{dx} + \frac{\partial}{\partial y}\left[(\nu + \epsilon)\frac{\partial \bar{u}}{\partial y}\right] \tag{6.124}$$

$$\bar{u}\frac{\partial \bar{T}}{\partial x} + \bar{v}\frac{\partial \bar{T}}{\partial y} = \frac{\partial}{\partial y}\left[(\alpha + \epsilon_H)\frac{\partial \bar{T}}{\partial y}\right] \tag{6.125}$$

282 Introduction to Convective Heat Transfer Analysis

Although not necessary, it is convenient to rewrite these equations in dimensionless form and the following variables are therefore introduced for this purpose:

$$U = \frac{\bar{u}}{u_r}, V = \frac{\bar{v}}{u_r}, P = \frac{\bar{p} - p_r}{\rho u_r^2}$$

$$X = \frac{u_r x}{\nu}, Y = \frac{u_r y}{\nu}, \theta = \frac{T - T_1}{T_{wr} - T_1} \qquad (6.126)$$

$$E = \frac{\epsilon}{\nu}, E_H = \frac{\epsilon_H}{\nu}$$

where u_r is some reference free-stream velocity, p_r is some reference pressure, T_{wr} is some reference wall temperature, and T_1 is the free-stream temperature. It is being assumed that the surface temperature distribution and not the surface heat flux distribution is being specified. The modifications to the procedure that are required to deal with the latter situation are, however, relatively minor and will be discussed later.

In terms of these variables, the governing equations are:

$$\frac{\partial U}{\partial X} + \frac{\partial V}{\partial Y} = 0 \qquad (6.127)$$

$$U\frac{\partial U}{\partial X} + V\frac{\partial U}{\partial Y} = -\frac{dP}{dX} + \frac{\partial}{\partial Y}\left[(1 + E)\frac{\partial U}{\partial Y}\right] \qquad (6.128)$$

$$U\frac{\partial \theta}{\partial X} + V\frac{\partial \theta}{\partial Y} = \frac{\partial}{\partial Y}\left[\left(\frac{1}{Pr} + \frac{E_H}{Pr_T}\right)\frac{\partial \theta}{\partial Y}\right] \qquad (6.129)$$

The boundary conditions on this set of equations are:

$$Y = 0: \quad U = 0, V = 0, \theta = \theta_w(X)$$

$$Y = \text{large}: \quad U \to U_1(X), \theta \to 0 \qquad (6.130)$$

The finite difference forms of the above set of equations and the method of solving them will first be discussed, it being assumed in this discussion that the variation of E and the value of Pr_T, which will here be assumed to be constant, are determined by the use of a turbulence model. After this discussion, one particular way of empirically describing E, i.e., one particular turbulence model, will be presented.

A series of grid lines running parallel to the X- and Y-axes are introduced as was done in dealing with laminar boundary layer flow. The nodal points used in obtaining the solution lie at the points of intersection of these grid lines. Because the gradients near the wall in a turbulent boundary layer are very high, a nonuniform grid spacing in the Y-direction will be used in the present finite difference solution, a smaller grid spacing being used near the wall than near the outer edge of the boundary layer.

First consider the finite difference form of the momentum equation, i.e., Eq. (6.128). As with laminar boundary layer flow, the four nodal points shown in Fig. 6.9 are used in deriving the finite difference form of this equation.

The following finite-difference approximations, which can be derived in the same way as that used in dealing with laminar boundary layer flow, to the various

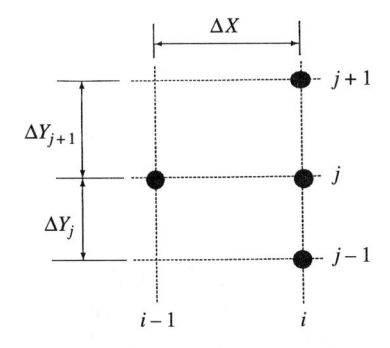

FIGURE 6.9
Nodal points used in obtaining the finite difference forms of the momentum and energy equations for turbulent boundary layer flow.

terms in the momentum equation are introduced:

$$U\frac{\partial U}{\partial X}\bigg|_{i,j} = U_{i-1,j}\left[\frac{U_{i,j} - U_{i-1,j}}{\Delta X}\right]$$

$$V\frac{\partial U}{\partial Y}\bigg|_{i,j} = V_{i-1,j}\left[\frac{\Delta Y_{j+1}}{\Delta Y_j(\Delta Y_{j+1} + \Delta Y_j)}(U_{i,j+1} - U_{i,j})\right.$$
$$\left.+ \frac{\Delta Y_j}{\Delta Y_{j+1}(\Delta Y_{j+1} + \Delta Y_j)}(U_{i,j} - U_{i,j-1})\right]$$

$$\frac{\partial}{\partial Y}\left[(1+E)\frac{\partial U}{\partial Y}\right]\bigg|_{i,j} = \frac{1}{\Delta Y_{j+1} + \Delta Y_j}\left[(2 + E_{i-1,j+1} + E_{i-1,j})\times\right.$$
$$\left.\left(\frac{U_{i,j+1} - U_{i,j}}{\Delta Y_{j+1}}\right) - (2 + E_{i-1,j} + E_{i-1,j-1})\left(\frac{U_{i,j} - U_{i,j-1}}{\Delta Y_j}\right)\right]$$

It will be seen that in treating the dimensionless eddy viscosity, it has been assumed that, like the coefficient terms, the values of the dimensionless eddy viscosity can be evaluated on the $(i-1)$-line, i.e., the line from which the solution is advancing.

Substituting the above finite-difference approximations into the momentum equation then leads to an equation of the form:

$$A_j U_{i,j} + B_j U_{i,j+1} + C_j U_{i,j-1} = D_j \tag{6.131}$$

where the coefficients are given by

$$A_j = \left(\frac{U_{i-1,j}}{\Delta X}\right) - V_{i-1,j}\left[\frac{\Delta Y_{j+1}}{\Delta Y_j(\Delta Y_{j+1} + \Delta Y_j)} + \frac{\Delta Y_j}{\Delta Y_{j+1}(\Delta Y_{j+1} + \Delta Y_j)}\right]$$
$$+ \frac{1}{\Delta Y_{j+1} + \Delta Y_j}\left[\frac{(2 + E_{i-1,j+1} + E_{i-1,j})}{\Delta Y_{j+1}} + \frac{(2 + E_{i-1,j} + E_{i-1,j-1})}{\Delta Y_j}\right]$$
$$\tag{6.132}$$

$$B_j = V_{i-1,j}\left[\frac{\Delta Y_{j+1}}{\Delta Y_j(\Delta Y_{j+1} + \Delta Y_j)}\right] + \frac{1}{\Delta Y_{j+1} + \Delta Y_j}\left[\frac{(2 + E_{i-1,j+1} + E_{i-1,j})}{\Delta Y_{j+1}}\right]$$
$$\tag{6.133}$$

284 Introduction to Convective Heat Transfer Analysis

$$C_j = -V_{i-1,j}\left[\frac{\Delta Y_j}{\Delta Y_{j+1}(\Delta Y_{j+1} + \Delta Y_j)}\right] - \frac{1}{\Delta Y_{j+1} + \Delta Y_j}\left[\frac{(2 + E_{i-1,j} + E_{i-1,j-1})}{\Delta Y_j}\right]$$

(6.134)

$$D_j = -\frac{U_{i-1,j}^2}{\Delta X} - \left.\frac{dP}{dX}\right|_i$$

(6.135)

The boundary conditions give $U_{i,1}$ the value of U at the wall, and $U_{i,N}$ the value of U at the outermost grid point which is always chosen to lie outside the boundary layer. Therefore, since $dP/dX|_i$ is a known quantity, the application of Eq. (6.131) to each of the points $j = 1, 2, 3, \ldots, N-1, N$ gives a set of N equations in the N unknown values of U, i.e., $U_1, U_2, U_3, U_4, \ldots U_{N-1}, U_N$. This set of equations has the following form because $U_{i,1}$ is zero

$$U_{i,1} = 0$$
$$A_2 U_{i,2} + B_2 U_{i,3} + C_2 U_{i,1} = D_2$$
$$A_3 U_{i,3} + B_3 U_{i,4} + C_3 U_{i,2} = D_3$$
$$\vdots$$
$$A_{N-1} U_{i,N-1} + B_{N-1} U_{i,N} + C_{N-1} U_{i,N-2} = D_{N-1}$$
$$U_{i,N} = U_1$$

(6.136)

This set of equations thus has the form:

$$\begin{bmatrix} 1 & 0 & 0 & 0 & 0 & . & 0 & 0 & 0 \\ C_2 & A_2 & B_2 & 0 & 0 & . & 0 & 0 & 0 \\ 0 & C_3 & A_3 & B_3 & 0 & . & 0 & 0 & 0 \\ 0 & 0 & C_4 & A_4 & B_4 & . & 0 & 0 & 0 \\ . & . & . & . & . & . & . & . & . \\ . & . & . & . & . & . & . & . & . \\ 0 & 0 & 0 & 0 & 0 & . & C_{N-1} & A_{N-1} & B_{N-1} \\ 0 & 0 & 0 & 0 & 0 & . & 0 & 0 & 1 \end{bmatrix} \begin{bmatrix} U_{i,1} \\ U_{i,2} \\ U_{i,3} \\ U_{i,4} \\ . \\ . \\ U_{i,N-1} \\ U_{i,N} \end{bmatrix} = \begin{bmatrix} 0 \\ D_2 \\ D_3 \\ D_4 \\ . \\ . \\ D_{N-1} \\ U_1 \end{bmatrix}$$

which has the form:

$$Q_U U_{i,j} = R_U$$

(6.137)

where Q_U is a tridiagonal matrix. This equation can be solved using the standard tridiagonal matrix solver algorithm discussed when considering laminar boundary layer flow.

Next, consider the energy equation (6.129). It will be assumed that the turbulent Prandtl number, Pr_T is constant in the flow. The following finite-difference approximations to the various terms in the energy equation are introduced:

$$\left.U\frac{\partial\theta}{\partial X}\right|_{i,j} = U_{i-1,j}\left[\frac{\theta_{i,j} - \theta_{i-1,j}}{\Delta X}\right]$$

(6.138)

CHAPTER 6: External Turbulent Flows   285

$$V\frac{\partial \theta}{\partial Y}\bigg|_{i,j} = V_{i-1,j}\left[\frac{\Delta Y_{j+1}}{\Delta Y_j(\Delta Y_{j+1} + \Delta Y_j)}(\theta_{i,j+1} - \theta_{i,j})\right.$$
$$\left. + \frac{\Delta Y_j}{\Delta Y_{j+1}(\Delta Y_{j+1} + \Delta Y_j)}(\theta_{i,j} - \theta_{i,j-1})\right] \quad (6.139)$$

$$\frac{\partial}{\partial Y}\left[\left(\frac{1}{Pr} + \frac{E_H}{Pr_T}\right)\frac{\partial \theta}{\partial Y}\right]\bigg|_{i,j} = \frac{1}{\Delta Y_{j+1} + \Delta Y_j}\left[\left(\frac{2}{Pr} + \frac{E_{i-1,j+1}}{Pr_T} + \frac{E_{i-1,j}}{Pr_T}\right)\left(\frac{\theta_{i,j+1} - \theta_{i,j}}{\Delta Y_{j+1}}\right)\right.$$
$$\left. - \left(\frac{2}{Pr} + \frac{E_{i-1,j}}{Pr_T} + \frac{E_{i-1,j-1}}{Pr_T}\right)\left(\frac{\theta_{i,j} - \theta_{i,j-1}}{\Delta Y_j}\right)\right] \quad (6.140)$$

Substituting these finite-difference approximations into the energy equation then leads to an equation of the form:

$$F_j \theta_{i,j} + G_j \theta_{i,j+1} + H_j \theta_{i,j-1} = L_j \quad (6.141)$$

where the coefficients are given by

$$F_j = \left(\frac{U_{i-1,j}}{\Delta X}\right) - V_{i-1,j}\left[\frac{\Delta Y_{j+1}}{\Delta Y_j(\Delta Y_{j+1} + \Delta Y_j)} + \frac{\Delta Y_j}{\Delta Y_{j+1}(\Delta Y_{j+1} + \Delta Y_j)}\right]$$
$$+ \frac{1}{\Delta Y_{j+1} + \Delta Y_j}\left[\frac{(2/Pr + E_{i-1,j+1}/Pr_T + E_{i-1,j}/Pr_T)}{\Delta Y_{j+1}}\right.$$
$$\left. + \frac{2/Pr + E_{i-1,j}/Pr_T + E_{i-1,j-1}/Pr_T}{\Delta Y_j}\right] \quad (6.142)$$

$$G_j = V_{i-1,j}\left[\frac{\Delta Y_{j+1}}{\Delta Y_j(\Delta Y_{j+1} + \Delta Y_j)}\right]$$
$$+ \frac{1}{\Delta Y_{j+1} + \Delta Y_j}\left[\frac{(2/Pr + E_{i-1,j+1}/Pr_T + E_{i-1,j}/Pr_T)}{\Delta Y_{j+1}}\right] \quad (6.143)$$

$$H_j = -V_{i-1,j}\left[\frac{\Delta Y_j}{\Delta Y_{j+1}(\Delta Y_{j+1} + \Delta Y_j)}\right]$$
$$- \frac{1}{\Delta Y_{j+1} + \Delta Y_j}\left[\frac{(2/Pr + E_{i-1,j}/Pr_T + E_{i-1,j-1}/Pr_T)}{\Delta Y_j}\right]$$
$$\quad (6.144)$$

$$L_j = -\frac{U_{i-1,j}\theta_{i-1,j}}{\Delta X} \quad (6.145)$$

The case where the wall temperature variation is known will be considered here. The boundary conditions then give $\theta_{i,1}$ and $\theta_{i,N}$, the outermost grid point being chosen to lie outside both the velocity and temperature boundary layers. The application of Eq. (6.141) to each of the internal points on the i-line, i.e., $j = 1, 2, 3, 4, \ldots,$

286   Introduction to Convective Heat Transfer Analysis

$N - 2$, $N - 1$, N again gives a set of N equations in the N unknown values of θ. Because $\theta_{i,N}$ is 0, this set of equations has the following form:

$$\theta_{i,1} = \theta_w$$
$$F_2\theta_{i,2} + G_2\theta_{i,3} + H_2\theta_{i,1} = L_2$$
$$F_3\theta_{i,3} + G_3\theta_{i,4} + H_3\theta_{i,2} = L_3$$
$$\vdots$$
$$F_{N-1}\theta_{i,N-1} + G_{N-1}\theta_{i,N} + H_{N-1}\theta_{i,N-2} = L_{N-1}$$
$$\theta_{i,N} = 0$$

(6.146)

This set of equations has the same form as that derived from the momentum equation, i.e.,

$$\begin{bmatrix} 1 & 0 & 0 & 0 & 0 & . & 0 & 0 & 0 \\ H_2 & F_2 & G_2 & 0 & 0 & . & 0 & 0 & 0 \\ 0 & H_3 & F_3 & G_3 & 0 & . & 0 & 0 & 0 \\ 0 & 0 & H_4 & F_4 & G_4 & . & 0 & 0 & 0 \\ . & . & . & . & . & . & . & . & . \\ . & . & . & . & . & . & . & . & . \\ 0 & 0 & 0 & 0 & 0 & . & H_{N-1} & F_{N-1} & G_{N-1} \\ 0 & 0 & 0 & 0 & 0 & . & 0 & 0 & 1 \end{bmatrix} \begin{bmatrix} \theta_{i,1} \\ \theta_{i,2} \\ \theta_{i,3} \\ \theta_{i,4} \\ . \\ . \\ \theta_{i,N-1} \\ \theta_{i,N} \end{bmatrix} = \begin{bmatrix} \theta_w \\ L_2 \\ L_3 \\ L_4 \\ . \\ . \\ L_{i,N-1} \\ 0 \end{bmatrix}$$

which has the form:

$$Q_T \theta_{i,j} = R_T \qquad (6.147)$$

where Q_T is again a tridiagonal matrix. Thus, the same form of equation is obtained from the energy equation as that obtained from the momentum equation. This equation can also be solved using the standard tridiagonal matrix-solver algorithm.

The continuity equation (6.127) has exactly the same form as that for laminar flow and it can therefore be treated using the same procedure as used in laminar flow. The grid points shown in Fig. 6.10 are therefore used in obtaining the finite difference approximation to the continuity equation.

The continuity equation is applied to the point (not a nodal point) denoted by $(i, j - 1/2)$ in Fig. 6.10 which lies on the i-line, halfway between the $(j - 1)$ and the

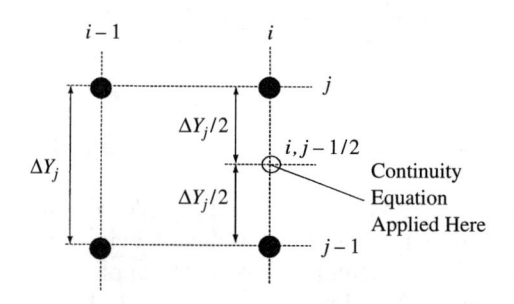

FIGURE 6.10
Nodal points used in finite-difference approximation to continuity equation.

j-points. The following is introduced:

$$\left.\frac{\partial V}{\partial Y}\right|_{i,j-1/2} = \frac{V_{i,j} - V_{i,j-1}}{\Delta Y_j} \qquad (6.148)$$

and it is assumed that the X-derivative at the point $(i, j - 1/2)$ is equal to the average of the X-derivatives at the points (i, j) and $(i, j - 1)$, i.e., it is assumed that:

$$\left.\frac{\partial U}{\partial X}\right|_{i,j-1/2} = \frac{1}{2}\left[\left.\frac{\partial U}{\partial X}\right|_{i,j} + \left.\frac{\partial U}{\partial X}\right|_{i,j-1}\right]$$

From this it follows that:

$$\left.\frac{\partial U}{\partial X}\right|_{i,j-1/2} = \frac{1}{2}\left[\frac{U_{i,j} - U_{i-1,j}}{\Delta X} + \frac{U_{i,j-1} - U_{i-1,j-1}}{\Delta X}\right]$$

i.e.:

$$\left.\frac{\partial U}{\partial X}\right|_{i,j-1/2} = \frac{1}{2\Delta X}(U_{i,j} - U_{i-1,j} + U_{i,j-1} - U_{i-1,j-1}) \qquad (6.149)$$

Therefore, since the continuity equation can be written as:

$$\frac{\partial V}{\partial Y} = -\frac{\partial U}{\partial X}$$

the following finite-difference approximation to this continuity equation is obtained:

$$\frac{V_{i,j} - V_{i,j-1}}{\Delta Y_j} = -\frac{1}{2\Delta X}(U_{i,j} - U_{i-1,j} + U_{i,j-1} - U_{i-1,j-1})$$

This equation can be rearranged to give:

$$V_{i,j} = V_{i,j-1} - \left(\frac{\Delta Y_j}{2\Delta X}\right)(U_{i,j} - U_{i-1,j} + U_{i,j-1} - U_{i-1,j-1}) \qquad (6.150)$$

Therefore, if the distribution of U across the i-line is first determined, this equation can be used to give the distribution of V across this line. This is possible because $V_{i,1}$ is given by the boundary conditions and the equation can, therefore, be applied progressively outward across the i-line starting at point $j = 2$ and then going to point $j = 3$ and so on across the line.

In order to apply the above equations to the calculation of turbulent boundary layer flows, the distribution of the eddy viscosity must be specified. The numerical procedure does not require this to be done in any particular way. For the purposes of illustration and because it leads to results that are in quite good agreement with experiment, this will be done here using an empirically derived [1],[2],[7],[13],[14],[15],[16],[17] distribution of mixing length, ℓ, it being recalled that the mixing length is related to the ϵ by:

$$\epsilon = \ell^2 \left|\frac{\partial \bar{u}}{\partial y}\right|$$

In terms of the dimensionless variables introduced above this becomes:

$$E = L_m^2 \left| \frac{\partial U}{\partial Y} \right| \tag{6.151}$$

where the dimensionless mixing length, L_m, is defined by:

$$L_m = \frac{\ell u_r}{\nu} \tag{6.152}$$

In a boundary layer flow it is usually adequate to assume that:

$$\ell = \text{function}(\tau_w, \rho, \mu, y, \delta) \tag{6.153}$$

where, as before, τ_w is the local wall shear stress and δ is the local boundary layer thickness. Applying dimensional analysis then gives after some rearrangement:

$$\frac{\ell}{\delta} = \text{function}\left[\frac{y}{\mu} \sqrt{\tau_w \rho}, \frac{y}{\delta} \right] \tag{6.154}$$

Defining:

$$\Delta = \frac{u_r \delta}{\nu} \tag{6.155}$$

and recalling that:

$$y^+ = \frac{y}{\mu} \sqrt{\tau_w \rho} = \frac{y}{\nu} \sqrt{\frac{\tau_w}{\rho}}$$

allows Eq. (6.154) to be written in terms of the dimensionless variables as:

$$\frac{L_m}{\Delta} = \text{function}\left[y^+, \frac{Y}{\Delta} \right] \tag{6.156}$$

Many experiments have established that, as mentioned before, there is a region near the wall where the local turbulent shear stress depends on the wall shear stress and the distance from the wall alone and is largely independent of the nature of the rest of the flow. In this region, the mixing length increases linearly with distance from the wall except that near the wall there is a "damping" of the turbulence due to viscosity. In the wall region, it is assumed therefore that:

$$\ell = 0.41 y [1 - \exp(-y^+/26)] \tag{6.157}$$

The term in the bracket is known as the van Driest damping factor.

In terms of the dimensionless variables introduced above this equation is:

$$L_m = 0.41 Y [1 - \exp(-y^+/26)] \tag{6.158}$$

Experimentally it has been found that Eq. (6.157) applies from the wall out to about 10% of the boundary layer thickness. Hence, it will here be assumed that:

$$\text{For } Y/\Delta < 0.1: L_m = 0.41 Y [1 - \exp(-y^+/26)] \tag{6.159}$$

In the outer portion of the boundary layer, experiment indicates that the mixing length has a constant value of about 0.089 of the boundary layer thickness, this result being applicable over approximately the outer 40% of the boundary layer. Thus, it will be assumed that:

$$\text{For } y/\delta > 0.6 : \ell = 0.089\delta \tag{6.160}$$

which can be written in terms of the dimensionless variables introduced above as:

$$\text{For } Y/\Delta > 0.6 : L_m = 0.089\Delta \tag{6.161}$$

Over the remaining 50% of the boundary layer, i.e., from $y = 0.1\delta$ to $y = 0.6\delta$, it will be assumed that ℓ varies smoothly from the value given by Eq. (6.157) to the value given by Eq. (6.160) and that the variation can be described by a fourth-order polynomial in y/δ, i.e., it will be assumed that:

$$\text{For } 0.1 \leq y/\delta \leq 0.6: \ell = 0.41y[1 - \exp(-y^+/26)]$$
$$- [1.54(y/\delta - 0.1)^2 - 2.76(y/\delta - 0.1)^3 + 1.88(y/\delta - 0.1)^4]\delta \tag{6.162}$$

which can be written in terms of the dimensionless variables introduced above as:

$$\text{For } 0.1 \leq Y/\Delta \leq 0.6: L_m = 0.41Y[1 - \exp(-y^+/26)] - [1.54(Y/\Delta - 0.1)^2$$
$$- 2.76(Y/\Delta - 0.1)^3 + 1.88(Y/\Delta - 0.1)^4]\Delta \tag{6.163}$$

Eqs. (6.159), (6.161), and (6.163) together describe the variation of the dimensionless mixing length, L_m, across the boundary layer, the variation being as shown in Fig. 6.11.

In order to utilize this set of equations to determine the mixing length distribution from the velocity distribution it is necessary, of course, to determine y^+. Now:

$$y^+ = \frac{y}{\nu}\sqrt{\frac{\tau_w}{\rho}} = Y\sqrt{\frac{\tau_w}{\rho u_r^2}} \tag{6.164}$$

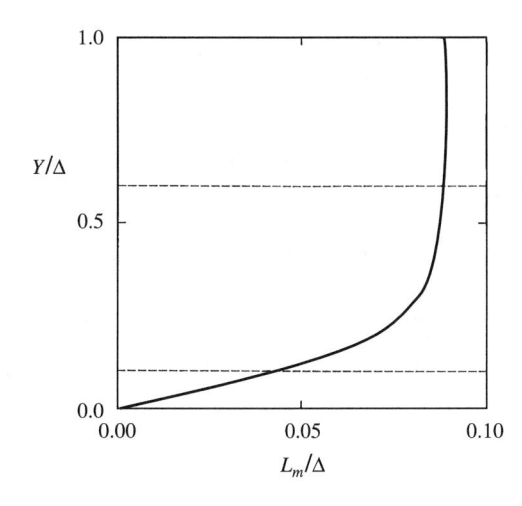

FIGURE 6.11
Assumed variation of dimensionless mixing length across boundary layer.

It is thus necessary to determine the value of $\tau_w/\rho u_r^2$. To do this, it is noted that the turbulent stress is zero at the wall and, therefore, that:

$$\tau_w = \mu \left.\frac{\partial \bar{u}}{\partial y}\right|_{y=0}$$

i.e.:

$$\frac{\tau_w}{\rho u_r^2} = \left.\frac{\partial U}{\partial Y}\right|_{Y=0} \tag{6.165}$$

Applying a Taylor expansion to give the velocity at the first nodal point away from the wall in terms of the velocity at the wall, i.e., zero, gives on neglecting the higher order terms:

$$U_{i,2} = 0 + \left.\frac{\partial U}{\partial Y}\right|_{i,1}(Y_2 - Y_1) + \left.\frac{\partial^2 U}{\partial Y^2}\right|_{i,1}\frac{(Y_2 - Y_1)^2}{2} \tag{6.166}$$

But $Y_1 = 0$ and applying the full momentum equation, i.e., Eq. (6.128), to conditions at the wall gives:

$$\left.\frac{\partial^2 U}{\partial Y^2}\right|_{i,1} = \left.\frac{dP}{dX}\right|_i$$

Hence, Eq. (6.166) gives:

$$\left.\frac{\partial U}{\partial Y}\right|_{i,1} = \frac{U_{i,2}}{Y_2} - \frac{Y_2}{2}\left.\frac{dP}{dX}\right|_i \tag{6.167}$$

Substituting Eq. (6.167) into Eq. (6.165) and the result into Eq. (6.164) then gives the following finite-difference approximation:

$$y^+ = Y_j \left[\frac{U_{i,2}}{Y_2} - \frac{Y_2}{2}\left.\frac{dP}{dX}\right|_i\right]^{1/2} \tag{6.168}$$

The second term in the bracket will usually be much smaller than the first and is often negligible.

It is also necessary to determine the local value of the dimensionless boundary layer thickness in order to find the mixing length distribution. It is usually adequate to take this as the value of Y at which U reaches 0.99 of its free-stream value.

Once the dimensionless mixing length distribution has been found using the above equations, the dimensionless eddy viscosity distribution, in terms of which the numerical method has been presented, can be found using the finite-difference approximation for the velocity derivative that was introduced earlier, i.e., using:

$$E = L_m^2 \left[\left|\frac{\Delta Y_{j+1}}{\Delta Y_j(\Delta Y_{j+1} + \Delta Y_j)}(U_{i,j+1} - U_{i,j}) + \frac{\Delta Y_j}{\Delta Y_{j+1}(\Delta Y_{j+1} + \Delta Y_j)}(U_{i,j} - U_{i,j-1})\right|\right]$$

$$\tag{6.169}$$

In many cases, a region of laminar flow will have to be allowed for over the initial part of the body. The calculation of this portion of the flow can be accomplished by using the same equations as for turbulent flow with E set equal to 0. This calculation can be started using initial conditions of the type discussed in Chapter 3 for purely laminar flows. From the point where transition is assumed to occur, the calculation can be continued using the equations presented above to describe E.

A computer program, TURBOUND, again written in FORTRAN, based on the numerical procedure described above is available as discussed in the Preface. It is, of course, essentially the same program as discussed earlier for the calculation of laminar boundary layer flows.

The program, as available, will calculate flow over a surface with a varying free-stream velocity and varying surface temperature. These variations are both assumed to be described by a third-order polynomial, i.e., by:

$$U_1 = A_U + B_U X_R + C_U X_R^2 + D_U X_R^3 \tag{6.170}$$

and:

$$\theta_w = A_T + B_T X_R + C_T X_R^2 + D_T X_R^3 \tag{6.171}$$

where:

$$X_R = \frac{X}{X_{max}}$$

It follows from the assumed form of the free-stream velocity distribution that since:

$$\frac{dp}{dx} = -\rho u_1 \frac{du_1}{dx}$$

the following applies:

$$\frac{dP}{dX} = -U_1 \frac{dU_1}{dX} = U_1(B_U + 2C_U X_R + 3D_U X_R^2)\frac{1}{X_{max}} \tag{6.172}$$

The program finds the distribution of the dimensionless temperature. The wall heat transfer rate is then found by noting that the turbulent heat transfer rate is 0 at the wall and, therefore, that:

$$q_w = -k \left.\frac{\partial \overline{T}}{\partial y}\right|_{y=0}$$

i.e.:

$$\frac{q_w x}{k(T_{wr} - T_1)} = X \left.\frac{\partial \theta}{\partial Y}\right|_{Y=0}$$

i.e.:

$$Nu_x = \frac{X}{\theta_w} \left.\frac{\partial \theta}{\partial Y}\right|_{Y=0} \tag{6.173}$$

where Nu_x is the local Nusselt number, i.e.:

$$Nu_x = \frac{q_w x}{k(T_w - T_1)} \tag{6.174}$$

Applying a Taylor expansion to give the temperature at the first nodal point away from the wall gives on neglecting the higher order terms:

$$\theta_{i,2} = \theta_{i,1} + \left.\frac{\partial \theta}{\partial Y}\right|_{i,1} (Y_2 - Y_1) + \left.\frac{\partial^2 \theta}{\partial Y^2}\right|_{i,1} \frac{(Y_2 - Y_1)^2}{2} \tag{6.175}$$

But $Y_1 = 0$ and applying the full energy equation, i.e., Eq. (6.128), to conditions at the wall gives:

$$\left.\frac{\partial^2 \theta}{\partial Y^2}\right|_{i,1} = 0$$

Hence Eq. (6.175) gives:

$$\left.\frac{\partial \theta}{\partial Y}\right|_{i,1} = \frac{\theta_{i,2} - \theta_{i,1}}{Y_2 - Y_1} \tag{6.176}$$

Substituting Eq. (6.176) into Eq. (6.173) then gives the following finite-difference approximation:

$$Nu_x = \frac{X}{\theta_w} \left[\frac{\theta_{i,2} - \theta_{i,1}}{Y_2 - Y_1}\right] \tag{6.177}$$

As with the program for laminar boundary layer flow discussed in Chapter 3, the turbulent boundary flow program calculates the velocity and temperature boundary layer thicknesses using, as discussed above, the assumption that Δ is the value of Y at which $U = 0.99 U_1$ and Δ_T is the value of Y at which $\theta = 0.01\theta_w$. If either Δ or Δ_T is greater than Y_{N-5}, the number of grid points is increased by ten.

The program is easily modified to utilize other turbulence models.

EXAMPLE 6.5. Numerically determine the dimensionless heat transfer rate variation with two-dimensional turbulent boundary layer air flow over an isothermal flat plate. Assume that the boundary layer is turbulent from the leading edge. Compare the numerical results with those given by the three-layer analogy solution discussed previously (see Eq. 6.79).

Solution. A solution for values of X up to 10^7 will be considered. Because flow over a flat plate with an isothermal surface is being considered, the free-stream velocity and surface temperature are constant. Therefore, because the boundary is assumed to be turbulent from the leading edge, the inputs to the program are:

$$XMAX = 10000000, \; XTRAN = 0, \; Pr = 0.7,$$

$$AT, BT, CT, DT = 1, 0, 0, 0 \quad AV, BV, CV, DV = 1, 0, 0, 0$$

The program, when run with these inputs, gives the values of Nu_x at various X values along the plate. These are shown in Fig. E6.5.

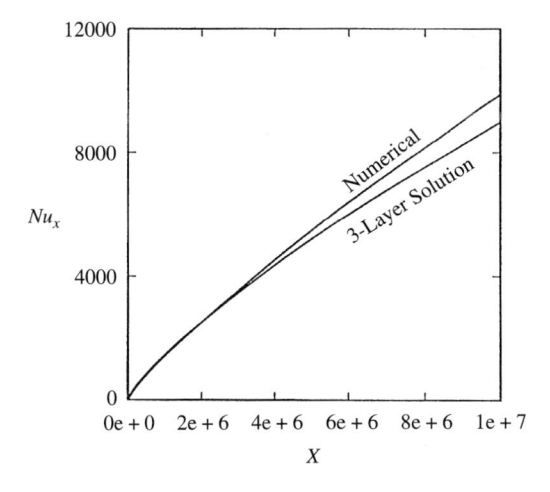

FIGURE E6.5

The variation given by using Eq. (6.79) is also shown in this figure. It will be seen that there is quite good agreement between the two results at Reynolds numbers (i.e., X values) up to about 5×10^6 but the numerically determined values of Nu_x are higher than those given by the three-layer equation at the higher Reynolds numbers.

EXAMPLE 6.6. Air flows over a flat plate which has a uniform surface temperature of 50°C, the temperature of the air ahead of the plate being 10°C. The air velocity is such that the Reynolds number based on the length of the plate is 6×10^6, the length of the plate being 2 m. Numerically determine the variation of the local heat transfer rate from the wall, q_w, with x assuming that (i) the boundary layer flow remains laminar, (ii) the boundary layer flow is turbulent from the leading edge of the plate, and (iii) boundary layer transition occurs at Re_T of 10^6. x is the distance from the leading edge of the plate.

Solution. The definition of dimensionless X is such that:

$$x = \frac{X}{X_{\max}} L = \frac{X}{6000000} \times 2 = \frac{X}{3000000} \text{ m} \qquad (i)$$

The local Nusselt number is defined by:

$$Nu_x = \frac{q_w x}{k(T_w - T_1)}$$

Therefore, since:

$$T_w - T_1 = 50 - 10 = 40°C$$

and, if the properties of the air are evaluated at $(50 + 10)/2 = 30°C$:

$$k = 0.02638 \text{ W/m-}°C, Pr = 0.7$$

Hence:

$$Nu_x = \frac{q_w x}{0.02638 \times 40}$$

i.e.:

$$q_w = 1.0552 \frac{Nu_x}{x} \qquad (ii)$$

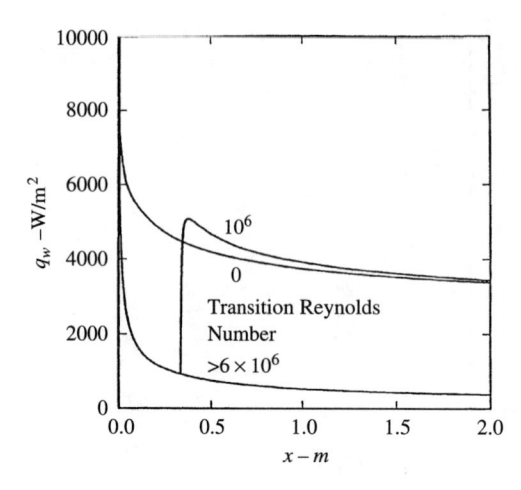

294 Introduction to Convective Heat Transfer Analysis

FIGURE E6.6

The program TURBOUND has been run with the following three sets of input values:

$$XMAX = 6000000, XTRAN = 6000000, Pr = 0.7,$$
$$AT, BT, CT, DT = 1, 0, 0, 0, AV, BV, CV, DV = 1, 0, 0, 0$$
$$XMAX = 6000000, XTRAN = 0, Pr = 0.7,$$
$$AT, BT, CT, DT = 1, 0, 0, 0, AV, BV, CV, DV = 1, 0, 0, 0$$
$$XMAX = 6000000, XTRAN = 1000000, Pr = 0.7,$$
$$AT, BT, CT, DT = 1, 0, 0, 0, AV, BV, CV, DV = 1, 0, 0, 0$$

The calculated variations of Nu_x with X have then been used in conjunction with Eqs. (i) and (ii) to derive the variations of q_w with x for the three cases. The variations so obtained are shown in Fig. E6.6.

In the above discussion it was assumed that the surface temperature variation was specified. The procedure is easily extended to deal with other thermal boundary conditions at the surface. For example, if the heat flux distribution at the surface is specified, it is convenient to define the following dimensionless temperature:

$$\theta^* = \frac{T - T_1}{(q_{wR}\nu/ku_r)} \tag{6.178}$$

where q_{wR} is some convenient reference wall heat flux. The energy equation in terms of this dimensionless temperature has the same form as that obtained in the specified surface temperature case; i.e., the dimensionless energy equation in this case is:

$$U\frac{\partial \theta^*}{\partial X} + V\frac{\partial \theta^*}{\partial Y} = \frac{1}{Pr}\frac{\partial^2 \theta^*}{\partial Y^2} \tag{6.179}$$

Using Fourier's law, the boundary condition on temperature at the wall in the specified heat flux case is:

$$y = 0: -k\frac{\partial T}{\partial y} = q_w(x) \tag{6.180}$$

i.e.,

$$Y = 0: \quad -\frac{\partial \theta^*}{\partial Y} = \frac{q_w(x)}{q_{wR}} = Q_R(X) \tag{6.181}$$

where $Q_R = q_w/q_{wR}$

Now, following the same procedure as used in deriving Eq. (6.176) gives the following finite-difference approximation:

$$\left.\frac{\partial \theta^*}{\partial Y}\right|_{i,1} = \frac{\theta^*_{i,2} - \theta^*_{i,1}}{Y_2 - Y_1} \tag{6.182}$$

Substituting this result into Eq. (6.181) then gives:

$$-\frac{\theta^*_{i,2} - \theta^*_{i,1}}{Y_2 - Y_1} = Q_R$$

i.e.:

$$\theta^*_{i,1} = \theta^*_{i,2} + Q_R \Delta Y_2 \tag{6.183}$$

This boundary condition is easily incorporated into the solution procedure that was outlined above, the set of equations governing the dimensionless temperature in this case having the form:

$$\begin{aligned}
\theta^*_{i,1} &= \theta^*_{i,2} + Q_R \Delta Y_2 \\
F_2 \theta^*_{i,2} + G_2 \theta^*_{i,3} + H_2 \theta^*_{i,1} &= L_2 \\
F_3 \theta^*_{i,3} + G_3 \theta^*_{i,4} + H_3 \theta^*_{i,2} &= L_3 \\
&\vdots \\
F_{N-1} \theta^*_{i,N-1} + G_{N-1} \theta^*_{i,N} + H_{N-1} \theta^*_{i,N-2} &= L_{N-1} \\
\theta^*_{i,N} &= 0
\end{aligned} \tag{6.184}$$

the coefficients having the same values as previously defined. The matrix equation that gives the dimensionless temperature therefore has the form.

$$\begin{bmatrix} 1 & 1 & 0 & 0 & 0 & . & 0 & 0 & 0 \\ H_2 & F_2 & G_2 & 0 & 0 & . & 0 & 0 & 0 \\ 0 & H_3 & F_3 & G_3 & 0 & . & 0 & 0 & 0 \\ 0 & 0 & H_4 & F_4 & G_4 & . & 0 & 0 & 0 \\ . & . & . & . & . & . & . & . & . \\ 0 & 0 & 0 & 0 & 0 & . & H_{N-1} & F_{N-1} & G_{N-1} \\ 0 & 0 & 0 & 0 & 0 & . & 0 & 0 & 1 \end{bmatrix} \begin{bmatrix} \theta^*_{i,1} \\ \theta^*_{i,2} \\ \theta^*_{i,3} \\ \theta^*_{i,4} \\ . \\ \theta^*_{i,N-1} \\ \theta^*_{i,N} \end{bmatrix} = \begin{bmatrix} Q_R \Delta Y \\ L_2 \\ L_3 \\ L_4 \\ . \\ L_{N-1} \\ 0 \end{bmatrix}$$

Thus, as before, a tridiagonal matrix is obtained.

The program discussed above is therefore easily modified to deal with the specified wall heat flux case. A program with this wall thermal boundary case,

TURBOUNQ, is also available as discussed in the Preface. In this program, the dimensionless wall heat flux variation is assumed to be of the form:

$$Q_R = A_Q + B_Q X + C_Q X^2 + D_Q X^3 \tag{6.185}$$

6.5 EFFECTS OF DISSIPATION ON TURBULENT BOUNDARY LAYER FLOW OVER A FLAT PLATE

Consider two-dimensional boundary layer flow over a flat plate as shown in Fig. 6.12.

If the effects of fluid property variations are neglected, the governing equations are:

$$\frac{\partial \bar{u}}{\partial x} + \frac{\partial \bar{v}}{\partial y} = 0 \tag{6.186}$$

$$\bar{u}\frac{\partial \bar{u}}{\partial x} + \bar{v}\frac{\partial \bar{u}}{\partial y} = \frac{\partial}{\partial y}\left[(\nu + \epsilon)\frac{\partial \bar{u}}{\partial y}\right] \tag{6.187}$$

$$\bar{u}\frac{\partial \bar{T}}{\partial x} + \bar{v}\frac{\partial \bar{T}}{\partial y} = \frac{\partial}{\partial y}\left[(\alpha + \epsilon_H)\frac{\partial \bar{T}}{\partial y}\right] + (\nu + \epsilon)\left[\frac{\partial \bar{u}}{\partial y}\right]^2 \tag{6.188}$$

where ϵ and ϵ_H are the eddy viscosity and eddy diffusivity as previously defined in Chapter 5. The last term on the right-hand side of the energy equation is, of course, the dissipation term, dissipation here arising both as a result of the presence of the viscous stress and as a result of the effective turbulent stress.

The above equations can be solved provided the turbulence terms in the momentum and energy equations are related to the other flow variables, i.e., provided a turbulence model is introduced. For example, a mixing-length model of the type introduced in Chapters 5 and 6 could be used, it being assumed that the viscous dissipation has no effect on the equation that describes the variation of the mixing length in the boundary layer. Using such a turbulence model, the adiabatic wall temperature and hence the recovery factor for turbulent boundary flow can be determined [18],[19],[20],[21],[22],[23],[24],[25],[26],[27]. This procedure gives:

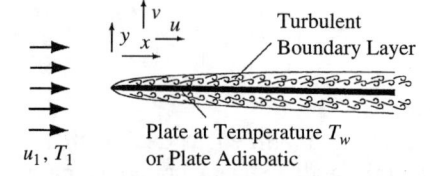

$$r = Pr^{1/3} \tag{6.189}$$

FIGURE 6.12
Turbulent boundary layer flow with viscous dissipation.

The effects of fluid property variations on heat transfer in turbulent boundary layer flow over a flat plate have also been numerically evaluated. This evaluation indicates that if the properties are as with laminar boundary layers evaluated at:

$$T_{prop} = T_1 + 0.5(T_w - T_1) + 0.22(T_{w_{ad}} - T_1)$$

the effects of these property variations can be neglected.

EXAMPLE 6.7. Air at a temperature of 0°C flows at a velocity of 600 m/s over a wide flat plate that has a length of 1 m. The pressure in the flow is 1 atm. The flow situation is therefore as shown in Fig. E6.7.

Find:

1. The wall temperature if the plate is adiabatic.
2. The heat transfer rate from the surface per unit span of the plate if the plate surface is maintained at a uniform temperature, T_w, of 60°C.

Solution

Part 1
In the free-stream:

$$a = \sqrt{\gamma R T} = \sqrt{1.4(287)(273)} = 332 \text{ m/s}$$

Hence:

$$M = \frac{V}{a} = \frac{600}{331} = 1.81$$

Now:

$$\frac{T_{w_{ad}}}{T_1} = 1 + r\left(\frac{\gamma - 1}{2}\right)M^2$$

In order to find the Reynolds number, the surface temperature must be known. However, in order to find the wall temperature, the recovery factor must be known and its value is different in laminar and turbulent flow. Therefore, an assumption as to the nature of the flow, i.e., laminar or turbulent, will be made. The wall temperature and then the gas properties will be found and then the Reynolds number, i.e.:

$$Re = \frac{\rho V L}{\mu}$$

will be evaluated and the initial assumption about the nature of the flow can be checked.

Here, it will be assumed that the flow in the boundary layer is turbulent. Experience suggests that this is very likely to be a correct assumption. Since the flow is assumed turbulent, it follows that since the Prandtl number of air can be assumed to be equal to 0.7:

$$r = Pr^{1/3} = 0.7^{1/3} = 0.89$$

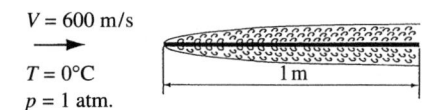

V = 600 m/s
T = 0°C
p = 1 atm.
1 m

FIGURE E6.7

Using this value then gives:

$$\frac{T_{w_{ad}}}{273} = 1 + 0.89 \times 0.2 \times 1.81^2$$

Hence:

$$T_{w_{ad}} = 432 \text{ K} = 159°C$$

Since the adiabatic surface case is being considered, the air properties are found at:

$$T_{prop} = T_1 + 0.5(T_w - T_1) + 0.22(T_{w_{ad}} - T_1)$$
$$= 0 + 0.5(159 - 0) + 0.22(159 - 0) = 114°C$$

Now at a temperature of 114°C, air has the following properties when the pressure is 1 atm:

$$\rho = 0.9 \text{ kg/m}^3$$
$$\mu = 225 \times 10^{-7} \text{ Ns/m}^2$$
$$k = 33 \times 10^{-3} \text{ W/mK}$$

If the pressure had not been 1 atm, the density would have had to be modified using the perfect gas law.

Using these values then gives:

$$Re = \frac{\rho V L}{\mu} = \frac{0.9 \times 600 \times 1}{225 \times 10^{-7}} = 2.4 \times 10^7$$

At this Reynolds number, the flow in the boundary layer will indeed be turbulent so the assumed recovery factor is, in fact, the correct value. Hence, the adiabatic wall temperature is 159°C.

Part 2

When the wall is at 60°C the air properties are found at the following temperature:

$$T_{prop} = T_1 + 0.5(T_w - T_1) + 0.22(T_{w_{ad}} - T_1)$$
$$= 0 + 0.5(60) + 0.22(159) = 65°C$$

Now at a temperature of 65°C, air has the following properties at a pressure of 1 atm:

$$\rho = 0.99 \text{ kg/m}^3$$
$$\mu = 208 \times 10^{-7} \text{ Ns/m}^2$$
$$k = 30 \times 10^{-3} \text{ W/m K}$$

In this case then:

$$Re = \frac{\rho V L}{\mu} = 4.3 \times 10^7$$

so the flow is again turbulent.

The same equation for the Nusselt number as derived for flow without dissipation can be used here. The Nusselt number is therefore given by:

$$Nu = 0.037 Re^{0.8} Pr^{1/3} = 0.37 \times (4.3 \times 10^7)^{0.8} \times (0.7)^{1/3} = 41980$$

From this it follows that:

$$\frac{hL}{k} = 41980$$

so:

$$h = \frac{41980 \times (30 \times 10^{-3})}{1} = 1259 \text{ W/m}^2\text{°C}$$

Therefore, considering both sides of the plate, the heat transfer rate from the plate is given by:

$$Q = hA(T_W - T_{W_{ad}}) = 2 \times 1259 \times 1 \times 1 \times (60 - 159) = -250,000 \text{ W}$$

The negative sign means that heat is transferred to the plate. Hence, the net rate of heat transfer **to** the plate is 250 kW.

6.6
SOLUTIONS TO THE FULL TURBULENT FLOW EQUATIONS

This chapter has mainly been devoted to the solution of the boundary layer form of the governing equations. While these boundary layer equations do adequately describe a number of problems of great practical importance, there are many other problems that can only be adequately modeled by using the full governing equations. In such cases, it is necessary to obtain the solution numerically and also almost always necessary to use a more advanced type of turbulence model [6],[12],[28],[29]. Such numerical solutions are most frequently obtained using the commercially available software based on the finite volume or the finite element method.

6.7
CONCLUDING REMARKS

While it may one day be possible to numerically solve, on a routine basis, the full unsteady form of the governing equations for turbulent flow, most solutions undertaken at the present time are based on the use of the time-averaged form of the governing equations together with a turbulence model. Such solutions for the flow over the outer surface of a body immersed in a fluid stream have been discussed in this chapter.

So-called analogy solutions for predicting the heat transfer rate from a knowledge of the wall shear stress distribution were first discussed. The initial discussion was of the simplest such analogy solution, the Reynolds analogy, which only really applies to fluids with a Prandtl number near one. Multi-layer analogy solutions which apply for all Prandtl numbers were then discussed. In order to use these analogy solutions for flow over bodies of complex shape it is necessary to solve for the surface shear stress distribution. The use of the integral equation method for this purpose was discussed.

PROBLEMS

6.1. Air flows over a wide 2-m long flat plate which has a uniform surface temperature of 80°C, the temperature of the air ahead of the plate being 20°C. The air velocity is such that the Reynolds number based on the length of the plate is 5×10^6. Derive an expression for the local wall heat flux variation along the plate. Use the Reynolds analogy and assume the boundary layer transition occurs at a Reynolds number of 10^6.

6.2. Air at a temperature of 50°C flows over a wide flat plate at a velocity of 60 m/s. The plate is kept at a uniform temperature of 10°C. If the plate is 3 m long, plot the variation of local heat transfer rate per unit area along the surface of the plate. Assume that transition occurs at a Reynolds number of 3×10^5.

6.3. In the discussion of the use of the Reynolds analogy for the prediction of the heat transfer rate from a flat plate it was assumed that when there was transition on the plate, the x-coordinate in the turbulent portion of the flow could be measured from the leading edge. Develop an alternative expression based on the assumption that the momentum thickness before and after transition is the same. This assumption allows an effective origin for the x-coordinate in the turbulent portion of the flow to be obtained.

6.4. Using the Taylor-Prandtl analogy, determine the relation between the velocity and temperature profiles in the boundary layer.

6.5. Derive a modified version of the Reynolds analogy assuming the Prandtl number and turbulent Prandtl number are equal but are not equal to one.

6.6. Air flows over a flat plate which has a uniform surface temperature of 50°C, the temperature of the air ahead of the plate being 30°C. The air velocity is such that the Reynolds number based on the length of the plate is 5×10^6, the length of the plate being 2 m. Using the Reynolds analogy, plot the variation of the local heat transfer rate from the wall, q_w, with x/L assuming that (i) the boundary layer flow remains laminar, (ii) the boundary layer flow is turbulent from the leading edge of the plate, and (iii) boundary layer transition occurs at Re_T of 10^6. x is the distance from the leading edge of the plate and L is the length of the plate.

6.7. Air flows through a large plane duct with isothermal walls. The Reynolds number based on the length of the duct and the inlet air velocity is 10^7. Using the integral equation method and assuming that the boundary layer is turbulent from the inlet and thin compared to the size of the duct, determine how the local Nusselt varies with distance along the duct if the duct cross-sectional area varies in such a way that the velocity increases linearly by 50% over the length of the duct.

6.8. Modify the integral equation computer program to use the Taylor-Prandtl analogy. Use this modified program to determine the local Nusselt number variation for the situation described in Problem 6.6.

6.9. Air at a temperature of 20°C flows at a velocity of 100 m/s over a 3-m long wide flat plate which is aligned with the flow. The first fifth of the plate is unheated and the remainder of the plate is maintained at a uniform wall temperature of 60°C. Plot the variation of the local heat transfer rate along the heated section of the plate. Evaluate the air properties at a temperature of 30°C and assume that the boundary layer is turbulent from the leading edge.

6.10. Modify the integral equation analysis of flow over a plate with an unheated leading edge section that was given in this chapter to apply to the case where the plate has a heated leading edge section followed by an adiabatic section.

6.11. Numerically determine the local Nusselt number variation with two-dimensional turbulent boundary layer air flow over an isothermal flat plate for a maximum Reynolds number of 10^7. Assume that transition occurs at a Reynolds number of 5×10^5. Compare the numerical results with those given by the Reynolds analogy.

6.12. In the numerical solution for boundary layer flow given in this chapter it was assumed that transition occurred at a point; i.e., the eddy viscosity was set equal to zero up to the transition point and then the full value given by the turbulence model was used. Show how this numerical method and the program based on it can be modified to allow for a transition zone in which the eddy viscosity increases linearly from zero at the beginning of the zone to the full value given by the turbulence model at the end of the zone.

6.13. Air flows over a 3-m long flat plate which has a uniform surface temperature of 60°C, the temperature of the air ahead of the plate being 20°C. The air velocity is 60 m/s. Numerically determine the variation of the local heat transfer rate from the wall, q_w, with x assuming that boundary layer transition occurs at Re_T of 10^6.

6.14. Numerically determine the local Nusselt number variation for the situation described in Problem 6.6.

6.15. Air flows over a wide flat plate which is aligned with the flow. The Reynolds number based on the length of the plate and the free-stream air velocity is 10^7. A specified heat flux is applied at the surface of the plate, the surface heat flux increasing linearly from 0.5 q_{wm} at the leading edge of the plate to 1.5 q_{wm} at the trailing edge of the plate, q_{wm} being the mean surface heat flux. Assuming that the boundary layer is turbulent from the leading edge of the plate, numerically determine how the dimensionless wall temperature varies with distance along the plate.

6.16. Consider air flow over a wide flat plate which is aligned with the flow. The Reynolds number based on the length of the plate and the free-stream air velocity is 6×10^6. The first third of the plate is adiabatic, the second third of the plate has a uniform heat flux applied at the surface, while the last third of the plate is again adiabatic. Assuming that the boundary layer is turbulent from the leading edge of the plate, numerically determine how the dimensionless wall temperature varies with distance along the plate.

6.17. Air at a temperature of 0°C and standard atmospheric pressure flows at a velocity of 50 m/s over a wide flat plate with a total length of 2 m. A uniform surface heat flux is applied over the first 0.7 m of the plate and the rest of the surface of the plate is adiabatic. Assuming that the boundary layer is turbulent from the leading edge, use the numerical solution to derive an expression for the plate temperature at the trailing edge of the plate in terms of the applied heat flux. What heat flux is required to ensure that the trailing edge temperature is 5°C?

6.18. Discuss how the computer program for calculating heat transfer from a surface with a specified surface temperature would have to be modified to incorporate the effect of suction at the surface.

6.19. Air at standard atmospheric pressure and a temperature of 30°C flows over a flat plate at a velocity of 20 m/s. The plate is 60 cm square and is maintained at uniform temperature of 90°C. The flow is normal to a side of the plate. Calculate the heat transfer from the plate assuming that the flow is two-dimensional.

6.20. The roof of a building is flat and is 20 m wide and long. If the wind speed over the roof is 10 m/s, determine the convective heat transfer rate to the roof (i) on a clear night when the roof temperature is 2°C and the air temperature is 12°C and (ii) on a hot, sunny day when the roof temperature is 46°C and the air temperature is 28°C. Assume two-dimensional turbulent boundary layer flow.

6.21. A rocket ascends vertically through the atmosphere with a velocity that can be assumed to increase linearly with altitude from zero at sea level to 1800 m/s at an altitude of 30,000 m. If the surface of this rocket is assumed to be adiabatic, estimate the variation of the skin temperature with altitude at a point on the surface of the rocket a distance of 3 m from the nose of the rocket. Use the flat plate equations given in this chapter and assume that at the distance from the nose considered, the Mach number and temperature outside the boundary layer are the same as those in the free-stream ahead of the rocket.

6.22. A flat plate with a length of 0.8 m and a width of 1.2 m is placed in the working section of a wind tunnel in which the Mach number is 4, the temperature is −70°C, and the pressure is 3 kPa. If the surface temperature of the plate is kept at 30°C by an internal cooling system, find the rate at which heat must be added to or removed from the plate. Consider both the top and the bottom of the plate.

6.23. At an altitude of 30,000 m the atmospheric pressure is approximately 1200 Pa and the temperature is approximately −45°C. Assuming a turbulent boundary layer flow over an adiabatic flat plate, plot the variation of the adiabatic wall temperature with Mach number for Mach numbers between 0 and 5.

6.24. Air at a pressure of 29 kPa and a temperature of −35°C flows at a Mach number of 4 over a flat plate. The plate is maintained at a uniform temperature of 90°C. If the plate is 0.5 m long, find the mean rate of heat transfer per unit surface area assuming a two-dimensional turbulent boundary layer flow.

REFERENCES

1. Schlichting, H., *Boundary Layer Theory*, 7th ed., McGraw-Hill, New York, 1979.
2. Hinze, J.O., *Turbulence*, 2nd ed., McGraw-Hill, New York, 1975.
3. Eckert, E.R.G. and Drake, R.M., Jr., *Analysis of Heat and Mass Transfer*, McGraw-Hill, New York, 1973.
4. Kestin, J. and Richardson, P.D., "Heat Transfer across Turbulent, Incompressible Boundary Layers", *J. Heat Mass Transfer*, Vol. 6, pp. 147–189, 1963.
5. Burmeister, L.C., *Convective Heat Transfer*, 2nd ed., Wiley-Interscience, New York, 1993.

6. Anderson, D., Tannehill, J.C., and Pletcher, R.H., *Computational Fluid Mechanics and Heat Transfer*, Hemisphere Publ., Washington, D.C., 1984.
7. Cebeci, T. and Smith, A.M.O., *Analysis of Turbulent Boundary Layers*, Academic Press, New York, 1974.
8. Christoph, G.H. and Pletcher, R.H., "Prediction of Rough-Wall Skin-Friction and Heat Transfer", *AIAA J.*, Vol. 21, No. 4, pp. 509–515, 1983.
9. Chow, C.Y., *Computational Fluid Mechanics*, Wiley, New York, 1979.
10. Patankar, S.V. and Spalding, D.B., *Heat and Mass Transfer in Boundary Layers*, 2nd ed., International Textbook Co., London, 1970.
11. Pletcher, R.H., "Prediction of Transpired Turbulent Boundary Layers", *J. Heat Transfer*, Vol. 96, pp. 89–94, 1974.
12. Roache, P.J., *Computational Fluid Dynamics*, Hermosa Publishers, Albuquerque, NM, 1976.
13. Bradshaw, P., "The Turbulent Structure of Equilibrium Turbulent Boundary Layers", *J. Fluid Mech.*, Vol. 29, pp. 625–645, 1967.
14. Galbraith, R.A. and Head, M.R., "Eddy Viscosity and Mixing Length from Measured Boundary Layer Developments", *Aeronautical Quarterly*, Vol. 26, pp. 133–154, 1975.
15. Kline, S.J., Reynolds, W.C., Schraub, F.A., and Runstadler, P.W., "The Structure of Turbulent Boundary Layers", *J. Fluid Mechanics,*Vol. 30, pp. 741–773, 1967.
16. Ng, K.H. and Spalding, D.B., "Turbulence Model for Boundary Layers Near Walls", *Phys. Fluids,* Vol. 15, pp. 20–30, 1972.
17. Notter, R.H. and Sleicher, C.A., "The Eddy Diffusivity in the Turbulent Boundary Layer near a Wall", *Eng. Sci.,* Vol. 26, pp. 161–171, 1971.
18. Beckwith, I.E. and Gallagher, J.J., "Local Heat Transfer and Recovery Temperatures on a Yawed Cylinder at a Mach Number of 4.15 and High Reynolds Numbers", NASA TR-R-104, Houston, TX, 1962.
19. Cary, A.M. and Bertram, M.H., "Engineering Prediction of Turbulent Skin Friction and Heat Transfer in High Speed Flow", NASA, TN D-7507, Houston, TX, 1974.
20. DeJarnette, F.R., Hamilton, H.H., Weilmuenster, K.L., and Cheatwook, F.M., "A Review of Some Approximate Methods Used in Aerodynamic Heating Analyses", *Thermophys. Heat Transfer,* Vol. 1, No. 1, pp. 5–12, 1987.
21. Eckert, E.R.G., "Engineering Relations for Friction and Heat Transfer to Surfaces in High Velocity Flow", *J. Aerosp. Sci.,* Vol. 22, p. 585, 1955.
22. Fischer, W.W. and Norris, R., "Supersonic Convective Heat Transfer Correlation from Skin Temperature Measurement on V-2 Rocket in Flight", *Trans. ASME,* Vol. 71, pp. 457–469, 1949.
23. Kaye, J., "Survey of Friction Coefficients, Recovery Factors, and Heat Transfer Coefficients for Supersonic Flow", *J. Aeronaut. Sci.,* Vol. 21, No. 2, pp. 117–229, 1954.
24. Rubesin, M.W. and Johnson, H.A., "Aerodynamic Heating and Convective Heat Transfer—Summary of Literature Survey", *Trans. ASME,* Vol. 71, pp. 383–388, 1949.
25. Truitt, R.W., *Fundamentals of Aerodynamic Heating*, Ronald Press, New York, 1960.
26. van Driest, E.R., "Turbulent Boundary Layer in Compressible Fluids", *J. Aeronaut. Sci.,* Vol. 18, No. 3, pp. 145–161, 1951.
27. van Driest, E.R., "The Problem of Aerodynamic Heating", *Aeronaut. Eng. Rev.,* Vol. 15, pp. 26–41, 1956.
28. Launder, B.W., "On the Computation of Convective Heat Transfer in Complex Turbulent Flows", *ASME J. Heat Transfer,* Vol. 110, pp. 1112–1128, 1988.
29. Patankar, S.V., *Numerical Heat Transfer and Fluid Flow*, Hemisphere Publ., Washington, D.C., 1980.

CHAPTER 7

Internal Turbulent Flows

7.1
INTRODUCTION

This chapter is concerned with the prediction of the heat transfer rate from the wall of a duct to a fluid flowing through the duct, the flow in the duct being turbulent. The majority of the attention will be given to axi-symmetric flow through pipes and two-dimensional flow through plane ducts, i.e., essentially to flow between parallel plates. These two types of flow are shown in Fig. 7.1.

Such flows effectively occur in many practical situations such as flows in heat exchangers. Attention will initially be given to fully developed flow, see References [1] to [12]. This will be followed by a discussion of developing duct flows. Lastly, a brief discussion of the numerical analysis of more complex duct flows will be presented.

7.2
ANALOGY SOLUTIONS FOR FULLY DEVELOPED PIPE FLOW

As discussed in the previous chapter, most early efforts at trying to theoretically predict heat transfer rates in turbulent flow concentrated on trying to relate the wall heat transfer rate to the wall shearing stress. In the present section an attempt will be made to outline some of the simpler such analogy solutions for duct flows [13], [14],[15],[16],[17].

The ideas used in the previous chapter to derive analogy solutions for boundary layer flows can easily be extended to obtain such analogy solutions for turbulent

Flow Through a Pipe

Flow Through a Plane Duct

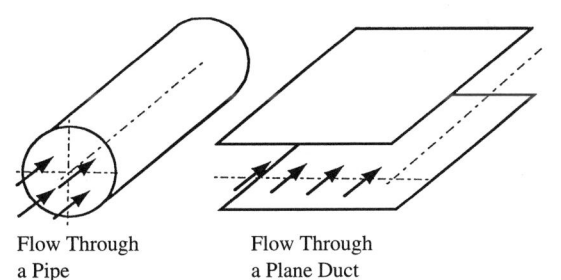

FIGURE 7.1
Flow in a pipe and in a plane duct.

duct flows. The main difference between the solutions for boundary layer flows and those for duct flows arises because in the latter case the Nusselt number is based on the difference between the wall temperature and the mean fluid temperature and the Reynolds number is based on the mean velocity; whereas in boundary layer flows the Nusselt number is based on the difference between the wall temperature and the fluid temperature outside the boundary layer and the Reynolds number is based on the velocity outside the boundary layer.

Consider, first, the simple Reynolds analogy for pipe flows. The pipe wall temperature will be assumed to be uniform. If y is the distance from the wall to the point in the flow being considered as shown in Fig. 7.2, the total shearing stress and heat transfer rate are again written as:

$$\tau = \tau_m + \tau_T = \mu \frac{\partial \bar{u}}{\partial y} - \rho \overline{v' u'} \tag{7.1}$$

and:

$$q = q_m + q_T = -k \frac{\partial \bar{T}}{\partial y} - \rho c_p \overline{v' T'} \tag{7.2}$$

The above two equations can be written as:

$$\tau = \rho(\nu + \epsilon) \frac{\partial \bar{u}}{\partial y} \tag{7.3}$$

and:

$$q = -\rho c_p (\alpha + \epsilon_H) \frac{\partial \bar{T}}{\partial y} \tag{7.4}$$

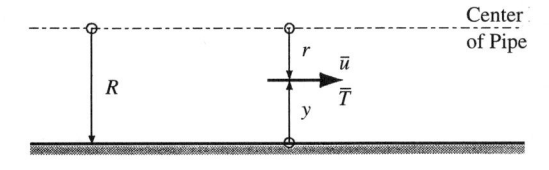

FIGURE 7.2
Coordinate system used.

306 Introduction to Convective Heat Transfer Analysis

These equations can be rearranged to give:

$$\frac{\partial \bar{u}}{\partial y} = \frac{\tau}{\rho(\nu + \epsilon)} \tag{7.5}$$

and:

$$\frac{\partial \bar{T}}{\partial y} = -\frac{q}{\rho c_p (\alpha + \epsilon_H)} \tag{7.6}$$

Integrating these two equations outward from the wall, where $\bar{u} = 0$ and $\bar{T} = T_w$, to a point at a distance of y from the wall where the mean velocity is $\bar{u}$ and the mean temperature is $\bar{T}$ then gives:

$$\bar{u} = \int_0^y \frac{\tau}{\rho(\nu + \epsilon)} dy \tag{7.7}$$

and:

$$T_w - \bar{T} = \int_0^y \frac{q}{\rho c_p (\alpha + \epsilon_H)} dy \tag{7.8}$$

The same assumptions as were used in deriving the Reynolds analogy for boundary layer flow are now introduced, i.e., it is assumed that:

$$\epsilon_H = \epsilon \tag{7.9}$$

this is, of course, equivalent to assuming that the turbulent Prandtl number, Pr_T, is equal to 1, and that

$$Pr = 1, \text{ i.e., } \alpha = \nu \tag{7.10}$$

No real fluids exist for which this is true, but, as mentioned earlier, most gases have Prandtl numbers that are close to unity.

It is also assumed that:

$$\frac{q}{q_w} = \frac{\tau}{\tau_w} \tag{7.11}$$

where τ_w and q_w are the values of the shear stress and the heat transfer rate at the wall.

Using the assumptions given in Eqs. (7.9) to (7.11), Eq. (7.8) can be rewritten as:

$$T_w - \bar{T} = \frac{q_w}{\tau_w} \int_0^y \frac{\tau}{\rho c_p (\nu + \epsilon)} dy \tag{7.12}$$

If the following is then defined:

$$G(y) = \int_0^y \frac{\tau}{\rho(\nu + \epsilon)} dy \tag{7.13}$$

Eqs. (7.7) and (7.12) can be written as:

$$\bar{u} = G(y) \tag{7.14}$$

and:

$$T_w - \overline{T} = \frac{q_w}{c_p T_w} G(y) \tag{7.15}$$

This shows that the mean velocity and temperature profiles are similar when the Reynolds analogy assumptions are adopted.

If the average velocity and average temperature across the pipe are defined as usual by:

$$\bar{u}_m = \frac{1}{\pi R^2} \int_0^R \bar{u} 2\pi r \, dr$$

and

$$\overline{T}_m = \frac{1}{\bar{u}_m \pi R^2} \int_0^R \bar{u}\overline{T} 2\pi r \, dr$$

then, because:

$$r = R - y$$

it follows that:

$$\bar{u}_m = \frac{2}{R^2} \int_0^R \bar{u}(R - y) \, dy \tag{7.16}$$

and:

$$\overline{T}_m = \frac{\int_0^R \bar{u}\overline{T}(R - y) \, dy}{\int_0^R \bar{u}(R - y) \, dy} \tag{7.17}$$

Using Eqs. (7.14) and (7.15) these give:

$$\bar{u}_m = \frac{2}{R^2} \int_0^R G(R - y) \, dy \tag{7.18}$$

and:

$$T_w - \overline{T}_m = \frac{q_w}{c_p T_w} \frac{\int_0^R G^2(R - y) \, dy}{\int_0^R G(R - y) \, dy} \tag{7.19}$$

Dividing Eq. (7.19) by Eq. (7.18) then gives:

$$\frac{T_w - \overline{T}_m}{\bar{u}_m} = \frac{q_w}{c_p T_w} \frac{\int_0^R G^2(R - y) \, dy}{\frac{2}{R^2} \left[\int_0^R G(R - y) \, dy \right]^2} \tag{7.20}$$

This equation can be written as:

$$\frac{T_w - \overline{T}_m}{\bar{u}_m} = \frac{q_w}{c_p T_w} F \tag{7.21}$$

where:

$$F = \frac{\int_0^R G^2(R-y)\,dy}{\frac{2}{R^2}\left[\int_0^R G(R-y)\,dy\right]^2} \tag{7.22}$$

This factor, F, can be evaluated approximately by noting that in pipe flow the mean velocity profile is approximately given by:

$$\frac{\bar{u}}{\bar{u}_c} = \left(\frac{y}{R}\right)^{1/7} \tag{7.23}$$

where $\bar{u}_c$ is the mean velocity on the center line of the pipe.

Comparing Eq. (7.23) with Eq. (7.14) then shows that the function G is approximately given by:

$$G = \bar{u}_c \left(\frac{y}{R}\right)^{1/7}$$

Substituting this into Eq. (7.22) then gives:

$$F = \frac{\int_0^1 \left(\frac{y}{R}\right)^{2/7}\left(1-\frac{y}{R}\right) d\left(\frac{y}{R}\right)}{2\left[\int_0^1 \left(\frac{y}{R}\right)^{1/7}\left(1-\frac{y}{R}\right) d\left(\frac{y}{R}\right)\right]^2}$$

from which it follows that:

$$F = \frac{150}{147}$$

i.e., F is effectively equal to 1 and Eq. (7.21) then gives:

$$\frac{T_w - \bar{T}_m}{\bar{u}_m} = \frac{q_w}{c_p \tau_w} \tag{7.24}$$

This is the basic Reynolds analogy equation for pipe flow. It is identical to that for boundary layer flow except that the mean velocity and temperature occur in place of the free-stream values.

Now, for fully developed pipe flow it is usual to work with the friction factor, f, rather than the wall shearing stress. The friction factor is defined in terms of the pressure gradient by:

$$f = -\frac{dp/dz}{\rho \bar{u}_m^2/2D} \tag{7.25}$$

$\bar{u}_m$ being, as before, the mean velocity in the pipe and D is its diameter.

In order to find the relation between the wall shearing stress, τ_w, and the friction factor, f, for the fully developed flow, the conservation of momentum principle is applied to the control volume shown in Fig. 7.3.

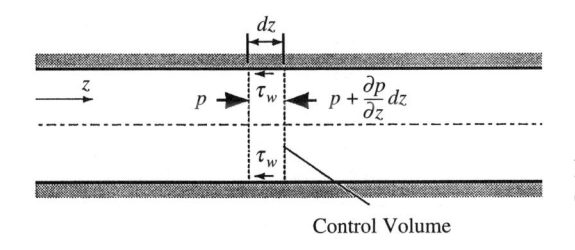

FIGURE 7.3
Control volume used in relating wall shear stress to the friction factor.

Because, in the fully developed flow, there is no change in the velocity profile with distance, z, along the pipe, there can be no change in fluid momentum through the control volume. The forces acting on the control must therefore balance, i.e.:

$$pA - \left(p + \frac{dp}{dz} dz\right)A = \tau_w C\, dz$$

where $C = \pi D$ is the circumference of the pipe and $A = \pi D^2/4$ is its cross-sectional area. This equation gives on rearrangement:

$$\frac{dp}{dz} = -\tau_w \frac{C}{A} = -\tau_w \frac{4}{D}$$

Substituting this equation into Eq. (7.25) then gives the following relation between the friction factor and the wall shearing stress:

$$\tau_w = \frac{f}{8}\rho \bar{u}_m^2 \qquad (7.26)$$

Substituting Eq. (7.26) into Eq. (7.24) then gives the following as the Reynolds analogy for pipe flow:

$$\frac{T_w - \bar{T}_m}{\bar{u}_m} = \frac{q_w}{c_p} \frac{8}{f \rho \bar{u}_m^2} \qquad (7.27)$$

This equation can be rearranged in terms of the Nusselt number, Nu_D, based on the mean temperature difference and on the pipe diameter, and the Reynolds number, Re_D, based on the mean velocity and the diameter, to give, on noting that Pr has already been assumed equal to 1:

$$Nu_D = \frac{f}{8} Re_D \qquad (7.28)$$

Charts and equations describing the variation of the friction factor, f, with the Reynolds number, Re_D, and wall roughness ratio, e/D, where e is a measure of the roughness of the walls, are available [18],[19], [20]. A Moody chart that gives the friction factor variation is shown in Fig. 7.4.

Alternatively, Colebrook gives the following equation for commercially rough pipes:

$$\frac{1}{\sqrt{f}} = 1.74 - 2\log_{10}\left[2\frac{e}{D} + \frac{18.7}{Re_D \sqrt{f}}\right] \qquad (7.29)$$

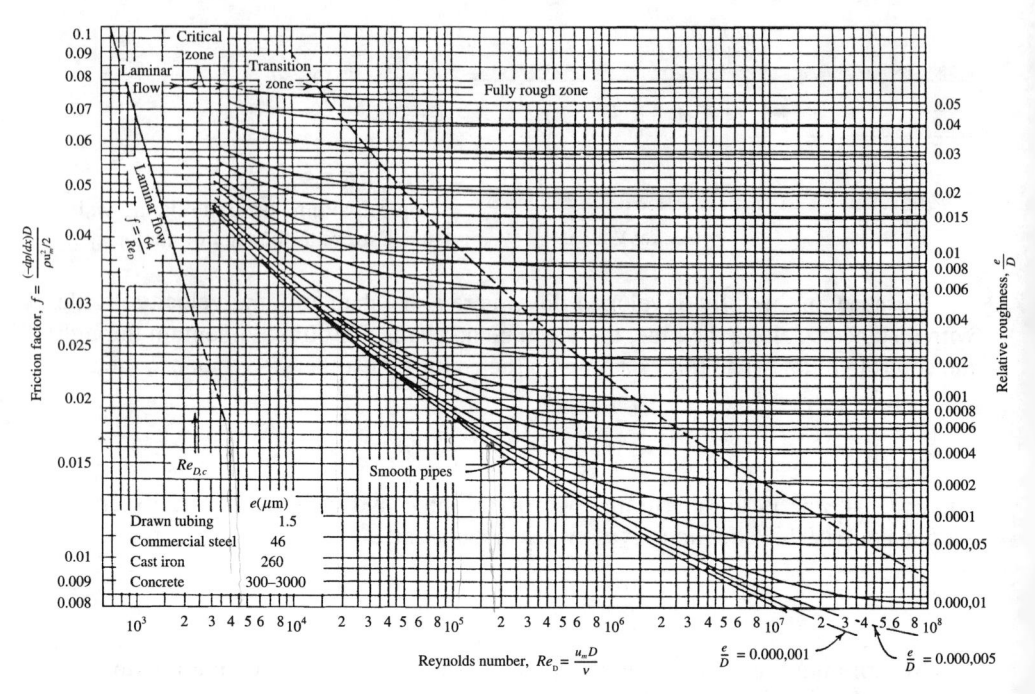

FIGURE 7.4
Moody chart.

Some typical values of the roughness, e, for pipes that are likely to be encountered in heat transfer practice are listed in Table 7.1.

For smooth pipes, i.e., for $e = 0$, Eq. (7.29) gives:

$$\frac{1}{\sqrt{f}} = 1.74 - 2\log_{10}\left[\frac{18.7}{Re_D \sqrt{f}}\right]$$

i.e.,

$$\frac{1}{\sqrt{f}} = 2\log_{10}\left[Re_D \sqrt{f}\right] - 0.8 \qquad (7.30)$$

Alternatively, the following simple equation can be used to obtain approximate values for the friction factor for flow in smooth pipes for Reynolds numbers less than

TABLE 7.1
Roughness values for commercial pipes

Pipe material	Wall roughness, e, mm
Drawn tubing	1.5×10^{-3}
Brass, glass	7.5×10^{-3}
Steel	4.6×10^{-2}
Galvanized iron	0.15

CHAPTER 7: Internal Turbulent Flows   311

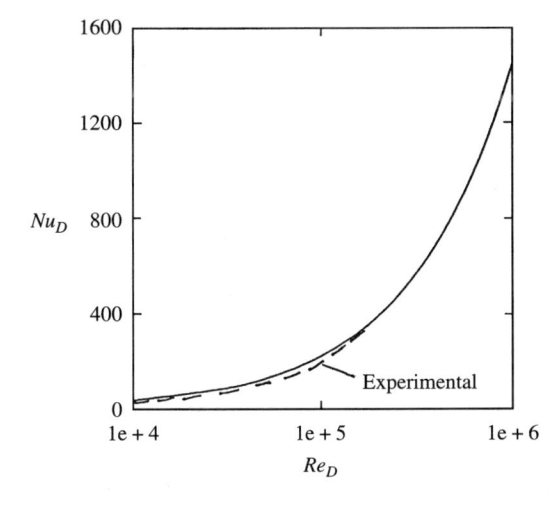

FIGURE 7.5
Variation of Nu_D with Re_D for smooth pipes.

about 10^5:

$$f = \frac{0.305}{Re_D^{0.25}} \tag{7.31}$$

The variation of Nu_D with Re_D for smooth pipes given by using Eq. (7.30) in conjunction with Eq. (7.28) is shown in Fig. 7.5 together with experimental results for air.

EXAMPLE 7.1. Air flows through a smooth pipe with a diameter of 50 mm at a mean velocity of 35 m/s. The walls of the pipe are kept at a temperature of 10°C. At a certain section of the pipe, the mean air temperature is 40°C. Find the value of the heat transfer coefficient and the rate of heat transfer per meter length of pipe from the air to the walls at this section.

Solution. At the mean temperature of 40°C assuming that the pressure is near ambient, air has the following properties:

$$\nu = 16.01 \times 10^{-6} \text{ m}^2/\text{s}, \quad k = 0.02638 \text{ W/m-°C}$$

Hence:

$$Re_D = \frac{u_m D}{\nu} = \frac{35 \times 0.05}{0.00001601} = 1.093 \times 10^5$$

The flow is therefore turbulent.

Because the pipe is smooth, f is given by Eq. (7.30), i.e., by:

$$\frac{1}{\sqrt{f}} = 2\log_{10}[Re_D \sqrt{f}] - 0.8 = 2\log_{10}[109300 \sqrt{f}] - 0.8$$

which can be solved to give:

$$f = 0.0177$$

Eq. (7.28) then gives:

$$Nu_D = \frac{f}{8} Re_D = \frac{0.0177}{8} \times 109300 = 241.8$$

Hence:

$$h = \frac{Nu_D k}{D} = \frac{241.8 \times 0.02638}{0.05} = 127.6 \text{ W/m}^2\text{°C}$$

The heat transfer rate per m length of the pipe is then given by:

$$Q_w = h\pi D(T_w - T_m) = 127.6 \times \pi \times 0.05 \times (10 - 40)$$

i.e.:

$$Q_w = -601.3 \text{ W/m}$$

This is negative because the heat transfer rate is taken as positive in the wall to air direction. The heat transfer rate from the air to the walls of the pipe is therefore 601.3 W/m.

Therefore, the heat transfer coefficient is 127.6 W/m²°C and the heat transfer from the air to the pipe walls per meter length is 601.3 W.

It will be seen from Fig. 7.5 that the agreement between the predicted values and experiment is quite good for the case considered because air has a Prandtl number near 1. When the Prandtl number of the fluid involved is very different from one, however, the results given by Eq. (7.28) are unsatisfactory. For this reason, more refined analogy solutions have also been developed for pipe flows just as in the case of boundary layer flows. Consideration will now be given to a three-layer analogy solution for pipe flows. In this discussion, attention will be restricted to flow through a pipe in which the wall heat flux is uniform. The analysis is, of course, similar to that given in Chapter 6 for boundary layer flow.

Now, if the longitudinal heat transfer in the fluid is neglected, the energy equation for fully developed constant fluid property turbulent pipe flow can be written as:

$$\bar{u}\frac{\partial \bar{T}}{\partial z} = \frac{1}{r}\frac{\partial}{\partial r}\left[r(\epsilon_H + \alpha)\frac{\partial \bar{T}}{\partial r}\right] \tag{7.32}$$

In writing this equation it has been noted that because the flow is fully developed, $\bar{v}$ is zero.

Because the flow is fully developed, the form of the mean temperature profile is not changing with distance along the pipe, i.e., defining:

$$\frac{T_w - \bar{T}}{T_w - \bar{T}_c} = F\left(\frac{r}{R}\right) \tag{7.33}$$

in fully developed flow the function, F, is independent of the distance along the pipe, z. In this equation, $\bar{T}_c$ is the mean temperature on the center line and T_w is the wall temperature.

Now, because F does not depend on z, it follows that:

$$\frac{\partial}{\partial z}\left[\frac{T_w - \bar{T}}{T_w - \bar{T}_c}\right] = 0$$

i.e., that:

$$\frac{\partial \overline{T}}{\partial z} = \frac{\partial T_w}{\partial z} - \left[\frac{T_w - \overline{T}}{T_w - \overline{T}_c}\right]\left[\frac{dT_w}{dz} - \frac{d\overline{T}_c}{dz}\right] \quad (7.34)$$

In the situation here being considered, the heat flux at the wall, q_w, is uniform and specified. Now, using Fourier's law, the heat transfer rate at the wall is given by:

$$q_w = +k\left.\frac{\partial \overline{T}}{\partial r}\right|_{r=R} \quad (7.35)$$

The positive sign arising because r is measured from the center toward the wall whereas the heat flux, q_w, is taken as positive in the inward direction, i.e., in the wall to fluid direction.

Eq. (7.35) can be written using the definition of the function, F, given in Eq. (7.33) as:

$$q_w = -\frac{k(T_w - \overline{T}_c)}{R}\left.\frac{dF}{d(r/R)}\right|_{r=R} \quad (7.36)$$

Because the case where q_w is uniform is being considered, this equation shows that $T_w - \overline{T}_c =$ constant from which it follows that:

$$\frac{dT_w}{dz} = \frac{d\overline{T}_c}{dz} \quad (7.37)$$

Substituting this result into Eq. (7.34) then shows that when the wall heat flux is uniform:

$$\frac{\partial \overline{T}}{\partial z} = \frac{dT_w}{dz} \quad (7.38)$$

Therefore, the gradient of temperature at all radii is the same. Hence, if $\overline{T}_m$ is the mean temperature at any section of the pipe it follows that:

$$\frac{\partial \overline{T}}{\partial z} = \frac{dT_w}{dz} = \frac{d\overline{T}_c}{dz} = \frac{d\overline{T}_m}{dz} \quad (7.39)$$

It should also be noted that an overall energy balance applied to the control volume shown in Fig. 7.6 gives:

$$q_w \pi D = \rho c_p \overline{u}_m \frac{\pi}{4} D^2 \left[\overline{T}_m + \frac{d\overline{T}_m}{dz} - \overline{T}_m\right]$$

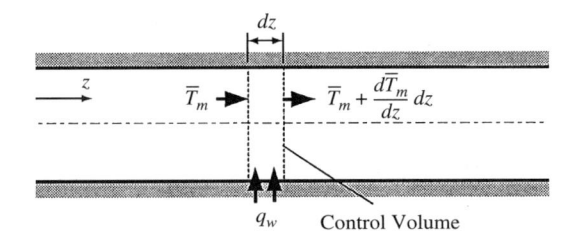

FIGURE 7.6
Control volume used in establishing energy balance.

which can be rearranged to give:

$$\frac{d\bar{T}_m}{dz} = \frac{4q_w \pi}{\rho c_p \bar{u}_m D} \tag{7.40}$$

Because q_w is constant, it follows from this that the gradient of temperature at all radii is constant.

Having established this result from the boundary conditions, attention can now be returned to the governing equation (7.32). Using Eq. (7.39), this equation gives:

$$u \frac{d\bar{T}_w}{dz} = \frac{1}{r} \frac{\partial}{\partial r}\left[r(\epsilon_H + \alpha) \frac{\partial \bar{T}}{\partial r}\right] \tag{7.41}$$

It is convenient to rewrite Eq. (7.41) in terms of the distance from the wall, y, to the point being considered. If, as above, R is the radius of the pipe, y is, as discussed before, related to r by:

$$y = R - r \tag{7.42}$$

In terms of y, Eq. (7.41) becomes:

$$u \frac{d\bar{T}_m}{dz} = \frac{1}{(R-y)} \frac{\partial}{\partial y}\left[(R-y)(\epsilon_H + \alpha) \frac{\partial \bar{T}}{\partial y}\right] \tag{7.43}$$

The boundary conditions on the solution to this equation are:

$$\text{When } r = 0 \text{ (i.e., } y = R\text{): } \frac{\partial \bar{T}}{\partial y} = 0 \tag{7.44}$$

$$\text{When } r = R \text{ (i.e., } y = 0\text{): } \bar{T} = T_w$$

The first of these conditions follows, of course, from the requirement that the profile be symmetrical about the center line.

Integrating Eq. (7.43) from the center line where $y = R$ to some point distance y from the wall and using the first boundary condition listed in Eq. (7.44) then gives:

$$(R-y)(\epsilon_H + \alpha) \frac{\partial \bar{T}}{\partial y} = \frac{d\bar{T}_m}{dz} \int_R^y u(R-y)\,dy$$

i.e.:

$$(R-y)(\epsilon_H + \alpha) \frac{\partial \bar{T}}{\partial y} = \frac{d\bar{T}_m}{dz}\left[\int_0^y u(R-y)\,dy - \int_0^R u(R-y)\,dy\right] \tag{7.45}$$

It is convenient to define:

$$I(y) = \int_0^y u(R-y)\,dy - \int_0^R u(R-y)\,dy \tag{7.46}$$

and in terms of this function, Eq. (7.45) becomes:

$$\frac{\partial \bar{T}}{\partial y} = \frac{I}{(R-y)(\epsilon_H + \alpha)} \frac{d\bar{T}_m}{dz} \tag{7.47}$$

CHAPTER 7: Internal Turbulent Flows 315

Integrating this equation out from the wall and using the second of the boundary conditions given in Eq. (7.44) then gives:

$$\overline{T} - T_w = \frac{d\overline{T}_m}{dz}\int_0^y \frac{I}{(R-y)(\epsilon_H + \alpha)}\,dy \tag{7.48}$$

Using Eq. (7.40) allows this equation to be written as:

$$\overline{T} - T_w = \frac{4q_w}{\rho c_p \overline{u}_m D}\int_0^y \frac{I}{(R-y)(\epsilon_H + \alpha)}\,dy \tag{7.49}$$

If the velocity profile is known together with the distribution of ϵ then, for any assumed relationship between the distributions of ϵ_H and ϵ, Eq. (7.49) can be used to deduce the temperature profile. Once this has been obtained the relation between the Nusselt and Reynolds numbers can be derived. Before illustrating this procedure, there is a simplifying assumption that can be introduced without any significant loss of accuracy. Because the velocity profile in turbulent pipe flow is relatively flat over a large portion of the pipe cross-section, it is usually sufficiently accurate to replace $\overline{u}$ in the integral in the expression for I by the constant value $\overline{u}_m$, i.e., to write:

$$\begin{aligned} I(y) &= \int_0^y \overline{u}(R-y)\,dy - \int_0^R \overline{u}(R-y)\,dy \\ &= \overline{u}_m\left[\int_0^y (R-y)\,dy - \int_0^R (R-y)\,dy\right] \\ &= -\frac{\overline{u}_m}{2}(R-y)^2 \end{aligned} \tag{7.50}$$

Substituting this approximate result into Eq. (7.49) then gives:

$$\overline{T} - T_w = -\frac{q_w}{\rho c_p}\int_0^y \frac{(1 - y/R)}{(\epsilon_H + \alpha)}\,dy \tag{7.51}$$

It should clearly be understood that the above approximation is not essential to the analysis. It is used because it leads to a considerable simplification and yet gives results that are of adequate accuracy for most purposes.

Because of the nature of the parameters that are used to describe the distribution of ϵ, it is convenient to write Eq. (7.51) as:

$$\overline{T} - T_w = -\left(\frac{q_w}{\rho c_p}\right)\sqrt{\frac{\rho}{\tau_w}}\int_0^{y^+} \frac{(1 - y/R)}{(E/Pr_T + 1/Pr)}\,dy^+ \tag{7.52}$$

where:

$$E = \frac{\epsilon}{\nu}, \quad y^+ = \frac{y}{\nu}\sqrt{\frac{\tau_w}{\rho}} \tag{7.53}$$

316 Introduction to Convective Heat Transfer Analysis

Now, because the flow is fully developed, the momentum equation can be written as:

$$\frac{1}{r}\frac{\partial}{\partial y}(r\tau) = \frac{dp}{dz} \tag{7.54}$$

where τ is the total shearing stress which is given as before by:

$$\tau = \rho(\nu + \epsilon)\frac{\partial \bar{u}}{\partial y} \tag{7.55}$$

Integrating Eq. (7.54) then gives since the shearing stress is zero on the center line:

$$r\tau = \frac{dp}{dz}\frac{r^2}{2}$$

i.e.:

$$\tau = \frac{dp}{dz}\frac{r}{2} = \frac{1}{2}\frac{dp}{dz}(R-y) \tag{7.56}$$

This shows that there is a linear variation of shear stress across the flow in fully developed pipe flow.

Applying this to conditions at the wall then gives:

$$\tau_w = \frac{R}{2}\frac{dp}{dz} \tag{7.57}$$

a result that was previously derived in a somewhat different way.

Dividing Eq. (7.56) by Eq. (7.57) then gives:

$$\frac{\tau}{\tau_w} = 1 - \frac{y}{R} \tag{7.58}$$

Now Eq. (7.55) can be written as:

$$1 + E = \frac{\tau/\rho\nu}{\partial \bar{u}/\partial y} \tag{7.59}$$

Substituting Eq. (7.58) into Eq. (7.59) then gives on rearrangement:

$$E = \frac{\frac{\tau_w}{\rho\nu}\left(1 - \frac{y}{R}\right)}{\frac{\partial \bar{u}}{\partial y}} - 1 \tag{7.60}$$

Defining as before:

$$u^+ = \bar{u}\sqrt{\frac{\rho}{\tau_w}}, \quad y^+ = \frac{y}{\nu}\sqrt{\frac{\tau_w}{\rho}} \tag{7.61}$$

Eq. (7.60) can be written as:

$$E = \frac{\left(1 - \dfrac{y}{R}\right)}{\dfrac{\partial u^+}{\partial y^+}} - 1 \tag{7.62}$$

Therefore if u^+ is a known function of y^+, the distribution of E can be found using Eq. (7.62) and if the value of Pr_T is assumed to be known, the temperature profile can be determined using Eq. (7.52) and this can then be used to find the Nusselt number. In carrying out this procedure it will here be assumed, based on experimental mean velocity measurements and intuitive reasoning, that the flow can be split into three regions. The three layers will be assumed to extend from $y^+ = 0$ to 5, from $y^+ = 5$ to 30, and from $y^+ = 30$ to the pipe center line. Each of these regions will now be separately discussed.

(i) Inner Layer: $0 < y^+ < 5$:

In this region which lies adjacent to the wall, the turbulent shearing stress and turbulent heat transfer rate will be assumed to be negligible, i.e., it will be assumed that in this region:

$$\epsilon = 0 \text{ and } \epsilon_H = 0$$

Therefore, for this region Eq. (7.52) gives:

$$\overline{T} - T_w = -\left(\frac{q_w}{\rho c_p}\right)\sqrt{\frac{\rho}{\tau_w}} Pr \int_0^{y^+} \left(1 - \frac{y}{R}\right) dy^+ \tag{7.63}$$

Also, because the extent of this region is small, y/R remains very small in this region and $1 - y/R$ is, therefore, effectively equal to 1 in this region. Eq. (7.63) therefore gives:

$$\overline{T} - T_w = -\left(\frac{q_w}{\rho c_p}\right)\sqrt{\frac{\rho}{\tau_w}} Pr \, y^+ \tag{7.64}$$

If $\overline{T}_s$ is the temperature at the outer edge of this layer, i.e., at $y^+ = 5$, then this equation gives:

$$\overline{T}_s - T_w = -5\left(\frac{q_w}{\rho c_p}\right)\sqrt{\frac{\rho}{\tau_w}} Pr \tag{7.65}$$

(ii) Buffer Layer: $5 < y^+ < 30$:

On the basis of available measurements, it will be assumed that in this layer the mean velocity profile is given by:

$$u^+ = 5 \ln y^+ - 3.05 \tag{7.66}$$

Differentiating this equation gives:

$$\frac{\partial u^+}{\partial y^+} = \frac{5}{y^+}$$

318 Introduction to Convective Heat Transfer Analysis

Substituting this result into Eq. (7.62) and assuming that in this region y/R also remains very small and $1 - y/R$, therefore, effectively equal to 1 in this region, gives:

$$E = \frac{y^+}{5} - 1 \tag{7.67}$$

Substituting this equation into Eq. (7.52) and assuming that the turbulent Prandtl number is 1 gives in the buffer layer:

$$\overline{T} - \overline{T}_s = -\left(\frac{q_w}{\rho c_p}\right)\sqrt{\frac{\rho}{\tau_w}} \int_5^{y^+} \frac{dy^+}{\frac{y^+}{5} - 1 + \frac{1}{Pr}} \tag{7.68}$$

it having already been assumed that in this region, region y/R remains very small and that $1 - y/R$ is, therefore, effectively equal to 1.

Carrying out the integration, this equation gives:

$$\overline{T} - \overline{T}_s = -\left(\frac{q_w}{\rho c_p}\right)\sqrt{\frac{\rho}{\tau_w}} \left[5 \ln\left(\frac{\frac{y^+}{5} - 1 + \frac{1}{Pr}}{\frac{1}{Pr}}\right)\right]$$

i.e.:

$$\overline{T} - \overline{T}_s = -5\left(\frac{q_w}{\rho c_p}\right)\sqrt{\frac{\rho}{\tau_w}} \ln\left(\frac{y^+}{5}Pr - Pr + 1\right) \tag{7.69}$$

If the mean temperature at the outer edge of the buffer layer, i.e., at $y^+ = 30$, is $\overline{T}_b$, Eq. (7.69) gives:

$$\overline{T}_b - \overline{T}_s = -5\left(\frac{q_w}{\rho c_p}\right)\sqrt{\frac{\rho}{\tau_w}} \ln(5Pr + 1) \tag{7.70}$$

(iii) Outer Layer: $y^+ > 30$:

In this outer region it is assumed that the velocity profile is given by:

$$u^+ = 2.5 \ln y^+ + 5.5 \tag{7.71}$$

Differentiating this equation gives for the outer layer:

$$\frac{\partial u^+}{\partial y^+} = \frac{2.5}{y^+} \tag{7.72}$$

In this region it can also be assumed that the molecular shearing stress and heat transfer rate are negligible compared to the turbulent components, i.e., that $\epsilon \gg \nu$ and $\epsilon_H \gg \alpha$ and that in this region therefore, because the turbulent Prandtl number is being assumed to be equal to 1, Eqs. (7.52) and (7.62) become:

$$\overline{T} - \overline{T}_b = -\left(\frac{q_w}{\rho c_p}\right)\sqrt{\frac{\rho}{\tau_w}} \int_{30}^{y^+} \frac{\left(1 - \frac{y}{R}\right)}{E} dy^+ \tag{7.73}$$

and

$$E = \frac{(1 - y/R)}{\dfrac{\partial u^+}{\partial y^+}} \tag{7.74}$$

respectively.
Using Eq. (7.72) in Eq. (7.74) then gives:

$$E = \left(1 - \frac{y}{R}\right)\frac{y^+}{2.5} \tag{7.75}$$

Substituting Eq. (7.75) into Eq. (7.73) then gives:

$$\overline{T} - \overline{T}_b = -\left(\frac{q_w}{\rho c_p}\right)\sqrt{\frac{\rho}{\tau_w}}\int_{30}^{y^+}\frac{2.5}{y^+}\,dy^+ \tag{7.76}$$

Carrying out the integration then gives:

$$\overline{T} - \overline{T}_b = -2.5\left(\frac{q_w}{\rho c_p}\right)\sqrt{\frac{\rho}{\tau_w}}\ln\left(\frac{y^+}{30}\right) \tag{7.77}$$

On the center line of the pipe, $y = R$. Hence, if y_1^+ is the value of y^+ on the center line, it follows that:

$$y_1^+ = \frac{R}{\nu}\sqrt{\frac{\tau_w}{\rho}} \tag{7.78}$$

Therefore, if $\overline{T}_c$ is the temperature on the center line, Eq. (7.77) gives:

$$\overline{T}_c - \overline{T}_b = -2.5\left(\frac{q_w}{\rho c_p}\right)\sqrt{\frac{\rho}{\tau_w}}\ln\left(\frac{R}{30\nu}\sqrt{\frac{\tau_w}{\rho}}\right) \tag{7.79}$$

This completes the analysis of the temperature distributions in the three layers. The overall temperature changes are obtained by adding Eqs. (7.65), (7.70), and (7.79) together to eliminate the intermediate temperature $\overline{T}_s$ and $\overline{T}_b$. This gives:

$$T_w - \overline{T}_c = \left(\frac{q_w}{\rho c_p}\right)\sqrt{\frac{\rho}{\tau_w}}\left[2.5\ln\left(\frac{R}{30\nu}\sqrt{\frac{\tau_w}{\rho}}\right) + 5\ln(5Pr + 1) + 5Pr\right] \tag{7.80}$$

In the discussion of the Reynolds analogy given earlier it was shown that:

$$\tau_w = \frac{f}{8}\rho \bar{u}_m^2 \tag{7.26}$$

Hence:

$$\frac{R}{\nu}\sqrt{\frac{\tau_w}{\rho}} = \frac{R\bar{u}_m}{\nu}\sqrt{\frac{f}{8}} = Re_D\sqrt{\frac{f}{32}} \tag{7.81}$$

where Re_D is as before the Reynolds number based on D and on $\bar{u}_m$.

320 Introduction to Convective Heat Transfer Analysis

Using Eqs. (7.26) and (7.81) allows Eq. (7.80) to be written as:

$$\overline{T}_w - \overline{T}_c = \left(\frac{q_w}{\rho c_p \overline{u}_m}\right)\sqrt{\frac{8}{f}}\left[2.5\ln\left(\frac{Re_D}{30}\sqrt{\frac{f}{32}}\right) + 5\ln(5Pr + 1) + 5Pr\right] \quad (7.82)$$

To proceed further, the relation between $(\overline{T}_w - \overline{T}_m)$ and $(\overline{T}_w - \overline{T}_c)$ must be obtained. This could be done by integrating the temperature distributions obtained for each of the three layers. However, sufficient accuracy for most purposes is obtained by using the approximate 1/7th power law, i.e., by assuming that the velocity and temperature distributions are effectively given by:

$$\frac{\overline{u}}{\overline{u}_c} = \left(\frac{y}{R}\right)^{1/7} \quad (7.83)$$

and:

$$\frac{\overline{T}_w - \overline{T}}{\overline{T}_w - \overline{T}_c} = \left(\frac{y}{R}\right)^{1/7} \quad (7.84)$$

where $\overline{u}_c$ is, as before, the mean velocity on the center line of the pipe.

Therefore using:

$$\overline{T}_w - \overline{T}_m = \frac{\int_0^R \overline{u}(\overline{T}_w - \overline{T})2\pi r\, dr}{\int_0^R \overline{u} 2\pi r\, dr}$$

it follows that:

$$\frac{\overline{T}_w - \overline{T}_m}{\overline{T}_w - \overline{T}_c} = \frac{\int_0^1 \left(\frac{y}{R}\right)^{2/7}\left(1 - \frac{y}{R}\right)d\left(\frac{y}{R}\right)}{\int_0^1 \left(\frac{y}{R}\right)^{1/7}\left(1 - \frac{y}{R}\right)\frac{y}{R}d\left(\frac{y}{R}\right)} \quad (7.85)$$

from which it follows that:

$$\frac{\overline{T}_w - \overline{T}_m}{\overline{T}_w - \overline{T}_c} = \frac{5}{6} \quad (7.86)$$

Substituting this into Eq. (7.82) then gives:

$$\overline{T}_w - \overline{T}_m = \frac{5}{6}\left(\frac{q_w}{\rho c_p \overline{u}_m}\right)\sqrt{\frac{8}{f}}\left[2.5\ln\left(\frac{Re_D}{30}\sqrt{\frac{f}{32}}\right) + \ln(5Pr + 1) + 5Pr\right] \quad (7.87)$$

This can be rearranged to give:

$$Nu_D = \frac{Re_D Pr\sqrt{\frac{f}{8}}}{\frac{5}{6}\left[2.5\ln\left(\frac{Re_D}{30}\sqrt{\frac{f}{32}}\right) + 5\ln(5Pr + 1) + 5Pr\right]} \quad (7.88)$$

For any value of Re_D, this equation allows Nu_D to be found by using the methods of finding f that were discussed before.

Eq. (7.88) gives results that are in good agreement with experiment for the flow of fluids with Prandtl numbers between about 0.5 and 30. For fluids with Prandtl numbers outside this range the effects of the various assumptions made in deriving Eq. (7.88), e.g., the neglect of the molecular heat transfer rate in the outer region, render it inaccurate. More refined analyses which overcome these deficiencies have been developed.

EXAMPLE 7.2. Air flows through a smooth pipe with a diameter of 65 mm at a mean velocity of 30 m/s. The walls of the pipe are kept at a temperature of 80°C. At a certain section of the pipe, the mean air temperature is 40°C. Using the three-layer solution, find the value of the heat transfer coefficient and the rate of heat transfer per m length of pipe from the air to the walls at this section.

Solution. At the mean temperature of 40°C assuming that the pressure is near ambient, air has the following properties:

$$\nu = 16.01 \times 10^{-6} \text{ m}^2/\text{s}, \quad k = 0.02638 \text{ W/m-°C}$$

Hence:

$$Re_D = \frac{u_m D}{\nu} = \frac{30 \times 0.065}{0.00001601} = 1.218 \times 10^5$$

The flow is therefore turbulent.

Because the pipe is smooth, f is given by Eq. (7.30), i.e., by:

$$\frac{1}{\sqrt{f}} = 2\log_{10}[Re_D \sqrt{f}] - 0.8 = 2\log_{10}[121800 \sqrt{f}] - 0.8$$

which can be solved to give:

$$f = 0.0173$$

Eq. (7.88) gives:

$$Nu_D = \frac{Re_D Pr \sqrt{\frac{f}{8}}}{\frac{5}{6}\left[2.5 \ln\left(\frac{Re_D}{30}\sqrt{\frac{f}{32}}\right) + 5\ln(5Pr + 1) + 5Pr\right]}$$

hence

$$Nu_D = \frac{121800 \times 0.7 \times \sqrt{\frac{0.0173}{8}}}{\frac{5}{6}\left[2.5 \ln\left(\frac{121800}{30}\sqrt{\frac{0.0173}{32}}\right) + 5\ln(5 \times 0.7 + 1) + 5 \times 0.7\right]} = 211.7$$

Hence:

$$h = \frac{Nu_D k}{D} = \frac{211.7 \times 0.02638}{0.065} = 85.9 \text{ W/m}^2\text{°C}$$

The heat transfer rate per m length of pipe is then given by:

$$Q_w = h\pi D(T_w - T_m) = 85.9 \times \pi \times 0.065 \times (80 - 40) = 701.6 \text{ W/m}$$

Therefore, the heat transfer coefficient is 85.9 W/m² °C and the heat transfer from the pipe walls to the air per m length is 701.6 W.

7.3
THERMALLY DEVELOPING PIPE FLOW

Here the concern is with flow in a long pipe with a long unheated initial section. The unheated section is long enough for the velocity field to become fully developed before the heating begins. When heating begins, the temperature field develops and it is this thermal development region that is considered in this section [21],[22],[23]. Fluid properties will be assumed to be constant and the velocity field will not change in the thermal development region. The flow situation being considered is thus as shown in Fig. 7.7.

Because the velocity profile is fully developed, the radial velocity component, i.e., $\bar{v}$, is zero. The energy equation for the situation here being considered is therefore as given in the previous section, i.e.:

$$\bar{u}\frac{\partial \bar{T}}{\partial z} = \frac{1}{r}\frac{\partial}{\partial r}\left[r(\epsilon_H + \alpha)\frac{\partial \bar{T}}{\partial r}\right] \tag{7.89}$$

the longitudinal flux of heat again having been ignored compared to the radial flux, i.e., the parabolic form of the governing equations having been used.

The mean velocity, $\bar{u}$, is here a function of r alone, i.e., does not vary with z. Eq. (7.89) can be written as:

$$\bar{u}\frac{\partial \bar{T}}{\partial z} = \frac{1}{r}\frac{\partial}{\partial r}\left[r\left(\frac{\epsilon}{Pr_T} + \frac{\nu}{Pr}\right)\frac{\partial \bar{T}}{\partial r}\right] \tag{7.90}$$

It is convenient to write this equation in dimensionless form. For this purpose the following dimensionless variables are introduced:

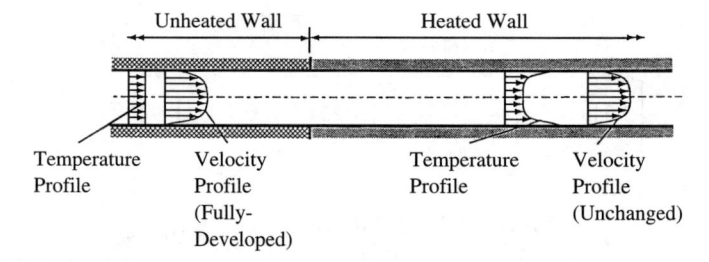

$$R = \frac{r}{D}, \quad U = \frac{\bar{u}}{\bar{u}_m}, \quad E = \frac{\epsilon}{\nu}, \quad Z = \frac{z/D}{Re_D Pr} \tag{7.91}$$

FIGURE 7.7
Thermally developing pipe flow.

Here, D is the diameter of the pipe and Re_D is as usual defined by:

$$Re_D = \frac{\bar{u}_m D}{\nu}$$

$\bar{u}_m$ being the average velocity across the pipe.

Attention will be focused on the case where the wall temperature is uniform and equal to T_w. The following dimensionless temperature is then introduced:

$$\theta = \frac{T - T_i}{T_w - T_i} \tag{7.92}$$

where T_i is the initial temperature of the fluid before the heating begins.

In terms of the variables defined in Eqs. (7.91) and (7.92), Eq. (7.90) becomes:

$$U \frac{\partial \theta}{\partial Z} = \frac{1}{R} \frac{\partial}{\partial R} \left[R \left(E \frac{Pr}{Pr_T} + 1 \right) \frac{\partial \theta}{\partial R} \right] \tag{7.93}$$

The boundary and initial conditions on the solution to this equation for θ are:

$$Z = 0: \theta = 0, \ R = 0.5: \theta = 1, \ R = 0: \frac{\partial \theta}{\partial R} = 0 \tag{7.94}$$

Because the velocity field is fully developed, the variations of U and E with R are known. The solution to Eq. (7.93) can therefore be obtained using a similar procedure to that used in Chapter 4 to solve for thermally developing laminar pipe flow, i.e., using separation of variables. Here, however, a numerical finite-difference solution procedure will be used because it is more easily adapted to the situation where the wall temperature is varying with Z.

The nodal points defined in Fig. 7.8 are used. The following finite-difference approximations are introduced, it being noted that U does not change in the i-direction:

$$U \left. \frac{\partial \theta}{\partial Z} \right|_{i,j} = U_j \left[\frac{\theta_{i,j} - \theta_{i-1,j}}{\Delta Z} \right] \tag{7.95}$$

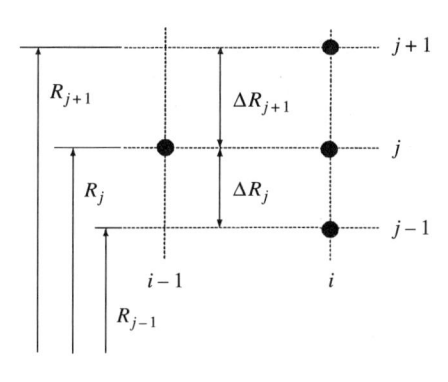

FIGURE 7.8
Nodal points used in obtaining finite-difference solution.

324 Introduction to Convective Heat Transfer Analysis

$$\frac{1}{R}\frac{\partial}{\partial R}\left[R\left(E\frac{Pr}{Pr_T}+1\right)\frac{\partial \theta}{\partial R}\right]\Big|_{i,j} = \frac{1}{R_j}\frac{2}{\Delta R_{j+1}+\Delta R_j}$$
$$\left\{\left[\frac{R_{j+1}(E_{j+1}Pr/Pr_T+1)+R_j(E_jPr/Pr_T+1)}{2}\right]\left(\frac{\theta_{i,j+1}-\theta_{i,j}}{\Delta R_{j+1}}\right)\right.$$
$$\left.-\left[\frac{R_j(E_jPr/Pr_T+1)+R_{j-1}(E_{j-1}Pr/Pr_T+1)}{2}\right]\left(\frac{\theta_{i,j}-\theta_{i,j-1}}{\Delta R_j}\right)\right\} \quad (7.96)$$

Substituting these finite-difference approximations into Eq. (7.93) gives an equation of the form:

$$A_j\theta_{i,j} + B_j\theta_{i,j+1} + C_j\theta_{i,j-1} = D_j \quad (7.97)$$

where the coefficients are given by

$$A_j = \frac{U_j}{\Delta Z} + \frac{1}{R_j}\frac{2}{\Delta R_{j+1}+\Delta R_j}\left\{\right.$$
$$\left[\frac{R_{j+1}(E_{j+1}Pr/Pr_T+1)+R_j(E_jPr/Pr_T+1)}{2}\right]\left(\frac{1}{\Delta R_{j+1}}\right)$$
$$+\left[\frac{R_j(E_jPr/Pr_T+1)+R_{j-1}(E_{j-1}Pr/Pr_T+1)}{2}\right]\left(\frac{1}{\Delta R_j}\right)\right\} \quad (7.98)$$

$$B_j = -\frac{1}{R_j}\frac{2}{\Delta R_{j+1}+\Delta R_j}\left\{\right.$$
$$\left.\left[\frac{R_{j+1}(E_{j+1}Pr/Pr_T+1)+R_j(E_jPr/Pr_T+1)}{2}\right]\left(\frac{1}{\Delta R_{j+1}}\right)\right\} \quad (7.99)$$

$$C_j = -\frac{1}{R_j}\frac{2}{\Delta R_{j+1}+\Delta R_j}\left\{\right.$$
$$\left.\left[\frac{R_j(E_jPr/Pr_T+1)+R_{j-1}(E_{j-1}Pr/Pr_T+1)}{2}\right]\left(\frac{1}{\Delta R_j}\right)\right\} \quad (7.100)$$

$$D_j = \frac{U_j\theta_{i-1,j}}{\Delta Z} \quad (7.101)$$

The boundary conditions give $\theta_{i,2} = \theta_{i,1}$ and $\theta_{i,N} = 1$. The use of these boundary conditions together with the application of Eq. (7.97) to all of the "internal" points on the i-line, i.e., $j = 2, 3, 4, \ldots, N-2, N-1$, gives a set of N equations in the N unknown values of θ. This set of equations has the following form:

$$\theta_{i,1} - \theta_{i,2} = 0$$
$$A_2\theta_{i,2} + B_2\theta_{i,3} + C_2\theta_{i,1} = D_2$$
$$A_3\theta_{i,3} + B_3\theta_{i,4} + C_3\theta_{i,2} = D_3$$
$$\vdots \quad (7.102)$$
$$A_{N-1}\theta_{i,N-1} + B_{N-1}\theta_{i,N} + C_{N-1}\theta_{i,N-2} = D_{N-1}$$
$$\theta_{i,N} = 1$$

i.e., has the form:

$$\begin{bmatrix} 1 & -1 & 0 & 0 & 0 & \cdots & 0 & 0 & 0 & \\ C_2 & A_2 & B_2 & 0 & 0 & \cdots & 0 & 0 & 0 & \\ 0 & C_3 & A_3 & B_3 & 0 & \cdots & 0 & 0 & 0 & 0 \\ 0 & 0 & C_4 & A_4 & B_4 & \cdots & 0 & 0 & 0 & \\ \vdots & \vdots & \vdots & \vdots & \vdots & \ddots & \vdots & \vdots & \vdots & \\ 0 & 0 & 0 & 0 & 0 & \cdots & C_{N-1} & A_{N-1} & B_{N-1} & \\ 0 & 0 & 0 & 0 & 0 & \cdots & 0 & 0 & 1 & \end{bmatrix} \begin{bmatrix} \theta_{i,1} \\ \theta_{i,2} \\ \theta_{i,3} \\ \theta_{i,4} \\ \vdots \\ \theta_{i,N-1} \\ \theta_{i,N} \end{bmatrix} = \begin{bmatrix} 0 \\ D_2 \\ D_3 \\ D_4 \\ \vdots \\ D_{N-1} \\ 1 \end{bmatrix}$$

i.e.,

$$Q\theta_{i,j} = R \tag{7.103}$$

where Q is a tridiagonal matrix.

This equation can be solved using the standard tridiagonal matrix solver algorithm.

The local heat transfer rate at any value of Z is obtained by noting that:

$$q_w = k \left. \frac{\partial \overline{T}}{\partial r} \right|_{r=R}$$

i.e.:

$$\frac{q_w D}{k(T_w - T_i)} = \left. \frac{\partial \theta}{\partial R} \right|_{R=0.5}$$

i.e.:

$$Nu_{iD} = \left. \frac{\partial \theta}{\partial R} \right|_{R=0.5} \tag{7.104}$$

where Nu_{iD} is the local Nusselt number based on the difference between the wall and the inlet temperatures, i.e.:

$$Nu_{iD} = \frac{q_w D}{k(T_w - T_i)} \tag{7.105}$$

But:

$$\left. \frac{\partial \theta}{\partial R} \right|_{i,1} = \frac{\theta_{i,N} - \theta_{1,N-1}}{\Delta R_N} \tag{7.106}$$

Substituting Eq. (7.106) into Eq. (7.104) then gives the following:

$$Nu_{iD} = \frac{\theta_{i,N} - \theta_{1,N-1}}{\Delta R_N} \tag{7.107}$$

Because pipe flow is being considered, it is more convenient to use a Nusselt number based on the difference between the wall and local average temperatures, i.e., to use:

$$Nu_D = \frac{q_w D}{k(T_w - \overline{T}_m)} \tag{7.108}$$

Comparing Eqs. (7.108) and (7.105) shows that:

$$Nu_D = Nu_{iD}\frac{T_w - T_i}{T_w - \bar{T}_m} = \frac{Nu_{iD}}{\theta_m} \qquad (7.109)$$

where:

$$\theta_m = \frac{T_w - T_i}{T_w - \bar{T}_m} \qquad (7.110)$$

Now:

$$\bar{T}_m = \frac{\int_0^{D/2} \bar{u}\bar{T}r\,dr}{\int_0^{D/2} \bar{u}r\,dr}$$

from which it follows that:

$$\theta_m = \int_0^{0.5} U\theta\,dR \bigg/ \int_0^{0.5} UR\,dR \qquad (7.111)$$

θ_m can therefore be obtained using the numerically determined variation of θ with R at any Z value. The value of Nu_D at this value of Z can then be determined. In this way the variation of Nu_D with Z can be obtained.

In order to use the above solution procedure, the variations of U and $E\,(=\epsilon/\nu)$ across the flow must be specified. Here, the distribution of E in the flow will be assumed to be described by the following set of equations:

$$y^+ < 5:\ E = 0$$

$$5 \leq y^+ \leq 30:\ E = \frac{y^+}{5} - 1 \qquad (7.112)$$

$$y^+ > 30:\ E = 1.6R(y^+)^{6/7} - 1$$

where:

$$y^+ = \frac{y}{\nu}\sqrt{\frac{\tau_w}{\rho}} = (0.5 - R)Re_D\sqrt{\frac{f}{8}}$$

where f is the friction factor. It will here be assumed, as discussed in the previous section, that:

$$f = \frac{0.305}{Re_D^{0.25}} \qquad (7.31)$$

this being applicable for flow in smooth pipes for Reynolds numbers less than about 10^5.

The above equations describe the variation of E with R.

It will be assumed that the mean velocity distribution is adequately described by:

$$\frac{\bar{u}}{\bar{u}_c} = \left(\frac{D - 2r}{D}\right)^{1/7} \qquad (7.113)$$

where $\bar{u}_c$ is the mean center-line velocity.

Now recalling that:

$$\bar{u}_m = \frac{8}{D^2} \int_0^{D/2} \bar{u} r \, dr$$

it follows that Eq. (7.32) gives:

$$\frac{\bar{u}_m}{\bar{u}_c} = 8 \int_0^{0.5} (1 - 2R)^{1/7} R \, dR = \frac{49}{60} \quad (7.114)$$

Using this result, Eq. (7.113) gives:

$$U = \frac{60}{49}(1 - 2R)^{1/7} \quad (7.115)$$

This equation describes the variation of U with R.

A computer program, TURDUCD, written in FORTRAN, that implements the procedure is available in the way discussed in the Preface. Some results obtained using this program are shown in Figs. 7.9 and 7.10.

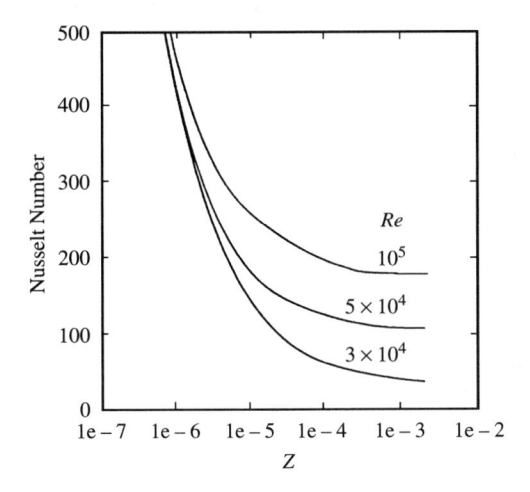

FIGURE 7.9
Variation of Nusselt number with Z in thermal entrance region for $Pr = 0.7$ for various values of Re_D.

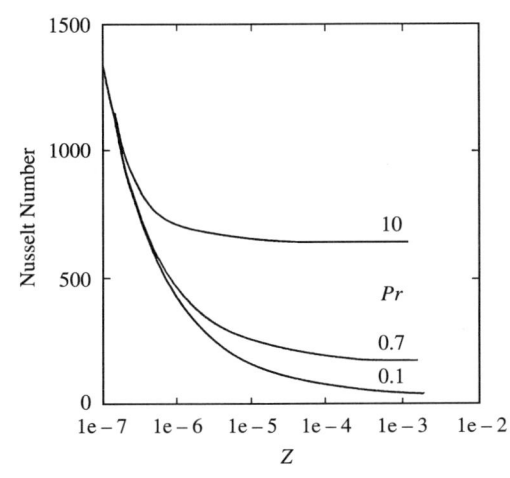

FIGURE 7.10
Variation of Nusselt number with Z in thermal entrance region for $Re_D = 10^5$ for various values of Pr.

EXAMPLE 7.3. Air flows through a long 5-cm diameter pipe at such a velocity that the Reynolds number is 10^5. The air enters the pipe at a temperature of 20°C. The first portion of the pipe is unheated and the velocity profile becomes fully developed in this portion of the pipe. The second portion of the pipe, which is heated to a uniform wall temperature of 40°C, has a length of 1.5 m. Determine how the wall heat transfer rate varies with z in the heated portion of the pipe.

Solution. The flow situation being considered is shown in Fig. E7.3a.

z is measured from the beginning of the heated section and the maximum value of Z is, assuming that $Pr = 0.7$, therefore given by:

$$Z_{max} = \frac{z_{max}/D}{Re_D Pr} = \frac{1.5/0.05}{10^5 \times 0.7} = 0.0004286$$

The computer program, TURDUCD, discussed above has therefore been run with the following inputs:

$$Z_{max} = 0.0004286, \; Pr = 0.7, \; Re_D = 100,000$$

and because the wall temperature is uniform, the wall boundary condition parameter has been set equal to 1 (the program also gives results for the case where the wall heat flux is uniform).

The program gives the variations of two Nusselt numbers with Z, these being $NuDi$, which is Nu_{iD}, and $NuDa$ which is Nu_D. Now:

$$Nu_{iD}(= NuDi) = \frac{q_w D}{k(T_w - T_i)}$$

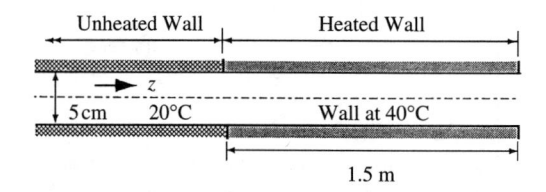

FIGURE E7.3a

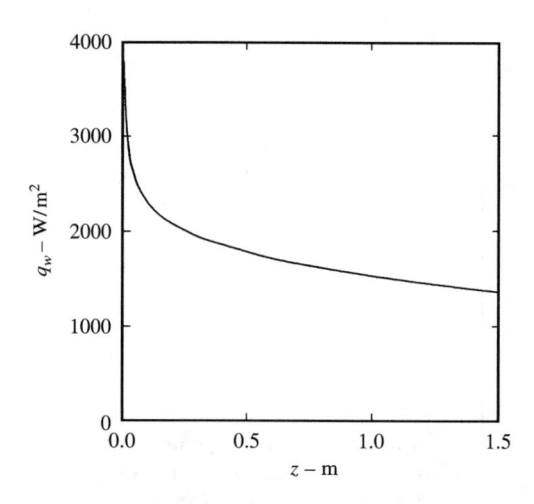

FIGURE E7.3b

If k is evaluated at $(20 + 40)/2 = 30°C$, it is equal to 0.02638 W/m-°C. Therefore, since $T_w = 40°C$ and $T_i = 20°C$, it follows that:

$$q_w = Nu_{iD}\frac{k(T_w - T_i)}{D} = Nu_{iD}\frac{0.02638(40 - 20)}{0.05} = 10.55 Nu_{iD} \text{W/m}^2$$

Also:

$$z = ZDRe_D Pr = Z \times 0.05 \times 100000 \times 0.7 = 3500Z \text{ m}$$

Therefore, using the calculated variation of Nu_{iD} with Z, the variation of q_w with z can be determined. The variation so found is shown in Fig. E7.3b.

7.4 DEVELOPING FLOW IN A PLANE DUCT

The previous section was concerned with a flow in which only the temperature field was developing, the velocity field having reached the fully developed state before the heating began. In general, however, both the velocity and temperature fields develop simultaneously [24],[25]. In order to illustrate the nature of such flows, developing two-dimensional flow in a plane duct will be considered here. The flow situation considered is shown in Fig. 7.11.

It will be assumed here that the flow enters the duct through a "shaped" unheated inlet section and that the velocity and temperature are therefore uniform across the inlet plane as illustrated in Fig. 7.12.

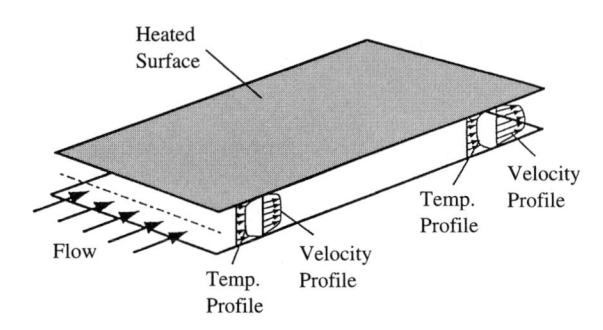

FIGURE 7.11
Flow in a plane duct.

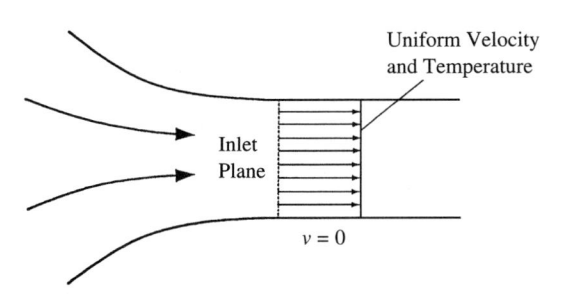

FIGURE 7.12
Assumed inlet plane conditions.

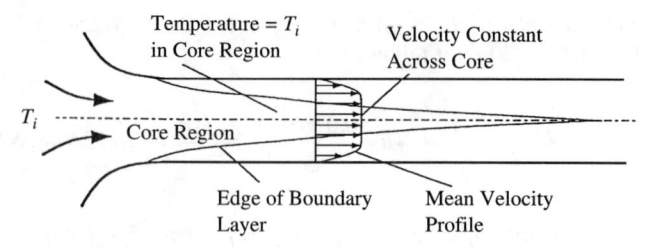

FIGURE 7.13
Assumed form of velocity distribution.

The flow in the development region initially consists essentially of a boundary layer on each wall with a constant velocity, uniform temperature core between these two boundary layers as illustrated in Fig. 7.13. These boundary layers grow until they meet on the center line of the duct. Following this, there is a region where the flow near the center line adjusts from that in the outer part of a boundary layer to that in a fully developed duct flow. However, the changes in this second region are relatively small and only the first boundary layer region will be considered here. The analysis presented here is based on the use of the integral equation method discussed in Chapter 6.

As the boundary layers grow on the wall, the velocity near the wall is decreased and, as a consequence, the velocity in the core region, u_1, increases. Because the velocity on the inlet plane is uniform, the velocity on the inlet plane must be equal to the mean velocity, $\bar{u}_m$, in the duct. Continuity therefore requires, assuming that the density is constant, that:

$$\rho \bar{u}_m W = \rho \int_0^W \bar{u}\, dy \tag{7.116}$$

where W is the width of the duct.

Assuming that the flow is symmetrical about the center line of the duct and that the velocity is uniform in the core region between the two boundary layers as shown in Fig. 7.13, Eq. (7.116) can be written as:

$$\rho \bar{u}_m W = 2\rho \left[\int_0^\delta \bar{u}\, dy + \left(\frac{W}{2} - \delta\right) u_1 \right]$$

i.e.:

$$\rho \bar{u}_m W = 2\rho \left[\frac{W}{2} u_1 - \int_0^\delta (u_1 - \bar{u})\, dy \right]$$

i.e.:

$$W u_1 - 2\int_0^\delta (u_1 - \bar{u})\, dy = \bar{u}_m W \tag{7.117}$$

Now the quantity:

$$\delta_1 = u_1 \int_0^\delta \left(1 - \frac{u}{u_1}\right) dy \qquad (7.118)$$

was defined in Chapter 6 as the boundary layer displacement thickness. Therefore, Eq. (7.117) can be written in terms of the displacement thickness as:

$$u_1(W - 2\delta_1) = \bar{u}_m W \qquad (7.119)$$

The right-hand side of this equation is a constant so this equation gives:

$$(W - 2\delta_1)\frac{du_1}{dz} - 2u_1 \frac{d\delta_1}{dz} = 0$$

i.e.:

$$\frac{1}{u_1}\frac{du_1}{dz} = \frac{2}{(W - 2\delta_1)} \frac{d\delta_1}{dz} \qquad (7.120)$$

This equation relates the gradient of the velocity in the core region to the rate of growth of the boundary layer displacement thickness.

The integral equation analysis given in Chapter 6 solved for the boundary layer momentum thickness, δ_2, which is related to the displacement thickness by the form factor, H, which is defined by:

$$H = \frac{\delta_1}{\delta_2} \qquad (7.121)$$

Eq. (7.120) can therefore be written as:

$$\frac{1}{u_1}\frac{du_1}{dz} = \frac{2}{(W - 2H\delta_2)} \frac{d}{dz}(H\delta_2)$$

i.e.:

$$\frac{1}{u_1}\frac{du_1}{dz} = \frac{2}{(W - 2H\delta_2)}\left[\delta_2 \frac{dH}{dz} + H \frac{d\delta_2}{dz}\right] \qquad (7.122)$$

Now the equations used in the integral equation analysis given in Chapter 6 can be written as:

$$\frac{d\delta_2}{dz} + \delta_2(2 + H)\frac{1}{u_1}\frac{du_1}{dz} = \frac{\tau_w}{\rho u_1^2} \qquad (7.123)$$

$$\frac{\tau_w}{\rho u_1^2} = \frac{0.123 \times 10^{-0.678H}}{(u_1\delta_2/\nu)^{0.268}} \qquad (7.124)$$

$$\left(\frac{u_1\delta_2}{\nu}\right)^{1/6} \delta_2 \frac{dH}{dz} = e^{5(H-1.4)}\left[-\left(\frac{u_1\delta_2}{\nu}\right)^{1/6} \delta_2 \frac{1}{u_1}\frac{du_1}{dz} - 0.0135(H - 1.4)\right] \qquad (7.125)$$

332 Introduction to Convective Heat Transfer Analysis

Eqs. (7.122) to (7.125) can be simultaneously solved to give the variations of δ_2, H, and u_1 with z. The analogy solutions can then be used to obtain the heat transfer rate, the Reynolds analogy, for example, giving:

$$\frac{T_w - T_i}{u_1} = \frac{q_w}{c_p \tau_w} \qquad (7.126)$$

it being noted that the temperature in the core region between boundary layer will be equal to the temperature on the inlet plane, i.e., T_i. It will be recalled that Eq. (7.126) was derived using the assumption that Pr is equal to 1.

The solution procedure outlined above applies until the boundary layer reaches the center line of the pipe, i.e., when:

$$\delta = \frac{W}{2} \qquad (7.127)$$

If a power law velocity distribution is assumed in the boundary layer, i.e., if it is assumed that:

$$\frac{\bar{u}}{u_1} = \left(\frac{y}{\delta}\right)^{1/n}$$

then using:

$$\delta_2 = \int_0^\delta \left(1 - \frac{\bar{u}}{u_1}\right)\frac{\bar{u}}{u_1} \, dy$$

it follows that:

$$\frac{\delta_2}{\delta} = \int_0^1 \left[\left(\frac{y}{\delta}\right)^{1/n} - \left(\frac{y}{\delta}\right)^{2/n}\right] d\frac{y}{\delta}$$

i.e., that:

$$\frac{\delta_2}{\delta} = \frac{n}{(n+1)(n+2)} \qquad (7.128)$$

Similarly, using the definition of the displacement thickness it follows that:

$$\frac{\delta_1}{\delta} = \int_0^1 \left[1 - \left(\frac{y}{\delta}\right)^{1/n}\right] d\frac{y}{\delta}$$

i.e., that:

$$\frac{\delta_1}{\delta} = \frac{1}{(n+1)} \qquad (7.129)$$

Dividing Eq. (7.129) by Eq. (7.128) gives:

$$H = \frac{(n+2)}{n}$$

i.e.:

$$n = \frac{2}{H-1}$$

Substituting this into Eq. (7.128) then gives:

$$\frac{\delta_2}{\delta} = \frac{H-1}{H(H+1)} \tag{7.130}$$

For any value of H, Eq. (7.130) allows δ to be found provided the value of δ_2 has been determined. Eq. (7.127) can then be used to determine if the boundary layer has reached the center line.

It is convenient to write the above equations in dimensionless form before obtaining the solution. For this purpose, the following dimensionless variables are introduced:

$$U = \frac{u_1}{u_m}, \quad Z = \frac{z}{W}, \quad \Delta_2 = \frac{\delta_2}{W}, \quad \Delta = \frac{\delta}{W} \tag{7.131}$$

Eqs. (7.122) to (7.125) can be written in terms of these variables as:

$$\frac{1}{U}\frac{dU}{dZ} = \frac{2}{(1-2H\Delta_2)}\left[\Delta_2 \frac{dH}{dZ} + H\frac{d\delta_2}{dz}\right] \tag{7.132}$$

$$\frac{d\Delta_2}{dZ} + \Delta_2(2+H)\frac{1}{U}\frac{dU}{dZ} = \frac{\tau_w}{\rho u_1^2} \tag{7.133}$$

$$\frac{\tau_w}{\rho u_1^2} = \frac{0.123 \times 10^{-0.678H}}{Re_w^{0.268}(U\Delta_2)^{0.268}} \tag{7.134}$$

$$Re_W^{1/6}(U\Delta_2)^{1/6}\Delta_2\frac{dH}{dZ} = e^{5(H-1.4)}\left[-Re_W^{1/6}(U\Delta_2)^{1/6}\frac{\Delta_2}{U}\frac{dU}{dZ} - 0.0135(H-1.4)\right] \tag{7.135}$$

where:

$$Re_W = \frac{u_m W}{\nu} \tag{7.136}$$

Eq. (7.127) indicates that the solution must be ended when:

$$\Delta = 0.5 \tag{7.137}$$

and Eq. (7.130) gives:

$$\Delta = \frac{H(H+1)}{(H-1)}\Delta_2 \tag{7.138}$$

Eq. (7.126) gives the heat transfer rate as:

$$Nu_W = \frac{\tau_w}{\rho u_1^2} U Re_W Pr^{0.4} \tag{7.139}$$

an approximate correction to account for the fact that Pr is not equal to 1 having been applied. Here:

$$Nu_W = \frac{q_w W}{k(T_w - T_i)} \text{ and } Re_W = \frac{u_m W}{\nu} \qquad (7.140)$$

The mean temperature, T_m, across any section of the duct is given by:

$$u_m(T_m - T_i)W = 2 \int_0^{W/2} \bar{u}(\bar{T} - T_i)\,dy$$

i.e., since T is equal to T_i outside the boundary layer:

$$u_m(T_m - T_i)W = 2 \int_0^{\delta} \bar{u}(\bar{T} - T_i)\,dy$$

i.e.:

$$\frac{T_m - T_i}{T_w - T_i} = 2\Delta \int_0^1 U\left(\frac{\bar{T} - T_i}{T_w - T_i}\right) d\left(\frac{y}{\delta}\right) \qquad (7.141)$$

Again assuming that:

$$\frac{\bar{u}}{u_1} = \left(\frac{y}{\delta}\right)^{1/n}$$

and also assuming:

$$\frac{\bar{T} - T_i}{T_w - T_i} = 1 - \left(\frac{y}{\delta}\right)^{1/n}$$

Eq. (7.141) gives:

$$\frac{T_m - T_i}{T_w - T_i} = \frac{2n}{(n+1)(n+2)} \Delta \qquad (7.142)$$

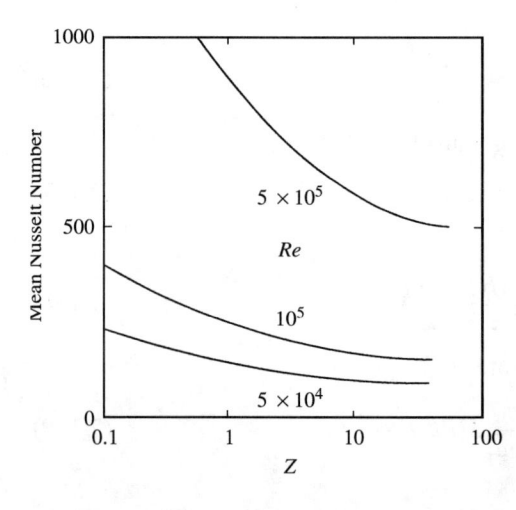

FIGURE 7.14
Variation of Nu_m with Z in entrance region for $Pr = 0.7$ for various values of Re_W.

Therefore, since:

$$n = \frac{2}{H-1}$$

Eq. (7.142) allows $(T_m - T_i)/(T_w - T_i)$ to be found. The Nusselt number based on $(T_w - T_m)$ can then be found using:

$$Nu_m = Nu_W \frac{T_w - T_i}{T_w - T_m} = \frac{Nu_W}{1 - (T_m - T_i)/(T_w - T_i)}$$

A computer program, TURINDEV, that implements this procedure is available as discussed in the Preface. This program is a modified version of that discussed in Chapter 6 for the calculation of external boundary layer flows.

Some typical results obtained using this procedure are shown in Fig. 7.14.

The values of du_1/dz are relatively small in the entrance region and results obtained, assuming that H is constant and equal to its flat plate value, agree quite closely with those obtained allowing for variations in H.

EXAMPLE 7.4. If the length of the entrance region is taken as the distance from the inlet at which the Nusselt number based on the difference between the wall and the mean fluid temperatures reaches within 1% of its fully developed value, use the program discussed above to determine how the length of the entrance region varies with Reynolds number for the flow of air in a plane duct.

Solution. The computer program discussed above has been run for Reynolds numbers between 10^4 and 3×10^5 for a Prandtl number of 0.7. The calculated variation of Nu_m with Z was then considered. The final value of Nu_m was multiplied by 1.01 and the results scanned to find the value of Z at which Nu_m was 1.01 times the final value. This value of Z was taken as z_{ent}/W, z_{ent} being the assumed entrance length, i.e., the distance from the inlet at which Nu_m is within 1% of its final value.

The results obtained in this way are shown in Fig. E7.4.

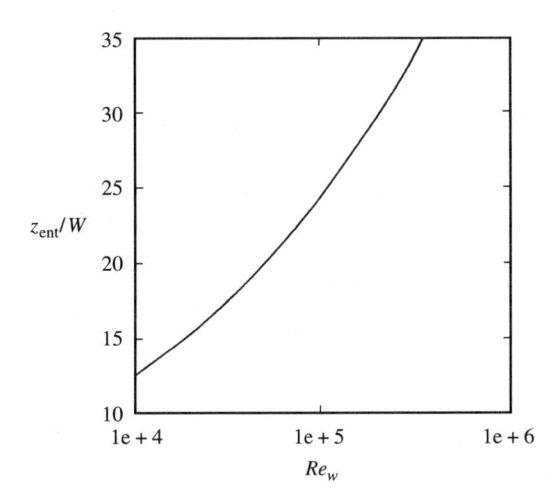

FIGURE E7.4
These results can be approximately represented by: $z_{ent}/W = 0.95 Re_w^{0.28}$

7.5
SOLUTIONS TO THE FULL GOVERNING EQUATIONS

Attention has in this chapter been devoted to fully developed turbulent duct flows or to flows in which the parabolic form of the governing equations can be used to obtain the solution. While the solutions obtained using these assumptions are applicable in many situations of great practical importance, there are many other important problems that cannot be adequately dealt with using these assumptions, see References [26] to [36]. For example, consider two-dimensional flow in a plane duct with a rectangular heated block mounted on one wall of the duct as shown in Fig. 7.15, this being a model of some situations that arise in electronic cooling. Because large regions of recirculating flow can occur in such a situation, this flow cannot be adequately treated using the equations given in this chapter. Instead, the heat transfer rates in such flows must be obtained by numerically solving the full

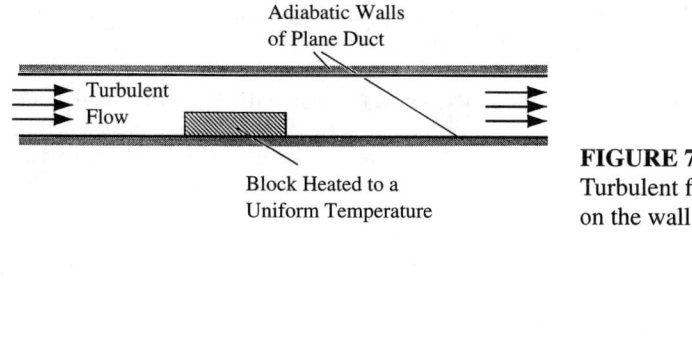

FIGURE 7.15
Turbulent flow over a heated block on the wall of a plane duct.

STREAMLINE CONTOUR PLOT

LEGEND

A - .5000E + 01
B - .1000E + 02
C - .1500E + 02
D - .2000E + 02
E - .2500E + 02
F - .3000E + 02
G - .3100E + 02

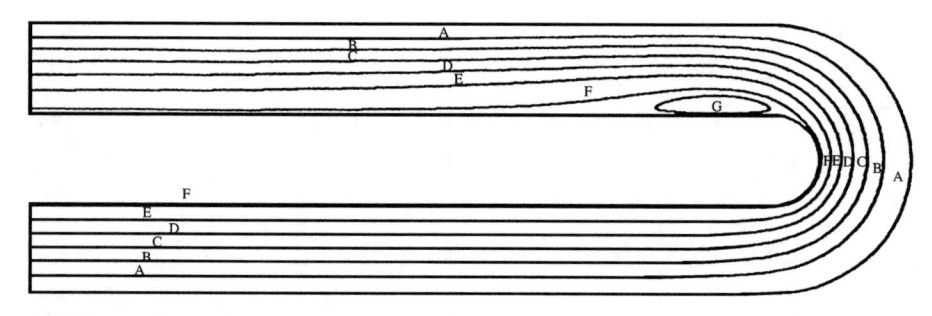

FIGURE 7.16
Streamlines for turbulent flow in an axi-symmetric turn-around duct. (Reproduced with permission of Fluid Dynamics International)

7.6
CONCLUDING REMARKS

Some simple methods of determining heat transfer rates to turbulent flows in a duct have been considered in this chapter. Fully developed flow in a pipe was first considered. Analogy solutions for this situation were discussed. In such solutions, the heat transfer rate is predicted from a knowledge of the wall shear stress. In fully developed pipe flow, the wall shear stress is conventionally expressed in terms of the friction factor and methods of finding the friction factor were discussed. The Reynolds analogy was first discussed. This solution really only applies to fluids with a Prandtl number of 1. A three-layer analogy solution which applies for all Prandtl numbers was then discussed.

Attention was then turned to developing duct flows. A numerical solution for thermally developing flow in a pipe was first considered. Attention was then turned to plane duct flow when both the velocity and temperature fields are simultaneously developing. An approximate solution based on the use of the boundary layer integral equations was discussed.

PROBLEMS

7.1. The walls of a smooth 75-mm diameter pipe are kept at a uniform temperature of 50°C. Air flows through the pipe at a mean velocity of 35 m/s. At a certain section of the pipe, the mean air temperature is 10°C. Using the Reynolds analogy and assuming that the flow in the pipe is fully developed, find the value of the heat transfer coefficient and the rate of heat transfer per m length of pipe from the walls of the pipe to the air at this section.

7.2. The walls of the 50-mm diameter pipe are kept at a uniform temperature of 70°C. Air flows through the pipe at a mean velocity of 40 m/s. At a certain section of the pipe, the mean air temperature is 30°C. Assuming that the flow in the pipe is fully developed, find the heat transfer coefficient and the rate of heat transfer per m length of pipe from the pipe to the air at this section using both the Reynolds analogy and the three-layer solution.

7.3. Assuming that if y is the distance from the wall of a pipe, the mean velocity and mean temperature profiles are given by:

$$\frac{\bar{u}}{\bar{u}_c} = \left(\frac{y}{R}\right)^{1/7}$$

and:

$$\frac{\overline{T} - \overline{T}_c}{\overline{T}_w - \overline{T}_c} = 1 - \left(\frac{y}{R}\right)^{1/7}$$

where R is the radius of the pipe and subscript c refers to conditions on the center line, derive expressions for $\bar{u}_m/u_c$ and $(\overline{T}_m - \overline{T}_c)/(\overline{T}_w - \overline{T}_c)$. Also recalling that in fully developed pipe flow:

$$\frac{\tau}{\tau_w} = 1 - \frac{y}{R}$$

find the variations of ϵ/ν and ϵ_H/ν across the flow. Assume that the turbulent Prandtl number is 0.9.

7.4. Use the Reynolds analogy to derive an expression for the Nusselt number for fully developed turbulent flow in an annulus in which the inner wall is heated to a uniform temperature and the outer wall is adiabatic. Assume that the friction factor can be derived by introducing the hydraulic diameter concept.

7.5. The so-called Taylor-Prandtl analogy was applied to boundary layer flow in Chapter 6. Use this analogy solution to derive an expression for the Nusselt number in fully developed turbulent pipe flow.

7.6. Consider the flow of water through a 65-mm diameter smooth pipe at a mean velocity of 4 m/s. The walls of the pipe are kept at a uniform temperature of 40°C and at a certain section of the pipe the mean water temperature is 30°C. Find the heat transfer coefficient for this situation using both the Reynolds analogy and the three-layer analogy solution.

7.7. Consider thermally developing flow in a smooth 60-mm diameter pipe. Air, at an initial temperature of 10°C, flows through this pipe, the mean air velocity being 30m/s. The first portion of the pipe is unheated and the velocity profile becomes fully developed in this portion of the pipe. The second portion of the pipe, which has a length of 2 m, is heated to a uniform temperature of 50°C. Determine how the wall heat transfer rate in W/m^2 varies with distance along the heated portion of the pipe.

7.8. If the length of the thermal entrance region in turbulent pipe flow is taken as the distance from the beginning of the heated section at which the Nusselt number based on the difference between the wall and the mean fluid temperatures reaches to within 1% of its fully developed value, use the computer program for thermally developing flow to determine how the length of the thermal entrance region varies with Reynolds number for the flow of air.

7.9. Air at a temperature of 10°C enters a plane duct with a distance of 6 cm between the two surfaces of the duct. The mean velocity of the air in the duct is 40 m/s. The walls of the duct are maintained at a uniform temperature of 40°C. Assuming that the velocity and temperature distributions are uniform at the entrance to the duct, determine how the mean heat transfer coefficient varies with distance along the duct.

7.10. Consider the fully developed flow at a mean velocity of 9 m/s through a 5-cm diameter smooth pipe. There is a uniform heat flux at the pipe wall. Find the heat-transfer coefficients for the following cases and discuss the reasons for the differences between the values obtained: (a) air at a mean temperature of 80°C and standard atmospheric pressure, (b) air at a mean temperature of 80°C and a pressure that is ten times the standard atmospheric pressure, (c) water at a mean temperature of 80°C.

7.11. Water flows at a rate of 1 kg/s through a 3-cm diameter smooth pipe. The water enters the pipe at a temperature of 15°C and must leave the the pipe at a mean temperature of 50°C. The pipe wall is heated in such a way that the wall temperature is 14°C higher than the local mean water temperature at all points along the pipe. What length of pipe is required? Ignore entrance region effects.

7.12. Water flows at a mean velocity of 5 m/s through a tube in a condenser that has a diameter of 3 cm and a length of 1.2 m. Because steam is condensing on the tube, its wall surface temperature is constant and equal to 90°C. If the water enters the tube at a temperature of 35°C, what will be its mean temperature at the tube exit? Assume that the tube is smooth and ignore entrance region effects.

7.13. Water flows at a mean velocity of 1 m/s through a 1-cm diameter pipe, the mean water temperature being 20°C. The flow can be assumed to be fully developed and turbulent. If the thickness of the sublayer is given by $y^+ = 12$, find the actual thickness of the sublayer in mm.

7.14. Water flows through a long smooth 2.7-cm diameter pipe. A uniform heat flux of 10 kW/m^2 is applied at the pipe walls and the difference between the wall and the mean water temperature is 5°C. If the mean water temperature is 40°C, find the mean water velocity in the pipe.

REFERENCES

1. Schlichting, H., *Boundary Layer Theory*, 7th ed., McGraw-Hill, New York, 1979.
2. Dipprey, D.F. and Sabersky, R.H., "Heat and Momentum Transfer in Smooth and Rough Tubes at Various Prandtl Numbers", *Int. J. Heat Mass Transfer*, Vol. 5, pp. 329–353, 1963.
3. Azer, N.Z., and Chao, B.T., "Turbulent Heat Transfer in Liquid Metals—Fully Developed Pipe Flow with Constant Wall Temperature", *Int. J. Heat Mass Transfer*, Vol. 3, p. 77, 1961.
4. Reynolds, W.C., "Turbulent Heat Transfer in a Circular Tube with Variable Circumferential Heat Flux", *Int. J. Heat Mass Transfer*, Vol. 6, pp. 445–454, 1963.
5. Schleicher, C.A. and Tribus, M., "Heat Transfer in a Pipe with Turbulent Flow and Arbitrary Wall-Temperature Distribution", *Trans. ASME*, vol. 79, pp. 789–797, 1957.
6. Cope, W.F., "Friction and Heat Transmission Coefficients of Rough Pipes", *Proc. Inst. Mech. Eng.*, Vol. 145, p. 99, 1941.
7. Gnielinski, V., "New Equations for Heat and Mass Transfer in Turbulent Pipe and Channel Flow", *Int. Chem. Eng.*, Vol. 16, pp. 359–368, 1976,
8. Lee, S., "Liquid Metal Heat Transfer in Turbulent Pipe Flow with Uniform Wall Flux", *Int. J. Heat Mass Transfer*, Vol. 26, pp. 349–356, 1983.
9. Mori, Y. and Futagami, K., "Forced Convective Heat Transfer in Uniformly Heated Horizontal Tubes (2nd Report, Theoretical Study)", *Int. J. Heat Mass Transfer*, Vol. 10, pp. 1801–1813, 1967.

10. Seban, R.A., "Heat Transfer to a Fluid Flowing Turbulently Between Parallel Walls with Asymmetric Wall Temperatures", *Trans. ASME,* Vol. 72, p. 789, 1950.
11. Deissler, R.G., "Investigation of Turbulent Flow and Heat Transfer in Smooth Tubes Including the Effect of Variable Properties", *Trans. ASME,* Vol. 73, p. 101, 1951.
12. Petukhov, B.S., "Heat Transfer and Friction in Turbulent Pipe Flow with Variable Physical Properties, in *Advances in Heat Transfer*", Hartnett, J.P., and Irvine, T.F., eds., Academic Press, New York, pp. 504–564, 1970.
13. Burmeister, L.C., *Convective Heat Transfer,* 2nd ed., Wiley-Interscience, New York, 1993.
14. Kays, W.M. and Perkins, K.R., "Forced Convection, Internal Flow in Ducts", in *Handbook of Heat Transfer Applications,* Vol. 1, Rohsenow, W.G., Hartnett, J.P., and Ganic, E.N., eds., McGraw-Hill, New York, chap. 7, 1985.
15. Sieder, E.N. and Tate, G.E., "Heat Transfer and Pressure Drop of Liquids in Tubes", *Ind. Eng. Chem.,* Vol. 28, pp. 1429–1453, 1936.
16. Seban, R.A. and Shimazaki, T.T., "Heat Transfer to Fluid Flowing Turbulently in a Smooth Pipe with Walls at Constant Temperature", *Trans. ASME,* Vol. 73, pp. 803–807, 1951.
17. Webb, R.L., "A Critical Evaluation of Analytical Solutions and Reynolds Analogy Equations for Turbulent Heat and Mass Transfer in Smooth Tubes", *Warme- und Stoffubertragung,* Vol. 4, pp. 197–204, 1971.
18. Haaland, S.E., "Simple and Explicit Formulas for the Friction Factor in Turbulent Flow", *Trans. ASME J. Fluid Eng.,* Vol.105, No. 3, pp. 89–90, 1983.
19. Moody, L.F., "Friction Factors for Pipe Flow", *Trans. ASME,* Vol. 66, pp. 671–684, 1944.
20. Rehme, K., "A Simple Method of Predicting Friction Factors of Turbulent Flow in Noncircular Channels", *Int. J. Heat Mass Transfer,* Vol. 16, pp. 933–950, 1973.
21. Notter, R.H. and Sleicher, C.A., "A Solution to the Turbulent Graetz Problem III. Fully Developed and Entry Region Heat Transfer Rates", *Chem. Eng. Sci.,* Vol. 27, pp. 2073–2093, 1972.
22. Azer, N.Z., "Thermal Entry Length for Turbulent Flow of Liquid Metals in Pipes with Constant Wall Heat Flux", *Trans. ASME Serv. C, J. Heat Transfer,* Vol. 90, pp. 483–485, 1968.
23. Sparrow, E.M., Hallman, T.M., and Siegel, R., "Turbulent Heat Transfer in the Thermal Entrance Region of a Pipe with Uniform Heat Flux", *Appl. Sci. Rest. Section A,* Vol. 7, p. 37, 1957.
24. Deissler, R.G., "Turbulent Heat Transfer and Friction in the Entrance Regions of Smooth Passages", Trans. ASME, Vol. 77, pp. 1211–1234, 1955.
25. Deissler, R.G., "Analysis of Turbulent Heat Transfer and Flow in the Entrance Regions of Smooth Passages", *NACA TN,* 3016, 1953.
26. Anderson, D., Tannehill, J.C., and Pletcher, R.H., *Computational Fluid Mechanics and Heat Transfer,* Hemisphere, Washington, D.C., 1984.
27. Choi, J.M., and Anand, N.K, "Turbulent Heat Transfer in a Serpentine Channel with a Series of Right-Angle Turns", *Inter. J. of Heat and Mass Transfer,* Vol. 38, No. 7, pp. 1225–1236, 1995.
28. Chow, C.Y., *Computational Fluid Mechanics,* Wiley, New York, 1979.
29. Garcia, A., and Sparrow, E.M., "Turbulent Heat Transfer Downstream of a Contraction-Related, Forward-Facing Step in a Duct", *J. Heat Trans.,* Vol. 109, pp. 621–626, 1987.
30. Gosman, A.D., Pun, W.M., Runchal, A.K., Spalding, D.B., and Wolfshtein, M., *Heat and Mass Transfer in Recirculating Flows,* Academic Press, New York, 1969.
31. Kim, S.H., and Anand, N.K., "Turbulent Heat Transfer Between a Series of Parallel Plates with Surface-Mounted Discrete Heat Sources", *J. Heat Transfer,* Vol. 116, pp. 577–587, 1994.

32. Lau, S.C., Kukreja, R.T., and McMillin, R.D., "Effects of V-Shaped Rib Arrays on Turbulent Heat Transfer and Friction of Fully Developed Flow in a Square Channel", *Int. J. Heat and Mass Transfer,* Vol. 34, pp. 1605–1616, 1991.
33. Liou, Tong Miin, Hwang, Jenn Jiang, Chen, Shih Hui, "Simulation and Measurement of Enhanced Turbulent Heat Transfer in a Channel with Periodic Ribs on One Principal Wall", *Int. J. Heat and Mass Transfer,* Vol. 36, No. 2, pp. 507–517, 1993.
34. Patankar, S.V., *Numerical Heat Transfer and Fluid Flow,* Hemisphere Publ. Corp., New York, 1980.
35. Roache, P.J., *Computational Fluid Dynamics,* Hermosa Publishers, Albuquerque, NM, 1976.
36. Launder, B.W., "On the Computation of Convective Heat Transfer in Complex Turbulent Flows", *J. Heat Transfer,* Vol. 110, pp. 1112–1128, 1988.

CHAPTER 8

Natural Convection

8.1 INTRODUCTION

As explained in Chapter 1, natural or free convective heat transfer is heat transfer between a surface and a fluid moving over it with the fluid motion caused entirely by the buoyancy forces that arise due to the density changes that result from the temperature variations in the flow, [1] to [5]. Natural convective flows, like all viscous flows, can be either laminar or turbulent as indicated in Fig. 8.1. However, because of the low velocities that usually exist in natural convective flows, laminar natural convective flows occur more frequently in practice than laminar forced convective flows. In this chapter attention will therefore be initially focused on laminar natural convective flows.

The majority of natural convective flows arise due to density changes in the presence of the gravitational force field. However, similar flows can arise in other force fields, e.g., very large buoyancy type forces can arise in a centrifugal force field as indicated in Fig. 8.2. However, in such cases, the flow is frequently not purely natural convective and such flows will not be considered here. A distinction is sometimes made between natural and free convection, the term natural convection then being applied to flows resulting due to the gravitational force field and the term free convection being applied to flows arising due to the presence of any force field. However, the terms are today both usually used to describe any flow arising due to temperature-induced density changes in a force field and they will be used interchangeably here.

8.2 BOUSSINESQ APPROXIMATION

Natural convective flows arise because the density of the fluid involved changes with temperature i.e., the flow arises because a property of the fluid, i.e., the density

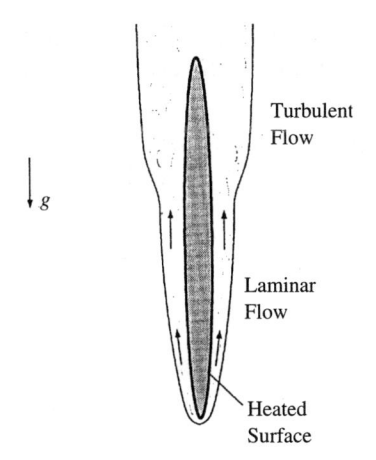

FIGURE 8.1
Laminar and turbulent free convective boundary layers on a vertical surface.

changes with temperature. The question that immediately then arises is that if a change in a fluid property causes the flow, must account be taken of the variations of all of the fluid properties with temperature in the analysis of such flows?

Now, the changes of density in most free convective flows are relatively small i.e., if ρ_0 is the initial density and $\Delta \rho$ is the overall change in density, then:

$$\frac{\Delta \rho}{\rho_0} \ll 1 \tag{8.1}$$

Consider a two-dimensional free convective flow. The momentum equation for the x-direction, which is assumed to be in the vertical direction, is:

$$\rho u \frac{\partial u}{\partial x} + \rho v \frac{\partial u}{\partial y} = -\frac{\partial p}{\partial x} + \mu \left(\frac{\partial^2 u}{\partial x^2} + \frac{\partial^2 u}{\partial y^2} \right) + g(\rho_0 - \rho) \tag{8.2}$$

the pressure being, of course, measured relative to the local hydrostatic pressure.

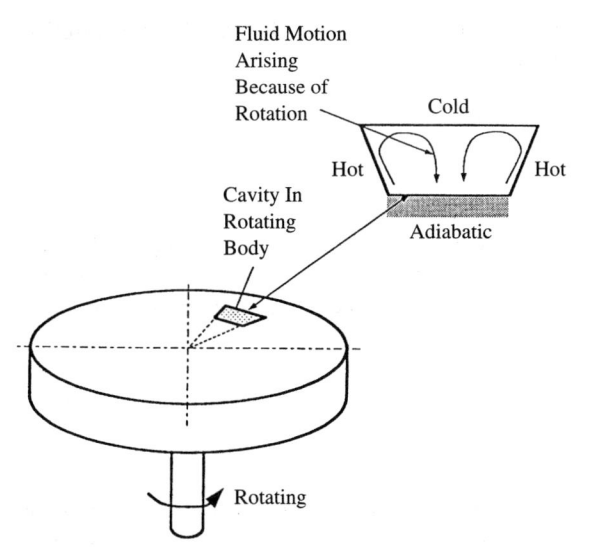

FIGURE 8.2
Free convective flow in a rotating enclosure.

The flow is induced by the buoyancy force, i.e., the terms in this equation must all be of the order of $\rho_0 - \rho$, i.e., of the order of $\Delta\rho$. Now, Eq. (8.2) can be written as:

$$\rho_0\left(u\frac{\partial u}{\partial x} + v\frac{\partial u}{\partial y}\right) + \Delta\rho\left(u\frac{\partial u}{\partial x} + v\frac{\partial u}{\partial y}\right) = -\frac{\partial p}{\partial x} + \mu_0\left(\frac{\partial^2 u}{\partial x^2} + \frac{\partial^2 u}{\partial y^2}\right)$$
$$+ \Delta\mu\left(\frac{\partial^2 u}{\partial x^2} + \frac{\partial^2 u}{\partial y^2}\right) + g(\Delta\rho) \quad (8.3)$$

where:

$$\mu = \mu_0 + \Delta\mu \quad (8.4)$$

$\Delta\mu$ will have the same order of magnitude as $\Delta\rho$. Hence, Eq. (8.3) has the following form:

$$\rho_0\left(u\frac{\partial u}{\partial x} + v\frac{\partial u}{\partial y}\right) + \text{term of order } \Delta\rho^2$$
$$= -\frac{\partial p}{\partial x} + \mu_0\left(\frac{\partial^2 u}{\partial x^2} + \frac{\partial^2 u}{\partial y^2}\right) + \text{term of order } \Delta\rho^2 + g(\Delta\rho) \quad (8.5)$$

But, since $\Delta\rho/\rho_0$ is small, terms of the order of $\Delta\rho^2$ can be neglected. Eq. (8.5) therefore becomes:

$$\rho_0\left(u\frac{\partial u}{\partial x} + v\frac{\partial u}{\partial y}\right) = -\frac{\partial p}{\partial x} + \mu_0\left(\frac{\partial^2 u}{\partial x^2} + \frac{\partial^2 u}{\partial y^2}\right) + g(\rho_0 - \rho) \quad (8.6)$$

i.e., the only effect of the change in density is the generation of the buoyancy term. In all of the other terms, the effect of the changes in density and other properties of the fluid due to the temperature changes can be neglected.

Now, by definition:

$$\rho_0 - \rho = \beta\rho_0(T - T_0) \quad (8.7)$$

where β is the bulk coefficient. It is a property of the fluid involved. Substituting Eq. (8.7) into Eq. (8.6) and rearranging gives:

$$u\frac{\partial u}{\partial x} + v\frac{\partial u}{\partial y} = \frac{1}{\rho}\frac{\partial p}{\partial x} + \nu\left(\frac{\partial^2 u}{\partial x^2} + \frac{\partial^2 u}{\partial y^2}\right) + \beta g(T - T_0) \quad (8.8)$$

The above discussion indicates that, in the analysis of free convective flows, the fluid properties can be assumed constant except for the density change with temperature which gives rise to the buoyancy force. This is, basically, the "Boussinesq approximation" [6]. This approximation will be adopted in all the analyses given in this chapter.

8.3
GOVERNING EQUATIONS

When the Boussinesq approximation discussed in the previous section is adopted, the equations governing steady laminar two-dimensional natural convective flow are:

CHAPTER 8: Natural Convection 345

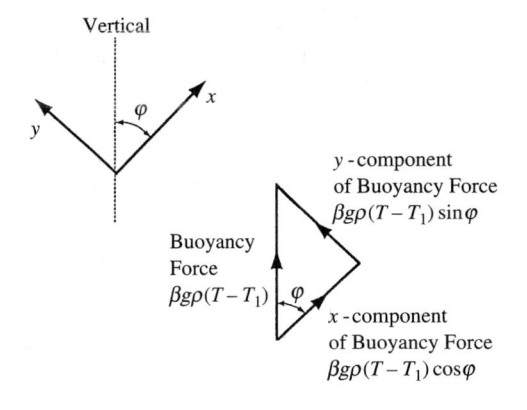

FIGURE 8.3
Buoyancy force components.

$$\frac{\partial u}{\partial x} + \frac{\partial v}{\partial y} = 0 \tag{8.9}$$

$$u\frac{\partial u}{\partial x} + v\frac{\partial u}{\partial y} = -\frac{1}{\rho}\frac{\partial p}{\partial x} + \nu\left(\frac{\partial^2 u}{\partial x^2} + \frac{\partial^2 u}{\partial y^2}\right) + \beta g(T - T_1)\cos\phi \tag{8.10}$$

$$u\frac{\partial v}{\partial x} + v\frac{\partial v}{\partial y} = -\frac{1}{\rho}\frac{\partial p}{\partial y} + \nu\left(\frac{\partial^2 v}{\partial x^2} + \frac{\partial^2 v}{\partial y^2}\right) + \beta g(T - T_1)\sin\phi \tag{8.11}$$

$$u\frac{\partial T}{\partial x} + v\frac{\partial T}{\partial y} = \left(\frac{k}{\rho c_p}\right)\left(\frac{\partial^2 T}{\partial x^2} + \frac{\partial^2 T}{\partial y^2}\right) \tag{8.12}$$

In these equations, T_1 is the temperature in the undisturbed fluid far from the surface and the pressure is, then, measured relative to the local hydrostatic pressure at this temperature of the undisturbed fluid. The dissipation term has been neglected in the energy equation (8.12) due to the low velocities involved.

The continuity equation (8.9) and the energy equation (8.12) are identical to those for forced convective flow. The x- and y-momentum equations, i.e., Eqs. (8.10) and (8.11), differ, however, from those for forced convective flow due to the presence of the buoyancy terms. The way in which these terms are derived was discussed in Chapter 1 when considering the application of dimensional analysis to convective heat transfer. In these buoyancy terms, ϕ is the angle that the x-axis makes to the vertical as shown in Fig. 8.3.

It should be noted that, in contrast to forced convective flows, in natural convective flows, due to the temperature-dependent buoyancy forces in the momentum equations, the velocity and temperature fields are interrelated even though the fluid properties are assumed to be constant except for the density change with temperature.

8.4
SIMILARITY IN FREE CONVECTIVE FLOWS

Consideration was given in Chapter 2 to the problem of determining the conditions under which the flow and temperature distributions about geometrically similar bodies are "similar", i.e., the conditions under which the velocity and temperature fields about the bodies will be "similar". By similar velocity and temperature fields, it is here meant that the velocity components and temperature when expressed in dimensionless form are related to the dimensionless coordinate system by the same functions in all of the flows considered. The same problem will now be briefly considered for free convective flows, i.e., the conditions under which the flow and temperature distributions about geometrically similar bodies are "similar" in natural convective flows will be investigated. Attention will be restricted to two-dimensional flow in the present discussion.

The governing equations are first expressed in terms of suitable dimensionless variables. In order to do this it is necessary to introduce a suitable characteristic velocity. Now, a measure of the buoyancy forces per unit volume existing in the flow will be $\beta g \rho (T_{wr} - T_1)$ where T_{wr} is a measure of the wall temperature. The buoyancy forces do work on the fluid as it flows over the surface, a measure of this being the product of the measure of the buoyancy forces and a measure of the distance over which these forces act. This distance will be the characteristic size of the body, D. Hence, a measure of the magnitude of the work being done by the buoyancy forces on the fluid per unit volume of the fluid is $\beta g \rho (T_{wr} - T_1) D$. As a result of this work, the fluid gains kinetic energy, a measure of this kinetic energy per unit volume of the fluid is $1/2 \rho u_r^2$, u_r being the characteristic velocity in the flow. Equating the measures of the work done and of the gain in kinetic energy in the fluid then gives:

$$\beta g \rho (T_{wr} - T_1) D = \frac{1}{2} \rho u_r^2$$

i.e.:

$$u_r = \sqrt{2 \beta g (T_{wr} - T_1) D}$$

Therefore, the characteristic velocity will be taken as $\sqrt{\beta g (T_{wr} - T_1) D}$.

The following dimensionless variables are then introduced:

$$X = x/D, \quad Y = y/D,$$
$$U = u/\sqrt{\beta g (T_{wr} - T_1) D}, \quad V = v/\sqrt{\beta g (T_{wr} - T_1) D} \quad (8.13)$$
$$P = p/\beta g \rho (T_{wr} - T_1) D, \quad \theta = (T - T_1)/(T_{wr} - T_1)$$

Substituting these variables into Eqs. (8.9) to (8.12) then gives:

$$\frac{\partial U}{\partial X} + \frac{\partial V}{\partial Y} = 0 \quad (8.14)$$

$$U \frac{\partial U}{\partial X} + V \frac{\partial U}{\partial Y} = -\frac{\partial P}{\partial X} + \left[\frac{\nu^2}{\beta g (T_{wr} - T_1) D^3} \right]^{0.5} \left(\frac{\partial^2 U}{\partial X^2} + \frac{\partial^2 U}{\partial Y^2} \right) + \theta \cos \phi \quad (8.15)$$

$$U \frac{\partial V}{\partial X} + V \frac{\partial V}{\partial Y} = -\frac{\partial P}{\partial Y} + \left[\frac{\nu^2}{\beta g (T_{wr} - T_1) D^3} \right]^{0.5} \left(\frac{\partial^2 V}{\partial X^2} + \frac{\partial^2 V}{\partial Y^2} \right) + \theta \sin \phi \quad (8.16)$$

$$U\frac{\partial \theta}{\partial X} + V\frac{\partial \theta}{\partial Y} = +\left[\frac{k}{\rho c_p D}\frac{1}{\sqrt{\beta g(T_{wr}-T_1)D}}\right]\left(\frac{\partial^2 \theta}{\partial X^2}+\frac{\partial^2 \theta}{\partial Y^2}\right) \quad (8.17)$$

From these equations it follows that the relations between U, V, θ, and P and X and Y will be the same in two flows if the following two parameters have the same values in the two flows:

$$\left[\frac{\beta g(T_{wr}-T_1)D^3}{\nu^2}\right], \left[\frac{\rho c_p D \sqrt{\beta g(T_{wr}-T_1)D}}{k}\right] \quad (8.18)$$

The first of these is, of course, the Grashof number, Gr, while the second can be written as:

$$\left(\frac{\mu c_p}{k}\right)\left[\frac{\beta g(T_{wr}-T_1)D^3}{\nu^2}\right]^{1/2} = Pr Gr^{1/2} \quad (8.19)$$

where Pr is the Prandtl number.

In view of these results and by considering the boundary conditions, it follows that two free convective flows over geometrically similar bodies will be similar if the following conditions exist:

i. The temperature distributions at the surface are similar, i.e., $(T_w-T_1)/(T_{wr}-T_1)$ is the same function of X and Y in the two flows.
ii. The Grashof number is the same in the two flows.
iii. The Prandtl number is the same in the two flows.

From the above discussion, it can be concluded that the dimensionless temperature field $\theta(X, Y)$ depends only on the dimensionless numbers Gr and Pr. In order to determine the implications this has for the local heat transfer rate at any point on the surface it is noted that this heat transfer rate is given by Fourier's Law as:

$$q_w = -k\left.\frac{\partial T}{\partial n}\right|_w \quad (8.20)$$

where $\partial T/\partial n|_w$ is the local temperature gradient at the surface measured normal to the surface. Now this derivative can be written as:

$$\left.\frac{\partial T}{\partial n}\right|_w = \left.\frac{\partial T}{\partial y}\right|_w \left.\frac{\partial y}{\partial n}\right|_w + \left.\frac{\partial T}{\partial x}\right|_w \left.\frac{\partial x}{\partial n}\right|_w \quad (8.21)$$

The derivatives $\partial y/\partial n|_w$ and $\partial x/\partial n|_w$ will depend only on the shape of the surface at the point considered.

Now the right-hand side of Eq. (8.21) is conveniently rewritten in dimensionless form to give:

$$\left.\frac{\partial T}{\partial n}\right|_w = \left\{\left.\frac{\partial \theta}{\partial Y}\right|_w \left.\frac{\partial Y}{\partial N}\right|_w + \left.\frac{\partial \theta}{\partial X}\right|_w \left.\frac{\partial X}{\partial N}\right|_w\right\}\left(\frac{T_{wr}-T_1}{D}\right) \quad (8.22)$$

where

$$N = \frac{n}{D} \tag{8.23}$$

Since $\partial X/\partial N|_w$ and $\partial Y/\partial N|_w$ will depend only on the shape of the surface and since the dimensionless temperature gradients $\partial \theta/\partial X|_w$ and $\partial \theta/\partial Y|_w$ must depend on the same variables as the dimensionless temperature field, i.e., on Gr and Pr, it follows that at any position on the surface:

$$\left.\frac{\partial T}{\partial n}\right|_w = \left(\frac{T_{wr} - T_1}{L}\right) \text{function}(Gr, Pr) \tag{8.24}$$

Substituting this result into Eq. (8.20) then gives:

$$\frac{q_w D}{(T_w - T_1)k} = \left(\frac{T_{wr} - T_1}{T_w - T_1}\right) \text{function}(Gr, Pr) \tag{8.25}$$

Now the left-hand side of this equation is the local Nusselt number, Nu, i.e.:

$$Nu = \frac{q_w D}{(T_w - T_1)k}$$

It follows, therefore that for a given surface temperature distribution, i.e., for a given distribution of $(T_w - T_1)/(T_{wr} - T_1)$:

$$Nu = \text{function}(Gr, Pr) \tag{8.26}$$

This result was, of course, previously derived by dimensional analysis.

EXAMPLE 8.1. Consider two vertical plane surfaces, both of which have a uniform surface temperature. These surfaces are exposed to stagnant air at a temperature of 15°C. One of the surfaces has a height of 0.1 m and a surface temperature of 65°C. The other surface has a height of 0.2 m. Under what conditions will the natural convective flows over the two surfaces be similar and what will be the ratio of the mean heat transfer rates from the two surfaces?

Solution. Because both surfaces are exposed to air, the Prandtl numbers will be essentially the same in the two flows. Hence, similar flows will exist when the Grashof numbers in the two flows are the same, i.e., when:

$$\left[\frac{\beta g(T_w - T_1)D^3}{\nu^2}\right]_A = \left[\frac{\beta g(T_w - T_1)D^3}{\nu^2}\right]_B$$

subscripts A and B denoting the two flows and D being the height of the surface.

Because the mean fluid temperatures in the two flows will be different, the fluid properties in the two flows will be somewhat different. This will, however, be neglected here, i.e., it will be assumed that:

$$\beta_A = \beta_B, \quad \nu_A = \nu_B$$

In this case, similar flows will exist when:

$$[(T_w - T_1)D^3]_A = [(T_w - T_1)D^3]_B$$

i.e.:

$$(65 - 15) \times 0.1^3 = (T_{wB} - 15) \times 0.2^3$$

i.e.:

$$T_{wB} = 21.25°C$$

Therefore, the flows over the two plates will be similar when the larger plate has a surface temperature of 21.25°C.

If the flows are similar, the Nusselt numbers in the two flows will be the same, i.e.:

$$\left[\frac{q_w D}{(T_w - T_1)k}\right]_A = \left[\frac{q_w D}{(T_w - T_1)k}\right]_B$$

Hence, because it is being assumed that:

$$k_A = k_B$$

it follows that:

$$\frac{q_{wA} 0.1}{(65 - 15)} = \frac{q_{wB} 0.2}{(21.25 - 15)}$$

i.e.:

$$\frac{q_{wB}}{q_{wA}} = 0.0625$$

Therefore, the mean heat transfer rate per unit surface area for the large plate will be 0.0625 times that for the smaller plate.

8.5
BOUNDARY LAYER EQUATIONS FOR NATURAL CONVECTIVE FLOWS

As in the case with forced convective flows, there are many free convective flows that can be analyzed with sufficient accuracy by adopting the boundary layer assumption. Essentially this boundary layer assumption is that the flow consists of two regions:

i. A thin region adjacent to the surface in which the effects of viscosity and heat transfer are important.
ii. An outer region in which the effects of viscosity and heat transfer are negligible.

Since the effects of heat transfer are negligible in the outer region, the temperature in this region will be effectively constant and, therefore, equal to the ambient temperature, T_1. There are, therefore, no buoyancy forces in this outer region.

The boundary layer equations for free convective flow will be deduced using essentially the same approach as was adopted in forced convective flow. Attention will, as discussed above, be restricted to the case of two-dimensional laminar boundary layer flow. Attention will initially be focused on a plane surface that is at an angle, ϕ, the vertical as shown in Fig. 8.4. The x-axis is chosen to be parallel to this surface as shown in Fig. 8.4.

In order to derive the boundary layer equations, it is necessary to introduce a velocity that is characteristic of that existing in the boundary layer. For the reasons given in the previous section this characteristic velocity will be taken as:

$$u_r = \sqrt{\beta g(T_{wr} - T_1)L} \qquad (8.27)$$

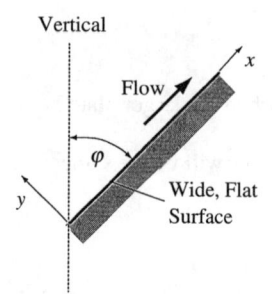

FIGURE 8.4
Free convective flow considered.

Here, L is a reference length that characterizes the size of the surface, e.g., its length. T_{wr} is again some reference wall temperature. If some measure of the boundary layer thickness, δ, is also introduced, then the governing equations can be written in terms of the following dimensionless variables:

$$X = x/L, \quad Y = y/\delta, \quad U = u/\sqrt{\beta g(T_{wr} - T_1)L}, \quad V = v/\sqrt{\beta g(T_{wr} - T_1)L}$$
$$P = p/\beta g \rho (T_{wr} - T_1)L, \quad \theta = (T - T_1)/(T_{wr} - T_1)$$
(8.28)

No distinction is made at this stage between the velocity and temperature boundary layer thicknesses, both being assumed to be of the same order of magnitude. δ is, therefore, a measure of the order of magnitude of both boundary layer thicknesses. It will be recalled that, by basic assumption, δ/L is small.

Because of the way in which they are defined, all of the above dimensionless variables except V have the order of magnitude of 1, i.e.:

$$X = o(1), \quad Y = o(1), \quad U = o(1), \quad P = o(1), \quad \theta = o(1)$$

The order of magnitude of V is determined from the continuity equation, i.e., Eq. (8.9), which can be written as:

$$\frac{\partial U}{\partial X} + \frac{\partial V}{\partial Y} \frac{1}{(\delta/L)} = 0 \qquad (8.29)$$

From this it follows that:

$$o(V) = \frac{o(U)}{o(X)} \times o(Y) \times o\left(\frac{\delta}{L}\right) \qquad (8.30)$$

Therefore, since, as discussed above, $o(U)$, $o(Y)$, and $o(X)$ are all 1, this equation indicates that:

$$o(V) = o\left(\frac{\delta}{L}\right) \qquad (8.31)$$

Since the basic assumption of boundary layer theory is that (δ/L) is small, it follows from this that V is also small. Thus, in a free convective boundary layer, as in the forced convective boundary layer, the lateral velocity component, v, is very much smaller than the longitudinal component, u, which is, of course, what is physically to be expected.

CHAPTER 8: Natural Convection   351

Now the x-momentum equation (8.10) can be written in terms of the dimensionless variables introduced above as:

$$U\frac{\partial U}{\partial X} + V\frac{\partial U}{\partial Y}\frac{1}{(\delta/L)} = -\frac{\partial P}{\partial X} + \left[\frac{\nu^2}{\beta g(T_{wr} - T_1)L^3}\right]^{0.5} \times \left(\frac{\partial^2 U}{\partial X^2} + \frac{\partial^2 U}{\partial Y^2}\frac{1}{(\delta/L)^2}\right) + \theta\cos\phi \quad (8.32)$$

i.e., introducing the orders of magnitude of the various terms in this equation gives the following:

$$o(1)\frac{o(1)}{o(1)} + o(\delta/L)\frac{o(1)}{o(1)}\frac{1}{(\delta/L)} = o(1) + \frac{1}{Gr_L^{0.5}}\left[\frac{o(1)}{o(1)} + \frac{o(1)}{o(1)}\frac{1}{(\delta/L)^2}\right] + o(1)$$

i.e.:

$$o(1) + o(1) = o(1) + \left[o\left(\frac{1}{Gr_L^{0.5}}\right) + \frac{o(1/Gr_L^{0.5})}{(\delta/L)^2}\right] + o(1) \quad (8.33)$$

The second term on the right-hand side of this equation shows that if (δ/L) is small, as it must be if the boundary layer approximations are to be applicable, then Gr_L must be large, its order of magnitude being such that:

$$\frac{o(1/Gr_L^{0.5})}{o[(\delta/L)^2]} = o(1)$$

i.e.:

$$o(Gr_L) = o\left[\frac{1}{(\delta/L)^4}\right] \quad (8.34)$$

From this it also follows that $\delta/L = o[1/Gr_L^{0.25}]$.

As a consequence of the above result, it can be seen that the first term within the square bracket on the right-hand side of Eq. (8.33), i.e., the term originating from the term $(\mu\partial^2 u/\partial x^2)$, will have $o(\delta/L)^2$ whereas, the other terms in the equation have $o(1)$. For this reason, when dealing with boundary layer flows, this term in the x-wise momentum equation is negligible. For such flows, therefore, the x-momentum equation is:

$$u\frac{\partial u}{\partial x} + v\frac{\partial u}{\partial y} = -\frac{1}{\rho}\frac{\partial p}{\partial x} + \left(\frac{\mu}{\rho}\right)\frac{\partial^2 u}{\partial y^2} + \beta g(T - T_1)\cos\phi \quad (8.35)$$

To find the order of magnitude of the pressure term, i.e., of the difference between the dimensionless pressure in the boundary layer and the pressure that would exist in the absence of the flow, attention is turned to the y-momentum equation. In terms of the dimensionless variables introduced above this equation becomes:

$$U\frac{\partial V}{\partial X} + V\frac{\partial V}{\partial Y}\frac{1}{(\delta/L)} = -\frac{\partial P}{\partial Y} + \left[\frac{\nu^2}{\beta g(T_{wr} - T_1)L^3}\right]^{0.5} \times \left(\frac{\partial^2 V}{\partial X^2} + \frac{\partial^2 V}{\partial Y^2}\frac{1}{(\delta/L)^2}\right) + \theta\sin\phi \quad (8.36)$$

352 Introduction to Convective Heat Transfer Analysis

i.e., introducing the orders of magnitude of the various terms in this equation gives the following:

$$o(1)\frac{o(\delta/L)}{o(1)} + o(\delta/L)\frac{o(\delta/L)}{o(1)}\frac{1}{(\delta/L)}$$
$$= \frac{o(\Delta P)}{o(\delta/L)} + \frac{1}{Gr_L^{0.5}}\left[\frac{o(\delta/L)}{o(1)} + \frac{o(\delta/L)}{o(1)}\frac{1}{(\delta/L)^2}\right] + o(1)$$

Here, ΔP is a measure of the dimensionless pressure change across the boundary layer.

But it was shown that $o(Gr_L) = o(\delta/L)^{-4}$. Hence, the above equation shows that:

$$o(\delta/L) + o(\delta/L) = \frac{o(\Delta P)}{o(\delta/L)} + [o(\delta/L)^3 + o(\delta/L)] + o(1)$$

From this it follows that:

$$o(\Delta P) = o(\delta/L) \tag{8.37}$$

Since terms of the order δ/L and less are being assumed to be negligible, this equation indicates that the changes in pressure in the y-direction across the boundary layer can be ignored, i.e., that the pressure in the boundary layer is equal to the pressure at the outer edge of the boundary layer. But in a free convective flow, the pressure outside the boundary layer is equal to the hydrostatic pressure in the stagnant surrounding fluid. Hence, since the pressure, p, being used is measured relative to the local hydrostatic pressure this means that p is equal to 0 everywhere in the boundary layer. It follows that the x-direction momentum equation for two-dimensional free convective boundary layer flows, i.e., Eq. (8.35), can be written as:

$$u\frac{\partial u}{\partial x} + v\frac{\partial u}{\partial y} = \left(\frac{\mu}{\rho}\right)\frac{\partial^2 u}{\partial y^2} + \beta g(T - T_1)\cos\phi \tag{8.38}$$

It should be noted that it was assumed in the above analysis that $\theta \cos\phi$ was of the order of 1. This will not be true when $\cos(\phi)$ is near 0, i.e., when ϕ is near 90°, i.e., the above equation will not apply to a near-horizontal surface. This is because in such a case there is essentially no component of the buoyancy force in the direction of flow, the flow actually being caused by the pressure differences in the flow that are induced by the buoyancy forces acting across the flow. These pressure differences were, of course, ignored in the above analysis.

Next, consider the energy equation. In terms of the dimensionless variables introduced above this becomes:

$$U\frac{\partial\theta}{\partial X} + V\frac{\partial\theta}{\partial Y}\frac{1}{(\delta/L)} = \left(\frac{1}{Pr}\right)\left[\frac{\nu^2}{\beta g(T_{wr} - T_1)L^3}\right]^{0.5}\left(\frac{\partial^2\theta}{\partial X^2} + \frac{\partial^2\theta}{\partial Y^2}\right) \tag{8.39}$$

Using the results concerning the orders of magnitude of V and Gr_L derived above gives the orders of magnitude of the various terms in Eq. (8.39) as:

$$o(1) + o(1) = \frac{1}{o(Pr)}[o(\delta/L)^2 + o(1)] \tag{8.40}$$

If terms of the order (δ/L) and less are again neglected and if it is assumed that the Prandtl number, Pr, has order 1 or greater, i.e., is not small, it follows that the x-wise diffusion term, i.e., $\partial^2 T/\partial x^2$, is negligible compared to the other terms. Hence, the energy equation for free convective laminar two-dimensional boundary layer flow becomes:

$$u\frac{\partial T}{\partial x} + v\frac{\partial T}{\partial y} = \left(\frac{k}{\rho c_p}\right)\frac{\partial^2 T}{\partial y^2} \tag{8.41}$$

To summarize, therefore, the equations for two-dimensional, constant fluid property boundary layer flow over a plane surface are:

$$\frac{\partial u}{\partial x} + \frac{\partial v}{\partial y} = 0$$

$$u\frac{\partial u}{\partial x} + v\frac{\partial u}{\partial y} = \left(\frac{\mu}{\rho}\right)\frac{\partial^2 u}{\partial y^2} + \beta g(T - T_1)\cos\phi$$

$$u\frac{\partial T}{\partial x} + v\frac{\partial T}{\partial y} = \left(\frac{k}{\rho c_p}\right)\frac{\partial^2 T}{\partial y^2}$$

These equations were derived for flow over a plane surface. They may be applied to flow over a curved surface provided that the boundary layer thickness remains small compared to the radius of curvature of the surface. When applied to flow over a curved surface, x is measured along the surface and y is measured normal to it at all points, i.e., body fitted coordinates are used. The value of ϕ is then dependent on x. This is illustrated in Fig. 8.5.

In order to solve the above set of boundary layer equations, the boundary conditions on the variables involved, i.e., u, v, and T, must be known. At the wall, the no-slip condition requires that the velocity be the same as that of the wall, i.e., usually zero, so that one such boundary condition is:

$$\text{At } y = 0: \quad u = v = 0 \tag{8.42}$$

If there is blowing or suction at the surface, the rate of blowing or sucking will determine v at the wall.

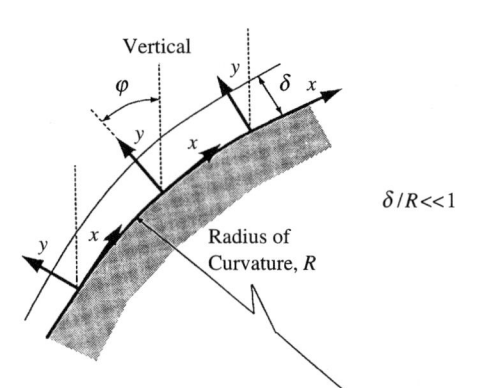

FIGURE 8.5
Free convective flow over a curved surface.

The boundary conditions on temperature at the wall depend on the thermal conditions specified at the wall. If the distribution of the temperature of the wall is specified then since the fluid in contact with the wall must be at the same temperature as the wall, the boundary condition on temperature is:

$$\text{At } y = 0: \quad T = T_w \tag{8.43}$$

where T_w is the temperature of the wall at the particular value of x being considered.

It is also possible for the distribution of the heat transfer rate at the wall to be specified and in this case the boundary condition becomes:

$$\text{At } y = 0: \quad \frac{\partial T}{\partial y} = -\frac{q_w}{k} \tag{8.44}$$

where q_w is the specified local heat transfer rate per unit area at the wall at the particular value of x being considered.

Because there are assumed to be no changes in temperature outside of the boundary layer, there is no flow in the x-direction outside of the boundary layer. Therefore, since the temperature is constant outside the boundary layer, the following boundary conditions apply:

$$\text{For large } y: \quad u \to 0, \quad T \to T_1 \tag{8.45}$$

It should again be noted that the velocity and temperature fields are interrelated by the buoyancy term in the momentum equations. Therefore, it is not possible, as in forced convection, to solve for the velocity components independently of the temperature field.

8.6
SIMILARITY SOLUTIONS FOR FREE CONVECTIVE LAMINAR BOUNDARY LAYER FLOWS

In order to illustrate how similarity solutions are obtained for free convective flows, see [7] to [23], consideration will initially be given to two-dimensional flow over a vertical flat plate with a uniform surface temperature. The situation being considered is thus, as shown in Fig. 8.6.

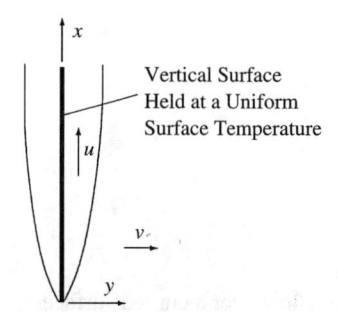

FIGURE 8.6
Flow situation considered.

Because the boundary layer assumptions are being adopted, the equations governing the flow are, as discussed above:

$$\frac{\partial u}{\partial x} + \frac{\partial v}{\partial y} = 0 \tag{8.46}$$

$$u\frac{\partial u}{\partial x} + v\frac{\partial u}{\partial y} = \left(\frac{\mu}{\rho}\right)\frac{\partial^2 u}{\partial y^2} + \beta g(T - T_1)\cos\phi \tag{8.47}$$

$$u\frac{\partial T}{\partial x} + v\frac{\partial T}{\partial y} = \left(\frac{k}{\rho c_p}\right)\frac{\partial^2 T}{\partial y^2} \tag{8.48}$$

$\cos\phi$ for a vertical surface being equal to 1.

The boundary conditions on the solution are:

At $y = 0$: $u = v = 0$, $T = T_w$

For large y: $u \to 0$, $T \to T_1$ \hfill (8.49)

As was done in dealing with forced convective flow over a uniform temperature plate, it is assumed that the velocity and temperature profiles are similar at all values of x, i.e., that:

$$\frac{u}{u_r} = \text{function}\left(\frac{y}{\delta}\right)$$

and that:

$$\frac{T - T_1}{T_w - T_1} = \text{function}\left(\frac{y}{\delta}\right)$$

where u_r is again a reference velocity and δ is a measure of both the local velocity and thermal boundary layer thicknesses.

The profiles are thus assumed to be similar in the sense that although δ varies with x and although the velocity and temperature at a fixed distance, y, from the wall vary with x, the velocity relative to the local reference value and temperature difference at fixed values of y/δ remain constant. The proof that such similar profiles do in fact exist follows from the analysis given below.

Now, if the distance, x, from the leading edge of the plate to the point under consideration is taken as the reference length, L, it follows from the derivation of the boundary layer equations presented earlier that:

$$\delta/x = o[1/Gr_x^{0.25}] \tag{8.50}$$

where Gr_x is the Grashof number based on x, i.e.:

$$Gr_x = \frac{\beta g(T_w - T_1)x^3}{\nu^2}$$

Therefore, a measure of the local value of y/δ is:

$$\eta = \frac{y}{x}Gr_x^{0.25} \tag{8.51}$$

This variable, η, is termed the similarity variable.

356 Introduction to Convective Heat Transfer Analysis

Therefore, it will be assumed that:

$$\frac{u}{u_r} = \text{function}(\eta)$$

and that:

$$\frac{T - T_1}{T_w - T_1} = \text{function}(\eta)$$

For the reasons given previously in this chapter, the reference velocity is taken as:

$$u_r = \sqrt{\beta g(T_w - T_1)x}$$

Therefore, it is assumed that the velocity and temperature profiles have the form:

$$\frac{u}{\sqrt{\beta g(T_w - T_1)x}} = F'(\eta) \tag{8.52}$$

and:

$$\frac{T - T_1}{T_w - T_1} = G(\eta) \tag{8.53}$$

The prime, of course, again denotes differentiation with respect to η. The differentiated function F' is, as in forced flow, used for convenience in describing the velocity profile.

The following dimensionless variables are now introduced:

$$U = \frac{u}{\sqrt{\beta g(T_w - T_1)x}} = \left(\frac{ux}{\nu}\right) Gr_x^{-0.5}$$

$$V = \frac{v}{\sqrt{\beta g(T_w - T_1)x}} = \left(\frac{vx}{\nu}\right) Gr_x^{-0.5} \tag{8.54}$$

$$\theta = (T - T_1)/(T_w - T_1)$$

Using these the continuity equation can be written as:

$$\frac{\partial}{\partial x}[U\sqrt{\beta g(T_w - T_1)x}] + \frac{\partial}{\partial y}[V\sqrt{\beta g(T_w - T_1)x}] = 0$$

i.e.:

$$\frac{\partial U}{\partial x} + \frac{U}{2x} + \frac{\partial V}{\partial y} = 0 \tag{8.55}$$

In terms of the similarity variable, η, this becomes:

$$\frac{\partial U}{\partial \eta}\frac{\partial \eta}{\partial x} + \frac{U}{2x} + \frac{\partial V}{\partial \eta}\frac{\partial \eta}{\partial y} = 0 \tag{8.56}$$

But using the definition of η given in Eq. (8.51):

$$\frac{\partial \eta}{\partial x} = -\frac{y}{4x^2} Gr_x^{0.25} = -\frac{\eta}{4x} \tag{8.57}$$

CHAPTER 8: Natural Convection   357

$$\frac{\partial \eta}{\partial y} = \frac{Gr_x^{0.25}}{x} \tag{8.58}$$

Substituting these two results into Eq. (8.56) then gives:

$$\frac{dF'}{d\eta}\left(-\frac{\eta}{4x}\right) + \frac{F'}{2x} + \frac{\partial V}{\partial \eta}\frac{Gr_x^{0.25}}{x} = 0$$

i.e.:

$$\frac{\partial V}{\partial \eta} = \frac{1}{2Gr_x^{0.25}}\left[\frac{\eta}{2}\frac{dF'}{d\eta} - F'\right] \tag{8.59}$$

The boundary conditions on the solution give:

At $y = 0$: $v = 0$

But when $y = 0$, $\eta = 0$, and when $v = 0$, $V = 0$ so this boundary condition implies:

At $\eta = 0$: $V = 0$

Eq. (8.59) can be integrated using this boundary condition to give:

$$V = \frac{1}{2Gr_x^{0.25}}\left[\int \frac{\eta}{2}\frac{dF'}{d\eta}d\eta - F\right]$$

i.e.:

$$V = \frac{1}{2Gr_x^{0.25}}\left[\frac{\eta F'}{2} - \frac{F}{2} - F\right] \qquad \frac{F \cdot 2F}{2} = \frac{3F}{2}$$

i.e.:

$$V = \frac{1}{4Gr_x^{0.25}}[\eta F' - 3F] \tag{8.60}$$

The first term in the bracket having been integrated by parts which gives:

$$\int \frac{\eta}{2}\frac{dF'}{d\eta}d\eta = \int \frac{d}{d\eta}\left(\frac{\eta}{2}F'\right)d\eta - \int \frac{F'}{2}d\eta$$

Consideration can now be given to the momentum equation which can be written in terms of the dimensionless variables introduced above as:

$$U\frac{\partial U}{\partial x} + \frac{U^2}{2x} + V\frac{\partial U}{\partial y} = \nu\frac{\partial^2 U}{\partial y^2}\frac{1}{\sqrt{\beta g(T_w - T_1)x}} + \frac{\theta}{x}$$

i.e.:

$$U\frac{\partial U}{\partial \eta}\frac{\partial \eta}{\partial x} + \frac{U^2}{2x} + V\frac{\partial U}{\partial \eta}\frac{\partial \eta}{\partial y} = \nu\frac{\partial^2 U}{\partial \eta^2}\left(\frac{\partial \eta}{\partial y}\right)\frac{1}{\sqrt{\beta g(T_w - T_1)x}} + \frac{\theta}{x}$$

i.e.:

$$F'F''\left(-\frac{\eta}{4x}\right) + \frac{F'^2}{2x} + \frac{1}{4Gr_x^{0.25}}(\eta F' - 3F)F''\frac{Gr_x^{0.25}}{x} = \frac{F'''}{x} + \frac{\theta}{x}$$

i.e.:

$$-\frac{\eta F'F''}{4} + \frac{F'^2}{2} + \frac{\eta F'F''}{4} - \frac{3FF''}{4} = F''' + G$$

i.e.:

$$F''' + \frac{3FF''}{4} - \frac{F'^2}{2} + G = 0 \qquad (8.61)$$

Lastly, consider the energy equation. Because T_w and T_1 are constants, the energy equation can be written as:

$$U\frac{\partial \theta}{\partial x} + V\frac{\partial \theta}{\partial y} = \left(\frac{k}{\rho c_p}\right)\frac{\partial^2 \theta}{\partial y^2}\frac{1}{\sqrt{\beta g(T_w - T_1)x}}$$

i.e.:

$$F'G'\left(-\frac{\eta}{4x}\right) + \frac{1}{4Gr_x^{0.25}}(\eta F' - 3F)G'\frac{Gr_x^{0.25}}{x} = \frac{G''}{Pr\,x}$$

i.e.:

$$G'' + \frac{3}{4}Pr\,FG' = 0 \qquad (8.62)$$

Eqs. (8.61) and (8.62) constitute a pair of simultaneous ordinary differential equations for the velocity and temperature functions, F and G. They must be solved subject to the following boundary conditions:

$$\begin{aligned}
&\text{At } y = 0: \quad u = 0 \text{ i.e., at } \eta = 0: \quad F' = 0 \\
&\text{At } y = 0: \quad v = 0 \text{ i.e., at } \eta = 0: \quad F = 0 \\
&\text{At } y = 0: \quad T = T_w \text{ i.e., at } \eta = 0: \quad G = 1 \\
&\text{For large } y: \quad u \to 0 \text{ i.e., for large } \eta: \quad F' \to 0 \\
&\text{For large } y: \quad T \to T_1 \text{ i.e., for large } \eta: \quad G \to 0
\end{aligned} \qquad (8.63)$$

The fact that the original partial differential equations have been reduced to a pair of ordinary differential equations confirms the assumption that similarity solutions do in fact exist.

Eq. (8.62) has the Prandtl number, Pr, as a parameter and a separate solution for the variation of F and G with η is, thus, obtained for each value of the Prandtl number, Pr. The heat transfer rate at the wall is then obtained by noting that:

$$q_w = -k\left.\frac{\partial T}{\partial y}\right|_{y=0} = -k(T_w - T_1)\left.\frac{dG}{d\eta}\right|_{\eta=0}\frac{Gr_x^{0.25}}{x}$$

i.e.:

$$\frac{Nu_x}{Gr_x^{0.25}} = -G'|_{\eta=0} \qquad (8.64)$$

where Nu_x is, as before, the local Nusselt number.

If the boundary conditions listed in Eq. (8.63) are considered, it will be seen that there are conditions at both $\eta = 0$ and at large η. Eqs. (8.61) and (8.62) cannot therefore, be directly integrated. One way of simultaneously solving these equations is to guess the values of F'' and G' at $\eta = 0$ and to then simultaneously numerically integrate Eqs. (8.61) and (8.62) to give the variations of F and G with η. The solutions so obtained will not generally satisfy the boundary conditions on F and G at large η. The solution is then obtained for other guessed values of F'' and G' at $\eta = 0$ and these results are used to deduce the values of these quantities that give solutions that satisfy the boundary conditions $F' = 0$ and $G = 0$ at large η. While there are some very elegant ways of obtaining the solution in this general manner, the simplest procedure is to guess F'' and G' at $\eta = 0$ and obtain the solution and then increase, first, the guessed value of F'' at $\eta = 0$ by a small amount and obtain the solution and, second, the guessed value of G' at $\eta = 0$ by a small amount and obtain the solution. From these solutions the effect of changes in the guessed values of F'' and G' at $\eta = 0$ on the differences between the values of F' and G at large η given by the solution and the required values can be determined. Improved guesses for the values of F'' and G' at $\eta = 0$ can then be deduced and the process repeated until converged values of F'' and G' at $\eta = 0$ are obtained. A computer program, SIMPLNAT, written in FORTRAN, that implements this procedure is available as discussed in the Preface. As set up, this program will give results for Prandtl number values up to about 30. The initial guessed values must be altered to obtain solutions at higher values of Prandtl number.

Some typical velocity and temperature profiles obtained using this procedure are shown in Figs. 8.7 and 8.8 while some typical values of $G'|_{\eta=0}$ are shown in Table 8.1.

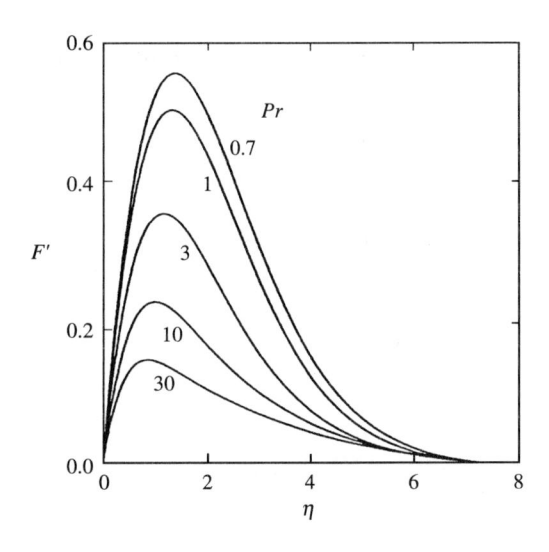

FIGURE 8.7
Dimensionless velocity profiles in natural convective boundary layer on a vertical plate for various values of Prandtl number.

FIGURE 8.8
Dimensionless temperature profiles in natural convective boundary layer on a vertical plate for various values of Prandtl number.

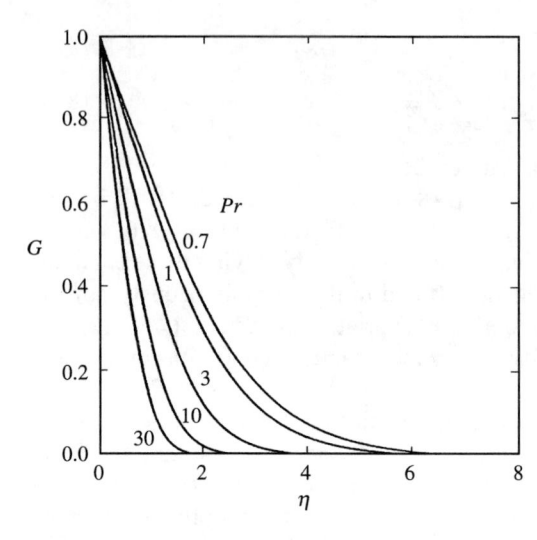

TABLE 8.1
Values of $G'|_{\eta=0}$ for various values of Pr

Pr	$-G'|_{\eta=0}$	$-G'|_{\eta=0}/Pr^{0.25}$
0.01	0.0570	0.1802
0.03	0.0962	0.2312
0.09	0.1549	0.2828
0.72	0.3568	0.3873
1	0.4010	0.4010
2	0.5066	0.4260
5	0.6746	0.4511
10	0.8259	0.4644
100	1.549	0.4898
1000	2.807	0.4992

It has been found that the values of $G'|_{\eta=0}$ can be approximately fitted by the following empirical equation:

$$\frac{Nu_x}{Gr_x^{0.25}} = -G'|_{\eta=0} = \left[\frac{0.316 Pr^{5/4}}{2.44 + 4.88 Pr^{1/2} + 4.95 Pr}\right]^{1/4} \quad (8.65)$$

The mean heat transfer rate for a plate of length, L (see Fig. 8.9) can be obtained by noting that:

$$\overline{q_w} = \frac{1}{L}\int_0^L q_w \, dx \quad (8.66)$$

Hence, using Eq. (8.64):

$$\overline{q_w} = \frac{1}{L}\int_0^L \left[-k(T_w - T_1)G'|_{\eta=0}\frac{Gr_x^{0.25}}{x}\, dx\right]$$

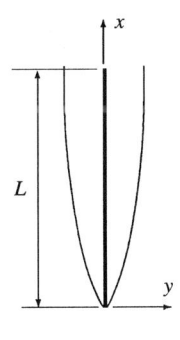

FIGURE 8.9
Flat plate being considered.

i.e.:

$$\frac{q_w L}{(T_w - T_1)k} = G'|_{\eta=0} \frac{4}{3} Gr_L^{0.25}$$

where Gr_L is the Grashof number based on the plate length, L. The mean Nusselt number for the whole plate $Nu_L (= \bar{h}L/k)$ is, therefore, given by:

> Derive this equation

$$\frac{Nu_L}{Gr_L^{0.25}} = \frac{4}{3} G'|_{\eta=0} \tag{8.67}$$

EXAMPLE 8.2. A 30-cm high vertical isothermal flat plate with a surface temperature of 50°C is exposed to stagnant air at 10°C and ambient pressure. Plot the velocity and temperature profiles in the boundary layer at the top of the plate and the variation of the local heat transfer rate along the surface. Assume the flow remains laminar.

Solution. The similarity solution given above indicates that the temperature and velocity distributions are given by:

$$\frac{u}{\sqrt{\beta g(T_w - T_1)x}} = F'(\eta) \tag{a}$$

and:

$$\frac{T - T_1}{T_w - T_1} = G(\eta) \tag{b}$$

where:

$$\eta = \frac{y}{x} Gr_x^{0.25}$$

The functions G and F' depend on the value of the Prandtl number, Pr.
The mean air temperature is:

$$T_{av} = \frac{(50 + 10)}{2} = 30°C$$

At this temperature air, at standard ambient pressure, has the following properties:

$$\beta = 0.0033 \text{ K}^{-1}, \quad \nu = 16 \times 10^{-6} \text{ m}^2/\text{s},$$
$$k = 0.02638 \text{ W/m-K}, \quad Pr = 0.7$$

Therefore, at the top of the plate where $x = 0.3$ m:

$$\eta = \frac{y}{x} \left[\frac{\beta g(T_w - T_\infty) x^3}{\nu^2} \right]^{0.25} = \frac{y}{0.3} \left[\frac{0.0033 \times 9.81 \times (50 - 10) 0.3^3}{(16 \times 10^{-6})^2} \right]^{0.25}$$

This equation gives on rearrangement:

$$y = \frac{\eta}{360.4} \quad (c)$$

Also:

$$\frac{u}{\sqrt{2\beta g(T_w - T_1)x}} = \frac{u}{\sqrt{2 \times 0.0033 \times 9.81 \times (50 - 10) \times 0.3}} = \frac{u}{0.8815} \quad (d)$$

and:

$$\frac{T - T_1}{T_w - T_1} = \frac{T - 10}{50 - 10} = \frac{T - 10}{40} \quad (e)$$

The similarity solution gives the variations of the functions G and F' with η for $Pr = 0.7$. For any value of η, Eq. (c) allows the corresponding value of y to be found while Eqs. (d) and (a) together allow the corresponding value of u to be found and Eqs. (e) and (b) together allow the corresponding value of T to be found. The variations of u and T with y obtained using this procedure are shown in Fig. E8.2a.

For a Prandtl number of 0.7, the similarity solution also gives:

$$\frac{Nu_x}{Gr_x^{0.25}} = 0.353$$

i.e.:

$$\frac{q_w x}{k(T_w - T_1)} = 0.353 \left[\frac{\beta g(T_w - T_\infty)x^3}{\nu^2} \right]^{0.25}$$

In the situation here being considered it therefore follows that:

$$\frac{q_w x}{0.02638 \times (50 - 10)} = 0.353 \left[\frac{0.0033 \times 9.81 \times (50 - 10)x^3}{(16 \times 10^{-6})^2} \right]^{0.25}$$

i.e.:

$$q_w = \frac{99.3}{x^{0.25}}$$

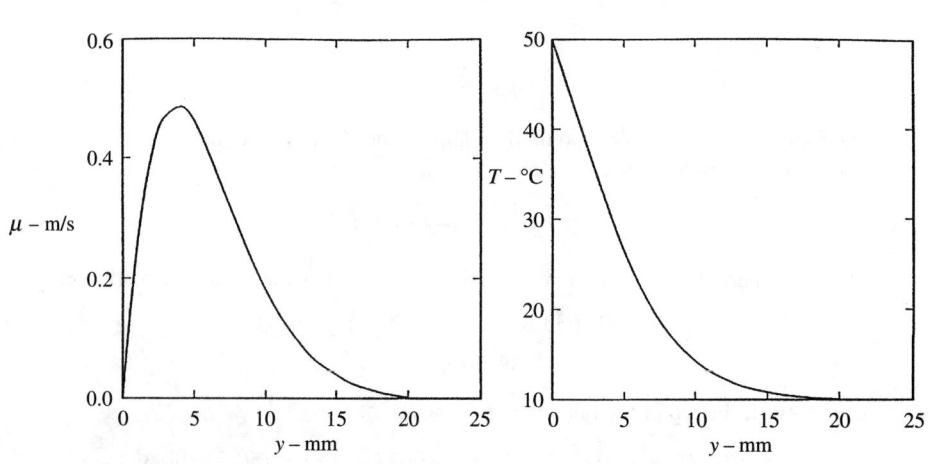

FIGURE E8.2a

CHAPTER 8: Natural Convection   363

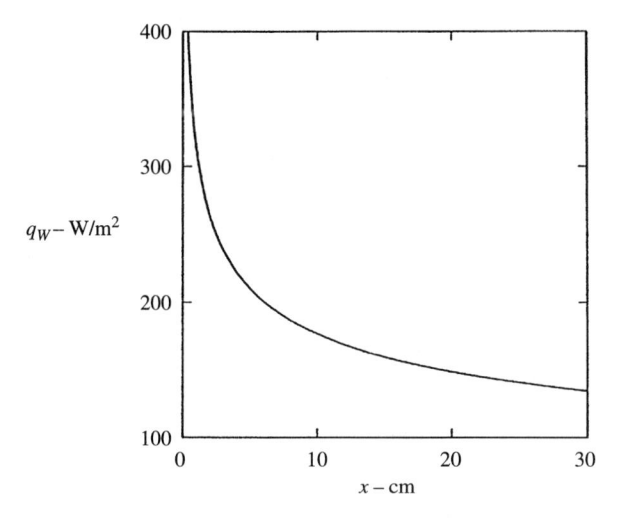

FIGURE E8.2b

The variation of the local heat transfer rate q_w with x as given by this equation is shown in Fig. E8.2b.

EXAMPLE 8.3. Plot the variation of the boundary layer thicknesses with distance along a 1-m high vertical plate which is maintained at a uniform surface temperature of 50°C and exposed to stagnant air at ambient pressure and a temperature of 10°C. Assume the flow remains laminar.

Solution. As indicated schematically in Fig. E8.3a, it can be deduced from Figs. 8.7 and 8.8 that the velocity decreases to 0.01 of its maximum value and the temperature difference decreases to 0.01 of its overall value for a Prandtl number of 0.7 (the assumed value for air) approximately when:

$$\eta = 7 \quad \text{and} \quad \eta = 5.5$$

respectively.

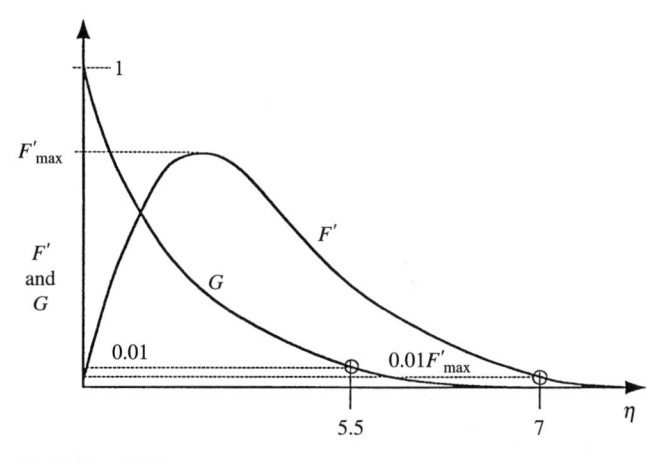

FIGURE E8.3a

Hence:

$$\frac{\delta}{x} Gr_x^{0.25} = 7$$

and:

$$\frac{\delta_T}{x} Gr_x^{0.25} = 5.5$$

where δ and δ_T are the effective thicknesses of the velocity and temperature boundary layers, respectively.

The average air temperature is given by:

$$T_{av} = \frac{(50 + 10)}{2} = 30°C$$

At this temperature, air at standard ambient pressure, has the following properties:

$$\beta = 0.0033 \text{ K}^{-1}, \quad \nu = 16 \times 10^{-6} \text{ m}^2/\text{s}$$

Hence:

$$Gr_x = \frac{\beta g(T_w - T_1)x^3}{\nu^2} = \frac{0.0033 \times 9.81 \times 40}{0.000016^2} x^3 = 5,058,280 x^3$$

where x is is the distance up the plate in m. Using this expression for Gr_x then gives:

$$\delta = \frac{7x}{(5,058,280 x^3)^{0.25}} = 0.1476 x^{0.25}$$

and:

$$\delta_T = \frac{5.5x}{(5,058,280 x^3)^{0.25}} = 0.1160 x^{0.25}$$

The variations of δ and δ_T with x as given by these two equations are shown in Fig. E8.3b.

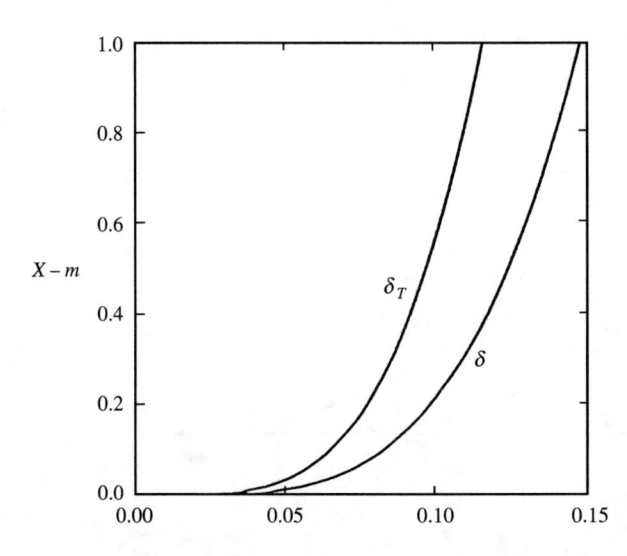

FIGURE E8.3b

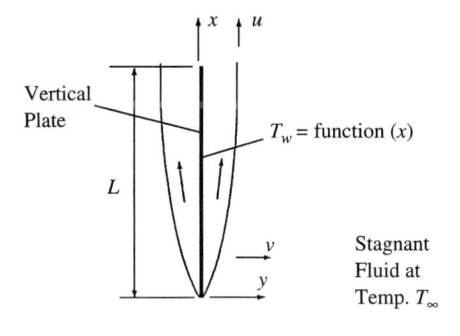

CHAPTER 8: Natural Convection 365

FIGURE 8.10
Flow situation considered in the discussion of the numerical solution of laminar natural convection.

8.7 NUMERICAL SOLUTION OF THE NATURAL CONVECTIVE BOUNDARY LAYER EQUATIONS

A numerical solution to the laminar boundary layer equations for natural convection can be obtained using basically the same method as applied to forced convection in Chapter 3. Because the details are similar to those given in Chapter 3, they will not be repeated here.

A computer program, LAMBNAT, written in FORTRAN that is based on this procedure is available as discussed in the Preface. This program applies to two-dimensional flow over a vertical flat plate with a surface temperature that in general is a function of the distance along the plate, x, i.e., to the flow situation shown in Fig. 8.10.

EXAMPLE 8.4. Use the computer program, LAMBNAT, for two-dimensional laminar boundary layer flow over a vertical plate to find the dimensionless heat transfer rate variation along a plate whose surface temperature varies linearly in such a way that T_w is equal to T_∞ at the leading edge and equal to $T_\infty + 40°C$ at the trailing edge.

Solution. The dimensionless temperature used in the numerical solution is given by:

$$\theta = (T - T_1)/(T_{wr} - T_1)$$

In the present situation $T_w - T_1$ varies from 0 at $X = 0$ to 40°C at $X = 1$. The reference wall temperature will therefore be taken to be such that:

$$T_{wr} - T_1 = 20$$

The dimensionless wall temperature will then vary from $\theta_w = 0$ at $X = 0$ to $\theta_w = 2$ at $X = 1$, i.e.:

$$\theta_w = 2X$$

The only change required to the program as available is in the subroutine TEMP which becomes:

```
*
      SUBROUTINE TEMP(X,TW)
*
*********** THIS DETERMINES THE WALL TEMPERATURE ********************
*
*
      TW = 2.0*X
      RETURN
```

FIGURE E8.4

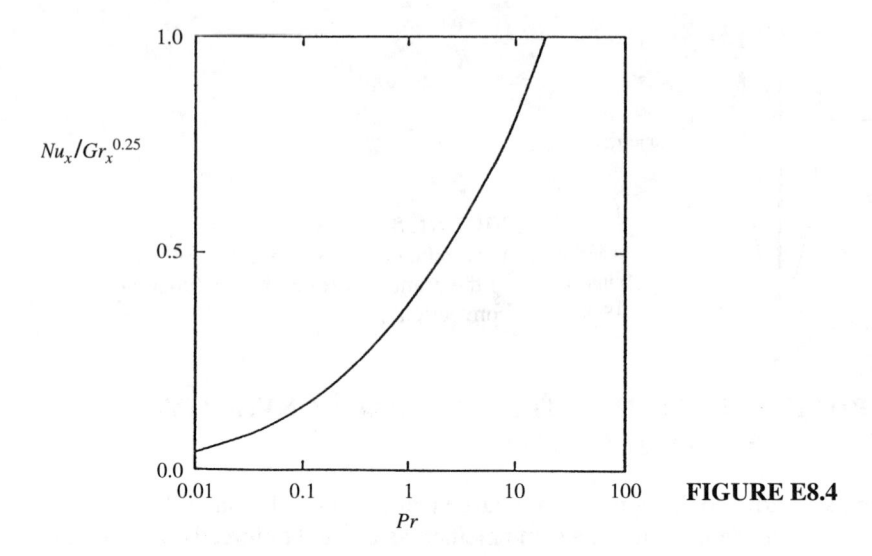

The program when run with this change gives the variation of $Nu_x/Gr_x^{0.25}$ with X where here:

$$Gr_x = \frac{\beta g(T_{wr} - T_1)x^3}{\nu^2}$$

The variation obtained in this way is shown in Fig. E8.4.

8.8
FREE CONVECTIVE FLOW THROUGH A VERTICAL CHANNEL

Attention will be given in this section to buoyancy-induced flows through a vertical channel in a large surrounding environment. The flow arises because all or part of the walls of the channel are at a different temperature from that of the fluid surrounding the channel, see [24] to [33]. It will be assumed in the present discussion that the walls are heated and that the flow is consequently upward through the channel. The same analysis can be used, of course, when the walls are cooled, the flow then being downward through the channel. The flow situation here being considered is, therefore, as shown in Fig. 8.11.

Flows of this general type occur, for example, in a number of situations involving the cooling of electrical and electronic equipment and in the flow through certain types of fin arrangement.

Here it will be assumed, as shown in Fig. 8.11, that the flow enters the channel through a smooth, convergent inlet in which the viscous losses are negligible. On the inlet plane of the channel, the velocity will consequently be uniform and equal to the mean velocity at any section of the channel, u_m. This is illustrated in Fig. 8.12.

Because viscous effects are assumed to be negligible in the inlet section, Bernoulli's equation gives across this inlet section:

$$p_\infty = p_i + \frac{\rho u_m^2}{2}$$

CHAPTER 8: Natural Convection   367

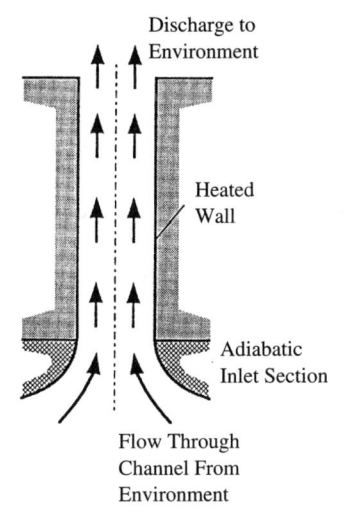

FIGURE 8.11
Free convective flow through a vertical channel.

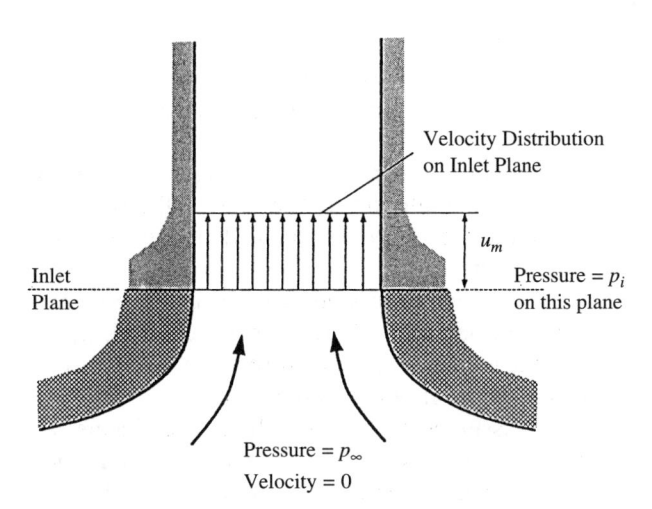

FIGURE 8.12
Assumed velocity distribution on inlet plane.

i.e.:
$$p_i - p_\infty = -\frac{\rho u_m^2}{2} \qquad (8.68)$$

The pressure therefore falls across the inlet section from p_∞ to $p_\infty - \rho u_m^2/2$, i.e., the pressure on the inlet plane is below ambient. As the fluid flows up the channel, viscous forces tend to cause the pressure to drop while the buoyancy forces tend to cause the pressure to rise. Because the flow leaving the channel is essentially parallel to the walls of the channel, the pressure on the exit plane is essentially uniform and equal to the ambient pressure. In the present analysis, it will therefore be assumed that across the exit plane the pressure is equal to the ambient pressure, p_∞. This is illustrated in Fig. 8.13.

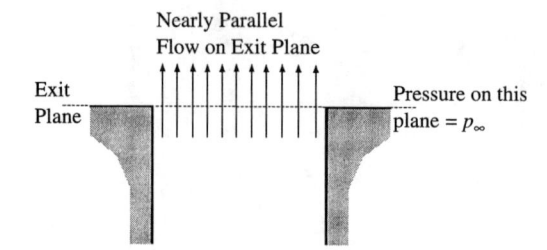

FIGURE 8.13
Assumed conditions on channel exit plane.

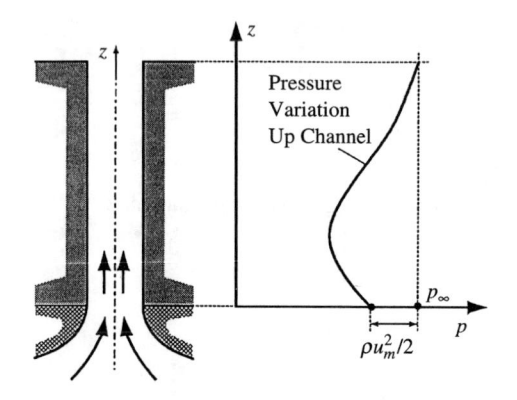

FIGURE 8.14
Pressure variation along channel.

The pressure variation up the channel therefore resembles that shown in Fig. 8.14. Near the inlet the viscous forces are high and the pressure falls. However, further up the channel, the buoyancy forces become dominant and the pressure rises, reaching the ambient value on the exit plane.

In order to illustrate how natural convection in a vertical channel can be analyzed, attention will be given to flow through a wide rectangular channel, i.e., to laminar, two-dimensional flow in a plane channel as shown in Fig. 8.15. This type of flow is a good model of a number of flows of practical importance.

In the present analysis it will be assumed that both walls of the channel are heated to the same uniform temperature and that the flow is therefore symmetrical about the channel center line. The coordinate system shown in Fig. 8.16 will therefore be used in the analysis and, because of the assumed symmetry, the solution will only be obtained for y values between 0 and $W/2$, W being, as indicated in Fig. 8.16, the full width of the channel.

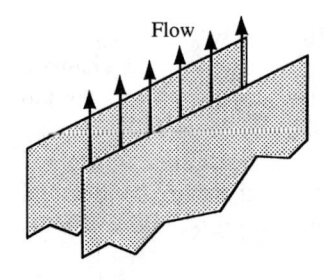

FIGURE 8.15
Type of flow considered.

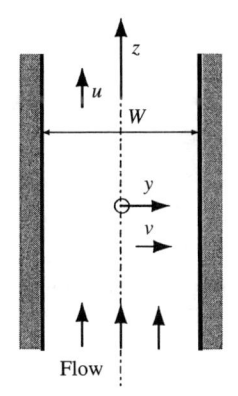

FIGURE 8.16
Coordinate system used.

If it is assumed that W is small compared to the height of the channel, H, which will mean that:

$$\frac{\partial^2 u}{\partial z^2} \ll \frac{\partial^2 u}{\partial y^2}, \quad v \ll u, \quad \frac{\partial p}{\partial y} \ll 1$$

i.e., if the "parabolic" flow assumptions are adopted, the momentum equation in the z-direction gives:

$$\rho u \frac{\partial u}{\partial z} + \rho v \frac{\partial u}{\partial y} = -\frac{dp}{dz} + \mu \left(\frac{\partial^2 u}{\partial y^2}\right) - \rho g \tag{8.69}$$

where p is the absolute pressure. The "parabolized" forms of the governing equation are thus being used.

It is convenient to work with the pressure relative to ambient pressure, i.e., with the "gauge" pressure. It is therefore noted that:

$$p = (p - p_\infty) + p_\infty$$

where p_∞ is the local ambient pressure at the value of z being considered.

The above equation gives:

$$\frac{dp}{dz} = \frac{d}{dz}(p - p_\infty) + \frac{dp_\infty}{dz} \tag{8.70}$$

But:

$$\frac{dp_\infty}{dz} = -\rho_\infty g$$

Hence, Eq. (8.70) gives:

$$\frac{dp}{dz} = \frac{d}{dz}(p - p_\infty) - \rho_\infty g \tag{8.71}$$

Substituting Eq. (8.71) into Eq. (8.69) and dividing through by ρ then gives:

$$u \frac{\partial u}{\partial z} + v \frac{\partial u}{\partial y} = -\frac{1}{\rho}\frac{d}{dz}(p - p_\infty) + \nu \left(\frac{\partial^2 u}{\partial y^2}\right) + \frac{(\rho_\infty - \rho)g}{\rho} \tag{8.72}$$

370 Introduction to Convective Heat Transfer Analysis

Hence, using:

$$\frac{\rho_\infty - \rho}{\rho} = \beta(T - T_\infty)$$

Eq. (8.72), i.e., the z-momentum equation, becomes:

$$u\frac{\partial u}{\partial z} + v\frac{\partial u}{\partial y} = -\frac{1}{\rho}\frac{d}{dz}(p - p_\infty) + \nu\left(\frac{\partial^2 u}{\partial y^2}\right) + \beta g(T - T_\infty) \quad (8.73)$$

The Boussinesq approximation has, of course, been used in deriving the above equation.

The continuity equation, as before, gives:

$$\frac{\partial u}{\partial z} + \frac{\partial v}{\partial y} = 0 \quad (8.74)$$

If it is assumed that:

$$\frac{\partial^2 T}{\partial z^2} \ll \frac{\partial^2 T}{\partial y^2}$$

and if the other assumptions used in deriving the z-momentum equation are adopted, the energy equation gives for the flow here being considered:

$$u\frac{\partial T}{\partial z} + v\frac{\partial T}{\partial y} = \left(\frac{k}{\rho c_p}\right)\left(\frac{\partial^2 T}{\partial y^2}\right) \quad (8.75)$$

The boundary conditions on the above set of equations are as follows, it being recalled that it is being assumed that the flow is symmetrical about the center line:

$$\text{At } y = 0: \quad \frac{\partial u}{\partial y} = 0, \quad v = 0, \quad \frac{\partial T}{\partial y} = 0 \quad (8.76)$$

and:

$$\text{At } y = W/2: \quad u = 0, \quad v = 0, \quad T = T_W \quad (8.77)$$

the y-coordinate, as discussed before and as shown in Fig. 8.17, being measured from the center line towards the wall. Eq. (8.76) follows from the assumed symmetry of the flow about the center line.

In addition, as discussed before, on the inlet and exit planes of the channel, it is assumed that:

$$\text{At } z = 0: \quad u = u_m, \quad p - p_\infty = -\frac{\rho u_m^2}{2} \quad (8.78)$$

and:

$$\text{At } z = \ell: \quad p = p_\infty \quad (8.79)$$

where ℓ is the vertical height of the channel.

It is also assumed that there is no heat transfer to the fluid in the inlet section so that on the inlet plane the fluid is at the ambient temperature, i.e.:

$$\text{At } z = 0: \quad T = T_\infty$$

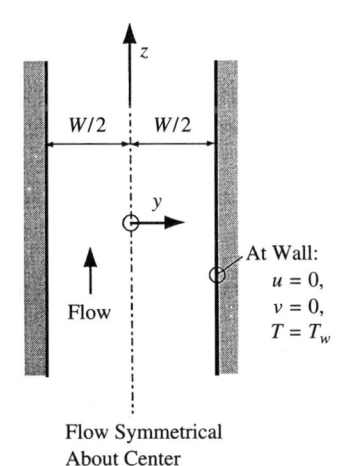

FIGURE 8.17
Boundary conditions on solution.

Two limiting types of flow can exist. If the channel is short and the Rayleigh number is high (see later), the flow will essentially consist of boundary layers on each wall with a uniform flow at temperature T_∞ between the boundary layers as shown in Fig. 8.18. Under these circumstances it is to be expected that a boundary-type relation for the heat transfer rate will apply.

The other limit would occur if the channel was long and the Rayleigh number was low. In this case, a type of fully developed flow would be expected to exist.

Approximate solutions for the two limiting cases discussed above can be obtained (see below). However, most real flows are not well described by either of these two limiting solutions. For this reason, a numerical solution of the governing equations must usually be obtained. To illustrate how such solutions can be obtained, a simple forward-marching, explicit finite-difference solution will be discussed here.

Although it is not a necessary step, the governing equations and the boundary conditions will be written in dimensionless form before obtaining the numerical

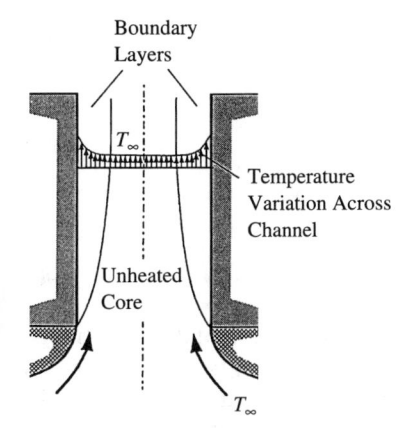

FIGURE 8.18
Boundary layer flow regime.

solution. For this purpose, the following is defined:

$$G = \frac{\beta g(T_w - T_\infty) W^4}{\nu^2 \ell} = Gr_w \frac{W}{\ell} \quad (8.80)$$

where Gr_w is the Grashof number based on the channel width, W. The following dimensionless variables are then defined:

$$Z = \frac{z}{\ell G}, \quad Y = \frac{y}{W}, \quad \theta = (T - T_\infty)/(T_w - T_\infty)$$

$$U = \left(\frac{uW}{\nu}\right)\left(\frac{W}{\ell G}\right), \quad V = \frac{vW}{\nu}, \quad P = \frac{(p - p_\infty)W^4}{\rho^2 \ell^2 G^2 \nu^2} \quad (8.81)$$

In terms of these dimensionless variables, the governing equations become:

$$\frac{\partial V}{\partial Y} + \frac{\partial U}{\partial Z} = 0 \quad (8.82)$$

$$U \frac{\partial U}{\partial Z} + V \frac{\partial U}{\partial Y} = -\frac{dP}{dZ} + \left(\frac{\partial^2 U}{\partial Y^2}\right) + \theta \quad (8.83)$$

$$U \frac{\partial \theta}{\partial Z} + V \frac{\partial \theta}{\partial Y} = \left(\frac{1}{Pr}\right)\left(\frac{\partial^2 \theta}{\partial Y^2}\right) \quad (8.84)$$

In terms of these dimensionless variables, the boundary conditions are:

$$\text{At } Y = 0: \quad \frac{\partial U}{\partial Y} = 0, \quad V = 0, \quad \frac{\partial \theta}{\partial Y} = 0 \quad (8.85)$$

and:

$$\text{At } Y = 0.5, \quad U = 0, \quad V = 0, \quad \theta = 1 \quad (8.86)$$

while the conditions on the inlet and exit planes are:

$$\text{At } Z = 0: \quad U = U_m, \quad P = -\frac{U_m^2}{2,} \quad \theta = 0 \quad (8.87)$$

and:

$$\text{At } Z = L: \quad P = 0 \quad (8.88)$$

where:

$$U_m = \left(\frac{u_m W}{\nu}\right)\left(\frac{W}{\ell G}\right) \quad (8.89)$$

and:

$$L = \frac{\ell}{\ell G} = \frac{1}{G} \quad (8.90)$$

are the dimensionless mean velocity and dimensionless channel length respectively.

Eqs. (8.82), (8.83), and (8.84) are the equations that must be solved. At first viewing, they may appear to be a set of three equations in four variables, i.e., U, V, P, and θ. However, if the boundary conditions given in Eqs. (8.85) and (8.86) are examined, it will be seen that there are more boundary conditions than are required to solve Eqs. (8.82) to (8.83). In fact, the requirement that V be equal to zero at both $Y = 0$ and at $Y = 0.5$, which by virtue of Eq. (8.82) is equivalent to:

$$\frac{\partial}{\partial Z} \int_0^{0.5} U \, dY = 0 \tag{8.91}$$

(this is equivalent to:

$$\int_0^{0.5} U \, dY = \text{constant}$$

i.e., is a statement of the fact that mass neither leaves or enters the duct) gives the extra relationship required to give four equations in four variables. Essentially, this equation is used to determine the pressure gradient which must be such that continuity of mass flow is ensured.

As previously mentioned, the solution to the above set of equations will here be obtained using a forward-marching, explicit finite-difference procedure. The solution starts with the known conditions on the inlet plane and marches forward in the z-direction from grid line to grid line as indicated in Fig. 8.19. Consider the nodal points shown in Fig. 8.20.

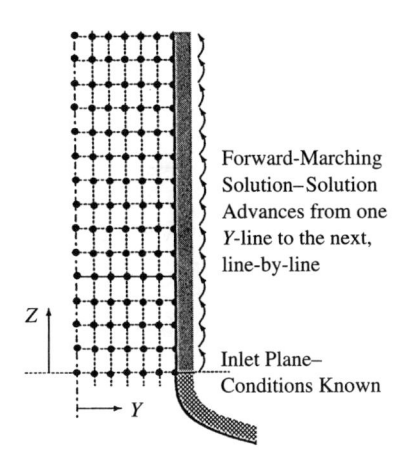

FIGURE 8.19
Forward-marching solution procedure used.

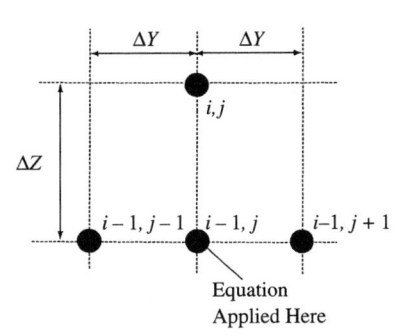

FIGURE 8.20
Nodal point considered.

In terms of the values of the variables at these points, the momentum equation, i.e., Eq. (8.83), gives the following, it being recalled that, since an explicit solution procedure is being used, all the Y derivatives are evaluated on the line from which the solution is advancing, i.e., on the $i - 1$ line:

$$U_{i-1,j}\left(\frac{U_{i,j} - U_{i-1,j}}{\Delta Z}\right) + V_{i-1,j}\left(\frac{U_{i-1,j+1} - U_{i-1,j-1}}{2\Delta Y}\right)$$

$$= -\frac{P_i - P_{i-1}}{\Delta Z} + \frac{(U_{i-1,j+1} + U_{i-1,j-1} - 2U_{i-1,j})}{\Delta Y^2} + \theta_{i-1,j}$$

which gives:

$$U_{i,j} = U_{i-1,j} - \frac{P_i}{U_{i-1,j}} + \frac{P_{i-1}}{U_{i-1,j}} - \frac{V_{i-1,j}}{U_{i-1,j}}\left(\frac{U_{i-1,j+1} - U_{i-1,j-1}}{2\Delta Y}\right)\Delta Z$$

$$+ \frac{(U_{i-1,j+1} + U_{i-1,j-1} - 2U_{i-1,j})}{\Delta Y^2}\frac{\Delta Z}{U_{i-1,j}} + \theta_{i-1,j}\frac{\Delta Z}{U_{i-1,j}}$$

i.e.:

$$U_{i,j} = U_{i-1,j} - \frac{P_i}{U_{i-1,j}} + A_j \qquad (8.92)$$

where:

$$A_j = \frac{P_{i-1}}{U_{i-1,j}} - \frac{V_{i-1,j}}{U_{i-1,j}}\left(\frac{U_{i-1,j+1} - U_{i-1,j-1}}{2\Delta Y}\right)\Delta Z$$

$$+ \frac{(U_{i-1,j+1} + U_{i-1,j-1} - 2U_{i-1,j})}{\Delta Y^2}\frac{\Delta Z}{U_{i-1,j}} + \theta_{i-1,j}\frac{\Delta Z}{U_{i-1,j}} \qquad (8.93)$$

This equation applies from $j = 2$ to $j = N - 1$. A_j is a known quantity at any stage of the calculation because it only involves the values of the variables on the $(i - 1)$ line which are known.

Attention will next be given to the continuity equation which can be written as:

$$\frac{\partial V}{\partial Y} = -\frac{\partial U}{\partial Z} \qquad (8.94)$$

This equation is again applied to the nodal points shown in Fig. 8.21. The equation is actually applied to the point O, shown in Fig. 8.21 which lies halfway between nodal points i, j and $i, j - 1$.

The value of $\partial U/\partial Z$ at point O is, as before, taken as the average of the values of $\partial U/\partial Z$ at these two points. Hence, the continuity equation gives:

$$\frac{V_{i,j} - V_{i,j-1}}{\Delta Y} = -\frac{1}{2}\left[\frac{U_{i,j} - U_{i-1,j}}{\Delta Z} + \frac{U_{i,j-1} - U_{i-1,j-1}}{\Delta Z}\right]$$

i.e.:

$$V_{i,j} - V_{i,j-1} = -\frac{\Delta Y}{2\Delta Z}[U_{i,j} - U_{i-1,j} + U_{i,j-1} - U_{i-1,j-1}] \qquad (8.95)$$

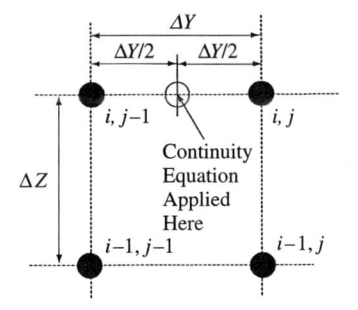

FIGURE 8.21
Nodal point used with continuity equation.

Substituting Eq. (8.92) into this equation then gives:

$$V_{i,j} - V_{i,j-1} = -\frac{\Delta Y}{2\Delta Z}\left[-\frac{P_i}{U_{i-1,j}} + A_j - \frac{P_i}{U_{i-1,j-1}} + A_{j-1}\right] \quad (8.96)$$

which can be written as:

$$V_{i,j} - V_{i,j-1} = -P_i\left[\frac{\Delta Y}{2\Delta Z}\left(\frac{1}{U_{i-1,j}} + \frac{1}{U_{i-1,j-1}}\right)\right] - \frac{\Delta Y}{2\Delta Z}(A_j + A_{j-1})$$

i.e., as:

$$V_{i,j} - V_{i,j-1} = -B_j P_i - C_j \quad (8.97)$$

where:

$$B_j = \frac{\Delta Y}{2\Delta Z}\left[\frac{1}{U_{i-1,j}} + \frac{1}{U_{i-1,j-1}}\right] \quad (8.98)$$

and:

$$C_j = \frac{\Delta Y}{2\Delta Z}(A_j + A_{j-1}) \quad (8.99)$$

It should be noted that Eq. (8.92) only applies for $j = 2$ to $N - 1$. The boundary conditions, however, give $U_{i,N} = U_{i-1,N} = 0$ so:

$$B_N = \frac{\Delta Y}{2\Delta Z} C_{N-1} \quad (8.100)$$

and:

$$C_N = \frac{\Delta Y}{2\Delta Z}\frac{1}{U_{i-1,N-1}} \quad (8.101)$$

It will be noted that since at any stage of the solution the conditions on the $(i-1)$ line are known, the coefficients A_j, A_{j-1}, B_j, and C_j are known quantities.

Now on the center line, i.e., at $j = 1$, $V_{i,j}$ is zero. Hence if Eq. (8.97) is applied sequentially outward from point $j = 2$ to the point $j = N$ which lies on the wall,

376 Introduction to Convective Heat Transfer Analysis

the following is obtained:
$$V_{i,2} - 0 = -B_2 P_i - C_2$$
$$V_{i,3} - V_{i,2} = -B_3 P_i - C_3$$
$$V_{i,4} - V_{i,3} = -B_4 P_i - C_4$$
$$V_{i,5} - V_{i,4} = -B_5 P_i - C_5$$
$$\vdots$$
$$V_{i,N} - V_{i,N-1} = -B_N P_i - C_N$$

Adding this set of equations together then gives:

$$V_{i,N} = -P_i \sum_{k=2}^{N} B_k - \sum_{k=2}^{N} C_k \tag{8.102}$$

But the boundary conditions give $V_{i,N} = 0$ so Eq. (8.102) gives:

$$P_i = \frac{\sum_{k=2}^{N} C_k}{\sum_{k=2}^{N} B_k} \tag{8.103}$$

Eq. (8.103) allows the dimensionless pressure, P_i, to be determined. Once this has been found, Eq. (8.92) can be used to find the $U_{i,j}$ values and Eq. (8.97) applied sequentially outward from $j = 2$ as discussed above allows the $V_{i,j}$ values to be found. The boundary conditions give $U_{i,N} = 0$ and since to first order of accuracy the zero gradient condition on the center line gives:

$$\frac{U_{i,2} - U_{i,1}}{\Delta Y} = 0$$

it follows that $U_{i,1} = U_{i,2}$.

Lastly, consider the finite-difference form of the dimensionless energy equation, i.e., Eq. (8.84), which is used to determine the distribution of $\theta_{i,j}$. The finite-difference form of Eq. (8.84) is obtained using the same nodal points as used in dealing with the momentum equation (see Fig. 8.20). The finite-difference form of Eq. (8.84) is:

$$U_{i-1,j}\left(\frac{\theta_{i,j} - \theta_{i-1,j}}{\Delta Z}\right) + V_{i-1,j}\left(\frac{\theta_{i-1,j+1} - \theta_{i-1,j-1}}{2\Delta Y}\right)$$
$$= \frac{1}{Pr}\left(\frac{\theta_{i-1,j+1} + \theta_{i-1,j-1} - 2\theta_{i-1,j}}{\Delta Y^2}\right)$$

which gives:

$$\theta_{i,j} = \theta_{i-1,j} + \left(\frac{\Delta Z}{U_{i-1,j}}\right)\left[\left(\frac{\theta_{i-1,j+1} + \theta_{i-1,j-1} - 2\theta_{i-1,j}}{Pr\Delta Y^2}\right) - V_{i-1,j}\left(\frac{\theta_{i-1,j+1} - \theta_{i-1,j-1}}{2\Delta Y}\right)\right]$$

$$\tag{8.104}$$

This equation allows the $\theta_{i,j}$ values for $j = 2$ to N to be found. The boundary conditions give $\theta_{i,N} = 1$ and, to first order accuracy, the symmetry condition gives $\theta_{i,1} = \theta_{i,2}$.

The above set of equations allows the values of the variables on the i-line to be determined from the known values of the variables on the $(i - 1)$ line. Therefore, starting with the known conditions on the inlet, the solution can be marched up the channel from i-line to i-line.

In general, the solution procedure adopted should allow the value of the dimensionless duct length, L, to be selected and the solution should give the value of the dimensionless mean velocity, U_m. It is, however, more convenient to select a value of U_m and use the solution procedure to calculate the value of L corresponding to this value of U_m. If the result for a particular value of L is required it can be obtained by interpolating between results obtained for several different values of U_m.

The actual solution procedure therefore involves the following steps:

1. A value of U_m is selected.
2. The values of the variables on the first i-line which lies on the inlet plane are specified, i.e.:

$$i = 1, \quad j = 1 \text{ to } N - 1: \quad U = U_m, \quad V = 0, \quad \theta = 0$$
$$i = 1, \quad j = N: \quad U = 0, \quad V = 0, \quad \theta = 1$$

   Also on the inlet plane:

$$i = 1: \quad P = -U_m^2/2$$

3. Using the values of $U_{1,j}$, $V_{1,j}$, and P_1 find the values of A_j, B_j, and C_j.
4. Use Eq. (8.103) to find the value of P_2.
5. Use Eq. (8.92) to find the values of $U_{2,j}$.
6. Use Eq. (8.97) to find the values of $V_{2,j}$.
7. Use Eq. (8.104) to find the values of $\theta_{2,j}$.
8. Having found the values of the variables in this way on the second i-line, the same procedure can be used to advance the solution to the next i-line and so on up the channel.
9. The process is continued until the value of the dimensionless pressure, P_i, becomes positive. The actual dimensionless duct length that corresponds to the chosen value of U_m, i.e., the Z value that gives $P = 0$ is then given by interpolation, i.e., by:

$$L = Z_{i-1} + \left(\frac{0 - P_{i-1}}{P_i - P_{i-1}}\right)(Z_i - Z_{i-1})$$

i.e.:

$$L = Z_{i-1} - \left(\frac{P_{i-1}}{P_i - P_{i-1}}\right)(Z_i - Z_{i-1})$$

Having discussed how the values of variables can be found, attention will be turned to the determination of the heat transfer rate. The mean heat transfer rate up

378   Introduction to Convective Heat Transfer Analysis

to any value of Z, i.e., $\overline{q_w}$, can be obtained by noting that:

$$\begin{matrix} \text{Total heat transfer} \\ \text{from inlet to } z \end{matrix} = \begin{matrix} \text{Rate enthalpy crosses} \\ \text{channel section at } z \end{matrix} - \begin{matrix} \text{Rate enthalpy} \\ \text{enters the channel} \end{matrix}$$

i.e., considering unit width of the duct and recalling that the flow is being assumed to be symmetrical about the center line:

$$2\overline{q_w}z = 2\int_0^{W/2} \rho u c_p T \, dy - \rho u_m c_p T_\infty W$$

$$= 2\int_0^{W/2} \rho u c_p (T - T_\infty) \, dy - 2\int_0^{W/2} \rho u c_p T_\infty \, dy - \rho u_m c_p T_\infty W$$

$$= 2\int_0^{W/2} \rho u c_p (T - T_\infty) \, dy - 2c_p T_\infty \int_0^{W/2} \rho u \, dy - \rho u_m c_p T_\infty W$$

(8.105)

But conservation of mass requires that:

$$2\int_0^{W/2} \rho u \, dy = \rho u_m W$$

so Eq. (8.105) becomes:

$$2\overline{q_w}z = 2\int_0^{W/2} \rho u c_p (T - T_\infty) \, dy$$

i.e.:

$$\overline{q_w} = \frac{1}{z}\int_0^{W/2} \rho u c_p (T - T_\infty) \, dy \qquad (8.106)$$

In terms of the dimensionless variables defined in Eq. (8.81) this equation becomes:

$$\frac{\overline{q_w}W}{k(T_w - T_\infty)} = \frac{1}{Z}\frac{\mu c_p}{k}\int_0^{0.5} U\theta \, dY$$

i.e.:

$$\frac{\overline{q_w}W}{k(T_w - T_\infty)Pr} = \frac{1}{Z}\int_0^{0.5} U\theta \, dY \qquad (8.107)$$

i.e., defining:

$$Q_a = \frac{\overline{q_w}W}{k(T_w - T_\infty)Pr} \qquad (8.108)$$

Eq. (8.107) can be written:

$$Q_a = \frac{1}{Z}\int_0^{0.5} U\theta \, dY \qquad (8.109)$$

Using the values of U and θ given by the numerical solution at any value of Z, the value of the right-hand side of this equation can be obtained by numerical integration so allowing Q_a to be found. For example, to the same order of accuracy as used in obtaining the solution for the flow, Eq. (8.109) gives:

$$Q_a = \frac{1}{Z_i}\left[\frac{U_{i,1}\theta_{i,1}}{2} + U_{i,2}\theta_{i,2} + U_{i,3}\theta_{i,3} + \ldots + U_{i,N-1}\theta_{i,N-1}\right]\Delta Y \quad (8.110)$$

it having been noted that the boundary conditions give $U_{i,N} = 0$.

The mean dimensionless heat transfer rate for the entire channel will be given by:

$$Q_L = \frac{1}{L}\int_0^{0.5} U\theta \, dY$$

the integral being evaluated on the exit plane of the channel.

A computer program, NATCHAN, written in FORTRAN that is based on the procedure that is outlined above is available as discussed in the Preface. Some results obtained using this program for $Pr = 0.7$ are shown in Figs. 8.22 and 8.23.

Because an explicit finite-difference procedure is being used to solve the momentum and energy equations, the solution can become unstable, i.e., as the solution proceeds it can diverge increasingly from the actual solution. The analysis of the conditions under which such an instability will develop that was given in Chapter 4 for the case of forced convection in a duct essentially applies here and shows that in order to avoid instability, ΔZ should be selected so that:

$$\Delta Z < 0.5\Delta Y^2 U_{i-1,N-1}$$

Because this criterion was obtained using an approximate analysis, it is usual to assume that for stability:

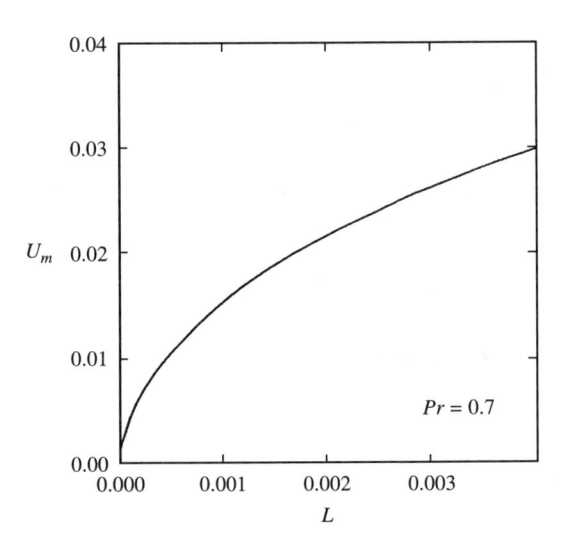

$$\Delta Z < K\Delta Y^2 U_{i-1,N-1}$$

FIGURE 8.22
Variation of U_m with L for $Pr = 0.7$.

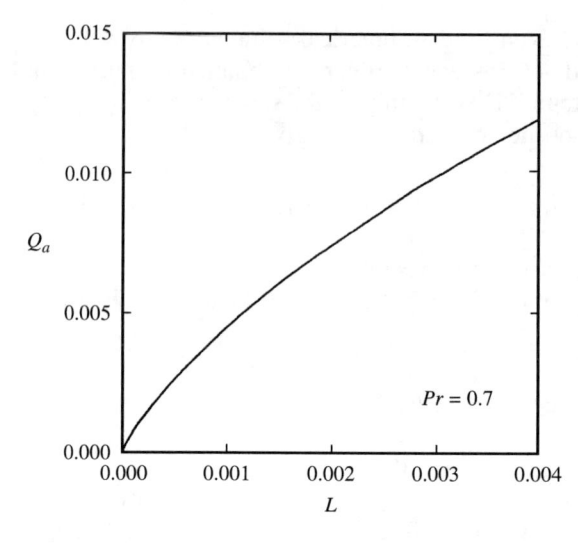

FIGURE 8.23
Variation of Q_a with L for $Pr = 0.7$.

where K is less than 0.5. This stability criterion is incorporated into the program NATCHAN.

As previously discussed, there are two limiting cases for natural convective flow through a vertical channel. One of these occurs when ℓ/W is large and the Rayleigh number is low. Under these circumstances all the fluid will be heated to very near the wall temperature within a relatively short distance up the channel and a type of fully developed flow will exist in which the velocity profile is not changing with Z and in which the dimensionless cross-stream velocity component, V, is essentially zero, i.e., in this limiting solution:

$$U = F(Y), \quad V = 0, \quad \theta = 1$$

the dimensionless wall temperature being 1. The velocity function F is independent of Z.

In this limiting case, then, Eq. (8.83) reduces to:

$$-\frac{dP}{dZ} + \left(\frac{d^2U}{dY^2}\right) + 1 = 0 \qquad (8.111)$$

The pressure gradient will be constant in fully developed flow so this equation shows that in this limiting case:

$$\frac{d^2U}{dY^2} = K$$

where K is a constant whose value has to be determined.

Integrating this equation once gives:

$$\frac{dU}{dY} = KY + C_1$$

where C_1 is a constant of integration. Integrating this equation in turn then gives:

$$U = \frac{KY^2}{2} + C_1Y + C_2Y^2 \tag{8.112}$$

C_2 being another constant of integration.

The two constants of integration are determined by applying the boundary conditions, i.e., by using:

$$Y = 0: \quad \frac{dU}{dY} = 0, \quad Y = 0.5: \quad U = 0$$

Applying these to Eq. (8.112) gives:

$$C_1 = 0, \quad \frac{K}{8} + 0.5C_1 + C_2 = 0$$

These equations give:

$$C_1 = 0, \quad C_2 = -\frac{K}{8}$$

Substituting these values back into Eq. (8.112) then gives the velocity profile as:

$$U = -\frac{K}{2}\left(\frac{1}{4} - Y^2\right) \tag{8.113}$$

The dimensionless mean velocity is such that:

$$U_m = \frac{1}{0.5}\int_0^{0.5} U\, dY$$

Substituting Eq. (8.113) into this equation then gives:

$$U_m = -K\int_0^{0.5}\left(\frac{1}{4} - Y^2\right) dY = -\frac{K}{12}$$

i.e.:

$$K = -12U_m \tag{8.114}$$

Substituting this result back into Eq. (8.113) then gives:

$$U = 6U_m\left(\frac{1}{4} - Y^2\right) \tag{8.115}$$

It follows that:

$$\left(\frac{d^2U}{dY^2}\right) = K = -12U_m \tag{8.116}$$

Because the effects of the developing region are being neglected in the present limiting case analysis and because the pressure gradient is constant in the flow being

382 Introduction to Convective Heat Transfer Analysis

considered, it follows that:

$$-\frac{dP}{dZ} = -\left(\frac{P_{exit} - P_{in}}{L}\right)$$

where P_{in} and P_{exit} are the pressures on the inlet and outlet planes of the duct. But, as discussed before:

$$P_{in} = -\frac{U_m^2}{2} \quad \text{and} \quad P_{exit} = 0$$

therefore:

$$-\frac{dP}{dZ} = -\left(\frac{U_m^2}{2L}\right) \tag{8.117}$$

Substituting Eqs. (8.116) and (8.117) into Eq. (8.111) then gives:

$$\frac{U_m^2}{2L} = 1 - 12U_m$$

i.e.:

$$L = \frac{U_m^2}{2(1 - 12U_m)} \tag{8.118}$$

It should be noted that according to this equation the highest possible value of U_m is equal to 1/12 and this is approached as $L \to \infty$.

Because in the limiting flow situation here being considered $\theta = 1$, Eq. (8.109) gives at the exit of the pipe:

$$Q_L = \frac{1}{L}\int_0^{0.5} U\theta \, dY = \frac{1}{L}\int_0^{0.5} U \, dY = 0.5\frac{U_m}{L}$$

Substituting Eq. (8.118) into this equation then gives:

$$Q_L = \frac{1 - 12U_m}{U_m} = \left(\frac{1}{U_m} - 12\right) \tag{8.119}$$

Eqs. (8.118) and (8.119) allow the variations of U_m and Q_L with L to be determined for the limiting case of large ℓ/W and low Rayleigh number to be found.

The other limiting solution is that in which the flow essentially consists of boundary layers on each wall of the duct, these boundary layers being so thin compared to W that there is no interaction between the flows in the two boundary layers, i.e., the boundary layer on each wall of the duct behaves as a boundary layer on a vertical plate in a large environment. Now, for free convective boundary layer flow over a vertical plate of height ℓ, it was shown earlier in this chapter that:

$$\frac{q_w \ell}{(T_w - T_\infty)k} = A\left[\frac{\beta g(T_w - T_\infty)\ell^3}{\nu^2}\right]^{0.25}$$

A being a function of Pr alone.

In terms of the dimensionless variables being used here in the analysis of duct flow, this equation can be written as:

$$\frac{\overline{q_w} W}{k(T_w - T_0) Pr} = \frac{A}{Pr} \left[\frac{\beta g(T_w - T_\infty) W^4}{\nu^2 \ell} \right]^{0.25} = \frac{A}{L^{0.25} Pr}$$

i.e.:

$$Q_L = \frac{A/Pr}{L^{0.25}} \tag{8.120}$$

In the limiting boundary layer situation here being considered, the total flow through the channel will be small, i.e., effectively:

$$U_m = 0 \tag{8.121}$$

Eqs. (8.120) and (8.121) represent the limiting boundary layer solution for natural convective flow through a vertical plane duct. For the particular case of $Pr = 0.7$, the similarity solution for natural convective boundary layer flow on a vertical plate was shown to give $A = 0.477$. Hence, for $Pr = 0.7$, Eq. (8.120) gives:

$$Q_L = \frac{0.681}{L^{0.25}}$$

The two limiting solutions, as defined by Eqs. (8.118) and (8.119) and by Eqs. (8.120) and (8.121) can be compared with the full numerical solutions shown in Figs. 8.22 and 8.23 and it will be seen that the numerical results do tend to these limiting solutions at high and at low values of L.

The analysis of natural convective flow through a vertical plane duct that was described above is easily extended to deal with other geometrical situations, such as natural convective flow through a vertical pipe, and to deal with other thermal boundary conditions at the wall.

EXAMPLE 8.5. Consider the natural convective flow of air though a plane vertical channel with isothermal walls whose temperature and height are such that the Grashof number based on the height of these walls is 10^5. Determine how the dimensionless heat transfer rate based on the height of these heated walls varies with the gap between the walls.

Solution. Here the height, ℓ, of the channel and the wall temperature and air temperature are fixed. Now:

$$G = \frac{\beta g(T_w - T_\infty)}{\nu^2} \frac{W^4}{\ell} = \frac{\beta g(T_w - T_\infty) \ell^3}{\nu^2} \left(\frac{W}{\ell}\right)^4 = Gr_\ell \left(\frac{W}{\ell}\right)^4$$

But:

$$Gr_\ell = 10^5$$

so:

$$G = 10^5 \left(\frac{W}{\ell}\right)^4$$

Hence:

$$L = \frac{1}{G} = \frac{0.00001}{(W/\ell)^4} \tag{a}$$

384 Introduction to Convective Heat Transfer Analysis

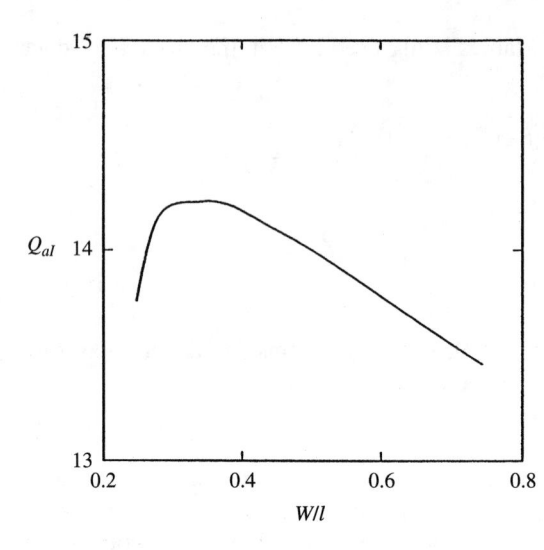

FIGURE E8.5

It is also noted that:

$$Q_a = \frac{\overline{q_w}W}{k(T_w - T_0)Pr} = \frac{\overline{q_w}\ell}{k(T_w - T_0)Pr}\frac{W}{\ell}$$

i.e.

$$Q_a = Q_\ell \frac{W}{\ell}$$

where:

$$Q_\ell = \frac{\overline{q_w}\ell}{k(T_w - T_0)Pr} \qquad (b)$$

The numerical solution discussed above gives the variation of Q_a with L. For any selected value of W/ℓ, Eq. (a) allows the value of L to be found. The numerical results then allow the value of Q_a corresponding to this value of L to be found. Eq. (b) then allows the value of Q_ℓ to be found. The variation of Q_ℓ with W/ℓ found using this procedure is shown in Fig. E8.5.

Because ℓ is constant in the situation being considered, Q_ℓ will be a measure of the heat transfer rate from the walls to the flow. It will be seen from Fig. E8.5 that there is therefore a value of W/ℓ that gives the highest heat transfer rate.

EXAMPLE 8.6. Air flows by natural convection through the channel formed between between 2 10-cm high plates which are kept at a temperature of 50°C. If the distance between the 2 plates is 2 cm and if the ambient air temperature is 10°C, find the rate of heat transfer from the 2 plates to the air. Also find the mean velocity of the air through the channel.

Solution. The mean air temperature will be taken as:

$$T_{av} = \frac{(50 + 10)}{2} = 30°C$$

At this temperature, air at standard ambient pressure, has the following properties:

$$\beta = 0.0033 \text{ K}^{-1}, \nu = 16 \times 10^{-6} \text{ m}^2/\text{s}, k = 0.02638 \text{ W/m-K}$$

Hence:

$$G = \frac{\beta g(T_w - T_\infty)}{\nu^2} \frac{W^4}{\ell} = \frac{0.0033 \times 9.81 \times (50 - 10)}{(16 \times 10^{-6})^2} \frac{0.02^4}{0.1} = 8093$$

From this it follows that:

$$L = \frac{1}{G} = \frac{1}{8093} = 0.0001236$$

From the numerically determined variations of U_m and Q_a with L for $Pr = 0.7$ it follows that for this value of L:

$$U_m = 0.00481 \text{ and } Q_a = 7.44$$

Hence:

$$Q_a = \frac{\overline{q_w} W}{k(T_w - T_0) Pr} = 7.44$$

and:

$$U_m = \left(\frac{u_m W}{\nu}\right)\left(\frac{W}{\ell G}\right) = 0.00481$$

Hence:

$$q_w = \frac{7.44 \times 0.02638 \times (50 - 10) \times 0.7}{0.02} = 274.8 \text{ W/m}^2$$

Therefore, considering unit width of the channel, the total heat transfer rate from the two walls will be:

$$Q = q_w A = 274.8 \times 2 \times 0.1 \times 1 = 54.96 \text{ W}$$

Also:

$$u_m = \frac{0.00481 \times 0.1 \times 8093 \times 16 \times 10^{-6}}{0.02} = 0.003114 \text{ m/s}$$

Therefore the heat transfer rate from the walls per unit channel width is 54.96 W and the mean velocity through the channel is 0.003114 m/s.

8.9
NATURAL CONVECTIVE HEAT TRANSFER ACROSS A RECTANGULAR ENCLOSURE

Heat transfer by natural convection across an enclosed space, called an enclosure or, sometimes, a cavity, occurs in many real situations, see [34] to [67]. For example, the heat transfer between the panes of glass in a double pane window, the heat transfer between the collector plate and the glass cover in a solar collector and in many electronic and electrical systems basically involves natural convective flow across an enclosure.

386 Introduction to Convective Heat Transfer Analysis

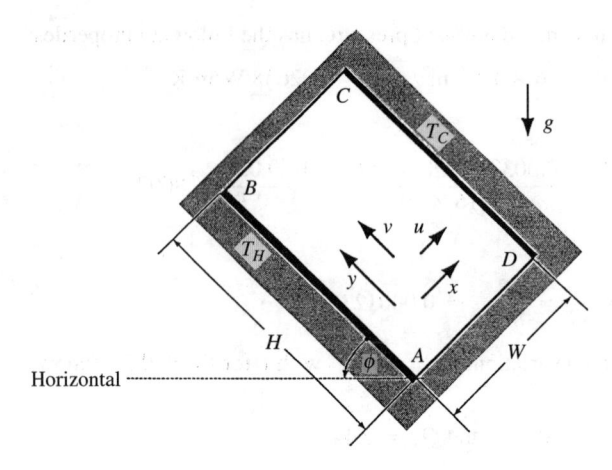

FIGURE 8.24
Type of enclosure considered.

Attention will here be restricted here to flow in a rectangular enclosure as shown in Fig. 8.24. In general, the enclosure is inclined to the vertical as illustrated in Fig. 8.24. For simplicity, in order to illustrate how enclosure flows can be analyzed, it will be assumed that one wall of the enclosure (AB in Fig. 8.24) is at a uniform high temperature, T_H, and that the opposite wall (CD in Fig. 8.24) is at a uniform low temperature, T_C. Two boundary conditions on temperature are usually considered on the two remaining "end" walls (BC and DA in Fig. 8.24). If these walls are made from a material that has a low thermal conductivity, it is usual to assume that these walls are adiabatic, i.e., that there is no net heat transfer to or from the wall at any point on the wall. Alternatively, if these walls are made from a material that has a relatively high thermal conductivity, it is usual to assume that these walls are "perfectly conducting" and that the temperature on these "end" walls varies linearly with distance from the hot wall from T_H to T_C.

It will be assumed here that the flow in the enclosure is steady and remains laminar. It will also be assumed that the flow is two-dimensional. With these assumptions and using the Boussinesq approximation that was discussed earlier, the equations governing the flow in the enclosure are:

$$\frac{\partial u}{\partial x} + \frac{\partial v}{\partial y} = 0 \tag{8.122}$$

$$u\frac{\partial u}{\partial x} + v\frac{\partial u}{\partial y} = -\frac{1}{\rho}\frac{\partial p}{\partial x} + \nu\left(\frac{\partial^2 u}{\partial x^2} + \frac{\partial^2 u}{\partial y^2}\right) + \beta g(T - T_1)\cos\phi \tag{8.123}$$

$$u\frac{\partial v}{\partial x} + v\frac{\partial v}{\partial y} = -\frac{1}{\rho}\frac{\partial p}{\partial y} + \nu\left(\frac{\partial^2 v}{\partial x^2} + \frac{\partial^2 v}{\partial y^2}\right) + \beta g(T - T_1)\sin\phi \tag{8.124}$$

$$u\frac{\partial T}{\partial x} + v\frac{\partial T}{\partial y} = \left(\frac{k}{\rho c_p}\right)\left(\frac{\partial^2 T}{\partial x^2} + \frac{\partial^2 T}{\partial y^2}\right) \tag{8.125}$$

In writing these equations, the density at the cold wall temperature, T_C, has been used as the reference relative to which the density changes are specified and p is then the gauge pressure measured relative to the ambient pressure that would exist at the point considered in the enclosure if all the fluid in the enclosure was at rest and at temperature, T_C.

The boundary conditions on the solution are:

On all walls, i.e., on AB, BC, CD, and DA: $u = v = 0$

On wall AB: $T = T_H$ (8.126)

On wall CD: $T = T_C$

The boundary conditions on temperature on walls BC and DA depend on whether these walls are assumed to be adiabatic or perfectly conducting. If they are assumed to be adiabatic, the boundary conditions on these walls are, using Fourier's law on these walls, i.e., using:

$$q = -k\frac{\partial T}{\partial y}$$

and noting that if the wall is adiabatic $q = 0$:

On walls BC and DA: $\dfrac{\partial T}{\partial y} = 0$ (8.127a)

On the other hand, if the end walls are perfectly conducting, the temperature is, as discussed above, assumed to vary linearly along these walls, i.e., the boundary conditions on these end walls then is:

On walls BC and DA: $T = T_H - \left(\dfrac{x}{w}\right)(T_H - T_C)$ (8.127b)

It should be noted that when the enclosure contains a gas, the convective heat transfer rates can be low and radiant heat transfer can be significant. Some gases, such as carbon dioxide and water vapor, absorb and emit radiation and in such cases the energy equation has to be modified to account for this. However, even when the gas in the enclosure is transparent to radiation, there can be an interaction between the radiant and convective heat transfer. For example, for the case where the end walls can be assumed to be adiabatic, if q_{rab} and q_{rem} are the rates at which radiant energy is being absorbed and emitted per unit wall area at any point on these end walls then the actual thermal boundary conditions on these walls are:

On wall BC: $q_{rab} - k\dfrac{\partial T}{\partial y} = q_{rem}$

and:

On wall DA: $q_{rab} = q_{rem} - k\dfrac{\partial T}{\partial y}$

The difference between these two expressions is due to the fact that the y-direction is from fluid-to-wall on BC and from wall-to-fluid on DA.

388 Introduction to Convective Heat Transfer Analysis

It will be assumed here that the convective heat transfer rates are high enough to allow the effects of radiation on the convective motion to be ignored, i.e., to assume that the convection and the radiation can be considered separately and that the total heat transfer rate will be the sum of the separately evaluated convective and radiant heat transfer rates.

Various approximate solutions to the governing equations for natural convection in an enclosure have been obtained. For example, the flow at high Rayleigh numbers can be assumed to consist of a boundary layer flow up the hot wall and a boundary layer flow down the cold wall, these two boundary layers being thin enough to ensure that there is no significant interaction between the two boundary layer flows. More consideration will be given to these approximate solutions later in this chapter. In general, however, the full governing equations must be numerically solved and this will be discussed here.

The governing equations given above, i.e., Eqs. (8.122) to (8.125), are given in terms of the so-called primitive variables, i.e., u, v, p, and T. The solution procedure discussed here is based on equations involving the stream function, ψ, the vorticity, ω, and the temperature, T, as variables. The stream function and vorticity are as before defined by:

$$u = \frac{\partial \psi}{\partial y}, \quad v = -\frac{\partial \psi}{\partial x} \tag{8.128}$$

$$\omega = \left(\frac{\partial v}{\partial x} - \frac{\partial u}{\partial y}\right) \tag{8.129}$$

As shown in Chapter 2, the stream function as so defined satisfies the continuity equation.

The vorticity equation is obtained by eliminating the pressure between the two momentum equations, i.e., by taking the y-derivative of Eq. (8.123) and subtracting from it the x-derivative of Eq. (8.124). This gives:

$$\frac{\partial u}{\partial y}\frac{\partial u}{\partial x} + u\frac{\partial^2 u}{\partial y \partial x} + \frac{\partial v}{\partial y}\frac{\partial u}{\partial y} + v\frac{\partial^2 u}{\partial y^2} - \frac{\partial u}{\partial x}\frac{\partial v}{\partial x} - u\frac{\partial^2 v}{\partial x^2} - \frac{\partial v}{\partial x}\frac{\partial v}{\partial y} - v\frac{\partial^2 v}{\partial x \partial y}$$

$$= \nu\left(\frac{\partial^3 u}{\partial y \partial x^2} + \frac{\partial^3 u}{\partial y^3} - \frac{\partial^3 v}{\partial x^3} - \frac{\partial^3 v}{\partial x \partial y^2}\right) + \beta g\left(\frac{\partial T}{\partial y}\cos\phi - \frac{\partial T}{\partial x}\sin\phi\right)$$

i.e.:

$$u\left(\frac{\partial^2 v}{\partial x^2} - \frac{\partial^2 u}{\partial x \partial y}\right) + v\left(\frac{\partial^2 v}{\partial y \partial x} - \frac{\partial^2 u}{\partial y^2}\right) + \frac{\partial v}{\partial x}\left(\frac{\partial u}{\partial x} + \frac{\partial v}{\partial y}\right) - \frac{\partial u}{\partial y}\left(\frac{\partial u}{\partial x} + \frac{\partial v}{\partial y}\right)$$

$$= \nu\left[\left(\frac{\partial^3 v}{\partial x^3} - \frac{\partial^3 u}{\partial x^2 \partial y}\right) + \left(\frac{\partial^3 v}{\partial y^2 \partial x} - \frac{\partial^3 u}{\partial y^3}\right)\right] - \beta g\left(\frac{\partial T}{\partial y}\cos\phi - \frac{\partial T}{\partial x}\sin\phi\right)$$

Using the definition of vorticity, i.e.:

$$\omega = \left(\frac{\partial v}{\partial x} - \frac{\partial u}{\partial y}\right)$$

CHAPTER 8: Natural Convection   389

and using the continuity equation, this equation can be written as:

$$u\frac{\partial \omega}{\partial x} + v\frac{\partial \omega}{\partial y} = \mu\left(\frac{\partial^2 \omega}{\partial x^2} + \frac{\partial^2 \omega}{\partial y^2}\right) - \beta g\left(\frac{\partial T}{\partial x}\sin\phi - \frac{\partial T}{\partial y}\cos\phi\right)$$

In terms of the stream function, this equation becomes:

$$\frac{\partial \psi}{\partial y}\frac{\partial \omega}{\partial x} - \frac{\partial \psi}{\partial x}\frac{\partial \omega}{\partial y} = \nu\left(\frac{\partial^2 \omega}{\partial x^2} + \frac{\partial^2 \omega}{\partial y^2}\right) - \beta g\left(\frac{\partial T}{\partial x}\sin\phi - \frac{\partial T}{\partial y}\cos\phi\right) \quad (8.130)$$

This equation, which basically expresses conservation of angular momentum, will be one of the equations on which the present numerical analysis of flow in an enclosure is based.

In terms of the stream function, the equation defining the vorticity, i.e.:

$$\left(\frac{\partial v}{\partial x} - \frac{\partial u}{\partial y}\right) = \omega$$

becomes:

$$\left(\frac{\partial^2 \psi}{\partial x^2} + \frac{\partial^2 \psi}{\partial y^2}\right) = -\omega \quad (8.131)$$

while in terms of the stream function the energy equation becomes:

$$\frac{\partial \psi}{\partial y}\frac{\partial T}{\partial x} - \frac{\partial \psi}{\partial x}\frac{\partial T}{\partial y} = \left(\frac{k}{\rho c_p}\right)\left(\frac{\partial^2 T}{\partial x^2} + \frac{\partial^2 T}{\partial y^2}\right) \quad (8.132)$$

Eqs. (8.130), (8.131), and (8.132) which involve the three variables ψ, ω, and T are the set of equations that are used in the present numerical solution. The boundary conditions on these equations are, as discussed before:

$$\text{On all walls, i.e., on AB, BC, CD, and DA: } \frac{\partial \psi}{\partial x} = \frac{\partial \psi}{\partial y} = 0$$

$$\text{On wall AB: } T = T_H$$

$$\text{On wall CD: } T = T_C \quad (8.133)$$

$$\text{On walls BC and DA: either } \frac{\partial T}{\partial y} = 0 \text{ or } T = T_H - \left(\frac{x}{w}\right)(T_H - T_C)$$

The actual value of the stream function is quite arbitrary because only the derivatives of ψ appear in the set of equations and in the boundary conditions discussed above. The value of ψ at point A in Fig. 8.24 will therefore arbitrarily be taken as 0. Because the boundary conditions give $\partial \psi/\partial y = 0$ along AB which indicates that ψ is not changing along AB, ψ will be 0 everywhere along AB. This means that ψ will be 0 at point B in Fig. 8.24. Because the boundary conditions give $\partial \psi/\partial x = 0$ along BC which indicates that ψ is not changing along BC, ψ will be 0 everywhere along BC. This means that ψ will be 0 at point C in Fig. 8.24. In the same way it follows because $\partial \psi/\partial y = 0$ on CD and $\partial \psi/\partial x = 0$ on DA that ψ will be 0 on CD and on DA. Therefore ψ will be 0 everywhere on ABCDA.

390 Introduction to Convective Heat Transfer Analysis

Hence the boundary conditions on stream function are:

$$\text{On walls on AB and CD: } \psi = 0, \frac{\partial \psi}{\partial x} = 0 \tag{8.134}$$

$$\text{On walls on BC and DA: } \psi = 0, \frac{\partial \psi}{\partial y} = 0$$

This can be written as:

$$\text{On walls on AB, BC, CD, and DA: } \psi = 0, \frac{\partial \psi}{\partial n} = 0 \tag{8.135}$$

where n is the coordinate measured normal to the surface being considered.

Before considering the numerical solution to the above set of equations, it is convenient to rewrite the equations in terms of dimensionless variables. This is not, it must be stressed, a necessary step in the solution process. The numerical solution can be obtained directly in terms of dimensional variables.

The following dimensionless variables will be used here:

$$\Psi = \frac{\psi Pr}{\nu}, \quad \Omega = \frac{\omega W^2 Pr}{\nu}$$

$$X = \frac{x}{W}, \quad Y = \frac{y}{W} \tag{8.136}$$

$$\theta = \frac{(T - T_C)}{(T_H - T_C)}$$

where Pr is the Prandtl number of the fluid.

In terms of these variables, the vorticity equation becomes:

$$\frac{\partial \Psi}{\partial Y}\frac{\partial \Omega}{\partial X} - \frac{\partial \Psi}{\partial X}\frac{\partial \Omega}{\partial Y} = Pr\left(\frac{\partial^2 \Omega}{\partial X^2} + \frac{\partial^2 \Omega}{\partial Y^2}\right)$$

$$- \frac{\beta g (T_H - T_C) W^3 Pr^2}{\nu^2}\left(\frac{\partial \theta}{\partial X}\sin\phi - \frac{\partial \theta}{\partial Y}\cos\phi\right)$$

But:

$$Ra = \frac{\beta g (T_H - T_C) W^3}{\nu^2} Pr = Gr_W Pr$$

is the Rayleigh number based on the enclosure width W. The dimensionless vorticity transport equation given above can therefore be written as:

$$\frac{\partial^2 \Omega}{\partial X^2} + \frac{\partial^2 \Omega}{\partial Y^2} = \frac{1}{Pr}\left(\frac{\partial \Psi}{\partial Y}\frac{\partial \Omega}{\partial X} - \frac{\partial \Psi}{\partial X}\frac{\partial \Omega}{\partial Y}\right) + Ra\left(\frac{\partial \theta}{\partial X}\sin\phi - \frac{\partial \theta}{\partial Y}\cos\phi\right) \tag{8.137}$$

In terms of the dimensionless variables, Eq. (8.131) becomes:

$$\frac{\partial^2 \Psi}{\partial X^2} + \frac{\partial^2 \Psi}{\partial Y^2} = -\Omega \tag{8.138}$$

Similarly, the energy equation, i.e., Eq. (8.132), becomes in terms of the dimensionless variables:

$$\frac{\partial^2 \theta}{\partial X^2} + \frac{\partial^2 \theta}{\partial Y^2} = \frac{\partial \Psi}{\partial Y}\frac{\partial \theta}{\partial X} - \frac{\partial \Psi}{\partial X}\frac{\partial \theta}{\partial Y} \tag{8.139}$$

In terms of these dimensionless variables, the boundary conditions are:

$$\text{On walls on AB, BC, CD and DA: } \Psi = 0, \; \frac{\partial \Psi}{\partial N} = 0$$

where $N = n/W$ is the dimensionless coordinate measured normal to the surface being considered, e.g., it is X for AB and Y for BC.

$$\text{On wall AB, i.e., at } X = 0: \theta = 1$$
$$\text{On wall CD, i.e., at } X = 1: \theta = 0$$
$$\text{On walls BC and DA, i.e., at } Y = 0 \text{ and } Y = A: \tag{8.140}$$
$$\text{either } \frac{\partial \theta}{\partial Y} = 0 \text{ or } \theta = 1 - X$$

where $A = H/W$ is the so-called aspect ratio of the enclosure.

An iterative finite-difference solution procedure will be discussed here. Many other more efficient solution procedures are available but the simple procedure considered here should serve to indicate the main features of such methods.

A series of evenly spaced nodal lines as shown in Fig. 8.25 will be used. The grid spacings in both the X- and Y-directions are thus constant and equal to ΔX and ΔY respectively as indicated in Fig. 8.25.

Consider the nodal points shown in Fig. 8.26. In terms of the values at these nodal points the finite-difference form of the dimensionless energy equation, i.e., Eq. (8.139), becomes:

$$\left(\frac{\theta_{i+1,j} + \theta_{i-1,j} - 2\theta_{i,j}}{\Delta X^2}\right) + \left(\frac{\theta_{i,j+1} + \theta_{i,j-1} - 2\theta_{i,j}}{\Delta Y^2}\right)$$
$$= \left(\frac{\Psi_{i,j+1} - \Psi_{i,j-1}}{2\Delta Y}\right)\left(\frac{\theta_{i+1,j} - \theta_{i-1,j}}{2\Delta X}\right) - \left(\frac{\Psi_{i+1,j} - \Psi_{i-1,j}}{2\Delta X}\right)\left(\frac{\theta_{i,j+1} - \theta_{i,j-1}}{2\Delta Y}\right)$$

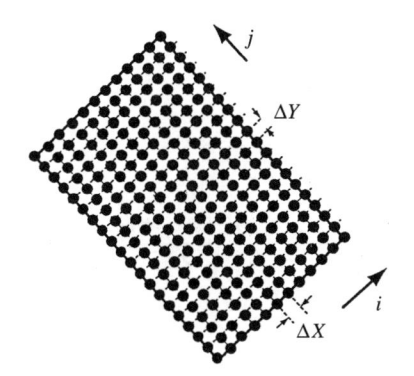

FIGURE 8.25
Grid system used.

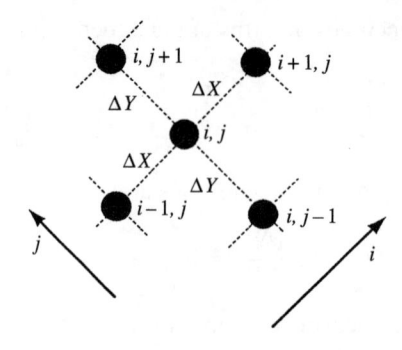

FIGURE 8.26
Nodal points used in obtaining finite-difference equations.

which can be rearranged to give:

$$\theta_{i,j} = \left\{ \left(\frac{-1}{4\Delta X \Delta Y}\right)[(\Psi_{i,j+1} - \Psi_{i,j-1})(\theta_{i+1,j} - \theta_{i-1,j}) \right.$$
$$- (\Psi_{i+1,j} - \Psi_{i-1,j})(\theta_{i,j+1} - \theta_{i,j-1})] + \left(\frac{\theta_{i+1,j} + \theta_{i-1,j}}{\Delta X^2}\right)$$
$$\left. + \left(\frac{\theta_{i,j+1} + \theta_{i,j-1}}{\Delta Y^2}\right) \right\} \bigg/ \left(\frac{2}{\Delta X^2} + \frac{2}{\Delta Y^2}\right) \quad (8.141)$$

Similarly, the finite-difference form of the vorticity transport equation, i.e., Eq. (8.137), gives:

$$\left(\frac{\Omega_{i+1,j} + \Omega_{i-1,j} - 2\Omega_{i,j}}{\Delta X^2}\right) + \left(\frac{\Omega_{i,j+1} + \Omega_{i,j-1} - 2\Omega_{i,j}}{\Delta Y^2}\right)$$
$$= \left(\frac{1}{Pr}\right)\left[\left(\frac{\Psi_{i,j+1} - \Psi_{i,j-1}}{2\Delta Y}\right)\left(\frac{\Omega_{i+1,j} - \Omega_{i-1,j}}{2\Delta X}\right) \right.$$
$$\left. - \left(\frac{\Psi_{i+1,j} - \Psi_{i-1,j}}{2\Delta X}\right)\left(\frac{\Omega_{i,j+1} - \Omega_{i,j-1}}{2\Delta Y}\right)\right]$$
$$+ Ra\left(\frac{\theta_{i+1,j} - \theta_{i-1,j}}{2\Delta X}\right)\sin\phi - \left(\frac{\theta_{i,j+1} - \theta_{i,j-1}}{2\Delta Y}\right)\cos\phi$$

which can be rearranged to give:

$$\Omega_{i,j} = \left\{ \left(\frac{-1}{4\Delta X \Delta Y Pr}\right)[(\Psi_{i,j+1} - \Psi_{i,j-1})(\Omega_{i+1,j} - \Omega_{i-1,j}) \right.$$
$$- (\Psi_{i+1,j} - \Psi_{i-1,j})(\Omega_{i,j+1} - \Omega_{i,j-1})] + \left(\frac{\Omega_{i+1,j} + \Omega_{i-1,j}}{\Delta X^2}\right)$$
$$+ \left(\frac{\Omega_{i,j+1} + \Omega_{i,j-1}}{\Delta Y^2}\right) - Ra\left[\left(\frac{\theta_{i+1,j} - \theta_{i-1,j}}{2\Delta X}\right)\sin\phi - \left(\frac{\theta_{i,j+1} - \theta_{i,j-1}}{2\Delta Y}\right)\cos\phi\right]\right\}$$
$$\bigg/ \left(\frac{2}{\Delta X^2} + \frac{2}{\Delta Y^2}\right) \quad (8.142)$$

Lastly, the finite-difference form of the equation relating the dimensionless vorticity to the dimensionless stream function, i.e., Eq. (8.138), is:

$$\frac{\Psi_{i+1,j} + \Psi_{i-1,j} - 2\Psi_{i,j}}{\Delta X^2} + \frac{\Psi_{i,j+1} + \Psi_{i,j-1} - 2\Psi_{i,j}}{\Delta Y^2} = -\Omega_{i,j}$$

which can be rearranged to give:

$$\Psi_{i,j} = \frac{\left[\left(\dfrac{\Psi_{i+1,j} + \Psi_{i-1,j}}{\Delta X^2}\right) + \left(\dfrac{\Psi_{i,j+1} + \Psi_{i,j-1}}{\Delta Y^2}\right)\right] + \Omega_{i,j}}{\left(\dfrac{2}{\Delta X^2} + \dfrac{2}{\Delta Y^2}\right)} \tag{8.143}$$

Eqs. (8.141), (8.142), and (8.143) apply to all "internal" points, i.e., to all points from $i = 2$ to $M - 1$ and for $j = 2$ to $N - 1$ where M and N are the number of nodal points in the X- and Y-directions, respectively. The boundary conditions give the values of the variables on the boundaries. These boundary conditions directly give:

$$\Psi_{1,j} = 0, \ \Psi_{M,j} = 0, \ \Psi_{i,1} = 0, \ \Psi_{i,N} = 0 \tag{8.144}$$

and:

$$\theta_{1,j} = 1, \ \theta_{M,j} = 0 \tag{8.145}$$

Consider the boundary conditions on the end walls. When the temperature gradient is zero on the end walls, i.e., when the end walls are adiabatic, since to the same order of accuracy as used in deriving the finite-difference equations:

$$\left.\frac{\partial \theta}{\partial y}\right|_{i,1} = \left[(\theta_{i,2} - \theta_{i,1}) - \frac{(\theta_{i,3} - \theta_{i,1})}{4}\right]\left(\frac{2}{\Delta Y}\right)$$

and:

$$\left.\frac{\partial \theta}{\partial y}\right|_{i,N} = \left[(\theta_{i,N} - \theta_{i,N-1}) - \frac{(\theta_{i,N} - \theta_{i,N-2})}{4}\right]\left(\frac{2}{\Delta Y}\right)$$

the boundary conditions give:

$$\theta_{i,1} = \frac{4}{3}\left(\theta_{i,2} - \frac{\theta_{i,3}}{4}\right) \tag{8.146}$$

and:

$$\theta_{i,N} = \frac{4}{3}\left(\theta_{i,N-1} - \frac{\theta_{i,N-2}}{4}\right) \tag{8.147}$$

Alternatively, if the end walls are perfectly conducting, the boundary conditions on these walls are:

$$\theta_{i,1} = 1 - X_i \tag{8.148}$$

and:

$$\theta_{i,N} = 1 - X_i \tag{8.149}$$

394 Introduction to Convective Heat Transfer Analysis

Now, a consideration of the boundary conditions indicates that there are no boundary conditions on dimensionless vorticity on the walls. However, it will be noted that the boundary condition:

$$\frac{\partial \Psi}{\partial N} = 0$$

where N is the dimensionless coordinate measured normal to the surface at the point considered, has not been used. To illustrate how this can be used to obtain a boundary condition on Ω, consider a point on wall surface AB such as that shown in Fig. 8.27.

Because $\Psi = 0$ at all points on AB, it follows that $\partial^2 \Psi / \partial Y^2 = 0$ at all points on AB. Hence when Eq. (8.138) is applied to any point on AB it gives:

$$\Omega_{X=0} = -\left.\frac{\partial^2 \Psi}{\partial X^2}\right|_{X=0} \tag{8.150}$$

Consider the two points adjacent to the surface AB that are shown in Fig. 8.27. A Taylor expansion gives to the same order of accuracy as used in obtaining the finite-difference approximations used in the rest of the analysis.

$$\Psi_{2,j} = \Psi_{1,j} + \left(\frac{\partial \Psi}{\partial X}\right)_{1,j} \Delta X + \left(\frac{\partial^2 \Psi}{\partial X^2}\right)_{1,j} \frac{\Delta X^2}{2!} \tag{8.151}$$

But the boundary conditions give:

$$\Psi_{1,j} = 0 \text{ and } \left(\frac{\partial \Psi}{\partial X}\right)_{1,j} = 0$$

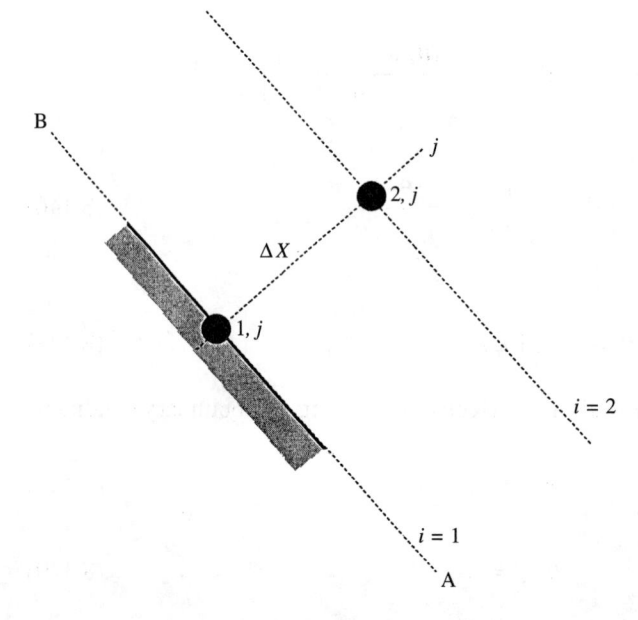

FIGURE 8.27
Wall points considered in obtaining boundary condition on vorticity.

CHAPTER 8: Natural Convection 395

so Eq. (8.151) gives:

$$\Psi_{2,j} = \left(\frac{\partial^2 \Psi}{\partial X^2}\right)_{1,j} \frac{\Delta X^2}{2}$$

i.e:

$$\left(\frac{\partial^2 \Psi}{\partial X^2}\right)_{1,j} = 2\frac{\Psi_{2,j}}{\Delta X^2}$$

Substituting this into Eq. (8.150) then gives for any point on AB:

$$\Omega_{1,j} = -2\frac{\Psi_{2,j}}{\Delta X^2} \tag{8.152}$$

Eq. (8.152) supplies a boundary condition on the dimensionless vorticity at the wall. Applying the same procedure to points on surfaces BC, CD, and DA gives:

$$\Omega_{i,N} = -2\frac{\Psi_{i,N-1}}{\Delta Y^2} \tag{8.153}$$

and:

$$\Omega_{M,j} = -2\frac{\Psi_{M-1,j}}{\Delta X^2} \tag{8.154}$$

and:

$$\Omega_{i,1} = -2\frac{\Psi_{1,2}}{\Delta Y^2} \tag{8.155}$$

The set of nonlinear dimensionless finite-difference equations with their associated boundary conditions that have been presented above are solved iteratively starting with guessed values of the variables at all points. The procedure, therefore, involves the following steps:

(1) Guess the values of $\Psi_{i,j}$, $\Omega_{i,j}$ and $\theta_{i,j}$ at all points. Typically, it could be assumed that initially:

$$\Psi_{i,j} = 0, \ \Omega_{i,j} = 0, \ \theta_{i,j} = 1 - X_{i,j}$$

the assumed dimensionless temperature variation stemming from the fact that if there is no fluid motion the temperature distribution will be that existing with pure conduction across the enclosure, i.e., in view of the boundary conditions, it will vary linearly with X.

(2) Eq. (8.142) is applied sequentially at all "internal" nodal points, i.e., all points that are not on a wall, to find new values of $\Omega_{i,j}$. The right-hand side of this equation is found using the assumed values of the variables. Under-relaxation is actually used so the "updated" values of $\Omega_{i,j}$ are actually taken as:

$$\Omega_{i,j}^1 = \Omega_{i,j}^0 + r(\Omega_{i,j}^{\text{calculated}} - \Omega_{i,j}^0)$$

where $\Omega_{i,j}^{\text{calculated}}$ is the value given directly by Eq. (8.142) and r is the relaxation factor (< 1). The superscripts 0 and 1 refer to conditions at the beginning and the end of the iteration step. The solution procedure, as mentioned above, is applied to all internal points, i.e., for $i = 2$ to $M - 1$ and for $j = 2$ to $N - 1$. The whole step is repeated before proceeding to step (3) of the solution procedure, this having been found to accelerate the convergence.

(3) Eq. (8.143) is applied sequentially at all internal nodal points to get updated values of $\Psi_{i,j}$ at these points, the right-hand side of this equation again being found using the assumed values of the variables. Under-relaxation is used so the "updated" values of $\Psi_{i,j}$ are actually taken as:

$$\Psi^1_{i,j} = \Psi^0_{i,j} + r(\Psi^{\text{calculated}}_{i,j} - \Psi^0_{i,j})$$

where $\Psi^{\text{calculated}}_{i,j}$ is the value given directly by Eq. (8.143) and r is the relaxation factor (< 1). The solution procedure, as mentioned above, is applied to all internal points, i.e., for $i = 2$ to $M - 1$ and for $j = 2$ to $N - 1$. The boundary conditions give the values of Ψ at the nodal points on the boundaries and these do not change. This whole step is also repeated before proceeding to step (4) of the solution procedure, this having been found to accelerate the convergence.

(4) Eqs. (8.152) to (8.155) are then used to get updated values of Ω on the boundary points. Under-relaxation is also used for these points so the "updated" values of $\Omega_{i,j}$ at the wall points are actually taken as:

$$\Omega^1_{i,j} = \Omega^0_{i,j} + r_b(\Omega^{\text{calculated}}_{i,j} - \Omega^0_{i,j})$$

where $\Omega^{\text{calculated}}_{i,j}$ is the value given directly by Eqs. (8.152) to (8.155) and r_b is the relaxation factor (< 1). The value of the relaxation factor for the wall points, as indicated by the subscript on the r, is often taken to be larger than that used for the internal points, this being found to accelerate convergence.

(5) Eq. (8.141) is applied sequentially at all internal nodal points to get updated values of $\theta_{i,j}$ at these points, the right-hand side of this equation again being found using the assumed values of the variables. Under-relaxation is also used so the "updated" values of $\theta_{i,j}$ are actually taken as:

$$\theta^1_{i,j} = \theta^0_{i,j} + r(\theta^{\text{calculated}}_{i,j} - \theta^0_{i,j})$$

where $\theta^{\text{calculated}}_{i,j}$ is the value given directly by Eq. (8.141) and r is the relaxation factor (< 1). The solution procedure, as mentioned above, is applied to all internal points, i.e., for $i = 2$ to $M - 1$ and for $j = 2$ to $N - 1$. The boundary conditions give the values of θ at the nodal points on the boundaries AB and CD and on BC and DA if perfectly conducting end walls are assumed and these values do not change. If adiabatic end walls are assumed the updated values of $\theta_{i,1}$ and $\theta_{i,N}$ are found using Eqs. (8.146) and (8.147). This whole step is also repeated before proceeding to step (6) of the solution procedure, this having been found to accelerate the convergence.

(6) Steps 2 to 5 are repeated over and over until the values of the variables cease to change from one iteration to the next by less than some prescribed value. A minimum number of iterations is usually prescribed in order to avoid false convergence and a maximum number of iterations also has to be prescribed because convergence will not always occur.

(7) Once the converged solution is obtained, the values of the heat transfer rate at the nodal points on the hot and cold walls can be found by applying Fourier's law

at these wall points. For example, consider a point on wall surface AB such as that shown in Fig. 8.27. Now:

$$\theta_{2,j} = \theta_{1,j} + \left(\frac{\partial \theta}{\partial X}\right)_{1,j} \Delta X + \left(\frac{\partial^2 \theta}{\partial X^2}\right)_{1,j} \frac{\Delta X^2}{2!} \tag{8.156}$$

But the velocity components are both 0 at the wall and $\theta_{i,j}$ is constant ($= 1$) everywhere on AB with the result that $\partial^2 \theta / \partial Y^2$ is 0 everywhere on AB. Using these results, it follows from the energy equation that on AB:

$$\left(\frac{\partial^2 \theta}{\partial X^2}\right)_{1,j} = 0$$

Hence, Eq. (8.156) gives:

$$-\left(\frac{\partial \theta}{\partial X}\right)_{1,j} = \frac{\theta_{1,j} - \theta_{2,j}}{\Delta X} = \frac{1 - \theta_{2,j}}{\Delta X} \tag{8.157}$$

But Fourier's law gives at the nodal points on AB:

$$q_w = -k \left.\frac{\partial T}{\partial x}\right|_{x=0}$$

which becomes in terms of the dimensionless variables:

$$\frac{q_w W}{k(T_H - T_C)} = -\left.\frac{\partial \theta}{\partial X}\right|_{1,j}$$

i.e., using Eq. (8.157):

$$Nu_{1,j} = \frac{q_w W}{k(T_H - T_C)} = -\left.\frac{\partial \theta}{\partial X}\right|_{1,j} = \frac{1 - \theta_{2,j}}{\Delta X} \tag{8.158}$$

For points on the wall surface CD, i.e., on the cold wall, the equivalent equation which can be derived using the same procedure is:

$$Nu_{M,j} = \frac{q_w W}{k(T_H - T_C)} = \frac{\theta_{M-1,j}}{\Delta X} \tag{8.159}$$

The mean heat transfer rates on AB and CD are then found by integrating the local distributions, i.e., by using:

$$\overline{q_w}|_1 = \frac{1}{H}\int_0^H q_w\,dx$$

Numerical integration using the values of the local dimensionless heat transfer rate at the points considered therefore gives:

$$Nu_H = \left.\frac{\overline{q_w} W}{k(T_H - T_C)}\right|_1 = \frac{\Delta X}{A}\left[\frac{Nu_{1,1}}{2} + Nu_{1,2} + Nu_{1,3} + \cdots + Nu_{1,N-1} + \frac{Nu_{1,N}}{2}\right] \tag{8.160}$$

Similarly, for the cold wall:

$$Nu_C = \frac{\overline{q_w}W}{k(T_H - T_C)}\bigg|_M = \frac{\Delta X}{A}\left[\frac{Nu_{M,1}}{2} + Nu_{M,2} + Nu_{M,3} + \cdots + Nu_{M,N-1} + \frac{Nu_{M,N}}{2}\right] \quad (8.161)$$

Nu_H and Nu_C being the mean Nusselt numbers, based on W, for the hot and the cold walls, respectively. With adiabatic end walls, because a steady state situation is being considered, these two values should have the same numerical value, i.e., the rate at which heat is transferred from the hot wall to the fluid should be equal to the rate at which heat is transferred from the fluid to the cold wall. Small differences usually exist between the values of Nu_H and Nu_C given by the numerical solution due to the small numerical round-off errors and due to the finite convergence criterion used in the numerical solution.

An examination of the governing equations and the boundary conditions on these equations indicates that the solution to the equations contains four parameters, i.e., the Rayleigh number, Ra, the Prandtl number, Pr, the angle of inclination, ϕ, and the aspect ratio $A = H/W$.

A computer program, ENCLREC, that is based on the procedure outlined above which is written in FORTRAN is available as discussed in the Preface. Besides the solution parameters, the inputs to the program are the number of nodal points in the X and Y directions M and N, the relaxation factors r and r_b, and the maximum and minimum number of iteration steps. The program and the numerical procedure discussed above are very simplistic, intended only to illustrate the main features of more advanced solution procedures.

Some results obtained using the program are presented in Figs. 8.28, 8.29, and 8.30. These results are all for the adiabatic end wall boundary condition. Figures

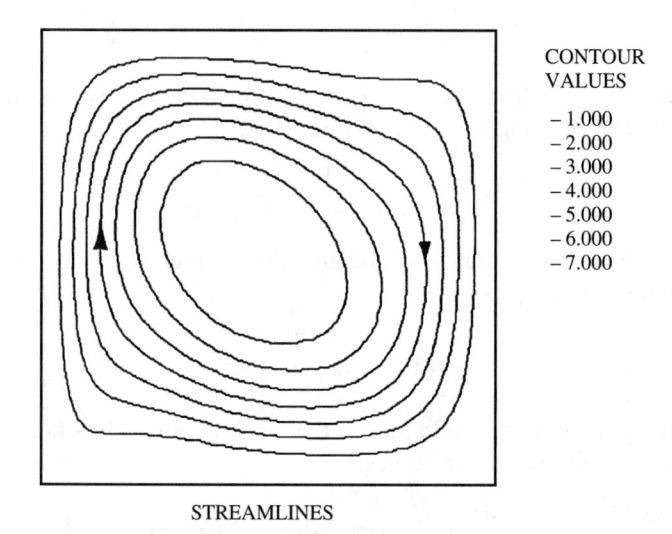

CONTOUR VALUES

- 1.000
- 2.000
- 3.000
- 4.000
- 5.000
- 6.000
- 7.000

STREAMLINES

FIGURE 8.28
Streamlines for $Ra = 3 \times 10^4$, $Pr = 0.7$, $\phi = 90°$, $A = 1$, and end walls adiabatic.

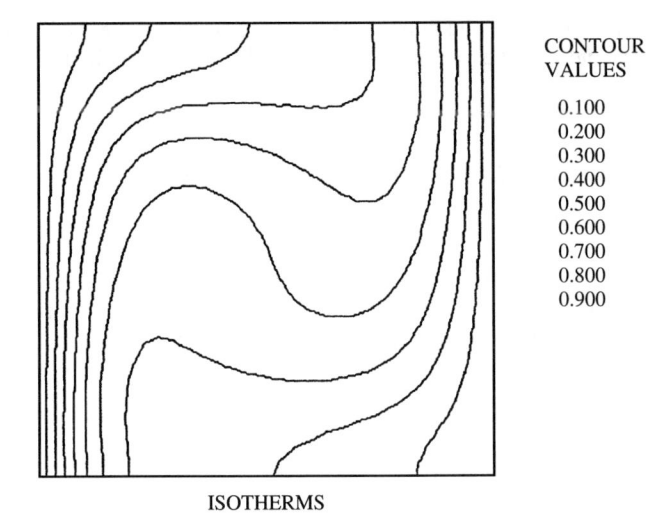

FIGURE 8.29
Isotherm for $Ra = 3 \times 10^4$, $Pr = 0.7$, $\phi = 90°$, $A = 1$, and end walls adiabatic.

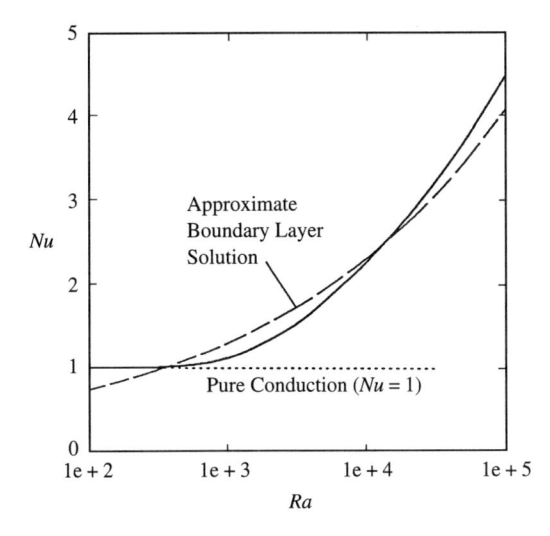

FIGURE 8.30
Variation of Nu with Ra for $Pr = 0.7$, $\phi = 90°$, $A = 1$, and end walls adiabatic.

8.28 and 8.29 show typical streamline and isotherm (lines of constant temperature) patterns in the enclosure while Fig. 8.30 shows the variation of the mean Nusselt number, Nu, with Rayleigh number, Ra, these results being for the particular case of $Pr = 0.7$ (air), $\phi = 90°$ (hot and cold walls vertical) and $A = 1$ (a square enclosure).

EXAMPLE 8.7. Use the computer program discussed above for natural convective flow in an enclosure to derive a relation between Nusselt number and Rayleigh number for a square enclosure with isothermal vertical walls and perfectly conducting horizontal walls. To do this, find the Nusselt numbers for Rayleigh numbers of between 10 and about 100,000 and plot the variation of Nu with Ra. Try to obtain a correlation of the form:

$$Nu = aRa^n$$

a and n being constants, for the results at the larger Rayleigh numbers.

Solution. The program has been run with the following input values:

`IEND=2 ; PR =0.7 ; RA=10, 30, 100, ........... , 30000, 100000, 300000`

The values of Nu obtained in this way are plotted against Ra in Fig. E8.7.

The values of Nu obtained at three of the larger values of Ra are listed in the following table:

Ra	Nu
10,000	1.751
30,000	2.395
100,000	3.400

Now if:

$$Nu = aRa^n$$

then:

$$\log Nu = \log a + n \log Ra$$

If the results at any two values in the above table are considered, these being indicated by subscripts 1 and 2, then the above equation gives:

$$n = \frac{\log Nu_2 - \log Nu_1}{\log Ra_2 - \log Ra_1}$$

Applying this equation to the results in the table gives:

$$n = 0.29$$

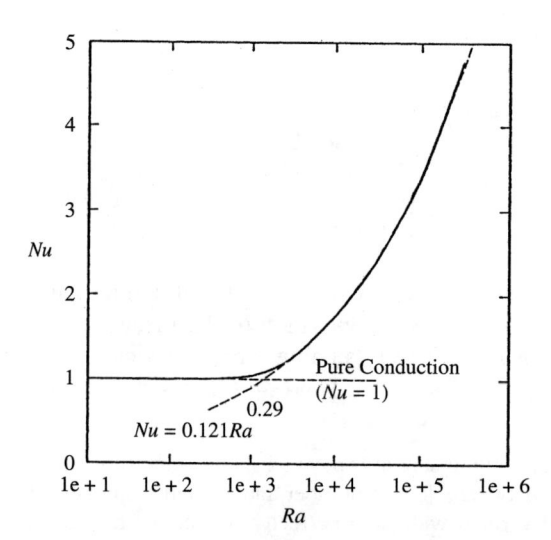

FIGURE E8.7

It then follows that:

$$a = \frac{Nu}{Ra^{0.29}}$$

Applying this to the results in the table gives:

$$a = 0.121$$

Hence, the results indicate that at the higher Rayleigh numbers:

$$Nu = 0.121 Ra^{0.29}$$

The values given by this equation are also shown in Fig. E8.7 and it will be seen that it describes the results quite well for Rayleigh numbers greater than about 3000.

As mentioned before, there are analytical solutions for the heat transfer that apply under certain limiting conditions. For example, if the Rayleigh number is very low, the convective motion will be so weak that it has no influence on the heat transfer rate and the heat transfer will effectively be by pure conduction. In this case the heat transfer rate is uniform on the hot and cold walls and is given by:

$$q_w = \frac{k(T_H - T_C)}{W}, \text{ i.e., } Nu = 1 \tag{8.162}$$

It will be seen from Fig. 8.30 that the numerically calculated values of Nu do indeed approach a value of 1 at low values of Ra.

Another approximate limiting solution for a vertical enclosure (i.e., $\phi < 90°$) is obtained, as mentioned before, by assuming that the flow consists of boundary layers on the hot and cold walls with an effectively stagnant layer between them and that the presence of these end walls has a negligible effect on the boundary layer flows. The assumed flow is therefore as shown in Fig. 8.31.

The boundary flow up the hot wall will be identical to the flow down the cold wall and the temperature of the fluid between the two layers will therefore be $(T_H + T_C)/2$. The temperature difference across both boundary layers will therefore be $(T_H - T_C)/2$.

Because the hot and cold walls are vertical and at a uniform temperature and because the fluid between the two boundary layers is assumed to be stagnant and at

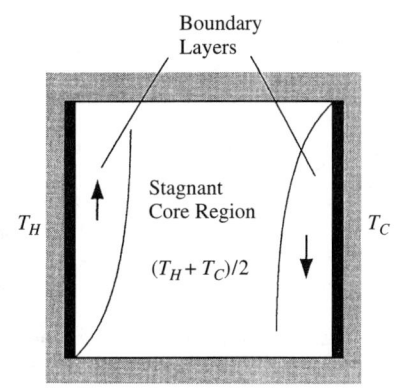

FIGURE 8.31
Assumed limiting form of flow.

a uniform temperature, the similarity solution for the boundary layer on a vertical plate derived earlier in this chapter will describe the flow in the boundary layers. This solution gives:

$$\frac{Nu_H^*}{Gr_H^{*0.25}} = \left[\frac{Pr^{5/4}}{2.44 + 4.88Pr^{1/2} + 4.95Pr}\right]^{1/4}$$

where Nu_H^* is the mean Nusselt number based on the height, H, of the plate and the temperature difference across the boundary layer and Gr_H^* is the Grashof number based on the height, H, of the plate and the temperature difference across the boundary layer, the temperature difference across the boundary layers being half the overall temperature difference. In terms of the variables being used here to describe the flow in the enclosure, this equation gives:

$$\frac{Nu}{Gr^{0.25}} A^{0.25} = \left[\frac{0.0313 Pr^{5/4}}{2.44 + 4.88Pr^{1/2} + 4.95Pr}\right]^{1/4}$$

where Gr is the Grashof number based on the width of the enclosure and the overall temperature difference. This equation can be written as:

$$Nu = \left[\frac{0.0313 Pr^{1/4}}{A(2.44 + 4.88Pr^{1/2} + 4.95Pr)}\right]^{1/4} Ra^{0.25} \quad (8.163)$$

The variation of of Nu with Ra given by this equation for $Pr = 0.7$ and $A = 1$ is compared with the numerical results in Fig. 8.30 and will be seen to compare quite well with the numerical results at higher values of Ra.

EXAMPLE 8.8. Consider the vorticity equation for flow in an enclosure, i.e.:

$$\frac{\partial^2 \Omega}{\partial X^2} + \frac{\partial^2 \Omega}{\partial Y^2} = \frac{1}{Pr}\left(\frac{\partial \Psi}{\partial Y}\frac{\partial \Omega}{\partial X} - \frac{\partial \Psi}{\partial X}\frac{\partial \Omega}{\partial Y}\right) - Ra\left(\frac{\partial \theta}{\partial X}\sin\phi - \frac{\partial \theta}{\partial Y}\cos\phi\right)$$

If the Prandtl number is large, the first term on the right-hand side of this equation, i.e., the inertia term, will be negligible, i.e., the equation will effectively be:

$$\frac{\partial^2 \Omega}{\partial X^2} + \frac{\partial^2 \Omega}{\partial Y^2} = Ra\left(\frac{\partial \theta}{\partial X}\sin\phi - \frac{\partial \theta}{\partial Y}\cos\phi\right)$$

With this approximation, the equations governing the flow in the enclosure are:

$$\frac{\partial^2 \Psi}{\partial X^2} + \frac{\partial^2 \Psi}{\partial Y^2} = -\Omega$$

$$\frac{\partial^2 \Omega}{\partial X^2} + \frac{\partial^2 \Omega}{\partial Y^2} = -Ra\left(\frac{\partial \theta}{\partial X}\sin\phi - \frac{\partial \theta}{\partial Y}\cos\phi\right)$$

$$\frac{\partial^2 \theta}{\partial X^2} + \frac{\partial^2 \theta}{\partial Y^2} = \frac{\partial \Psi}{\partial Y}\frac{\partial \theta}{\partial X} - \frac{\partial \Psi}{\partial X}\frac{\partial \theta}{\partial Y}$$

The only parameter in these equations is the Rayleigh number, Ra, i.e., the solution depends only on the Rayleigh number and not on the Prandtl number. From this it follows

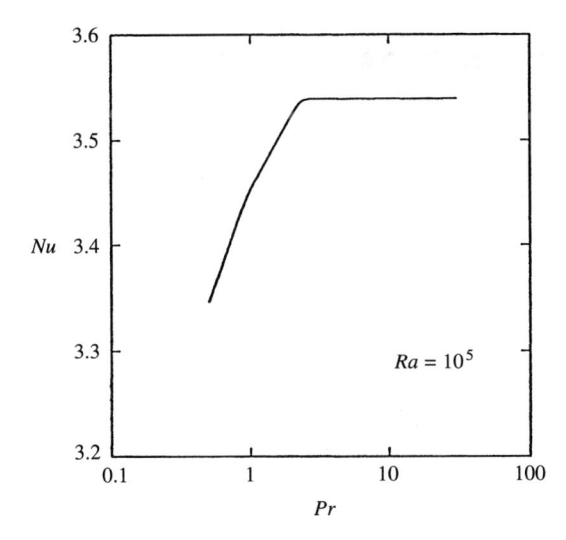

FIGURE E8.8

that:

$$Nu = \text{function}(Ra)$$

Verify this conclusion by calculating the Nusselt number for a fixed Rayleigh number of 10^5 for the perfectly conducting wall case for Prandtl numbers of between 0.5 and 30 and verify that the Nusselt number tends to a constant value at the larger values of Pr.

Solution. The program has been run with the following input values:

```
IEND=2 ; RA=100000.0 ; PR=0.5, 0.7, 1, 2, 3, 10, 30
```

The variation of Nu with Pr obtained in this way is shown in Fig. E8.8.

It will be seen from the results given in Fig. E8.8 that Nu does appear to be essentially independent of Pr when $Pr > 2$. It will also be seen that the change in Nu over the entire Pr range considered is small. This indicates that the inertia term in the vorticity equation, i.e.:

$$\frac{1}{Pr}\left(\frac{\partial \Psi}{\partial Y}\frac{\partial \Omega}{\partial X} - \frac{\partial \Psi}{\partial X}\frac{\partial \Omega}{\partial Y}\right)$$

is small compared to the viscosity and buoyancy terms because the inertia term is the only term containing Pr as a separate parameter.

8.10
HORIZONTAL ENCLOSURES HEATED FROM BELOW

The above discussion was concerned with enclosures in which the heated and cooled walls were at an angle to the horizontal. In the present section, the concern is with rectangular enclosures in which the hot and cold walls are horizontal and in which the hot wall is at the bottom, see [68] to [83]. The flow situation being considered is therefore as shown in Fig. 8.32.

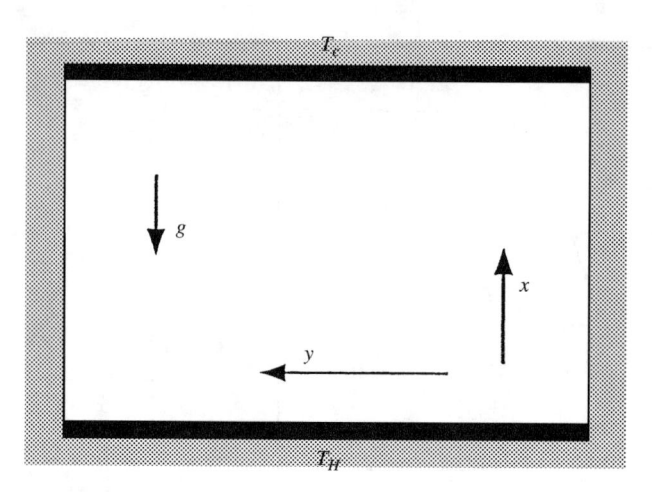

FIGURE 8.32
Horizontal enclosure heated from below.

Because the hot wall is at the bottom, the buoyancy forces will tend to cause a fluid motion, hot fluid flowing upwards from the lower surface and cold fluid flowing downward from the upper surface. This motion will only develop however, when the Rayleigh number based on the difference between the temperatures of the upper and lower surfaces and on the size of the gap between the lower and upper surfaces exceeds a certain critical value. The value of this critical Rayleigh number can be determined by imposing a small disturbance on the initially stationary fluid between the two walls. At low Rayleigh numbers this disturbance will decay with time. However, at high Rayleigh numbers, the disturbance will grow with time. The Rayleigh number at which the disturbances neither grow nor decay is the required critical Rayleigh number. This process is illustrated in Chapter 10 where it is applied to the case where the gap between the two surfaces is filled with a porous medium. In the present case, an attempt will be made to find the critical Rayleigh number by using a numerical approach. The computer program, ENCLREC, discussed above is easily applied to a horizontal enclosure, i.e., an enclosure with horizontal hot and cold walls as shown in Fig. 8.32, and with perfectly conducting end walls, by setting ϕ equal to 0. This modified program has been used to determine the variation of mean Nusselt with Rayleigh number for a horizontal enclosure for various enclosure aspect ratios for a Prandtl number of 0.7. A typical such variation, this being for an aspect ratio of 5, is shown in Fig. 8.33.

It will be seen that at low Rayleigh numbers the Nusselt number has a value of 1, the pure conduction value. At these Rayleigh numbers there is no fluid motion in the enclosure. However it will be seen that at a Rayleigh number of approximately between 1700 and 1750, the Nusselt number starts to rise above the pure conduction value of 1, i.e., fluid motion develops in the enclosure. This is further illustrated by the results given in Fig. 8.34. These results are for an enclosure with an aspect ratio of 10. Figure 8.34 shows the variation of the maximum positive value of the stream function with Rayleigh number for Rayleigh numbers near 1700. Because of the nature of the flow in the enclosure (see below) there are both positive and negative

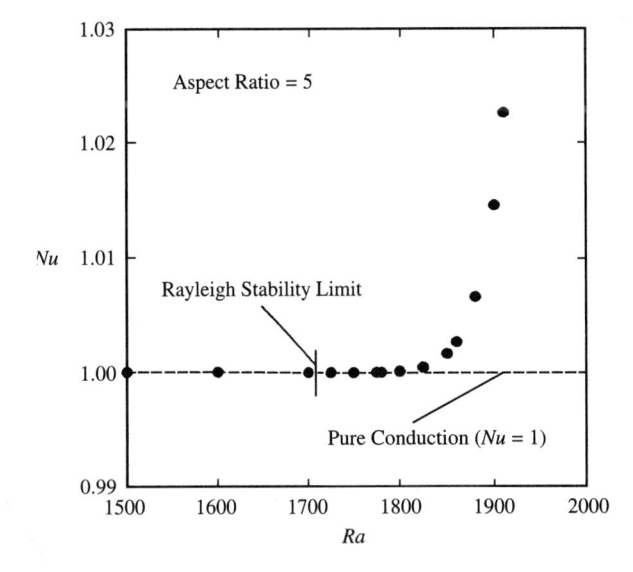

FIGURE 8.33
Variation of mean Nusselt number with Rayleigh number for an enclosure with an aspect ratio of 5 for $Pr = 0.7$.

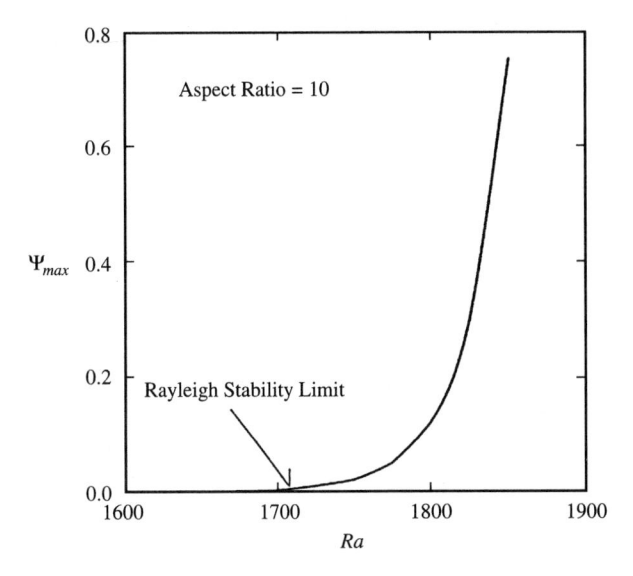

FIGURE 8.34
Variation of maximum value of the stream function with Rayleigh number for an enclosure with an aspect ratio of 10 for $Pr = 0.7$.

values of the stream function in the enclosure flow. The maximum negative value of the stream function varies with Rayleigh number in basically the same way as the maximum positive value. The maximum value of the stream function is an indication of the intensity of the fluid motion in the enclosure. The results given in Fig. 8.34 therefore also indicate that a significant fluid motion develops in the enclosure when the Rayleigh number reaches a value of approximately 1700 to 1720.

In running the program discussed above, the small errors introduced by numerical round-off are sufficient to trigger the instability. In some related situations, however, it may be necessary to introduce very small random disturbances in order to trigger the instability.

(a)

Two-dimensional Rolls

Three-dimensional
Hexagonal Cells

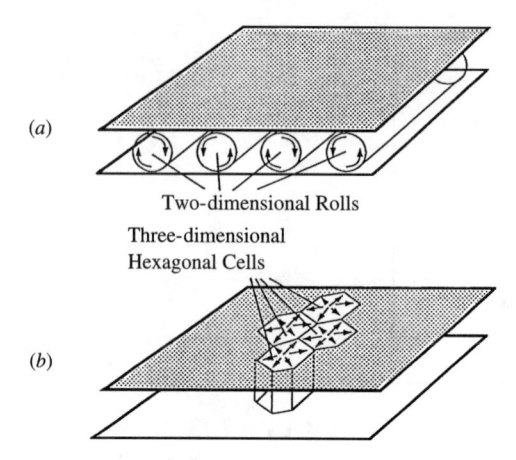

(b)

FIGURE 8.35
Flow patterns in a horizontal enclosure.

Using the small disturbance approach gives a value of 1708 for the Rayleigh number at which disturbances start to grow, i.e., at which fluid motion will develop in the enclosure. This value will be independent of Prandtl number because the fluid motion for Rayleigh numbers near the critical value is very weak and the effect of the convective terms in the momentum equation is then negligible and the governing equations, i.e., Eqs. (8.137) to (8.139) then are:

$$\frac{\partial^2 \Omega}{\partial X^2} + \frac{\partial^2 \Omega}{\partial Y^2} = -Ra\left(\frac{\partial \theta}{\partial X}\sin\phi - \frac{\partial \theta}{\partial Y}\cos\phi\right)$$

$$\frac{\partial^2 \Psi}{\partial X^2} + \frac{\partial^2 \Psi}{\partial Y^2} = -\Omega$$

$$\frac{\partial^2 \theta}{\partial X^2} + \frac{\partial^2 \theta}{\partial Y^2} = \frac{\partial \Psi}{\partial Y}\frac{\partial \theta}{\partial X} - \frac{\partial \Psi}{\partial X}\frac{\partial \theta}{\partial Y}$$

The only parameter in these equations is the Rayleigh number. Therefore, the Rayleigh number at which the flow starts to develop in a horizontal enclosure will be independent of the Prandtl number.

When the Rayleigh number exceeds the critical value, fluid motion develops. Initially, this consists of a series of parallel two-dimensional vortices as indicated in Fig. 8.35a. However at higher Rayleigh numbers a three-dimensional cellular flow of the type indicated in Fig. 8.35b develops. These three-dimensional cells have a hexagonal shape as indicated in the figure. This type of flow is termed Bénard cells or Bénard convection.

EXAMPLE 8.9. Two large horizontal plane surfaces are separated by a layer of air. The lowest surface is at a uniform temperature of 40°C and the upper surface is at a uniform temperature of 20°C. Find the gap between the two plates at which significant convective motion in the air layer can be expected to first occur.

Solution. Convective motion will be assumed to begin when:

$$Ra = \frac{\beta g(T_H - T_C)W^3}{\nu^2}Pr = 1708$$

The mean air temperature is:

$$T_{av} = \frac{(40 + 20)}{2} = 30°C$$

At this temperature, air at standard ambient pressure, has the following properties:

$$\beta = 0.0033 \text{K}^{-1}, \nu = 16 \times 10^{-6} \text{ m}^2/\text{s}, Pr = 0.7$$

Hence, convective motion will be assumed to begin when:

$$\frac{0.0033 \times 9.81 \times (40 - 20)W^3}{(16 \times 10^{-6})^2} \times 0.7 = 1708$$

From which it follows that:

$$W = 0.00988 \text{ m} = 9.88 \text{ mm}$$

Therefore, convective motion will first occur when the gap between the plates is 9.88 mm.

8.11
TURBULENT NATURAL CONVECTIVE FLOWS

In the discussions of natural convective flows presented so far in this chapter it has been assumed that the flow is laminar. Turbulent flow can, however, as discussed before, occur in natural convective flows, see [84] to [95], this being illustrated schematically in Fig. 8.36.

For flow over a vertical surface, it has often been assumed that transition occurs when the local Rayleigh number is equal to 10^8 or when the local Grashof number is equal to 10^8. For fluids with Prandtl numbers near one, the difference between these two criteria is small. For fluids with Prandtl numbers that are very different from one, available analyses suggest that the Grashof number-based criterion should be used.

EXAMPLE 8.10. A vertical isothermal flat plate with a surface temperature of 50°C is exposed to stagnant air at 10°C and ambient pressure. Find the distance from the bottom of the plate at which transition to turbulence can be expected to occur.

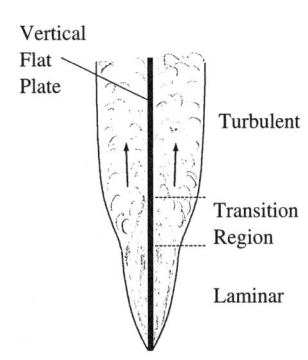

FIGURE 8.36
Transition to turbulence in the natural convective flow over a vertical plate.

Solution. Transition will be assumed to occur when:

$$Gr_x = \frac{\beta g(T_H - T_C)x^3}{\nu^2} = 10^8$$

The mean air temperature is:

$$T_{av} = \frac{(50 + 10)}{2} = 30°C$$

At this temperature air, at standard ambient pressure, has the following properties:

$$\beta = 0.0033 \text{K}^{-1}, \quad \nu = 16 \times 10^{-6} \text{m}^2/\text{s}$$

Hence, transition is assumed to occur when:

$$\frac{0.0033 \times 9.81 \times (50 - 10)x^3}{(16 \times 10^{-6})^2} = 10^8$$

From which it follows that transition occurs when:

$$x = 0.2704 \text{ m}$$

Therefore, transition will occur at a distance of 0.2704 m from the bottom of the plate.

Available analyses of turbulent natural convection mostly rely in some way on the assumption that the turbulence structure is similar to that which exists in turbulent forced convection, see [96] to [105]. In fact, the buoyancy forces influence the turbulence and the direct use of empirical information obtained from studies of forced convection to the analysis of natural convection is not always appropriate. This will be discussed further in Chapter 9. Here, however, a discussion of one of the earliest analyses of turbulent natural convective boundary layer flow on a flat plate will be presented. This analysis involves assumptions that are typical of those used in the majority of available analyses of turbulent natural convection.

This analysis is based on the use of the momentum and energy integral equations which for natural convective flow over a vertical plate are:

$$\frac{d}{dx}\left[\int_0^\delta u^2 \, dy\right] = \beta g \int_0^{\delta_T} (T - T_1) \, dy - \frac{\tau_w}{\rho} \quad (8.164)$$

and:

$$\frac{d}{dx}\left[\int_0^\delta u(T - T_1) \, dy\right] = \frac{q_w}{\rho c_p} \quad (8.165)$$

These equations are the same as for forced convection except for the presence of the integrated buoyancy force term on the right-hand side of Eq. (8.164).

An overbar here will not be used to denote time-averaged values of the variables, all variables used in the present section being time-averaged.

In using these equations, the forms of the velocity and temperature profiles in the boundary layer are assumed. Now, in turbulent forced convective boundary layer flows it has been found that the velocity profile is well described by:

$$\frac{u}{u_1} = \eta^{\frac{1}{n}} \tag{8.166}$$

where:

$$\eta = \frac{y}{\delta} \tag{8.167}$$

δ being the local boundary layer thickness and u_1 being the local free-stream velocity; n is an integer with a value of near 7.

Equation (8.166) cannot be directly applied to natural convective boundary layer flows because in such flows the velocity is zero at the outer edge of the boundary layer. However, Eq. (8.166) should give a good description of the velocity distribution near the wall. It is therefore assumed that in a turbulent natural convective boundary layer:

$$\frac{u}{u_1} = \eta^{\frac{1}{n}}(1-\eta)^2 \tag{8.168}$$

where u_1 is a characteristic velocity for the near wall flow. It will have a value that is close to the maximum velocity in the boundary layer. Eq. (8.168) satisfies the boundary conditions:

At $y = 0$: $u = 0$

At $y = \delta$: $u = 0$

At $y = \delta$: $\dfrac{\partial u}{\partial y} = 0$

and ensures that the velocity distribution near the wall (i.e., for small values of η) is effectively described by Eq. (8.166).

The form of the temperature profile must also be assumed. It will be assumed that the velocity and temperature boundary layers have the same thickness, δ. This is a relatively good assumption in turbulent boundary layer flows. Now, in turbulent forced convective boundary layer flows it has been found that the temperature profile is well described by:

$$\frac{T - T_1}{T_w - T_1} = 1 - \eta^{\frac{1}{n}} \tag{8.169}$$

where T_1 is the free-stream temperature and T_w is the wall temperature. This equation satisfies the boundary conditions:

At $y = 0$: $T = T_w$

At $y = \delta$: $T = T_1$

these boundary conditions applying in both forced and natural convective boundary layer flows. Eq. (8.169) is really only expected to describe the temperature distribution near the wall in natural convective flows but because it does satisfy the boundary conditions it will be assumed here to apply over the whole boundary layer.

Now, Eqs. (8.164) and (8.165) can be written as:

$$\frac{d}{dx}\left(u_1^2 \delta \left[\int_0^1 \left(\frac{u}{u_1}\right)^2 d\eta\right]\right) = \beta g(T_w - T_1)\delta \int_0^1 \left(\frac{T - T_1}{T_w - T_1}\right) d\eta - \frac{\tau_w}{\rho} \qquad (8.170)$$

and:

$$\frac{d}{dx}\left(u_1(T_w - T_1)\delta \left[\int_0^1 \left(\frac{u}{u_1}\right)\left(\frac{T - T_1}{T_w - T_1}\right) d\eta\right]\right) = \frac{q_w}{\rho c_p} \qquad (8.171)$$

If the following are then defined:

$$I_1 = \int_0^1 \left(\frac{u}{u_1}\right)^2 d\eta = \int_0^1 \eta^{\frac{2}{n}}(1-\eta)^4 \, d\eta \qquad (8.172)$$

$$I_2 = \int_0^1 \left(\frac{T - T_1}{T_w - T_1}\right) dy = \int_0^1 (1-\eta^{\frac{1}{n}}) d\eta \qquad (8.173)$$

$$I_3 = \int_0^1 \left(\frac{u}{u_1}\right)\left(\frac{T - T_1}{T_w - T_1}\right) d\eta = \int_0^1 \eta^{\frac{1}{n}}(1-\eta)^2 (1-\eta^{\frac{1}{n}}) d\eta \qquad (8.174)$$

The values of these integrals can be determined for any chosen value of n.

With the integrals defined in this way, Eqs. (8.170) and (8.171) can be written as:

$$I_1 \frac{d}{dx}(u_1^2 \delta) = I_2 \beta g(T_w - T_1)\delta - \frac{\tau_w}{\rho} \qquad (8.175)$$

and:

$$I_3 \frac{d}{dx}[u_1(T_w - T_1)\delta] = \frac{q_w}{\rho c_p} \qquad (8.176)$$

To proceed further, relationships for the wall shear stress, τ_w, and the wall heat transfer rate, q_w, must be assumed. It is consistent with the assumption that the flow near the wall in a turbulent natural convective boundary layer is similar to that in a turbulent forced convective boundary layer to assume that the expressions for τ_w and q_w that have been found to apply in forced convection should apply in natural convection. It will therefore be assumed here that the following apply in a natural convective boundary layer:

$$\frac{\tau_w}{\rho u_1^2} = \frac{0.023}{(\rho u_1 \delta / \mu)^{0.25}}$$

i.e.:

$$\tau_w = 0.023 \frac{\rho^{0.75} u_1^{1.75} \mu^{0.25}}{\delta^{0.25}} \qquad (8.177)$$

and that:

$$\frac{q_w \delta}{k(T_w - T_1)} = 0.023 Pr^{0.33} \left(\frac{\rho u_1 \delta}{\mu}\right)^{0.75}$$

i.e.:

$$q_w = 0.023 Pr^{0.33} k(T_w - T_1) \frac{(\rho u_1/\mu)^{0.75}}{\delta^{0.25}} \quad (8.178)$$

it being assumed here that the characteristic forced velocity in these equations is the same as that used in defining the velocity profile.

Using these equations, Eqs. (8.175) and (8.176) become:

$$I_1 \frac{d}{dx}(u_1^2 \delta) = I_2 \beta g (T_w - T_1) \delta - 0.023 \left(\frac{\mu}{\rho}\right)^{0.25} \frac{u_1^{1.75}}{\delta^{0.25}} \quad (8.179)$$

and:

$$I_3 \frac{d}{dx}[u_1(T_w - T_1)\delta] = \frac{q_w}{\rho c_p} 0.023 Pr^{-0.67} \nu^{0.25} (T_w - T_1) \frac{u_1^{0.75}}{\delta^{0.25}} \quad (8.180)$$

These two equations together describe the variations of δ and u_1. An inspection of the form of these two equations indicates that the solutions are of the form:

$$\delta = Rx^r, \quad u_1 = Sx^s \quad (8.181)$$

This form of solution implies that the boundary layer can be assumed to be turbulent from the leading edge of the surface, i.e., from $x = 0$, or that x is measured from some artificial origin.

Substituting Eq. (8.181) into Eqs. (8.179) and (8.180) and equating the indices of x in the various terms gives:

$$2s + r - 1 = r$$

and:

$$2s + r - 1 = 1.75s - 0.25r$$

and:

$$s + r - 1 = 0.75s - 0.25r$$

Adding s to both sides of the third equation shows that it is the same as the second equation. Solving between the first and second equation then gives:

$$s = 0.5, \quad r = 0.7 \quad (8.182)$$

Substituting these results back into the governing equations and rearranging gives:

$$1.7 S^2 R I_1 = I_2 \beta g (T_w - T_1) R - 0.023 \nu^{0.25} \frac{S^{1.75}}{R^{0.25}}$$

and:

$$1.2 I_3 SR = 0.023 Pr^{-0.67} \nu^{0.25} \frac{S^{0.75}}{R^{0.25}}$$

Solving between these two equations gives:

$$S = \left[\frac{I_2}{1.7 I_1 + 1.2 I_3 Pr^{0.67}}\right]^{0.5} [\beta g (T_w - T_1)]^{0.5} \quad (8.183)$$

and:
$$R = \left[\frac{0.023}{1.2I_3 Pr^{0.67}}\right]^{0.8} \left[\frac{1.7I_1 + 1.2I_3 Pr^{0.67}}{I_2}\right]^{0.1} \left[\frac{\nu^2}{\beta g(T_w - T_1)}\right]^{0.1} \quad (8.184)$$

These two equations can be written as:
$$S = F_1[\beta g(T_w - T_1)]^{0.5} \quad (8.185)$$

and:
$$R = F_2 \left[\frac{\nu^2}{\beta g(T_w - T_1)}\right]^{0.1} \quad (8.186)$$

where:
$$F_1 = \left[\frac{I_2}{1.7I_1 + 1.2I_3 Pr^{0.67}}\right]^{0.5} \quad (8.187)$$

and:
$$F_2 = \left[\frac{0.023}{1.2I_3 Pr^{0.67}}\right]^{0.8} \left[\frac{1.7I_1 + 1.2I_3 Pr^{0.67}}{I_2}\right]^{0.1} \quad (8.188)$$

Combining the results derived above then gives:
$$u_1 = F_1[\beta g(T_w - T_1)]^{0.5} x^{0.5} \quad (8.189)$$

and:
$$\delta = F_2 \left[\frac{\nu^2}{\beta g(T_w - T_1)}\right]^{0.1} x^{0.7} \quad (8.190)$$

Therefore, because Eq. (8.178) can be written as:
$$\frac{q_w x}{k(T_w - T_1)} = 0.023 Pr^{0.33} \left(\frac{u_1 x}{\nu}\right)^{0.75} \left(\frac{x}{\delta}\right)^{0.25}$$

It follows that:
$$Nu_x = 0.023 Pr^{0.33} \frac{F_1^{0.75}}{F_2^{0.25}} Gr_x^{0.4} \quad (8.191)$$

The present analysis therefore predicts that in turbulent boundary layer flow over a vertical surface:
$$\frac{Nu_x}{Gr_x^{0.4}} = \text{function}(Pr) = A \quad (8.192)$$

If n is assumed to be 7, the above equation gives for $Pr = 0.7$:
$$Nu_x = 0.0185 Gr_x^{0.4} \quad (8.193)$$

The mean heat transfer rate for a plate of length, L, can be obtained by noting that:

$$\overline{q_w} = \frac{1}{L}\int_0^L q_w\,dx$$

Hence, using Eq. (8.192):

$$\overline{q_w} = \frac{1}{L}\int_0^L \left[-k(T_w - T_1)A\frac{Gr_x^{0.4}}{x}\right]dx$$

i.e.:

$$\frac{\overline{q_w}L}{(T_w - T_1)k} = \frac{A}{1.2}Gr_L^{0.4}$$

where Gr_L is the Grashof number based on the plate length, L. The mean Nusselt number for the whole plate, $Nu_L(=\overline{h}L/k)$ is, therefore, given by:

$$\frac{Nu_L}{Gr_L^{0.25}} = \frac{A}{1.2} \qquad (8.194)$$

If n is assumed to be 7, the above equation gives for $Pr = 0.7$:

$$Nu_L = 0.0154 Gr_L^{0.4} \qquad (8.195)$$

A comparison of Eq. (8.195) with some experimental results is shown in Fig. 8.37. Quite good agreement will be seen to be obtained.

Turbulent natural convective flows can also be analyzed by numerically solving the governing equations together with some form of turbulence model. This is

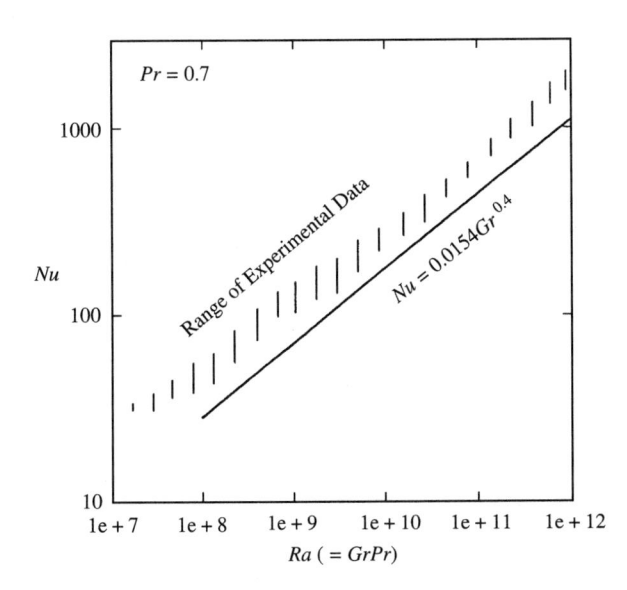

FIGURE 8.37
Comparison of measured and predicted Nusselt number variations for turbulent natural convective flow over a vertical plate.

however, hampered by the fact that many turbulence models do not correctly describe the effect of the buoyancy forces on the turbulence structure.

EXAMPLE 8.11. Plot the variations of boundary layer thickness, maximum velocity, and local heat transfer rate along a 10-m high vertical plate which is maintained at a uniform surface temperature of 50°C and exposed to stagnant air at ambient pressure and a temperature of 10°C. Assume the flow is turbulent from the leading edge.

Solution. The following integrals arise in the approximate solution for turbulent natural convective boundary layer flow over a flat plate discussed above:

$$I_1 = \int_0^1 \eta^{\frac{2}{n}}(1-\eta)^4 \, d\eta$$

$$I_2 = \int_0^1 (1-\eta^{\frac{1}{n}}) \, d\eta$$

$$I_3 = \int_0^1 \eta^{\frac{1}{n}}(1-\eta)^2(1-\eta^{\frac{1}{n}}) \, d\eta$$

These integrals are easily evaluated for any chosen value of the index, n. It will here be assumed that:

$$n = 7$$

and in this case:

$$I_1 = 0.10972, I_2 = 0.125, I_3 = 0.052723$$

The following functions of these integrals and of the Prandtl number were also defined in the analysis:

$$F_1 = \left[\frac{I_2}{1.7I_1 + 1.2I_3 Pr^{0.67}}\right]^{0.5}$$

and:

$$F_2 = \left[\frac{0.023}{1.2I_3 Pr^{0.67}}\right]^{0.8} \left[\frac{1.7I_1 + 1.2I_3 Pr^{0.67}}{I_2}\right]^{0.1}$$

Because air flow is being considered it will be assumed that $Pr = 0.7$. In this case, using the values of the integrals given above, the following are obtained:

$$F_1 = 0.72856, F_2 = 0.57408$$

The analysis then gives:

$$u_1 = F_1[\beta g(T_w - T_1)]^{0.5} x^{0.5}$$

$$\delta = F_2 \left[\frac{\nu^2}{\beta g(T_w - T_1)}\right]^{0.1} x^{0.7}$$

$$\frac{q_w x}{k(T_w - T_1)} = 0.023 Pr^{0.33} \frac{F_1^{0.75}}{F_2^{0.25}} Gr_x^{0.4}$$

CHAPTER 8: Natural Convection **415**

The mean air temperature in the boundary layer is:

$$T_{av} = \frac{(50 + 10)}{2} = 30°C$$

At this temperature, air at standard ambient pressure has the following properties:

$$\beta = 0.0033 \text{ K}^{-1}, \nu = 16 \times 10^{-6} \text{ m}^2/\text{s}, k = 0.02638 \text{ W/m-K}$$

Hence:

$$u_1 = 0.72856[0.0033 \times 9.81 \times (50 - 10)]^{0.5} x^{0.5} = 0.8292 x^{0.5}$$

$$\delta = 0.57408 \left[\frac{(16 \times 10^{-6})^2}{0.0033 \times 9.81 \times (50 - 10)}\right]^{0.1} x^{0.7} = 0.04882 x^{0.7}$$

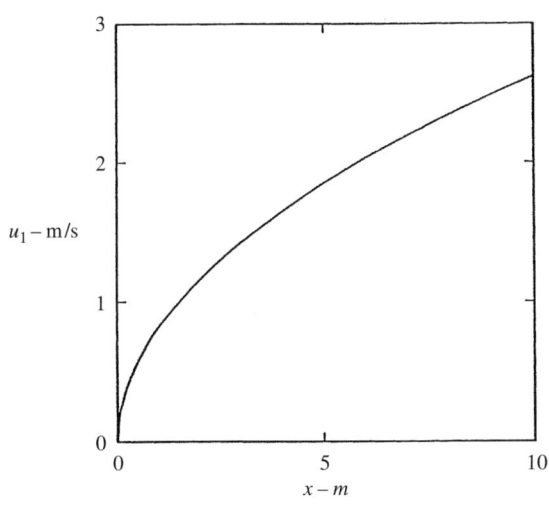

FIGURE E8.11a

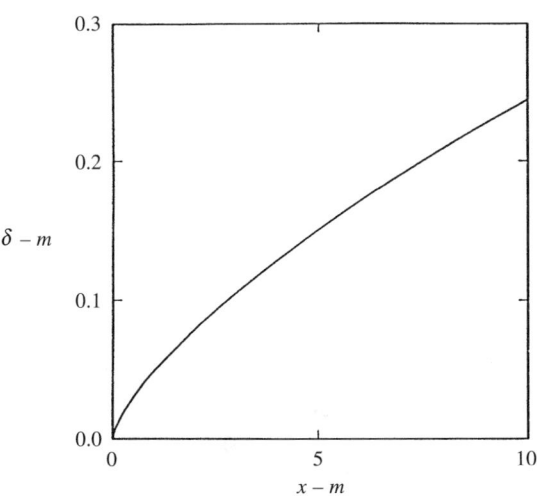

FIGURE E8.11b

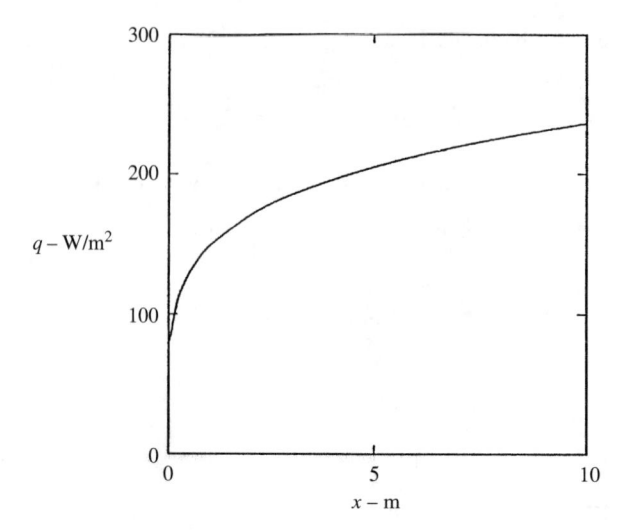

FIGURE E8.11c

and:

$$\frac{q_w x}{0.02638 \times (50 - 10)} = 0.023 \times 0.7^{0.33} \times \frac{0.72856^{0.75}}{0.57408^{0.25}} \times \left[\frac{0.0033 \times 9.81 \times (50 - 10)x^3}{(16 \times 10^{-6})^2} \right]^{0.4}$$

i.e.:

$$q_w = 148.8 x^{0.2}$$

The variations of u_1, δ, and q_w with x as given by these equations is shown in Figs. E8.11a, E8.11b, and E8.11c respectively.

8.12 CONCLUDING REMARKS

Some of the more commonly used methods of obtaining solutions to problems involving natural convective flow have been discussed in this chapter. Attention has been given to laminar natural convective flows over the outside of bodies, to laminar natural convection through vertical open-ended channels, to laminar natural convection in a rectangular enclosure, and to turbulent natural convective boundary layer flow. Solutions to the boundary layer forms of the governing equations and to the full governing equations have been discussed.

PROBLEMS

8.1. A vertical isothermal flat plate with a surface temperature of 50°C is exposed to stagnant air at 20°C and standard ambient pressure. Plot the velocity and temperature profiles in the boundary layer at distances of 15, 30, and 45 cm from the leading edge.

8.2. A 0.3-m vertical plate is maintained at a surface temperature of 65°C and is exposed to stagnant air at a temperature of 15°C and standard ambient pressure. Compare the natural convective heat transfer rate from this plate with that which would result from forcing air over the plate at a velocity equal to the maximum velocity that occurs in the natural convective boundary layer.

8.3. Compare the heat-transfer coefficients for laminar forced and free convection over vertical flat plates. Develop an approximate relation between the Reynolds and Grashof numbers such that the heat-transfer coefficients for pure forced convection and pure free convection are equal.

8.4. A 0.2-m square vertical plate is heated to 400°C and placed in room air at 25°C. Calculate the heat loss from one side of the plate.

8.5. A vertical plate 10 cm high is immersed in a stagnant fluid. The plate is maintained at a temperature of 50°C and the fluid temperature is 10°C. Determine the average heat transfer coefficient for this situation if the fluid is:
(i) air at standard atmospheric pressure
(ii) air at $0.01 \times$ standard atmospheric pressure
(iii) water

8.6. Plot the free-convection boundary-layer thickness along a 0.3-m high vertical plate which is maintained at a uniform surface temperature of 50°C and exposed to stagnant air at ambient pressure and a temperature of 10°C. Assume the flow remains laminar.

8.7. Using the similarity solution results, derive an expression for the maximum velocity in the natural convective boundary layer on a vertical flat plate. At what position in the boundary layer does this maximum velocity occur?

8.8. Two vertical flat plates held at a uniform surface temperature of at 40°C are placed in a tank of water which is at a temperature of 20°C. If the plates are 10 cm high, estimate the minimum spacing between the plates if there is to be no interference between the boundary layers on the two plates.

8.9. A vertical flat plate is maintained at a uniform surface temperature and is exposed to air at standard ambient pressure. At a distance of 10 cm from the leading edge of the plate the boundary layer thickness is 2 cm. Estimate the thickness of the boundary layer at a distance of 25 cm from the leading edge. Assume a laminar boundary layer flow.

8.10. Consider laminar free-convective flow over a vertical flat plate at whose surface the heat transfer rate per unit area, q_w, is constant. Show that a similarity solution to the two-dimensional laminar boundary layer equations can be derived for this case.

8.11. It will be seen from the results given by the similarity solution that the velocities are very low in natural convective boundary layers in fluids with high Prandtl numbers. In such circumstances, the inertia terms (i.e., the convective terms) in the momentum equation are negligible and the boundary layer momentum equation for a vertical surface effectively is:

$$\left(\frac{\mu}{\rho}\right)\frac{\partial^2 u}{\partial y^2} + \beta g(T - T_1) = 0$$

i.e., there is a balance between the buoyancy and the viscous forces. If the following are assumed:

$$\frac{ux/\alpha}{Ra_x^{1/2}} = F'(\eta)$$

and:

$$\frac{T - T_1}{T_w - T_1} = G(\eta)$$

where:

$$\eta = \frac{y}{x} Ra_x^{1/4}$$

Ra_x being the Rayleigh number based on x, i.e.:

$$Ra_x = \frac{\beta g(T - T_1)x^3}{\nu\alpha} = Gr_x Pr$$

the prime, of course, again denotes differentiation with respect to η, show that the boundary layer equations give:

$$F''' + G = 0$$

$$G'' + \frac{3}{4} FG' = 0$$

with boundary conditions:

$$\text{At } \eta = 0 : F' = 0$$
$$\text{At } \eta = 0 : F = 0$$
$$\text{At } \eta = 0 : G = 1$$
$$\text{For large } \eta : F' \to 0$$
$$\text{For large } \eta : G \to 0$$

Show from these results that this solution indicates that

$$Nu_x = \text{function}(Ra_x)$$

i.e., show that at large values of the Prandtl number the local Nusselt number depends only on the local Rayleigh number and not separately on the Prandtl number and the local Grashof number.

8.12. A 30-cm high vertical plate has a surface temperature that varies linearly from 15°C at the lower edge to 45°C at the upper edge. This plate is exposed to air at 15°C and ambient pressure. Use the computer program for natural convective boundary layer flow to determine how the local heat transfer rate varies with distance up the plate from the lower edge.

8.13. Use the computer program for two-dimensional laminar free convective flow to find the Nusselt number variation along a vertical plate whose surface temperature varies in such a way that $T_w - T_\infty$ is equal to 10°C over the lower half of the plate and equal to 30°C over the upper half of the plate.

8.14. When the density variation with temperature of water is considered, a maximum is found to occur near 4°C as indicated schematically in the following figure.

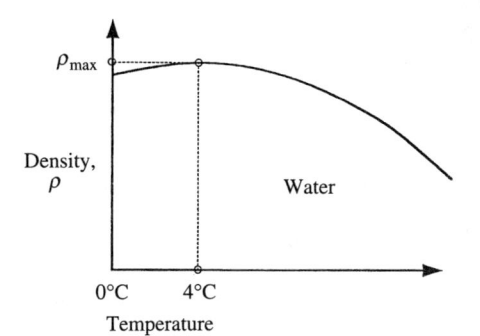

FIGURE P8.14

Near this point of maximum density, the density is approximately given by:

$$\rho_{max} - \rho = A(T - 4)^2$$

where A is a constant. Modify the program for solving the boundary layer equation for natural convective flow discussed in this text to apply to this situation. In this case define:

$$\theta = \frac{T - 4}{T_w - 4}$$

T_w being the uniform wall temperature which is between 0°C and 4°C.

8.15. Air flows by natural convection through the channel formed between 2 20-cm high plates kept at a temperature of 50°C. If the distance between the 2 plates is 3 cm and if the ambient air temperature is 20°C, find the rate of heat transfer from the 2 plates to the air and the mean velocity of the air through the channel.

8.16. Consider the natural convective flow of air at 10°C though a plane vertical channel with isothermal walls whose temperature is 40°C and whose height is 10 cm. Determine how the mean heat transfer rate from the heated walls varies with the gap between the walls.

8.17. Consider heat transfer across an air-filled inclined square enclosure with one wall heated to a uniform temperature and the parallel wall cooled to a uniform temperature with the remaining two walls being perfectly conducting. The wall temperature and enclosure size are such that the Rayleigh number based on the enclosure size and on the difference between the temperatures of the hot and cold walls is 10^5. Examine the effect of the angle of the heated wall to the vertical on the heat transfer rate by numerically solving the governing equations. Consider angles of inclination to the vertical of between $+60°$ and $-60°$.

8.18. When the density variation with temperature of water is considered, a maximum is found to occur near 4°C as discussed in Problem 8.14. Near this point of maximum, density is approximately given by:

$$\rho_{max} - \rho = A(T - 4)^2$$

where A is a constant. Modify the program for solving for the natural convective flow in an enclosure to apply to this situation. Consider the case of a vertical square enclosure with perfectly conducting top and bottom walls whose cold wall is at 0°C and determine how the dimensionless hot wall temperature affects the mean Nusselt number. Use a constant suitably modified Rayleigh number based on the difference between the temperatures of the hot and cold walls of 10^5. Define:

$$\theta = \frac{T - 4}{T_H - T_C}$$

8.19. The discussion of free convective flows given in this chapter has implicitly been concerned with flows that arise due to the presence of temperature gradients in a fluid acted on by gravity. Free convective flows can be caused by the presence of temperature gradients in a fluid acted upon by centrifugal forces. Consider a 5-cm high plane fin kept at a uniform temperature of 60°C and exposed to air at 20°C. This fin is part of a rotating electrical machine and the center of the fin is rotating on a radius of 0.5 m at a speed of 600 rpm. Estimate the mean heat transfer rate from the plate assuming that the air to which the plate is exposed is rotating with the machine.

8.20. Plot the variations of boundary layer thickness and local heat transfer rate along a 2-m high vertical plate which is maintained at a uniform surface temperature of 60°C and exposed to stagnant air at ambient pressure and a temperature of 20°C assuming that (i) the flow is laminar from the leading edge and (ii) the flow is turbulent from the leading edge.

8.21. A vertical isothermal flat plate with a surface temperature of 20°C is exposed to stagnant water at 10°C. Find the distance from the bottom of the plate at which transition to turbulence can be expected to occur.

8.22. A wide vertical plate 1 m high is immersed in a stagnant air which has a temperature of 30°C. The plate is maintained at a uniform temperature of 50°C. Plot the variation of average heat transfer rate from plate with air pressure for air pressures between 0.1 and 10 times standard atmospheric pressure.

REFERENCES

1. Ede, A.J., "Advances in Natural Convection", *Adv. Heat Transfer*, Vol. 4, pp. 1-64, 1967.
2. Gebhart, B., *Heat Transfer*, 2nd ed., McGraw-Hill, New York, 1970.
3. Jaluria, Y., *Natural Convection Heat and Mass Transfer*, Pergamon Press, New York, 1980.
4. Raithby, G.D. and Hollands, K.G.T., "Natural Convection", in *Handbook of Heat Transfer Fundamentals*, 2nd ed., Rohsenow, W.M., Hartnett, J.P., and Ganic, E.N., Eds., McGraw-Hill, New York, 1985.
5. Gebhart, B., Jaluria, Y., Mahajan, R.L., and Sammakia, B., *Buoyancy-Induced Flows and Transport*, Hemisphere Publ., Washington, DC, 1988.
6. Gray, D.D. and Giorgini, A., "The Validity of the Boussinesq Approximation for Liquids and Gases", *Int. J. Heat Mass Transfer*, Vol. 19, pp. 545-551, 1976.
7. LeFevre, E.J., "Laminar Free Convection from a Vertical Plane Surface", Proc. Ninth Int. Congr. Appl. Mech., Brussels, Vol. 4, p. 168, 1956.

8. Martin, B.W., "An Appreciation of Advances in Natural Convection Along an Isothermal Vertical Surface", *Int. J. Heat Mass Transfer,* Vol. 27, pp.1583-1586, 1984.
9. Ostrach, S., "An Analysis of Laminar-Free-Convection Flow and Heat Transfer About a Flat Plate Parallel to the Direction of the Generating Body Force", NACA Tech. Rep 1111, 1953.
10. Schlichting, H., *Boundary Layer Theory,* 7th ed., McGraw-Hill, New York, 1979.
11. Touloukian, Y.S., Hawkins, G.A., and Jakob, M., "Heat Transfer by Free Convection from Heated Vertical Surfaces to Liquids," *Trans. ASME,* Vol. 70, pp. 13-23, 1948.
12. Sparrow, E.M. and Gregg, J.L., "Laminar Free Convection from a Vertical Flat Plate", *Trans. ASME,* Vol. 78, p. 435, 1956.
13. Sparrow, E.M. and Gregg, J.L., "Laminar Free Convection From a Vertical Flat Plate with Uniform Surface Heat Flux", *Trans. ASME,* Vol. 78, pp. 435-440, 1956.
14. Sparrow, E.M. and Gregg, J.L., "Similar Solutions for Free Convection from a Nonisothermal Vertical Plate", *Trans. ASME,* Vol. 80, pp. 379-386, 1958.
15. Sparrow, E.M., "Laminar Free Convection on a Vertical Plate with Prescribed Nonuniform Wall Heat Flux or Prescribed Nonuniform Wall Temperature", NACA TN 3508, July 1955.
16. Sparrow, E.M. and Yu, H.S., "Local Nonsimilarity Thermal Boundary Layer Solutions", *J. Heat Transfer,* Vol. 93, pp. 328-334, 1971.
17. Sparrow, E.M. and Gregg, J.L.,"The Variable Fluid Property Problem in Free Convection", *Trans. ASME,* Vol. 80, pp. 879-886, 1958.
18. Yang, K.T., "Possible Similarity Solutions for Laminar Free Convection on Vertical Plates and Cylinders", *J. Appl. Mech.,* Vol. 27, pp. 230-236, 1960.
19. Clausing, A.M., "Natural Convection Correlations for Vertical Surfaces Including Influences of Variable Properties", *J. Heat Transfer,* Vol. 105, No. 1, pp. 138-143, 1983.
20. Gryzagoridis, J., "Natural Convection from a Vertical Flat Plate in the Low Grashof Number Range", *Int. J. Heat Mass Transfer,* Vol. 14, pp. 162-164, 1971.
21. Kuiken, H.K., "An Asymptotic Solution for Large Prandtl Number Free Convection," *J. Eng. Math.,* Vol. 2, pp. 355-371, 1968.
22. Fujii, T. and Imura, H., "Natural Convection from a Plate with Arbitrary Inclination", *Int. J. Heat Mass Transfer,* Vol. 15, pp. 755-767, 1972.
23. Vliet, G.C., "Natural Convection Local Heat Transfer on Constant Heat Flux Inclined Surfaces", *J. Heat Transfer,* Vol. 9, pp. 511-516, 1969.
24. Bodoia, J.R. and Osterle, J.F., "The Development of Free Convection Between Heated Vertical Plates", *Trans. ASME, J. Heat Transfer,* Vol. 84, pp. 40-44, 1962.
25. Elenbaas, W., "Heat Dissipation of Parallel Plates by Free Convection", *Physica IX,* No. 1, pp. 2-28, 1942.
26. Lighthill, M.J., "Theoretical Consideration on Free Convection in Tubes", *J. Mech. Appl. Math,* Vol. 6, pp. 398-439, 1953.
27. Oosthuizen, P.H., "A Numerical Study of Laminar Free Convective Flow Through a Vertical Open Partially Heated Plane Duct", Fundamentals of Natural Convection/Electronic Equipment Cooling, ASME HTD, Vol. 32, pp. 41-48, 22nd National Heat Transf. Conf. and Exhibition, Niagara Falls, New York, 1984.
28. Oosthuizen, P.H. and Paul, J.T., "A Numerical Study of Free Convective Flow through a Vertical Annular Duct", ASME Paper 86-WA/HT-81, ASME Winter Annual Meeting, Anaheim, CA, December 7-12, 1986.
29. Azevedo, L.F.A. and Sparrow, E.M., "Natural Convection in Open-Ended Inclined Channels", *J. Heat Transfer,* Vol. 107, p. 893, 1985.
30. Bahrani, P.A. and Sparrow, E.M., "Experiments on Natural Convection from Vertical Parallel Plates with Either Open or Closed Edges", *ASME J. Heat Transfer,* Vol. 102, pp. 221-227, 1980.

31. Bar-Cohen, A. and Rohsenow, W.M., "Thermally Optimum Spacing of Vertical Natural Convection Cooled, Parallel Plates", *J. Heat Transfer*, Vol. 106, p. 116, 1984.
32. Eckert, E. and Diaguila, A.J., "Convective Heat Transfer for Mixed, Free, and Forced Flow Through Tubes", *Trans. ASME*, Vol. 76, pp. 497-504, 1954.
33. Yan, W.M. and Lin, T.F., "Theoretical and Experimental Study of Natural Convection Pipe Flows at High Rayleigh Numbers", *Int. J. Heat Mass Transfer*, Vol. 34, pp. 291-302, 1991.
34. Arnold, J.N., Catton, I., and Edwards, D.K., "Experimental Investigation of Natural Convection in Inclined Rectangular Regions of Differing Aspects Ratios", *J. Heat Transfer*, Vol. 98, pp. 67-71, 1976.
35. Ayyaswamy, P.S. and Catton, I., "The Boundary-Layer Regime for Natural Convection in a Differentially Heated, Tilted Rectangular Cavity ", *J. Heat Transfer*, Vol. 95, pp. 543-545, 1973.
36. Batchelor, G.K., "Heat Transfer by Free Convection Across a Closed Cavity between Vertical Boundaries at Different Temperatures", *J. Appl. Math*, Vol. 12, pp. 209-233, 1954.
37. Blythe, P.A. and Simpkins, P.G., "Thermal Convection in a Rectangular Cavity", *Physiochemical Hydrodynamics*, Vol. 2, pp. 511-524, 1977.
38. Buchberg, H., Catton, I., and Edwards, D.K., "Natural Convection in Enclosed Spaces—A Review of Application to Solar Energy", *J. Heat Transfer*, 98, p. 182, 1976.
39. Catton, I., "Natural Convection in Enclosures", Proc. 6th. International Heat Trans. Conf., Toronto, Vol. 6, pp. 13-31, Hemisphere Publ, Washington, DC, 1978.
40. de Vahl Davis, G. and Jones, I.P., "Natural Convection in a Square Cavity - A Comparison Exercise", *Int. J. Numer. Mech. Fluids*, Vol. 3, pp. 227-249, 1983.
41. de Vahl Davis, G., "Laminar Natural Convection in an Enclosed Rectangular Cavity", *Int. J. Heat Mass Trans.*, Vol. 11, pp. 1675-1693, 1968.
42. Eckert, E.R.G. and Carlson, W.O., "Natural Convection in an Air Layer Enclosed Between Two Vertical Plates with Different Temperatures", *Int. J. Heat Mass Transfer*, Vol. 2, pp. 106-120, 1961.
43. El Sherbing, S.M., Raithby, G.D., and Hollands, K.G.T., "Heat Transfer by Natural Convection Across Vertical and Inclined Air Layers", *J. Heat Transfer, Trans. ASME*, Vol. 104, p. 96, 1982.
44. Emery, A. and Chu, N.C., "Heat Transfer Across Vertical Layers", *J. Heat Transfer, Trans. ASME*, Ser. C, Vol. 87, pp. 110-116, 1965.
45. Gill, A.E., "The Boundary Layer Regime for Convection in a Rectangular Cavity", *J. Fluid Mech.*, Vol. 26, pp. 515-536, 1966.
46. Graebel, W.P., "The Influence of Prandtl Number on Free Convection in a Rectangular Cavity", *Int. J. Heat Mass Transfer*, Vol. 24, pp. 125-131, 1981.
47. Hollands, K.G.T., Unny, T.E., Raithby, G.D., and Konicek, L., "Free Convection Heat Transfer Across Inclined Air Layers", *J. Heat Transfer*, Vol. 98, pp. 189-193, 1976.
48. Jakob, M., "Free Convection through Enclosed Plane Gas Layers", *Trans. ASME*, Vol. 68, p. 189, 1946.
49. Kamotani, Y., Wang, L.W., and Ostrach, S., "Experiments on Natural Convection Heat Transfer in Low Aspect Ratio Enclosures", Paper No. AIAA-81-1066, AIAA 16th Thermophysics Conference, Palo Alto, California, June 23-25, 1981.
50. Kimura, S. and Bejan, A., "The Boundary Layer Natural Convection Regime in a Rectangular Cavity with Uniform Heat Flux From the Side", *J. Heat Transfer*, Vol. 106, pp. 98-103, 1984.
51. Lee, E.I. and Sernas, V., "Numerical Study of Heat Transfer in Rectangular Air Enclosures of Aspect Ratios Less Than One", ASME Paper No. 80-WA/HT-43, Am. Soc. Mech. Eng., New York, 1980.

52. MacGregor, R.K. and Emery, A.P., "Free Convection Vertical Plane Layers—Moderate and High Prandtl Number Fluids", *J. Heat Transfer,* Vol. 91, p. 391, 1969.
53. Markatos, N.C. and Pericleous, K.A., "Laminar and Turbulent Natural Convection in an Enclosed Cavity", *Int. J. Heat Mass Transfer,* Vol. 27, pp. 755-772, 1984.
54. Naylor, D. and Oosthuizen, P.H., "A Numerical Study of Free Convective Heat Transfer in a Parallelogram-Shaped Enclosure", *Heat & Fluid Flow,* Vol. 4, No. 6, pp. 553-559, 1994.
55. Newell, M.E. and Schmidt, F.W., "Heat Transfer by Laminar Natural Convection Within Rectangular Enclosures", *J. Heat Transfer,* Vol. 92, pp. 159-168, 1970.
56. Oosthuizen, P.H. and Paul, J.T., "Natural Convective Flow in a Cavity with Conducting Top and Bottom Walls", Proc. 9th International Heat Trans. Conf., Vol. 2, Hemisphere Publisher, New York, pp. 263-268, 1990.
57. Oosthuizen, P.H. and Paul, J.T., "Free Convection in a Square Cavity with a Partially Heated Wall and a Cooled Top", *Thermophysics and Heat Transfer,* Vol. 5, No. 4, pp. 583-588, 1991.
58. Oosthuizen, P.H., "Free Convective Flow in an Enclosure with a Cooled Inclined Upper Surface", *Computational Mechanics,* Vol. 14, No. 5, pp. 420-430, 1994.
59. Ostrach, S., "Natural Convection in Enclosures", in *Advances in Heat Transfer,* Vol. 8, Hartnett, J.P. and Irvine, T.F., Eds., Academic Press, New York, pp. 161-227, 1972.
60. Pepper, D.W. and Harris, S.D., "Numerical Simulation of Natural Convection in Closed Containers by a Fully Implicit Method", *J. Fluids Eng.,* Vol. 99, pp. 649-656, 1977.
61. Quon, C., "Free Convection in an Enclosure Revisited", *J. Heat Transfer,* Vol. 99, pp. 340-342, 1977.
62. Quon, C., "High Rayleigh Number Convection in an Enclosure: A Numerical Study", *Phys. Fluids,* Vol. 15-I, pp. 12-19, 1972.
63. Rubel, A. and Landis, R., "Numerical Study of Natural Convection in a Vertical Rectangular Enclosure", *Phys. Fluids,* Suppl. II, Vol. 12-II, pp. 208-213, 1969.
64. Tichy, J. and Gadgil, A., "High Rayleigh Number Laminar Convection in Low Aspect Ratio Enclosures with Adiabatic Horizontal Walls and Differentially Heated Vertical Walls", *J. Heat Transfer,* Vol. 104, pp. 103-110, 1982.
65. Wirtz, R.A. and Tseng, W.F., "Natural Convection Across Tilted, Rectangular Enclosures of Small Aspect Ratio", *Natural Convection in Enclosures,* ASME publication HTD-Vol. 8, pp. 47-54, 1980.
66. Zhong, Z.Y., Lloyd, J.R., and Yang, K.T., "Variable-Property Natural Convection in Tilted Square Cavities", *Numerical Methods in Thermal Problems,* Vol. III, Lewis, R.W., Johnson, J.A., and Smith, W.R., Eds., Pineridge Press, Swansea, U.K., pp. 968-979, 1983.
67. Oosthuizen, P.H. and Paul, J.T., "Free Convective Heat Transfer in a Rotating Square Cavity", Paper AIAA-88-2621, Presented at the AIAA Thermophysics, Plasmadynamics and Lasers Conference, San Antonio, 1988.
68. Lord Rayleigh, "On Convection Currents in a Horizontal Layer of Fluid when the Higher Temperature is on the Underside", *Philos. Mag.,* Vol. 6, No. 32, pp. 529-546, 1916.
69. Jeffreys, H., "The Stability of a Layer of Fluid Heated Below", *Philos. Mag.,* Vol. 2, pp. 833-844, 1926.
70. Jeffreys, H., "Some Cases of Instability in Fluid Motion", *Proc. R. Soc. London Ser. A.,* Vol. 118, pp. 195-208, 1928.
71. Pellew, A. and Southwell, R.V., "On Maintained Convection Motion in a Fluid Heated from Below", *Proc. R. Soc. London,* Vol. A176, pp. 312-343, 1940.
72. Platten, J.K. and Legros, J.C., Convection in Liquid, Springer-Verlag, Berlin, 1984.
73. Turner, J.S., *Buoyancy Effects in Fluids,* Cambridge University Press, London, 1973.

74. Zimmerman, G. and Muller, U., "Bernard Convection in a Two-Component System with Soret Effects", *Int. J. Heat Mass Transfer,* Vol. 35, pp. 2245-2256, 1992.
75. Clever, R.M., "Finite Amplitude Longitudinal Convection Rolls in an Inclined Layer", *J. Heat Trans.,* Vol. 95, p. 407, 1973.
76. Globe, S. and Dropkin, D., "Natural Convection Heat Transfer in Liquids Confined by Two Horizontal Plates and Heated From Below", *J. Heat Transfer,* Vol. 81, pp. 24-28, 1959.
77. Goldstein, R.J. and Chu, T.Y., "Thermal Convection in a Horizontal Layer of Air", *Prog. Heat Mass Transfer,* Vol. 2, p. 55, 1969.
78. Goldstein, R.J., Sparrow, E.M., and Jones, D.C., "Natural Convection Mass Transfer Adjacent to Horizontal Plates", *Int. J. Heat Mass Transfer,* Vol. 16, pp. 1025-1035, 1973.
79. Husar, R.B. and Sparrow, E.M., "Patterns of Free Convection Flow Adjacent to Horizontal Heated Surfaces", *Int. J. Heat Mass Transfer,* Vol. 11, p. 1206, 1968.
80. Schmidt, E., "Free Convection in Horizontal Fluid Spaces Heated from Below", Proc. Int. Heat Trans. Conf., Boulder, Colo., ASME, 1961.
81. Sparrow, E.M., Husar, R.B., and Goldstein, R.J., "Observations and Other Characteristics of Thermals", *J. Fluid Mechanics,* Vol. 41, pp. 793-800, 1970.
82. Hollands, K.G.T., Raithby, G.D., and Konicek, L., "Correlation Equations for Free Convection Heat Transfer in Horizontal Layers of Air and Water", *Int. J. Heat Mass Transfer,* Vol. 18, pp. 879-884, 1975.
83. Pera, L. and Gebhart, B., "Natural Convection Boundary Layer Flow Over Horizontal and Slightly Inclined Surfaces", *Int. J. Heat Mass Transfer,* Vol. 16, p. 1131, 1973.
84. Bejan, A. and Cunnington, G.R., "Theoretical Consideration of Transition to Turbulence in Natural Convection Near a Vertical Wall", *Int. J. Heat Fluid Flow,* Vol. 4, No. 3, pp. 131-139, 1983.
85. Gebhart, B., "Natural Convection Flows and Stability", in *Advances in Heat Transfer,* Academic Press, New York, 1973.
86. Gebhart, B. and Mahajan, R.L., "Characteristic Disturbance Frequency in Vertical Natural Convection", *Int. J. Heat Mass Transfer,* Vol. 18, pp. 1143-1148, 1975.
87. Godaux, R. and Gebhart, B., "An Experimental Study of the Transition of Natural Convection Flow Adjacent to a Vertical Surface", *Int. J. Heat Mass Transfer,* Vol. 17, pp. 93-107, 1974.
88. Lloyd, J.R. and Sparrow, E.M., "On the Instability of Natural Convection Flow on Inclined Plates", *J. Fluid Mech.,* Vol. 42, pp. 465-470, 1970.
89. Mahajan, R.L. and Gebhart, B., "An Experimental Determination of Transition Limits in a Vertical Natural Convection Flow Adjacent to a Surface", *J. Fluid Mech.,* Vol. 91, pp. 131-154, 1979.
90. Oosthuizen, P.H., "A Note on the Transition Point in a Free Convective Boundary Layer on an Isothermal Vertical Plane Surface", *J.S.A. Inst. Mech. Engs.,* Vol. 13, No. 10, pp. 265-268, 1964.
91. Polymeropoulos, C.E. and Gebhart, B., "Incipient Instability in Free Convection Laminar Boundary Layers", *J. Fluid Mechanics,* Vol. 30, pp. 225-239, 1967.
92. Szewczyk, A.A., "Stability and Transition of the Free Convection Layer Along a Vertical Flat Plate", *Int. J. Heat Mass Transfer,* Vol. 5, pp. 903-914, 1962.
93. Eckert, E.R.G. and Soehnghen, E., "Interferometric Studies on the Stability and Transition to Turbulence of a Free-Convection Boundary Layer", *Proc. of the General Discussion on Heat Transfer,* pp. 321-323, ASME-IME, London, 1951.
94. Reshotkao, E., "Boundary Layer Stability and Transition", *Annu. Rev. Fluid Mech.,* Vol. 8, pp. 311-349, 1976.
95. Bejan, A. and Lage, J.L., "The Prandtl Number Effect on the Transition in Natural Convection Along a Vertical Surface", *J. Heat Transfer,* Vol. 112, pp. 787-790, 1990.

96. Bayley, F.J., "An Analysis of Turbulent Free Convection Heat Transfer", *Proc. Inst. Mech. Eng.,* Vol. 169, No. 20, p. 361, 1955.
97. Cheesewright, R., "Turbulent Natural Convection from a Vertical Plane Surface", *J. Heat Transfer,* Vol. 90, p. 1, 1968.
98. Churchill, S.W. and Chu, H.H.S., "Correlating Equations for Laminar and Turbulent Free Convection from a Vertical Plate", *Int. J. Heat Mass Transfer,* Vol. 18, pp. 1323-1329, 1975.
99. Eckert, E.R.G. and Jackson, T.W., "Analysis of Turbulent Free Convection Boundary Layer on a Flat Plate", NACA Rept. 1015, 1951.
100. Henkes, R.A.W.M. and Hoogendoorn, C.J., "Comparison of Turbulence Models for the Natural Convection Boundary Layer Along a Heated Vertical Plate", *Int. J. Heat Mass Transfer,* Vol. 32, pp. 157-169, 1989.
101. Lin, S.-J. and Churchill, S.W., "Turbulent Free Convection from a Vertical, Isothermal Plate", *Num. Heat Trans.,* Vol. 1, pp. 129-145, 1978.
102. Ruckstein, E. and Felski, J.D., "Turbulent Natural Convection at High Prandtl Numbers", *J. Heat Transfer,* Vol. 102, pp. 773-775, 1980.
103. Vliet, G.C. and Liu, D.C., "An Experimental Study of Turbulent Natural Convection", *J. Heat Transfer,* Vol. 91, pp. 517-531, 1969.
104. Warner, C.Y. and Arpaci, V.S., "An Experimental Investigation of Turbulent Natural Convection in Air at Low Pressure along a Vertical Heated Plate", *Int. J. Heat Mass Transfer,* Vol. 11, pp. 397-406, 1968.
105. To, W.M. and Humphrey, J.A.C., "Numerical Simulation of Buoyant, Turbulent Flow. I. Free Convection Along a Heated, Vertical, Flat Plate", *Int. J. Heat Mass Transfer,* Vol. 29, pp. 573-592, 1986.

CHAPTER 9

Combined Convection

9.1
INTRODUCTION

The buoyancy forces that arise as the result of the temperature differences and which cause the fluid flow in free convection also exist when there is a forced flow. The effects of these buoyancy forces are, however, usually negligible when there is a forced flow. In some cases, however, these buoyancy forces do have a significant influence on the flow and consequently on the heat transfer rate. In such cases, the flow about the body is a combination or mixture of forced and free convection as indicated in Fig. 9.1 and such flows are referred to as combined or mixed forced and free (or natural) convection.

Such combined convective flows are normally associated with low forced velocities. They can occur, for example, in some electronic cooling situations and in some heat exchangers.

9.2
GOVERNING PARAMETERS

Consider combined convective flow over a series of geometrically identical bodies as indicated in Fig. 9.2. In such a case, as discussed in Chapter 1, the mean heat transfer coefficient, h, will depend on:

- The conductivity, k, of the fluid with which the body is exchanging heat.
- The viscosity, μ, of the fluid with which the body is exchanging heat.
- The specific heat, c_p, of the fluid with which the body is exchanging heat.

CHAPTER 9: Combined Convection 427

FIGURE 9.1
Combined convective flow.

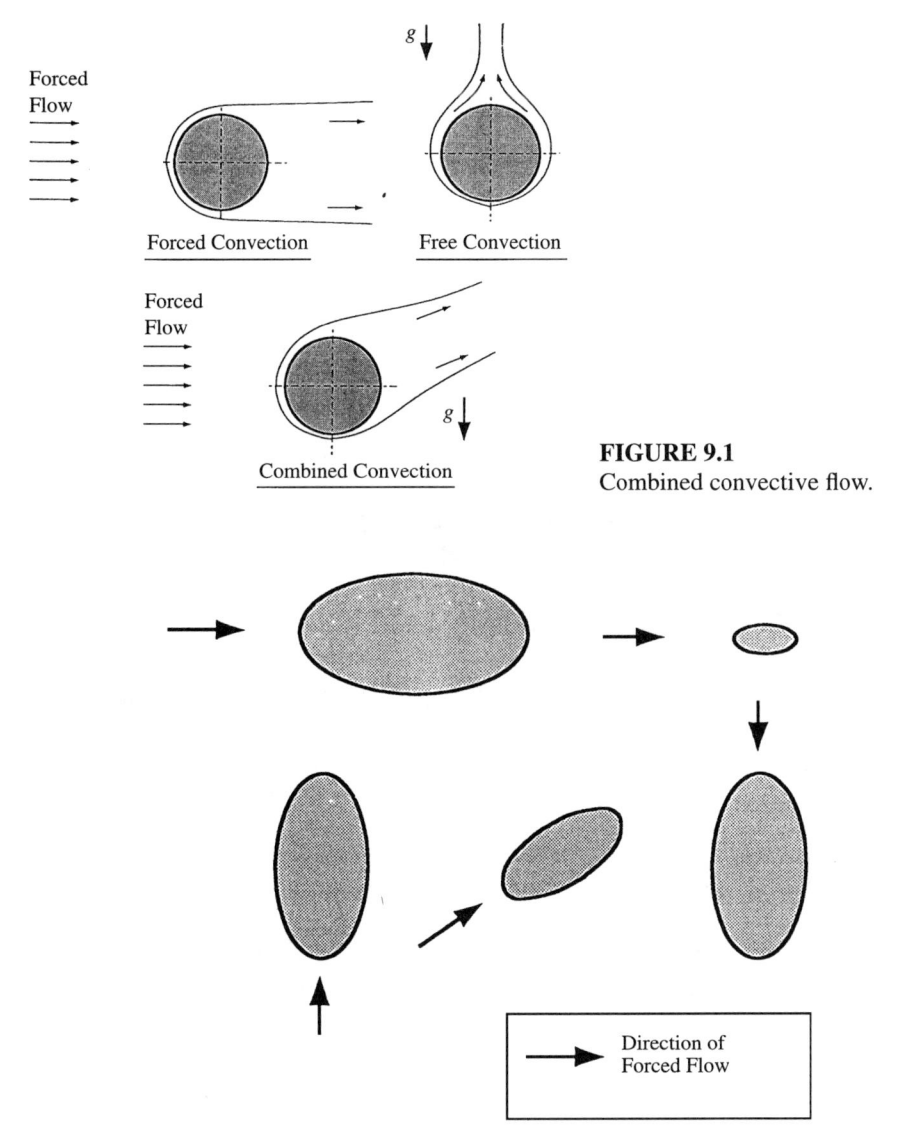

FIGURE 9.2
Flow situation considered.

- The density, ρ, of the fluid with which the body is exchanging heat.
- The size of the body as specified by some characteristic dimension, ℓ.
- The magnitude of the forced fluid velocity, U, relative to the body.
- The buoyancy force parameter, $\beta g(T_w - T_f)$.
- The angle, ϕ, between the direction of the forced velocity and the direction of the gravity vector, i.e., between the direction of the forced velocity and the direction in which the buoyancy forces act (see Fig. 9.3).

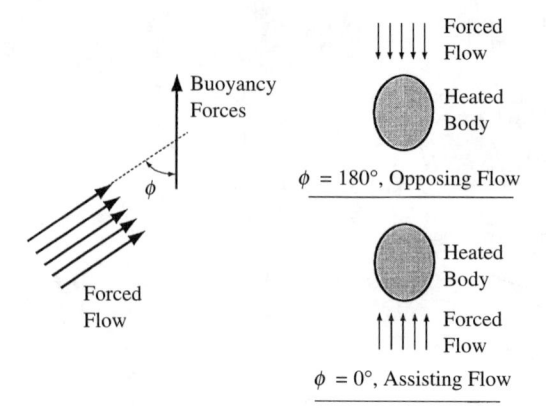

FIGURE 9.3
Angle between direction of forced flow and that of the buoyancy force.

It is assumed, therefore, that:

$$h = \text{function}[k, \mu, c_p, \rho, \ell, u, \beta g(T_w - T_f), \phi] \quad (9.1)$$

which can be written as:

$$f[h, k, \mu, c_p, \rho, \ell, U, \beta g(T_w - T_f), \phi] = 0 \quad (9.2)$$

where f is some function.

There are, thus, nine dimensional variables involved in this type of convection. As a result, there are five dimensionless variables involved in describing combined convective heat transfer. Since the angle, ϕ, is dimensionless, these dimensionless variables are:

$$\pi_1 = (h\ell/k) = Nu, \text{ the Nusselt number}$$

$$\pi_2 = (U\ell\rho/\mu) = (U\ell/\nu) = Re, \text{ the Reynolds number}$$

$$\pi_3 = \beta g(T_w - T_o)\rho^2 \ell^3/\mu^2 = \beta g(T_w - T_o)\ell^3/\nu^2 = Gr, \text{ the Grashof number}$$

$$\pi_4 = (c_p \mu/k) = Pr, \text{ the Prandtl number}$$

$$\pi_5 = \phi$$

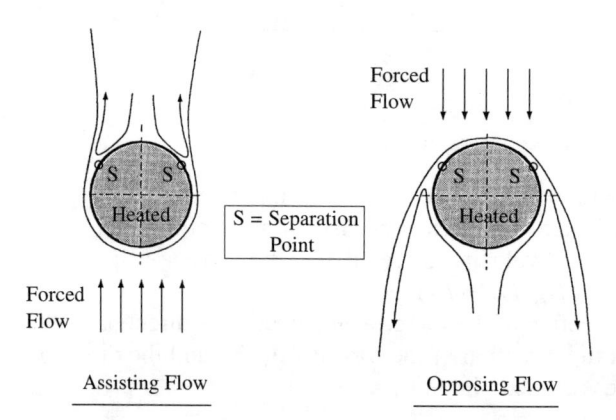

FIGURE 9.4
Assisting and opposing combined convective flow.

Therefore, in combined convection, in general

$$\text{function}(Nu, Re, Gr, Pr, \phi) = 0 \tag{9.3}$$

which can be rewritten as

$$Nu = \text{function}(Re, Gr, Pr, \phi) \tag{9.4}$$

When the angle ϕ is 0°, the forced flow is in the same direction as the buoyancy forces. In this case, the flow is referred to as "aiding" or "assisting" combined convective flow. When ϕ is 180°, the forced flow is in the opposite direction to the buoyancy forces. In this case the flow is referred to as "opposing" combined convective flow. The flows in these two cases can be very different particularly if the flow involves significant regions of separated flow. In assisting flow, the buoyancy forces tend to delay the separation, i.e., to move the separation point rearward on the body whereas in opposing flow they tend to move the separation point forward on the body. This is illustrated in Fig. 9.4.

The effect of the flow direction on the flow field in mixed convection is further illustrated by the experimental results given in Fig. 9.5. This figure essentially shows

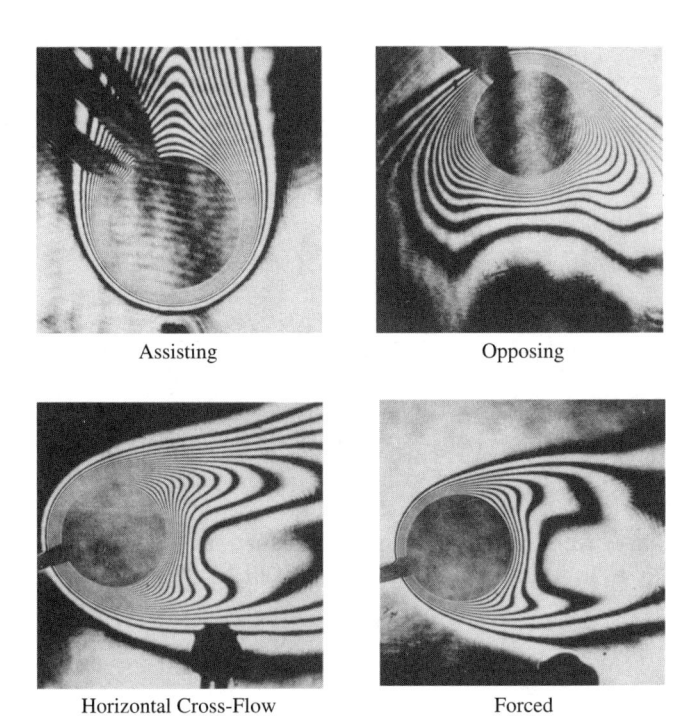

FIGURE 9.5
Effect of flow direction on the temperature field around a heated cylinder. (From Krause, J.R., "An Interferometric Study of Mixed Convection from a Horizontal Cylinder to a Cross Flow of Air", M.E.Sc. Thesis, The University of Western Ontario, London, Ontario, Canada, 1985. By permission.)

the distribution of the isotherms (lines of constant temperature) near a heated cylinder in a vertically upward, a vertically downward, and a horizontal air flow under such conditions that the buoyancy forces are important, i.e., in combined convective flow. For comparison, the isotherms in an effectively forced convective flow are also shown.

9.3 GOVERNING EQUATIONS

Attention will initially be restricted to two-dimensional steady laminar flow. The Boussinesq assumptions will again be used and, consistent with this assumption, dissipation effects will be neglected. The governing equations, expressed in Cartesian coordinates, are then [1]:

Continuity:

$$\frac{\partial u}{\partial x} + \frac{\partial v}{\partial y} = 0 \tag{9.5}$$

x-momentum:

$$u\frac{\partial u}{\partial x} + v\frac{\partial u}{\partial y} = -\frac{1}{\rho}\frac{\partial p}{\partial x} + \nu\left(\frac{\partial^2 u}{\partial x^2} + \frac{\partial^2 u}{\partial y^2}\right) + \beta g(T - T_1)\cos\phi \tag{9.6}$$

y-momentum:

$$u\frac{\partial v}{\partial x} + v\frac{\partial v}{\partial y} = -\frac{1}{\rho}\frac{\partial p}{\partial y} + \nu\left(\frac{\partial^2 v}{\partial x^2} + \frac{\partial^2 v}{\partial y^2}\right) + \beta g(T - T_1)\sin\phi \tag{9.7}$$

Energy:

$$u\frac{\partial T}{\partial x} + v\frac{\partial T}{\partial y} = \left(\frac{k}{\rho c_p}\right)\left(\frac{\partial^2 T}{\partial x^2} + \frac{\partial^2 T}{\partial y^2}\right) \tag{9.8}$$

In these equations, p is the pressure relative to the local ambient pressure and, as before, ϕ is the angle between the direction of the forced velocity and the direction of the buoyancy forces as defined in Fig. 9.3. The x-axis is in the direction of the undisturbed forced flow.

The boundary layer forms of the governing equations given above can be derived in the same way as in forced convection. The resultant equations are:

$$\frac{\partial u}{\partial x} + \frac{\partial v}{\partial y} = 0 \tag{9.9}$$

$$u\frac{\partial u}{\partial x} + v\frac{\partial u}{\partial y} = -\frac{1}{\rho}\frac{dp}{dx} + \nu\frac{\partial^2 u}{\partial y^2} + \beta g(T - T_1)\cos\phi \tag{9.10}$$

$$u\frac{\partial T}{\partial x} + v\frac{\partial T}{\partial y} = \left(\frac{k}{\rho c_p}\right)\frac{\partial^2 T}{\partial y^2} \tag{9.11}$$

where, as in forced convection, dp/dx is the pressure gradient in the free-stream which is imposed on the boundary layer.

CHAPTER 9: Combined Convection   431

A consideration of the orders of magnitude of the terms in the momentum equation for boundary layer flow indicates that if $u = o(u_1)$, where u_1 is a characteristic free-stream velocity, then the buoyancy force term will be important if:

$$G = \frac{\beta g(T_{wr} - T_1)\cos\phi L}{u_1^2} \qquad (9.12)$$

is of the order of magnitude, 1. Here L is some characteristic dimension of the body and T_{wr} is a measure of the surface temperature. If G is of a significantly lower order than one, the buoyancy force term will be negligible and forced convective flow will exist. On the other hand, if G is of a significantly greater order of magnitude than one, the buoyancy force effects will predominate and the flow will essentially be free convective. Hence, combined convective flow exists when:

$$G = o(1) \qquad (9.13)$$

G is usually termed the "buoyancy force parameter". It is a form of Richardson number.

Now it will be noted that:

$$G = \left(\frac{\beta g(T_{wr} - T_1)\cos\phi L^3}{\nu^2}\right)\left(\frac{\nu^2}{u_1^2 L^2}\right) = \frac{Gr}{Re^2} \qquad (9.14)$$

Therefore, the parameter Gr/Re^2 will be important in determining whether a boundary layer flow can be treated as a forced convective flow or as a free convective flow or as a combined convective flow.

9.4
LAMINAR BOUNDARY LAYER FLOW OVER AN ISOTHERMAL VERTICAL FLAT PLATE

The flow situation being considered here is shown in Fig. 9.6.

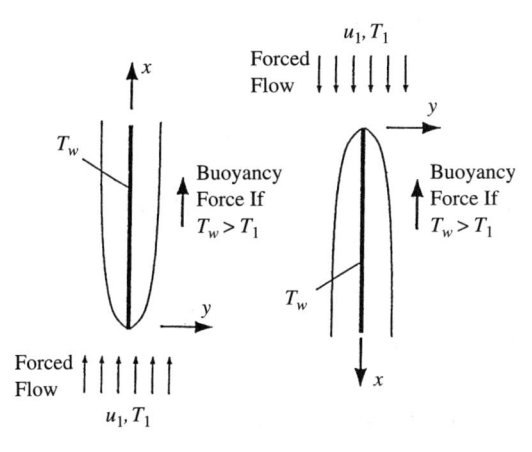

FIGURE 9.6
Flow situation considered.

Studies of this type of flow are described in [2] to [14]. The governing equations are:

$$\frac{\partial u}{\partial x} + \frac{\partial v}{\partial y} = 0 \tag{9.15}$$

$$u\frac{\partial u}{\partial x} + v\frac{\partial u}{\partial y} = \nu\frac{\partial^2 u}{\partial y^2} \pm \beta g(T - T_1) \tag{9.16}$$

$$u\frac{\partial T}{\partial x} + v\frac{\partial T}{\partial y} = \left(\frac{k}{\rho c_p}\right)\frac{\partial^2 T}{\partial y^2} \tag{9.17}$$

It has been noted that because flow over flat plate is being considered, the pressure gradient dp/dx is 0. The $(+)$ and $(-)$ signs on the buoyancy term refer to assisting and opposing flow, respectively.

In seeking a solution to the above equations, it seems worth first investigating whether similarity solutions can be obtained in the same way as in forced convection. Therefore, the similarity variable used in forced convection, i.e.:

$$\eta = y\sqrt{\frac{u_1}{\nu x}} = \frac{y}{x}Re_x^{0.5} \tag{9.18}$$

is again introduced and it is assumed that similar velocity and temperature profiles exist in the boundary layer, i.e., that:

$$\frac{u}{u_1} = F'(\eta), \quad \frac{T - T_1}{T_w - T_1} = H(\eta) \tag{9.19}$$

The prime, of course, again denotes differentiation with respect to η.

As in forced convection, it follows from the continuity equation by using the boundary condition $\eta = 0$, $v = 0$ that:

$$\frac{v}{u_1} = \frac{1}{2}\sqrt{\frac{\nu}{xu_1}}(\eta F' - F) \tag{9.20}$$

Substituting this into the momentum equation and using:

$$\frac{\partial \eta}{\partial x} = -\frac{\eta}{2x}, \quad \frac{\partial \eta}{\partial y} = \frac{1}{x}\sqrt{\frac{u_1}{\nu x}} \tag{9.21}$$

gives the following equation:

$$F'F''\left(-\frac{\eta}{2x}\right) + \left[\frac{1}{2}\sqrt{\frac{\nu}{u_1 x}}(\eta F' - F)\right]\frac{F''}{x}\sqrt{\frac{u_1 x}{\nu}} = \frac{\nu}{u_1}F'''\left(\frac{u_1}{x\nu}\right) \pm \frac{\beta g(T_w - T_1)}{u_1^2}H$$

i.e.:

$$F''' + \frac{FF''}{2} \pm G_x H = 0 \tag{9.22}$$

The primes, of course, denote differentiation with respect to η and:

$$G_x = \frac{\beta g(T_w - T_1)x}{u_1^2} = \frac{Gr_x}{Re_x^2} \tag{9.23}$$

and, as discussed before, the (+) and (−) signs on the buoyancy term refer to assisting and opposing flow, respectively.

Similarly, the energy equation gives as in forced convection:

$$H'' + \frac{Pr}{2}H'F = 0 \tag{9.24}$$

The boundary conditions on these equations are:

When $y = 0$, $u = 0$, i.e., when $\eta = 0$, $F' = 0$
When $y = 0$, $v = 0$, i.e., when $\eta = 0$, $F = 0$
When $y = 0$, $T = T_w$, i.e., when $\eta = 0$, $H = 1$ (9.25)
When y is large, $u \to u_1$, i.e., when η is large, $F' \to 1$
When y is large, $T \to T_1$, i.e., when η is large, $H \to 0$

A consideration of Eq. (9.22) shows that because G_x is a function of x, similarity solutions do not exist, i.e., F and H cannot be expressed as functions of η alone.

To examine the conditions under which the buoyancy forces can be neglected, consideration is given to the case where G_x is small and the solution differs by only a small amount from that existing in purely forced convection. If terms of the order of G_x^2 and higher are neglected because G_x is small, the solution will have the form:

$$F = F_0(\eta) + G_x F_1(\eta) \tag{9.26}$$

$$H = H_0(\eta) + G_x H_1(\eta) \tag{9.27}$$

Here F_0 and H_0 are the functions that apply in purely forced convection. Substituting these equations into Eq. (9.22) then gives, since terms of the order G_x^2 are being neglected:

$$F_0''' + G_x F_1''' + \frac{F_0 F_0''}{2} + G_x \frac{F_0 F_1''}{2} + G_x \frac{F_1 F_0''}{2} \pm G_x H_0 = 0 \tag{9.28}$$

But the analysis of forced convective gives:

$$F_0''' + \frac{F_0 F_0''}{2} = 0 \tag{9.29}$$

Hence, Eq. (9.28) gives on canceling the G_x in all the remaining terms:

$$F_1''' + \frac{F_0 F_0''}{2} + \frac{F_0 F_1''}{2} + \frac{F_1 F_0''}{2} \pm H_0 = 0 \tag{9.30}$$

This equation allows $F_1(\eta)$ to be found, the boundary conditions being:

$\eta = 0$, $F_1' = 0$
$\eta = 0$, $F_1 = 0$
$\eta = 0$, $H_1 = 1$ (9.31)
η large, $F_1' \to 0$
η large, $H_1 \to 0$

Hence, using the known variation of F_0 and H_0 with η, Eq. (9.30) can be used to solve for F_1.

Similarly, substituting Eqs. (9.26) and (9.27) into Eq. (9.24) gives when terms of the order of G_x^2 and larger are neglected:

$$\frac{H_0''}{Pr} + \frac{H_1''}{Pr}G_x + \frac{F_0 H_0'}{2} + \frac{F_1 H_0'}{2}G_x + \frac{F_0 H_1'}{2}G_x = 0 \tag{9.32}$$

But the forced convection solution gives:

$$\frac{H_0''}{Pr} + \frac{F_0 H_0'}{2} = 0$$

so this equation gives:

$$\frac{H_1''}{Pr} + \frac{F_1 H_0'}{2} + \frac{F_0 H_1'}{2} = 0 \tag{9.33}$$

Because $F_0(\eta)$, $H_0(\eta)$, and $F_1(\eta)$ are known, this equation can be used to solve for H_1. The solution will depend on the value of Pr. Once the values of F_1 and H_1 have been determined, the heat transfer rate at the wall can be found by applying Fourier's law which gives:

$$q_w = -k\frac{\partial T}{\partial y}\bigg|_{y=0} = -k(T_w - T_1)\left[H_0'(0) + G_x H_1'(0)\right]\frac{\sqrt{Re_x}}{x}$$

i.e.:

$$\frac{Nu_x}{\sqrt{Re_x}} = -H_0'(0) - G_x H_1'(0) \tag{9.34}$$

If the buoyancy forces are negligible, this equation gives:

$$\frac{Nu_{xF}}{\sqrt{Re_x}} = -H_0'(0) \tag{9.35}$$

Nu_{xF} being the local Nusselt number in forced convection.

Dividing the above two equations then gives:

$$\frac{Nu_x}{Nu_{xF}} = 1 + \frac{H_1'(0)}{H_0'(0)}G_x \tag{9.36}$$

where Nu_x is the local Nusselt number in the actual combined convective flow.

Similarly, the wall shear stress is given by:

$$\tau_w = \mu\frac{\partial u}{\partial y}\bigg|_{y=0} = \mu u_1[F_0''(0) + G_x F_1''(0)]\frac{\sqrt{Re_x}}{x}$$

i.e.:

$$\frac{\tau_w}{\rho u_1^2}\sqrt{Re_x} = F_0''(0) - G_x F_1''(0) \tag{9.37}$$

If the buoyancy forces are negligible, this gives:

$$\frac{\tau_{wF}}{\rho u_1^2}\sqrt{Re_x} = F_0''(0)$$

τ_{wF} being the local wall shear stress in forced convection.

Dividing the above two equations then gives:

$$\frac{\tau_w}{\tau_{wF}} = 1 + \frac{F_1''(0)}{F_0''(0)}G_x \qquad (9.38)$$

where τ_w is the actual local wall shear stress in the combined convective flow.

A "shooting" technique can be used in solving Eq. (9.30), it being assumed that the functions F_0 and H_0 are known. Basically, a type of iterative solution procedure is used in which the value of $F_1''(0)$ is guessed and the value of F_1' at large η is then calculated by numerically integrating Eq. (9.30). The value of $F_1''(0)$ that gives $F_1' = 0$ at large η is then iteratively determined. Once $F_1(\eta)$ is found in this way, Eq. (9.33) can be integrated to give $H_1(\eta)$ and the value of $H_1'(0)$ can then be determined. The solution depends on the value of the Prandtl number, Pr. A computer program, SIMPLCOM, that implements this procedure in a very simple, basic manner is available as discussed in the Preface and some results given by this program are given in Fig. 9.7.

A consideration of the results given in Fig. 9.7 in conjunction with Eqs. (9.35) and (9.36) shows that the buoyancy forces increase the heat transfer rate and wall shear stress in assisting flow and decrease these quantities in opposing flow. If it is assumed that the effect of the buoyancy forces on the heat transfer rate can be neglected, i.e., that the flow can be assumed to be a purely forced convective flow, when:

$$\left|\frac{Nu_x}{Nu_{xF}} - 1\right| < 0.01$$

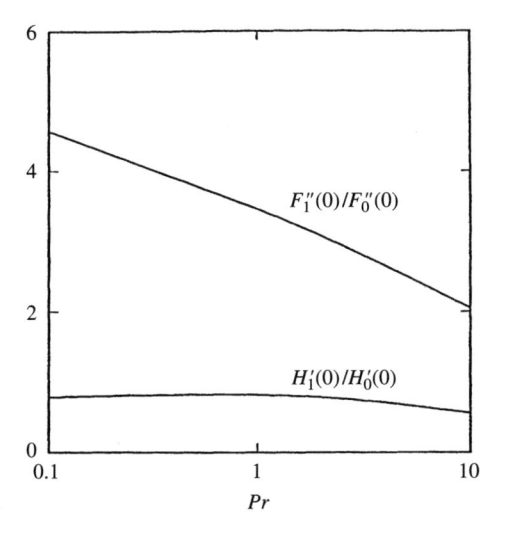

FIGURE 9.7
Variation of $H_1'(0)/H_0'(0)$ and $F_1''(0)/F_0''(0)$ with Prandtl number.

436 Introduction to Convective Heat Transfer Analysis

it will be seen that, according to Eq. (9.35), purely forced convection will exist if:

$$G_x < \frac{0.01}{|H_1'(0)/H_0'(0)|} \tag{9.39}$$

EXAMPLE 9.1. Air at a temperature of 20°C flows upward over a wide vertical 10-cm high flat plate which is maintained at a uniform surface temperature of 60°C. Below what velocity will the buoyancy forces have an effect on the heat transfer rate at the trailing edge of the plate?

Solution. Here, the buoyancy forces will act in the same direction as the forced flow, i.e., assisting flow will exist. In this case, buoyancy force effects on the heat transfer rate will be important if:

$$G_x > \frac{0.01}{|H_1'(0)/H_0'(0)|}$$

For air for which $Pr = 0.7$, the solution discussed above gives:

$$\frac{H_1'(0)}{H_0'(0)} = 0.8$$

Hence, in this case, buoyancy force effects will be important if:

$$G_x > \frac{0.01}{0.8}$$

i.e.:

$$G_x > 0.0125$$

Now, for a vertical plate

$$G_x = \left[\frac{\beta g(T_w - T_1)x^3}{\nu^2}\right]\left(\frac{\nu^2}{u_1^2 x^2}\right) = \frac{\beta g(T_w - T_1)x}{u_1^2}$$

At the trailing edge, i.e., the uppermost edge of the plate, $x = 10$ cm $= 0.1$ m. Hence, buoyancy forces will be important at the trailing edge if:

$$\frac{\beta g(T_w - T_1)0.1}{u_1^2} > 0.0125$$

i.e., if:

$$u_1^2 < \frac{\beta g(T_w - T_1)0.1}{0.0125}$$

The mean air temperature in the boundary layer is:

$$\frac{20 + 60}{2} = 40°C$$

At this temperature:

$$\beta = \frac{1}{273 + 40} = \frac{1}{313} \text{ K}^{-1}$$

Hence, buoyancy forces will be important if:

$$u_1^2 < \frac{(1/313) \times 9.81 \times (60 - 20) \times 0.1}{0.0125}$$

i.e.:

$$u_1 < 3.17 \text{ m/s}$$

Therefore, the buoyancy forces can be expected to influence the heat transfer rate when the forced velocity is less than about 3.2 m/s.

The above analysis applies only for very small values of G_x because only the first-order terms, i.e., terms of the order of G_x, were considered. The analysis could have been expanded to include some of the higher order terms, e.g., Eqs. (9.26) and (9.27) could have been written as:

$$F = F_0 + G_x F_1 + G_x^2 F_2 + G_x^3 F_3 + G_x^4 F_4$$

and:

$$H = H_0 + G_x H_1 + G_x^2 H_2 + G_x^3 H_3 + G_x^4 H_4$$

and the analysis could have been carried out retaining terms up to and including those involving G_x^4 but ignoring higher order terms in G_x. The procedure used is basically an extension of that described above to deal with the first-order analysis. However, the analysis becomes quite involved and if results for larger values of G_x are required it is easier to apply a numerical solution procedure of the type discussed in the next section.

Before turning to a discussion of other methods of solving the laminar boundary layer equations for combined convection, a series-type solution aimed at determining the effects of small forced velocities on a free convective flow will be considered. In the analysis given above to determine the effect of weak buoyancy forces on a forced flow, the similarity variables for forced convection were applied to the equations for combined convection. Here, the similarity variables that were previously used in obtaining a solution for free convection will be applied to these equations for combined convection. Therefore, the following similarity variable is introduced:

$$\eta = \frac{y}{x} Gr_x^{0.25} = y \left[\frac{\beta g (T_w - T_1)}{\nu^2 x} \right]^{0.25} \tag{9.40}$$

where Gr_x is the local Grashof number.

As in free convection it is assumed that:

$$\frac{u}{\sqrt{\beta g (T_w - T_1) x}} = F'(\eta) \tag{9.41}$$

and:

$$\frac{T - T_1}{T_w - T_1} = H(\eta) \tag{9.42}$$

The prime, of course, again denotes differentiation with respect to η. The differentiated function F' is, as in forced flow, used for convenience in describing the velocity profile. The functions F' and H are, of course, different from those in forced convective flows.

As with purely free convective flow, the continuity equation gives:

$$\frac{v}{\sqrt{\beta g(T_w - T_1)x}} = \frac{1}{4Gr_x^{0.25}}(\eta F' - 3F) \tag{9.43}$$

Substituting this into the momentum equation and using:

$$\frac{\partial \eta}{\partial x} = -\frac{\eta}{4x}, \quad \frac{\partial \eta}{\partial y} = \frac{Gr_x^{0.25}}{x} \tag{9.44}$$

gives:

$$F'F''\left(-\frac{\eta}{4x}\right) + \frac{F'^2}{2x} + \frac{1}{4Gr_x^{0.25}}(\eta F' - 3F)F''\frac{Gr_x^{0.25}}{x} = \frac{F'''}{x} \pm \frac{H}{x}$$

i.e.:

$$F''' + \frac{3FF''}{4} - \frac{F'^2}{2} \pm H = 0 \tag{9.45}$$

Similarly, the energy equation gives:

$$F'H'\left(-\frac{\eta}{4x}\right) + \frac{1}{4Gr_x^{0.25}}(\eta F' - 3F)H'\frac{Gr_x^{0.25}}{x} = \frac{H''}{Prx}$$

i.e.:

$$H'' + \frac{3}{4}PrFH' = 0 \tag{9.46}$$

A consideration of Eqs. (9.45) and (9.46) might, at first sight, appear to indicate that a similarity solution can be obtained in this case. However, consider the boundary conditions:

$$\begin{aligned}
&\text{At } y = 0: u = 0, \text{ i.e., at } \eta = 0: F' = 0 \\
&\text{At } y = 0: v = 0, \text{ i.e., at } \eta = 0: F = 0 \\
&\text{At } y = 0: T = T_w, \text{ i.e., at } \eta = 0: H = 1 \\
&\text{For large } y: u \to u_1, \text{ i.e., for large } \eta: F' \to G_x^{-0.5} \\
&\text{For large } y: T \to T_1, \text{ i.e., for large } \eta: H \to 0
\end{aligned} \tag{9.47}$$

it having been noted that:

$$\text{for large } \eta: F' \to \frac{u_1}{[\beta g(T_w - T_1)x]^{0.5}} = \frac{Re_x}{Gr_x^{0.5}} = \frac{1}{G_x^{0.5}}$$

The solution thus depends on G_x so a similarity solution cannot be obtained.

As G_x tends to infinity, the flow tends towards a purely free convective flow. To examine the conditions under which the effects of the forced velocity on a free convective flow can be neglected, a series solution in terms of $1/G_x^{0.5}$ will be con-

CHAPTER 9: Combined Convection   439

sidered and terms of the order of $(1/G_x^{0.5})^2$ and higher will be neglected, i.e., it will be assumed that:

$$F = F_0(\eta) + \frac{1}{G_x^{0.25}} F_1(\eta) \tag{9.48}$$

$$H = H_0(\eta) + \frac{1}{G_x^{0.25}} H_1(\eta) \tag{9.49}$$

In this case, F_0 and H_0 are the functions that apply in purely free convection. Substituting these equations into Eq. (9.45) and ignoring terms involving $(1/G_x^{0.5})^2$ then gives:

$$F_0''' + \frac{F_1'''}{G_x^{0.5}} + \frac{3}{4} F_0 F_0'' + \frac{3}{4 G_x^{0.5}} F_0 F_1''$$
$$+ \frac{3}{4 G_x^{0.5}} F_1 F_0'' - \frac{F_0'^2}{2} - \frac{F_0' F_1'}{G_x^{0.5}} + H_0 + \frac{H_1}{G_x^{0.5}} = 0 \tag{9.50}$$

Only the (+) sign has been accounted for on the buoyancy term, i.e., only the case where the forced flow is in the same direction as the buoyancy forces has been considered. This is because in opposing flow the outer forced flow would be in the opposite direction to the buoyancy-driven flow near the surface. In this case the boundary layer equations would not apply.

It follows from the analysis of free convection that:

$$F_0''' + \frac{3}{4} F_0 F_0'' - \frac{F_0'^2}{2} + H_0 = 0 \tag{9.51}$$

so, Eq. (9.50) gives:

$$F_1''' + \frac{3}{4} F_0 F_1'' + \frac{3}{4} \frac{F_1 F_0''}{2} - F_0' F_1' + H_1 = 0 \tag{9.52}$$

The boundary conditions give:

$$\eta = 0, \; F_0' = 0, \; F_1' = 0$$
$$\eta = 0, \; F_0 = 0, \; F_1 = 0$$
$$\eta = 0, \; H_0 = 1, \; H_1 = 0 \tag{9.53}$$
$$\eta \text{ large}, \; F_0' \to 0, \; F_1' \to 1$$
$$\eta \text{ large}, \; H_0 \to 0, \; H_1 \to 0$$

The boundary condition on F' at large η is obtained by noting that:

$$\eta \text{ large}, \; F_0' + \frac{F_1'}{G_x^{0.5}} \to \frac{1}{G_x^{0.5}}$$

But the free convective flow solution requires that $F_0' \to 0$ at large η so the above equation requires that:

$$\eta \text{ large}, \; F_1' \to 1$$

Next, Eqs. (9.48) and (9.49) are substituted into the energy equation, i.e., Eq. (9.46). This gives to first-order accuracy in $1/G_x^{0.5}$:

$$H_0'' + \frac{H_1''}{G_x^{0.5}} + \frac{3}{4}PrF_0H_0' + \frac{3}{4G_x^{0.5}}PrF_0H_1' + \frac{3}{4G_x^{0.5}}PrF_1H_0' = 0 \quad (9.54)$$

But the free convective flow solution requires:

$$H_0'' + \frac{3}{4}PrF_0H_0' = 0 \quad (9.55)$$

so Eq. (9.54) gives:

$$H_1'' + \frac{3}{4}PrF_0H_1' + \frac{3}{4}PrF_1H_0' = 0 \quad (9.56)$$

Therefore, since the purely free convection solution gives F_0 and H_0, Eqs. (9.52) and (9.56) can be simultaneously solved to give F_1 and H_1 using the boundary conditions given in Eq. (9.53). The heat transfer rate is then given by applying Fourier's law at the wall. This gives:

$$q_w = -k\left.\frac{\partial T}{\partial y}\right|_{y=0} = -k(T_w - T_1)\left[H_0'(0) + \frac{H_1'(0)}{G_x^{0.5}}\right]\frac{Gr_x^{0.25}}{x}$$

i.e.:

$$\frac{Nu_x}{Gr_x^{0.25}} = -H_0'(0) - H_1'(0)G_x^{0.5} \quad (9.57)$$

If the effects of the forced velocity are negligible, i.e., if purely free convection can be assumed, Eq. (9.57) gives:

$$\frac{Nu_{xN}}{Gr_x^{0.25}} = -H_0'(0) \quad (9.58)$$

Nu_{xN} being the local Nusselt number in purely free or natural convection.

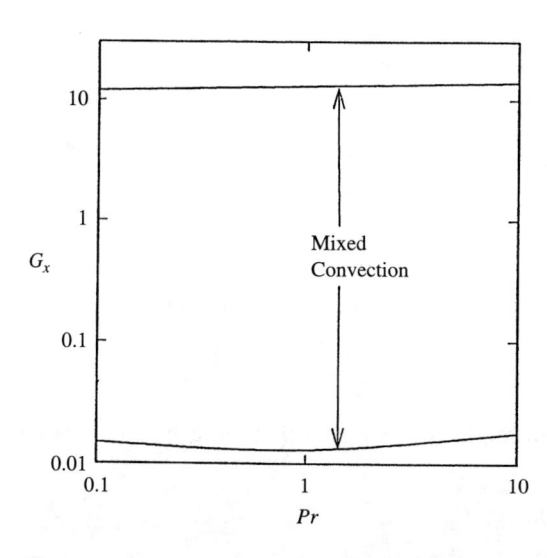

FIGURE 9.8
Limiting values of G_x for purely forced and purely free convection.

Dividing Eq. (9.57) by Eq. (9.58) then gives:

$$\frac{Nu_x}{Nu_{xN}} = 1 + \frac{H_1'(0)}{H_0'(0)} \frac{1}{G_x^{0.5}}$$

This allows the value of G_x above which:

$$\left|\frac{Nu_x}{Nu_{xN}} - 1\right| < 0.01$$

i.e., above which the Nusselt number is within 1% of the purely free convective value, to be found. This will be given by:

$$G_x > 10^4 |H_1'(0)/H_0'(0)|^2 \qquad (9.59)$$

The value of G_x so found will depend on the value of Pr. Some typical values are shown in Fig. 9.8 which also shows the values of G_x below which the flow can be assumed to be purely forced convective. These values are given by Eq. (9.39).

EXAMPLE 9.2. Consider free convective flow over a wide vertical plate which is held at a uniform surface temperature of 60°C placed in air at a temperature of 20°C. If a forced flow is introduced over the plate, at what air velocity will the forced flow start to affect the heat transfer rate at a distance of 20 cm from the leading edge of the plate?

Solution. Using the results given in Fig. 9.8, it will be seen that for air, i.e., for $Pr = 0.7$, the forced flow will be important approximately when:

$$G_x < 12$$

i.e., when:

$$\frac{\beta g(T_w - T_1)x}{u_1^2} < 12$$

i.e., when:

$$u_1^2 > \frac{\beta g(T_w - T_1)x}{12}$$

The mean temperature in the boundary layer is $(60 + 20)/2 = 40°C$ and x at the point being considered is 0.2 m. Hence, since:

$$\beta = \frac{1}{273 + 40} = \frac{1}{313} \text{ K}^{-1}$$

it follows that forced flow effects will be important when:

$$u_1^2 > \frac{(1/313) \times 9.81 \times (60 - 20) \times 0.2}{12}$$

i.e., when:

$$u_1 > 0.12 \text{ m/s}$$

Therefore, the forced flow will affect the heat transfer at the point considered if the forced velocity is greater than 0.12 m/s.

9.5
NUMERICAL SOLUTION OF BOUNDARY LAYER EQUATIONS

The numerical procedure for solving the laminar boundary layer equations for forced convection that was described in Chapter 3 is easily extended to deal with combined convection. The details of the procedure are basically the same as those for forced convection and the details will not be repeated here [16]. A computer program, LAMBMIX, based on the procedure is available in the way discussed in the Preface. This program can actually allow the wall temperature or wall heat flux to vary with X but as available, the program is set for the case of a uniform wall temperature or a uniform wall heat flux.

This program allows the variation of $Nu_x/\sqrt{Re_x}$ with G_x to be found in the uniform surface temperature case and the variation of $Nu_x/\sqrt{Re_x}$ with:

$$G_x^* = \frac{\beta g q_w x^4}{k\nu^2 Re_x^{5/2}} = \frac{Gr_x^*}{Re_x^{5/2}} \qquad (9.60)$$

to be obtained in the uniform heat flux case. Here Gr_x^* is the heat flux Grashof number, i.e.:

$$Gr_x^* = \frac{\beta g q_w x^4}{k\nu^2} \qquad (9.61)$$

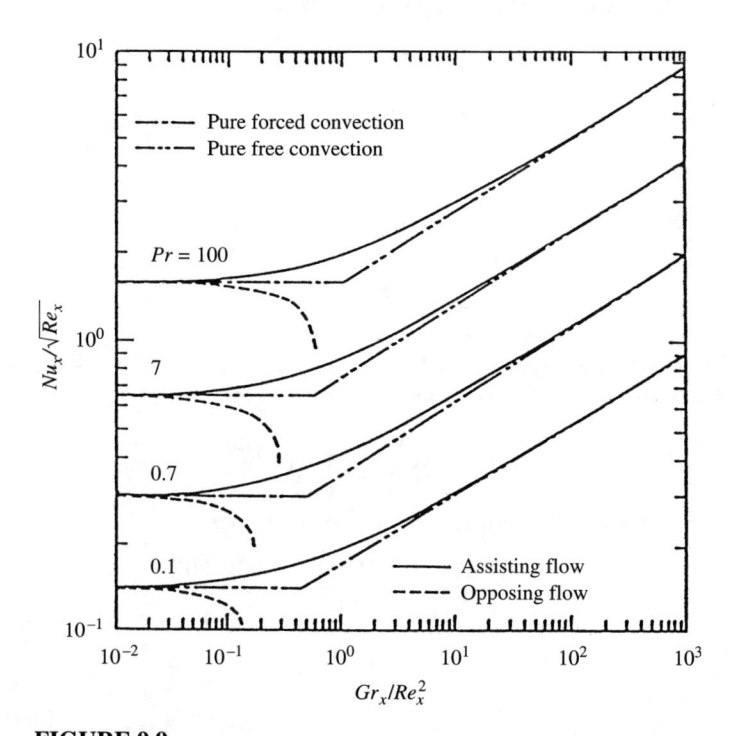

FIGURE 9.9
Variations of $Nu_x/\sqrt{Re_x}$ with G_x for mixed convective flow over a vertical plate with a uniform surface temperature.

CHAPTER 9: Combined Convection

The variations of $Nu_x/\sqrt{Re_x}$ with G_x for the uniform surface temperature case and of $Nu_x/\sqrt{Re_x}$ with G_x^* for the uniform wall heat flux case for various values of Pr as given by the computer program are shown in Figs. 9.9 and 9.10, respectively.

Of course, if attention was to have been restricted to the case where the wall temperature or the wall heat flux was uniform, it would have been better to write the governing equations in terms of dimensionless variables that directly gave the variations shown in Figs. 9.9 and 9.10, and not to have used the same dimensionless variables as in forced convection. However, with the variables used, the computer program based on this procedure can easily be modified to allow for a variable wall temperature or a variable wall heat flux.

If it is again assumed that forced convection effectively exists when:

$$\left| \frac{Nu_x}{Nu_{xF}} - 1 \right| < 0.01$$

and that free convection effectively exists when:

$$\left| \frac{Nu_x}{Nu_{xN}} - 1 \right| < 0.01$$

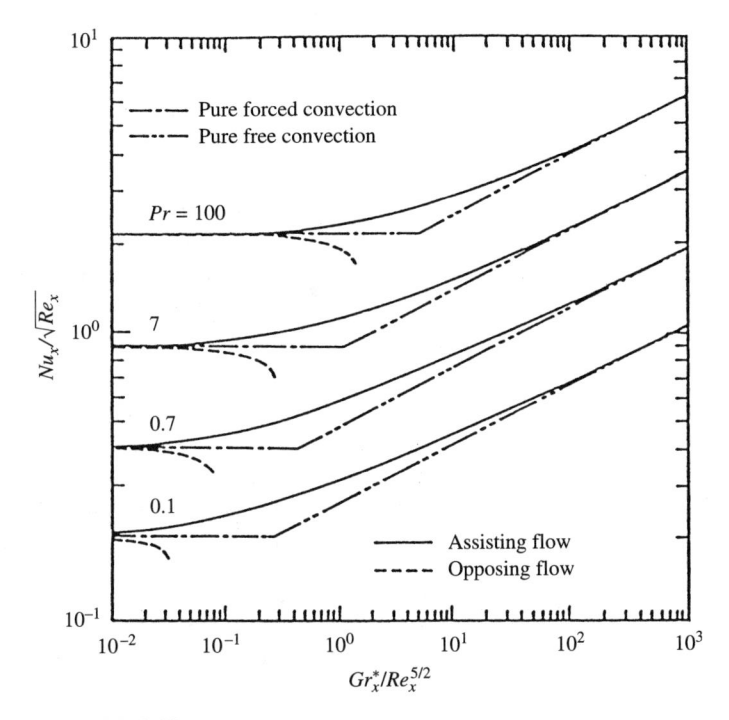

FIGURE 9.10
Variations of $Nu_x/\sqrt{Re_x}$ with G_x^* for mixed convective flow over a vertical plate with a uniform surface heat flux.

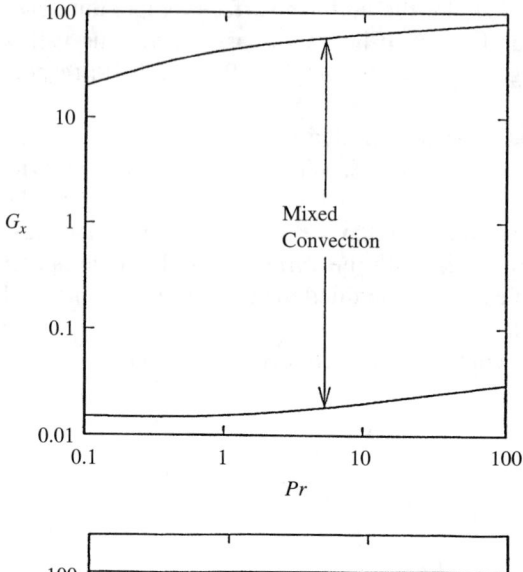

FIGURE 9.11
Values of G_x for effectively purely forced and purely free convection for a plate with a uniform surface temperature.

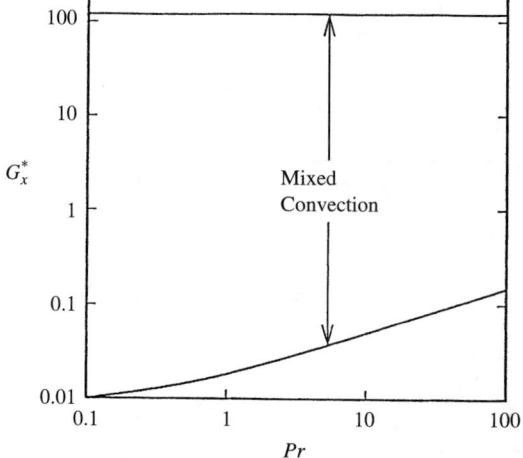

FIGURE 9.12
Values of G_x^* for effectively purely forced and purely free convection for a plate with a uniform surface heat flux.

then the results given in Figs. 9.9 and 9.10 can be used to deduce the variations of G_x with Pr for the uniform wall temperature case for which effectively purely forced convection and purely free convection exist and of G_x^* with Pr for the uniform wall heat flux case for which effectively purely forced convection and purely free convection exist. These variations are shown in Figs. 9.11 and 9.12.

EXAMPLE 9.3. Air at a temperature of 15°C flows upward over a 0.2-m high vertical plate which is kept at a uniform surface temperature of 45°C. Plot the variation of the local heat transfer rate along the plate for air velocities of 2 and 0.3 m/s. Assume two-dimensional flow.

Solution. The mean temperature of the air in the boundary layer is $(15 + 45)/2 = 30$°C. At this temperature for air at standard ambient pressure:

$$\beta = \frac{1}{303} \text{ K}^{-1}, \quad \nu = 16.01 \times 10^{-6} \text{ m}^2/\text{s}, \quad k = 0.02638 \text{ W/m-}°\text{C}$$

Now:

$$G_x = \frac{\beta g(T_w - T_1)x}{u_1^2} = \frac{(1/303) \times 9.81 \times 30 \times x}{u_1^2} = 0.9713 \frac{x}{u_1^2}$$

i.e.:

$$x = 1.030 G_x u_1^2$$

where x is in m and u_1 is in m/s.

For any value of G_x and for either value of u_1 that is to be considered, this equation allows the corresponding value of x to be determined. For $u_1 = 2$ m/s, this equation gives:

$$x = 4.12 G_x$$

while for $u_1 = 0.3$ m/s it gives:

$$x = 0.0927 G_x$$

Because the maximum value that x can have is 0.2 m, this shows that the highest value of G_x is 0.0486 for $u_1 = 2$ m/s and 2.158 for $u_1 = 0.3$ m/s.

For any value of G_x, the numerical solution discussed above gives the corresponding value of $Nu_x/Re_x^{0.5}$. But:

$$\frac{Nu_x}{Re_x^{0.5}} = \frac{q_w x/k(T_w - T_1)}{(u_1 x/\nu)^{0.5}} = \frac{q_w x/30 \times 0.02638}{(u_1 x/0.00001601)^{0.5}} = \frac{q_w x^{0.5}}{197.8 u_1^{0.5}}$$

i.e.:

$$q_w = \frac{197.8 u_1^{0.5}}{x^{0.5}} \frac{Nu_x}{Re_x^{0.5}}$$

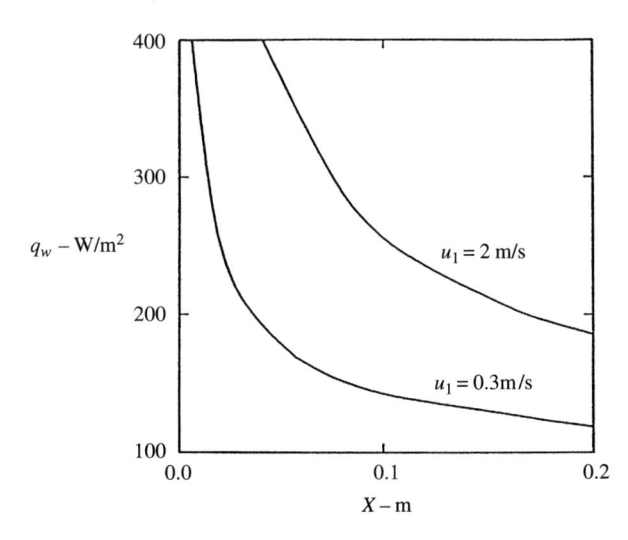

FIGURE E9.3

Hence, since for any value of G_x the numerical solution discussed above gives the corresponding value of $Nu_x/Re_x^{0.5}$ and since, for any value of G_x and for either value of u_1 that is to be considered, the corresponding value of x can be determined, the variation of q_w with x for either value of u_1 that is to be considered can be determined. For example, the numerical solution gives for $G_x = 0.5$, $Nu_x/Re_x^{0.5} = 0.430$. If the case of $u_1 = 0.3$ m/s is considered, this value of G_x corresponds to a value of x that is given by:

$$x = 0.0927 G_x = 0.0927 \times 0.5 = 0.0464 \text{ m}$$

The corresponding value of q_w is given by:

$$q_w = \frac{197.8 \times 0.3^{0.5}}{0.0464^{0.5}} \times 0.430 = 503 \text{ W}$$

Using this procedure, the variation of q_w with x for the two velocities considered can be found, the results being shown in Fig. E9.3.

The numerical procedure described above for solving the laminar boundary layer equations is easily extended to deal with situations in which the free-stream velocity is varying with x, i.e., to deal with situations involving flow over bodies of arbitrary shape.

9.6
COMBINED CONVECTION OVER A HORIZONTAL PLATE

In combined convective flow over a horizontal flat surface, the buoyancy forces are at right angles to the flow direction and lead to pressure changes across the boundary layer, i.e., there is an induced pressure gradient in the boundary layer despite the fact that flow over a flat plate is involved. Under some circumstances, this can lead to complex three-dimensional flow in the boundary layer. This type of flow will not be considered here, more information being available in [17] to [23].

9.7
SOLUTIONS TO THE FULL GOVERNING EQUATIONS

The boundary layer equations can, as previously discussed, only be applied to flows in which the Reynolds number is relatively large and in which there is no significant areas of reversed flow. This, in particular, severely limits the applicability of these equations in situations involving opposing flow. When these conditions are not satisfied, the solution must be obtained using the full governing equations. For example, if the flow can be assumed to be two-dimensional and if the Boussinesq approximations are applicable, the equations governing the flow are Eqs. (9.5) to (9.7). If the x-axis is vertical, these equations become:

$$\frac{\partial u}{\partial x} + \frac{\partial v}{\partial y} = 0 \qquad (9.62)$$

$$u\frac{\partial u}{\partial x} + v\frac{\partial u}{\partial y} = -\frac{1}{\rho}\frac{\partial p}{\partial x} + \nu\left(\frac{\partial^2 u}{\partial x^2} + \frac{\partial^2 u}{\partial y^2}\right) \pm \beta g(T - T_1) \qquad (9.63)$$

CHAPTER 9: Combined Convection   447

$$u\frac{\partial v}{\partial x} + v\frac{\partial v}{\partial y} = -\frac{1}{\rho}\frac{\partial p}{\partial y} + \nu\left(\frac{\partial^2 v}{\partial x^2} + \frac{\partial^2 v}{\partial y^2}\right) \quad (9.64)$$

$$u\frac{\partial T}{\partial x} + v\frac{\partial T}{\partial y} = \left(\frac{k}{\rho c_p}\right)\left(\frac{\partial^2 T}{\partial x^2} + \frac{\partial^2 T}{\partial y^2}\right) \quad (9.65)$$

The ± sign on the buoyancy force term in Eq. (9.64) arises because the x-axis is in the direction of the forced flow and can thus be either vertically upwards or vertically downwards, i.e., either in the same direction as the buoyancy forces or in the opposite direction to the buoyancy forces.

The above equations can be solved using numerical methods, i.e., using the same basic procedures as used with forced convection. There is, however, one major difference between the procedures used in forced convection and in mixed convection. In forced convection, the velocity field is independent of the temperature field because fluid properties are here being assumed constant. Thus, in forced convection it is possible to first solve for the momentum and continuity equations and then, once this solution is obtained, to solve for the temperature distribution in the flow. However, in combined convection, because of the presence of the temperature-dependent buoyancy force term in the momentum equation, all of the equations must be solved simultaneously. Studies of flows for which the boundary layer equations are not applicable are described in [24] to [43].

To illustrate the form of results obtained, consider assisting and opposing combined convective flow over a square cylinder at low Reynolds numbers (Fig. 9.13). A computer program, MIXSQCYL, that finds this solution can be obtained in the way discussed in the Preface. The program assumes that the flow over the cylinder is symmetrical about the vertical center-line of the cylinder.

Some results obtained using this program are shown in Figs. 9.14 and 9.15. Figure 9.14 shows some typical streamline patterns. Because the flow is assumed to be symmetrical, only half of the patterns are shown in these figures. It will be seen from Fig. 9.14 that, as discussed earlier in this chapter, the buoyancy forces tend to reduce the size of the separation region downstream of the cylinder in assisting flow whereas they tend to increase the size of this separation region in opposing flow. Figure 9.15

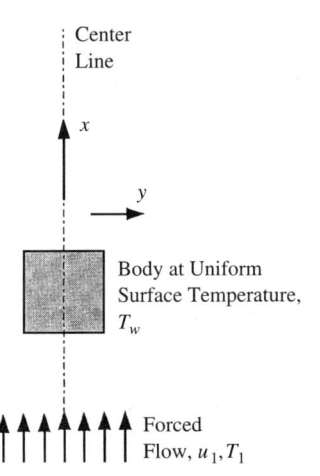

FIGURE 9.13
Flow situation considered.

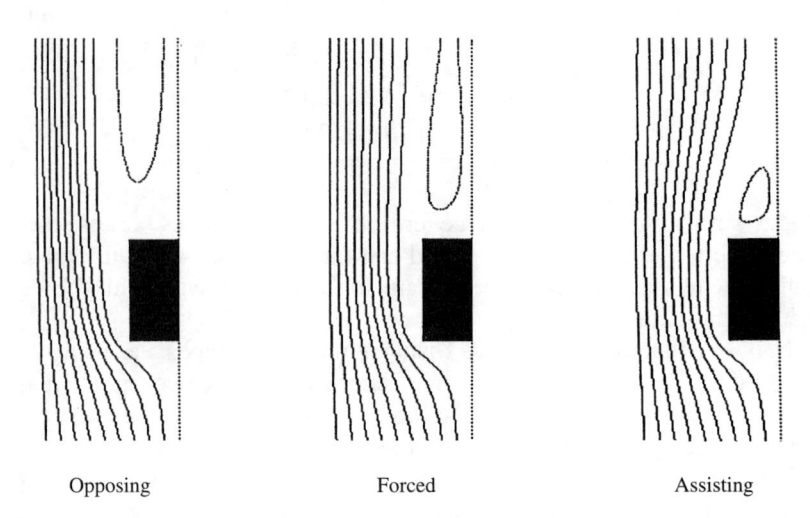

FIGURE 9.14
Typical streamline patterns for combined convective flow over a square cylinder at a Reynolds number of 50 and a Prandtl number of 0.7.

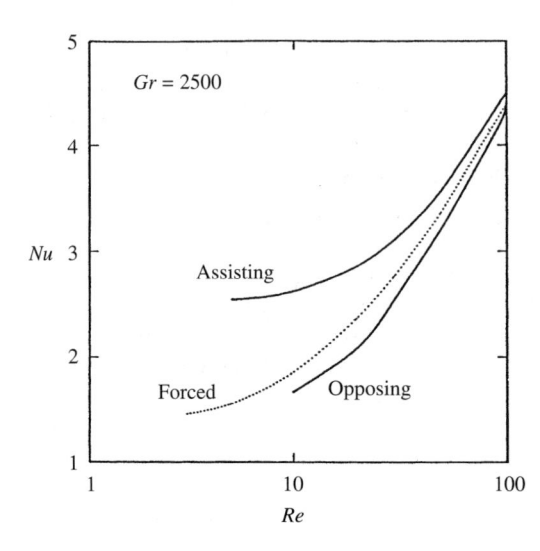

FIGURE 9.15
Variation of Nusselt number with Reynolds number in forced convection and in assisting and opposing convection for a Grashof number of 2500.

shows some typical changes in the Nusselt number produced by the buoyancy forces in mixed convection. As with flow over a flat plate, the buoyancy forces increase the Nusselt number in assisting flow and decrease it in opposing flow. These Nusselt number changes are much less with flow over a square cylinder than they are for flow over a circular cylinder because, in the latter case, the buoyancy forces cause large movements in the point of separation. These changes do not occur with a square cylinder due to the presence of the sharp corners. This is shown schematically in Fig. 9.16.

CHAPTER 9: Combined Convection   449

| Opposing Flow | Forced Flow | Assisting Flow |

S = Separation Point

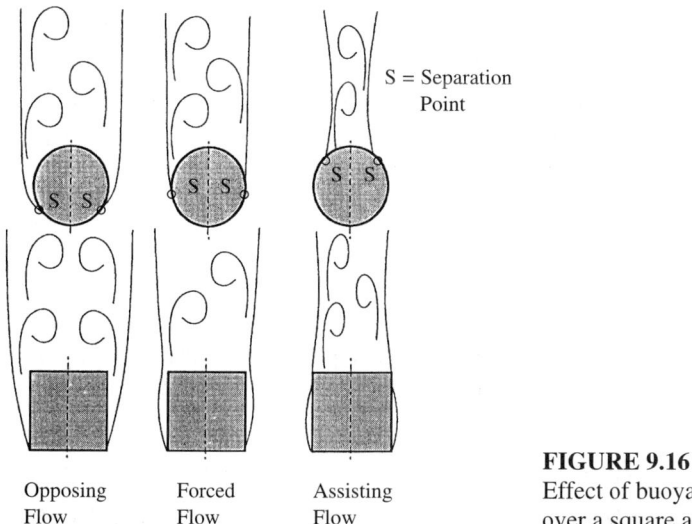

FIGURE 9.16
Effect of buoyancy forces on flow over a square and a circular cylinder.

9.8
CORRELATION OF HEAT TRANSFER RESULTS FOR MIXED CONVECTION

Consider assisting combined convective flow over a body. If the Grashof number is kept constant and the Reynolds number is varied, the variation of Nu with Re would resemble that shown in Fig. 9.17. This type of result could be obtained by considering a body of fixed size, kept at a fixed temperature (this would keep the Grashof number constant), that is placed in a fluid flow in which the velocity could be varied (this would allow the Reynolds number to be varied).

At low Reynolds numbers, the Nusselt number will tend to the constant value that would exist in purely free convection, this being designated as Nu_N, whereas at high Reynolds numbers, when the effects of the buoyancy forces are small, the Nusselt numbers will tend to the values that would exist in purely forced convection at the same Reynolds number as that being considered. These forced convection Nusselt numbers are here designated as Nu_F. In the combined convection regions between these two limits, the Nusselt number variation can be approximately

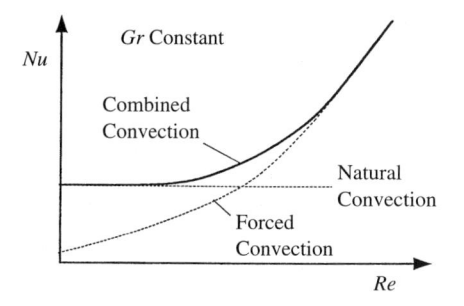

FIGURE 9.17
Typical variation of Nusselt number with Reynolds number in assisting combined convective flow.

represented by:

$$Nu^n = Nu_N^n + Nu_F^n \tag{9.66}$$

where the value of the index n depends on the geometrical situation being considered. For example, consider flow over a vertical isothermal plate. The Nusselt numbers in the limiting purely forced and purely free convective flow cases are given by:

$$Nu_F = C_F Re_x^{0.5} \tag{9.67}$$

and:

$$Nu_N = C_N Gr_x^{0.25} \tag{9.68}$$

The values of the coefficients, C_F and C_N, are dependent on the Prandtl number of the fluid being considered. For a Prandtl number of 0.7, they are, as discussed in Chapters 3 and 8, given by:

$$\text{For } Pr = 0.7: C_F = 0.29, \; C_N = 0.36 \tag{9.69}$$

Substituting Eqs. (9.67) and (9.68) into (9.66) and dividing through by $Re_x^{0.5}$ gives:

$$\left(\frac{Nu_x}{Re_x^{0.5}}\right)^n = C_F^n + \left(C_N Gx^{0.25}\right)^n \tag{9.70}$$

The results given by this equation with n set equal to 3.5 are compared in Fig. 9.18 with the numerically calculated results for assisting flow for a Prandtl number of 0.7 that were given earlier in Fig. 9.9.

It will be seen that Eq. (9.70) does, in fact, describe the variation in the mixed convection region with assisting flow to an accuracy that is quite acceptable for most purposes. Eq. (9.66) does, therefore, apply to assisting mixed convective flow over a flat plate. It has been shown that it does, in fact, also describe experimental results for other more complex situations to a good degree of accuracy provided the value

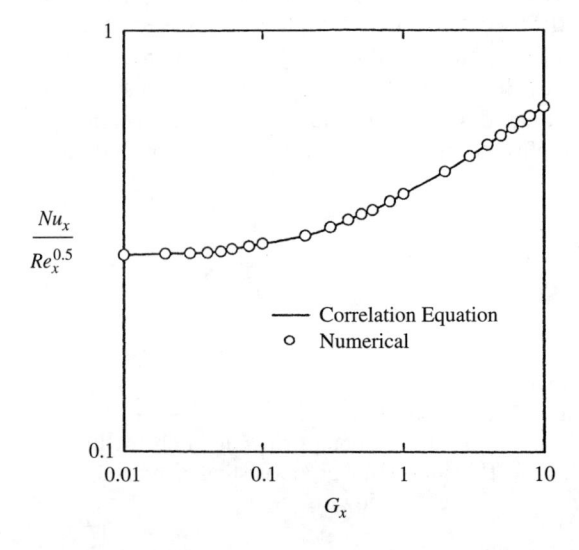

FIGURE 9.18
Comparison with variation of $Nu_x/Re_x^{0.5}$ with Gx given by Eq. (9.70) with actual calculated variation.

TABLE 9.1
Values of n for various geometries

Situation	Index, n
Vertical plate, assisting flow	3.5
Horizontal cylinder, assisting flow, uniform surface temperature	3.5
Horizontal cylinder, assisting flow, uniform surface heat flux	4
Horizontal cylinder, horizontal cross flow	7
Sphere, assisting flow, uniform surface temperature	3.5

of n is properly chosen. Some values of n deduced from experimental results for various geometrical situations are shown in Table 9.1, see [44] to [54]. Cross flow involves a horizontal forced flow which is thus at right angles to the buoyancy forces.

The conditions under which the heat transfer rate can be assumed to be equal to that in purely forced convection and under which it can be assumed to be equal to that in purely free convective can be deduced from Eq. (9.66). For this purpose, this equation can be written as:

$$\frac{Nu}{Nu_F} = \left[\left(\frac{Nu_N}{Nu_F}\right)^n + 1\right]^{1/n} \tag{9.71}$$

If it is again assumed that the flow can be taken to be forced convective if:

$$\frac{Nu}{Nu_F} - 1 < 0.01$$

then it will be seen that Eq. (9.66) indicates that forced convection can be assumed to exist if:

$$\left(\frac{Nu_N}{Nu_F}\right)^n + 1 < 1.01^n \tag{9.72}$$

i.e., if:

$$\frac{Nu_N}{Nu_F} < (1.01^n - 1)^{1/n} \tag{9.73}$$

This defines the conditions under which the flow can be assumed to be purely forced convective.

Similarly, if it is again assumed that the flow can be taken to be free convective if:

$$\frac{Nu}{Nu_N} - 1 < 0.01$$

then it will be seen that since Eq. (9.66) indicates that:

$$\frac{Nu}{Nu_N} = \left[\left(\frac{Nu_F}{Nu_N}\right)^n - 1\right]^{1/n} \tag{9.74}$$

FIGURE 9.19
Correlation of results for opposing flow over form of cylinder indicated [47].

free convection can be assumed to exist if:

$$\left[\left(\frac{Nu_F}{Nu_N}\right)^n - 1\right]^{1/n} < 1.01$$

i.e., if:

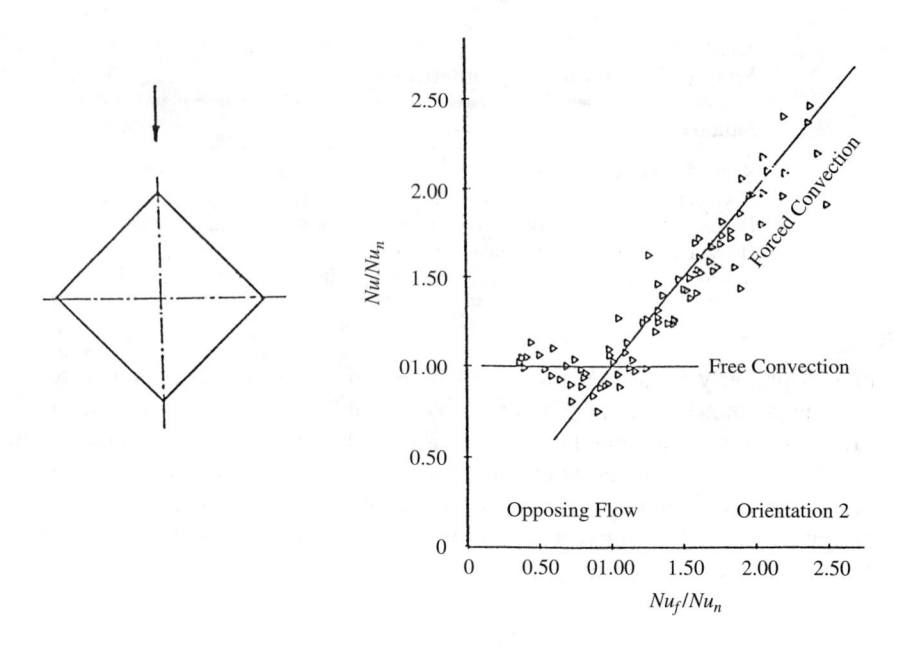

$$\frac{Nu_F}{Nu_N} < (1.01^n - 1)^{1/n} \qquad (9.75)$$

The above discussion applied only to assisting flow and cross flow. Now, it will be noted that Eq. (9.74) indicates that:

$$\frac{Nu}{Nu_N} = \text{function}\left(\frac{Nu_F}{Nu_N}\right) \qquad (9.76)$$

Attempts have been made to correlate results for other geometrical situations by assuming that this type of relation applies. Some success has been achieved by using this approach. For example, some experimental results for opposing flow over a diamond-shaped cylinder are correlated in this way in Fig. 9.19.

EXAMPLE 9.4. Air at a temperature of 15°C flows over a 25-mm diameter horizontal cylinder which has a uniform surface temperature of 45°C. If the air flow is also horizontal and at right angles to the axis of the cylinder, determine how the heat transfer rate per meter length from the cylinder varies with air velocity. Determine the air velocity above which the flow can be assumed to be purely forced convective.

Solution. The mean temperature of the air in the flow about the cylinder is: (15 + 45)/2 = 30°C. At this temperature for air at standard ambient pressure:

$$\beta = \frac{1}{303} \text{ K}^{-1}, \ \nu = 16.01 \times 10^{-6} \text{ m}^2/\text{s}, \ k = 0.02638 \text{ W/m-°C}$$

Because cross flow over a horizontal cylinder is being considered, the results given in Table 9.1 indicate that:

$$n = 7$$

i.e., that:

$$Nu^7 = Nu_N^7 + Nu_F^7$$

Hence:

$$Nu = (Nu_N^7 + Nu_F^7)^{1/7}$$

The Grashof number for the situation being considered is given by:

$$Gr = \frac{\beta g (T_w - T_1) D^3}{\nu^2} = \frac{(1/303) \times 9.81 \times (45 - 15) \times 0.025^3}{(16.01 \times 10^{-6})^2} = 59{,}210$$

For this value of the Grashof number, standard correlation equations for natural convection from a horizontal cylinder indicate that for a Prandtl number of 0.7:

$$Nu_N = 5.98$$

Hence:

$$Nu^7 = 5.98^7 + Nu_F^7 = 273470 + Nu_F^7$$

The Reynolds number for the situation being considered is given by:

$$Re = \frac{VD}{\nu} = \frac{0.025 V}{16.01 \times 10^{-6}} = 1562 V$$

where the velocity, V, is in m/s. Thus, for any value of V, the Reynolds number can be found. For any value of the Reynolds number and for a Prandtl number of 0.7, standard correlation equations for forced convection from a cylinder allow Nu_F to be found which then, using the above equation, allows Nu to be found. But:

$$Nu = \frac{q_w D}{k(T_w - T_1)}$$

from which it follows that:

$$q_w = \frac{Nu \, k (T_w - T_1)}{D} = \frac{Nu \times 0.02638 \times (45 - 15)}{0.025} = 31.66 Nu \text{ W/m}^2$$

The heat transfer rate per m length of the cylinder is given by:

$$Q = q_w \pi D = 31.66 \times Nu \times \pi \times 0.025 = 2.487 Nu \text{ W}$$

Therefore, for any value of the velocity, V, the corresponding value of Q can be found. Some values of Q obtained in this way are shown in the following table. If the effect of the buoyancy forces can be neglected, Q is given by:

$$Q_F = 2.487 Nu_F \text{ W}$$

Values of Q_F are also shown in the table. Alternatively, if forced flow effects are negligible, Q is given by:

$$Q_N = 2.487 Nu_N = 2.487 \times 5.98 = 14.87 \text{ W}$$

V – m/s	Re	Nu_F	Nu	Q_F – W	Q – W
0	0	0	5.98	0	14.87
0.02	31.24	3.000	5.9875	7.460	14.89
0.05	78.100	4.569	6.102	11.36	15.18
0.1	156.2	6.337	6.816	15.76	16.95
0.15	234.3	7.693	7.869	19.13	19.57
0.2	312.4	8.837	9.917	21.98	22.18
0.4	624.8	12.37	12.38	30.77	30.80
0.6	937.2	15.09	15.09	37.52	37.53
0.8	1249.6	17.37	17.38	43.21	43.21
1.0	1562.0	18.60	18.60	46.26	46.26
1.5	2343.0	23.68	23.68	58.89	58.89

The heat transfer rates are shown plotted against the velocity in Fig. E9.4.

It will be seen from the results given in the above table that the heat transfer rate is within 1% of its forced convective value when the velocity is greater than approximately 0.2 m/s.

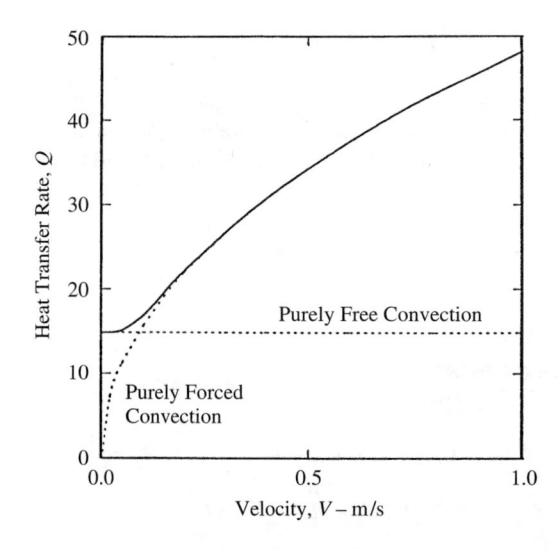

FIGURE E9.4

9.9
EFFECT OF BUOYANCY FORCES ON TURBULENT FLOWS

In a turbulent flow, the buoyancy forces have two effects, [55] to [60]:

i. The momentum equation must be modified to include the buoyancy force.
ii. The turbulence model employed must be modified to account for the effect of the buoyancy forces on the turbulence quantities.

Consider, for example, assisting turbulent mixed convection over a vertical flat plate. This situation is schematically shown in Fig. 9.20. If it is assumed that the boundary layer assumptions apply, the governing equations for the mean velocity components and temperature are:

$$\frac{\partial \bar{u}}{\partial x} + \frac{\partial \bar{v}}{\partial y} = 0 \tag{9.77}$$

$$\bar{u}\frac{\partial \bar{u}}{\partial x} + \bar{v}\frac{\partial \bar{u}}{\partial y} = \frac{\partial}{\partial y}\left[(\nu + \epsilon)\frac{\partial \bar{u}}{\partial y}\right] + \beta g(\bar{T} - T_1) \tag{9.78}$$

$$\bar{u}\frac{\partial \bar{T}}{\partial x} + \bar{v}\frac{\partial \bar{T}}{\partial y} = \frac{\partial}{\partial y}\left[(\alpha + \epsilon_H)\frac{\partial \bar{T}}{\partial y}\right] \tag{9.79}$$

The last term in the momentum equation, i.e., Eq. (9.78), represents the affect of the buoyancy forces on the mean momentum balance. However, these buoyancy forces also affect the variation of ϵ and ϵ_H in the flow. To illustrate how the buoyancy forces can effect ϵ and ϵ_H, consider again the simple mixing length model discussed in Chapter 5. "Lumps" or "eddies" of fluid are assumed to move across the flow through a transverse distance, ℓ_m, while retaining their initial velocity and temperature. They then interact with the local fluid layer giving rise to the fluctuations in velocity and temperature that occur in turbulent flow.

Thus, to recap the derivation, consider the situation shown in Fig. 9.21. As discussed in Chapter 5, it is assumed that "lumps" from layer 2 arrive at layer 1 with a velocity deficit of:

$$\Delta u_2 = \ell_m \frac{\partial \bar{u}}{\partial y}$$

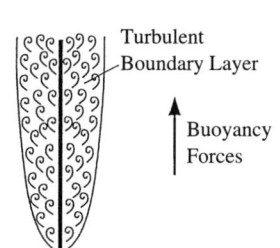

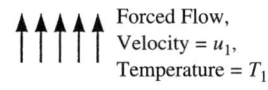

FIGURE 9.20
Flow situation being considered.

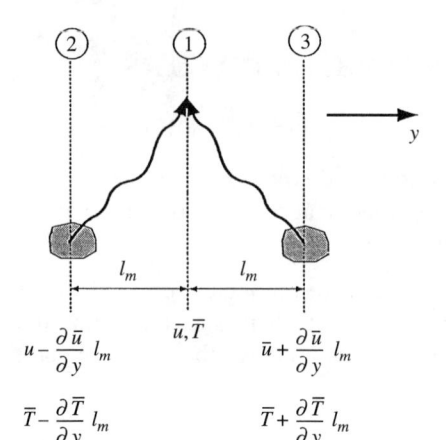

456 Introduction to Convective Heat Transfer Analysis

FIGURE 9.21
Mixing length model.

while lumps from layer 3 arrive at layer 1 with a velocity excess of:

$$\Delta u_3 = \ell_m \frac{\partial \bar{u}}{\partial y}$$

It is then assumed that the magnitude of the velocity fluctuation at 1 is proportional to the average of $|\Delta u_2|$ and $|\Delta u_3|$, i.e.:

$$|u'| \propto \tfrac{1}{2}[|\Delta u_2| + |\Delta u_3|]$$

i.e.:

$$|u'| \propto \ell_m \left|\frac{\partial \bar{u}}{\partial y}\right|$$

It is further assumed that the transverse velocity fluctuations arise from the longitudinal velocity fluctuations in order to satisfy continuity requirements, i.e.:

$$|v'| \propto |u'|$$

Lastly, it is assumed that turbulent stress is proportional to $|v'||u'|$, i.e., that:

$$\tau_T = -\overline{\rho v' u'} \propto \rho |v'||u'|$$

Hence:

$$\tau_T = \rho C_u \ell_m^2 \left|\frac{\partial \bar{u}}{\partial y}\right| \frac{\partial \bar{u}}{\partial y}$$

i.e.:

$$\tau_T = \rho \ell^2 \left|\frac{\partial \bar{u}}{\partial y}\right| \frac{\partial \bar{u}}{\partial y} \tag{9.80}$$

where C_u is a constant of proportionality that has been combined with ℓ_m to give ℓ. In writing Eq. (9.80), account has been taken of the fact that the sign of τ_T depends on the sign of $\partial \bar{u}/\partial y$. Hence, $|\partial \bar{u}/\partial y|$ has been used instead of $(\partial \bar{u}/\partial y)^2$.

Next, consider the temperature fluctuations. The lumps arriving at 1 from 3 and arriving at 1 from 2 do so with a temperature excess and deficit of:

$$\Delta T_3 = \ell_m \frac{\partial \overline{T}}{\partial y}$$

and

$$\Delta T_2 = \ell_m \frac{\partial \overline{T}}{\partial y}$$

respectively.

As with the velocity fluctuation, it is assumed that the magnitude of the temperature fluctuation at 1 is proportional to the average of $|\Delta T_2|$ and $|\Delta T_3|$, i.e.:

$$|T'| \propto \tfrac{1}{2} [|\Delta T_2| + |\Delta T_3|]$$

i.e.:

$$|T'| \propto \ell_m \left|\frac{\partial \overline{T}}{\partial y}\right| \tag{9.81}$$

It is also assumed that the turbulent heat transfer rate is proportional to $|v'||T'|$, i.e., that:

$$q_T = -\rho c_p \overline{v'T'} \propto \rho c_p |v'||T'|$$

Hence:

$$q_T = \rho c_p C_T \ell_m^2 \left|\frac{\partial \overline{u}}{\partial y}\right| \frac{\partial \overline{T}}{\partial y}$$

i.e.:

$$q_T = \rho c_p C \ell^2 \left|\frac{\partial \overline{u}}{\partial y}\right| \frac{\partial \overline{T}}{\partial y} \tag{9.82}$$

where C_T and C are constants of proportionality. In writing Eq. (9.82), account has been taken of the fact that the sign of q_T depends on the sign of $\partial \overline{T}/\partial y$.

In the derivation of Eqs. (9.80) and (9.82), it was assumed that the buoyancy forces had no effect on the flow. However, by assumption, as the fluid "lumps" move across the flow from 2 to 1 and from 3 to 1 they are at a different temperature than the surrounding fluid and, as a result, they are acted on by buoyancy forces. For example, consider a fluid "lump" moving from 2 to 1. If $\partial \overline{u}/\partial y$ is taken as positive and $\partial \overline{T}/\partial y$ is taken as negative, then as the "lump" moves from 2 to 1 the temperature difference between the "lump" and the surrounding fluid will increase. When the "lump" is at 2, it is at the same temperature as the surrounding fluid and there is no buoyancy force acting on it. When the lump arrives at 1 it is at a temperature that is $\ell_m(\partial \overline{T}/\partial y)$ above that of the surrounding fluid and the buoyancy force acting on it per unit volume is $-\beta g \ell_m (\partial \overline{T}/\partial y)$. The average buoyancy force per unit volume acting on the lump between 2 and 1 is, therefore:

$$-\tfrac{1}{2} \beta g \rho \ell_m \frac{\partial \overline{T}}{\partial y}$$

458 Introduction to Convective Heat Transfer Analysis

This buoyancy force will increase the velocity of the "lump" of fluid in the direction of the buoyancy force, i.e., in the x-direction, by an amount Δu_B that is given by Newton's law as:

$$\rho \frac{\Delta u_B}{\Delta t} = -\frac{1}{2}\beta g \rho \ell_m \frac{\partial \overline{T}}{\partial y}$$

because a unit volume is being considered. Δt is the time taken by the lump to move from 2 to 1.

Hence:

$$\Delta u_B = -\frac{1}{2}\beta g \ell_m \frac{\partial \overline{T}}{\partial y}\Delta t \tag{9.83}$$

But, because the "lump" of fluid moves a distance, ℓ_m, in going in the y-direction from 2 to 1 and because the movement of the "lump" is caused by the presence of the transverse velocity fluctuation v', it follows that:

$$\Delta t = \frac{\ell_m}{|v'|}$$

Substituting this into Eq. (9.83) then gives:

$$\Delta u_B = -\frac{1}{2}\beta g \frac{\ell_m^2}{|v'|} \frac{\partial \overline{T}}{\partial y} \tag{9.84}$$

But it has been assumed that $|v'|$ is proportional to $|u'|$, i.e., that:

$$|v'| = K|u'|$$

Hence, Eq. (9.84) gives:

$$\Delta u_B = -\frac{1}{2}\beta g \frac{\ell_m^2}{K|u'|} \frac{\partial \overline{T}}{\partial y} \tag{9.85}$$

Now, the "lump" of fluid from 2 will arrive at 1 with a velocity deficit (as mentioned before, $\partial \overline{u}/\partial y$ is taken as positive) that is given by:

$$\Delta u_2 = \ell_m \frac{\partial \overline{u}}{\partial y} - \Delta u_B \tag{9.86}$$

the buoyancy forces having increased the velocity of the "lump", thus causing it to arrive at 1 with a reduced velocity deficit.

Substituting Eq. (9.85) into Eq. (9.86) then gives:

$$\Delta u_2 = \ell_m \frac{\partial \overline{u}}{\partial y} + \frac{\beta g \ell_m^2}{2K|u'|} \frac{\partial \overline{T}}{\partial y} \tag{9.87}$$

Similarly, a "lump" of fluid moving across the flow from 3 to 1 will arrive with a velocity excess of:

$$\Delta u_3 = \ell_m \frac{\partial \overline{u}}{\partial y} + \frac{\beta g \ell_m^2}{2K|u'|} \frac{\partial \overline{T}}{\partial y} \tag{9.88}$$

In this case, the fluid "lump" moving from 3 to 1 will have a lower temperature than the surrounding fluid so the buoyancy forces will reduce the velocity component in the x-direction and thus reduce the velocity difference between that of the "lump" and that of the surrounding fluid when the lump arrives at 1.

It is now again assumed that:

$$|u'| \propto \tfrac{1}{2}\left[|\Delta u_2| + |\Delta u_3|\right]$$

which gives:

$$|u'| \propto \ell_m \frac{\partial \bar{u}}{\partial y} + \frac{\beta g \ell_m^2}{2K|u|} \frac{\partial \bar{T}}{\partial y} \quad (9.89)$$

It will be assumed that the effect of the buoyancy forces on the turbulence structure is relatively small, i.e., that the magnitude of the second term in Eq. (9.89) is much smaller than the magnitude of the first. Now Eq. (9.89) can be written as:

$$|u'| \propto \ell_m \frac{\partial \bar{u}}{\partial y}\left[1 + \frac{\beta g}{2K}\frac{\partial \bar{T}/\partial y}{\partial \bar{u}/\partial y|\partial \bar{u}/\partial y|}\right] \quad (9.90)$$

But, as discussed above, it is also assumed that:

$$|v'| \propto |u'|$$

hence:

$$|v'| \propto \ell_m \frac{\partial \bar{u}}{\partial y}\left[1 + \frac{\beta g}{2K}\frac{\partial \bar{T}/\partial y}{\partial \bar{u}/\partial y|\partial \bar{u}/\partial y|}\right] \quad (9.91)$$

It is also, as before, assumed that:

$$\tau_T = -\overline{\rho v'u'} \propto \rho|v'||u'|$$

from which it follows that:

$$\tau_T \propto \rho \ell_m^2 \left|\frac{\partial \bar{u}}{\partial y}\right| \frac{\partial \bar{u}}{\partial y}\left[1 + \frac{\beta g}{2K}\frac{\partial \bar{T}/\partial y}{\partial \bar{u}/\partial y|\partial \bar{u}/\partial y|}\right]^2 \quad (9.92)$$

But, as discussed above, the buoyancy force effect is assumed to be small, i.e., the second term in the bracket is assumed to be much less than 1. Hence, Eq. (9.92) can be written as:

$$\tau_T = \rho \ell^2 \left|\frac{\partial \bar{u}}{\partial y}\right| \frac{\partial \bar{u}}{\partial y}\left[1 + J\beta g \frac{\partial \bar{T}/\partial y}{\partial \bar{u}/\partial y|\partial \bar{u}/\partial y|}\right] \quad (9.93)$$

where:

$$J = \frac{1}{K}$$

Next, consider the turbulent heat transfer. The buoyancy forces have no effect on the temperature fluctuations so Eq. (9.81) still applies. Hence, again assuming:

$$q_T = -\rho c_p \overline{v'T'} \propto \rho |v'||T'|$$

and using Eq. (9.91), the following is obtained:

$$q_T \propto \rho c_p \ell_m^2 \left|\frac{\partial \overline{u}}{\partial y}\right| \frac{\partial \overline{T}}{\partial y}\left[1 + \frac{\beta g}{2K}\frac{\partial \overline{T}/\partial y}{\partial \overline{u}/\partial y|\partial \overline{u}/\partial y|}\right]$$

i.e.:

$$q_T = \rho c_p C \ell^2 \left|\frac{\partial \overline{u}}{\partial y}\right| \frac{\partial \overline{T}}{\partial y}\left[1 + \frac{J\beta g}{2}\frac{\partial \overline{T}/\partial y}{\partial \overline{u}/\partial y|\partial \overline{u}/\partial y|}\right] \quad (9.94)$$

Eqs. (9.93) and (9.94) give:

$$\epsilon = \ell^2 \left|\frac{\partial \overline{u}}{\partial y}\right|\left[1 + J\beta g\frac{\partial \overline{T}/\partial y}{\partial \overline{u}/\partial y|\partial \overline{u}/\partial y|}\right] \quad (9.95)$$

and:

$$\epsilon_H = C\ell^2 \left|\frac{\partial \overline{u}}{\partial y}\right|\left[1 + \frac{J\beta g}{2}\frac{\partial \overline{T}/\partial y}{\partial \overline{u}/\partial y|\partial \overline{u}/\partial y|}\right] \quad (9.96)$$

Hence:

$$Pr_T = \frac{\epsilon}{\epsilon_H} = \frac{1}{C}\left[1 + \frac{J}{2}\beta g\frac{\partial \overline{T}/\partial y}{\partial \overline{u}/\partial y|\partial \overline{u}/\partial y|}\right] \quad (9.97)$$

A comparison of results given by using these equations with experiment indicates that J is approximately equal to 2.

The effect of the buoyancy forces on the turbulence therefore depends on the value of the dimensionless quantity:

$$S = \beta g \frac{\partial \overline{T}/\partial y}{\partial \overline{u}/\partial y|\partial \overline{u}/\partial y|} \quad (9.98)$$

which is a form of Richardson number.

It will be noted that the above derivation was based on the assumption that $\partial \overline{T}/\partial y$ was negative and that $\partial \overline{u}/\partial y$ was positive and that the buoyancy forces acted in the x-direction, i.e., they applied to assisting flow. In opposing flow, the buoyancy forces act in the opposite direction to the x-axis, this x-axis conventionally being taken in the direction of the forced flow. Thus, in general, Eqs. (9.95), (9.96), and (9.97) can be written as:

$$\epsilon = \ell^2 \left|\frac{\partial \overline{u}}{\partial y}\right|[1 \pm JS] \quad (9.99)$$

$$\epsilon_H = C\ell^2 \left|\frac{\partial \overline{u}}{\partial y}\right|\left[1 \pm \frac{J}{2}S\right] \quad (9.100)$$

CHAPTER 9: Combined Convection 461

$$Pr_T = \frac{1}{C}\left[1 \pm \frac{J}{2}S\right] \qquad (9.101)$$

where the upper sign on J applies in assisting flow and the lower sign on J applies in opposing flow.

Because, for flow over a heated surface, $\partial \bar{u}/\partial y$ is positive and $\partial \bar{T}/\partial y$ is negative, S will normally be a negative. Hence, in assisting flow, the buoyancy forces will tend to decrease ϵ and ϵ_H, i.e., to damp the turbulence, and thus to decrease the heat transfer rate below the purely forced convective flow value. However, the buoyancy force in the momentum equation tends to increase the mean velocity and, therefore, to increase the heat transfer rate. In turbulent assisting flow over a flat plate, this can lead to a Nusselt number variation with Reynolds number that resembles that shown in Fig. 9.22.

At higher Reynolds numbers, i.e., between points A and B in Fig. 9.22, the effect of the buoyancy forces on the turbulence quantities is the dominant effect and the Nusselt number is, therefore, decreased below its forced convective value. At the lower Reynolds number, i.e., between B and C in Fig. 9.22, the direct effect of the buoyancy forces on the mean momentum balance becomes the dominant effect and the Nusselt number rises above its forced convective value. The changes are displayed by the numerical results shown in Fig. 9.23. These results were obtained using a more advanced turbulence model than that discussed here.

Based on results similar to those shown in Fig. 9.23, the conditions under which mixed convection effects are important in assisting flow over a vertical plate can be derived. These conditions are indicated in Fig. 9.24.

EXAMPLE 9.5. In forced convective turbulent boundary layer flow over a plate the velocity and temperature distributions are approximately given by:

$$\frac{\bar{u}}{u_1} = \left[\frac{y}{\delta}\right]^{\frac{1}{7}}$$

and:

$$\frac{\bar{T} - T_1}{T_w - T_1} = 1 - \left[\frac{y}{\delta}\right]^{\frac{1}{7}}$$

where u_1 is the free-stream velocity, T_1 is the free-stream temperature, T_w is the wall temperature, and δ is the local boundary layer thickness, the thermal and velocity boundary

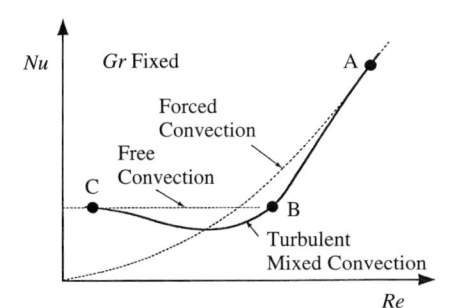

FIGURE 9.22
Variation of Nusselt number variation with Reynolds number in turbulent assisting mixed convective flow.

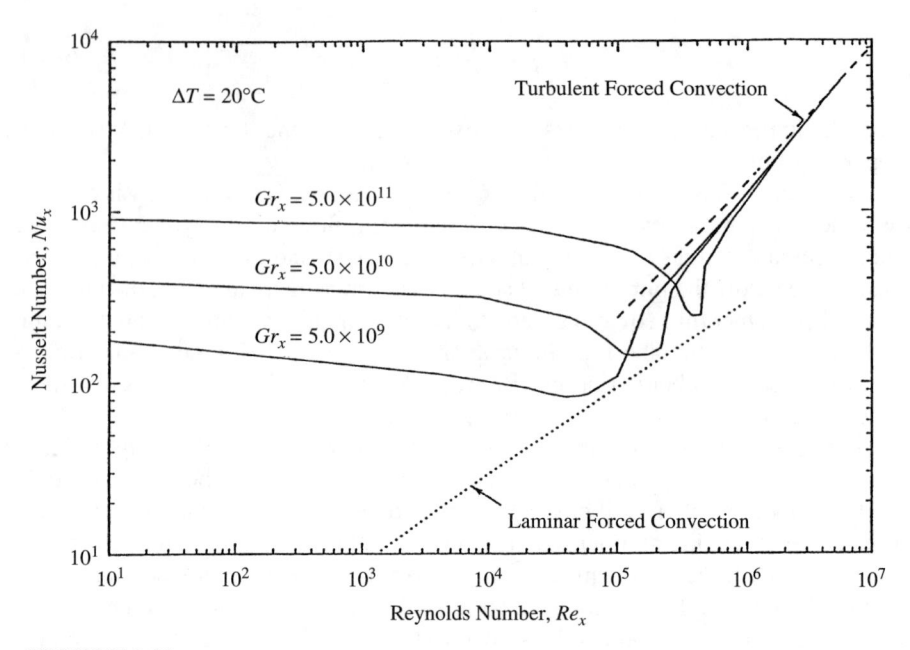

FIGURE 9.23
Numerically predicted variation of Nusselt number variation with Reynolds number in turbulent assisting mixed convective flow over a vertical plate. (Based on results obtained by Patel K., Armaly B.F., and Chen T.S., "Transition from Turbulent Natural to Turbulent Forced Convection Adjacent to an Isothermal Vertical Plate", ASME HTD, Vol. 324, pp. 51–56, 1996. With permission.)

layers being assumed to have the same thickness. Using these relations derive expressions for the ratio of the turbulent shear stress in assisting mixed convective flow to the turbulent shear in forced convective flow and for the ratio of the turbulent heat transfer rate in mixed convective flow to the turbulent heat transfer rate in forced convective flow.

Solution. In forced convective flow the turbulent shear stress is given by:

$$\tau_{T0} = \rho \ell^2 \left| \frac{\partial \bar{u}}{\partial y} \right| \frac{\partial \bar{u}}{\partial y}$$

while in mixed convective flow it is given by:

$$\tau_T = \rho \ell^2 \left| \frac{\partial \bar{u}}{\partial y} \right| \frac{\partial \bar{u}}{\partial y} \left[1 + J\beta g \frac{\partial \bar{T}/\partial y}{\partial \bar{u}/\partial y |\partial \bar{u}/\partial y|} \right]$$

Hence:

$$\frac{\tau_T}{\tau_{T0}} = \left[1 + J\beta g \frac{\partial \bar{T}/\partial y}{\partial \bar{u}/\partial y |\partial \bar{u}/\partial y|} \right]$$

But:

$$\frac{\partial \bar{u}}{\partial y} = \frac{u_1}{7} \left[\frac{y}{\delta} \right]^{\frac{1}{7}-1} \frac{1}{\delta}$$

CHAPTER 9: Combined Convection   463

FIGURE 9.24
Purely forced, purely free, and mixed convective regions in assisting flow over a vertical plate. (Based on results obtained by Patel K., Armaly B.F., and Chen T.S., "Transition from Turbulent Natural to Turbulent Forced Convection Adjacent to an Isothermal Vertical Plate", ASME HTD, Vol. 324, pp. 51–56, 1996. With permission.)

and:

$$\frac{\partial T}{\partial y} = -\frac{(T_w - T_1)}{7}\left[\frac{y}{\delta}\right]^{\frac{1}{7}-1}\frac{1}{\delta}$$

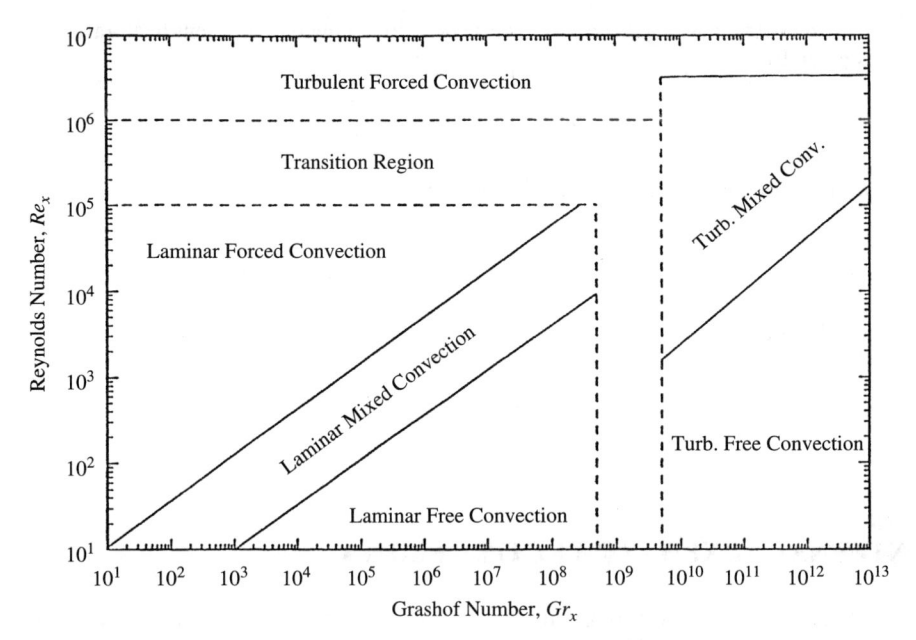

Therefore:

$$\frac{\tau_T}{\tau_{T0}} = 1 - 7J\frac{\beta g(T_w - T_1)\delta}{u_1^2}\left[\frac{y}{\delta}\right]^{\frac{6}{7}}$$

i.e.:

$$\frac{\tau_T}{\tau_{T0}} = 1 - 7J\frac{Gr_x}{Re_x^2}\frac{\delta}{x}\left[\frac{y}{\delta}\right]^{\frac{6}{7}}$$

where Gr_x and Re_x are the local Grashof number and the local Reynolds number, respectively.

The boundary layer thickness is approximately given by:

$$\frac{\delta}{x} = \frac{0.036}{Re_x^{0.2}}$$

Using this gives:

$$\frac{\tau_T}{\tau_{T0}} = 1 - 0.252J\frac{Gr_x}{Re_x^{2.2}}\left[\frac{y}{\delta}\right]^{\frac{6}{7}}$$

In a similar way, because in forced convective flow the turbulent heat transfer rate is given by:

$$q_{T0} = \rho c_p C \ell^2 \left| \frac{\partial \bar{u}}{\partial y} \right| \frac{\partial \bar{T}}{\partial y}$$

while in mixed convective flow it is given by:

$$q_T = \rho c_p C \ell^2 \left| \frac{\partial \bar{u}}{\partial y} \right| \frac{\partial \bar{T}}{\partial y} \left[1 + \frac{J \beta g}{2} \frac{\partial \bar{T}/\partial y}{\partial \bar{u}/\partial y |\partial \bar{u}/\partial y|} \right]$$

it follows that:

$$\frac{q_T}{q_{T0}} = \frac{\tau_T}{\tau_{T0}} = 1 - 0.252 J \frac{Gr_x}{Re_x^{2.2}} \left[\frac{y}{\delta} \right]^{\frac{6}{7}}$$

9.10
INTERNAL MIXED CONVECTIVE FLOWS

The discussion up to this point in this chapter has been concerned with external mixed convective flows. However, buoyancy forces can also sometimes have a significant influence on internal flows, see [61] to [67]. One example is with the internal cooling of rotor blades in gas turbine engines. In this application, the blade rotation generates large centrifugal forces that cause high buoyancy forces to exist in the internal cooling channels and these buoyancy forces can cause the heat transfer rates to be very different from those that would exist with purely forced convective flow. Mixed convection effects are also often encountered in nuclear reactor cooling problems, particularly during shutdown and emergency situations.

As with external mixed convection, the influence of buoyancy forces on the flow depends on the angle that the buoyancy forces makes to the direction of the forced flow. The heat transfer rate also, of course, depends on the duct cross-sectional shape as well as on whether the flow is laminar or turbulent.

For laminar flow in a horizontal pipe, the buoyancy forces can cause a secondary flow that can significantly enhance the heat transfer rate. The nature of the buoyancy-induced secondary motion is illustrated in Fig. 9.25. In a horizontal pipe the buoyancy force acts perpendicular to the direction of the forced flow. As a result, in a heated pipe, the fluid rises on the outside of the pipe wall and

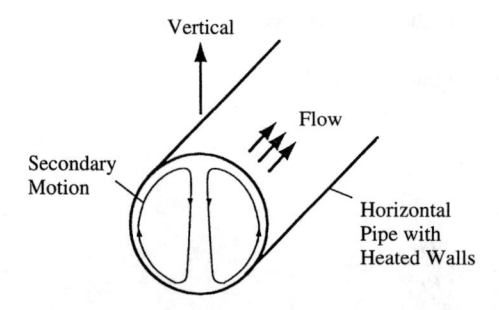

FIGURE 9.25
Secondary flow in mixed convection in a horizontal pipe.

descends near the vertical center line of the pipe forming a pair of counter-rotating spirals that travel down the pipe. This buoyancy-driven secondary flow causes the heat transfer rate to increase significantly on the lower surface of the pipe.

There have been many studies of internal mixed convection, particularly in circular pipes. The conditions under which flow in a circular pipe can be assumed to be purely forced convective, purely free, and mixed convective have been presented in graphical form by Metais and Eckert [62], the form of these graphs being given in Figs. 9.26 and 9.27. Figure 9.26 applies to flow in a vertical pipe while Fig. 9.27 is for flow in a horizontal pipe.

In these figures, the mixed convection regime is defined as the conditions for which the convective heat transfer deviates by more than 10% from either the purely forced or purely free convective value. Qualitatively, the regions shown in these figures are as is to be expected. At low Reynolds and high Rayleigh numbers, free convection is the dominant mode. Conversely, at high Reynolds and low Rayleigh numbers, forced convection dominates. In these figures, the Rayleigh number is based on the tube diameter and L/D is the pipe length-to-diameter ratio. Note that only the boundary between forced and mixed convection is shown in Fig. 9.27.

The results given in Figs. 9.26 and 9.27 are largely based on experimental results. Here, however, the main interest is in the analysis of internal mixed convective

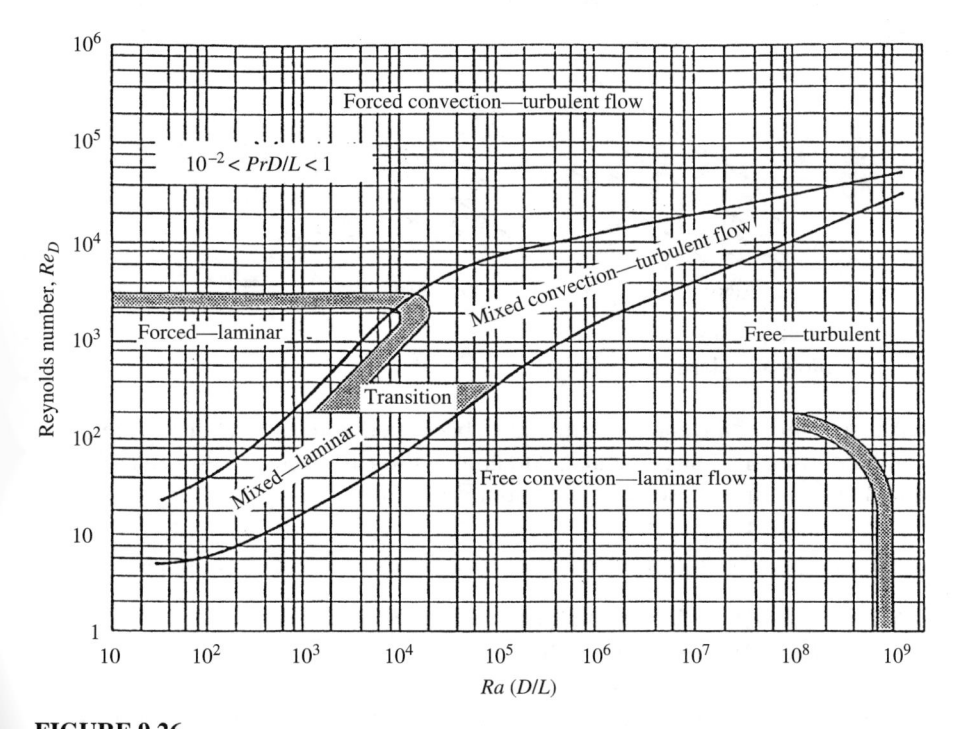

FIGURE 9.26
Regimes for purely forced, purely free, and mixed convection for flow through vertical pipes [62].

FIGURE 9.27
Regimes for purely forced, purely free, and mixed convection for flow through horizontal pipes [62].

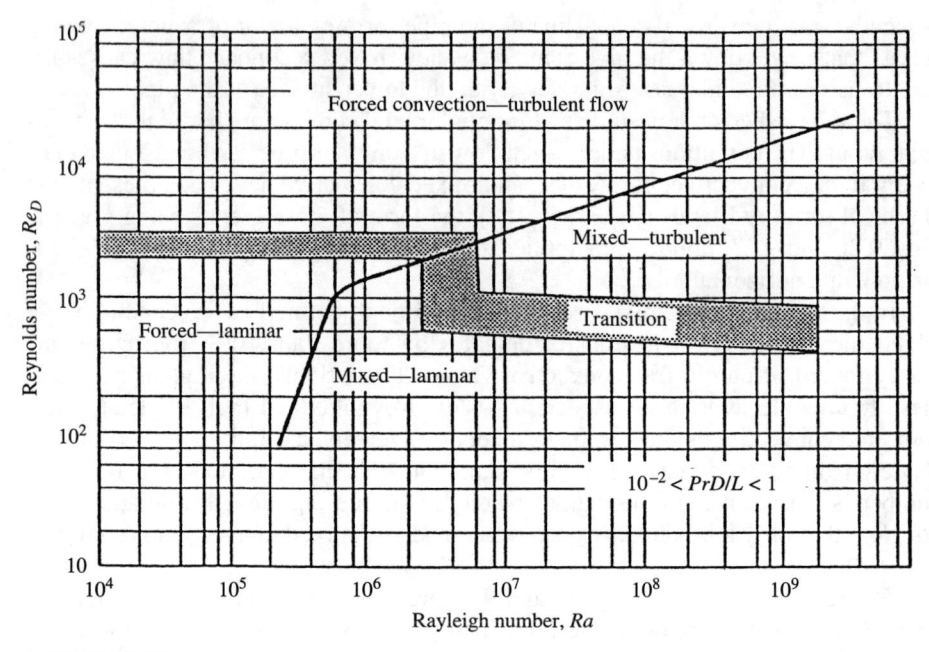

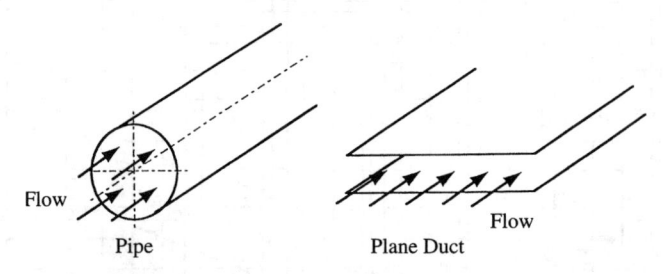

FIGURE 9.28
Flow in a plane duct and a pipe.

flows. A discussion of the equations that have been used in most analysis of internal flows will therefore now be presented. Attention will be restricted to flow in a plane duct and flow in a pipe, these being shown in Fig. 9.28.

9.11
FULLY DEVELOPED MIXED CONVECTIVE FLOW IN A VERTICAL PLANE CHANNEL

In order to illustrate some of the characteristics of internal mixed convective flows, consider internal mixed convection between two parallel plates held at different uniform temperatures as shown in Fig. 9.29.

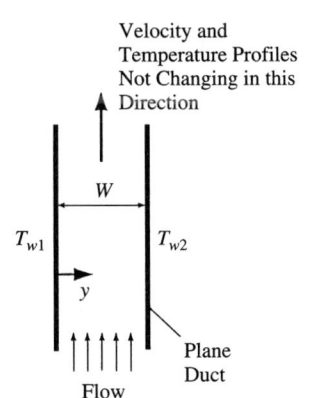

FIGURE 9.29
Flow in a differentially heated plane duct.

In a sufficiently long channel, the velocity and temperature profiles will cease to change with distance along the channel, i.e., a fully developed flow will exist, see [68] to [91]. Defining:

$$\theta = \frac{T - T_0}{T_{w1} - T_0}, U = \frac{u}{u_m}, Y = \frac{y}{W} \qquad (9.102)$$

where u_m is the mean velocity across the duct, T_0 is a reference temperature typically taken as the fluid temperature at the inlet to the channel, and T_{w1} is the temperature of one of the walls of the channel; then in fully developed flow:

$$U = F(Y), \theta = G(Y) \qquad (9.103)$$

with F and G being functions that do not vary with distance, z, along the duct. Because the mass flow rate is constant, the fact that F is not changing with z implies that the value of u at any y is not changing with z. As a consequence of this it follows that in fully developed flow, v is everywhere 0 which means that the pressure will be the same everywhere across the flow. Further, since the function G is not varying with z and since the wall temperatures T_{w1} and T_{w2} are also not varying with z, it follows that the value of T at any y is not changing with z.

Using the above results, it follows that the equations governing the flow are, using the Buossinesq approximation and ignoring viscous dissipation:

$$0 = -\frac{dp}{dz} + \mu \frac{\partial^2 u}{\partial y^2} \pm \beta g \rho (T - T_0) \qquad (9.104)$$

$$0 = \left(\frac{k}{\rho c_p}\right) \frac{\partial^2 T}{\partial y^2}, \text{ i.e., } \frac{\partial^2 T}{\partial y^2} = 0 \qquad (9.105)$$

The pressure, p, is thus being measured relative to that which would exist at the same elevation in the stagnant fluid if it were at a uniform temperature of T_0. The positive sign in front of the buoyancy term applies to buoyancy-assisted flow and the negative sign again applies to buoyancy-opposed flow.

468 Introduction to Convective Heat Transfer Analysis

The boundary conditions on these equations are:

$$\begin{aligned} \text{When } y &= 0, u = 0 \\ \text{When } y &= W, u = 0 \\ \text{When } y &= 0, T = T_{w1} \\ \text{When } y &= W, T = T_{w2} \end{aligned} \quad (9.106)$$

Eqs. (9.104) and (9.105) can be written as:

$$0 = -\frac{dP}{dZ} + \frac{\partial^2 U}{\partial Y^2} \pm \frac{Gr}{Re}\theta \quad (9.107)$$

and:

$$\frac{\partial^2 \theta}{\partial Y^2} = 0 \quad (9.108)$$

where:

$$Z = \frac{z}{ReW}, \quad P = \frac{p}{\rho u_1^2}$$

$$Gr = \frac{\beta g(T_{w1} - T_0)W^3}{\nu^2}, \quad Re = \frac{u_m W}{\nu} \quad (9.109)$$

In terms of the dimensionless variables introduced above, the boundary conditions on the above two equations are:

$$\begin{aligned} \text{When } Y &= 0, U = 0 \\ \text{When } Y &= 1, U = 0 \\ \text{When } Y &= 0, \theta = 1 \\ \text{When } Y &= 1, \theta = r_T \end{aligned} \quad (9.110)$$

where:

$$r_T = \frac{T_{w2} - T_0}{T_{w1} - T_0}$$

Eq. (9.108) indicates that the dimensionless temperature varies linearly across the duct. Integrating Eq. (9.108) and applying the boundary conditions gives:

$$\theta = 1 + (r_T - 1)Y \quad (9.111)$$

This equation can alternatively be written in more conventional form as:

$$\frac{T - T_0}{T_{w1} - T_0} = 1 + \left[\frac{T_{w2} - T_0}{T_{w1} - T_0} - 1\right]Y$$

i.e.:

$$\frac{T - T_{w1}}{T_{w2} - T_{w1}} = Y \quad (9.112)$$

CHAPTER 9: Combined Convection 469

This is the same temperature distribution as that which would exist with pure conduction across the channel and with purely forced convection, i.e., the buoyancy forces have no effect on the temperature distribution and thus on the heat transfer rate in the situation considered.

Substituting Eq. (9.111) into the momentum equation, Eq. (9.107) gives:

$$0 = -\frac{dP}{dZ} + \frac{\partial^2 U}{\partial Y^2} \pm \frac{Gr}{Re}[1 + (r_T - 1)Y]$$

i.e.:

$$\frac{\partial^2 U}{\partial Y^2} = \frac{dP}{dZ} \mp \frac{Gr}{Re}[1 + (r_T - 1)Y] \tag{9.113}$$

Integrating this equation gives:

$$U = \frac{dP}{dZ}\left(\frac{Y^2}{2}\right) \mp \frac{Gr}{Re}\left[\frac{Y^2}{2} + (r_T - 1)\frac{Y^3}{6}\right] + C_1 Y + C_2 \tag{9.114}$$

where C_1 and C_2 are constants of integration.

Applying the boundary conditions gives:

$$C_2 = 0, \quad C_1 = -\frac{dP}{dZ}\left(\frac{1}{2}\right) \pm \frac{Gr}{Re}\left[\frac{1}{2} + (r_T - 1)\frac{1}{6}\right] \tag{9.115}$$

Substituting Eq. (9.115) into Eq. (9.114) then gives:

$$U = \frac{dP}{dZ}\left(\frac{Y^2}{2}\right) \mp \frac{Gr}{Re}\left[\frac{Y^2}{2} + (r_{T-1})\frac{Y^3}{6}\right] - \frac{dP}{dZ}\left(\frac{1}{2}\right) \mp \frac{Gr}{Re}\left[\frac{1}{2} + (r_T - 1)\frac{1}{6}\right]Y$$

i.e.:

$$U = \frac{dP}{dZ}\left(\frac{Y - Y^2}{2}\right) \mp \frac{Gr}{Re}\left[\left(\frac{Y^2 - Y}{2}\right) + (r_{T-1})\left(\frac{Y^3 - Y}{6}\right)\right] \tag{9.116}$$

But, because:

$$u_m W = \int_0^W u\, dy$$

it follows that:

$$\int_0^1 U\, dY = 1$$

Substituting Eq. (9.116) into this equation then gives:

$$-\frac{dP}{dZ}\left(\frac{1}{12}\right) \mp \frac{Gr}{Re}\left[-\frac{1}{12} - (r_T - 1)\frac{1}{24}\right] = 1$$

i.e.:

$$-\frac{dP}{dZ} = 12 \mp \frac{Gr}{Re}\left[1 + \frac{(r_T - 1)}{2}\right] \tag{9.117}$$

Substituting Eq. (9.117) back into Eq. (9.116) then gives:

$$U = \left\{12 \mp \frac{Gr}{Re}\left[1 + (r_T - 1)\frac{1}{2}\right]\right\}\left(\frac{Y - Y^2}{2}\right) \pm \frac{Gr}{Re}\left[\left(\frac{Y^2 - Y}{2}\right) + (r_T - 1)\left(\frac{Y^3 - Y}{6}\right)\right]$$

i.e.:

$$U = 6(Y - Y^2) \pm \frac{Gr}{Re}(r_T - 1)\left(\frac{1}{12}\right)(Y - 3Y^2 + 2Y^3) \tag{9.118}$$

If the buoyancy forces are negligible, i.e., if $Gr/Re = 0$, the velocity distribution is given by:

$$U = 6(Y - Y^2)$$

with the maximum dimensionless velocity occurring at $Y = 0.5$ and being equal to 1.5.

It will be seen from Eq. (9.118) that the effect of the buoyancy forces on the velocity profile is characterized by the parameter:

$$\frac{Gr}{Re}(r_T - 1) = \frac{\beta g(T_{w1} - T_0)W^3/\nu^2}{u_m W/\nu}\left[\frac{T_{w2} - T_0}{T_{w1} - T_0} - 1\right]$$

This buoyancy force effect parameter can be written as:

$$\frac{\beta g(T_{w1} - T_{w2})W^3/\nu^2}{(u_m W/\nu)} = \frac{Gr_T}{Re} \tag{9.119}$$

where:

$$Gr_T = \frac{\beta g(T_{w1} - T_{w2})W^3}{\nu^2} \tag{9.120}$$

is the Grashof number based on $T_{w1} - T_{w2}$. The reference temperature, T_0, therefore has no effect on the form of the velocity profile, i.e., as is to be expected, it is only the buoyancy forces that arise due to temperature differences across the flow that have an effect on the velocity profile which can be written as:

$$U = 6(Y - Y^2) \pm \frac{Gr_T}{Re}\left(\frac{1}{12}\right)(Y - 3Y^2 + 2Y^3) \tag{9.121}$$

The variation of U with Y for various values of Gr_T/Re is shown in Fig. 9.30. Results are only given in this figure for assisting flow, i.e., for the $-$ sign on the buoyancy term in Eq. (9.121).

It will be seen from Fig. 9.30 that the buoyancy forces increase the velocity near the hotter wall (at $Y = 1$). Since the total mass flow rate is fixed, the increase in velocity near the hot wall is associated with a decrease in velocity near the cooler wall (at $Y = 0$). As the parameter Gr_T/Re increases, the velocity profiles become increasingly distorted and at high values of Gr_T/Re flow reversal can occur adjacent to the cooler wall, i.e., a downward flow can occur near the cooler wall. The condition under which such a reverse flow occurs can be deduced by considering the shear

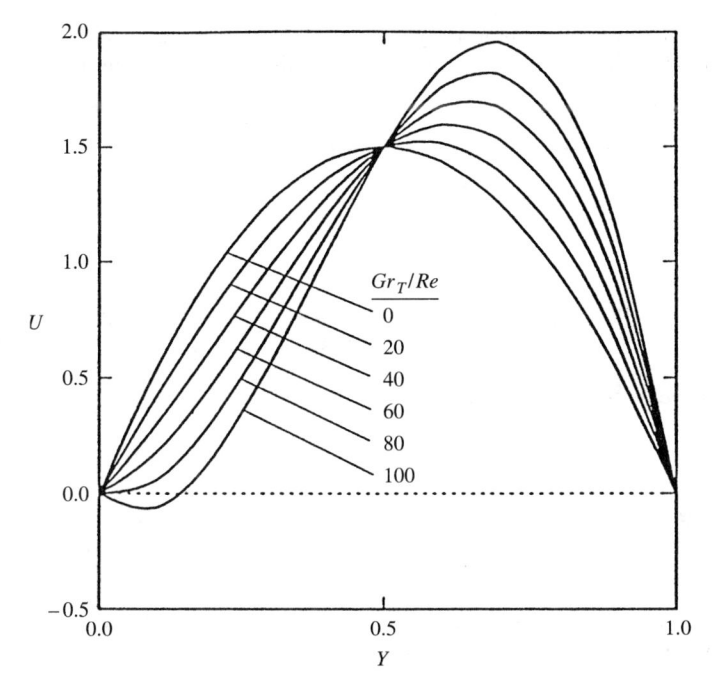

FIGURE 9.30
Velocity profiles in fully developed mixed convective flow in a vertical plane channel. Results are for assisting flow.

stress at $Y = 0$. Now the shear stress on the cold wall is given by:

$$\tau_C = \mu \left.\frac{\partial u}{\partial y}\right|_{y=W} = \frac{\mu u_m}{W} \left.\frac{\partial U}{\partial Y}\right|_{Y=1}$$

Using Eq. (9.121), this gives:

$$\frac{\tau_C}{\rho u_m^2} \frac{u_m W}{\nu} = 6(1 - 2Y) \mp \frac{Gr_T}{Re}\left(\frac{1}{12}\right)(1 - 6Y + 6Y^2)\Big|_{Y=1} = 6 \mp \left(\frac{1}{12}\right)\frac{Gr_T}{Re}$$

(9.122)

This equation indicates that in assisting flow this shear stress will be zero when:

$$\frac{Gr_T}{Re} = 72 \tag{9.123}$$

i.e., flow reversal will occur at the cooler wall if $Gr_T/Re > 72$.

The above results were for assisting flow, i.e., where the buoyancy forces act in the direction of the forced flow through the duct. In opposing flow, the same results are obtained except that the velocity decreases, and ultimately flow reversal, occurs at $Y = 1$ instead of at $Y = 0$.

EXAMPLE 9.6. Air flows vertically upward at a mean velocity of 1.5 m/s through a vertical plane channel in which the distance between the the two walls is 3 cm. The air enters the channel at a temperature of 10°C. The walls of the channel are at temperatures of 20°C and 40°C. Assuming fully developed flow, plot the velocity and temperature distribution in the pipe and find the pressure gradient in the channel.

Solution. The mean air temperature of the air in the flow is $(20 + 40)/2 = 30°C$. At this temperature for air at standard ambient pressure:

$$\beta = \frac{1}{303} \text{ K}^{-1}, \nu = 16.01 \times 10^{-6} \text{ m}^2/\text{s}, \rho = 1.164 \text{ kg/m}^3$$

Hence:

$$\frac{Gr_T}{Re} = \frac{\beta g(T_{w1} - T_{w2})W^2}{u_m \nu} = \frac{(1/303) \times 9.81 \times (40 - 20) \times 0.03^2}{1.5 \times 16.01 \times 10^{-6}} = 24.26$$

The velocity profile is then given by Eq. (9.121)

$$U = 6(Y - Y^2) - 24.26\left(\frac{1}{12}\right)(Y - 3Y^2 + 2Y^3)$$

where:

$$Y = \frac{y}{0.03}, U = \frac{u}{1.5}$$

where y is in m and u is in m/s and y is measured from the cooler wall.

The temperature distribution is given by Eq. (9.112) as:

$$\frac{T - T_{w1}}{T_{w2} - T_{w1}} = Y$$

i.e.:

$$\frac{T - 20}{40 - 20} = Y$$

i.e.:

$$T = 20 + 20Y$$

where T is in °C. The variations of u and T with y as given by these equations are shown in Figs. E9.6a and E9.6b.

The pressure gradient in the fully developed flow is given by Eq. (9.117) as:

$$-\frac{dP}{dZ} = 12 \mp \frac{Gr}{Re}\left[1 + \frac{(r_T - 1)}{2}\right]$$

But:

$$\frac{Gr}{Re} = \frac{\beta g(T_{w1} - T_0)W^2}{u_m \nu} = \frac{(1/303) \times 9.81 \times (20 - 10) \times 0.03^2}{1.5 \times 16.01 \times 10^{-6}} = 12.13$$

and:

$$r_T = \frac{T_{w2} - T_0}{T_{w1} - T_0} = \frac{40 - 10}{20 - 10} = 3$$

Hence, for the situation being considered:

$$-\frac{dP}{dZ} = 12 - 12.13\left[1 + \frac{(3 - 1)}{2}\right] = -12.26$$

FIGURE E9.6a

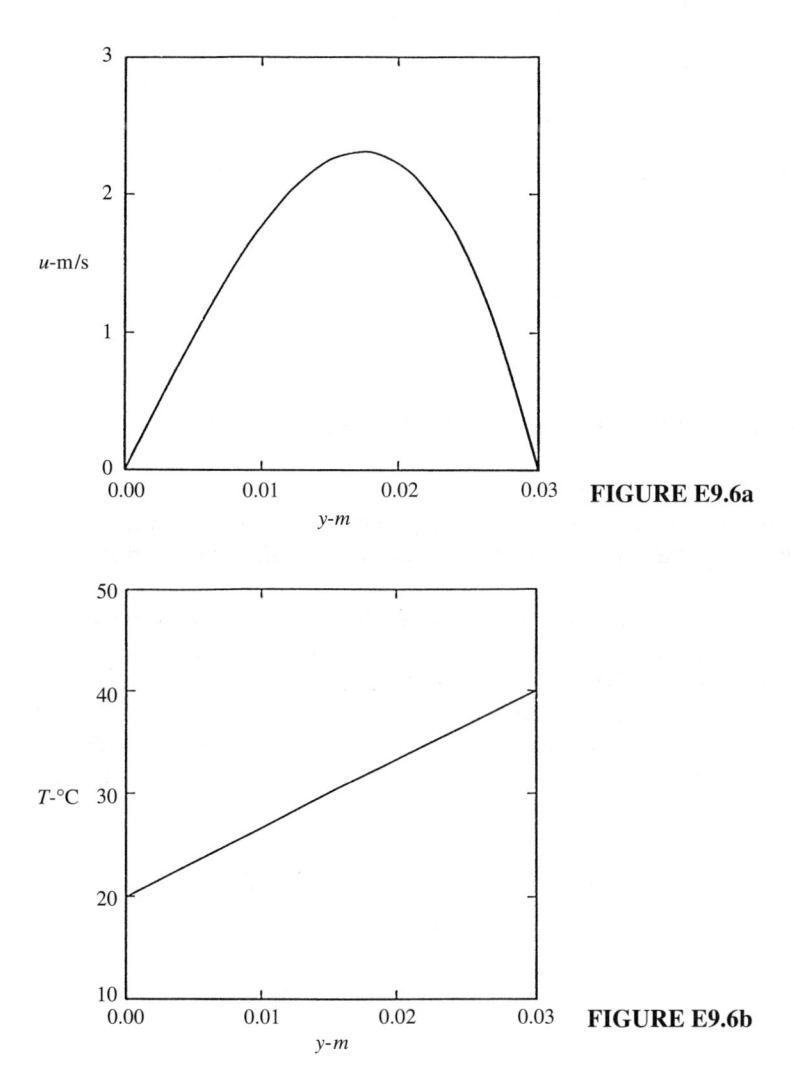

FIGURE E9.6b

Therefore, the pressure is increasing with distance along the duct as a result of the buoyancy forces. But:

$$Z = \frac{z}{ReW}, \quad P = \frac{p}{\rho u_1^2}$$

i.e.:

$$Z = \frac{z}{u_m W^2/\nu} = \frac{z \times 16.01 \times 10^{-6}}{1.5 \times 0.03^2} = 0.01186z$$

and:

$$P = \frac{p}{1.164 \times 1.5^2} = 0.3818p$$

Hence:

$$\frac{dP}{dZ} = 32.19 \frac{dp}{dz}$$

Using the above results then gives:

$$\frac{dp}{dz} = +0.380 \text{ Pa/m}$$

Therefore, the pressure gradient is +0.3809 Pa/m.

9.12
MIXED CONVECTIVE FLOW IN A HORIZONTAL DUCT

With mixed convective flow in a horizontal pipe the buoyancy forces act at right angles to the direction of forced flow leading to the generation of a secondary motion as discussed earlier. The equations governing this type of flow will be briefly discussed in this section.

Attention will be restricted to fully developed flow, i.e., to flow in which all the flow variables except temperature are not changing with distance, z, along the pipe. It will also be assumed that the wall heat flux is axially constant and the wall temperature is constant around the periphery although it of course varies with axial distance. Using the coordinate system shown in Fig. 9.31, the equations governing the flow are, if the Boussinesq approximation is adopted and if viscous dissipation

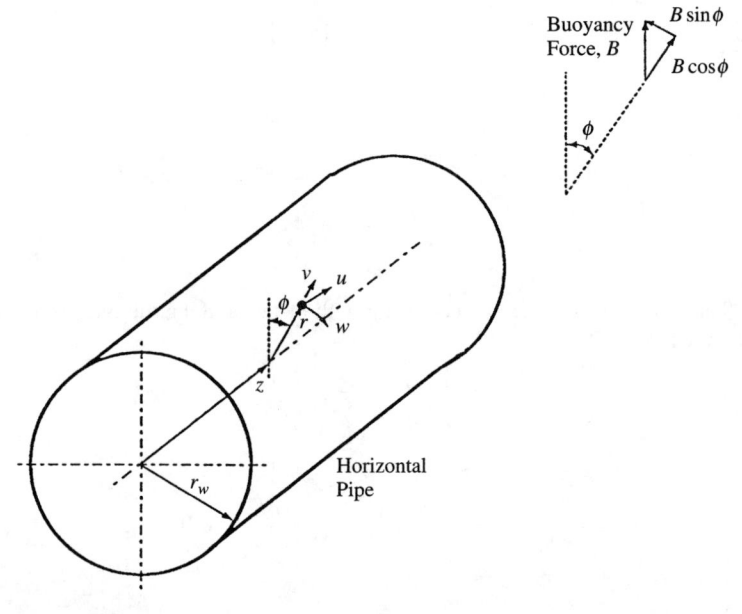

FIGURE 9.31
Coordinate system used.

is neglected:
$$\frac{1}{r}\frac{\partial}{\partial r}(vr) + \frac{\partial w}{\partial \phi} = 0 \tag{9.124}$$

$$v\frac{\partial v}{\partial r} + \frac{w}{r}\frac{\partial v}{\partial \phi} - \frac{w^2}{r} = -\frac{1}{\rho}\frac{\partial p}{\partial r} + \nu\left(\frac{\partial^2 v}{\partial r^2} + \frac{1}{r}\frac{\partial v}{\partial r} + \frac{1}{r^2}\frac{\partial^2 v}{\partial \phi^2} - \frac{v}{r^2} + \frac{2}{r^2}\frac{\partial w}{\partial \phi}\right)$$
$$- \beta g(T_w - T)\cos\phi \tag{9.125}$$

$$v\frac{\partial w}{\partial r} + \frac{w}{r}\frac{\partial w}{\partial \phi} - \frac{vw}{r} = -\frac{1}{\rho r}\frac{\partial p}{\partial \phi} + \nu\left(\frac{\partial^2 w}{\partial r^2} + \frac{1}{r}\frac{\partial w}{\partial r} + \frac{1}{r^2}\frac{\partial^2 w}{\partial \phi^2} - \frac{w}{r^2} + \frac{2}{r^2}\frac{\partial v}{\partial \phi}\right)$$
$$+ \beta g(T_w - T)\sin\phi \tag{9.126}$$

$$v\frac{\partial u}{\partial r} + \frac{w}{r}\frac{\partial u}{\partial \phi} = -\frac{1}{\rho}\frac{\partial p}{\partial z} + \nu\left(\frac{\partial^2 u}{\partial r^2} + \frac{1}{r}\frac{\partial u}{\partial r} + \frac{1}{r^2}\frac{\partial^2 u}{\partial \phi^2}\right) \tag{9.127}$$

$$u\frac{\partial T}{\partial z} + v\frac{\partial T}{\partial r} + \frac{w}{r}\frac{\partial T}{\partial \phi} = \left(\frac{\nu}{\text{Pr}}\right)\left(\frac{\partial^2 T}{\partial r^2} + \frac{1}{r}\frac{\partial T}{\partial r} + \frac{1}{r^2}\frac{\partial^2 T}{\partial \phi^2}\right) \tag{9.128}$$

These equations express conservation of mass, conservation of momentum in the r, ϕ, and z directions, and conservation of energy, respectively. The reference temperature, T_w, is taken as the temperature of the wall of the pipe at the particular value of z being considered.

The pressure can be expressed as:
$$p = P(z) + F(r, \phi)$$
and Eqs. (9.125), (9.126), and (9.127) can then be written as:

$$v\frac{\partial v}{\partial r} + \frac{w}{r}\frac{\partial v}{\partial \phi} - \frac{w^2}{r} = -\frac{1}{\rho}\frac{\partial F}{\partial r}\nu\left(\frac{\partial^2 v}{\partial r^2} + +\frac{1}{r}\frac{\partial v}{\partial r} + \frac{1}{r^2}\frac{\partial^2 v}{\partial \phi^2} - \frac{v}{r^2} + \frac{2}{r^2}\frac{\partial w}{\partial \phi}\right)$$
$$- \beta g(T_w - T)\cos\phi \tag{9.129}$$

$$v\frac{\partial w}{\partial r} + \frac{w}{r}\frac{\partial w}{\partial \phi} - \frac{vw}{r} = -\frac{1}{\rho r}\frac{\partial F}{\partial \phi} + \nu\left(\frac{\partial^2 w}{\partial r^2} + \frac{1}{r}\frac{\partial w}{\partial r} + \frac{1}{r^2}\frac{\partial^2 w}{\partial \phi^2} - \frac{w}{r^2} + \frac{2}{r^2}\frac{\partial v}{\partial \phi}\right)$$
$$+ \beta g(T_w - T)\sin\phi \tag{9.130}$$

$$v\frac{\partial u}{\partial r} + \frac{w}{r}\frac{\partial u}{\partial \phi} = -\frac{1}{\rho}\frac{dP}{dz} + \nu\left(\frac{\partial^2 u}{\partial r^2} + +\frac{1}{r}\frac{\partial u}{\partial r} + \frac{1}{r^2}\frac{\partial^2 u}{\partial \phi^2}\right) \tag{9.131}$$

A dimensionless temperature defined as follows is next introduced:
$$\theta = \frac{T_w - T}{T_w - T_m} \tag{9.132}$$

where T_m is the mean temperature at the particular value of z being considered. Because fully developed flow is being considered, the temperature profile will not be changing with distance along the pipe, i.e.:
$$\theta = \text{function}(r, \phi)$$

476 Introduction to Convective Heat Transfer Analysis

The energy equation can therefore be written as:

$$u\frac{dT_w}{dz} - \theta\frac{d}{dz}(T_w - T_m) + v\frac{\partial\theta}{\partial r} + \frac{w}{r}\frac{\partial\theta}{\partial\phi} = \left(\frac{\nu}{Pr}\right)\left(\frac{\partial^2\theta}{\partial r^2} + \frac{1}{r}\frac{\partial\theta}{\partial r} + \frac{1}{r^2}\frac{\partial^2\theta}{\partial\phi^2}\right)$$
(9.133)

But because a uniform axial heat flux is being considered and because:

$$q_w\frac{\pi D^2}{4} = \rho c_p \int_0^R\int_0^{2\pi} ruT\,d\phi\,dr = \rho c_p \int_0^R\int_0^{2\pi} ru[T_w - \theta(T_w - T_m)]\,d\phi\,dr$$

it follows that T_w is increasing linearly with z and that $T_w - T_m$ is a constant. The energy equation therefore can be written as:

$$u\frac{dT_w}{dz} + v\frac{\partial\theta}{\partial r} + \frac{w}{r}\frac{\partial\theta}{\partial\phi} = \left(\frac{\nu}{Pr}\right)\left(\frac{\partial^2\theta}{\partial r^2} + \frac{1}{r}\frac{\partial\theta}{\partial r} + \frac{1}{r^2}\frac{\partial^2\theta}{\partial\phi^2}\right) \quad (9.134)$$

and the r- and ϕ-momentum equations can be written as:

$$v\frac{\partial v}{\partial r} + \frac{w}{r}\frac{\partial v}{\partial\phi} - \frac{w^2}{r} = -\frac{1}{\rho}\frac{\partial F}{\partial r} + \nu\left(\frac{\partial^2 v}{\partial r^2} + \frac{1}{r}\frac{\partial v}{\partial r} + \frac{1}{r^2}\frac{\partial^2 v}{\partial\phi^2} - \frac{v}{r^2} + \frac{2}{r^2}\frac{\partial w}{\partial\phi}\right)$$
$$- \beta g(T_w - T_m)\theta\cos\phi \quad (9.135)$$

and:

$$v\frac{\partial w}{\partial r} + \frac{w}{r}\frac{\partial w}{\partial\phi} - \frac{vw}{r} = -\frac{1}{\rho r}\frac{\partial F}{\partial\phi} + \nu\left(\frac{\partial^2 w}{\partial r^2} + \frac{1}{r}\frac{\partial w}{\partial r} + \frac{1}{r^2}\frac{\partial^2 w}{\partial\phi^2} - \frac{w}{r^2} + \frac{2}{r^2}\frac{\partial v}{\partial\phi}\right)$$
$$+ \beta g(T_w - T_m)\theta\sin\phi \quad (9.136)$$

In addition, conservation of total mass requires that:

$$\frac{d}{dz}\int_0^R\int_0^{2\pi} ru\,d\phi\,dr = 0 \quad (9.137)$$

Eqs. (9.124), (9.135), (9.136), (9.131), (9.137), and (9.134) can be simultaneously solved to give the solution. For example, the effect of the buoyancy forces can initially be ignored, i.e., v, w, and F can be set equal to zero, Eqs. (9.137) and (9.131) can be solved to give the variation of u with r, and Eq. (9.134) can be used to give the variation of θ with r. These solutions are, of course, the same as those for forced convective flow in a pipe that were discussed in Chapter 4. These solutions can then be used in Eqs. (9.135) and (9.136) together with Eq. (9.124) to solve for first approximations for the variations of v, w, and F with r and ϕ. These results can then be used in Eqs. (9.137) and (9.131) to solve for a second approximation to the variation of u with r and ϕ and in Eq. (9.134) to give a second approximation to the variation of θ with r and ϕ. These second approximations to the variations of u and θ can then be used in Eqs. (9.135), (9.136), and (9.124) to solve for second approximations to the variations of v, w, and F with r and ϕ. The procedure can be repeated until convergence to an acceptable degree is achieved.

Studies of flow in horizontal ducts are discussed in [96] to [112].

9.13
CONCLUDING REMARKS

This chapter has been concerned with flows in which the buoyancy forces that arise due to the temperature difference have an influence on the flow and heat transfer values despite the presence of a forced velocity. In external flows it was shown that the deviation of the heat transfer rate from that which would exist in purely forced convection was dependent on the ratio of the Grashof number to the square of the Reynolds number. It was also shown that in such flows the Nusselt number can often be expressed in terms of the Nusselt numbers that would exist under the same conditions in purely forced and purely free convective flows. It was also shown that in turbulent flows, the buoyancy forces can affect the turbulence structure as well as the momentum balance and that in turbulent flows the heat transfer rate can be decreased by the buoyancy forces in assisting flows whereas in laminar flows the buoyancy forces essentially always increase the heat transfer rate in assisting flow. Some consideration was also given to the effect of buoyancy forces on internal flows.

PROBLEMS

9.1. Water at a temperature of 60°F flows horizontally at a velocity of 8 ft/sec perpendicular to the axis of a 1-inch diameter horizontal tube that is kept at a uniform surface temperature of 80°F. Determine whether the buoyancy forces have an effect on the heat transfer rate.

9.2. Consider flow over a wide 20-cm high vertical plate which is held at a uniform surface temperature of 40°C placed in air at a temperature of 10°C. If a vertical forced air flow is induced over the plate, find the air velocity at which the forced flow starts to affect the mean heat transfer rate from the plate.

9.3. A solution that was accurate to first order in the buoyancy parameter, G_x, for near-forced convective laminar two-dimensional boundary layer flow over an isothermal vertical plate was discussed in this chapter. Derive the equations that would allow a solution that was second order accurate in G_x to be obtained. Clearly state the boundary conditions on the solution.

9.4. A solution that was accurate to first order in $1/G_x^{0.5}$ for near free convective laminar two-dimensional boundary layer flow over an isothermal vertical plate was discussed in this chapter. Derive the equations that would allow a solution that was second order accurate in $1/G_x^{0.5}$, i.e., accurate to order $(1/G_x^{0.5})^2$ to be obtained. Clearly state the boundary conditions on the solution.

9.5. Air at a temperature of 10°C flows upward at a velocity of 0.8 m/s over a wide vertical 15-cm high flat plate which is maintained at a uniform surface temperature of 50°C. Plot the variation of the local heat transfer rate with distance along the plate from the leading edge. Also show the variations that would exist in purely forced and purely free convective flow.

9.6. Consider water flow over a vertical flat plate with a height of 30 cm and at a velocity that is such that the Reynolds number based on the height of the plate is 5×10^4. If the water temperature is 15°C and if the plate temperature is 35°C, can the effect of the buoyancy forces on the mean heat transfer rate be neglected? What is the mean heat transfer rate from the plate?

9.7. The side of a small laboratory furnace can be idealized as a vertical plate 0.6 m high and 2.5 m wide. The furnace sides are at 40°C and the surrounding air is at 25°C. If air is blown vertically over the side of the furnace, estimate the lowest forced air velocity that would cause the heat-transfer coefficient to depart noticeably from its natural convection value.

9.8. Air at a temperature of 30°C flows at a velocity of 1 m/s vertically downwards over a wide vertical flat plate which is held at a uniform surface temperature of 10°C. Plot the variations of the velocity and temperature in the boundary layer at a distance of 20 cm from the leading edge of the plate. Also plot the variations that would exist if the buoyancy force effects were negligible.

9.9. Air at a temperature of 10°C flows upward over a 0.25 m high vertical plate which is kept at a uniform surface temperature of 40°C. Plot the variation of the velocity boundary layer thickness and local heat transfer rate along the plate for air velocities of between 0.2 and 1.5 m/s. Assume two-dimensional flow.

9.10. Air at a temperature of 10°C flows vertically upwards over a 0.1 m high vertical plate whose surface temperature increases linearly from 10°C to 40°C with distance along the plate. Numerically determine how the heat transfer rate varies along the plate for various forced velocities between that which gives effectively forced convection and that which gives effectively free convection. Assume two-dimensional flow.

9.11. Consider mixed convective laminar boundary layer flow over a horizontal flat plate that is heated to a uniform surface temperature. In such a flow there will be a pressure change across the boundary induced by the buoyancy forces, i.e.:

$$-\frac{\partial p}{\partial y} = \beta g \rho (T - T_\infty)$$

where T_∞ is the uniform temperature outside the boundary layer. Integrating this equation gives, because the pressure is expressed relative to the uniform pressure outside the boundary layer:

$$p = \int_0^\delta \beta g \rho (T - T_\infty) dy - \int_0^y \beta g \rho (T - T_\infty) dy$$

Using this equation, derive the boundary layer momentum integral equation for this type of flow.

9.12. Consider two-dimensional air flow over a square cylinder. How does the Nusselt number vary with Reynolds number at a fixed Grashof number of 10,000?

9.13. Air at a temperature of 10°C flows over a 30-mm diameter horizontal cylinder which has a uniform surface temperature of 50°C. The air flow is at right angles to the axis of the cylinder and in the vertically upwards direction. Determine the air velocity above which the flow can be assumed to be purely forced convective and below which it can be assumed to be purely free convective.

9.14. Consider assisting air flow normal to the axis of a horizontal circular cylinder. Using the equation of the form:

$$Nu^n = Nu_N^n + Nu_F^n$$

that was discussed in this chapter, together with standard equations for Nu_N and Nu_F, derive equations that define when the flow can be assumed to be purely forced convective and when it can be assumed to be purely free convective. Using these equations, plot the variations of Grashof number with Reynolds number that define these limits.

9.15. In forced convective turbulent boundary layer flow over a plate the velocity and temperature distributions are approximately given by:

$$\frac{u}{u_1} = \left[\frac{y}{\delta}\right]^{\frac{1}{n}}$$

and:

$$\frac{\overline{T} - T_1}{T_w - T_1} = 1 - \left[\frac{y}{\delta}\right]^{\frac{1}{n}}$$

where u_1 is the free-stream velocity, T_1 is the free-stream temperature, T_w is the wall temperature, and δ is the local boundary layer thickness, the thermal and velocity boundary layers being assumed to have the same thickness, and n being a constant with a value of approximately between 5 and 7. Derive expressions for the ratio of the turbulent shear stress in assisting mixed convective flow to the turbulent shear in forced convective flow and for the ratio of the turbulent heat transfer rate in mixed convective flow to the turbulent heat transfer rate in forced convective flow.

9.16. The turbulent kinetic energy equation was derived in Chapter 5 using the momentum equations and assuming buoyancy force effects were negligible. Re-derive this equation starting with momentum equations in which the buoyancy terms are retained. Assume a vertically upward flow and use the Boussinesq approximation.

9.17. A numerical procedure for calculating the heat transfer rate with turbulent boundary layer flow was discussed in Chapter 5. This procedure used a mixing length-based turbulence model. Discuss the modifications that must be made to this procedure to apply it to mixed convective flow over a vertical plate.

9.18. Air at mean temperature of 40°C flows through a horizontal pipe that is 1.8 m long with a diameter of 25 mm. The air velocity is such that the Reynolds number is 150. If the wall of the pipe is kept at a uniform temperature of 100°C determine if the flow can be assumed to be purely forced convective.

9.19. Air at a mean temperature of 60°C flows vertically upward through a 20-cm diameter pipe that is 5 m long. The wall of the pipe is kept at a uniform temperature of 30°C. Estimate the flow velocity at which buoyancy force effects will become important.

9.20. Air flows vertically upward at a mean velocity of 1 m/s through a vertical plane channel whose walls are temperatures of 30°C and 40°C, the distance between the walls being 4 cm. The air enters the channel at a temperature of 15°C. Plot the velocity and temperature distribution in the channel assuming fully developed flow.

9.21. Derive an expression for the velocity distribution in fully developed mixed convective flow in a vertical annulus. The inner and outer surfaces have diameters of D_i and D_o respectively and are kept at uniform temperatures of T_{wi} and T_{wo} respectively.

9.22. Consider developing flow in a vertical wide channel when there is the same uniform heat flux, q_w, applied at each wall and where the flow enters the channel at a temperature of T_i. Write out the governing equations, clearly stating the assumptions on which these equations are based. Express the governing equations in dimensionless form, defining the dimensionless temperature as $(T - T_i)k/q_w W$ where W is the width of the channel. Discuss how these equations can be numerically solved.

REFERENCES

1. Kakac, S., Shah, R.K., and Aung, W., Eds., *"Handbook of Single-Phase Convective Heat Transfer"*, Wiley, New York, 1987.
2. Sparrow, E.M. and Gregg, J.L., "Buoyancy Effects in Forced Convection Flow and Heat Transfer", *Trans. ASME, J. Appl. Mech.*, Sect. E, Vol. 81, pp. 133–135, 1959.
3. Szewczyk, A.A., "Combined Forced and Free-Convection Laminar Flow", *J. Heat Transfer*, Vol. 86, pp. 501–507, 1964.
4. Lloyd, J.R. and Sparrow, E.M., "Combined Forced and Free Convection Flow on Vertical Surfaces", *Int. J. Heat Mass Transfer*, Vol. 13, pp. 434–438, 1970.
5. Merkin, J.H., "The Effect of Buoyancy Forces on the Boundary-Layer Flow over a Semi-Infinite Vertical Flat Plate in a Uniform Free Stream", *J. Fluid Mech.*, Vol. 35, pp. 439–450, 1969.
6. Wickern, G., "Mixed Convection from an Arbitrarily Inclined Semi-Infinite Flat Plate", *Int. J. of Heat and Mass Transfer*, Vol. 34, pp. 1935–57, 1991.
7. Yao, L.S., "Two-Dimensional Mixed Convection Along a Flat Plate", *J. of Heat Transfer*, Vol. 109, pp. 440–445, 1987.
8. Ramachandran, N., Armaly, B.F., and Chen, T.S., "Measurements and Predictions of Laminar Mixed Convection Flow Adjacent to a Vertical Surface", *J. Heat Transfer*, Vol. 107, pp. 636–641, 1985.
9. Oosthuizen, P.H. and Bassey, M., "An Experimental Study of Combined Forced and Free Convective Heat Transfer from Flat Plates to Air at Low Reynolds Numbers", *J. Heat Transfer*, Vol. 96, pp. 120–121, 1973.
10. Manning, K.S. and Qureshi, Z.H., "Transport Correlations for Laminar Aiding Mixed Convection over a Vertical Isothermal Surface", *J. Heat Transfer*, Vol. 116, pp. 777–80, 1994.
11. Cebeci, T., Broniewski, D., and Joubert, C., "Mixed Convection on a Vertical Flat Plate with Transition and Separation", *J. Heat Transfer*, Vol. 112, Feb. '90, pp. 144–50, 1990.
12. Chen, T.S., Armaly, B.F., and Ramachandran, N., "Correlations for Laminar Mixed Convection Flows on Vertical, Inclined, and Horizontal Flat Plates", *J. Heat Transfer*, Vol. 108, pp. 835–840, 1986.
13. Abu Mulaweh, H.I., Armaly, B.F., and Chen, T.S., "Measurements of Laminar Mixed Convection Adjacent to a Vertical Plate-Uniform Wall Heat Flux Case", *J. Heat Transfer*, Vol. 114, Nov. '92, pp. 1057–1059, 1992.
14. Lin, Hsiao Tsung and Chen, Yao Han, "The Analogy Between Fluid Friction and Heat Transfer of Laminar Mixed Convection Flat Plates", *Int. J. Heat and Mass Transfer*, Vol. 37, No. 11, pp. 1683–1686, 1994.

15. Oosthuizen, P.H., "A Note on the Combined Free and Forced Convective Laminar Flow Over a Vertical Isothermal Plate", *J.S.A. Inst. Mech. Engs.*, Vol. 15, No. 1, August, pp. 8–13, 1965.
16. Oosthuizen, P.H. and Hart, R., "A Numerical Study of Laminar Combined Convective Flow Over Flat Plates", *J. Heat Trans.*, Feb., 1973, pp. 60–63.
17. Mori, Y., "Buoyancy Effects in Forced Laminar Convection Flow over a Horizontal Flat Plate", *J. Heat Transfer*, Sect. C, Vol. 83, pp. 479–482, 1961.
18. Afzal, N. and Hussain, T., "Mixed Convection over a Horizontal Plate", *J. of Heat Transfer*, Vol. 106, Feb. '84, pp. 240–241, 1984.
19. Aldoss, T.K., Jarrah, M.A., and Duwairi, H.M., "Wall Effect on Mixed Convection from Horizontal Surfaces with a Variable Surface Heat Flux", *Canadian J. of Chem. Eng.*, Vol. 72, pp. 35–42, 1994.
20. Lee, H.R., Chen, T.S., and Armaly, B.F., "Nonparallel Thermal Instability of Mixed Convection Flow on Nonisothermal Horizontal and Inclined Flat Plates", *Int. J. Heat and Mass Transfer*, Vol. 35, Aug. '92, pp. 1913–1925, 1992.
21. Risbeck, W.R, Chen, T.S., and Armaly, B.F., "Laminar Mixed Convection on Horizontal Flat Plates with Variable Surface Heat Flux", *Int. J. of Heat and Mass Transfer*, Vol. 37, No. 4, March '94, pp. 699–704, 1994.
22. Risbeck, W.R., Chen, T.S., and Armaly, B.F., "Laminar Mixed Convection over Horizontal Flat Plates with Power-Law Variation in Surface Temperature", *Int. J. of Heat and Mass Transfer*, Vol. 36, No. 7, May '93, pp. 1859–1866, 1993.
23. Steinruck, H., "Mixed Convection over a Cooled Horizontal Plate: Non-uniqueness and Numerical Instabilities of the Boundary-layer Equations", *J. of Fluid Mechanics*, Vol. 278, Nov. 10, '94, pp. 251–265, 1994.
24. Hong, B., Armaly, B.F., and Chen, T.S., "Laminar Mixed Convection in a Duct with a Backward-facing Step: the Effects of Inclination Angle and Prandtl Number", *Int. J. of Heat and Mass Transfer*, Vol. 36, Aug. '93, pp. 3059–3067, 1993.
25. Huang, C.C. and Lin, T.F., "Vortex Flow and Thermal Characteristics in Mixed Convection of Air in a Horizontal Rectangular Duct: Effects of the Reynolds and Grashof Numbers", *Int. J. of Heat and Mass Transfer*, Vol. 38, June '95, pp. 1661–1674, 1995.
26. Kang, B.H, Jaluria, Y., and Tewari, S.S., "Mixed Convection Transport from an Isolated Heat Source Module on a Horizontal Plate", *J. of Heat Transfer*, Vol. 112, Aug. '90, pp. 653–61, 1990.
27. Lei, Q.M. and Trupp, A.C., "Experimental Study of Laminar Mixed Convection in the Entrance Region of a Horizontal Semicircular Duct", *Int. J. of Heat and Mass Transfer*, Vol. 34, Sept. '91, pp. 2361–2372, 1991.
28. Lin, W.L., Ker, Y.T., and Lin, T.F., "Experimental Observation and Conjugated Heat Transfer Analysis of Vortex Flow Development in Mixed Convection of Air in a Horizontal Rectangular Duct", *Int. J. of Heat and Mass Transfer*, Vol. 39, No.17, Nov. '96, pp. 3667–3683, 1996.
29. Lin, J.T., Armaly, B.F., and Chen, T.S., "Mixed Convection Heat Transfer in Inclined Backward-facing Step Flows", *Int. J. of Heat and Mass Transfer*, Vol. 34, June '91, pp. 1568–1571, 1991.
30. Lin, Jenn Nan, Yen, Chen Shyong, and Chou, Fu Chu, "Laminar Mixed Convection in the Entrance Region of Shrouded Arrays of Heated Rectangular Blocks", *Canadian J. of Chem. Eng.*, Vol. 66, June '88, pp. 361–366, 1988.
31. Oosthuizen, P.H., "Numerical Study of Combined Convective Heat Transfer from a Vertical Cylinder in a Horizontal Flow", Paper MC-4, 6th Int. Heat Transfer Conference, Toronto, August 1978, pp. 19–24.

32. Oosthuizen, P.H., "Combined Forced and Free Convective Heat Transfer from a Horizontal Cylinder in an Axial Stream", Proc. 3rd Int. Conf. Num. Methods in Thermal Problems, Vol. 3, Pineridge Press, Swansea, U.K., pp. 529–539, 1983.
33. Oosthuizen, P.H., "Mixed Convective Heat Transfer in a Plane Duct Containing Multiple Heated Plates", ASME Proc. 1988 National Heat Transfer Conference, HTD-96, Vol. 2, Houston, TX, July 24–27, 1988, pp. 79–85.
34. Oosthuizen, P.H., "Mixed Convective Heat Transfer From Inclined Circular Cylinders", Experimental Heat Transfer, Fluid Mechanics, and Thermodynamics 1989, Proc. First World Conference on Experimental Heat Transfer, Fluid Mechanics and Thermodynamics, Dubrovnik, Yugoslavia, Sept. 1988, pp. 200–207.
35. Oosthuizen, P.H. and de Champlain, A., "Combined Convective Heat Transfer in an Inclined Cavity with a Heated Rectangular Element on the Wall", Paper AIAA-88-2620, presented at the AIAA Thermophysics, Plasma-dynamics and Lasers Conference, San Antonio, TX, June 27–29, 1988.
36. Oosthuizen, P.H. and de Champlain, A., "Mixed Convective Heat Transfer in an Inclined Cavity with Multiple Heated Elements on One Wall", Paper AIAA-89-1687, AIAA 24th Thermophysics Conference, Buffalo, NY, June 12–14, 1989.
37. Oosthuizen, P.H. and de Champlain, A., "A Numerical and Experimental Study of Laminar Mixed Convective Flow Through an Enclosure with a Heated Element on One Wall", Computational Methods and Experimental Measurements V, July 23–26, 1991, Montreal, pp. 177–188.
38. Oosthuizen, P.H. and Leung, R.K., "Combined Convective Heat Transfer from Vertical Cylinders in a Horizontal Air-Stream", *Trans. CSME*, Vol. 5, No. 2, pp. 115–117, 1978.
39. Oosthuizen, P.H. and Taralis, D.N., "Combined Convective Heat Transfer from Vertical Cylinders in a Horizontal Fluid Flow", ASME Paper 76-HT-41, ASME-AIChE Heat Trans. Conf., St. Louis, MO, Aug. 1976.
40. Papanicolaou, E. and Jaluria, Y., "Mixed Convection from Simulated Electronic Components at Varying Relative Positions in a Cavity", *J. Heat Transfer*, Vol. 116, pp. 960–970, 1994.
41. Sankar, R., Mees, P.A.J., and Nandakumar, K., "Development of Three-Dimensional, Streamwise-Periodic Flows in Mixed-Convection Heat Transfer", *J. Fluid Mechanics*, Vol. 255, pp. 683–705, 1993.
42. Subhashis, R. and Srinivasan, J., "Analysis of Conjugate Laminar Mixed Convection Cooling in a Shrouded Array of Electronic Components", *Int. J. Heat and Mass Transfer*, Vol. 35, April '92, pp. 815–822, 1992.
43. Yan,-Wei-Mon, "Transport Phenomena of Developing Laminar Mixed Convection Heat and Mass Transfer in Inclined Rectangular Ducts", *Int. J. Heat and Mass Transfer*, Vol. 38, Oct. '95, pp. 2905–2914, 1995.
44. Churchill, S.W., "A Comprehensive Correlation Equation for Laminar, Assisting Forced and Free Convection", *AIChE J.*, Vol. 23, pp. 10–16, 1977.
45. Morgan, V.T., "The Overall Convective Heat Transfer from Smooth Circular Cylinders", *Adv. Heat Transfer*, Vol. 11, p. 199, 1975.
46. Oosthuizen, P.H. and Barnes, G.B., "Experimental Study of Combined Convective Heat Transfer from Oblong Cylinders to Air", ASME Paper 76-HT-40, ASME-AIChE Heat Trans. Conf., St. Louis, MO, Aug. 1976.
47. Oosthuizen, P.H. and Bishop, M., "An Experimental Study of Mixed Convective Heat Transfer from Square Cylinders", Paper No. AIAA-87-1592, presented at the AIAA 22nd Thermophysics Conf., Honolulu, June 8–10, 1987.
48. Oosthuizen, P.H. and Madan, S., "Combined Convective Heat Transfer from Horizontal Cylinders in Air", *J. Heat Transfer*, Vol. 92, 1970, pp. 194–196, 1970.

49. Oosthuizen, P.H. and Madan, S., "The Effect of Flow Direction on Combined Convective Heat Transfer from Cylinders to Air", *J. Heat Transfer,* May, pp. 240–242, 1971.
50. Oosthuizen, P.H. and Rangarajan, N., "Experimental Study of Combined Convective Heat Transfer from Horizontal Cylinders in an Axial Flow", *Trans. CSME,* Vol. 6, No. 2, pp. 103–105, 1980.
51. Yuge, T., "Experiments on Heat Transfer from Spheres Including Combined Natural and Forced Convection", *J. Heat Transfer,* Vol. 82, pp. 214–220, 1960.
52. Armaly, B. F., Chen, T.S., and Ramachandran, N., "Correlations for Mixed Convection Flows Across Horizontal Cylinders and Spheres", *J. Heat Transfer,* Vol. 110, pp. 511–514, 1988.
53. Fand, R.M. and Keswani, K.K., "Combined Natural and Forced Convection Heat Transfer from Horizontal Cylinders to Water", *Int. J. Heat Mass Transfer,* Vol. 16, p. 175, 1973.
54. Oosthuizen, P.H., "Laminar Combined Convection from an Isothermal Circular Cylinder to Air", *Trans. Inst. of Chem. Engs.,* Vol. 48, Nos. 7–10, pp. T227–T231, 1970.
55. Oosthuizen, P.H., "Turbulent Combined Convective Flow Over a Vertical Plane Surface", Proc. 5th Int. Heat Transfer Conf., 1974, pp. 129–133.
56. Krishnamurthy, R. and Gebhart, B., "An Experimental Study of Transition to Turbulence in Vertical Mixed Convection Flows", *J. Heat Transfer,* Vol. 111, pp. 121–130, 1989.
57. Chen, T.S, Armaly, B.F., and Ali, M.M., "Turbulent Mixed Convection along a Vertical Plate", *J. Heat Transfer,* Vol. 109, pp. 251–253, 1987.
58. Ramachandran, N., Armaly, B. F., and Chen, T.S., "Turbulent Mixed Convection over an Isothermal Horizontal Flat Plate", *J. Heat Transfer,* Vol. 112, pp. 124–129, 1990.
59. Siebers, D.L., Schwind, R.G., and Moffat, R.J., "Experimental Mixed Convection Heat Transfer from a Large, Vertical Surface in Horizontal Flow", Sandia Rept. SAND 83-8225, Sandia National Laboratories, Albuquerque, NM, 1983.
60. Brewster, R.A. and Gebhart, B., "Instability and Disturbance Amplification in a Mixed-convection Boundary Layer", *J. Fluid Mechanics,* Vol. 229, pp. 115–133, 1991.
61. Eckert, R.G. and Diaguila, A.J., "Convective Heat Transfer for Mixed Free and Forced Flow through Tubes", *Trans. ASME,* Vol. 76, pp. 497–504, 1954.
62. Metais, B. and Eckert, E.R.G., "Forced Mixed and Free Convection Regimes", *J. Heat Transfer,* Vol. 86, pp. 295–296, 1964.
63. Jackson, T.W., Harrison, W.B., and Boteler, W.C., "Combined Free and Forced Convection in a Constant-Temperature Vertical Tube", *Trans. ASME,* Vol. 80, pp. 739–745, 1958.
64. Buhr, H.D., Carr, A.D., and Balzhiser, R.R., "Temperature Profiles in Liquid Metals and the Effects of Superimposed Free Convection in Turbulent Flow", *Int. J. Heat Mass Transfer,* Vol. 11, p. 641, 1968.
65. Buhr, H.A., Horsten, E.A., and Carr, A.D., "The Distribution of Turbulent Velocity and Temperature Profiles on Heating, for Mercury in a Vertical Pipe", Natl. Heat Transfer Conf., Denver, CO, ASME 72-HT-21, 1972.
66. Hallman, T.M., "Combined Forced and Free-Laminar Heat Transfer in a Vertical Tube with Uniform Internal Heat Generation", *Trans. ASME,* Vol. 78, No. 8, pp. 1831–1841, 1956.
67. Scheele, G.F., Rosen, E.M., and Hanratty, T.J., "Effect of Natural Convection on Transition to Turbulence in Vertical Pipes", *Can. J. Chem. Eng.,* Vol. 38, pp. 67–73, 1960.
68. Cebeci, T., Khattab, A.A., and LaMont, R., "Combined Natural and Forced Convection in Vertical Ducts", *J. Heat Transfer,* Vol. 2, pp. 419–424, 1982.
69. Chen, Yen-Cho and Chung, J.N., "The Linear Stability of Mixed Convection in a Vertical Channel Flow", *J. Fluid Mechanics,* Vol. 325, pp. 29–51, 1996.

70. Gau, C., Yih, K.A., and Aung, W., "Reversed Flow Structure and Heat Transfer Measurements for Buoyancy-Assisted Convection in a Heated Vertical Duct", *J. Heat Transfer*, Vol. 114, pp. 928–935, 1992.
71. Hamadah, T.T. and Wirtz, R.A., "Analysis of Laminar Fully Developed Mixed Convection in a Vertical Channel with Opposing Buoyancy", *J. Heat Transfer*, Vol. 113, pp. 507–510, 1991.
72. Joye, D.D., "Correlation for Opposing Flow, Mixed Convection Heat Transfer in a Vertical Tube with Constant Wall Temperature", *J. Heat Transfer*, Vol. 118, pp. 787–789, 1996.
73. Joye, D.D., "Design Criterion for the Heat-Transfer Coefficient in Opposing Flow, Mixed Convection Heat Transfer in a Vertical Tube", *Ind. Eng. Chem. Res.*, Vol. 35, pp. 2399–2403, 1996.
74. Joye, D.D., "Comparison of Correlations and Experiment in Opposing Flow, Mixed Convection Heat Transfer in a Vertical Tube with Grashof Number Variation", *Int. J. Heat and Mass Transfer*, Vol. 39, No. 5, pp. 1033–1038, 1996.
75. Joye, D.D., Bushinsky, J.P., and Saylor, P.E., "Mixed Convection Heat Transfer at High Grashof Number in a Vertical Tube", *Ind. Eng. Chem. Res.*, Vol. 28, pp. 1899–1903, 1989.
76. Lin, Tsing Fa, Chang, Tsai Shou, and Chen, Yu Feng, "Development of Oscillatory Asymmetric Recirculating Flow in Transient Laminar Opposing Mixed Convection in a Symmetrically Heated Vertical Channel", *J. Heat Transfer*, Vol. 115, pp. 342–352, 1993.
77. Maitra, D. and Sabba Raju, K., "Combined Free and Forced Convection Laminar Heat Transfer in a Vertical Annulus", *J. Heat Transfer*, Vol. 97, pp. 135–137, 1975.
78. Marner, W.J. and McMillan, H.K., "Combined Free and Forced Laminar Convection in a Vertical Tube with Constant Wall Temperature", *J. Heat Transfer*, Vol. 92, pp. 559–562, 1970.
79. Rogers, B.B. and Yao, L.S., "Finite-Amplitude Instability of Mixed-Convection in a Heated Vertical Pipe", *Int. J. Heat and Mass Transfer*, Vol. 36, pp. 2305–2315, 1993.
80. Sparrow, E.M., Chrysler, G.M., and Azevedo, L.F., "Observed Flow Reversals and Measured-Predicted Nusselt Numbers for Natural Convection in a One-Sided Heated Vertical Channel", *J. Heat Transfer*, Vol. 106, No. 2, pp. 325–332, 1984.
81. Suslov, S.A and Paolucci, S., "Stability of Mixed-Convection Flow in a Tall Vertical Channel under Non-Boussinesq Conditions", *J. Fluid-Mechanics*, Vol. 302, pp. 91–115, 1995.
82. Velusamy, K. and Garg, V.K. "Laminar Mixed Convection in Vertical Elliptic Ducts", *Int. J. Heat and Mass Transfer*, Vol. 39, No. 4, pp. 745–752, 1996.
83. Vilemas, J.V., Poskas, P.S., and Kaupas, V.E., "Local Heat Transfer in a Vertical Gas-Cooled Tube with Turbulent Mixed Convection and Different Heat Fluxes", *Int. J. Heat and Mass Transfer*, Vol. 35, pp. 2421–2428, 1992.
84. Watson, J.C., Anand, N.K., and Fletcher, L.S., "Mixed Convective Heat Transfer Between a Series of Vertical Parallel Plates with Planar Heat Sources", *J. Heat Transfer*, Vol. 118, pp. 984–990, 1996.
85. Yan, Wei Mon and Tsay, Hsin Chuen, "Mixed Convection Heat and Mass Transfer in Vertical Annuli with Asymmetric Heating", *Int. J. of Heat and Mass Transfer*, Vol. 34, pp. 1309–1313, Apr.-May 1991.
86. Aung, W. and Worku, G., "Theory of Fully Developed Combined Convection Including Flow Reversal", *J. Heat Transfer*, Vol. 108, pp. 485–488, 1986.
87. Aung, Win and Worku, G., "Mixed Convection in Ducts with Asymmetric Wall Heat Fluxes", *J. Heat Transfer*, Vol. 109, pp. 947–951, 1987.

88. Badr, H.M., "Mixed Convection from a Straight Isothermal Tube of Elliptic Cross-Section", *Int. J. Heat and Mass Transfer,* Vol. 37, p. 2343–2365, 1994.
89. Law, Hin Sum, Masliyah, J.H., and Nandakumar, K., "Effect of Nonuniform Heating on Laminar Mixed Convection in Ducts", *J. Heat Transfer,* Vol. 109, pp. 131–137, 1987.
90. Oosthuizen, P.H. and Paul, J.T., "Mixed Convective Flow in an Asymmetrically Heated Plane Duct with Buoyancy Induced Flow Separation", *Mixed Convection Heat Transfer-1987,* ASME HTD-Vol. 84, Am. Soc. Mech. Eng., New York, 1987, pp. 21–28.
91. Yao, L.S. and Rogers, B.B., "Mixed Convection in an Annulus of Large Aspect Ratio", *J. Heat Transfer,* Vol. 111, pp. 683–689, 1989.
92. Aung, W., Moghadam, H.E., and Tsou, F.K., "Simultaneous Hydrodynamic and Thermal Development in Mixed Convection in a Vertical Annulus with Fluid Property Variations", *J. Heat Transfer,* Vol. 113, pp. 926–931, 1991.
93. Wang, M., Tsuji, T., and Nagano, Y., "Mixed Convection with Flow Reversal in the Thermal Entrance Region of Horizontal and Vertical Pipes", *Int. J. Heat Mass Transfer,* Vol. 37, pp. 2305–2319, 1994.
94. Yao, L.S., "Free and Forced Convection in the Entry Region of a Heated Vertical Channel", *Int. J. Heat Mass Transfer,* Vol. 26, No. 1, pp. 65–72, 1983.
95. Zeldin, B. and Schmidt, F.W., "Developing Flow with Combined Forced-Free Convection in an Isothermal Vertical Tube", *J. Heat Transfer,* Vol. 94, pp. 211–223, 1972.
96. Choi, D.K. and Choi, D.H., "Dual Solution for Mixed Convection in a Horizontal Tube under Circumferentially Non-Uniform Heating", *Int. J. Heat and Mass Transfer,* Vol. 35, pp. 2053–2056, 1992.
97. Chou, Fu Chu, "Laminar Mixed Convection in the Thermal Entrance Region of Horizontal Rectangular Channels with Uniform Heat Input Axially and Uniform Wall Temperature Circumferentially", *Canadian J. Chem. Eng.,* Vol. 68, pp. 577–584, 1990.
98. Depew, C.A. and August, S.E., "Heat Transfer Due to Combined Free and Forced Convection in a Horizontal and Isothermal Tube", *J. Heat Transfer,* Vol. 93, pp. 380–384, 1971.
99. Hwang, G.J. and Lai, H.C., "Laminar Convective Heat Transfer in a Horizontal Isothermal Tube for High Rayleigh Numbers", *Int. J. Heat and Mass Transfer,* Vol. 37, No. 11, pp. 1631–1640, 1994.
100. Incropera, F.P., Knox, A.J., and Maughan, J.R., "Mixed Convection Flow and Heat Transfer in the Entry Region of a Horizontal Rectangular Duct", *J. Heat Transfer,* Vol. 109, pp. 434–439, 1987.
101. Lin, Jenn Nan and Chou, Fu Chu, "Laminar Mixed Convection in the Thermal Entrance Region of Horizontal Isothermal Rectangular Channels", *Canadian J. Chemical Eng.,* Vol. 67, pp. 361–367, 1989.
102. Maughan, J.R. and Incropera, F.P., "Mixed Convection Heat Transfer for Airflow in a Horizontal and Inclined Channel", *Int. J. Heat Mass Transfer,* Vol. 30, p. 1307, 1987.
103. Morcos, S.M. and Bergles, A.E., "Experimental Investigation of Combined Forced and Free Laminar Convection in Horizontal Tubes", *J. Heat Transfer,* Vol. 97, pp. 212–219, 1975.
104. Nandakumar, K. and Weinitschke, H.J., "A Bifurcation Study of Mixed-Convection Heat Transfer in Horizontal Ducts", *J. Fluid Mech.,* Vol. 231, pp. 157–187, 1991.
105. Narusawa, U., "Numerical Analysis of Mixed Convection at the Entrance Region of a Rectangular Duct Heated from Below", *Int. J. Heat and Mass Transfer,* Vol. 36, pp. 2375–2384, 1993.
106. Nieckele, A.O. and Patankar, S.V., "Laminar Mixed Convection in a Concentric Annulus with Horizontal Axis", *J. Heat Transfer,* Vol. 107, pp. 902–909, 1985.

107. Nyce, T.A., Ouazzani, J., and Durand Daubin, A., "Mixed Convection in a Horizontal Rectangular Channel-Experimental and Numerical Velocity Distributions", *Int. J. Heat and Mass Transfer,* Vol. 35, pp. 1481–1494, 1992.
108. Osborne, D.G. and Incropera, F.P., "Laminar, Mixed Convection Heat Transfer for Flow between Horizontal Parallel Plates with Asymmetric Heating", *Int. J. Heat Mass Transfer,* Vol. 28, No. 1, pp. 207–217, 1985.
109. Shome, B. and Jensen, M.K., "Mixed Convection Laminar Flow and Heat Transfer of Liquids in Isothermal Horizontal Circular Ducts", *Int. J. of Heat and Mass Transfer,* Vol. 38, No. 11, pp. 1945–1956, 1995.
110. Yousef, W.W. and Tarasuk, J.D., "Free Convection Effects on Laminar Forced Convective Heat Transfer in a Horizontal Isothermal Tube", *J. Heat Transfer,* Vol. 104, pp. 145–152, 1982.
111. Choi, C.Y and Ortega, A., "Mixed Convection in an Inclined Channel with a Discrete Heat Source", *Int. J. Heat and Mass Transfer,* Vol. 36, pp. 3119–3134, 1993.
112. Lavine, A.S., Kim, M.Y. and Shores, C.N., "Flow Reversal in Opposing Mixed Convection Flow in Inclined Pipes", *J. Heat Transfer,* Vol. 111, pp. 114–120, 1989.

CHAPTER 10

Convective Heat Transfer Through Porous Media

10.1 INTRODUCTION

A porous medium basically consists of a bed of many relatively closely packed particles or some other form of solid matrix which remains at rest and through which a fluid flows. If the fluid fills all the gaps between the particles, the porous medium is said to be saturated with the fluid, i.e., with a saturated porous medium it is not possible to add more fluid to the porous medium without changing the conditions at which the fluid exists, e.g., its density if the fluid is a gas. A porous medium is shown schematically in Fig. 10.1.

An example of heat transfer through a porous medium is heat transfer through a layer of granular insulating material. This material will be saturated with air, i.e., the space between the granules of insulating material is entirely filled with air, and this air will flow through the insulation material as a result of the temperature difference imposed on the material, i.e., there will be a free convective flow in the porous material. Even when a fibrous insulation is used, the flow in the insulation can be

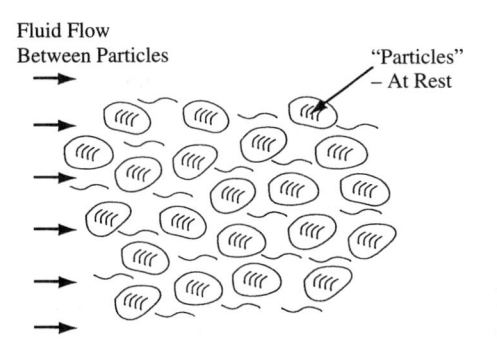

FIGURE 10.1
A porous medium.

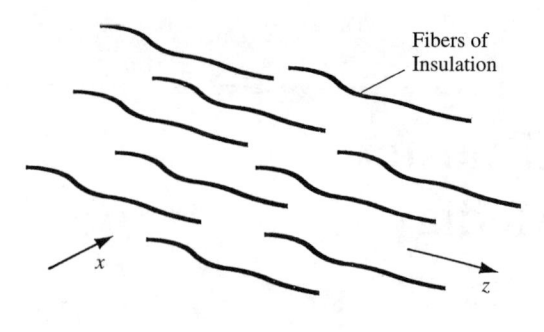

488 Introduction to Convective Heat Transfer Analysis

FIGURE 10.2
Possible directional dependency of properties of a fibrous insulation material.

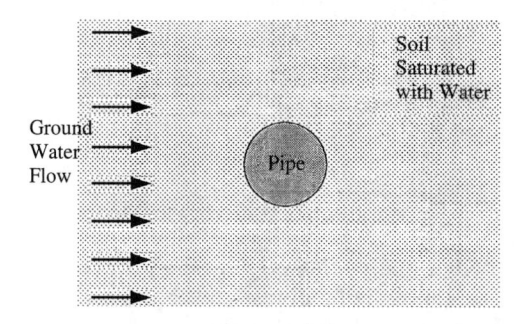

Resistance to Flow in x-direction > Resistance to Flow in z-direction.

FIGURE 10.3
Heat transfer to a pipe buried in water saturated soil.

treated as flow in a porous medium although, in this case, the fibers are often aligned with each other and as a result the properties of the porous medium will vary with the direction considered. This is illustrated in Fig. 10.2.

Another example of heat transfer involving a porous medium is heat transfer from a pipe or cable buried in soil or in a bed of crushed stones which is saturated with ground water which is flowing through the soil or stones. This is illustrated in Fig. 10.3.

Many geological flows such as flow in an oil reservoir or in a geothermal power system involve convective heat transfer in a porous medium.

The present chapter gives no more than a brief introduction to convective heat transfer in a porous medium. It is an area of considerable practical importance and there is a large body of literature on the topic to which the reader is referred for more detail, for example see [1] to [12].

Attention will be restricted to steady flows in this chapter and multiphase flows will not be considered.

10.2
AREA AVERAGED VELOCITY

If a plane drawn in a porous media flow is considered, the velocity will not be uniform over the plane. There will be no flow where this plane intersects the solid particles, i.e., the velocity is zero over these areas, and even where the plane passes between particles, the velocity will not, in general, be uniform. This is illustrated schematically in Fig. 10.4 in which the velocity variation along a line drawn in a porous media flow is shown.

Because the size of the particles considered is small compared to the overall size of the system, the concern is usually not with the details of the local velocity variation but only with the mean velocity over an area of the plane that is small compared with the overall size of the system but large compared to the size of the particles [12].

The mean velocity over a rectangular area $\Delta y \Delta z$ in size on the plane is then given by:

$$u = \frac{1}{\Delta y \Delta z} \int_0^{\Delta z} \int_0^{\Delta y} u_p \, dy \, dz \qquad (10.1)$$

where u_p is the local velocity in the porous material.

Because Δy and Δz are small compared to the size of the system, u may be regarded as the area-averaged velocity component at a point in the flow. The area-averaged velocity components in the other coordinate directions at a point in the flow are defined in the same way, i.e., by:

$$v = \frac{1}{\Delta x \Delta z} \int_0^{\Delta x} \int_0^{\Delta z} v_p \, dz \, dx$$

and:

$$w = \frac{1}{\Delta x \Delta y} \int_0^{\Delta y} \int_0^{\Delta x} w_p \, dx \, dy$$

The analyses of convection in porous media presented in this chapter will all be based on the use of these area-averaged velocity components. Their use is, of course, similar to the use of time-averaged velocity components in turbulent flows.

By definition, with the area-averaged velocity defined as above, the continuity equation for flow in a porous medium will have the same form as that for the flow of a pure fluid, i.e., the continuity equation for flow through a porous medium is, if density variations are negligible:

$$\frac{\partial u}{\partial x} + \frac{\partial v}{\partial y} + \frac{\partial w}{\partial z} = 0 \qquad (10.2)$$

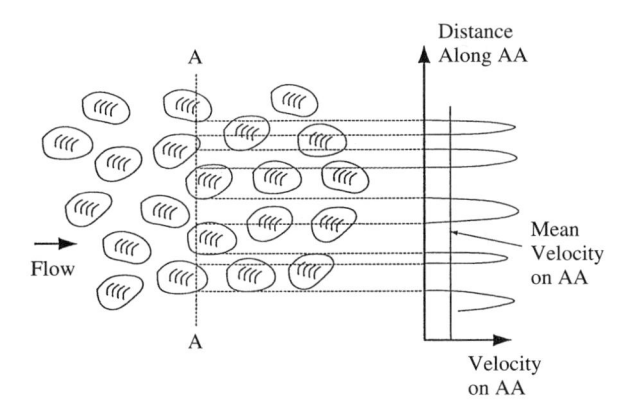

FIGURE 10.4
Velocity distribution in a porous medium flow.

10.3
DARCY FLOW MODEL

When the fluid flows through a porous medium, the solid particles exert a force on the fluid equal and opposite to the drag force on the solid particles. This force must be balanced by the pressure gradient in the flow, i.e., for flow through a control volume for any chosen direction:

Difference between rate fluid momentum leaves and the rate fluid momentum enters control volume = Net viscous force on surface of control volume

+ Net pressure force on control volume

− Drag force on particles in control volume

+ Net buoyancy force   (10.3)

In the Darcy model of flow through a porous medium, it is assumed that the flow velocities are low and that momentum changes and viscous forces in the fluid are consequently negligible compared to the drag force on the particles, i.e., if flow through a control volume of the type shown in Fig. 10.5 is considered, then:

Net pressure force on control volume in any direction = Drag force on particles in control volume in direction considered + Buoyancy force   (10.4)

Thus, in the Darcy model, the rate of change of momentum through the control volume and the viscous forces acting on the surfaces of the control volume are assumed to be negligible compared to the drag force and the buoyancy force [13],[14].

Buoyancy forces will, for the present, be neglected. In this case, if the x-direction is considered, Eq. (10.4) gives:

$$p\,dy\,dz - \left(p + \frac{\partial p}{\partial x}dx\right)dy\,dz = D_{Fx}$$

where D_{Fx} is the net drag force. Hence:

$$-\frac{\partial p}{\partial x} = \frac{D_{Fx}}{dx\,dy\,dz} = D_x \qquad (10.5)$$

D_x being the drag force per unit volume in the x-direction.

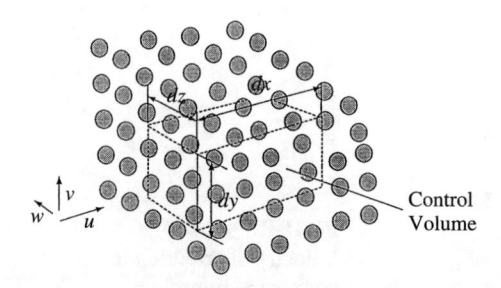

FIGURE 10.5
Control volume considered.

CHAPTER 10: Convective Heat Transfer Through Porous Media   491

Now because, by the basic assumptions being made, the fluid velocity over the particles is small, the Reynolds number based on particle size will be very small and the flow over the particles will be a "creeping" or "Stokes" type flow. In such flows, the drag force on a body is proportional to the velocity over the body and to the viscosity of the fluid, μ_f. Hence, D_x in Eq. (10.4) will be proportional to $u\mu_f$. This means that Eq. (10.5) can be written as:

$$u\mu_f = -K\frac{\partial p}{\partial x} \tag{10.6}$$

where K is the constant of proportionality. This is often referred to as Darcy's law or the Darcy model. K is called the "permeability" of the porous medium. It will depend on the number of particles per unit volume of the medium and on the shape and size of the particles. K has the dimensions of (Length)^2, e.g., the units of m^2.

Eq. (10.6) can be written as

$$u = -\frac{K}{\mu_f}\frac{\partial p}{\partial x} \tag{10.7}$$

Similarly, in the other flow directions, the Darcy model gives:

$$v = -\frac{K}{\mu_f}\frac{\partial p}{\partial y} \tag{10.8}$$

and

$$w = -\frac{K}{\mu_f}\frac{\partial p}{\partial z} \tag{10.9}$$

In writing Eqs. (10.7) to (10.9), it has been assumed that K has the same value in each direction, i.e., that the porous medium is isotropic. In general, as discussed earlier, because the particles are not spherical and because they can all be aligned in some way as shown in Fig. 10.6 (see also Fig. 10.2), this will not be the case. In such a case, i.e., in the case of anisotropic porous materials, Eqs. (10.7) to (10.9) must be written:

$$u = -\frac{K_x}{\mu_f}\frac{\partial p}{\partial x}, \quad v = -\frac{K_y}{\mu_f}\frac{\partial p}{\partial y}, \quad w = -\frac{K_z}{\mu_f}\frac{\partial p}{\partial z} \tag{10.10}$$

where K_x, K_y, and K_z are the permeabilities in the x, y, and z directions, respectively.

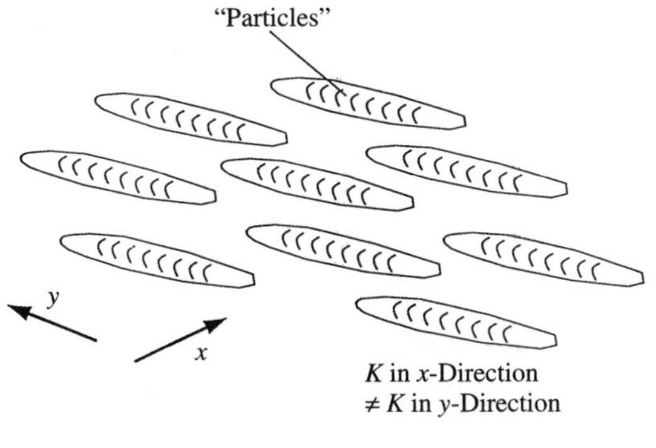

K in x-Direction
$\neq K$ in y-Direction

FIGURE 10.6
An anisotropic porous material.

492 Introduction to Convective Heat Transfer Analysis

It will be assumed in the present chapter that the material being considered isotropic, i.e., that $K_x = K_y = K_z = K$.

Hence, for a flow in which the Darcy assumptions can be used, the governing equations for the velocity components are, if buoyancy forces are neglected:

$$\frac{\partial u}{\partial x} + \frac{\partial v}{\partial y} + \frac{\partial w}{\partial z} = 0 \quad (10.11)$$

$$u = -\frac{K}{\mu_f}\frac{\partial p}{\partial x} \quad (10.12)$$

$$v = -\frac{K}{\mu_f}\frac{\partial p}{\partial y} \quad (10.13)$$

$$w = -\frac{K}{\mu_f}\frac{\partial p}{\partial z} \quad (10.14)$$

This represents a set of four equations in the four variables u, v, w, and p. In solving these equations, it must be noted that because the effects of viscosity are being ignored, the no-slip boundary condition cannot be applied, i.e., the only boundary condition on velocity at a solid surface is that the normal component of the velocity at the surface is 0. This is illustrated in Fig. 10.7.

When the Darcy assumptions are adopted and the effect of buoyancy forces are neglected, no vorticity can be generated in the flow because the only forces on the surface of a control volume being considered are pressure forces which act normal to the surface of the control volume. Forces with a component that is tangential to the surface are required in order to generate vorticity. Therefore, the velocity distribution in Darcy flow will be that for irrotational flow, i.e., exactly the same as that that would exist in inviscid or potential flow in the situation being considered. To illustrate that this is the case, consider two-dimensional flow. The governing equations for the velocity components in this case are:

$$\frac{\partial u}{\partial x} + \frac{\partial v}{\partial y} = 0 \quad (10.15)$$

$$u = -\frac{K}{\mu_f}\frac{\partial p}{\partial x} \quad (10.16)$$

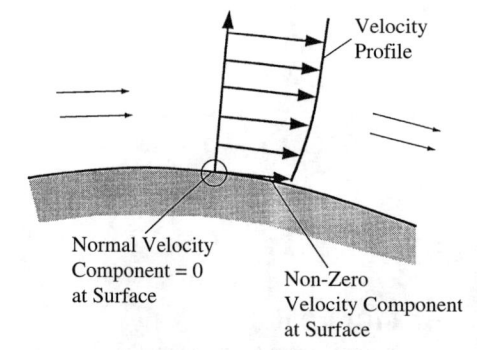

FIGURE 10.7
Boundary condition on velocity at surface.

CHAPTER 10: Convective Heat Transfer Through Porous Media   493

$$v = -\frac{K}{\mu_f}\frac{\partial p}{\partial y} \tag{10.17}$$

The pressure can be eliminated by differentiating Eq. (10.16) with respect to y and Eq. (10.17) with respect to x and subtracting the first result from the second. This process gives:

$$\frac{\partial v}{\partial x} - \frac{\partial u}{\partial y} = 0 \tag{10.18}$$

Therefore, the equations governing the velocity components in Darcy flow are:

$$\frac{\partial u}{\partial x} + \frac{\partial v}{\partial y} = 0 \tag{10.15}$$

and:

$$\frac{\partial v}{\partial x} - \frac{\partial u}{\partial y} = 0 \tag{10.18}$$

If the stream function, ψ, is introduced, it being defined as before by:

$$u = \frac{\partial \psi}{\partial y}, \quad v = -\frac{\partial \psi}{\partial x}$$

then it will be seen that Eq. (10.15) is always satisfied and Eq. (10.18) becomes in terms of ψ:

$$\frac{\partial^2 \psi}{\partial x^2} + \frac{\partial^2 \psi}{\partial y^2} = 0$$

This is the equation that has traditionally been used to obtain solutions for two-dimensional potential flows. The methods used to obtain such solutions can therefore be used to find the velocity distributions in Darcy flows.

The above equations for the pressure variation in Darcy flow involve the permeability, K. This quantity is mainly dependent on the porosity of the material, ϕ, i.e., the ratio of the pore volume, V_p, to the total volume, V, of the porous material, and on the mean diameter of the particles. The relation between these quantities depends on the shape of the particles and how they are arranged. If, for example, the particles

TABLE 10.1
Typical Values of Permeability and Porosity

Material	Permeability - m²	Porosity - percent
Brick	4.8×10^{-14} to 2.2×10^{-11}	12 to 35
Fiberglass	2.4×10^{-11} to 5.1×10^{-11}	88 to 93
Limestone	2.0×10^{-15} to 4.5×10^{-14}	4 to 10
Sand (Loose Bed)	4.8×10^{-11} to 1.8×10^{-10}	35 to 50
Sandstone	5.0×10^{-16} to 3.0×10^{-12}	8 to 38
Silica Powder	1.3×10^{-14} to 5.1×10^{-14}	37 to 49
Soil	2.9×10^{-13} to 1.4×10^{-11}	43 to 55

are spherical, it has been proposed that:

$$K = \frac{\phi^3 d^2}{150(1-\phi)^2} \tag{10.19}$$

Some typical values of permeability and porosity for a few materials are shown in Table 10.1.

Deviations from the Darcy model will be considered in a later section.

EXAMPLE 10.1. In some cases, unidirectional flow through a porous medium can be approximately modeled as flow through a bundle of small diameter parallel tubes of diameter, D, as shown in Fig. E10.1. Assuming that the pressure gradient in the tubes is given by the Hagen-Poiseuille equation, i.e., by:

$$\frac{dp}{dx} = -\frac{64\mu_f U}{D^2}$$

where U is the mean velocity in the tubes, derive an expression for the permeability, K.

Solution. If u is the mean velocity through the porous medium, then the mean velocity in the tubes is given by continuity considerations as:

$$U = \frac{u}{\phi}$$

Hence, using the Hagen-Poiseuille equation it follows that:

$$\frac{dp}{dx} = -\frac{64\mu_f u}{\phi D^2}$$

i.e.:

$$\frac{\mu_f u}{-(dp/dx)} = \frac{\phi D^2}{64}$$

But the definition of the permeability is such that:

$$K = \frac{\mu_f u}{-(dp/dx)}$$

so the model here being considered gives:

$$K = \frac{\phi D^2}{64}$$

Flow Through Series of Parallel Tubes

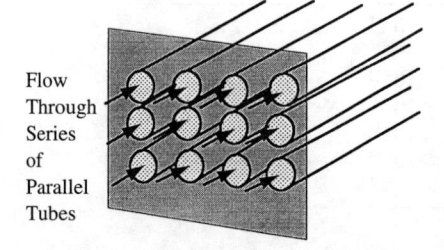

FIGURE E10.1

10.4
ENERGY EQUATION

The energy equation is derived, as before, by applying conservation of energy considerations to flow through a control volume of the type shown in Fig. 10.8.

Viscous dissipation effects are neglected in line with the neglect of viscous stresses in the consideration of momentum conservation that was given in the previous section. Hence, since steady flow is being assumed, conservation of energy requires that:

$$\begin{matrix} \text{Rate enthalpy leaves} \\ \text{control volume} \end{matrix} - \begin{matrix} \text{Rate enthalpy enters} \\ \text{control volume} \end{matrix} =$$
$$\begin{matrix} \text{Rate of heat transfer} \\ \text{into control volume} \end{matrix} - \begin{matrix} \text{Rate of heat transfer} \\ \text{out of control volume} \end{matrix} \quad (10.20)$$

It will be assumed that the fluid and the particulate material are in thermal equilibrium, i.e., that locally the fluid and the particulate matter in contact with it are at the same temperature.

The enthalpy changes will first be considered. Consider the x-direction. Because the fluid properties are being assumed to be constant, the rate enthalpy enters the left-hand face of the control volume is given by:

$$H_x = \dot{m}_x c_{pf} T = \rho_f u \, dy \, dz \, c_{pf} T \quad (10.21)$$

Here, ρ_f and c_{pf} are the density and specific heat of the fluid. The fluid values are used because only the fluid is in motion.

The rate that enthalpy leaves the right-hand face of the control volume is given by:

$$H_{x+dx} = H_x + \frac{\partial}{\partial x}(H_x) dx$$

Hence, the difference between the rate that enthalpy leaves and the rate it enters the control volume in the x-direction is given by:

$$H_{x+dx} - H_x = \frac{\partial}{\partial x}(\rho_f c_{pf} u T) dx \, dy \, dz \quad (10.22)$$

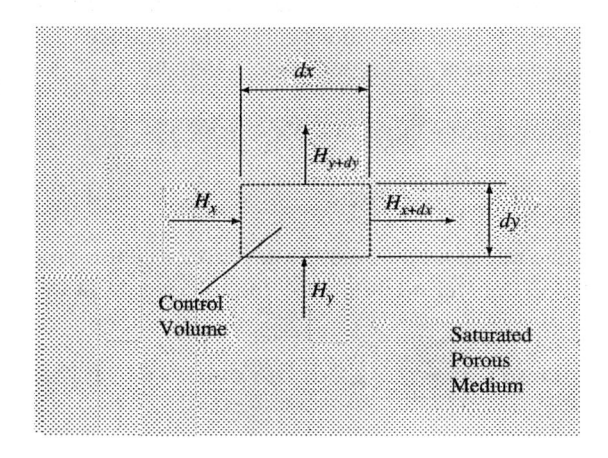

FIGURE 10.8
Control volume used in deriving energy equations.

Since the properties of the fluid are being assumed constant, this gives:

$$H_{x+dx} - H_x = \rho_f c_{pf} \frac{\partial}{\partial x}(uT)\,dx\,dy\,dz \qquad (10.23)$$

Similarly, in the y- and z-directions, the differences between the rates that enthalpy leaves and enters the control volume are:

$$H_{y+dy} - H_y = \rho_f c_{pf} \frac{\partial}{\partial y}(vT)\,dx\,dy\,dz$$

and

$$H_{z+dz} - H_z = \rho_f c_{pf} \frac{\partial}{\partial z}(wT)\,dx\,dy\,dz$$

Hence, the net difference between the rate that enthalpy leaves and the rate at which it enters the control volume is given by:

$$\rho_f c_{pf}\left[\frac{\partial}{\partial x}(uT) + \frac{\partial}{\partial y}(vT) + \frac{\partial}{\partial z}(wT)\right]dx\,dy\,dz \qquad (10.24)$$

This can be written as:

$$\rho_f c_f \left\{ T\left[\frac{\partial u}{\partial x} + \frac{\partial v}{\partial y} + \frac{\partial w}{\partial z}\right] + \left[u\frac{\partial T}{\partial x} + v\frac{\partial T}{\partial y} + w\frac{\partial w}{\partial z}\right] \right\} dx\,dy\,dz \qquad (10.25)$$

The first term in this equation is 0 by virtue of the continuity equation so Eq. (10.24) can be written as:

$$\rho_f c_f \left[u\frac{\partial T}{\partial x} + v\frac{\partial T}{\partial y} + w\frac{\partial T}{\partial z} \right] dx\,dy\,dz \qquad (10.26)$$

Next, consider the heat transfer rate into the control volume. The coordinate directions will again be separately considered. In the x-direction, the rate of heat transfer into the control volume through the left-hand face is given by:

$$Q_x = -k_a \frac{\partial T}{\partial x} dy\,dz \qquad (10.27)$$

The thermal conductivity has to be carefully defined. The fluid and the particulate matter will, in general, have different thermal conductivities. The conductivity, k_a, in the above equation is an area averaged or apparent thermal conductivity of the porous material.

The rate of heat transfer out of the control volume in the x-direction is given by:

$$Q_{x+dx} = Q_x + \frac{\partial}{\partial x}(Q_x)\,dx \qquad (10.28)$$

Hence, the difference between the rate at which heat is transferred into the control volume and the rate at which it is transferred out of the control volume in the x-direction is given by:

$$Q_x - Q_{x+dx} = -\frac{\partial}{\partial x}(Q_x)\,dx = +\frac{\partial}{\partial x}\left[k_a \frac{\partial T}{\partial x}\right] dx\,dy\,dz \qquad (10.29)$$

CHAPTER 10: Convective Heat Transfer Through Porous Media 497

If the apparent thermal conductivity is assumed to be constant, this gives:

$$Q_x - Q_{x+dx} = k_a \frac{\partial^2 T}{\partial x^2} dx\,dy\,dz \qquad (10.30)$$

Similarly, in the y- and z-directions, the differences between the heat transfer rate into the control volume and the heat transfer rate out of the control volume are given by:

$$Q_y - Q_{y+dy} = k_a \frac{\partial^2 T}{\partial y^2} dx\,dy\,dz \qquad (10.31)$$

and:

$$Q_z - Q_{z+dz} = k_a \frac{\partial^2 T}{\partial z^2} dx\,dy\,dz \qquad (10.32)$$

Hence, the net difference between the rate that heat is transferred into the control volume and the rate at which it is transferred out of the control volume is given by:

$$k_a \left[\frac{\partial^2 T}{\partial x^2} + \frac{\partial^2 T}{\partial y^2} + \frac{\partial^2 T}{\partial y^2} \right] dx\,dy\,dz \qquad (10.33)$$

Substituting Eqs.(10.26) and (10.33) into Eq. (10.20) and canceling $(dx\,dy\,dz)$ gives, after rearrangement:

$$u\frac{\partial T}{\partial x} + v\frac{\partial T}{\partial y} + w\frac{\partial T}{\partial z} = \left(\frac{k_a}{\rho_f c_{pf}}\right)\left(\frac{\partial^2 T}{\partial x^2} + \frac{\partial^2 T}{\partial y^2} + \frac{\partial^2 T}{\partial z^2}\right) \qquad (10.34)$$

The quantity $(k_a/\rho_f c_{pf})$ is termed the "apparent thermal diffusivity" and will here be given the symbol α_a. It is not the thermal diffusivity of the fluid since it is the ratio of the apparent conductivity of the porous medium to the product of the density and specific heat at constant pressure of the fluid.

Eq. (10.34) together with Eqs. (10.11) to (10.14) constitutes the set of equations governing forced convective flow through a porous medium. As discussed in the previous section, the distribution of the velocity components is the same as would exist with potential flow in the same geometrical situation. This potential flow solution gives the values of u, v, and w which can then be used in Eq. (10.34) to give the temperature distribution.

The apparent conductivity that occurs in the energy equation is a result of conduction in the solid material and in the fluid. If a simple "parallel" path model is assumed, k_a will be given by:

$$k_a = \phi k_f + (1 - \phi)k_s \qquad (10.35)$$

k_f and k_s being the conductivities of the fluid and solid material respectively and ϕ is the porosity as previously defined. Other more complex models that relate k_a to ϕ, k_f, and k_s usually must be used or k_a has to be measured experimentally.

10.5
BOUNDARY LAYER SOLUTIONS FOR TWO-DIMENSIONAL FORCED CONVECTIVE HEAT TRANSFER

If the Darcy assumptions are used then with forced convective flow over a surface in a porous medium, because the velocity is not assumed to be 0 at the surface, there is no velocity change induced by viscosity near the surface and there is therefore no velocity boundary layer in the flow over the surface. There will, however, be a region adjacent to the surface in which heat transfer is important and in which there are significant temperature changes in the direction normal to the surface. Under many circumstances, the normal distance over which such significant temperature changes occur is relatively small, i.e., a thermal boundary layer can be assumed to exist around the surface as shown in Fig. 10.9, the ratio of the boundary layer thickness, δ_T, to the size of the body as measured by some dimension, L, being small [15],[16].

The velocity component in the x-direction shown in Fig. 10.9 can, because the boundary layer is assumed to be thin, be taken as equal to the velocity at the surface, i.e., as equal to the velocity that would exist at the surface at the value of x considered in inviscid flow over the surface (see discussion in Section 10.3 above). The boundary layer form of the full energy equation for porous media flow is derived using the same procedure as used in dealing with pure fluid flows, this procedure having been discussed in Chapter 2. Attention will be restricted to two-dimensional flow.

Now the continuity equation gives:

$$\frac{\partial v}{\partial y} = -\frac{\partial u}{\partial x}$$

Therefore, if u_∞ is a measure of the value of u equal, for example, to the velocity in the free-stream ahead of the body and if L is some measure of the size of the body in the x-direction then because y is of the order of δ_T in the boundary layer, the above equation shows that:

$$o\left(\frac{v}{u_\infty}\right) = o\left(\frac{\delta_T}{L}\right) \tag{10.36}$$

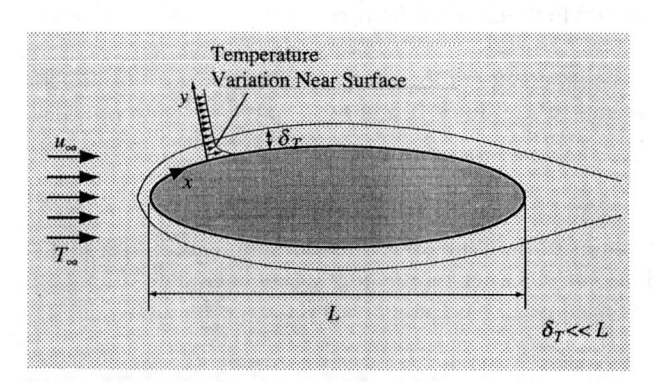

FIGURE 10.9
Thermal boundary layer in a porous medium.

Defining the following for convenience:

$$X = \frac{x}{L}, \quad Y = \frac{y}{\delta_T}, \quad U = \frac{u}{u_\infty}, \quad V = \frac{v}{u_\infty}, \quad \theta = \frac{T - T_\infty}{T_{wr} - T_\infty} \tag{10.37}$$

where T_∞ is the temperature in the free-stream temperature outside the boundary layer and T_{wr} is a measure of the wall temperature. X, Y, and θ will all be of order 1 in the boundary layer.

The energy equation for the two-dimensional flow which is being considered here can be written as:

$$u\frac{\partial T}{\partial x} + v\frac{\partial T}{\partial y} = \left(\frac{k_a}{\rho_f c_{pf}}\right)\left(\frac{\partial^2 T}{\partial x^2} + \frac{\partial^2 T}{\partial y^2}\right) \tag{10.38}$$

In terms of the variables defined above, this equation gives:

$$U\frac{\partial \theta}{\partial X} + V\frac{\partial \theta}{\partial Y}\left(\frac{\delta_T}{L}\right) = \left(\frac{\alpha_a}{u_1 L}\right)\left[\frac{\partial^2 \theta}{\partial X^2} + \frac{1}{(\delta_T/L)^2}\frac{\partial^2 \theta}{\partial Y^2}\right] \tag{10.39}$$

The left-hand side of this equation is of order 1. Considering the right-hand side, δ_T/L is by assumption small so, since $\partial^2\theta/\partial X^2$ and $\partial^2\theta/\partial Y^2$ are both of order 1, the first term on the right-hand side of the above equation will be much less than the second and the first term is, therefore, negligible compared to the second. Hence, the following form of the energy equation applies in two-dimensional boundary layer flow in a porous medium:

$$u\frac{\partial T}{\partial x} + v\frac{\partial T}{\partial y} = \left(\frac{k_a}{\rho_f c_{pf}}\right)\frac{\partial^2 T}{\partial y^2} \tag{10.40}$$

Further, since the left-hand side of Eq. (10.39) has order of magnitude 1, the right-hand side of this equation must also be of order 1. Since $\partial^2\theta/\partial Y^2$ is of order 1, this requires that:

$$\left(\frac{\alpha_a}{u_1 L}\right)\frac{1}{(\delta_T/L)^2} = o(1)$$

i.e.:

$$\frac{\delta_T}{L} = o\left(\frac{1}{\sqrt{u_\infty L/\alpha_a}}\right) = o(Pe^{-1/2}) \tag{10.41}$$

where:

$$Pe = \frac{u_\infty L}{\alpha_a} \tag{10.42}$$

is the Peclet number. For the boundary layer assumptions to apply, i.e., for δ_T/L to be small, it is necessary, therefore, that the Peclet number be large.

In order to illustrate the use of the boundary layer equations, consider, first, two-dimensional forced convection flow over a flat plate that is buried in a porous material in such a way that it is aligned with the fluid flow. The situation being considered is thus as shown in Fig. 10.10.

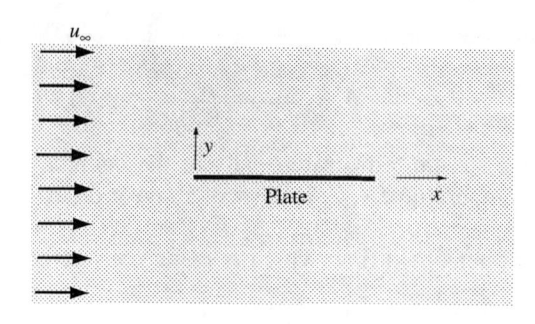

FIGURE 10.10
Flow over a flat plate in a porous medium.

Because the plate is aligned with the flow and because the effects of viscosity are being ignored, the plate does not disturb the flow, and the velocity distribution is given by:

$$u = u_\infty, \quad v = 0$$

everywhere, this being the solution for inviscid flow in the situation.

With this solution for the velocity components, the energy equation gives if a boundary layer-type flow is assumed:

$$u_\infty \frac{\partial T}{\partial x} = \alpha_a \frac{\partial^2 T}{\partial y^2} \tag{10.43}$$

It will first be assumed that the plate is at a uniform temperature, T_w. The following is then defined for this case:

$$\theta = \frac{T_w - T}{T_w - T_\infty} \tag{10.44}$$

The governing equation, Eq. (10.43), then becomes:

$$u_\infty \frac{\partial \theta}{\partial x} = \alpha_a \frac{\partial^2 \theta}{\partial y^2} \tag{10.45}$$

Now it seems reasonable to assume that the temperature profiles in the boundary layer at all values of x will be similar, i.e., that:

$$\theta = f\left(\frac{y}{\delta_T}\right) \tag{10.46}$$

where the function f does not vary with x. This is illustrated in Fig. 10.11.

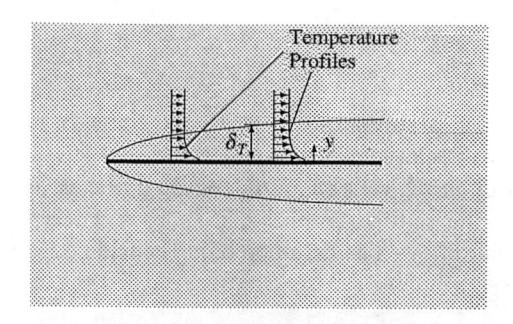

FIGURE 10.11
Similar temperature profiles in boundary layer.

CHAPTER 10: Convective Heat Transfer Through Porous Media   501

Now δ_T is basically not a clearly defined quantity. Hence, in view of Eq. (10.41) and noting that δ_T is here the boundary layer thickness at any value of x, δ_T is assumed to be proportional to $x/\sqrt{u_\infty x/\alpha_a}$. Eq. (10.46) is, therefore, written as:

$$\theta = \text{function}\left(\frac{y}{x/\sqrt{u_\infty x/\alpha_a}}\right) = \theta(\eta) \tag{10.47}$$

where η is here given by:

$$\eta = \frac{y}{x}\sqrt{\frac{u_\infty x}{\alpha_a}} = \frac{y}{x}Pe_x^{1/2} \tag{10.48}$$

Here, Pe_x is the Peclet number based on x.

Substituting Eq. (10.47) into Eq. (10.45) and noting that, by assumption, θ does not depend on x, gives the following:

$$u_\infty \theta' \frac{\partial \eta}{\partial x} = \alpha_a \theta'' \left(\frac{\partial \eta}{\partial y}\right)^2 \tag{10.49}$$

where the primes denote differentiation with respect to η, i.e:

$$\theta' = \frac{d\theta}{d\eta}, \quad \theta'' = \frac{d^2\theta}{d\eta^2}$$

Using Eq. (10.48), Eq. (10.49) becomes

$$-u_\infty \theta' \frac{\eta}{2x} = \alpha_a \theta'' \frac{Pe_x}{x^2}$$

i.e.:

$$-\left(\frac{u_\infty}{\alpha_a}\right)\frac{\eta \theta'}{2x} = \theta'' \frac{Pe_x}{x^2}$$

i.e.:

$$-\eta \frac{\theta'}{2} = \theta''$$

i.e.:

$$\theta'' + \frac{\eta \theta'}{2} = 0 \tag{10.50}$$

The fact that an ordinary differential equation has been obtained indicates that the assumption of similar temperature profiles is valid.

Now, the boundary conditions on temperature are:

$$\begin{aligned} y = 0, T &= T_w \\ y \text{ large}, T &\to T_\infty \end{aligned} \tag{10.51}$$

y large again being meant to indicate y values outside the boundary layer.

In terms of θ and η, these boundary conditions are:

$$\begin{aligned} \eta = 0, \theta &= 0 \\ \eta \text{ large}, \theta &\to 1 \end{aligned} \tag{10.52}$$

502 Introduction to Convective Heat Transfer Analysis

The temperature distribution is therefore given by solving Eq. (10.50) subject to the boundary conditions given in Eq. (10.52). Now Eq. (10.50) can be written as:

$$\frac{(\theta')'}{\theta'} = -\frac{\eta}{2} \tag{10.53}$$

Integrating this equation once gives:

$$\theta' = C_1 \exp\left(-\frac{\eta^2}{4}\right)$$

and then integrating again gives:

$$\theta = C_1 \int_0^\eta \exp\left(-\frac{\eta^2}{4}\right) d\eta + C_2$$

where C_1 and C_2 are constants of integration. Applying the boundary conditions gives:

$$C_2 = 0, \; C_1 = \frac{1}{\int_0^\infty \exp\left(-\frac{\eta^2}{4}\right) d\eta}$$

Hence the solution can be written as:

$$\theta = \frac{\int_0^\eta \exp\left(\frac{\eta^2}{4}\right) d\eta}{\int_0^\infty \exp\left(-\frac{\eta^2}{4}\right) d\eta} \tag{10.54}$$

This can be written as:

$$\theta = \mathrm{erf}\left(\frac{\eta}{2}\right) \tag{10.55}$$

where erf is the so-called error function. The form of this solution is shown in Fig. 10.12.

The local heat transfer rate at any value of x is given by:

$$q_w = -k_a \left.\frac{\partial T}{\partial y}\right|_{y=0} = +k_a(T_w - T_\infty)\frac{Pe_x^{1/2}}{x}\theta'_0$$

where:

$$\theta'_0 = \left.\frac{d\theta}{d\eta}\right|_{\eta=0}$$

This equation can be rearranged to give:

$$\frac{q_w x}{k_a(T_w - T_\infty)} = \theta'_0 Pe_x^{1/2}, \; \text{i.e., } Nu_x = \theta'_0 Pe_x^{1/2} \tag{10.56}$$

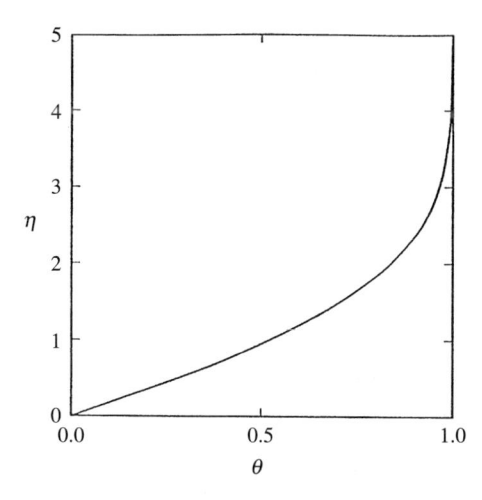

CHAPTER 10: Convective Heat Transfer Through Porous Media   503

FIGURE 10.12
Variation of θ with η in boundary layer on an isothermal flat plate.

Now the error function is such that:

$$\theta'_0 = \frac{1}{\pi^{1/2}} = 0.564$$

Hence, Eq. (10.56) gives:

$$Nu_x = 0.564 Pe_x^{1/2} \qquad (10.57)$$

This equation allows the determination of the variation of the local heat transfer rate along the plate. The mean heat transfer rate is given, as before, by:

$$\overline{q_w} = \frac{1}{L}\int_0^L q_w\,dx$$

i.e., using Eq. (10.56)

$$\overline{q_w} = k_a(T_w - T_a)\theta'_0\left(\frac{u_\infty}{\alpha_a}\right)^{1/2}\int_0^L \frac{dx}{x^{1/2}} = k_a(T_w - T_a)\theta'_0\left(\frac{u_\infty}{\alpha_a}\right)^{1/2} 2L^{1/2}$$

which, using Eq.(10.57), can be rearranged to give:

$$Nu_L = 1.128 Pe_L^{1/2} \qquad (10.58)$$

where Nu_L is the mean Nusselt number for the plate, i.e., $\overline{h}L/k_a$, $\overline{h}$ being the mean heat transfer coefficient.

Similarity solutions to the boundary layer equations for certain other thermal boundary conditions at the surface of the plate can be obtained, e.g., such a solution can be obtained for a plate with a uniform heat flux at the surface.

EXAMPLE 10.2. A heat pump system rejects heat to a water flow that trickles through a bed of small pebbles in the ground. The water velocity in the bed is estimated to be 1 cm/s. The heat exchanger can be approximately modeled as a series of 10-cm wide flat plates that are aligned with the water flow. If the plates are at a temperature of 20°C and if the ground water has a temperature of 10°C, find the heat transfer rate from the plates to the water per meter length of plate considering both sides of the plate. The pebble bed has a porosity of 0.35.

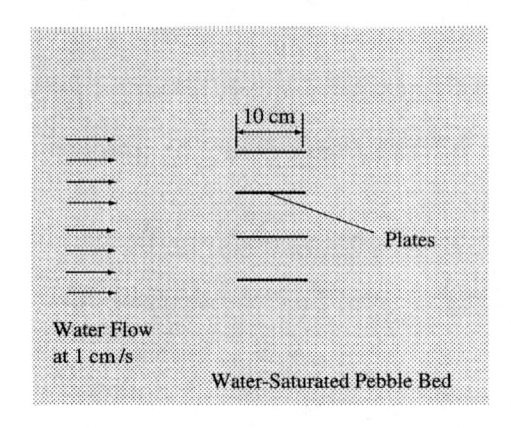

504 Introduction to Convective Heat Transfer Analysis

FIGURE E10.2

Solution. The water properties will be evaluated at the mean temperature of the water in the boundary layer, i.e., at $(20 + 10)/2 = 15°C$. At this temperature, for water:

$$\rho_f = 999.1 \text{ kg/m}^3$$
$$c_{pf} = 4186 \text{ J/kg K}$$
$$k_f = 0.59 \text{ W/m K}$$

The thermal conductivity of the pebbles will be assumed to be given by:

$$k_s = 1.8 \text{ W/m K}$$

In the absence of other information, the thermal conductivity of the porous material will be assumed to be given by:

$$k_a = \phi k_f + (1 - \phi)k_s = 0.35 \times 0.59 + 0.65 \times 1.8 = 1.38 \text{ W/m K}$$

Hence:

$$\alpha_a = \frac{k_s}{\rho_f c_{pf}} = \frac{1.38}{999.1 \times 4186} = 0.33 \times 10^{-6} \text{ m}^2/\text{s}$$

The Peclet number is therefore given by:

$$Pe_L = \frac{0.01 \times 0.1}{0.33 \times 10^{-6}} = 3030$$

Eq. (10.58) then gives:

$$\frac{\bar{h} \times 0.1}{1.38} = 1.128 \times (3030)^{1/2}$$

Hence:

$$\bar{h} = 857 \text{ W/m}^2\text{-}°C$$

Therefore, considering both sides of the plate:

$$Q = 2 \times 857 \times 0.1 \times 1 \times (20 - 10) = 1.71 \text{ kW}$$

Hence, the heat transfer rate from the two sides of the plate per meter length is 1.71 kW.

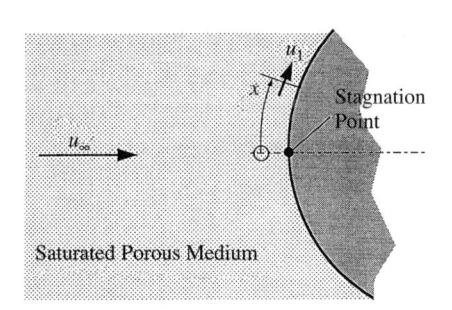

CHAPTER 10: Convective Heat Transfer Through Porous Media   505

FIGURE 10.13
Flow in the region of a stagnation point.

As another example of a situation in which a similarity-type solution can be obtained, consider flow in the region of a stagnation point of an isothermal body as shown in Fig. 10.13.

The inviscid flow solution gives for this region:

$$u_1 = 4u_\infty(x/D) \tag{10.59}$$

where D is twice the radius of curvature of the leading edge. The velocity given by this equation will be taken as the x-component of the velocity in the boundary layer in the stagnation point region. The y-component of velocity in the boundary layer is then given by using the continuity equation to give:

$$\frac{\partial v}{\partial y} = -\frac{\partial u}{\partial x}$$

i.e., because u is assumed to be equal to u_1 as given by Eq. (10.59) it follows that:

$$\frac{\partial v}{\partial y} = -4\frac{u_\infty}{D}$$

Because v is zero at the wall, i.e., at $y = 0$, this can be integrated to give:

$$v = -4u_\infty(y/D) \tag{10.60}$$

The boundary layer energy equation therefore gives for the stagnation point region:

$$4u_\infty(x/D)\frac{\partial T}{\partial x} - 4u_\infty(y/D)\frac{\partial T}{\partial y} = \left(\frac{k_a}{\rho_f c_{pf}}\right)\frac{\partial^2 T}{\partial y^2} \tag{10.61}$$

Again defining:

$$\theta = \frac{T_w - T}{T_w - T_\infty}$$

and assuming that:

$$\theta = \theta(\eta) \tag{10.62}$$

where, because it is to be expected from stagnation region solutions for pure fluids that the boundary layer thickness will be constant in this stagnation point region, η is defined by:

$$\eta = \frac{y}{D}\sqrt{\frac{u_\infty D}{\alpha_a}} = \frac{y}{D}Pe_D^{1/2} \tag{10.63}$$

Because this quantity is independent of x, the use of Eq. (10.62) implies that in the stagnation point region, $\partial \theta / \partial x = 0$. Substituting Eq. (10.62) into Eq. (10.61) therefore gives:

$$-4u_\infty \left(\frac{y}{D}\right) \frac{\partial \theta}{\partial \eta} \frac{d\eta}{dy} = \left(\frac{k_a}{\rho_f c_p f}\right) \frac{\partial^2 \theta}{\partial \eta^2} \left[\frac{d\eta}{dy}\right]^2$$

i.e., because $d\eta/dy = Pe_D^{1/2}/D$:

$$-4\eta \theta' = \theta''$$

i.e.:

$$\theta'' + 4\eta \theta' = 0 \tag{10.64}$$

The boundary conditions on temperature are:

$$\begin{aligned} y = 0, \ T &= T_w \\ y \text{ large}, \ T &\to T_\infty \end{aligned} \tag{10.65}$$

y large again being meant to indicate y values outside the boundary layer.

In terms of θ and η, these boundary conditions are:

$$\begin{aligned} \eta = 0, \ \theta &= 0 \\ \eta \text{ large}, \ \theta &\to 1 \end{aligned} \tag{10.66}$$

The temperature distribution is therefore given by solving Eq. (10.64) subject to the boundary conditions given in Eq. (10.66). Now Eq. (10.64) can be written as:

$$\frac{(\theta')'}{\theta'} = -4\eta \tag{10.67}$$

Integrating this equation once gives:

$$\theta' = C_1 \exp\left(-2\eta^2\right)$$

and then integrating again gives:

$$\theta = C_1 \int_0^\eta \exp\left(-2\eta^2\right) d\eta + C_2 \tag{10.68}$$

where C_1 and C_2 are constants of integration. Applying the boundary conditions gives:

$$C_2 = 0, \ C_1 = \frac{1}{\int_0^\infty \exp\left(-2\eta^2\right) d\eta} \tag{10.69}$$

Hence, the solution can be written as:

$$\theta = \frac{\int_0^\eta \exp\left(-2\eta^2\right) d\eta}{\int_0^\infty \exp\left(-2\eta^2\right) d\eta} \tag{10.70}$$

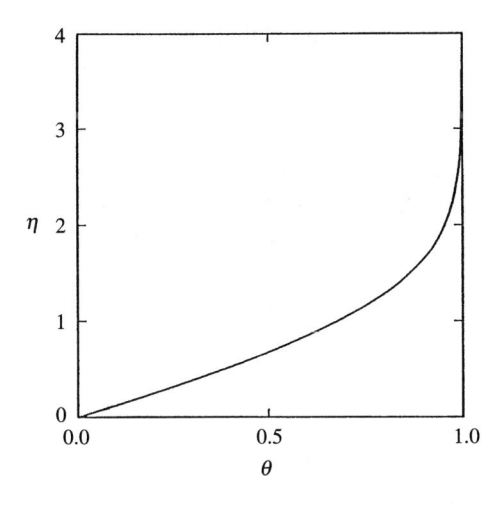

FIGURE 10.14
Variation of θ with η in boundary layer in stagnation point region.

This can be written as:

$$\theta = \text{erf}\left(2^{1/2}\eta\right) \qquad (10.71)$$

The form of this solution is shown in Fig. 10.14.

The heat transfer rate is given by:

$$q_w = -k_a \left.\frac{\partial T}{\partial y}\right|_{y=0} = +k_a(T_w - T_\infty)\frac{Pe_D^{1/2}}{D}\theta'_0 \qquad (10.72)$$

where:

$$\theta'_0 = \left.\frac{d\theta}{d\eta}\right|_{\eta=0} \qquad (10.73)$$

Eq. (10.72) can be rearranged to give:

$$\frac{q_w D}{k_a(T_w - T_\infty)} = \theta'_0 Pe_D^{1/2}, \text{ i.e., } Nu_D = \theta'_0 Pe_D^{1/2} \qquad (10.74)$$

Now the error function is such that:

$$\theta'_0 = \frac{2^{3/2}}{\pi^{1/2}} = 1.596 \qquad (10.75)$$

Hence, Eq. (10.74) gives:

$$Nu_D = 1.596\, Pe_D^{1/2} \qquad (10.76)$$

There are many situations in which similarity-type solutions to the boundary layer equations cannot be obtained. Numerical solutions to these equations can be obtained in such cases. In general, such solutions first involve numerically solving for the surface velocity distribution and then using the energy equation to obtain the temperature distribution. Here, in order to illustrate how the energy equation can be numerically solved, it will be assumed that the variation of the surface velocity with distance x about the surface is known, i.e., it will be assumed that:

$$u = u_\infty F\left(\frac{x}{D}\right) \tag{10.77}$$

where u_∞ is some characteristic velocity such as the free-stream velocity ahead of the body and D is some characteristic dimension of the body involved. The function, F, is assumed to be known.

The continuity equation is again used to obtain an expression for the velocity component in the y-direction in the boundary layer, i.e., using:

$$\frac{\partial v}{\partial y} = -\frac{\partial u}{\partial x}$$

and noting that Eq. (10.77) gives:

$$\frac{du}{dx} = \frac{u_\infty}{D} F'$$

where $F' = dF/d(x/D)$, it follows that because v is zero at the wall, i.e., at $y = 0$:

$$v = -u_\infty (y/D) F' \tag{10.78}$$

Because F is assumed to be a known function, this equation allows the value of v at any value of y in the boundary layer to be found.

Attention will be given to flow over a surface with a specified temperature and the following dimensionless variables are introduced for convenience:

$$X = \frac{x}{D},\ Y = \frac{y}{D} Pe_D^{1/2},\ U = \frac{u}{u_\infty},\ V = \frac{v}{u_\infty} Pe_D^{1/2},\ \theta = \frac{T - T_\infty}{T_{wr} - T_\infty} \tag{10.79}$$

where T_∞ is the temperature in the free-stream, T_{wr} is a measure of the wall temperature, and Pe_D is the Peclet number based on u_∞ and D.

Eqs. (10.77) and (10.78) then give:

$$U = F(X),\ V = -Y \frac{dF}{dX} \tag{10.80}$$

In terms of the above dimensionless variables, the energy equation for the boundary layer becomes:

$$U \frac{\partial \theta}{\partial X} + V \frac{\partial \theta}{\partial Y} = \frac{\partial^2 \theta}{\partial Y^2} \tag{10.81}$$

The boundary conditions on the solution are:

$$Y = 0 : \theta = \theta_w(X),\ Y \text{ large} : \theta \to 0 \tag{10.82}$$

A forward-marching implicit finite-difference solution of the energy equation will again be considered. In order to obtain this solution, a series of nodal lines running parallel to the x- and y-axes are again introduced as shown in Fig. 10.15.

A uniform grid spacing in the Y-direction will be used here. Consider, as was done in Chapter 3, the nodal points shown in Fig. 10.16.

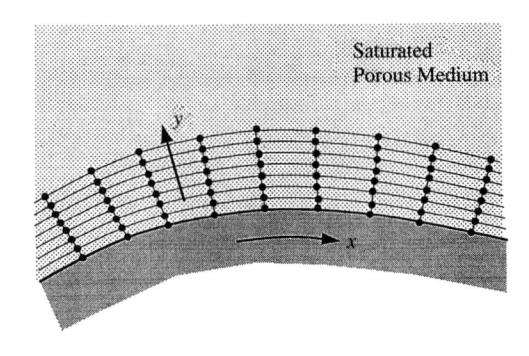

CHAPTER 10: Convective Heat Transfer Through Porous Media   509

FIGURE 10.15
Nodal lines used in numerically solving the energy equation in the boundary layer.

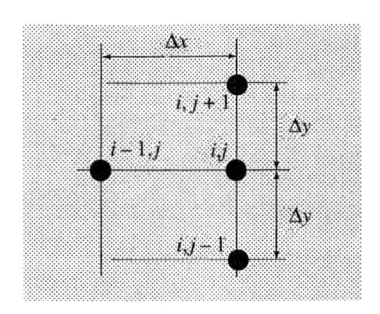

FIGURE 10.16
Nodal lines used in deriving finite-difference approximations.

At any stage of the solution, θ is known at all points on the $(i-1)$ line and must be determined at all points on the i-line. Because U and V are known at all points in the boundary layer and because U is assumed to depend only on X, the following finite-difference approximations are introduced:

$$\left(U\frac{\partial \theta}{\partial X}\right)\bigg|_{i,j} = U_i \frac{(\theta_{i,j} - \theta_{i-1,j})}{\Delta X} \tag{10.83}$$

$$\left(V\frac{\partial \theta}{\partial Y}\right)\bigg|_{i,j} = V_{i,j} \frac{(\theta_{i,j+1} - \theta_{i,j-1})}{2\Delta Y} \tag{10.84}$$

$$\frac{\partial^2 \theta}{\partial Y^2}\bigg|_{i,j} = \frac{(\theta_{i,j+1} + \theta_{i,j-1} - 2\theta_{i,j})}{\Delta Y^2} \tag{10.85}$$

Substituting these into the energy equation then gives, on rearrangement, an equation that has the form:

$$E_j \theta_{i,j} + F_j \theta_{i,j+1} + G_j \theta_{i,j-1} = H_j \tag{10.86}$$

where the coefficients in this equation are given by:

$$E_j = \left(\frac{U_i}{\Delta X}\right) + \left(\frac{2}{\Delta Y^2}\right) \tag{10.87}$$

$$F_j = \left(\frac{V_{i,j}}{\Delta Y}\right) - \left(\frac{1}{\Delta Y^2}\right) \tag{10.88}$$

510 Introduction to Convective Heat Transfer Analysis

$$G_j = -\left(\frac{V_{i,j}}{\Delta Y}\right) - \left(\frac{1}{\Delta Y^2}\right) \qquad (10.89)$$

$$H_j = \left(\frac{U_i \theta_{i-1,j}}{\Delta X}\right) \qquad (10.90)$$

Because the boundary conditions give $\theta_{i,1}$ and $\theta_{i,N}$, the outermost grid point being chosen to lie outside both the boundary layers, the application of Eq.(10.86) to each of the internal points on the i-line, i.e., $j = 2, 3, 4, \ldots, N - 2, N - 1$ again gives a set of N equations in the N unknown values of θ. Because $\theta_{i,N}$ is 0, this set of equations has the following form:

$$\theta_{i,1} = \theta_w$$
$$E_2 \theta_{i,2} + F_2 \theta_{i,3} + G_2 \theta_{i,1} = H_2$$
$$E_3 \theta_{i,3} + F_3 \theta_{i,4} + G_3 \theta_{i,2} = H_3$$
$$\cdot \qquad (10.91)$$
$$\cdot$$
$$E_{N-1} \theta_{i,N-1} + F_{N-1} \theta_{i,N} + G_{N-1} \theta_{i,N-2} = H_{N-1}$$
$$\theta_{i,N} = 0$$

i.e., has the form:

$$\begin{bmatrix} 1 & 0 & 0 & 0 & 0 & \cdot & 0 & 0 & 0 \\ G_2 & E_2 & F_2 & 0 & 0 & \cdot & 0 & 0 & 0 \\ 0 & G_3 & E_3 & F_3 & 0 & \cdot & 0 & 0 & 0 \\ 0 & 0 & G_4 & E_4 & F_4 & \cdot & 0 & 0 & 0 \\ \cdot & \cdot & \cdot & \cdot & \cdot & \cdot & \cdot & \cdot & \cdot \\ \cdot & \cdot & \cdot & \cdot & \cdot & \cdot & \cdot & \cdot & \cdot \\ 0 & 0 & 0 & 0 & 0 & \cdot & G_{N-1} & E_{N-1} & F_{N-1} \\ 0 & 0 & 0 & 0 & 0 & \cdot & 0 & 0 & 1 \end{bmatrix} \begin{bmatrix} \theta_{i,1} \\ \theta_{i,2} \\ \theta_{i,3} \\ \theta_{i,4} \\ \cdot \\ \cdot \\ \theta_{i,N-1} \\ \theta_{i,N} \end{bmatrix} = \begin{bmatrix} \theta_w \\ H_2 \\ H_3 \\ H_4 \\ \cdot \\ \cdot \\ H_{N-1} \\ 0 \end{bmatrix}$$

i.e., has the form:

$$Q_T \theta_{i,j} = R_T \qquad (10.92)$$

where Q_T is a tridiagonal matrix. This equation can be solved, as discussed in Chapter 3, using the standard tridiagonal matrix solver algorithm often termed the Thomas algorithm. It should be noted that at any stage of the procedure, it is only necessary to know the values of the variables on two adjacent grid lines.

In the above procedure, it was assumed that the outermost grid point, i.e., the N-point, was always outside the boundary layer. To ensure that this is always the case, the solution starts with a relatively small number of j-grid lines and the boundary layer growth is monitored. When the boundary layer has reached to within a few nodal points of the outermost point, the number of j-lines is increased. Since the additional points so generated initially lie outside the boundary layer, the values of the variables on these points are initially known. The procedure was discussed in Chapter 3.

Once the distribution of θ has been determined using this procedure, any other property of the flow can be determined. In the present discussion, the heat transfer

CHAPTER 10: Convective Heat Transfer Through Porous Media **511**

rate at the wall, q_w, is the most important such property. This is, of course, given by Fourier's law as:

$$q_w = -k_a \left.\frac{\partial T}{\partial y}\right|_{y=0} \tag{10.93}$$

In terms of the dimensionless variables being used here this gives:

$$\frac{q_w D}{(T_{wr} - T_\infty)k_a} = -\left.\frac{\partial \theta}{\partial Y}\right|_{Y=0} Pe_D^{1/2} \tag{10.94}$$

This can be rearranged as:

$$\frac{Nu_D}{\sqrt{Pe_D}} = -\left.\frac{\partial \theta}{\partial Y}\right|_{Y=0} \left(\frac{1}{\theta_w}\right) \tag{10.95}$$

where Nu_D is the local Nusselt number and θ_w is the local dimensionless wall temperature. Now since point $j = 1$ lies on the wall:

$$\left.\frac{\partial \theta}{\partial Y}\right|_{Y=0} = \left(\frac{\partial \theta}{\partial Y}\right)_{i,1} \tag{10.96}$$

In order to determine this from the values of θ calculated at the nodal points, it is noted that, to the same degree of approximation as previously used:

$$\theta_{i,2} = \theta_{i,1} + \left(\frac{\partial \theta}{\partial Y}\right)_{i,1} \Delta Y + \left(\frac{\partial^2 \theta}{\partial Y^2}\right)_{i,1} \frac{\Delta Y^2}{2!} \tag{10.97}$$

But the application of the boundary layer energy equation to conditions at the wall gives:

$$\left.\frac{\partial^2 \theta}{\partial Y^2}\right|_{Y=0} = \left(\frac{\partial^2 \theta}{\partial Y^2}\right)_{i,1} = 0 \tag{10.98}$$

Therefore, Eq.(10.97) can be rearranged to give:

$$\left.\frac{\partial \theta}{\partial Y}\right|_{i,1} = \left(\frac{\theta_{i,2} - \theta_{i,1}}{\Delta Y}\right) \tag{10.99}$$

and Eq.(10.95) then gives:

$$\frac{Nu_D}{\sqrt{Pe_D}} = \left(\frac{\theta_{i,1} - \theta_{i,2}}{\Delta Y}\right)\left(\frac{1}{\theta_w}\right) \tag{10.100}$$

θ_w being, of course, the same as $\theta_{i,1}$.

In order to illustrate the use of this procedure, consider flow over an isothermal cylinder of diameter D. In this case:

$$U = 2\sin(2X) \tag{10.101}$$

Eq. (10.80) then gives:

$$V = -4Y\cos(2X) \tag{10.102}$$

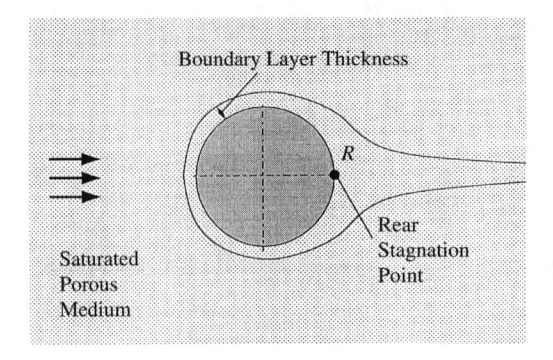

FIGURE 10.17
Boundary layer thickness variation around cylinder.

Using these equations, the numerical procedure outlined above can be used to find the variation of the local heat transfer rate in the form of $Nu_D/Pe_D^{1/2}$ around the cylinder. The solution procedure does not really apply near the rear stagnation point (R in Fig. 10.17) because the effective boundary thickness becomes very large in this area as indicated in Fig. 10.17. This is however, of little practical importance because the heat transfer rate is very low in the region of the rear stagnation point.

Once the distribution of the local heat transfer rate about the the cylinder has been found, the mean heat transfer can be found using:

$$\overline{q_w} = \frac{1}{\pi} \int_0^\pi q_w \, dX \tag{10.103}$$

i.e., if $\overline{Nu_D}$ is the mean Nusselt number:

$$\frac{\overline{Nu_D}}{\sqrt{Pe_D}} = \frac{1}{\pi} \int_0^\pi \frac{Nu_D}{\sqrt{Pe_D}} \, dX \tag{10.104}$$

A simple computer program, PORCYL, written in FORTRAN and based on the procedure outlined above is available as discussed in the Preface. The variation of $Nu_D/\sqrt{Pe_D}$ with X given by this program is shown in Fig. 10.18.

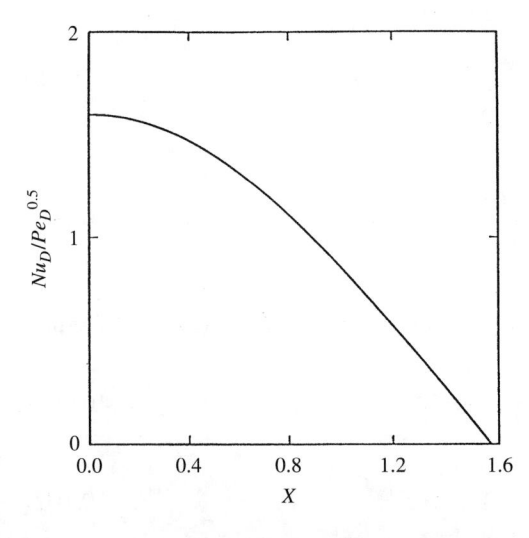

FIGURE 10.18
Variation of local dimensionless heat transfer rate around cylinder.

CHAPTER 10: Convective Heat Transfer Through Porous Media 513

It will be seen from Fig. 10.18 that in the forward stagnation point region the heat transfer rate obtained is in agreement with that previously given for the stagnation point region, i.e., $Nu_D/Pe_D^{1/2} = 1.596$.

The program gives the mean heat transfer rate from the cylinder as:

$$\frac{\overline{Nu_D}}{\sqrt{Pe_D}} = 1.015 \qquad (10.105)$$

EXAMPLE 10.3. Consider the same situation as dealt with in Example 10.2, i.e., in which a heat pump system rejects heat to a water flow that trickles through a bed of small pebbles in the ground, the water velocity in the bed being estimated to be 1 cm/s. In the present case, the heat exchanger consists of a series of long 2.5-cm diameter pipes that are arranged with their axis normal to the water flow. If the pipes are again at a temperature of 20°C and if the ground water again has a temperature of 10°C, find the heat transfer from the pipes to the water per meter length of pipe. As in Example 10.2, assume that the pebble bed has a porosity of 0.35.

Solution. As discussed in Example 10.2, the water properties are evaluated at the mean temperature of the water in the boundary layer, i.e., at 15°C. At this temperature, for water:

$$\rho_f = 999.1 \text{ kg/m}^3$$
$$c_{pf} = 4186 \text{ J/kg K}$$
$$k_f = 0.59 \text{ W/m K}$$

As was also discussed in Example 10.2, the thermal conductivity of the porous material will be assumed to be given by:

$$k_a = \phi k_f + (1 - \phi)k_s = 1.38 \text{ W/m K}$$

Hence:

$$\alpha_a = \frac{k_s}{\rho_f c_{pf}} = \frac{1.38}{999.1 \times 4186} = 0.33 \times 10^{-6} \text{ m}^2/\text{s}$$

The Peclet number based on the pipe diameter is therefore given by:

$$Pe_D = \frac{0.01 \times 0.025}{0.33 \times 10^{-6}} = 757.6$$

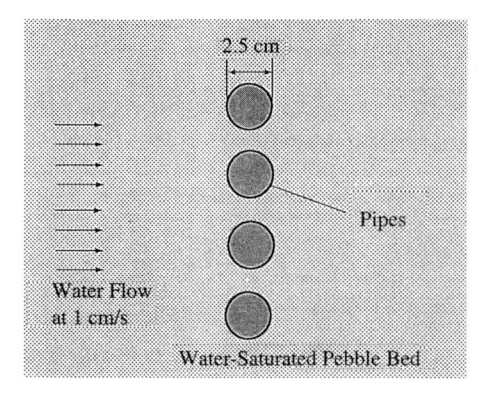

FIGURE E10.3

This is assumed to be large enough for the boundary layer assumptions to apply. It is also assumed that the pipes are far enough apart to avoid interaction between the flows over the pipes. Eq. (10.105) is therefore used and gives:

$$\frac{\bar{h} \times 0.025}{1.38} = 1.015 \times (757.6)^{1/2}$$

Therefore:

$$\bar{h} = 1542 \text{ W/m}^2$$

Considering a 1-m length of 1 pipe:

$$Q = 1542 \times \pi \times 0.025 \times 1 \times (20 - 10) = 1.21 \text{ kW}$$

Hence, the heat transfer rate per meter length of the pipes is 1.21 kW.

The approximate integral equation method that was discussed in Chapters 2 and 3 can also be applied to the boundary layer flows on surfaces in a porous medium. As discussed in Chapters 2 and 3, this integral equation method has largely been superceded by purely numerical methods of the type discussed above. However, integral equation methods are still sometimes used and it therefore appears to be appropriate to briefly discuss the use of the method here. Attention will continue to be restricted to two-dimensional constant fluid property forced flow.

As discussed in Chapters 2 and 3, in the integral method it is assumed that the boundary layer has a definite thickness and the overall or integrated momentum and thermal energy balances across the boundary layer are considered. In the case of flow over a body in a porous medium, if the Darcy assumptions are used, there is, as discussed before, no velocity boundary layer, the velocity parallel to the surface near the surface being essentially equal to the surface velocity given by the potential flow solution. For flow over a body in a porous medium, therefore, only the energy integral equation need be considered. This equation was shown in Chapter 2 to be:

$$\frac{d}{dx}\left[\int_0^{\delta_T} u(T - T_\infty)\,dy\right] = \frac{q_w}{\rho_f c_{pf}} \qquad (10.106)$$

where δ_T is the boundary layer thickness as shown in Fig. 10.19.

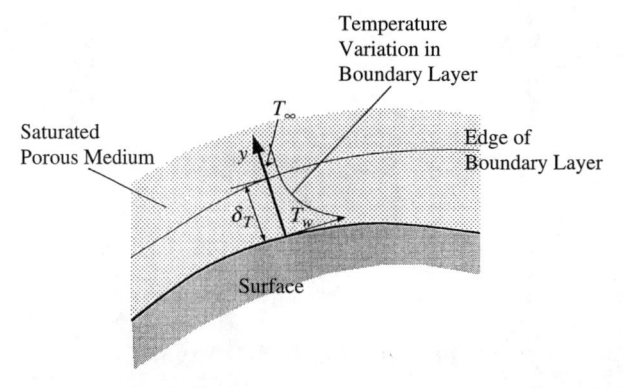

FIGURE 10.19
Boundary layer on a surface in a porous medium.

Because δ_T is assumed to be small, the velocity u can, as discussed before, be assumed to be independent of y and equal to the value of u at the surface at the particular value of x being considered that is given by the potential flow solution, this value here being designated by $u_1(x)$. Eq. (10.106) can therefore be written as:

$$\frac{d}{dx}\left[u_1\int_0^{\delta_T}(T-T_\infty)dy\right] = \frac{q_w}{\rho_f c_{pf}} \tag{10.107}$$

In the integral method, the form of the temperature profile is assumed. For example, if a situation in which the wall temperature variation is specified is considered then it is assumed that:

$$\frac{T-T_\infty}{T_w-T_\infty} = f\left(\frac{y}{\delta_T}\right) \tag{10.108}$$

where the form of the function f is assumed, any coefficients in this function being determined by boundary conditions on temperature. Using Eq. (10.108), the integral equation as given in Eq. (10.107) can be written as:

$$\left[\int_0^1 f d\left(\frac{y}{\delta_T}\right)\right]\frac{d}{dx}[u_1(T_w-T_\infty)\delta_T] = \frac{q_w}{\rho_f c_{pf}} \tag{10.109}$$

The heat transfer rate at the wall is given by Fourier's Law as:

$$q_w = -k_a\left.\frac{\partial T}{\partial y}\right|_{y=0} = \frac{-k_a(T_w-T_\infty)}{\delta_T}\left.\frac{df}{d(y/\delta_T)}\right|_{y/\delta_T=0} = \frac{-k_a(T_w-T_\infty)}{\delta_T}f_0' \tag{10.110}$$

where k_a is, as before, the apparent thermal conductivity of the porous medium.
Substituting this into Eq. (10.109) then gives:

$$\left[\int_0^1 f d\left(\frac{y}{\delta_T}\right)\right]\frac{d}{dx}[u_1(T_w-T_\infty)\delta_T] = -\alpha_a\frac{(T_w-T_\infty)}{\delta_T}f_0' \tag{10.111}$$

where α_a is, as before, the apparent thermal diffusivity, i.e., $k_a/\rho_f c_{pf}$.

In order to illustrate the procedure, the temperature profile will here be assumed to be described by a second-order polynomial, i.e., it will be assumed that in the boundary layer:

$$T = a + by + cy^2 \tag{10.112}$$

The coefficients in this equation, i.e., a, b, and c, are determined by applying the boundary conditions on temperature at the inner and outer edges of the thermal boundary layer. Because the case where the wall temperature variation is specified is being considered, these boundary conditions are:

$$\begin{array}{l}\text{At } y = 0: T = T_w\\ \text{At } y = \delta_T: T = T_\infty\\ \text{At } y = \delta_T: \partial T/\partial y = 0\end{array} \tag{10.113}$$

The first of these conditions follows from the requirement that the fluid in contact with the wall must attain the same temperature as the wall. The other two conditions follow from the requirement that the boundary layer temperature profile must blend smoothly into the free-stream temperature distribution at the outer edge of the boundary layer.

Applying the boundary conditions given in Eqs. (10.113) to Eq. (10.112) then gives:

$$T_w = a$$
$$T_1 = a + b\delta_T + c\delta_T^2 \quad (10.114)$$
$$0 = b + 2c\delta_T$$

This is a set of three equations in the three unknown coefficients. Solving between them then gives:

$$a = T_w, \quad b = -2(T_w - T_1)/\delta_T, \quad c = (T_w - T_1)/\delta_T^2 \quad (10.115)$$

Substituting these values into Eq. (10.112) and rearranging gives the temperature profile as:

$$\frac{T - T_\infty}{T_w - T_\infty} = 1 - 2\left(\frac{y}{\delta_T}\right) + \left(\frac{y}{\delta_T}\right)^2 \quad (10.116)$$

Comparing Eq. (10.116) with Eq. (10.108) shows that:

$$f = 1 - 2\left(\frac{y}{\delta_T}\right) + \left(\frac{y}{\delta_T}\right)^2 \quad (10.117)$$

Hence:

$$\int_0^1 f\, d\left(\frac{y}{\delta_T}\right) = \frac{1}{3}, \quad f_0' = \left.\frac{df}{d(y/\delta_T)}\right|_{y/\delta_T = 0} = -2 \quad (10.118)$$

Substituting these values into Eq. (10.111) gives:

$$\frac{d}{dx}\left[u_1(T_w - T_\infty)\delta_T\right] = 6\alpha_a \frac{(T_w - T_\infty)}{\delta_T} \quad (10.119)$$

If attention is restricted to a surface that has a uniform temperature, i.e., for which $T_w - T_\infty$ is constant, Eq. (10.119) becomes:

$$\frac{d}{dx}(u_1 \delta_T) = \frac{6\alpha_a}{\delta_T} \quad (10.120)$$

Consider, first, the case of flow over a flat plate for which, as discussed before, u_1 is a constant, equal say to u_∞. In this case, Eq. (10.120) gives:

$$\delta_T \frac{d\delta_T}{dx} = \frac{6\alpha_a}{u_\infty} \quad (10.121)$$

This equation can be directly integrated, because δ_T is by assumption 0 at $x = 0$, (see Fig. 10.20), to give:

$$\frac{\delta_T^2}{2} = \frac{6\alpha_a x}{u_\infty} \quad (10.122)$$

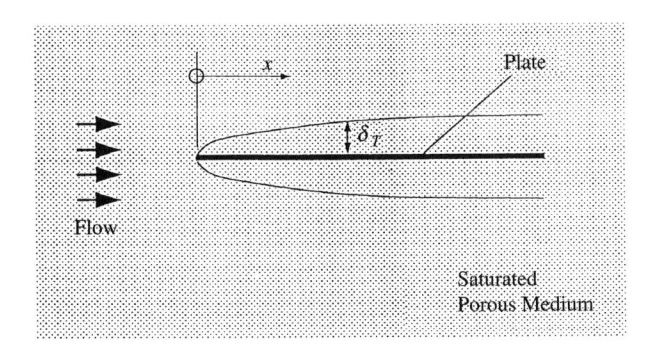

CHAPTER 10: Convective Heat Transfer Through Porous Media 517

FIGURE 10.20
Boundary layer growth on a flat plate in a porous medium.

i.e.:

$$\frac{\delta_T}{x} = \frac{\sqrt{12}}{Pe_x^{0.5}} = \frac{3.464}{Pe_x^{0.5}} \tag{10.123}$$

where Pe_x is, as before, the Peclet number based on x.

Now Eq. (10.110) gives using Eq. (10.118):

$$q_w = 2k_a \frac{(T_w - T_\infty)}{\delta_T} \tag{10.124}$$

which when combined with Eq. (10.123) gives for flow over a flat plate in a porous medium:

$$\frac{q_w x}{k_a(T_w - T_\infty)} = \frac{2}{3.464} Pe_x^{0.5}, \text{ i.e., } Nu_x = 0.577\, Pe_x^{0.5} \tag{10.125}$$

This equation gives the variation of the local heat transfer rate along the plate. The mean heat transfer rate is given by:

$$\overline{q_w} = \frac{1}{L} \int_0^L q_w\, dx$$

i.e., using Eq. (10.125):

$$\overline{q_w} = \frac{0.577 k_a(T_w - T_\infty)}{L} \left(\frac{u_\infty}{\alpha_a}\right)^{1/2} \int_0^L \frac{dx}{x^{1/2}} = \frac{0.577 k_a(T_w - T_\infty)}{L} \left(\frac{u_\infty}{\alpha_a}\right)^{1/2} 2L^{1/2}$$

which can be rearranged to give:

$$Nu_L = 1.154 Pe_L^{1/2} \tag{10.126}$$

where Nu_L is the mean Nusselt number for the plate, i.e., $\overline{h}L/k_a$, $\overline{h}$ being the mean heat transfer coefficient.

The results obtained using the integral equation method as given in Eqs. (10.125) and (10.126) agree to within about 2% with the exact result derived earlier.

EXAMPLE 10.4. Consider forced convective boundary layer flow over a flat plate in a porous medium, the plate being aligned with the forced flow. The surface temperature of the plate is given by $T_w - T_\infty = Cx^m$. Use the integral equation method to determine an expression for the variation of the local heat transfer rate along the plate.

Solution. The following applies to flow over a flat plate whatever the form of the wall temperature variation:

$$u_\infty \frac{d}{dx}[(T_w - T_\infty)\delta_T] = 6\alpha_a \frac{(T_w - T_\infty)}{\delta_T}$$

which can be written as:

$$\frac{d\delta_T}{dx} + \frac{\delta_T}{(T_w - T_\infty)}\frac{d}{dx}(T_w - T_\infty) = \frac{6\alpha_a}{u_\infty \delta_T}$$

Substituting:

$$T_w - T_\infty = Cx^m$$

into this equation gives:

$$\frac{d\delta_T}{dx} + \frac{m\delta_T}{x} = \frac{6\alpha_a}{u_\infty \delta_T}$$

i.e.:

$$\delta_T \frac{d\delta_T}{dx} + \frac{m\delta_T^2}{x} = \frac{6\alpha_a}{u_\infty} \quad (i)$$

An inspection of the form of this equation shows that the solution for δ_T must have the form:

$$\delta_T = Ax^{1/2} \quad (ii)$$

where A is a constant. Substituting this into Eq. (i) then gives:

$$\frac{A^2}{2} + mA^2 = \frac{6\alpha_a}{u_\infty}$$

i.e.:

$$A^2 = \frac{12\alpha_a}{u_\infty(1 + 2m)}$$

Substituting this into Eq. (ii) then gives:

$$\frac{\delta_T}{x} = \sqrt{\frac{12}{1 + 2m}} \frac{1}{Pe_x^{0.5}} \quad (iii)$$

But the heat transfer rate is given by:

$$q_w = 2k_a \frac{(T_w - T_\infty)}{\delta_T}, \quad \text{i.e., } Nu_x = \frac{2}{\delta_T/x}$$

Substituting Eq. (iii) into this equation then gives:

$$Nu_x = \left[\frac{2(1 + 2m)}{\sqrt{12}}\right]Pe_x^{0.5} = 0.577(1 + 2m)Pe_x^{0.5}$$

As another example of the use of the energy integral equation, consider again the flow in the region of a stagnation point on an isothermal body as shown in Fig. 10.21.

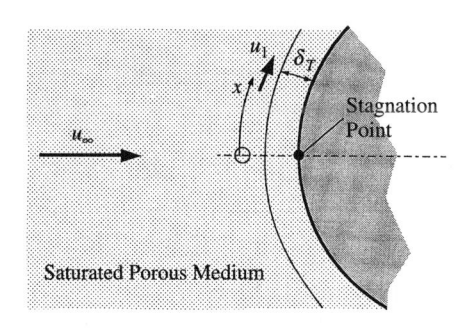

FIGURE 10.21
Flow considered in integral equation solution for stagnation region.

In this case, as discussed before, for small x:

$$u = 4u_\infty \left(\frac{x}{D}\right) \tag{10.127}$$

where D is twice the radius of curvature of the leading edge and u_∞ is the velocity in the free-stream ahead of the body.

Substituting this expression for u into Eq. (10.120) gives:

$$\frac{4u_\infty}{D}\frac{d}{dx}(x\delta_T) = \frac{6\alpha_a}{\delta_T}$$

i.e.:

$$\delta_T \frac{d}{dx}(x\delta_T) = \frac{1.5\alpha_a D}{u_\infty}$$

i.e.:

$$\delta_T^2 + x\delta_T \frac{d\delta_T}{dx} = \frac{1.5\alpha_a D}{u_\infty} \tag{10.128}$$

Because the right-hand side of this equation is independent of x, it follows that in the stagnation point region $d\delta_T/dx = 0$ and in this region therefore:

$$\delta_T^2 = \frac{1.5\alpha_a D}{u_\infty} \tag{10.129}$$

Substituting this result into Eq. (10.124) gives:

$$\frac{q_w}{k_a(T_w - T_\infty)} = \frac{2}{1.5^{1/2}}\left(\frac{u_\infty}{\alpha_a D}\right)^{0.5}$$

i.e.:

$$Nu_D = 1.633 Pe_D^{0.5} \tag{10.130}$$

Comparing this result with the exact result given in Eq. (10.76) shows that the integral equation method agrees in this case to within 3% of the exact solution.

As a last example of the use of the integral equation method consider again two-dimensional flow about an isothermal cylinder in a porous medium. The situation considered is shown in Fig. 10.22.

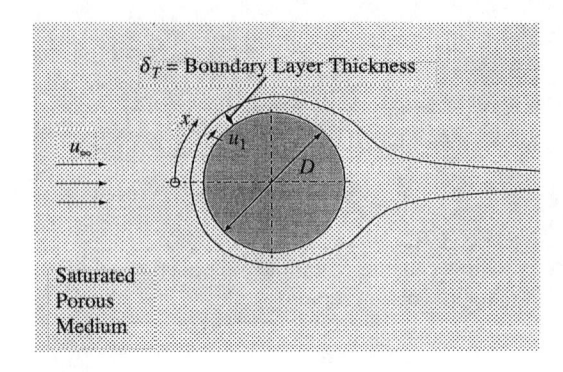

520 Introduction to Convective Heat Transfer Analysis

FIGURE 10.22
Flow considered in integral equation solution for flow over a cylinder.

In this case, as previously discussed:

$$u = 2u_\infty \sin\left(\frac{2x}{D}\right) \tag{10.131}$$

Eq. (10.120) therefore gives for this case:

$$4\cos\left(\frac{2x}{D}\right)\left(\frac{\delta_T}{D}\right) + 2\sin\left(\frac{2x}{D}\right)\frac{d\delta_T}{dx} = \frac{6\alpha_a}{\delta_T u_\infty} \tag{10.132}$$

The following are then defined for convenience:

$$X = \frac{x}{D}, \quad \Delta = \frac{\delta_T}{D} Pe_D^{1/2}, \quad Pe_D = \frac{u_\infty D}{\alpha_a} \tag{10.133}$$

In terms of these variables, Eq. (10.132) gives:

$$2\Delta^2 \cos(2X) + \sin(2X)\Delta \frac{d\Delta}{dX} = 3 \tag{10.134}$$

For small values of X, i.e., in the region of the forward stagnation point, Δ will, as discussed before, be a constant and is given by Eq. (10.134) as:

$$\Delta_0^2 = \frac{3}{2}, \quad \text{i.e.,} \quad \Delta_0 = 1.225 \tag{10.135}$$

Starting with this value of Δ at $X = 0$, Eq. (10.134) can be integrated to give the variation of Δ with X around the cylinder. For this purpose, Eq. (10.134) is written as:

$$\frac{d\Delta}{dX} = \frac{3 - 2\Delta^2 \cos(2X)}{\Delta \sin(2X)} \tag{10.136}$$

The variation of Δ with X obtained by the integration of this equation is shown in Fig. 10.23.

Once the variation of Δ has been obtained in this way, the local heat transfer rate distribution can be obtained by using Eq. (10.124) which can be rearranged to give:

$$\frac{q_w D}{k_a(T_w - T_\infty)} = \frac{2Pe_D^{0.5}}{\Delta}, \quad \text{i.e.,} \quad \frac{Nu_D}{Pe_D^{0.5}} = \frac{2}{\Delta} \tag{10.137}$$

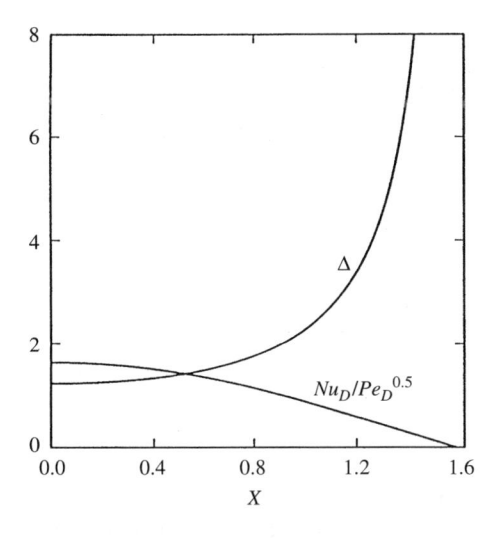

FIGURE 10.23
Variation of Δ and $Nu_D/Pe_D^{0.5}$ with X around cylinder.

Using the calculated variation of Δ with X, this equation allows the variation of $Nu_D/Pe_D^{0.5}$ with X to be determined. The variation so obtained is also shown in Fig. 10.23. With the variation of the local dimensionless heat transfer rate obtained in this way, the mean heat transfer rate for the cylinder can be obtained by using:

$$\overline{q_w} = \frac{1}{\pi/2} \int_0^{\pi/2} q_w \, dX$$

i.e., if $\overline{Nu_D}$ is the mean Nusselt number:

$$\frac{\overline{Nu_D}}{Pe_D^{0.5}} = \frac{1}{\pi/2} \int_0^{\pi/2} \frac{Nu_D}{Pe_D^{0.5}} \, dX \qquad (10.138)$$

Using this with the calculated variation of $Nu_D/Pe_D^{0.5}$ with X gives:

$$\frac{\overline{Nu_D}}{Pe_D^{0.5}} = 1.039 \qquad (10.139)$$

This value agrees to within 3% with that obtained by numerically solving the boundary layer energy equation.

10.6 FULLY DEVELOPED DUCT FLOW

While flow in a duct containing a porous medium [19] does not occur extensively in practice there are a few important situations that can be represented by this type of flow. For example, it has been suggested that filling a duct with a porous medium can be used as a method of enhancing the heat transfer rate from the walls of the duct. The disadvantage of this is, of course, that a much higher pressure drop will be incurred. Some geological flows can also be adequately modeled as flow through a porous medium-filled duct with heated or cooled walls.

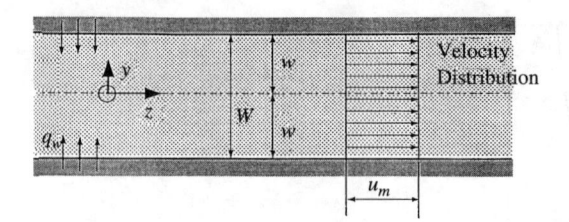

522 Introduction to Convective Heat Transfer Analysis

FIGURE 10.24
Flow situation considered.

In order to illustrate how fully developed flow through a duct filled with a porous medium can be analyzed, consider flow through a wide duct with plane walls, i.e., flow between parallel plates, with a uniform heat flux at the wall. The flow situation is thus as shown in Fig. 10.24.

Because in the porous media flow model being used, the effects of viscosity are assumed to be negligible, the velocity will be uniform across the duct, i.e., the velocity will be equal to u_m everywhere and the cross-stream velocity component, v, will, therefore, be zero everywhere. It will further be assumed that the temperature gradients across the flow, i.e., in the y-direction, will be much greater than those in the z-direction because $W < L$, L being the length of the duct. With these assumptions, the governing equation, i.e., Eq. (10.34), reduces to:

$$u_m \frac{\partial T}{\partial z} = \alpha_a \frac{\partial^2 T}{\partial y^2} \tag{10.140}$$

Now it is being assumed that the flow is fully developed, i.e., that if T_c is the centerline temperature:

$$\frac{T_w - T}{T_w - T_c} = f\left(\frac{y}{w}\right) \tag{10.141}$$

the function f not being dependent on z, i.e.:

$$\frac{\partial}{\partial z}\left(\frac{T_w - T}{T_w - T_c}\right) = 0 \tag{10.142}$$

From this it follows that:

$$\left(\frac{dT_w}{dz} - \frac{\partial T}{\partial z}\right)\left(\frac{T_w - T}{T_w - T_c}\right)\left(\frac{dT_w}{dz} - \frac{dT_c}{dz}\right) = 0 \tag{10.143}$$

But the heat transfer rate at the wall is given by:

$$q_w = +k_a \left.\frac{\partial T}{\partial y}\right|_{y=w} \tag{10.144}$$

the (+) sign occurring because the heat flux is taken as positive from the wall to the fluid, i.e., in the negative y-direction.

Substituting Eq. (10.141) into Eq. (10.144) then gives:

$$q_w = -\frac{k_a(T_w - T_c)}{w} \left.\frac{df}{d(y/w)}\right|_{y=w} \tag{10.145}$$

CHAPTER 10: Convective Heat Transfer Through Porous Media   523

Because q_w is constant and because f does not depend on z, this equation shows that $(T_w - T_c) =$ constant, i.e., that

$$\frac{dT_w}{dz} = \frac{dT_c}{dz} \tag{10.146}$$

Substituting this into Eq. (10.143) then gives:

$$\frac{dT_w}{dz} = \frac{\partial T}{\partial z} \tag{10.147}$$

Hence, Eq. (10.140), the governing equation, can be written as:

$$u_m \frac{dT_w}{dz} = \alpha_a \frac{\partial^2 T}{\partial y^2} \tag{10.148}$$

Integrating this once and noting that the left-hand side does not depend on y gives:

$$\frac{\partial T}{\partial y} = \frac{u_m}{\alpha_a} \frac{dT_w}{dz} y + C_1 \tag{10.149}$$

where C_1 is a constant of integration. But, with the y-coordinate defined as in Fig. 10.24, the following boundary condition applies because the temperature profile is symmetrical about the center line:

$$y = 0: \quad \frac{\partial T}{\partial y} = 0 \tag{10.150}$$

This gives $C_1 = 0$.

Integrating again then gives:

$$T = \frac{u_m}{\alpha_a} \frac{dT_w}{dz} \frac{y^2}{2} + C_2 \tag{10.151}$$

where C_2 is a second constant of integration. But the other boundary condition is:

$$y = w(= W/2): \quad T = T_w \tag{10.152}$$

Hence:

$$T_w = \frac{u_m}{\alpha_a} \frac{dT_w}{dz} \frac{w^2}{2} + C_2 \tag{10.153}$$

i.e.:

$$C_2 = T_w - \frac{u_m}{\alpha_a} \frac{dT_w}{dz} \frac{w^2}{2} \tag{10.154}$$

Substituting this back into Eq. (10.151) then gives:

$$T = T_w - \frac{u_m}{\alpha_a} \frac{dT_w}{dz} \left(\frac{w^2}{2} - \frac{y^2}{2} \right) \tag{10.155}$$

524 Introduction to Convective Heat Transfer Analysis

The wall temperature gradient is now related to the specified uniform wall heat flux q_w by substituting Eq. (10.155) into Eq. (10.144). This gives:

$$q_w = k_a \frac{u_m}{\alpha_a} \frac{dT_w}{dz} w$$

i.e.:

$$\frac{u_m}{\alpha_a} \frac{dT_w}{dz} = \frac{q_w}{k_a w} \qquad (10.156)$$

Substituting this back into Eq. (10.155) gives the temperature profile as:

$$T_w - T = \left(\frac{q_w}{k_a w}\right)\left(\frac{w^2}{2} - \frac{y^2}{2}\right) \qquad (10.157)$$

From this it follows that:

$$T_w - T_c = \left(\frac{q_w}{k_a w}\right) \frac{w^2}{2} \qquad (10.158)$$

Hence:

$$f = \frac{T_w - T}{T_w - T_c} = 1 - \left(\frac{y}{w}\right)^2 \qquad (10.159)$$

Now in duct flow, the Nusselt number is defined in terms of the difference between the wall temperature and the mean fluid temperature, this mean temperature being defined by:

$$T_m = \frac{1}{u_m w} \int_0^w uT\, dy = \frac{1}{w} \int_0^w T\, dy \qquad (10.160)$$

because $u = u_m$ at all y. Using Eq. (10.157), this gives:

$$T_m = \frac{1}{w}\left[T_w w - \left(\frac{q_w}{k_a w}\right)\left(\frac{w^3}{2} - \frac{w^3}{6}\right)\right] = T_w - \frac{q_w}{k_a}\frac{w}{3}$$

i.e.:

$$\frac{q_w w}{(T_w - T_m) k_a} = 3 \qquad (10.161)$$

But the mean Nusselt number is conventionally defined in terms of the full duct width $W (= 2w)$ so Eq. (10.161) is written as:

$$\frac{q_w W}{(T_w - T_m) k_a} = 6$$

i.e.:

$$Nu = 6 \qquad (10.162)$$

Therefore, with fully developed flow in the gap between parallel plates, when this gap is filled with a porous medium and when there is a uniform heat flux at the plates, the Nusselt number is 6. Using the same approach it can be shown that, if the wall temperature is constant, $Nu = 5$. For flow through a pipe filled with a porous medium it can be shown that $Nu = 8$ for the uniform heat flux case and $Nu = 5.78$ for the uniform temperature case. The Nusselt number, Nu, is here based on the pipe diameter, D. Notice that these Nusselt numbers are all much higher than those that would exist with the flow of a pure fluid in the same geometrical situation and that they are based on the effective conductivity. This is basically the reason that has led to the suggestion that filling a channel with a porous medium be used as a means of increasing, i.e., enhancing, the heat transfer rate.

EXAMPLE 10.5. In a heat exchanger, air flows between parallel plates that are separated by 3 mm. The plates are effectively kept at a uniform temperature of 40°C and the air is heated from 10°C to 20°C as it passes through the heat exchanger. The air velocity between the plates is 1 m/s. It has been proposed that the size of the heat exchanger can be reduced by incorporating a loosely packed porous medium into the gap between the plates. Evaluate this proposal by finding the plate lengths with and without the porous medium and by finding the pressure drop across the heat exchanger with and without the porous medium. Neglect any entry region effects.

Solution. First consider the flow without the porous medium. At the mean temperature of 15°C, air has the following properties:

$$\rho = 1.22 \text{ kg/m}^3, \ c_p = 1007 \text{ J/kg K}$$

$$k = 0.026 \text{ W/m K}, \ \mu = 18 \times 10^{-6} \text{ kg/ms}$$

The Reynolds number based on the gap between the plates is therefore:

$$Re = \rho u W/\mu = 1.22 \times 1 \times 0.003/18 \times 10^{-6} = 203$$

The flow is therefore laminar and as a consequence, as shown in Chapter 4, $Nu = 4.12$. Hence:

$$h = Nuk/W = 4.12 \times 0.026/0.003 = 35.7 \text{ W/m}^2 \text{ K}$$

Consider a unit width of a passage.

$$Q = \dot{m}c_p(T_{out} - T_{in}) = \rho u A c_p(T_{out} - T_{in})$$
$$= 1.22 \times 1 \times (0.03 \times 1) \times 1007 \times (20 - 10) = 368.6 \text{ W}$$

But:

$$Q = hA_{wall}(T_w - T_{air}) = h(2 \times L \times 1)(40 - 15)$$

where L is the length of the channels. It has been recalled that there is heat transfer from the top and bottom walls. Using the above results then gives:

$$L = Q/(50h) = 368.6/(50 \times 35.7) = 0.207 \text{ m}$$

Therefore, the length of the heat exchanger is 207 mm. The pressure drop through one channel and therefore across the heat exchanger is given by:

$$\Delta p = 12 \times \frac{\mu u L}{W^2} = 12 \frac{18 \times 10^{-6} \times 1 \times 0.207}{0.003^2} = 4.97 \text{ Pa}$$

Consideration will next be given to the case where the porous medium is incorporated into the system. It will be assumed that:

$$K = 10^{-9} \text{ m}^2, \; k_a = 0.04 \text{ W/m K}$$

In this case, $Nu = 5$. Hence:

$$h = Nuk/W = 5 \times 0.026/0.04 = 81.3 \text{ W/m}^2\text{K}$$

In this case then:

$$L = Q/(50h) = 368.6/(50 \times 81.3) = 0.091 \text{ m}$$

Thus, the required length of the heat exchanger is reduced by more than 50%.

The pressure drop through one channel and therefore across the heat exchanger when the porous medium is in the channel is given by:

$$\Delta p = \frac{\mu u L}{K} = \frac{18 \times 10^{-6} \times 1 \times 0.091}{10^{-9}} = 1638 \text{ Pa}$$

While this is much higher than the pressure drop when there is no porous medium, it may be acceptable under many circumstances.

10.7
NATURAL CONVECTIVE BOUNDARY LAYER FLOWS

Natural convective flows in porous media occur in a number of important practical situations, e.g., in air-saturated fibrous insulation material surrounding a heated body and about pipes buried in water-saturated soils. To illustrate how such flows can be analyzed, e.g., see [20] to [22], attention will be given in this section to flow over the outer surface of a body in a porous medium, the flow being caused purely by the buoyancy forces resulting from the temperature differences in the flow. The simplest such situation is two-dimensional flow over an isothermal vertical flat surface imbedded in a porous medium, this situation being shown schematically in Fig. 10.25.

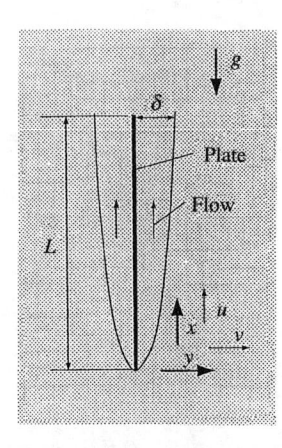

FIGURE 10.25
Natural convective flow over a vertical plate.

It is assumed that the thickness of the layer, i.e., the boundary layer, in which there are significant temperature differences and velocities parallel to the surface is small, i.e., that $\delta/L \ll 1$. In such a flow, as discussed previously, the pressure changes across the boundary layer are negligible and the momentum equation for the x-direction, i.e., the vertical direction, is

$$\frac{\mu_f u}{K} = \beta g \rho_f (T - T_\infty) \tag{10.163}$$

T_∞ being the temperature far from the plate, i.e., outside the boundary layer.

Eq. (10.163) can be written as:

$$u = \frac{\beta g K \rho_f (T - T_\infty)}{\mu_f} \tag{10.164}$$

Because a boundary layer type flow is being assumed, the gradients in the y-direction are much greater than those in the x-direction and the energy equation can, therefore, as before, be approximated by:

$$u \frac{\partial T}{\partial x} + v \frac{\partial T}{\partial y} = \alpha_a \frac{\partial^2 T}{\partial y^2} \tag{10.165}$$

It is also noted that the continuity equation gives:

$$\frac{\partial u}{\partial x} + \frac{\partial v}{\partial y} = 0 \tag{10.166}$$

Eq. (10.166) is, as previously discussed, satisfied by a stream function, ψ, which is defined, as before, by:

$$u = \frac{\partial \psi}{\partial y}, \quad v = -\frac{\partial \psi}{\partial x} \tag{10.167}$$

Eqs. (10.164) and (10.165) can be written in terms of the stream function as follows:

$$\frac{\partial \psi}{\partial y} = \frac{\beta g K \rho_f (T - T_\infty)}{\mu_f} \tag{10.168}$$

$$\frac{\partial \psi}{\partial y} \frac{\partial T}{\partial x} - \frac{\partial \psi}{\partial x} \frac{\partial T}{\partial y} = \alpha_a \frac{\partial^2 T}{\partial y^2} \tag{10.169}$$

The first of these equations can be written as:

$$\frac{\partial^2 \psi}{\partial y^2} = \frac{\beta g K \rho_f}{\mu_f} \frac{\partial T}{\partial y} \tag{10.170}$$

The order of magnitude of ψ in the boundary layer must, therefore, be such that:

$$\frac{\psi}{\delta^2} = o\left(\frac{\beta g K \rho_f}{\mu_f} \frac{\Delta T}{\delta}\right) \tag{10.171}$$

where ΔT is a measure of the temperature change across the boundary layer. Hence:

$$\psi = o\left(\frac{\beta g K \Delta T \delta}{\mu_f}\right) \tag{10.172}$$

Similarly, if the first and last terms in Eq. (10.169) are considered, the orders of magnitude are such that:

$$\frac{\psi \Delta T}{\delta \, x} = o\left(\alpha_a \frac{\Delta T}{\delta^2}\right) \qquad (10.173)$$

i.e.:

$$\psi = o\left(\frac{\alpha_a x}{\delta}\right) \qquad (10.174)$$

Eqs. (10.172) and (10.174) together give:

$$\frac{\alpha_a x}{\delta} = o\left(\frac{\beta g K \rho_f \Delta T \delta}{\mu_f}\right)$$

i.e.:

$$\frac{\delta^2}{x^2} = o\left(\frac{\alpha_a \mu_f}{\beta_g K \rho_f \Delta T x}\right) \qquad (10.175)$$

i.e., defining the following as the Darcy-modified Rayleigh number:

$$Ra = \frac{\beta g K \rho_f \Delta T x}{\alpha_a \mu_f} = \frac{\beta g K \Delta T x}{\alpha_a \nu_f} \qquad (10.176)$$

it follows from Eq. (10.175) that:

$$\frac{\delta}{x} = o(Ra^{-0.5}) \qquad (10.177)$$

and substituting this result into Eq. (10.174) gives:

$$\psi = o(\alpha_a Ra^{0.5}) \qquad (10.178)$$

Now it is to be expected that the stream function and temperature profiles in the boundary layer will be similar, i.e., that in view of the results given in Eqs. (10.177) and (10.178) that:

$$\frac{T - T_\infty}{T_w - T_\infty} = \theta\left(\frac{y}{\delta}\right) = \theta\left(\frac{y}{x} Ra^{0.5}\right) = \theta(\eta) \qquad (10.179)$$

and

$$\frac{\psi}{\alpha_a Ra^{0.5}} = f(\eta) \qquad (10.180)$$

where T_w is the wall temperature which is assumed to be uniform and η is the similarity variable which is given by:

$$\eta = \frac{y}{x} Ra^{0.5} \qquad (10.181)$$

and the Darcy-modified Rayleigh number is given as before by:

$$Ra = \frac{\beta g K (T_w - T_\infty) x}{\alpha_a \nu_f}$$

In terms of the variables introduced in Eqs. (10.179) and (10.180), Eqs. (10.170) and (10.169) become, i.e.:

$$f'' = \theta' \qquad (10.182)$$

and

$$\theta'' = -\frac{f\theta'}{2} \qquad (10.183)$$

respectively.

The boundary conditions on the solution are:

$$y = 0, v = 0, T = T_w, \text{ i.e., } \eta = 0, f = 0, \theta = 1 \qquad (10.184)$$

$$y \text{ large}, u \to 0, T \to \infty, \text{ i.e., } \eta \text{ large } f' \to 0, \theta \to 0$$

Eqs. (10.182) and (10.183), which are a pair of simultaneous ordinary differential equations, can be integrated simultaneously to give the variations of f and θ with η. For this purpose it is convenient to note that Eq. (10.183) can be written as:

$$\frac{\theta''}{\theta'} = -\frac{f}{2} \qquad (10.185)$$

Integrating this gives:

$$\frac{\theta'}{\theta'_0} = \exp\left[-\int_0^\eta \frac{f}{2} d\eta\right] \qquad (10.186)$$

where θ'_0 is the value of θ' at $\eta = 0$.

Integrating this equation again then gives:

$$\theta = \theta'_0 \int_0^\eta \exp\left[-\int_0^\eta \frac{f}{2} d\eta\right] d\eta + 1 \qquad (10.187)$$

the fact that $\theta = 1$ at $\eta = 0$ having been used. If the fact that θ tends to zero at large η is next used, the above equation gives:

$$\theta'_0 = \frac{-1}{\int_0^\infty \exp\left[-\int_0^\eta \frac{f}{2} d\eta\right] d\eta} \qquad (10.188)$$

Substituting this back into the original equation then gives:

$$\theta = 1 - \frac{\int_0^\eta \exp\left[-\int_0^\eta \frac{f}{2} d\eta\right] d\eta}{\int_0^\infty \exp\left[-\int_0^\eta \frac{f}{2} d\eta\right] d\eta} \qquad (10.189)$$

Integrating Eq. (10.182) gives:

$$f' - f'_0 = \int_0^\eta \theta' d\eta = \theta - 1 \qquad (10.190)$$

where f'_0 is the value of f' at $\eta = 0$. But f' and θ both tend to 0 as η tends to infinity. It follows therefore that $f'_0 = 1$ so the above equation gives:

$$f' = \theta \tag{10.191}$$

Integrating this and recalling that $f = 0$ at $\eta = 0$ then gives:

$$f = \int_0^\eta \theta \, d\eta \tag{10.192}$$

These equations are easily solved to give the variations of f and θ with η. The variations so obtained are shown in Fig. 10.26.

The heat transfer rate at the wall is given by:

$$q_w = -k_a \left.\frac{\partial T}{\partial y}\right|_{y=0}$$

i.e., terms of the dimensionless temperature.

$$\frac{q_w x}{k_a(T_w - T_\infty)} = -Ra^{0.5}\theta'|_0 \tag{10.193}$$

But the calculated variation of θ with η gives:

$$\theta'|_0 = 0.444$$

so Eq. (10.193) gives:

$$Nu_x = 0.444 Ra^{0.5} \tag{10.194}$$

where Nu_x is the local Nusselt number based on x. This gives the variation of the local heat transfer rate with x. The mean heat transfer rate is then given by using:

$$\overline{q_w} = \frac{1}{L}\int_0^L q_w \, dx$$

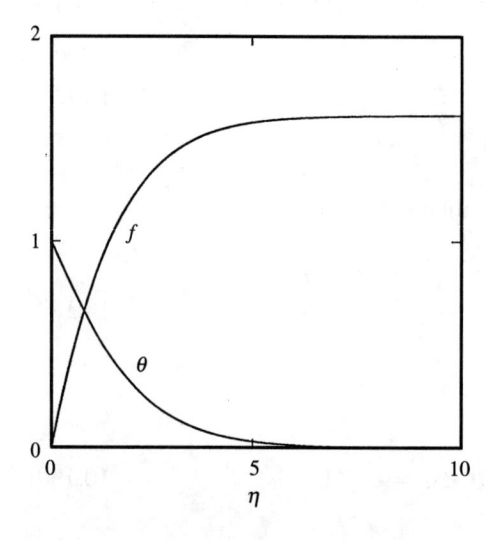

FIGURE 10.26
Variations of dimensionless stream function and temperature with similarity variable, η.

Using Eq. (10.194) in this equation and integrating gives:

$$Nu_L = 0.888 Ra_L^{0.5} \qquad (10.195)$$

where Nu_L is the mean Nusselt number based on the plate length L and Ra_L is the Darcy-modified Rayleigh number based on L.

Natural convective boundary layer-type solutions have been obtained for a number of other geometrical configurations. A number of studies of mixed convective flows in porous media are also available, e.g., [23] to [35].

EXAMPLE 10.6. In order to keep it cool during a journey, a bottle of a medication is kept in the center of a large box containing a loosely packed low quality granular insulating material that has a permeability of 10^{-7} m². The bottle can be approximately modeled as a 20-cm high by 10-cm wide flat plate placed vertically in the insulating material. The initial temperature of the medication is 5°C and the temperature of the insulation is equal to the ambient temperature which is 30°C. Find the initial rate at which heat is transferred to the medication.

Solution. It will be assumed that the insulation material and air have the following properties:

$$K = 10^{-7} \text{ m}^2, \; \alpha_a = 37 \times 10^{-6} \text{ m}^2/\text{s}, \; k_a = 0.046 \text{ W/m K}$$

$$\nu_f = 14.8 \times 10^{-6} \text{ m}^2/\text{s}, \; \beta = 3.5 \times 10^{-3} \text{ K}^{-1}$$

Using these values the Darcy-modified Rayleigh number based on the height of the bottle is:

$$Ra = \frac{\beta g K \Delta T x}{\alpha_a \nu_f} = \frac{0.0035 \times 9.81 \times 0.0000001 \times 0.2 \times (30 - 5)}{0.000037 \times 0.0000148} = 31.3$$

While this is rather small to apply the boundary layer assumptions it will, nevertheless, be assumed that the mean Nusselt number is given by:

$$Nu_L = 0.888 Ra_L^{0.5} = 0.888 \times (31.3)^{0.5} = 4.97$$

The mean heat transfer coefficient is therefore given by:

$$\bar{h} = \frac{Nu_L k_a}{L} = \frac{4.97 \times 0.046}{0.2} = 1.14 \text{ W/m}^2\text{-K}$$

The initial rate of heat transfer from the bottle is therefore given, since there are two sides to the bottle, by:

$$Q = \bar{h} A (T_w - T_\infty) = 1.14 \times (2 \times 0.2 \times 0.1) \times (5 - 30) = -1.14 \text{ W}$$

The negative sign indicates that the heat transfer is to the bottle.

10.8 NATURAL CONVECTION IN POROUS MEDIA-FILLED ENCLOSURES

Heat transfer by natural convection across porous media-filled enclosures occurs in a number of practical situations and will be considered in this section, [36] to [51].

532 Introduction to Convective Heat Transfer Analysis

Situations of this type arise for example in some building applications in which heat is transferred across an insulation-filled enclosure.

Attention will here be restricted to two-dimensional steady flow in a rectangular porous medium-filled enclosure which is, in general, inclined at an angle to the vertical. One wall of the enclosure is kept at a uniform high temperature and the opposite wall is kept at a uniform low temperature. The other two walls of the enclosure are assumed to be adiabatic, i.e., it is assumed that no heat is transferred into or out of these walls. The situation is, therefore, as shown in Fig. 10.27.

Using the Darcy flow model and the Boussinesq approximation, the governing equations are:

$$\frac{\partial y}{\partial x} + \frac{\partial v}{\partial y} = 0 \tag{10.196}$$

$$\frac{\mu_f u}{K} = -\frac{\partial p}{\partial x} + \beta g \rho_f (T - T_C) \cos \phi \tag{10.197}$$

$$\frac{\mu_f v}{K} = -\frac{\partial p}{\partial y} + \beta g \rho_f (T - T_C) \sin \phi \tag{10.198}$$

$$u \frac{\partial T}{\partial x} + v \frac{\partial T}{\partial y} = \left(\frac{k_a}{\rho_f c_{pf}}\right)\left(\frac{\partial^2 T}{\partial x^2} + \frac{\partial^2 T}{\partial y^2}\right) \tag{10.199}$$

The first of these equations expresses, as before, conservation of mass, the second and third express conservation of momentum in the x- and y-directions respectively while the last expresses conservation of energy. The cold wall temperature, T_C, has been taken as the reference temperature.

The solution will, here, be obtained in terms of the stream function. It must be stressed that this is not necessary, solutions in terms of the so-called primitive variables, i.e., u, v, and p, being in fact more widely used. The stream function is, as before, defined by:

$$u = \frac{\partial \psi}{\partial y}, v = -\frac{\partial \psi}{\partial x} \tag{10.200}$$

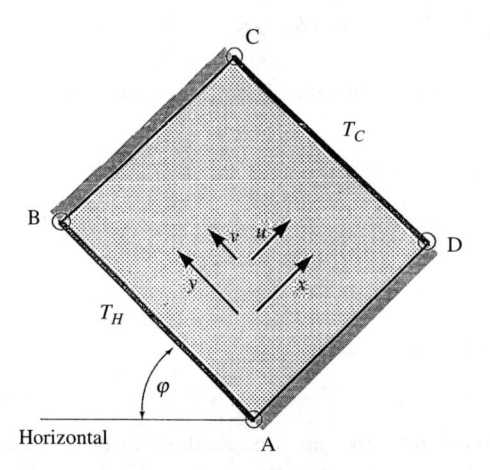

FIGURE 10.27
Situation being analyzed.

CHAPTER 10: Convective Heat Transfer Through Porous Media 533

The stream function, so defined, satisfies the continuity equation.

Now, if the x-derivative is taken of Eq. (10.197) and the y-derivative is taken of Eq. (10.198) and the two results are subtracted, the pressure is eliminated and the following is obtained:

$$\frac{\mu_f}{K}\left(\frac{\partial u}{\partial y} - \frac{\partial v}{\partial x}\right) = \beta g \rho_f \left(\frac{\partial T}{\partial y}\cos\phi - \frac{\partial T}{\partial x}\sin\phi\right) \quad (10.201)$$

i.e., using the definition of the stream function:

$$\frac{\partial^2 \psi}{\partial x^2} + \frac{\partial^2 \psi}{\partial y^2} = \frac{\beta g \rho_f K}{\mu_f}\left(\frac{\partial T}{\partial y}\cos\phi - \frac{\partial T}{\partial x}\sin\phi\right) \quad (10.202)$$

In terms of the stream function, the energy equation, i.e., Eq. (10.199), is:

$$\frac{\partial \psi}{\partial y}\frac{\partial T}{\partial x} - \frac{\partial \psi}{\partial x}\frac{\partial T}{\partial y} = \left(\frac{k_a}{\rho_f c_{pf}}\right)\left(\frac{\partial^2 T}{\partial x^2} + \frac{\partial^2 T}{\partial y^2}\right) \quad (10.203)$$

Before discussing the solution to the above pair of equations, they will be written in dimensionless form. For this purpose, the following dimensionless variables are defined:

$$X = x/W, Y = y/W$$
$$\Psi = \psi/\alpha_a, \theta = (T - T_C)/(T_H - T_C) \quad (10.204)$$

where, as before, the apparent thermal diffusivity of the porous material, α_a, is equal to $k_a/\rho_f c_{pf}$

In terms of the variables defined in Eq. (10.204), Eqs. (10.202) and (10.203) become:

$$\frac{\partial^2 \Psi}{\partial X^2} + \frac{\partial^2 \Psi}{\partial Y^2} = Ra_W\left(\frac{\partial \theta}{\partial Y}\cos\phi - \frac{\partial \theta}{\partial X}\sin\phi\right) \quad (10.205)$$

$$\frac{\partial^2 \theta}{\partial X^2} + \frac{\partial^2 \theta}{\partial Y^2} = \frac{\partial \Psi}{\partial Y}\frac{\partial \theta}{\partial X} - \frac{\partial \Psi}{\partial X}\frac{\partial \theta}{\partial Y} \quad (10.206)$$

where Ra_W is the Darcy-modified Rayleigh number based on the enclosure width W, i.e.:

$$Ra_W = \frac{\beta g K (T_H - T_C) W}{\alpha_a \nu_f} \quad (10.207)$$

Now, the boundary conditions on the solution for flow in the enclosure are:

Velocity component normal to wall = 0 on all walls

$$T = T_H \text{ at } x = 0$$
$$T = T_C \text{ at } x = W \quad (10.208)$$
$$\frac{\partial T}{\partial y} = 0 \text{ at } y = 0 \text{ and } y = H$$

534 Introduction to Convective Heat Transfer Analysis

In terms of the stream function, this means that:

$$\frac{\partial \psi}{\partial y} = 0 \text{ at } x = 0 \text{ and } x = W$$

$$\frac{\partial \psi}{\partial x} = 0 \text{ at } y = 0 \text{ and } y = W \tag{10.209}$$

But the absolute value of the stream function is quite arbitrary because it is only its derivatives that occur in the governing equations. Hence, it will arbitrarily be assumed that the stream function has a value of 0 at the point A shown in Fig. 10.27. But the boundary conditions indicate that $\partial \psi / \partial y$ is 0 along AB. Hence, since ψ is 0 at point A, it is zero everywhere along AB. Along BC the boundary conditions indicate that $\partial \psi / \partial x = 0$. Hence, since ψ is zero at B, it is 0 everwhere along BC. In a similar way, it can be deduced that ψ is 0 everywhere along CD and DA. Hence, on all walls, i.e., along all of ABCD, ψ is 0 if it is arbitrarily taken as 0 at point A. Hence, the boundary condition on the stream function is:

$$\text{All wall surfaces: } \psi = 0 \tag{10.210}$$

In terms of the dimensionless variables defined in Eq. (10.204), the boundary conditions are:

$$\begin{aligned}
\text{On all walls:} \quad & \Psi = 0 \\
X = 0: \quad & \theta = 1 \\
X = 1: \quad & \theta = 0 \\
Y = 0: \quad & \frac{\partial \theta}{\partial y} = 0 \\
Y = A: \quad & \frac{\partial \theta}{\partial Y} = 0
\end{aligned} \tag{10.211}$$

Here $A = H/W$ is the so-called aspect ratio of the enclosure.

A number of approximate analytical solutions to Eqs. (10.205) and (10.206) have been obtained. However, those solutions are of limited applicability and it is now usual to obtain a numerical solution to the equations. A very simple finite-difference numerical procedure, basically identical to that used before to solve for the flow in a fluid-filled enclosure, will be discussed here.

If a uniformly spaced grid is used and if attention is directed to the grid points shown in Fig. 10.28, the following finite-difference form of Eq. (10.206) is obtained:

$$\left[\frac{\theta_{i+1,j} + \theta_{i-1,j} - 2\theta_{i,j}}{\Delta X^2}\right] + \left[\frac{\theta_{i,j+1} + \theta_{i,j-1} - 2\theta_{i,j}}{\Delta Y^2}\right]$$
$$= \left[\frac{\Psi_{i,j+1} - \Psi_{i,j-1}}{2\Delta Y}\right]\left[\frac{\theta_{i+1,j} - \theta_{i-1,j}}{2\Delta X}\right] - \left[\frac{\Psi_{i+1,j} - \Psi_{i-1,j}}{2\Delta X}\right]\left[\frac{\theta_{i,j+1} - \theta_{i,j-1}}{2Y}\right]$$
$$\tag{10.212}$$

An iterative procedure is actually used in which the values of the variables at all nodal points are first guessed. Updated values are then obtained by applying the

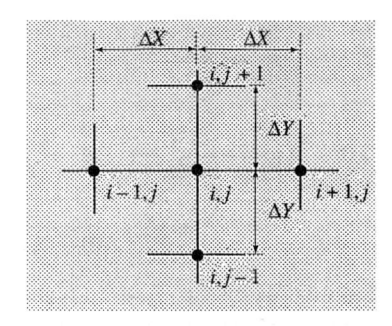

FIGURE 10.28
Nodal points used.

governing equations and the process is repeated until convergence is attained. For this reason, Eq. (10.212) is written as:

$$\theta_{i,j} = \left[\left(\frac{\theta_{i+1,j} + \theta_{i-1,j}}{\Delta X^2}\right) + \left(\frac{\theta_{i,j+1} + \theta_{i,j-1}}{\Delta Y^2}\right) - \left(\frac{\Psi_{i,j+1} - \Psi_{i,j-1}}{2\Delta Y}\right)\left(\frac{\theta_{i+1,j} + \theta_{i-1,j}}{2\Delta X}\right)\right.$$
$$\left. + \left(\frac{\Psi_{i+1,j} - \Psi_{i-1,j}}{2\Delta X}\right)\left(\frac{\theta_{i,j+1} - \theta_{i,j-1}}{2\Delta Y}\right)\right] / \left(\frac{2}{\Delta X^2} + \frac{2}{\Delta Y^2}\right) \quad (10.213)$$

The right-hand side of this equation is calculated using the "most recent" values of the variables. Under-relaxation is actually used so the updated value of θ is given by:

$$\theta_{i,j}^1 = \theta_{i,j}^0 + r(\theta_{i,j}^{\text{calc}} - \theta_{i,j}^0) \quad (10.214)$$

where $\theta_{i,j}^{\text{calc}}$ is the value given by Eq. (10.213) and $\theta_{i,j}^0$ is the value of $\theta_{i,j}$ at the previous iteration.

Eq. (10.213) is applied at all "internal" nodal points, i.e., at all points within the enclosure. The boundary conditions determine the dimensionless temperatures on the walls. These give:

$$J = 1, N: \theta_{i,j} = 1.0, \theta_{M,j} = 0 \quad (10.215)$$

there being N nodal points in the Y-direction and M in the X-direction.

On the other two walls, since the gradient in the Y-direction is zero, to first order accuracy:

$$i = i, M: \theta_{i,1} = \theta_{i,2}, \theta_{i,N} = \theta_{i,N-1} \quad (10.216)$$

The stream function equation, i.e., Eq. (10.205), is treated in the same way as the energy equation. The following finite-difference form of Eq. (10.205) is, therefore, obtained:

$$\left(\frac{\Psi_{i+1,j} + \Psi_{i-1,j} - 2\Psi_{i,j}}{\Delta X^2}\right) + \left(\frac{\Psi_{i,j+1} + \Psi_{i,j-1} - 2\Psi_{i,j}}{\Delta Y^2}\right)$$
$$= Ra_W\left[\left(\frac{\theta_{i,j+1} - \theta_{i,j-1}}{2\Delta Y}\right)\cos\phi - \left(\frac{\theta_{i+1,j} - \theta_{i-1,j}}{2\Delta X}\right)\sin\phi\right] \quad (10.217)$$

This can be rearranged to give:

$$\Psi_{i,j} = \left\{ \left(\frac{\Psi_{i+1,j} + \Psi_{i-1,j}}{\Delta X^2} \right) + \left(\frac{\Psi_{i,j+1} + \Psi_{i,j-1}}{\Delta Y^2} \right) \right.$$
$$\left. - Ra_W \left[\left(\frac{\theta_{i,j+1} - \theta_{i,j-1}}{2\Delta Y} \right) \cos\phi - \left(\frac{\theta_{i+1,j} - \theta_{i-1,j}}{2\Delta X} \right) \sin\phi \right] \right\} \bigg/ \left(\frac{2}{\Delta X^2} + \frac{2}{\Delta Y^2} \right)$$
(10.218)

The right-hand side of this equation is again calculated using the "most" recent values of the variables. Under-relaxation is also again used so the updated value of $\Psi_{i,j}$ is actually given by:

$$\Psi_{i,j}^1 = \Psi_{i,j}^0 + r(\Psi_{i,j}^{\text{calc}} - \Psi_{i,j}^0)$$
(10.219)

where $\Psi_{i,j}^{\text{calc}}$ is the value given by Eq. (10.218) and $\Psi_{i,j}^0$ is the value of $\Psi_{i,j}$ at the previous iteration. $r(<1)$ is again the under-relaxation parameter.

Eq. (10.218) is applied at all "internal" nodal points. The boundary conditions give the value of $\Psi = 0$ on all boundary points, i.e.:

$$j = 1, N: \Psi_{i,j} = 0, \ \Psi_{M,j} = 0$$
$$i = 1, M: \Psi_{i,j} = 0, \ \Psi_{i,N} = 0$$
(10.220)

The above procedure is actually implemented in the following way:

1. The values of $\Psi_{i,j}$ and $\theta_{i,j}$ at all nodal points are first set equal to arbitrary initial values. Typically, the following are used:

$$\Psi_{i,j} = 0$$
$$\theta_{i,j} = 1.0 - X_{i,j}$$
(10.221)

The assumed θ distribution is that which would exist if there was no connective motion, i.e., if conduction alone existed. Its use is consistent with the assumed distribution of Ψ which implies that there is no flow in the enclosure.

2. Eq. (10.218) in conjunction with Eq. (10.219) is used to obtain updated values of $\Psi_{i,j}$. Because iteration is being used, this process should really be repeated over and over until the values of $\Psi_{i,j}$ corresponding to the initially assumed distribution of θ are obtained. Experience suggests, however, that it is quite adequate to undertake this step just twice.
3. Eq. (10.213) in conjunction with Eq. (10.214) is used to obtain updated values of $\theta_{i,j}$. This step is also undertaken twice.
4. Steps (2) and (3) are repeated over and over until convergence is obtained to a specified degree.
5. The heat transfer rate distribution is obtained by applying Fourier's law at the heated and cooled walls, i.e.:

$$q_{w_{1,j}} = \frac{k_a(T_{1,j} - T_{2,j})}{\Delta y}, \quad q_{w_{M,j}} = \frac{k_a(T_{M-1,j} - T_{M,j})}{\Delta y}$$
(10.222)

CHAPTER 10: Convective Heat Transfer Through Porous Media 537

In terms of the dimensionless variables, this equation becomes:

$$Nu_{1,j} = \frac{1 - \theta_{2,j}}{\Delta Y}, \quad Nu_{M,j} = \frac{\theta_{M-1,j}}{\Delta Y} \qquad (10.223)$$

where Nu is the local Nusselt number based on W and where it has been noted that $\theta_{i,j} = 1.0$ and $\theta_{M,j} = 0$.

Because uniform grid spacing has been used, the mean Nusselt numbers are given by:

$$\overline{Nu_H} = \frac{\Delta Y}{A}\left(\frac{Nu_{1,1}}{2} + Nu_{1,2} + Nu_{1,3} + \cdots + Nu_{M,N-1} + \frac{Nu_{1,N}}{2}\right)$$

$$\overline{Nu_C} = \frac{\Delta Y}{A}\left(\frac{Nu_{M,i}}{2} + Nu_{M,2} + Nu_{M,3} + \cdots + Nu_{M,N-1} + \frac{Nu_{M,N}}{2}\right) \qquad (10.224)$$

Because a steady-state situation is being considered, $\overline{Nu_H}$ and $\overline{Nu_C}$ will be equal. Numerically, slightly different values will normally be obtained because of the finite degree of convergence used.

A computer program, ENCLPOR, written in FORTRAN that is based on this procedure is available as discussed in the Preface. This program applies to a "vertical" enclosure, i.e., to the case $\phi = 90°$.

The solution discussed above has, as parameters, Ra_W, the Darcy-modified Rayleigh number based on the enclosure width, and A, the enclosure aspect ratio H/W. Some typical results computed with the above program are shown in Figs. 10.29 and 10.30.

It will be seen from Fig. 10.29 that for small Ra_W, when the convective motion is weak, the Nusselt number tends to the conduction limit value of 1. However, when Ra_W is large, the flow will consist mainly of a boundary layer flow up the hot wall and a boundary layer flow down the cold wall. It is to be expected, therefore, that in

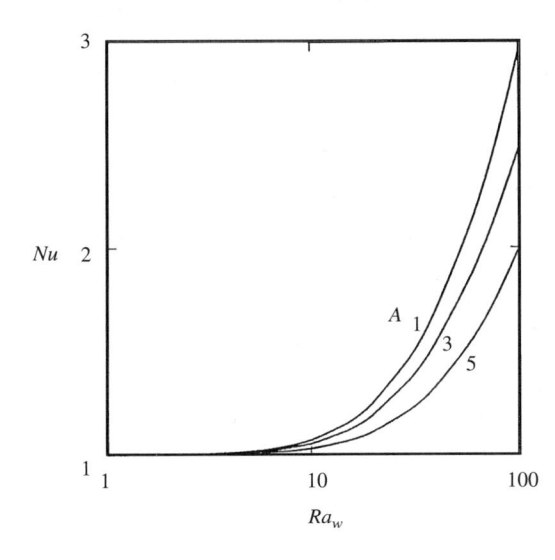

FIGURE 10.29
Variation of mean Nusselt number with Ra_W for various A.

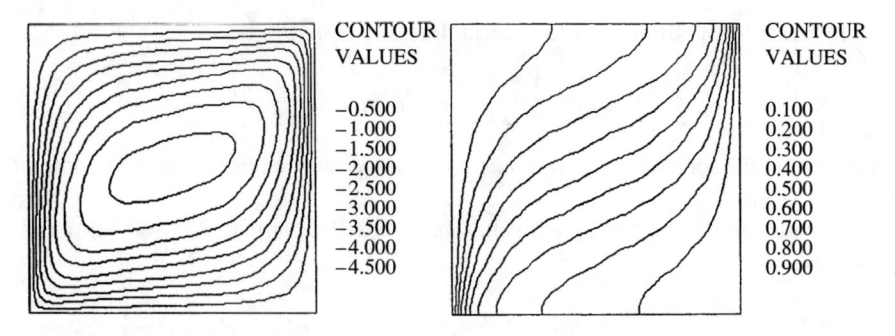

FIGURE 10.30
Typical streamline and isotherm patterns ($Ra_W = 50$, $A = 1$).

view of the form of the boundary layer solution obtained earlier, i.e:

$$\frac{\overline{q_w}H}{(T_H - T_C)k_a} \propto Ra_H^{1/2}$$

that:

$$\overline{Nu} \propto Ra_W^{1/2}/A^{1/2} \tag{10.225}$$

and, indeed, it will be seen from the results that at large values of Ra_W:

$$\overline{Nu} = 0.508\frac{Ra_W^{1/2}}{A^{1/2}} \tag{10.226}$$

This equation will apply as long as $Ra_W^{1/2}/A^{1/2}$ is large, i.e., as long as $A \ll Ra_W$.

EXAMPLE 10.7. The wall of an insulated shipping container can be modeled as a 10 cm thick by 50 cm high vertical enclosure. The inner and outer surfaces can be assumed to be at 50°C and 10°C respectively. The enclosure is filled with an air-saturated loosely packed granular insulating material that has a permeability of 2×10^{-7} m². Estimate the heat transfer rate across the wall per m width.

Solution. It will be assumed that the insulation material and air have the following properties:

$$K = 2 \times 10^{-7} \text{ m}^2, \quad \alpha_a = 40 \times 10^{-6} \text{ m}^2/\text{s}, \quad k_a = 0.048 \text{ W/m K}$$
$$\nu_f = 15 \times 10^{-6} \text{ m}^2/\text{s}, \quad \beta = 3.3 \times 10^{-3} \text{ K}^{-1}$$

Using these values the Darcy-modified Rayleigh number based on the width of the enclosure is:

$$Ra_W = \frac{\beta g K(T_H - T_C)W}{\alpha_a \nu_f} = \frac{0.0033 \times 9.81 \times 0.0000002 \times (50 - 10) \times 0.1}{0.00004 \times 0.000015} = 43$$

The aspect ratio of the enclosure is $50/10 = 5$. Hence $Ra_W = 43$ and $A = 5$. For these values of Ra_W and A, Fig. 10.29 indicates that because $Ra_W^{1/2}/A^{1/2} = 2.9$, Eq. (10.226) applies, i.e.:

$$\overline{Nu} = 0.508\frac{Ra_W^{1/2}}{A^{1/2}} = 0.508\frac{43^{1/2}}{5^{1/2}} = 1.49$$

CHAPTER 10: Convective Heat Transfer Through Porous Media   539

The mean heat transfer coefficient is therefore given by:

$$\bar{h} = \frac{\overline{Nu} k_a}{L} = \frac{1.49 \times 0.048}{0.1} = 0.72 \text{ W/m}^2\text{K}$$

Therefore the rate of heat transfer across the enclosure per unit length is given by:

$$Q = \bar{h} A (T_H - T_C) = 0.72 \times (0.5 \times 1) \times (50 - 10) = 14.3 \text{ W}$$

The computer program for the calculation of flow in an enclosure discussed above is easily modified to deal with nonvertical enclosures. Some results obtained using such a modified program are shown in Fig. 10.31. These results are for a square enclosure, i.e., for an enclosure with an aspect ratio of 1 and a Darcy-modified Rayleigh number of 100. This figure shows the variations of mean Nusselt number and of the stream function at the center of the enclosure, i.e., Ψ_{center}, with angle of inclination. The magnitude of Ψ_{center} is a measure of the strength of the fluid motion in the enclosure.

It will be seen that the heat transfer rate is a maximum when ϕ is about 50°. At this angle, the flow along both the heated wall and the adiabatic "top" wall is acted upon by buoyancy forces that are near parallel to the wall as shown in Fig. 10.32. It will also be noted that the highest magnitude of Ψ_{center} occurs near this value of ϕ.

When $\phi = 180°$, the hot wall is horizontal and at the top there is no fluid motion and the Nusselt number has its pure conduction value. When $\phi = 0°$, the hot wall is horizontal and at the bottom and the flow is unstable. However, the Rayleigh number for the aspect ratio considered is too low for this to occur and the conduction value for the Nusselt number is obtained at $\phi = 0°$. However, inclining the enclosure by a small amount provides this "trigger" leading to a value of Nusselt number that is much higher than the conduction value, e.g., $\overline{Nu}$ is calculated to be 1 when $\phi = 0°$ and to be approximately 2.6 when when $\phi = 1°$.

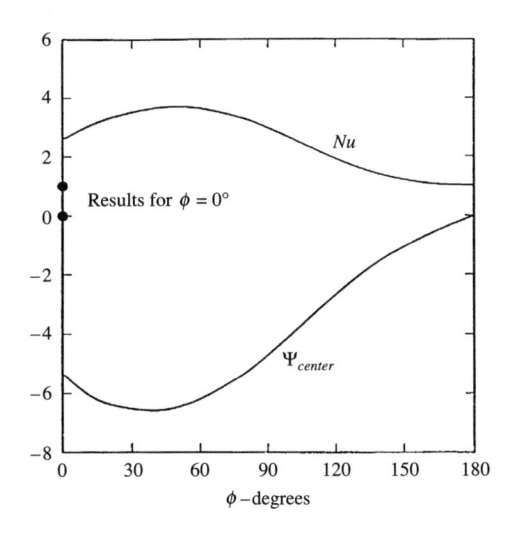

FIGURE 10.31
Variation of mean Nusselt number and dimensionless center-stream function with angle of inclination for an enclosure with an aspect ratio of 1.

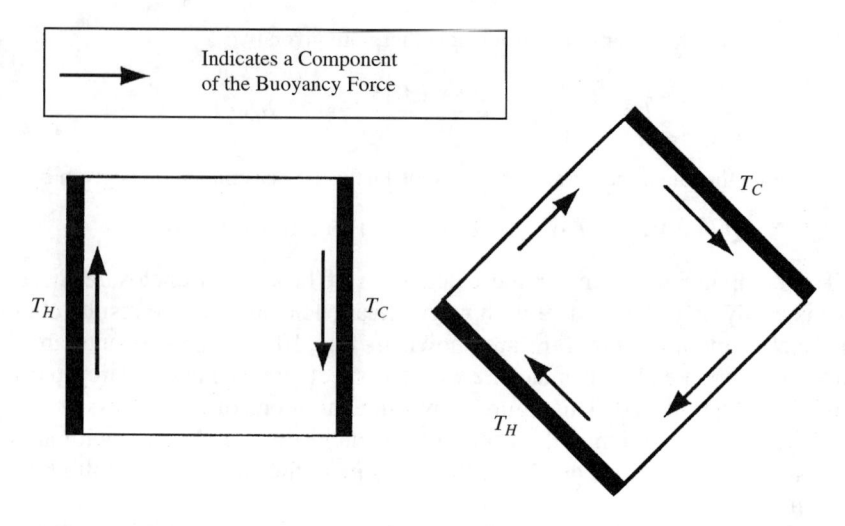

FIGURE 10.32
Buoyancy force components in an inclined enclosure.

10.9
STABILITY OF HORIZONTAL POROUS LAYERS HEATED FROM BELOW

Consider an essentially infinite horizontal layer of saturated porous medium that is heated from below and cooled from the top. The situation is shown in Fig. 10.33.

If the temperature difference is very small no fluid motion will occur and the heat transfer from the hot surface to the cold surface will be by conduction alone. In this case, the temperature varies linearly through the material and the temperature distribution is given by:

$$T = T_H - \left(\frac{y}{H}\right)\Delta T \tag{10.227}$$

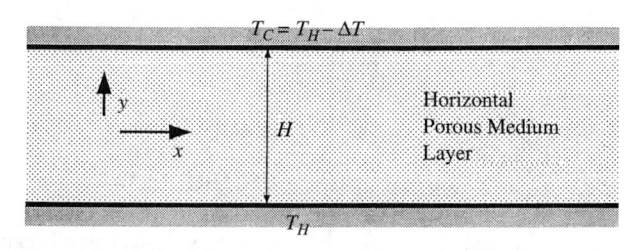

FIGURE 10.33
Porous layer heated from below.

where $\Delta T = T_H - T_C$ is the temperature difference across the layer. As the temperature difference is increased, a condition will be reached at which fluid motion begins, i.e., the fluid layer becomes "unstable", e.g., see [12], [52] to [56]. The fluid motion is, of course, caused by the buoyancy forces, the least dense fluid being adjacent to the lower hot surface which tends to rise while the fluid with the highest density is adjacent to the upper cold surface and tends to fall.

Here, an attempt will be made to determine the conditions under which the instability develops, i.e., under what conditions fluid motion first occurs.

The equations governing the fluid motion are:

$$\frac{\partial u}{\partial x} + \frac{\partial v}{\partial y} = 0 \tag{10.228}$$

$$u = -\frac{K}{\mu_f}\frac{\partial p}{\partial x} \tag{10.229}$$

$$v = -\frac{K}{\mu_f}\frac{\partial p}{\partial y} + \frac{\beta g \rho_f K (T - T_c)}{\mu_f} \tag{10.230}$$

$$\sigma\frac{\partial T}{\partial t} + u\frac{\partial T}{\partial x} + v\frac{\partial T}{\partial y} = \alpha_a\left(\frac{\partial^2 T}{\partial x^2} + \frac{\partial^2 T}{\partial y^2}\right) \tag{10.231}$$

Because the growth of the disturbances must be considered, the unsteady form of the governing equations has been used. The term σ is given by:

$$\sigma = \frac{\rho c}{\rho_f c_f} = \frac{\phi \rho_f c_f + (1-\phi)\rho_s c_s}{\rho_f c_f} \tag{10.232}$$

the subscripts f and s referring to the fluid and the solid matrix respectively, ϕ being the porosity. This term arises because the energy required to produce a given change in temperature will depend on ρc whereas the enthalpy changes will depend on $\rho_f c_f$.

It is again convenient to eliminate the pressure, p, by differentiating Eq. (10.229) with respect to y and Eq. (10.230) with respect to x and subtracting the two results. This gives:

$$\frac{\partial u}{\partial y} - \frac{\partial v}{\partial x} = -\frac{\beta \rho_f g K}{\mu_f}\frac{\partial T}{\partial x} \tag{10.233}$$

If the flow is stable, there is no fluid motion and under these circumstances:

$$u = v = 0, \ T = T_H - \left(\frac{y}{H}\right)\Delta T \tag{10.234}$$

The interest here is with the conditions under which the flow just begins, i.e., under which the flow deviates only slightly from that existing under stable conditions.

542 Introduction to Convective Heat Transfer Analysis

Hence, the following solution is assumed:

$$T = \left(T_H - \frac{y}{H}\Delta T\right) + T' \tag{10.235}$$

$$u = u' \tag{10.236}$$

$$v = v' \tag{10.237}$$

where u', v', and T' are functions of x, y, and t and are small.

In terms of these variables the governing equations become:

$$\frac{\partial u'}{\partial x} + \frac{\partial v'}{\partial y} = 0 \tag{10.238}$$

$$\frac{\partial u'}{\partial y} - \frac{\partial v'}{\partial x} = \frac{\beta g \rho_f K}{\mu_f} \frac{\partial T'}{\partial x} \tag{10.239}$$

$$\sigma \frac{\partial T'}{\partial t} + u' \frac{\partial T'}{\partial x} + v' \left[\frac{\partial T'}{\partial y} - \frac{\Delta T}{H}\right] = \alpha_a \left(\frac{\partial^2 T'}{\partial x^2} + \frac{\partial^2 T'}{\partial y^2}\right) \tag{10.240}$$

Now u', v', and T' are, by assumption, small so products of such terms will be very small and will be neglected in the present analysis. As a result, Eq. (10.240) can be written:

$$\sigma \frac{\partial T'}{\partial t} - v' \frac{\Delta T}{H} = \alpha_a \left(\frac{\partial^2 T'}{\partial x^2} + \frac{\partial^2 T'}{\partial y^2}\right) \tag{10.241}$$

Because this equation does not contain u', it is convenient to eliminate u' between Eqs. (10.238) and (10.239) by differentiating Eq. (10.238) with respect to y and Eq. (10.239) with respect to x and subtracting the two results. This gives:

$$\frac{\partial^2 v'}{\partial x^2} + \frac{\partial^2 v'}{\partial y^2} = \frac{\beta g K}{\mu_f} \frac{\partial^2 T'}{\partial x^2} \tag{10.242}$$

It is convenient to write Eqs. (10.241) and (10.242) in dimensionless form by introducing the following dimensionless variables:

$$X = \frac{x}{H}, \quad Y = \frac{y}{H}, \quad \tau = \frac{\alpha_a t}{H^2 \sigma}$$

$$V' = \frac{v' H}{\alpha_a} \quad \theta' = \frac{T'}{\Delta T} \tag{10.243}$$

In terms of these variables, Eqs. (10.241) and (10.242) become:

$$\frac{\partial \theta'}{\partial \tau} - V' = \frac{\partial^2 \theta'}{\partial X^2} + \frac{\partial^2 \theta'}{\partial Y^2} \tag{10.244}$$

and

$$\frac{\partial^2 V'}{\partial X^2} + \frac{\partial^2 V'}{\partial Y^2} = Ra \frac{\partial^2 \theta'}{\partial X^2} \tag{10.245}$$

CHAPTER 10: Convective Heat Transfer Through Porous Media 543

where as before

$$Ra = \frac{\beta g K \rho_f \Delta T H}{\alpha_a \mu_f} \qquad (10.246)$$

is the Darcy-modified Rayleigh number based on H. The boundary conditions on the above pair of equations are:

$$Y = 0 \text{ and } Y = 1: \quad V' = 0, \quad T' = 0 \qquad (10.247)$$

Now the conditions under which the disturbances grow are being sought and physical observation indicates that the disturbances grow periodically with X. It will, therefore, be assumed that:

$$\theta' = A(Y)e^{r\tau + i\alpha X} \qquad (10.248)$$

$$V' = B(Y)e^{r\tau + i\alpha X} \qquad (10.249)$$

Substituting these into Eqs. (10.244) and (10.245) gives:

$$rA - B = -\alpha^2 A + \frac{d^2 A}{dY^2} \qquad (10.250)$$

and

$$-\alpha^2 B + \frac{d^2 B}{dY^2} = -\alpha^2 Ra A \qquad (10.251)$$

the boundary conditions being:

$$Y = 0 \text{ and } Y = 1: \quad A = 0, \quad B = 0 \qquad (10.252)$$

Now, consideration of Eqs. (10.248) and (10.249) indicates that if $r < 0$, the disturbance will decrease with time, i.e., will die out. However, if $r > 0$, the disturbance will increase with time and flow will develop. The condition $r = 0$ represents neutral stability, the disturbances, therefore, neither increasing or decreasing with time under these conditions. This neutral stability condition will determine the conditions beyond which instability develops. If $r = 0$, Eq. (10.250) gives

$$B = \alpha^2 A - \frac{d^2 A}{dY^2} \qquad (10.253)$$

Using this value of B to evaluate the left-hand side of Eq. (10.251) then gives:

$$-\alpha^4 A + \alpha^2 \frac{d^2 A}{dY^2} + \alpha^2 \frac{d^2 A}{dY^2} - \frac{d^4 A}{dY^4} = -\alpha^2 Ra A$$

i.e.:

$$\frac{d^4 A}{dY^4} - 2\alpha^2 \frac{d^2 A}{dY^2} + \alpha^4 A = \alpha^2 Ra A \qquad (10.254)$$

Now consider the following solution to this equation:

$$A = C \sin(n\pi Y) \qquad (10.255)$$

544 Introduction to Convective Heat Transfer Analysis

Substituting this into Eq. (10.254) gives

$$Cn^4\pi^4 \sin(n\pi Y) + 2\alpha^2 Cn^2\pi^2 \sin(n\pi Y) + \alpha^4 C \sin(n\pi Y)$$
$$= \alpha^2 RaC \sin(n\pi Y) \quad (10.256)$$

In order for the boundary conditions to be satisfied, n must be an integer. Eq. (10.256) then gives:

$$(n^4\pi^4 + 2\alpha^2 n^2\pi^2 + \alpha^4) = \alpha^2 R_A \quad (10.257)$$

i.e.:

$$R_A = \frac{(n^2\pi^2 + \alpha^2)^2}{\alpha^2}$$

For any value of n, this allows the value of R_A at which instability starts to develop to be determined as a function of α. Some typical results are shown in Fig. 10.34.

It will be seen that for each value of n there is a minimum Rayleigh number in the variation with α. It will further be seen that the lowest of these minimum Rayleigh numbers occurs when $n = 1$. This minimum Rayleigh number with $n = 1$ is, therefore, the lowest Rayleigh number at which instability could be expected to occur. Now, when $n = 1$, Eq. (10.257) gives:

$$Ra = \frac{(\pi^2 + \alpha^2)^2}{\alpha^2} \quad (10.258)$$

The minimum in the curve will occur when:

$$\frac{d\,Ra}{d\alpha} = \frac{4(\pi^2 + \alpha^2)\alpha}{\alpha^2} - \frac{2(\pi^2 + d^2)^2}{\alpha^3} = 0$$

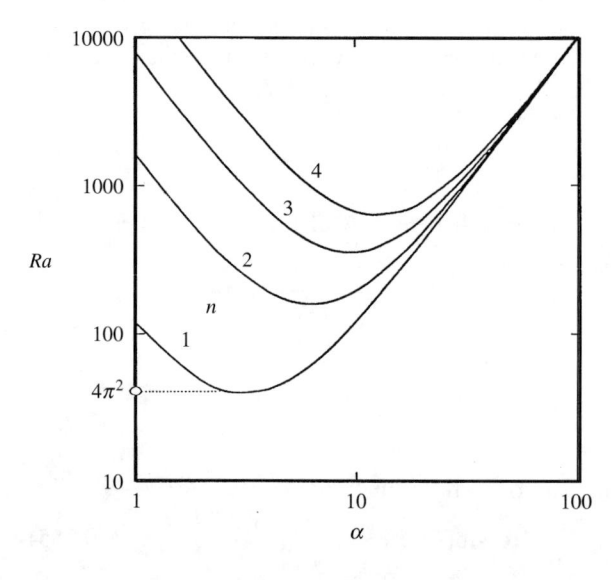

FIGURE 10.34
Variation of minimum Rayleigh number for stability.

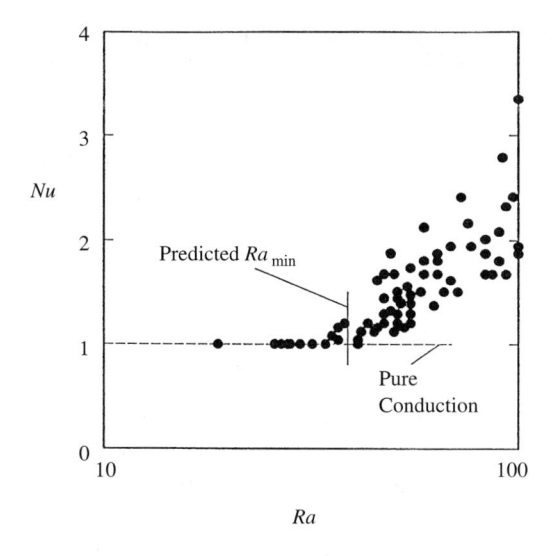

CHAPTER 10: Convective Heat Transfer Through Porous Media

FIGURE 10.35
Heat transfer rate across a horizontal porous layer.

i.e., when:
$$\pi^2 + \alpha^2 = 2\alpha^2$$

i.e., when:
$$\alpha^2 = \pi^2 \tag{10.259}$$

Substituting this result back into Eq. (10.258) then gives:
$$Ra_{min} = 4\pi^2 = 39.5 \tag{10.260}$$

Hence, this analysis indicates that instability will first occur, i.e., convective motions will first occur, when the Darcy-modified Rayleigh number based on the thickness of the layer reaches 39.5. If Ra is greater than this value, convective motion will occur. This value for Ra_{min} is in good agreement with experiment, some experimental results being shown in Fig. 10.35.

10.10 NON-DARCY AND OTHER EFFECTS

The Darcy model is based, as discussed earlier, on the assumptions that:

1. The inertia terms in the governing equations are negligible.
2. The viscous terms in the governing equations are negligible.
3. The losses associated with the flow over the particles of the porous material are proportional to the volume-averaged velocity.

The use of these assumptions is, however, not always justified, e.g., see [57] to [71].

If the "pore" Reynolds number, i.e., the Reynolds number based on the local velocity vector and on $K^{1/2}$, i.e.:
$$Re_p = \frac{(u^2 + v^2 + w^2)^{1/2} K^{1/2}}{\nu} \tag{10.261}$$

546 Introduction to Convective Heat Transfer Analysis

is greater than 1, the drag on the particles associated with the square of the velocity starts to become important, i.e., Eq. (10.7) and its equivalent in the other coordinate directions must be written as:

$$-\frac{\partial p}{\partial x} = \frac{\mu_f}{K} u + b\rho_f |u|u \qquad (10.262)$$

$$-\frac{\partial p}{\partial y} = \frac{\mu_f}{K} v + b\rho_f |v|v \qquad (10.263)$$

$$-\frac{\partial p}{\partial z} = \frac{\mu_f}{K} w + b\rho_f |w|w \qquad (10.264)$$

The last term, which is written as $|V|V$ to ensure that the direction of the loss is correctly specified, is termed the Forchheimer extension of the Darcy model. The factor b is a constant that depends on the characteristics of the porous material. If the material is assumed to be made up of tightly packed spheres, it can be shown that

$$b = \frac{1.75(1-\phi)}{\phi^3 d} \qquad (10.265)$$

where d is again the diameter of the particles.

The above equations apply to forced convection. Their extension and the extension of the rest of the equations given in this section to natural convective flows is straight-forward and will not be discussed here.

If the permeability of the material is very high, it may not be possible to ignore the viscous forces in setting up the momentum equation. In this case, if the effective viscosity is taken to be the viscosity of the fluid, the Darcy equations, i.e., Eqs. (10.12) to (10.14) become:

$$u = -\frac{K}{\mu_f} \frac{\partial p}{\partial x} + K \left(\frac{\partial^2 u}{\partial x^2} + \frac{\partial^2 u}{\partial y^2} + \frac{\partial^2 u}{\partial z^2} \right) \qquad (10.266)$$

$$v = -\frac{K}{\mu_f} \frac{\partial p}{\partial y} + K \left(\frac{\partial^2 v}{\partial x^2} + \frac{\partial^2 v}{\partial y^2} + \frac{\partial^2 v}{\partial z^2} \right) \qquad (10.267)$$

$$w = -\frac{K}{\mu_f} \frac{\partial p}{\partial z} + K \left(\frac{\partial^2 w}{\partial x^2} + \frac{\partial^2 w}{\partial y^2} + \frac{\partial^2 w}{\partial z^2} \right) \qquad (10.268)$$

These equations represent the Brinkman extension of the Darcy model. The Forchheimer and Brinkman extensions of the basic Darcy model often must be simultaneously used.

While it is easy to extend the governing equations to include the inertia terms, these terms are seldom important in studying the flow through a porous medium.

In addition to the modifications to the Darcy flow model, it is also sometimes necessary to note that the apparent thermal conductivity, k_a, as used above arose from conduction through the fluid and solid media. This was the basis for Eq. (10.36). However, in fact, as the fluid flows through the tortuous paths between the solid

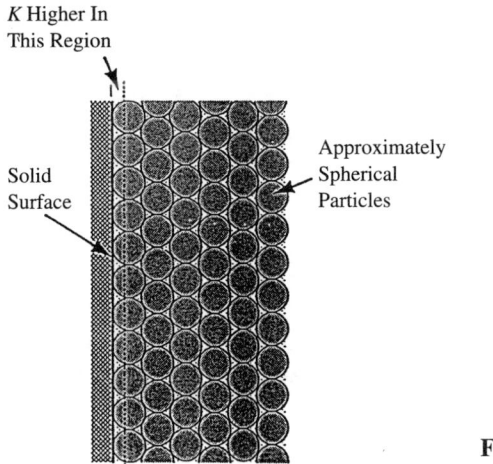

FIGURE 10.36
Conditions near a wall.

matrix, additional mixing occurs which effectively causes an increase in the thermal conductivity. This effect is referred to as "thermal dispersion". The effective thermal conductivity is thus higher than the value calculated by only accounting for the molecular level conduction in the fluid and in the solid matrix.

Near a wall it is possible for K to vary with distance from the wall. This can cause so-called tunnelling near the wall and so affect the heat transfer rate at the wall (see Fig. 10.36).

10.11
CONCLUDING REMARKS

Flow through a solid matrix which is saturated with a fluid and through which the fluid is flowing occurs in many practical situations. In many such cases, temperature differences exist and heat transfer, therefore, occurs. The extension of the methods of analyzing convective heat transfer rates that were discussed in the earlier chapters of this book to deal with heat transfer in porous media flows have been discussed in this chapter. Both forced and natural convective flows have been discussed.

PROBLEMS

10.1. Unidirectional flow through a porous medium can often be approximately modeled as flow through a series of parallel plane channels as shown in Fig. P10.1. Using this model derive an expression for the permeability, K, in terms of the channel size, W, and the porosity, ϕ.

10.2. Show that a similarity-type solution can be obtained for the case of two-dimensional flow over a flat plate in a porous medium, the plate being aligned with a forced flow

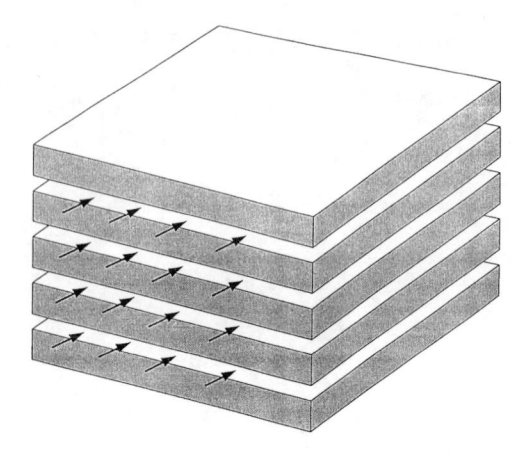

FIGURE P10.1

through the porous medium, for the case where there is a uniform heat flux at the surface of the plate. In treating this problem define:

$$\theta = \frac{T - T_\infty}{q_w x / k_a}$$

where q_w is the uniform heat flux at the surface of the plate.

10.3. A wide flat plate with a length of 20 cm kept at a uniform surface temperature of 30°C is buried in a bed of approximately spherical pebbles with a mean diameter of 7 mm with a porosity of 0.35. Water trickles through the bed at a velocity of 0.03 m/s, the plate being aligned with the water flow. If the water has a temperature of 20°C, find the variation of the heat transfer rate along the plate.

10.4. In Chapter 3 it was shown that similarity-type solutions could be found for fluid flow over isothermal surfaces whose shape is such that the the velocity distribution outside the boundary is described by $u_1 = Ax^m$. Investigate whether this is also true when such bodies are placed in a flow through a porous medium.

10.5. A long 5-cm diameter pipe kept at a surface temperature of 15°C is buried in a bed of approximately spherical pebbles with a mean diameter of 5 mm with a porosity of 0.4. Water trickles through the bed at a mean velocity of 0.015 m/s, the axis of the pipe being normal to the water flow. If the water has a temperature of 10°C, find the heat transfer from the pipe to the water per meter length of pipe.

10.6. A cylindrical container kept at a surface temperature of 40°C is buried in a bed of approximately spherical pebbles with a mean diameter of 5 mm with a porosity of 0.3. Water at a temperature of 15°C trickles through the bed at a mean velocity of 0.01 m/s, the axis of the cylinder being normal to the water flow. Find the distribution of the local heat transfer rate around the container assuming two-dimensional flow.

10.7. Use the integral equation method to derive an expression for the variation of the local heat transfer rate along a wide flat plate buried in a porous medium through which there is a forced flow parallel to the plate surface. The surface temperature of the plate is given by $T_\infty + A(x/L)^{0.5}$, L being the length of the plate in the flow direction.

10.8. Derive the value for the Nusselt number for fully developed flow through a porous medium-filled pipe with a uniform heat flux at the wall.

10.9. A heat pump system utilizes a heat exchanger buried in water-saturated soil as a heat source. The heat exchanger basically consists of a series of vertical plates with height of 30 cm and a width of 10 cm. These plates are effectively at a uniform temperature of 5°C. The soil can be assumed to have a permeability of 10^{-10} m² and an apparent thermal conductivity of 0.1 W/m-K. The temperature of the saturated soil far from the heat exchanger is 30°C. Assuming natural convective flow and that there is no interference between the flows over the individual plates, find the mean heat transfer rate to a plate.

10.10. Discuss how the analysis of natural convective flow over a vertical flat plate in a saturated porous medium must be modified if there is a uniform heat flux rather than a uniform temperature at the surface.

10.11. Explain how the computer program for the calculation of flow in a porous medium-filled vertical enclosure given in this chapter must be modified to deal with the case where the heat flux rather than the temperature at the hot wall is specified. For this purpose, define the following dimensionless temperature:

$$\theta = \frac{T - T_C}{q_w W/k_a}$$

where q_w is the uniform heat flux at the hot wall. Also introduce a suitably defined heat flux Rayleigh number. Use the modified program to determine the variation of $\overline{Nu}$ with Rayleigh number for the particular case of a "vertical" enclosure with $A = 1$.

10.12. The wall of an insulated container can be modeled as a 5-cm wide vertical enclosure with an aspect ratio of 5. The enclosure is filled with a low grade insulating material that has a permeability of 10^{-8} m² and an apparent thermal conductivity of 0.04 W/m-K. The inner wall of the enclosure is at 5°C. Plot a curve showing how the heat transfer rate across the enclosure varies with outer wall temperature.

10.13. An approximate model of the flow in a vertical porous medium-filled enclosure assumes that the flow consists of boundary layers on the hot and cold walls with a stagnant layer between the two boundary layers, this layer being at a temperature that is the average of the hot and cold wall temperatures. Use this model to find an expression for the heat transfer rate across the enclosure and discuss the conditions under which this model is likely to be applicable.

10.14. Explain how the computer program for the calculation of flow in an enclosure discussed in this chapter must be modified to deal with nonisothermal walls.

10.15. Using the procedure outlined in this chapter for using the boundary layer equations to find the forced convective heat transfer rate from a circular cylinder buried in a saturated porous medium, investigate the heat transfer rate from cylinders with an elliptical cross-section with their major axes aligned with the forced flow. The surface velocity distribution should be obtained from a suitable book on fluid mechanics.

10.16. An insulated wall can be modeled as a 6-cm wide vertical enclosure with an aspect ratio of 3. The inner and outer surfaces are at temperatures of 50°C and 10°C respectively. This wall is divided into two 3-cm wide sections by a vertical impermeable barrier. The inner section is filled with an insulating material that has a permeability of 10^{-10} m^2 and an apparent thermal conductivity of 0.02 W/m-K while the outer section is filled with an insulating material that has a permeability of 10^{-8} m^2 and an apparent thermal conductivity of 0.05 W/m-K. Assuming that the barrier between the two sections is at a uniform temperature and that the flow in the insulation is two-dimesional, find the mean heat transfer rate across the wall per m^2 of wall area.

10.17. Consider air flow through a 1000-mm long, 5-mm diameter pipe whose wall is kept at a uniform temperature of 50°C. The air is heated from 20°C to 30°C as it passes through the pipe and the mean air velocity in the pipe is 0.6 m/s. Using results given in Chapter 4, find the mean wall heat flux assuming full-developed flow, i.e., neglect any entry region effects. Also find the mean heat transfer rate if the pipe is filled with a loosely packed porous medium which has a permeability of 10^{-9} m^2 and an apparent thermal conductivity of 0.04 W/m-K. Again assume fully developed flow. Compare and discuss the values of the mean heat transfer rates obtained for the two cases. As part of this discussion, the pressure drop across the pipe in the two cases should be calculated.

REFERENCES

1. Kaviany, M., *Principles of Heat Transfer in Porous Media,* Springer-Verlag, New York, 1991.
2. Scheidegger, A.E., *The Physics of Flow Through Porous Media,* 3rd ed., University of Toronto Press, Toronto, 1974.
3. Bear, J., *Dynamics of Fluids in Porous Media,* American Elsevier, New York, 1972.
4. Bear, J. and Bachmat, Y., *Introduction to Modeling of Transport Phenomena in Porous Media,* Kluwer Academic Publishers, Dordrecht, 1991.
5. Carbonell, R.G. and Whitaker, S., "Heat and Mass Transfer in Porous Media", in *Fundamntals of Transport Phenomena in Porous Media,* Bear, J. and Corapciolglu, M.Y., Eds., Martinus Nijhoff Publishers, The Hague, 1984.
6. Lapwood, E.R., "Convection of a Fluid in a Porous Medium", *Proc. Cambridge Philos. Soc.,* Vol. 44, pp. 508-521, 1948.
7. Combarnous, M.A. and Bories, S.A., "Hydrothermal Convection in Saturated Porous Media", *Adv. Hydrosci.,* Vol. 10, pp. 231-307, 1975.
8. Cheng, P., "Heat Transfer in Geothermal Systems", *Adv. Heat Transfer,* Vol. 14, pp. 1-105, 1979.
9. Dybbs, A. and Edwards, R.V., "A New Look at Porous Media Fluid Mechanics: Darcy to Turbulent", in *Fundamentals of Transport Phenomena in Porous Media,* Bear, J. and Corapcigoln, M.Y., Eds., Martinus Nijhoff Publishers, The Hague, 1984.
10. Muskat, M., *The Flow of Homogeneous Fluids through Porous Media,* McGraw-Hill, New York, 1937; 2nd printing by Edwards Brothers, Ann Arbor, MI, 1946.
11. Nield, D.A. and Bejan, A., *Convection in Porous Media,* Springer-Verlag, New York, 1992.
11a. Simpkins, P.G. and Blythe, P.A., "Convection in a Porous Layer", *Int. J. Heat Mass Transfer,* Vol. 23, pp. 881-887, 1980.

12. Bejan, A., *Convection Heat Transfer,* 2nd. ed., Wiley, New York, 1995.
13. Whitaker, S., "Flow in Porous Media. I: A Theoretical Derivation of Darcy's Law", *Transp. Porous Media,* Vol. 1, pp. 3-25, 1986.
14. Dagan, G., "The Generalization of Darcy's Law for Nonuniform Flows", *Water Resour. Res.,* Vol. 15, pp. 1-7, 1979.
15. Kaviany, M., "Boundary Layer Treatment of Forced Convection Heat Transfer from a Semi-Infinite Flat Plate Embedded in Porous Media", *J. Heat Transfer,* Vol. 109, pp. 345-349, 1987.
16. Whitaker, S., "Forced Convection Heat-Transfer Correlations for Flow in Pipes, Past Flat Plates, Single Cylinders, Single Spheres, and Flow in Packed Beds and Tube Bundles", *AIChE J.,* Vol. 18, pp. 361-371, 1972.
17. Henderson, C. and Oosthuizen, P.H., "Heat Transfer by Mixed Convection from Tandem Cylinders in a Porous Medium", *Proc. AIAA/ASME Thermophysics and Heat Transfer Conference,* HTD-Vol. 129, Am. Soc. Mech. Eng., New York, 1990, pp. 1-9.
18. Oosthuizen, P.H. and Paul, J.T., "Forced Convective Heat Transfer from a Flat Plate Embedded Near an Impermeable Surface in a Porous Medium", *Symp. on Fundamentals of Forced Convection Heat Transfer,* ASME HTD-Vol. 101, Am. Soc. Mech. Eng., New York, 1988, pp. 105-111.
19. Kaviany, M., "Laminar Flow Through a Porous Channel Bounded by Isothermal Parallel Plates", *Int. J. Heat Mass Transfer,* Vol. 28, pp. 851-858, 1985.
20. Cheng, P. and Minkowycz, W.J., "Free Convection About a Vertical Flat Plate Embedded in a Porous Medium with Application to Heat Transfer from a Dike", *J. Geophs. Res.,* Vol. 82, pp. 2040-2044, 1977.
21. Kaviany, M. and Miltal, M., "Natural Convection Heat Transfer from a Vertical Plate to High Permeability Porous Media: An Experiment and an Approximate Solution", *Int. J. Heat Mass Transfer,* Vol. 30, pp. 967-977, 1987.
22. Nakayama, A. and Pop, I., "A Unified Similarity Transformation for Free, Forced and Mixed Convection in Darcy and Non-Darcy Porous Media", *Int. J. Heat and Mass Transfer,* Vol. 34, pp. 357-367, 1991.
23. Aldoss, T.K., Chen, T.S., and Armaly, B.F., "Nonsimilarity Solutions for Mixed Convection from Horizontal Surfaces in a Porous Medium", *Int. J. Heat and Mass Transfer,* Vol. 36, No. 2, pp. 463-477, 1993.
24. Choi, C.Y and Kulacki, F.A., "Mixed Convection Through Vertical Porous Annuli Locally Heated from the Inner Cylinder", *J. Heat Transfer,* Vol. 114, pp. 143-151, 1992.
25. Cheng, P., "Combined Free and Forced Convection Flow About Inclined Surfaces in Porous Media", *Int. J. Heat Mass Transfer,* Vol. 20, pp. 807-814, 1977.
26. Combarnous, M.A. and Bia, P., "Combined Free and Forced Convection in Porous Media", *Soc. Petrol. Eng. J.,* Vol. 11, pp. 399-405, 1971.
27. Hsieh, J.C., Chen, T.S., and Armaly, B.F., "Nonsimilarity Solutions for Mixed Convection from Vertical Surfaces in Porous Media: Variable Surface Temperature or Heat Flux", *Int. J. of Heat and Mass Transfer,* Vol. 36, pp. 1485-1493, 1993.
28. Hsieh, J.C., Chen, T.S., and Armaly, B.F., "Mixed Convection Along a Nonisothermal Vertical Flat Plate Embedded in a Porous Medium: the Entire Regime", *Int. J. Heat Mass Transfer,* Vol. 36, No. 7, pp. 1819-1825, 1993.
29. Huang, Ming Jer, Yih, Kuo, and Chou, You Li, "Mixed Convection Flow over a Horizontal Cylinder or a Sphere Embedded in a Saturated Porous Medium", *J. Heat Transfer,* Vol. 108, pp. 469-471, 1986.
30. Kwendakwema, N.J. and Boehm, R.F., "Parametric Study of Mixed Convection in a Porous Medium Between Vertical Concentric Cylinders", *J. Heat Transfer,* Vol. 113, pp. 128-134, 1991.

31. Oosthuizen, P.H., "Mixed Convective Heat Transfer from a Cylinder in a Porous Medium Near an Impermeable Surface", *Mixed Convection Heat Transfer–1987*, ASME HTD-Vol. 84, Am. Soc. Mech. Eng., New York, 1987, pp. 75-82.
32. Oosthuizen, P.H., "Mixed Convective Heat Transfer from a Heated Horizontal Plate in a Porous Medium Near an Impermeable Surface", *ASME J. Heat Transfer*, Vol. 110, No. 2, pp. 390-394, 1988.
33. Oosthuizen, P.H. and Paul, J.T., "Mixed Convective Heat Transfer from a Cylinder in a Porous Medium Including Maximum Density Effects", Proc. ASME Winter Annual Meeting, HTD-Vol. 180, *Fundamentals of Forced and Mixed Convection and Transport Phenomena*, Am. Soc. Mech. Eng., New York, 1991, pp. 19-27.
34. Parang, M. and Keyhani, M., "Boundary Effects in Laminar Mixed Convection Flow Through an Annular Porous Medium", *J. Heat Transfer*, Vol. 109, pp. 1039-1041, 1987.
35. Reda, D.C., "Mixed Convection in a Liquid-Saturated Porous Medium", *J. of Heat Transfer*, Vol. 110., pp. 147-154, 1988.
36. Bejan, A. and Tien, C.L., "Natural Convection in a Horizontal Porous Medium Subjected to an End-to-End Temperature Difference", *J. Heat Transfer*, Vol. 100, pp. 191-198, 1978.
37. Bejan, A., "On the Boundary Layer Regime in a Vertical Enclosure Filled with a Porous Medium", *Lett. Heat Mass Transfer*, Vol. 6, pp. 93-102, 1979.
38. Chan, B.K.C., Ivey, C.M., and Barry, J.M., "Natural Convection in Enclosed Porous Media with Rectangular Boundaries", *J. Heat Transfer*, Vol. 92, pp. 21-27, 1970.
39. Hickox, C.E. and Gartling, D.K., "A Numerical Study of Natural Convection in a Horizontal Porous Layer Subjected to an End-to-End Temperature Difference", *J. Heat Transfer*, Vol. 103, pp. 797-802, 1981.
40. Naylor, D. and Oosthuizen, P.H., "Free Convection in a Horizontal Enclosure Partly Filled with a Porous Medium", *J. of Thermophysics and Heat Transfer*, Vol. 9, No. 4, pp. 797-800, 1995.
41. Naylor, D. and Oosthuizen, P.H., "Free Convection in an Enclosure Partly Filled with a Porous Medium and Partially Heated from Below", *Proc. 10th Int. Heat Trans. Conf.*, Brighton, UK, Vol. 5, pp. 351-356, 1994.
42. Oosthuizen, P.H. and Paul, J.T., "Natural Convection in a Square Enclosure Partly Filled with a Centrally Positioned Porous Layer and with a Partially Heated Wall", *Proc. 2nd Thermal-Sciences and 14th UIT National Heat Trans. Conf.*, Rome, Italy, Vol. 2, 1996, pp. 851-856.
43. Oosthuizen, P.H. and Paul, J.T., "Free Convective Flow in a Cavity Filled with a Vertically Layered Porous Medium", *Natural Convection in Porous Media*, ASME HTD-Vol. 56, AIAA/ASME 4th Thermophysics and Heat Trans. Conf., Am. Soc. Mech. Eng., New York, 1986, pp. 75-84.
44. Oosthuizen, P.H. and Paul, J.T., "Natural Convective Flow in a Square Cavity Partly Filled with a Porous Medium", *Proc. 1987 ASME/JSME Thermal Engineering Joint Conf.*, Vol. 2, Am. Soc. Mech. Eng., New York, 1987, pp. 407-412.
45. Oosthuizen, P.H. and Paul, J.T., "Natural Convection in a Rectangular Enclosure with a Partially Heated Wall and Partly Filled with a Porous Medium", *Proc. Eighth Int. Conf. on Numerical Methods in Thermal Problems*, Vol. VIII, Part 1, Pineridge Press, Swansea, U.K., 1993, pp. 467-478.
46. Oosthuizen, P.H., "Natural Convection in a Square Enclosure Partly Filled with Two Layers of Porous Material", Proc. 4th Int. Conf. on Advanced Computational Methods in Heat Trans., Advanced Computational Methods in Heat Transfer IV, Udine, Italy, 1996, pp. 63-72.

47. Oosthuizen, P.H., "Natural Convection in an Inclined Square Enclosure Partly Filled with a Porous Medium and with a Partially Heated Wall", *Heat Transfer in Porous Media and Two-Phase Flow,* ASME HTD-Vol. 302, Energy- Sources Technology Conference and Exhibition, Houston, TX, 1995, pp. 29-42.
48. Oosthuizen, P.H. and Naylor, D., "Natural Convective Heat Transfer from a Cylinder in an Enclosure Partly Filled with a Porous Medium", *Int. J. Numer. Methods Heat and Fluid Flow,* Vol. 6, No. 6, pp. 51-63, 1996.
49. Poulikakos, D. and Bejan, A., "Numerical Study of Transient High Rayleigh Number Convection in an Attic-Shaped Porous Layer", *J. Heat Transfer,* Vol. 105, pp. 476-484, 1983.
50. Walker, K.L. and Homsy, G.M., "Convection in a Porous Cavity", *J. Fluid Mech.,* Vol. 87, pp. 449-474, 1978.
51. Weber, J.E., "The Boundary Layer Regime for Convection in a Vertical Porous Layer", *Int. J. Heat Mass Transfer,* Vol. 18, pp. 569-573, 1975.
52. Elder, J.W., "Steady Free Convection in a Porous Medium Heated from Below", *J. Fluid Mech.,* Vol. 27, pp. 29-48, 1967.
53. Yen, Y.C., "Effects of Density Inversion on Free Convective Heat Transfer in Porous Layer Heated from Below", *Int. J. Heat Mass Transfer,* Vol. 17, pp. 1349-1356, 1974.
54. Lai, F.C. and Kulacki, F.A., "Oscillatory Mixed Convection in Horizontal Porous Layers Locally Heated from Below", *Int. J. Heat and Mass Transfer,* Vol. 34, pp. 887-890, 1991.
55. Lai, F.C and Kulacki, F.A., "Experimental Study of Free and Mixed Convection in Horizontal Porous Layers Locally Heated from Below", *Int. J. Heat and Mass Trans.,* Vol. 34, pp. 525-541, 1991.
56. Prasad, V., Lai, F.C., and Kulacki, F.A., "Mixed Convection in Horizontal Porous Layers Heated from Below", *J. Heat Transfer,* Vol. 110, pp. 395-402, 1988.
57. Bejan, A. and Poulikakos, D., "The NonDarcy Regime for Vertical Boundary Layer Natural Convection in a Porous Medium", *Int. J. Heat Mass Transfer,* Vol. 27, pp. 717-722, 1984.
58. Vafai, K. and Tien, C.-L., "Boundary and Inertia Effects on Flow and Heat Transfer in Porous Media", *Int. J. Heat Mass Transfer,* Vol. 24, pp. 195-203, 1981.
59. Ward, J.C., "Turbulent Flow in Porous Media", *J. Hydraul. Div. ASCE,* Vol. 90, No. HY5, pp. 1-12, 1964.
60. Chou, F.C., Su, J.H., and Lien, S.S., 1994 "A Re-evaluation of Non-Darcian Forced and Mixed Convection in Cylindrical Packed Tubes", *J. of Heat Transfer,* Vol. 116, pp. 513-516, 1994.
61. Hong, J.-T., Tien, C.-L. and Kaviany, M., "Non-Darcean Effects on Vertical Plate Natural Convection in Porous Media with High Porosity", *Int. J. Heat Mass Transfer,* Vol. 28, pp. 2149-2157, 1985.
62. Islam, Rafiqul M. and Nandakumar, K., "Mixed Convection Heat Transfer in Porous Media in the Non-Darcy Regime", *Can. J. of Chem. Eng.,* Vol. 66, pp. 68-74, 1988.
63. Jang, Jiin Yuh and Chen, Jiing Lin, "Thermal Dispersion and Inertia Effects on Vortex Instability of a Horizontal Mixed Convection Flow in a Saturated Porous Medium", *Int. J. Heat and Mass Transfer,* Vol. 36, No. 2, pp. 383-389, 1993.
64. Kumari, M. and Nath, G., "Non-Darcy Mixed Convection Flow over a Nonisothermal Cylinder and Sphere Embedded in a Saturated Porous Medium", *J. Heat Transfer,* Vol. 112, pp. 518-521, 1990.
65. Lai, F.C. and Kulacki, F.A., "Non-Darcy Mixed Convection Along a Vertical Wall in a Saturated Porous Medium", *J. Heat Transfer,* Vol 113, pp. 252-255, 1991.

66. Aldoss, T.K., Jarrah, M.A. and Al Sha'er, B.J., "Mixed Convection from a Vertical Cylinder Embedded in a Porous Medium: Non-Darcy Model", *Int. J. Heat and Mass Transfer,* Vol. 39, No. 6, pp. 1141-1148, 1996.
67. Nakayama, A., "A Unified Theory for Non-Darcy Free, Forced, and Mixed Convection Problems Associated with a Horizontal Line Heat Source in a Porous Medium", *J. Heat Transfer,* Vol. 116, pp. 508-513, 1994.
68. Choi, C.Y. and Kulacki, F.A., "Non-Darcian Effects on Mixed Convection in a Vertical Packed-Sphere Annulus", *J. of Heat Transfer,* Vol. 115, pp. 506-510, 1993.
69. Chen, Chien Hsin; Chen, T.S., and Chen, Cha'o Kuang, "Non-Darcy Mixed Convection Along Nonisothermal Vertical Surfaces in Porous Media", *Int. J. Heat and Mass Transfer,* Vol. 39, No. 6, pp. 1157-1164, 1996.
70. Chen, Cha'o-Kuang, Chen, Chien Hsin, and Minkowycz, W.J., "Non-Darcian Effects on Mixed Convection About a Vertical Cylinder Embedded in a Saturated Porous Medium", *Int. J. Heat and Mass Transfer,* Vol. 35, pp. 3041-3046, 1992.
71. Yu, W.S, Lin, H.T., and Lu, C.S., "Universal Formulations and Comprehensive Correlations for Non-Darcy Natural Convection and Mixed Convection in Porous Media", *Int. J. Heat and Mass Transfer,* Vol. 34, pp. 2859-2868, 1991.

CHAPTER 11

Condensation

11.1 INTRODUCTION

Condensation occurs when a vapor is exposed to a surface that is at a temperature below the saturation temperature of the vapor. The latent heat of the vapor is transferred to the surface and condensate, i.e., a liquid, is formed on the surface. Condensation thus involves a change in phase from vapor to liquid at a surface. The process is shown schematically in Fig. 11.1, see [1] to [12].

The prediction of the heat transfer rates associated with condensation is an essential stage in the design of a wide range of industrial devices and processes. For example, in a modern steam-powered electricity generating plant, steam that has expanded through the turbine is condensed before being pumped back to the boiler via the feedwater heating system. Similarly, in a refrigeration plant, the refrigerant is condensed, in this way providing the cooling. The design of the condenser is critical to the proper operation and to the achievement of a high thermal efficiency in both of these situations. Condensation is also involved in many chemical process plants, e.g., in an oil refinery.

It has been observed that when a vapor condenses on a surface the condensate can form in two possible ways or modes, these two modes of condensation being termed Film (or Filmwise) Condensation and Dropwise Condensation. In film condensation, the condensate forms as a continuous layer (or film) on the surface, this layer draining continuously off the surface under the influence of gravity. In dropwise condensation, the vapor condenses in the form of individual droplets. These droplets grow and sometimes coalesce with other nearby droplets to form larger droplets. Sporadically, as individual droplets become large enough, they drain off the surface under the influence of gravity. The two modes of condensation are shown in Fig. 11.2.

The way in which the condensate forms on the surface, i.e., in the form of a continuous film or in the form of discrete droplets, has a strong influence on the heat transfer rate. In film condensation, the latent heat released by the vapor must be

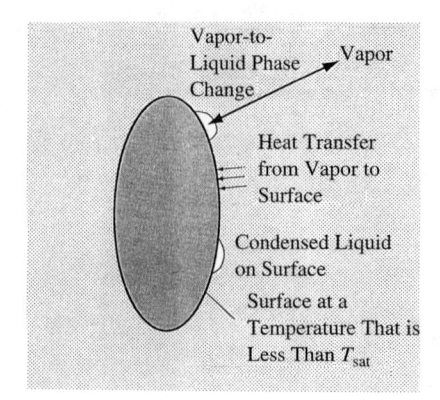

FIGURE 11.1
Condensation of a vapor on a surface.

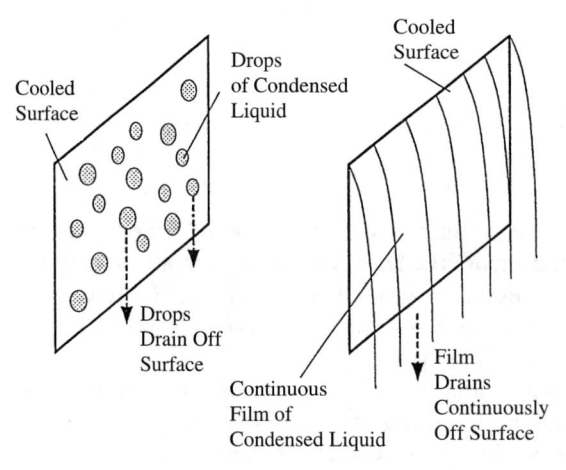

FIGURE 11.2
Film and dropwise condensation on a vertical surface.

transferred to the surface through the liquid film. The film forms a thermal "barrier" between the warm vapor and the cool surface, i.e., it offers substantial resistance to heat transfer. In contrast, during dropwise condensation, parts of the surface are directly exposed to the vapor. For example, as a large droplet runs down the surface under the influence of gravity, it coalesces with other droplets in its path, leaving a dry surface in its wake. Many new droplets form and grow rapidly on the newly exposed surface, producing a high heat transfer rate. Also, the droplets tend to form at distinct sites and parts of the surface may as a result always be directly exposed to the vapor. As a result of these differences, heat transfer rates with dropwise condensation can be more than ten times higher than those with film condensation under the same conditions [13]. Typical values for film and dropwise condensation are shown in Fig. 11.3.

The high heat transfer rates associated with dropwise condensation make it very attractive for engineering applications. High heat transfer rates lead to high condensation rates allowing the use of smaller condensers. The controlling factor for

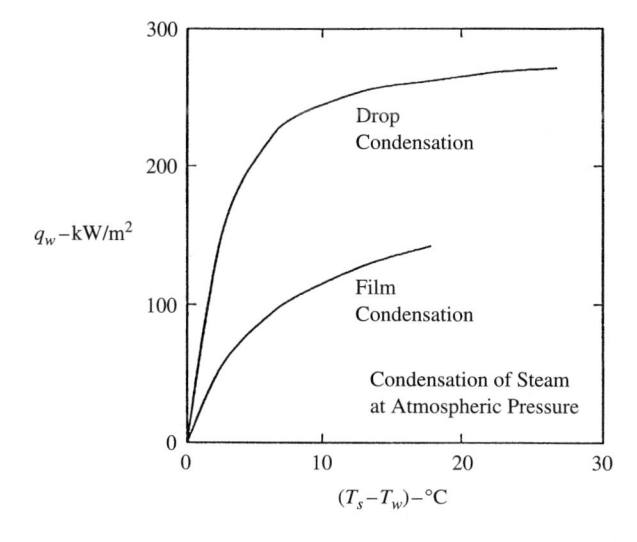

FIGURE 11.3
Heat transfer rates with film and dropwise condensation of steam on a vertical surface.

promoting dropwise condensation is the wettability of the surface as determined by the contact angle of a droplet. As shown in Fig. 11.4, a liquid/surface combination is considered nonwetting if the contact angle (θ) is greater than 90°. If $\theta > 90°$, the condensate will tend to form as discrete droplets, i.e., dropwise condensation will tend to occur, while if $\theta < 90°$, the droplets will tend to merge into a continuous film, i.e., film condensation will tend to occur, e.g. see [14] to [29].

Considerable research on creating long-lasting nonwettable surface-liquid combinations that can be used to produce dropwise condensation in industrial devices has been undertaken. Surface coatings (such as grease, Teflon®, and nonoxidizing "noble" metals) and additives to the working fluid have been used to promote dropwise condensation. A detailed review of much of this work can be found in reference 13. Although some success in sustaining dropwise condensation has been obtained using thin Teflon® and gold coatings, in most practical situations the process eventually reverts to film condensation as unavoidable contaminants build up on the surface. For this reason, dropwise condensation will not be considered further here and this chapter will, therefore, focus on the analysis of film condensation. Condensers are most often designed based on the assumption of film condensation, this design practice being adopted because of the difficulty of sustaining dropwise condensation and because of the lower heat transfer rates associated with film condensation. These lower rates yield a more conservative, safer design.

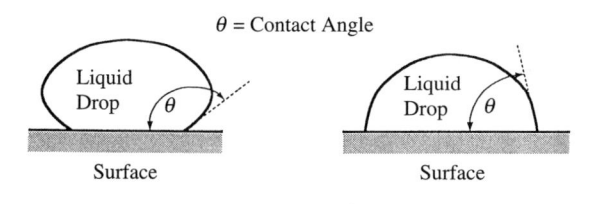

FIGURE 11.4
Wettability of a surface.

11.2
LAMINAR FILM CONDENSATION ON A VERTICAL PLATE

The situation shown in Fig. 11.5 is considered here, i.e., consideration is given to film condensation on a cold isothermal vertical plate, held at temperature T_w, which is exposed to a reservoir of saturated vapor at saturation temperature, T_s.

It should be noted that the x-coordinate is measured vertically downward along the plate surface and the y-coordinate is measured perpendicular to the plate surface. For condensation to occur, the wall temperature, T_w, must be lower than the saturation temperature, T_s, corresponding to the vapor reservoir pressure. Vapor condenses on the plate forming a thin film of liquid that flows down the plate under the influence of gravity. The thickness of the film, δ, and the local mass flow rate increase with distance down the plate as condensate forms continuously along the entire film/vapor interface.

Steady film condensation on an isothermal surface can be analyzed using an approximate analysis first proposed by Nusselt [30], see also [31] to [40]. To simplify the problem, the following assumptions are made:

1. The flow in the liquid film is laminar and the condensed liquid has constant properties.
2. The vapor reservoir is at rest and is everywhere at the saturation temperature, T_s.
3. Momentum changes (i.e., inertia forces) within the film are negligible, i.e., in writing the momentum equation for the film, the viscous, gravitational, and pressure forces acting on the film are assumed to be far greater than the rate of change of liquid momentum within the film.
4. The temperature of the film/vapor interface is the saturation temperature, T_s, corresponding to the vapor reservoir pressure.
5. The temperature within the liquid film layer varies linearly from the wall temperature, T_w, at $y = 0$ to the saturation temperature, T_s, at $y = \delta$.
6. The viscous shear force of the vapor on the film at $y = \delta$ is negligible, i.e.,

$$\partial u/\partial y\big|_{y=\delta} = 0$$

7. Pressure changes in the y-direction across the film are negligible.

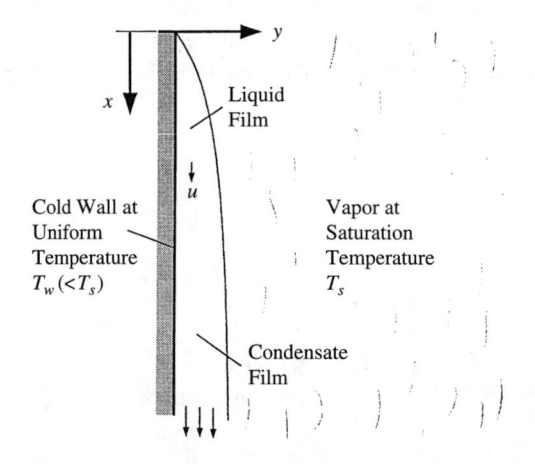

FIGURE 11.5
Film condensation on a vertical isothermal surface.

CHAPTER 11: Condensation   559

While this is a very simplified model of the process, it does allow much insight to be gained into factors that govern film condensation and does, in fact, give results that are, in many situations, in good agreement with experiment. More sophisticated models that relax some of the assumptions on which the Nusselt analysis is based will be discussed later in this chapter.

The analysis is, of course, based on the application of the conservation of mass, momentum, and energy principles. These conservation principles are applied to a control volume spanning all or part of the liquid film such as that shown in Fig. 11.6 and, because two-dimensional flow is being assumed in the following analysis, a control volume with unit width will be considered.

Conservation of momentum is first applied. Consider the forces on the control volume. Because the changes of momentum within the film are being neglected (assumption 3 above), conservation of momentum requires that the forces on the control volume must balance. Thus, considering the control volume shown in Fig. 11.7, the sum of the net pressure, and shear and gravitational forces in the x-direction on the element must be 0, i.e.:

$$g\,dm - \mu_\ell \frac{\partial u}{\partial y} dx - \left(\frac{dp}{dx} dx\right)(\delta - y) = 0 \quad (11.1)$$

where dm is the mass of the liquid in the control volume which is given by:

$$dm = \rho_\ell \, dV = \rho_\ell (\delta - y)\, dx \quad (11.2)$$

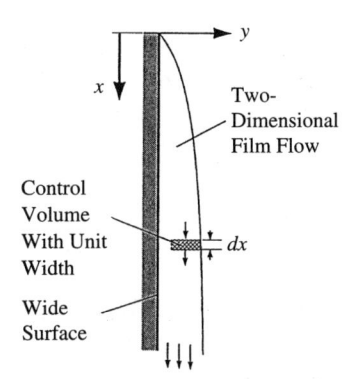

FIGURE 11.6
Type of control volume across condensed film used in Nusselt analysis.

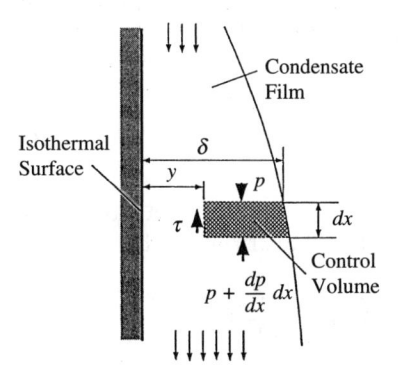

FIGURE 11.7
Control value used in applying conservation of momentum.

560 Introduction to Convective Heat Transfer Analysis

where dV is the volume of the liquid in the control volume and where it has been noted that a unit width of the control volume is being considered.

The subscript, ℓ, in the above equations refers to the properties of the liquid. The subscript, v, will be used below to denote the properties of the vapor in the reservoir.

Because the vapor in the reservoir is assumed to be at rest, i.e., to be stagnant, the pressure distribution in the vapor reservoir will be given by the hydrostatic pressure equation. The vertical pressure gradient in the vapor is therefore given by:

$$\frac{dp}{dx} = \rho_v g \tag{11.3}$$

In writing this equation it has been noted that the coordinate, x, is being measured in the vertically downwards direction, i.e., in the direction of increasing hydrostatic pressure.

This pressure gradient is imposed on the outer edge of the liquid film. However, the liquid film on the surface is assumed to remain thin and the velocity component normal to the surface in the film is therefore assumed to be very small. As a result, as mentioned above, pressure changes across the liquid film are neglected. This is why a total derivative was used for the pressure in Eq. (11.1). Eq. (11.3) therefore gives the pressure gradient everywhere in the condensed film.

Substituting Eqs. (11.2) and (11.3) into Eq. (11.1) gives:

$$g\rho_\ell(\delta - y)\,dx - \mu_\ell \frac{\partial u}{\partial y}\,dx - \rho_v g\,dx(\delta - y) = 0 \tag{11.4}$$

This first term in this equation is the result of the gravitational force on the liquid in the control volume, the second term is due to the viscous shear force, while the third represents the pressure exerted on the control volume due to the gravitational force on vapor.

Dividing Eq. (11.4) by dx and rearranging then gives:

$$\mu_\ell \frac{\partial u}{\partial y} = g(\delta - y)(\rho_\ell - \rho_v) \tag{11.5}$$

This equation indicates that the shear force is equal to the net gravitational force, i.e., the gravitational force on the liquid less the gravitational force that would have acted on the control volume had it contained vapor and not liquid.

Integrating Eq. (11.5) with respect to y, i.e., across the liquid film and applying the no-slip condition at the surface, i.e, setting $u = 0$ at $y = 0$, gives the velocity distribution in the film as:

$$u = \frac{(\rho_\ell - \rho_v)g}{\mu_\ell}\left(\delta y - \frac{y^2}{2}\right) \tag{11.6}$$

The mass flow rate per unit width of the plate, denoted here by Γ, at any x location can be then be calculated by integrating this velocity distribution. This gives:

$$\Gamma = \int_0^\delta \rho_\ell u\,dy = \int_0^\delta \rho_\ell \left[\frac{(\rho_\ell - \rho_v)g}{\mu_\ell}\left(\delta y - \frac{y^2}{2}\right)\right]dy \tag{11.7}$$

CHAPTER 11: Condensation   561

from which it follows that:

$$\Gamma = \frac{\rho_\ell(\rho_\ell - \rho_v)g\delta^3}{3\mu_\ell} \tag{11.8}$$

It will be noted that both Eqs. (11.6) and (11.8) contain the film thickness δ which at this stage of the analysis is not known.

Conservation of mass and conservation of energy are next considered. Consider the control volume shown in Fig. 11.8. Between x and $x + dx$, i.e., over the length of the control volume, the mass flow rate in the film (Γ) increases because vapor condenses at the rate $d\dot{m}_v$ on the film and is absorbed into it. Applying conservation of mass to the control volume gives:

Rate mass is condensed on the outer surface of control volume = Rate mass leaves control volume − Rate mass enters control volume

i.e.:

$$d\dot{m}_v = \left(\Gamma + \frac{d\Gamma}{dx}dx\right) - \Gamma$$

i.e.:

$$d\dot{m}_v = \frac{d\Gamma}{dx}dx \tag{11.9}$$

As the vapor condenses onto the film, its latent heat (h_{fg} per unit mass) must be transferred to the liquid film. The latent heat plus the sensible heat associated with the cooling of the liquid film below the saturation temperature is then transferred to the wall. In most situations, the amount of sensible heat associated with the subcooling, i.e., the cooling of the liquid below the saturation temperature of the liquid in the condensed film, is very small compared to the released latent heat. Therefore, to simplify the analysis, the effects of this subcooling on the energy balance will be neglected. This will be discussed in more detail later in this section. This assumption that subcooling effects are negligible implies that at any value of x, the rate at which heat is transferred into the outer surface of the liquid layer is equal to the rate heat is transferred to the wall. The assumption of a linear temperature profile in the film is consistent with this assumption.

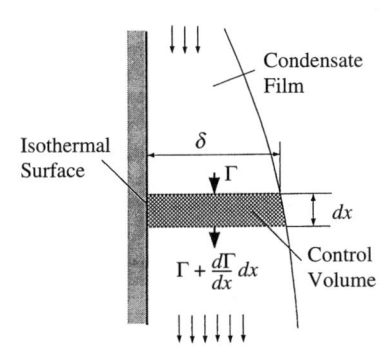

FIGURE 11.8
Control volume used in applying conservation of mass.

562 Introduction to Convective Heat Transfer Analysis

Because the temperature profile is assumed to be linear (assumption 5 above), the heat transfer rate, which is given by Fourier's law, is the same at all points across the film. It follows therefore that the heat transfer rate to the wall from the liquid film is given by:

$$q_x = k_\ell \left.\frac{\partial T}{\partial y}\right|_{y=0} = k_\ell \frac{(T_s - T_w)}{\delta} \tag{11.10}$$

Because sensible heat effects are being neglected, the rate of latent heat release by the condensing vapor can be equated to heat transfer rate through the film to the wall, i.e., considering the control volume of length dx:

$$h_{fg}\, d\dot{m}_v = k_\ell \frac{(T_s - T_w)}{\delta}\, dx \tag{11.11}$$

Substituting Eq. (11.9) into Eq. (11.11) then gives:

$$h_{fg}\frac{d\Gamma}{dx}\, dx = k_\ell \frac{(T_s - T_w)}{\delta}\, dx \tag{11.12}$$

Dividing this equation through by dx and substituting the expression for Γ given by Eq. (11.8) then gives:

$$h_{fg}\frac{d}{dx}\left[\frac{\rho_\ell(\rho_\ell - \rho_v)g\delta^3}{3\mu_\ell}\right] = k_\ell \frac{(T_s - T_w)}{\delta} \tag{11.13}$$

Because the liquid properties in the film are being assumed constant, this equation can be written as:

$$h_{fg}\left[\frac{\rho_\ell(\rho_\ell - \rho_v)g}{\mu_\ell}\right]\delta^2 \frac{d\delta}{dx} = k_\ell \frac{(T_s - T_w)}{\delta} \tag{11.14}$$

i.e.:

$$\left[\frac{h_{fg}\rho_\ell(\rho_\ell - \rho_v)g}{\mu_\ell}\right]\delta^3 \frac{d\delta}{dx} = k_\ell(T_s - T_w) \tag{11.15}$$

Integrating this equation and noting that the film thickness must be 0 at the top of the plate, i.e., that $\delta = 0$ at $x = 0$, gives:

$$\frac{\delta^4}{4} = \left[\frac{\mu_\ell k_\ell(T_s - T_w)}{h_{fg}\rho_\ell(\rho_\ell - \rho_v)g}\right]x$$

i.e.:

$$\delta = \left[\frac{4\mu_\ell k_\ell(T_s - T_w)x}{gh_{fg}\rho_\ell(\rho_\ell - \rho_v)}\right]^{1/4} \tag{11.16}$$

Using this expression for the film thickness, δ, Eq. (11.10) gives the local heat transfer rate at the wall as:

$$q_x = k_\ell \left[\frac{gh_{fg}\rho_\ell(\rho_\ell - \rho_v)}{4\mu_\ell k_\ell(T_s - T_w)x}\right]^{1/4}(T_s - T_w) \tag{11.17}$$

This can be written in terms of a local heat transfer coefficient which, since q_x is the local heat rate from the film to the wall, is conventionally defined by:

$$-q_x = h(T_s - T_w)$$

Using Eq. (11.17), this gives:

$$h = \left[\frac{gh_{fg}\rho_\ell(\rho_\ell - \rho_v)k_\ell^3}{4\mu_\ell(T_s - T_w)x}\right]^{1/4} \tag{11.18}$$

This can be written in terms of the local Nusselt number as:

$$Nu_x = \frac{hx}{k_\ell} = \left[\frac{gh_{fg}\rho_\ell(\rho_\ell - \rho_v)x^3}{4\mu_\ell k_\ell(T_s - T_w)}\right]^{1/4} \tag{11.19}$$

The mean heat transfer rate from a wall of height L is given by:

$$\bar{q} = \frac{1}{L}\int_0^L q_x\,dx$$

i.e., using Eq. (11.17):

$$\bar{q} = \frac{4}{3}k_\ell\left[\frac{gh_{fg}\rho_\ell(\rho_\ell - \rho_v)}{4\mu_\ell k_\ell(T_s - T_w)L}\right]^{1/4}(T_s - T_w) \tag{11.20}$$

which can be rewritten in terms of the mean Nusselt number as:

$$\overline{Nu} = \frac{\bar{h}L}{k_\ell} = 0.943\left[\frac{gh_{fg}\rho_\ell(\rho_\ell - \rho_v)L^3}{\mu_\ell k_\ell(T_s - T_w)}\right]^{1/4} \tag{11.21}$$

When applying Eq. (11.19) or Eq. (11.21) the effects of temperature dependent liquid properties can normally be adequately accounted for by evaluating these liquid properties at the mean temperature in the film, i.e., at $(T_s + T_w)/2$. The latent heat, h_{fg}, is evaluated at the saturation temperature. In some cases, in order to achieve greater accuracy, it may be necessary to account for the effect of variable liquid properties by using a more complex procedure to find the temperature at which the properties are evaluated. The appropriate temperature will then depend on the specific liquid involved. This is discussed in references [41] to [43].

EXAMPLE 11.1. Stagnant saturated steam at 100°C condenses on a 0.5 m high vertical plate with a surface temperature of 95°C. Assuming steady laminar flow, calculate the heat transfer rate and condensation rate per m width of the plate. Also find the maximum film thickness.

Solution. The water properties in the film are evaluated at the mean film temperature, i.e., at:

$$T_f = \frac{(T_s + T_w)}{2} = 97.5°C$$

At this temperature for water:

$$\rho_\ell = 960 \text{ kg/m}^3,\ \mu_\ell = 2.89 \times 10^{-4} \text{ kg/m s},\ k_\ell = 0.680 \text{ W/m K}$$

The vapor density and latent heat are evaluated at the saturation temperature, T_s which is 100°C and are:

$$\rho_v = 0.598 \text{ kg/m}^3, \quad h_{fg} = 2257 \text{ kJ/kg}$$

In the situation being considered it will be seen that the vapor density is very small compared to the liquid density so: $\rho_\ell(\rho_\ell - \rho_v) = \rho_\ell^2$ is a good approximation. This is usually true for the condensation of steam.

The average Nusselt number is calculated by applying Eq. (11.21), i.e.:

$$\overline{Nu} = 0.943 \left[\frac{gh_{fg}\rho_\ell(\rho_\ell - \rho_v)L^3}{\mu_\ell k_\ell (T_s - T_w)} \right]^{1/4}$$

i.e.:

$$\overline{Nu} = 0.943 \left[\frac{9.81 \times 2{,}257{,}000 \times 960^2 \times 0.5^3}{0.000289 \times 0.68 \times (100 - 95)} \right]^{1/4} = 6730$$

The average heat transfer coefficient is then given by:

$$\bar{h} = \frac{\overline{Nu} k_\ell}{L} = \frac{6730 \times 0.68}{0.5} = 9152 \text{ W/m}^2 \text{ °C}$$

Considering both sides of the plate, the total heat transfer rate per m plate width is then given by:

$$Q = \bar{h}A(T_s - T_w) = 9152 \times (0.5 \times 2) \times (100 - 95) = 45{,}760 \text{ W} = 45.76 \text{ kW}$$

The condensation rate will be equal to the total heat transfer rate divided by the latent heat, i.e.:

$$\dot{m} = \frac{Q}{h_{fg}} = \frac{45{,}760}{2{,}257{,}000} = 0.0203 \text{ kg/s} = 73.0 \text{ kg/h}$$

The maximum film thickness is found using Eq. (11.16), it being noted that the maximum film thickness will occur at the bottom of the plate, i.e., at $x = L$. Hence, the maximum film thickness is given by:

$$\delta = \left[\frac{4\mu_\ell k_\ell (T_s - T_w)L}{gh_{fg}\rho_\ell(\rho_\ell - \rho_v)} \right]^{1/4} = \left[\frac{4\mu_\ell k_\ell (T_s - T_w)L}{gh_{fg}\rho_\ell^2} \right]^{1/4}$$

$$= \left[\frac{4 \times 0.000289 \times 0.68 \times (100 - 95) \times 0.5}{9.81 \times 2{,}257{,}000 \times 960^2} \right]^{1/4}$$

$$= 0.0000991 \text{ m} \approx 0.1 \text{ mm}$$

This illustrates the fact that condensate films are usually very thin.

It was assumed in this example that the flow in the liquid film remained laminar. This may not be true, however, as will be discussed later.

Although not necessary, some insight can be gained by writing Eq. (11.21) in terms of commonly used dimensionless products, i.e., to write Eq. (11.21) as:

$$\overline{Nu} = 0.943 \left[\frac{Gr_L Pr_\ell}{Ja} \right]^{1/4} \qquad (11.22)$$

where Gr_L is the Grashof number based on the plate length and Pr_ℓ is the fluid Prandtl number. The dimensionless group, Ja, is termed the Jakob number and is defined as:

$$Ja = \frac{c_{p\ell}(T_s - T_w)}{h_{fg}} = \frac{\text{sensible heat}}{\text{latent heat}} \qquad (11.23)$$

The Jakob number is basically a measure of the importance of subcooling expressing, as it does, the change in the sensible heat per unit mass of condensed liquid in the film relative to the enthalpy associated with the phase change. The Jakob number is small for many problems, i.e the sensible heat change across the liquid film is small compared to the latent heat release. For example, for cases involving the condensation of steam, Ja, is typically of the order of 0.01.

The definition of the Grashof number, Gr_L, in Eq. (11.22) is:

$$Gr_L = \frac{\rho_\ell(\rho_\ell - \rho_v)gL^3}{\mu_\ell^2} \qquad (11.24)$$

i.e., the density change in this Grashof number is associated with the density change that results from the change in phase whereas in the conventional Grashof number used in the analysis of natural convection, the density change is associated purely with the temperature changes in the fluid. The density changes associated with a liquid-vapor phase change are usually much greater than those associated purely with a temperature change in a fluid.

It should be noted that the Eqs. (11.19) and (11.21) are also valid for condensation on the outside surface of vertical tubes [44], provided the radius of the tube is large compared to the film thickness as indicated in Fig. 11.9. Since condensate films are typically very thin in practical situations, this condition is usually met.

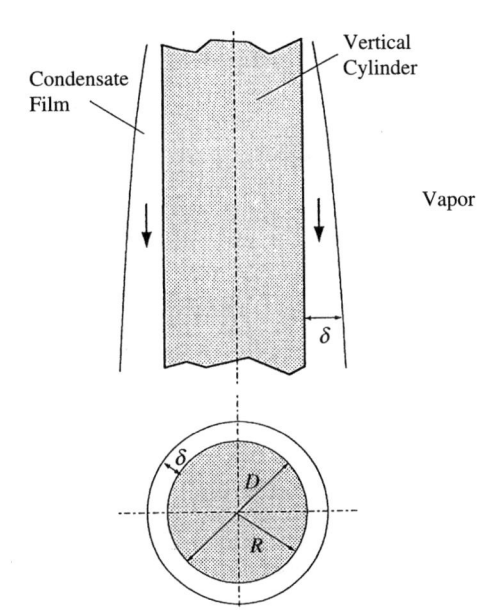

FIGURE 11.9
Condensation on the outer surface of a vertical cylinder.

566 Introduction to Convective Heat Transfer Analysis

The above analysis of condensation on a vertical plate can be easily extended to condensation on an inclined plate. Consider a plate inclined at an angle, θ, with respect to the gravity vector as shown in Fig. 11.10.

The x-coordinate is again measured in the flow direction along the plate and the y-coordinate is measured normal to the plate surface as indicated in Fig. 11.10. When the plate is inclined, there are components of the gravitational force both parallel to and normal to the direction of film flow. The normal, i.e., y-direction, component of the gravitational force will, in general, produce a pressure change across the film. However, because the film is by assumption thin, this pressure change across the film is here assumed to be negligible. Hence, as shown previously, the net body force on a fluid element will be $g(\rho_\ell - \rho_v)\,dV$. However, only the x component of the net body force acts in the direction of the flow, so, in this case, a force balance in the x-direction (see Fig. 11.11) gives:

$$\mu_\ell \frac{\partial u}{\partial y} = g(\delta - y)(\rho_\ell - \rho_v)\cos\theta \tag{11.25}$$

This is the same as Eq. (11.5) except that $g\cos\theta$ occurs in place of g. The remainder of the analysis is identical to that given above for a vertical plate and the same result therefore applies to an inclined plate except that g is replaced by $g\cos\theta$.

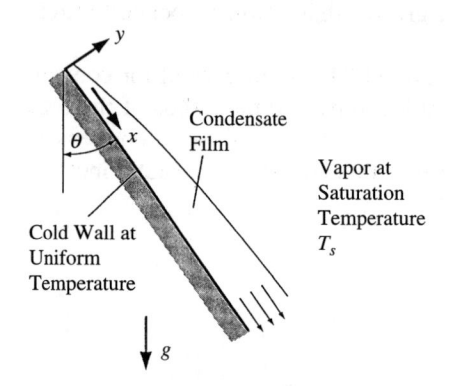

FIGURE 11.10
Condensation on an inclined plate.

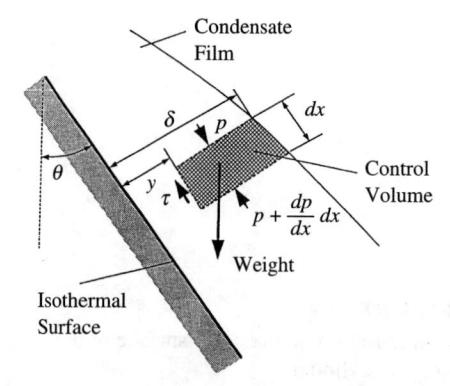

FIGURE 11.11
Control volume used in setting up force balance for a condensed film on an inclined plate.

CHAPTER 11: Condensation **567**

For example, for an inclined plate, Eq. (11.21) becomes:

$$\overline{Nu} = 0.943 \left[\frac{g \cos \theta h_{fg} \rho_\ell (\rho_\ell - \rho_v) L^3}{\mu_\ell k_\ell (T_s - T_w)} \right]^{1/4} \tag{11.26}$$

The effect of subcooling in the film will next be considered. As indicated previously, the sensible heat transfer associated with the subcooling of the liquid film is often small compared to the latent heat release and has been neglected in the above analysis. However, for working fluids with low values of latent heat, such as most refrigerants, a correction to the above analysis to account for subcooling may sometimes be needed.

To illustrate how the effect of subcooling is accounted for, consider an energy balance for an elemental control volume such as that shown in Fig. 11.8. When subcooling of the liquid film is allowed for, the total enthalpy flowing into and out of the control volume must be included in the energy balance. In this case, conservation of energy requires for the control volume:

$$\begin{array}{c}\text{Heat transfer}\\\text{rate to wall}\end{array} = \begin{array}{c}\text{Rate of mass}\\\text{influx at the outer}\\\text{edge of the film}\end{array} \times \begin{array}{c}\text{Latent heat}\\\text{per unit mass}\end{array}$$

$$- \left(\begin{array}{c}\text{Rate enthalpy leaves}\\\text{the control volume}\end{array} - \begin{array}{c}\text{Rate enthalpy enters}\\\text{control volume}\end{array} \right)$$

the enthalpy being that of the liquid. This equation gives:

$$q_x \, dx = d\dot{m}_v h_{fg} - \left[\int_0^\delta \rho_\ell u h_\ell \, dy + \frac{d}{dx} \left(\int_0^\delta \rho_\ell u h_\ell \, dy \right) dx - \int_0^\delta \rho_\ell u h_\ell \, dy \right]$$

Hence:

$$k_\ell \left. \frac{\partial T}{\partial y} \right|_{y=0} dx = d\dot{m}_v h_{fg} - \frac{d}{dx} \left(\int_0^\delta \rho_\ell u h_\ell \, dy \right) dx \tag{11.27}$$

The enthalpy of the subcooled liquid is measured relative to that existing at the saturation temperature, i.e., assuming that the liquid specific heat is constant, the enthalpy of the subcooled liquid film h_ℓ can be expressed as:

$$h_\ell = c_{p\ell}(T - T_s) \tag{11.28}$$

Now, it will be recalled that it was shown [see Eq. (11.9)] that the mass flux of vapor into the control volume, $d\dot{m}_v$, is given by:

$$d\dot{m}_v = \frac{d\Gamma}{dx} dx = \frac{d}{dx} \left(\int_0^\delta \rho_\ell u \, dy \right) dx \tag{11.29}$$

Substituting Eqs. (11.28) and (11.29) into Eq. (11.27) gives:

$$k_\ell \left. \frac{\partial T}{\partial y} \right|_{y=0} = h_{fg} \frac{d\Gamma}{dx} + c_{p\ell} \frac{d}{dx} \left(\int_0^\delta \rho_\ell u (T_s - T) \, dy \right) \tag{11.30}$$

568 Introduction to Convective Heat Transfer Analysis

As discussed before, the temperature distribution in the film is being assumed to be linear and can therefore be expressed as:

$$(T_s - T) = (T_s - T_w)\left(1 - \frac{y}{\delta}\right)$$

Hence, substituting this expression for the temperature profile and the expression for the velocity profile given in Eq. (11.6) into the integral into Eq. (11.30) gives:

$$c_{p\ell}\rho_\ell \frac{d}{dx}\left(\int_0^\delta u(T_s - T)\,dy\right)dx = \frac{c_{p\ell}\rho_\ell(\rho_\ell - \rho_v)(T_s - T_w)g}{\mu_\ell}\frac{d}{dx}\left\{\delta^3 \int_0^1 \left[\left(\frac{y}{\delta}\right)\right.\right.$$
$$\left.\left. - \frac{1}{2}\left(\frac{y}{\delta}\right)^2\right]\left[1 - \left(\frac{y}{\delta}\right)\right]d\left(\frac{y}{\delta}\right)\right\}dx \quad (11.31)$$

Carrying out the integration then gives:

$$c_{p\ell}\rho_\ell \frac{d}{dx}\left(\int_0^\delta u(T_s - T)\,dy\right)dx = \frac{c_{p\ell}\rho_\ell(\rho_\ell - \rho_v)(T_s - T_w)g}{\mu_\ell}\frac{1}{8}\frac{d\delta^3}{dx}dx$$

But Eq. (11.8) gives:

$$\Gamma = \frac{\rho_\ell(\rho_\ell - \rho_v)g\delta^3}{3\mu_\ell}$$

from which it follows that:

$$\frac{d\Gamma}{dx} = \frac{\rho_\ell(\rho_\ell - \rho_v)g}{3\mu_\ell}\frac{d\delta^3}{dx}$$

Hence:

$$c_{p\ell}\rho_\ell \frac{d}{dx}\left[\int_0^\delta u(T_s - T)\,dy\right]dx = c_{p\ell}(T_s - T_w)\frac{3}{8}\frac{d\Gamma}{dx}dx \quad (11.32)$$

Substituting this back into Eq. (11.30) then gives:

$$\frac{k_\ell(T_s - T_w)}{\delta} = \left[h_{fg} + \frac{3}{8}c_{p\ell}(T_s - T_w)\right]\frac{d\Gamma}{dx} \quad (11.33)$$

This result is the same as Eq. (11.12) which was obtained by ignoring subcooling except that h_{fg} has been replaced by:

$$h_{fg} + \frac{3}{8}c_{p\ell}(T_s - T_w)$$

Therefore, to include subcooling effects, h_{fg} is simply replaced by a modified latent heat, h_{fg}^*, which is given by:

$$h_{fg}^* = h_{fg} + \frac{3}{8}c_{p\ell}(T_s - T_w) = h_{fg}\left(1 + \frac{3}{8}Ja\right) \quad (11.34)$$

Rohsenow [32] has improved on the above analysis by performing an integral analysis that drops the assumption of a linear temperature profile. Based on this analysis, the following equation for the modified latent heat is recommended:

$$h_{fg}^* = h_{fg}(1 + 0.68 Ja) \tag{11.35}$$

Now the heat transfer rate depends on $h_{fg}^{*1/4}$. If it is assumed, therefore, that subcooling effects are negligible if the term $(1 + 0.68Ja)^{1/4}$ is 1.01 or less, it will be seen that subcooling effects are negligible if Ja is approximately less than 0.06, i.e., if:

$$c_{p\ell}(T_s - T_w) < 0.06 h_{fg}$$

EXAMPLE 11.2. Consider the situation described in Example 11.1 in which stagnant saturated steam at 100°C condenses on a 0.5-m high vertical plate with a surface temperature of 95°C. Determine whether subcooling effects have a significant influence on the condensation rate.

Solution. Using the water and vapor properties given in Example 11.1, the following value for the Jakob number is obtained:

$$Ja = \frac{c_{p\ell}(T_s - T_w)}{h_{fg}} = \frac{4217 \times (100 - 95)}{2257000} = 0.00934$$

Eq. (11.35) therefore indicates that the effect of film subcooling will be small. The modified latent heat is given by:

$$h_{fg}^* = h_{fg}(1 + 0.68 Ja) = 2257 \times (1 + 0.68 \times 0.00934) = 2271 \text{ kJ/kg}$$

The average Nusselt number is then given by:

$$\overline{Nu} = 0.943 \left[\frac{g h_{fg}^* \rho_\ell (\rho_\ell - \rho_v) L^3}{\mu_\ell k_\ell (T_s - T_w)} \right]^{1/4}$$

i.e.:

$$\overline{Nu} = 0.943 \left[\frac{9.81 \times 2{,}271{,}000 \times 960^2 \times 0.5^3}{0.000289 \times 0.68 \times (100 - 95)} \right]^{1/4} = 6740$$

The average heat transfer coefficient is then given by:

$$\bar{h} = \frac{\overline{Nu} k_\ell}{L} = \frac{6740 \times 0.68}{0.5} = 9166 \text{ W/m}^2 \text{ °C}$$

The total heat transfer rate per meter width of the plate is then given by:

$$Q = \bar{h} A (T_s - T_w) = 9166 \times (0.5 \times 2) \times (100 - 95) = 45{,}800 \text{ W} = 45.8 \text{ kW}$$

The condensation rate per meter plate width is therefore given by:

$$\dot{m} = \frac{Q}{h_{fg}} = \frac{45{,}800}{2{,}271{,}000} = 0.0202 \text{ kg/s} = 72.7 \text{ kg/hr}$$

When undercooling effects were ignored in Example 11.1, the condensation rate was found to be 73.0 kg/hr which differs by approximately 0.5% from the above value. In the situation here being considered, therefore, the effects of subcooling are very small.

The effect of superheat of the vapor will next be considered. In the previous analysis it was assumed that the vapor reservoir was at the saturation temperature. If the vapor reservoir is superheated to temperature, T_∞, the enthalpy of the vapor, h_v, condensing onto the film will be:

$$h_v = h_{fg} + c_{pv}(T_\infty - T_s) \tag{11.36}$$

where c_{pv} is the constant pressure specific heat of the vapor. Using the above equation, the energy balance for the control volume in the Nusselt analysis when both superheat and subcooling effects are included is:

$$k_\ell \left. \frac{\partial T}{\partial y} \right|_{y=0} dx = \frac{d}{dx} \left\{ \int_0^\delta \rho_\ell u \left[h_{fg} + c_{pv}(T_\infty - T_s) + c_{p\ell}(T_s - T) \right] dy \right\} dx \tag{11.37}$$

Comparing Eqs. (11.30) and (11.37) shows that the effect of superheat can be accounted for by replacing h_{fg} with $h_{fg} + c_{pv}(T_\infty - T_s)$. In most applications, this correction will be very small.

11.3
WAVY AND TURBULENT FILM CONDENSATION ON A VERTICAL SURFACE

Transition from laminar to turbulent flow within the condensed film can occur when the vapor is condensed on a tall surface or on a tall vertical bank of horizontal tubes [45] to [47]. It has been found that the film Reynolds number, based on the mean velocity in the film, u_m, and the hydraulic diameter, D_h, can be used to characterize the conditions under which transition from laminar flow occurs. The mean velocity in the film is given by definition as:

$$u_m = \frac{\Gamma}{\rho_\ell \delta} \tag{11.38}$$

The hydraulic diameter is defined by:

$$D_h = \frac{4A}{P} \tag{11.39}$$

where A is the flow area and P is the wetted perimeter, i.e., the length of the line of contact between the fluid and the solid surfaces over which it is flowing. Consider a portion of a plane film as shown in Fig. 11.12.

The only wetted surface is that in contact with the wall. Hence, for a unit width of film, the wetted perimeter is unity and the hydraulic diameter is given by:

$$D_h = \frac{4A}{P} = \frac{4 \times (\delta \times 1)}{1} = 4\delta \tag{11.40}$$

Therefore, the film Reynolds number is:

$$Re = \frac{u_m D_h \rho_\ell}{\mu_\ell} = \frac{(\Gamma/\rho_\ell \delta)(4\delta \rho_\ell)}{\mu_\ell} = \frac{4\Gamma}{\mu_\ell} \tag{11.41}$$

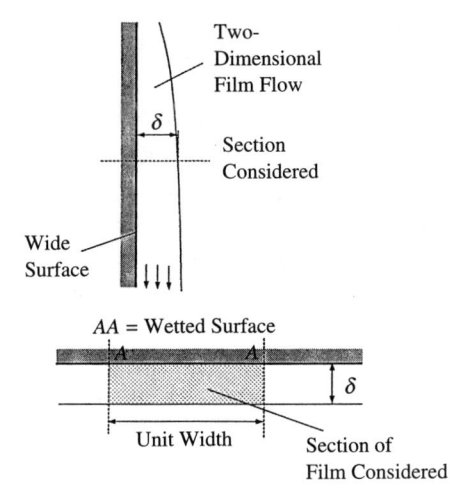

FIGURE 11.12
Quantities used in defining the hydraulic diameter of a condensed film.

Consider the condensed film on a vertical plate as shown in Fig. 11.13. There are three possible distinct flow regimes, as shown in Fig. 11.13. Near the top of the plate the mass flow rate is low and the flow is laminar. In this regime, corresponding to approximately $Re < 30$, the surface of the film is smooth. Further down the plate, as the mass flow rate of condensate increases, a series of regular ripples appear on the surface of the film. These ripples promote mixing in the film, enhancing the local heat transfer and condensation rates. The wavy region has been found to correspond approximately to $30 < Re < 1800$. Eventually, if the wall is sufficiently high such that $Re > 1800$, the flow in the liquid film becomes fully turbulent.

The analysis of the wavy and turbulent regions is very difficult because of the random nature of the flow. For engineering design purposes, it is usually necessary to resort to empirical correlations. Many correlation equations can be found in the literature. After an extensive review, Chen, Gerner, and Tien [4] recommend the

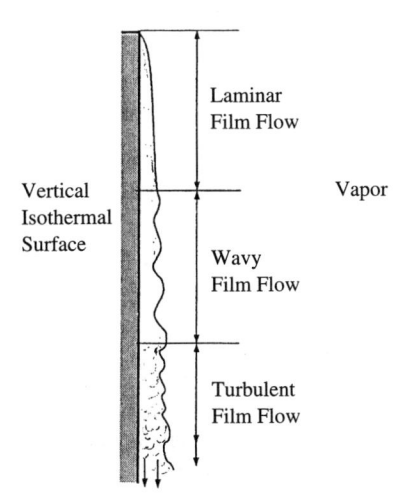

FIGURE 11.13
Flow regimes in a condensed film on a vertical flat plate.

following correlation for the wavy and turbulent regions:

$$\frac{\bar{h}}{k_\ell}\left(\frac{\nu_\ell^2}{g}\right)^{1/3} = (Re^{-0.44} + 5.82 \times 10^{-6} Re^{0.8} Pr_\ell^{1/3})^{1/2}, Re > 30 \qquad (11.42)$$

The quantity $(\nu_\ell^2/g)^{1/3}$ in this equation has the dimensions of length. Therefore, the left-hand side of Eq. (11.42) has the form of a modified average Nusselt number with characteristic length $(\nu_\ell^2/g)^{1/3}$. The heat transfer rate for the laminar region can also be cast in terms of this modified average Nusselt number, the equation having the form:

$$\frac{\bar{h}}{k_\ell}\left(\frac{\nu_\ell^2}{g}\right)^{1/3} = 1.468 Re^{-1/3}, Re < 30 \qquad (11.43)$$

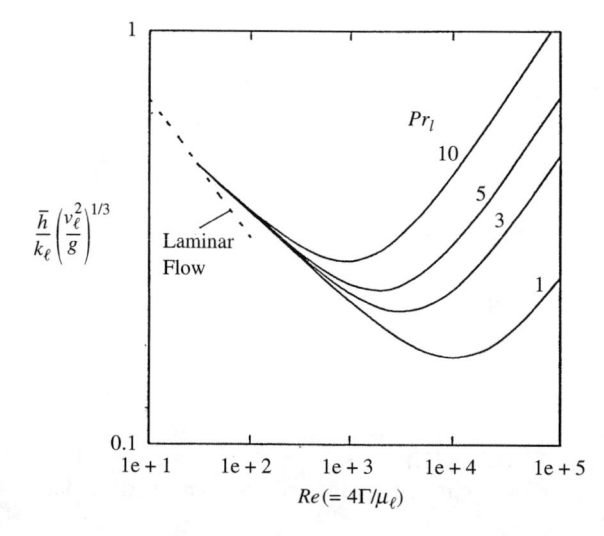

Figure 11.14 shows the variation of modified average Nusselt number with Reynolds number for the laminar, wavy, and turbulent regimes. In the laminar region, the average Nusselt number decreases with Reynolds number because of the thermal resistance of the thickening film. In the turbulent regime, however, the Nusselt number increases as Reynolds number increases despite the thickening of the film. In this regime, turbulent "mixing" reduces the thermal resistance of the film and enhances the heat transfer and condensation rates.

Eqs. (11.42) and (11.43) are very convenient for design calculations when the mass flow rate of condensate is specified and the required temperature difference is to be determined. However, when the condensation rate is not specified, the solution of Eq. (11.42) requires an iterative procedure since the Reynolds number cannot be calculated a priori. A simple iterative approach is described in Example 11.3. In the laminar regime, if the condensation rate is not known and the temperature difference is specified, iteration can be avoided by using Eq. (11.21) instead of Eq. (11.43).

FIGURE 11.14
Variation of modified average Nusselt number with Reynolds number for the laminar, wavy, and turbulent regimes.

EXAMPLE 11.3. Stagnant saturated steam at 100°C condenses on a vertical plate with surface temperature of 95°C. The plate is 0.5 m high and 2.0 m wide. Find the condensation rate on the plate.

Solution. The water properties in the film are evaluated at the mean film temperature $T_f = (T_s + T_w)/2 = 97.5°C$. At this temperature for water:

$$\rho_\ell = 960 \text{ kg/m}^3, \quad \mu_\ell = 2.8910 \times 10^{-4} \text{ kg/ms}$$

$$k_\ell = 0.680 \text{ W/mK}, \quad c_{p,\ell} = 4217 \text{ J/kg}$$

The vapor density and latent heat are evaluated at the saturation temperature, $T_s = 100°C$ and are:

$$\rho_v = 0.598 \text{ kg/m}^3, \quad h_{fg} = 2257.0 \text{ kJ/kg}$$

In the situation being considered, the vapor density is very small compared to the liquid density so $\rho_\ell(\rho_\ell - \rho_v) = \rho_\ell^2$ is a good approximation. This is usually true for the condensation of steam. The Jakob number is given by:

$$Ja = \frac{c_{p,\ell}(T_s - T_w)}{h_{fg}} = \frac{(4217)(100 - 95)}{2257 \times 10^3} = 0.00934$$

Hence:

$$h_{fg}^* = h_{fg}(1 + 0.68 Ja) = 1.0063 h_{fg} = 2271 \text{ kJ/kg}$$

The average Nusselt number is then calculated using Eq. (11.21),

$$\overline{Nu} = \frac{\bar{h}L}{k_\ell} = 0.943 \left[\frac{g h_{fg}^* \rho_\ell^2 L^3}{\mu_\ell k_\ell (T_s - T_w)} \right]^{1/4} = \left[\frac{(9.81)(2271 \times 10^3)(960)^2 (0.5)^3}{(2.89 \times 10^{-4})(0.680)(100 - 95)} \right]^{1/4} = 6740$$

The average heat transfer coefficient is then given by:

$$\bar{h} = \frac{\overline{Nu} k_\ell}{L} = \frac{6740 \times 0.680}{0.5} = 9166 \text{ W/m}^2 \text{ °C}$$

From which it follows that the total heat transfer rate is given by:

$$Q = \bar{h}A(T_s - T_w) = (9166)(0.5 \times 2.0)(100 - 95) = 45.8 \text{ kW}$$

The condensation rate is again found using the total heat transfer rate divided by the latent heat, i.e.:

$$m = \frac{Q}{h_{fg}^*} = \frac{45.8 \times 10^3}{2271 \times 10^3} = 2.02 \times 10^{-2} \text{ kg/s} = 72.7 \text{ kg/h}$$

Therefore, the condensation rate per unit width is $\Gamma = (2.02 \times 10^{-2})/2 = 1.01 \times 10^{-2}$ kg/s and the Reynolds number of the film at the bottom of the plate is:

$$Re = \frac{4\Gamma}{\mu_\ell} = \frac{4(1.01 \times 10^{-2})}{2.89 \times 10^{-4}} = 140$$

This Reynolds number is greater than 30 so the film will not be entirely laminar and it is therefore necessary to use Eq. (11.42), i.e.:

$$\frac{\bar{h}}{k_\ell} \left(\frac{v_\ell^2}{g} \right)^{1/3} = \left(Re^{-0.44} + 5.82 \times 10^{-6} Re^{0.8} Pr_\ell^{1/3} \right)^{1/2}$$

While the actual Reynolds number will be greater than 140 the value $Re = 140$ is a good initial value to use in the above equation. Using the liquid properties given above, the liquid kinematic viscosity and Prandtl number are:

$$\nu_\ell = \frac{\mu_\ell}{\rho_\ell} = \frac{2.89 \times 10^{-4}}{960} = 3.01 \times 10^{-7} \text{ m}^2/\text{s}$$

$$Pr_\ell = \frac{\mu_\ell c_{p\ell}}{k_\ell} = \frac{2.89 \times 10^{-4}(4217)}{0.680} = 1.79$$

Hence, Eq. (11.42), gives:

$$\bar{h} = k_\ell \left(\frac{g}{\nu_\ell^2}\right)^{1/3} \left(Re^{-0.44} + 5.82 \times 10^{-6} Re^{0.8} Pr_\ell^{1/3}\right)^{1/2}$$

$$= 0.68 \left[\frac{9.81}{(3.01 \times 10^{-0.7})^2}\right]^{1/3} [(140)^{-0.44} + (5.82 \times 10^{-6})(140)^{0.8}(1.79)^{1/3}]^{1/2}$$

$$= 10{,}950 \text{ W/m}^2\text{K}$$

Now the Reynolds number can be recalculated using the following relationship:

$$\bar{h}L(T_s - T_w) = \Gamma h_{fg}^* = \frac{\mu_\ell Re}{4} h_{fg}^*$$

i.e.

$$Re = \frac{4\bar{h}L(T_s - T_w)}{\mu_\ell h_{fg}^*} = \frac{4(10{,}950)(0.5)(100 - 95)}{(2.89 \times 10^{-4})(2271 \times 10^3)} = 167$$

This improved estimate of the Reynolds number can now be used in Eq. (11.42) to recalculate the average heat transfer coefficient,

$$\bar{h} = 0.68 \left[\frac{9.81}{(3.01 \times 10^{-7})^2}\right]^{1/3} [(167)^{-0.44} + (5.82 \times 10^{-6})(167)^{0.8}(1.79)^{1/3}]^{1/2}$$

$$= 10{,}520 \text{ W/m}^2\text{K}$$

The Reynolds number can then be recalculated using this value of the heat transfer coefficient which then allows a new value of the condensation rate and therefore a new value of the Reynolds number to be found and so on. Repeating this iterative procedure gives converged values for the Reynolds number and heat transfer coefficient of 162 and 10,590 W/m²K respectively. Using these values, the total heat transfer rate is,

$$Q = \bar{h}A(T_s - T_w) = (10{,}590)(0.5 \times 2.0)(100 - 95) = 52.6 \text{ kW}$$

and the condensation rate is,

$$\dot{m} = \frac{Q}{h_{fg}^*} = \frac{5.26 \times 10^3}{2271 \times 10^3} = 2.32 \times 10^{-2} \text{ kg/s} = 83.4 \text{ kg/h}$$

11.4 FILM CONDENSATION ON HORIZONTAL TUBES

There are a number of very important practical situations in which film condensation occurs on the outside of horizontal tubes. An obvious example is a shell-and-tube

CHAPTER 11: Condensation 575

condenser used in steam power plants. The analysis of film condensation on a horizontal tube proceeds along the same lines as that for a vertical plate. Figure 11.15 shows the coordinate system used and the control volume considered in the analysis. In this case, then, the x-coordinate is measured circumferentially around the tube and the y-coordinate is measured normal to the tube surface at all points. The tube has radius, R, and is assumed to be maintained at a uniform surface temperature T_w.

The gravitational force on the liquid in the control volume has components in both the x- and y-directions. The y-direction component will, in general, produce a pressure change across the film. However, it will again be assumed that the film is thin and this pressure change across the film will therefore again be neglected. As shown previously, the net body force on the liquid in the control volume is $g(\rho_\ell - \rho_v)\,dV$. The x-direction (or tangential) component of this force is $g(\rho_\ell - \rho_v)\,dV \sin\theta$. Therefore, because $dV = (\delta - y)R\,d\theta$, a tangential force balance for the control volume gives:

$$g(\delta - y)R\,d\theta(\rho_\ell - \rho_v)\sin\theta = \mu_\ell \frac{\partial u}{\partial y} R\,d\theta \qquad (11.44)$$

Integrating this equation with respect to y across the liquid film and applying the no-slip condition on velocity at the surface, i.e., setting $u = 0$ at $y = 0$, gives the velocity distribution in the film as:

$$u = \frac{(\rho_\ell - \rho_v)\sin\theta\, g}{\mu_\ell}\left(\delta y - \frac{y^2}{2}\right) \qquad (11.45)$$

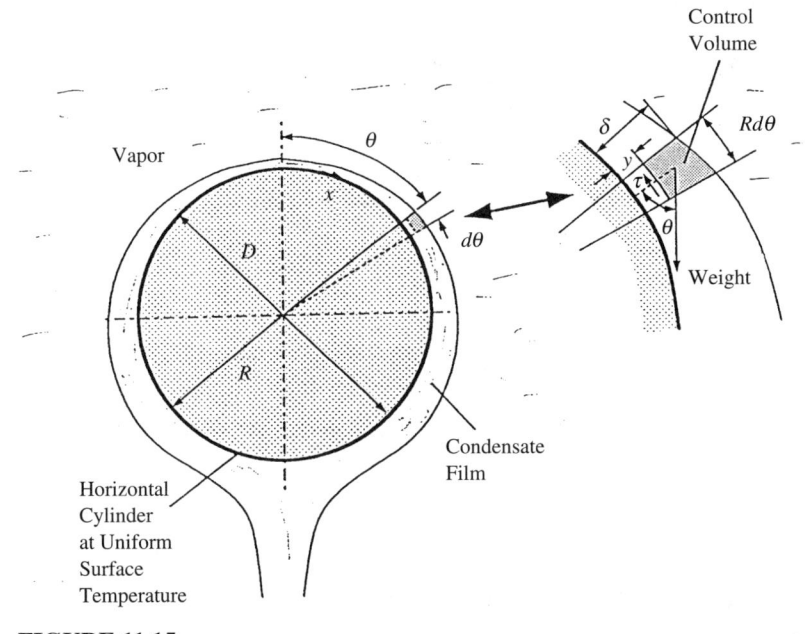

FIGURE 11.15
Condensation on a horizontal tube.

576 Introduction to Convective Heat Transfer Analysis

The mass flow rate in the film can then be found by again using:

$$\Gamma = \int_0^\delta \rho_\ell u\,dy = \int_0^\delta \rho_\ell \left[\frac{(\rho_\ell - \rho_v)\sin\theta\, g}{\mu_\ell}\left(\delta y - \frac{y^2}{2}\right)\right] dy \quad (11.46)$$

Carrying out the integration gives:

$$\Gamma_{1/2} = \frac{\rho_\ell(\rho_\ell - \rho_v)g\sin\theta\,\delta^3}{3\mu_\ell} \quad (11.47)$$

Because only one side of the tube is being considered, the local mass flow rate is denoted by $\Gamma_{1/2}$.

Next, noting that $x = R\theta$, it will be seen that an energy balance for the control volume gives:

$$h_{fg}\frac{d\Gamma_{1/2}}{dx} = \frac{h_{fg}}{R}\frac{d\Gamma_{1/2}}{d\theta} = k_\ell\frac{(T_s - T_w)}{\delta} \quad (11.48)$$

Solving Eq. (11.47) for δ and substituting the result into Eq. (11.48) gives:

$$\frac{\Gamma_{1/2}^{1/3}}{R}\frac{d\Gamma_{1/2}}{d\theta} = \left[\frac{g\rho_\ell(\rho_\ell - \rho_v)}{3\mu_\ell}\right]^{1/3}\frac{k_\ell(T_s - T_w)}{h_{fg}}\sin^{1/3}\theta \quad (11.49)$$

The analysis here differs from that given previously for condensation on a flat plate. Eq. (11.49) is expressed in terms of the mass flow, $\Gamma_{1/2}$, rather than the film thickness, δ, because the film thickness is not known at $\theta = 0$. However, $\Gamma_{1/2}$ is known at $x = 0$, i.e., by symmetry $\Gamma_{1/2} = 0$ at $x = 0$. Therefore, Eq. (11.49) can be integrated from $\theta = 0$ to $\theta = \pi$ to give:

$$\frac{3}{4}\Gamma_{1/2,\text{total}}^{4/3} = \left[\frac{g\rho_\ell(\rho_\ell - \rho_v)R^3}{3\mu_\ell}\right]^{1/3}\frac{k_\ell(T_s - T_w)}{h_{fg}}\int_0^\pi \sin^{1/3}\theta\, d\theta \quad (11.50)$$

Hence, noting that:

$$\int_0^\pi \sin^{1/3}\theta\, d\theta = 2\int_0^{\pi/2}\sin^{1/3}\theta\, d\theta \quad (11.51)$$

and noting that:

$$\int_0^{\pi/2}\sin^{1/3}\theta\, d\theta = 1.2936 \quad (11.52)$$

it follows from Eq. (11.50) that the total condensation rate for one side of the tube is given by:

$$\Gamma_{1/2,\text{total}} = 1.923\left[g\rho_\ell(\rho_\ell - \rho_v)R^3\frac{k_\ell^3(T_s - T_w)^3}{\mu_\ell h_{fg}^3}\right]^{1/4} \quad (11.53)$$

Now, the average heat transfer coefficient for the entire tube can be evaluated by noting that:

$$2\Gamma_{1/2,\text{total}}h_{fg} = 2\pi R\bar{h}(T_s - T_w) \quad (11.54)$$

Hence, using Eq. (11.53) and Eq. (11.54), the average Nusselt number based on the tube diameter (D) becomes:

$$\overline{Nu} = \frac{\bar{h}D}{k_\ell} = 0.729 \left[\frac{gh_{fg}\rho_\ell(\rho_\ell - \rho_v)D^3}{\mu_\ell k_\ell(T_s - T_w)} \right]^{1/4} \quad (11.55)$$

For problems in which the temperature difference is unknown, it is useful to cast Eq. (11.55) in terms of the condensation Reynolds number. It will be recalled that the Reynolds number is defined as $Re = 4\Gamma/\mu_\ell$, where Γ is the total condensation rate from the entire tube. Using this definition and noting that $\Gamma = 2\Gamma_{1/2,\text{total}}$, Eq. (11.54) can be written as:

$$(T_s - T_w) = \frac{\mu_\ell h_{fg} Re}{8\pi R \bar{h}} \quad (11.56)$$

Solving Eq. (11.53) for $(T_s - T_w)$ and substituting the result into Eq. (11.56) gives:

$$\frac{\bar{h}}{k_\ell}\left(\frac{\nu^2}{g}\right)^{1/3} = 1.52 Re^{-1/3} \quad (11.57)$$

In this equation it has been assumed that $\rho_\ell(\rho_\ell - \rho_v)$ is effectively equal to ρ_ℓ^2.

Eqs. (11.55) and (11.57) are valid for laminar flow. If flat plate flow is used as a rough guide, it will be seen that these equations can be expected to yield accurate results for about $Re < 60$. This Reynolds number is twice the value for a flat plate because condensation occurs on both sides of the tube.

In many practical applications, condensation occurs on a column of vertically aligned horizontal tubes. In such a case, as illustrated in Fig. 11.16, the condensate cascades from tube to tube down the column of tubes. If the flow remains laminar, the heat transfer rate from lower tubes will decrease because of the thickening condensate film.

Referring to Fig. 11.16, Eq. (11.49) can be applied to the nth tube in the vertical array. Now the mass flow rate at $\theta = 0$ for the nth tube is the total mass flow rate

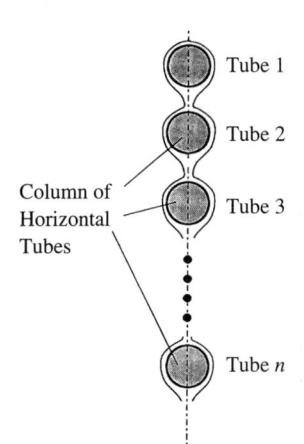

FIGURE 11.16
Condensation on a column of horizontal tubes.

from the $(n-1)^{th}$ tube, denoted $\Gamma_{1/2,n-1}$. So, Eq. (11.53) gives:

$$\Gamma_{1/2,n}^{4/3} - \Gamma_{1/2,n-1}^{4/3} = B \tag{11.58}$$

where:

$$B = 2.394 \frac{k_\ell R(T_s - T_w)}{h_{fg}} \left[\frac{g\rho_\ell(\rho_\ell - \rho_v)}{\mu_\ell} \right]^{1/3}$$

For tube 1, $\Gamma_{1/2,n-1}$ is 0, so:

$$\Gamma_{1/2,1}^{4/3} = B \tag{11.59}$$

The following is then obtained for tube 2:

$$\Gamma_{1/2,2}^{4/3} = \Gamma_{1/2,1}^{4/3} + B = 2B \tag{11.60}$$

Continuing down the column of cylinders, it can be seen that for the nth tube the following will be obtained:

$$\Gamma_{1/2,n}^{4/3} = nB \tag{11.61}$$

The average heat transfer coefficient for the tube bank can then be obtained by noting that:

$$2\Gamma_{1/2,n} h_{fg} = n 2\pi R \bar{h}(T_s - T_w) \tag{11.62}$$

Using Eq. (11.58) and Eq. (11.62) the average Nusselt number for the entire column of tubes is therefore given by:

$$\overline{Nu} = \frac{\bar{h}(nD)}{k_\ell} = 0.729 \left[\frac{g h_{fg} \rho_\ell (\rho_\ell - \rho_v)(nD)^3}{\mu_\ell k_\ell (T_s - T_w)} \right]^{1/4} \tag{11.63}$$

It will be noted that the above equation is identical to the equation for condensation on a single horizontal cylinder, i.e., Eq. (11.55), except that D is replaced with nD.

Experiments have shown that Eq. (11.63) tends to underestimate the heat transfer rate from a column of tubes, and is thus conservative for design purposes. The actual heat transfer rates are enhanced by several factors not accounted for in the theoretical model discussed above. These factors include such effects as splashing of the film when it impinges on a lower tube, additional condensation on the subcooled film as it falls between tubes, and uneven run-off because of bowed or slightly inclined tubes.

EXAMPLE 11.4. A compact condenser has 100 horizontal tubes arranged in a square array. Saturated steam at 30°C condenses onto the tubes. Each tube has an outside diameter of 1.5 cm and has a wall temperature of 15°C. Assuming laminar flow, calculate the condensation rate per unit length of the tubes.

Solution. The fluid properties are evaluated at the film temperature $T_f = (T_s + T_w)/2 = 22.5°C$ and the latent heat is evaluated at the saturation temperature $T_s = 30°C$. Hence:

$$\rho_\ell = 997 \text{ kg/m}^3 \quad \mu_\ell = 9.82 \times 10^{-4} \text{ kg/ms} \quad k = 0.602 \text{ W/mK}$$

$$c_{p\ell} = 4181 \text{ J/kg K}, \quad h_{fg} = 2430.0 \text{ kJ/kg}$$

For this problem the vapor density is very small, so it is assumed that $\rho_\ell(\rho_\ell - \rho_v)$ is equal to ρ_ℓ^2. The effect of film subcooling is small, but is included here for illustration purposes.

$$h_{fg}^* = h_{fg} + 0.68 c_{p\ell}(T_s - T_w) = (2430 \times 10^3) + 0.68(4181)(30 - 15) = 2473 \text{ kJ/kg}$$

There are 10 tubes in each column. So, using $n = 10$ in Eq. (11.63), the following is obtained:

$$\overline{Nu} = \frac{\bar{h}nD}{k_\ell} = 0.729 \left[\frac{g h_{fg}^* \rho_\ell^2 (nD)^3}{\mu_\ell k_\ell (T_s - T_w)} \right]^{1/4}$$

$$= 0.729 \left[\frac{(9.81)(2473 \times 10^3)(997)^2(10 \times 0.015)^3}{(9.82 \times 10^{-4})(0.602)(30 - 15)} \right]^{1/4} = 1270$$

The average heat transfer coefficient is then:

$$\bar{h} = \frac{\overline{Nu} k_\ell}{nD} = \frac{(1270)(0.6024)}{10(0.015)} = 5096 \text{ W/m}^2 \text{ °C}$$

So, the total heat transfer rate (per unit length) is:

$$Q = \bar{h} A (T_s - T_w) = (5096)(100)\pi(0.015)(30 - 15) = 36.0 \text{ kW}$$

and the total condensation rate (per unit length) is therefore:

$$m = \frac{Q}{h_{fg}^*} = \frac{360 \times 10^3}{2473 \times 10^3} = 0.145 \text{ kg/s} = 524 \text{ kg/h}$$

The Reynolds number must now be checked to ensure that the flow is laminar as assumed:

$$Re = \frac{4\Gamma}{\mu_\ell} = \frac{4(0.145/10)}{9.82 \times 10^{-4}} = 59$$

Because this value is less than 60, the assumption that the flow is laminar is reasonable.

11.5
EFFECT OF SURFACE SHEAR STRESS ON FILM CONDENSATION ON A VERTICAL PLATE

In the analysis of film condensation given in the previous sections it was assumed that the shear stress at the outer edge of the film was negligible. In some situations, however, particularly when the vapor velocity is high, this assumption may not be justified, i.e., the shear stress exerted on the outer surface of the condensed liquid film may have a significant influence on the heat transfer rate. The action of the shear stress on the surface of the liquid film is illustrated in Fig. 11.17.

If τ_v is the shear stress acting on the outer surface of the liquid film at any distance down the surface, then the momentum balance for the control volume shown in Fig. 11.18 (it is the same as the control volume shown before in Fig. 11.7) gives for the situation here being considered:

$$g \, dm - \mu_\ell \frac{\partial u}{\partial y} dx - \left(\frac{dp}{dx} dx \right) (\delta - y) + \tau_v \, dx = 0 \quad (11.64)$$

580 Introduction to Convective Heat Transfer Analysis

FIGURE 11.17
Shear stress on outer surface of liquid film on a vertical plate.

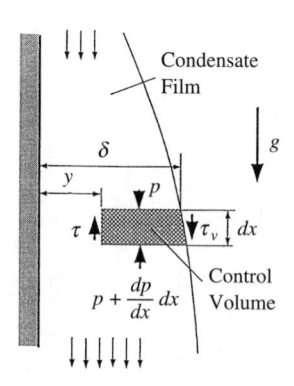

FIGURE 11.18
Control volume used in analysis.

Here, τ_v has been taken as positive in the downward direction, i.e., in the direction in which gravity acts.

Using Eqs. (11.2) and (11.3) and dividing by dx, Eq. (11.64) gives after rearrangement:

$$\mu_\ell \frac{\partial u}{\partial y} = g(\delta - y)(\rho_\ell - \rho_v) + \tau_v \tag{11.65}$$

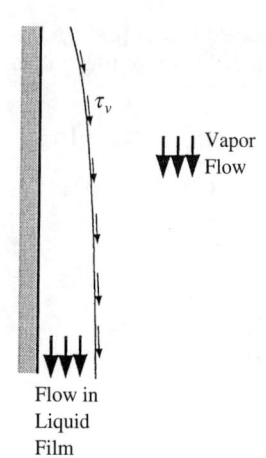

Integrating this equation across the film gives, since $u = 0$ at $y = 0$:

$$u = \frac{(\rho_\ell - \rho_v)g}{\mu_\ell}\left(\delta y - \frac{y^2}{2}\right) + \frac{\tau_v}{\mu_\ell} y \tag{11.66}$$

The mass flow rate in the film is then given by:

$$\Gamma = \int_0^\delta \rho_\ell u \, dy = \frac{\rho_\ell(\rho_\ell - \rho_v)g\delta^3}{3\mu_\ell} + \frac{\tau_v}{\mu_\ell}\frac{\delta^2}{2}\rho_\ell \tag{11.67}$$

But Eq. (11.12) gives:

$$h_{fg}\frac{d\Gamma}{dx} dx = k_\ell \frac{(T_s - T_w)}{\delta} dx$$

Hence, using Eq. (11.67), the following is obtained:

$$h_{fg}\left[\frac{\rho_\ell(\rho_\ell - \rho_v)g\delta^2}{\mu_\ell} + \rho_\ell \frac{\tau_v}{\mu_\ell}\delta\right]\frac{d\delta}{dx} = k_\ell\frac{(T_s - T_w)}{\delta} \quad (11.68)$$

Integrating this equation and again noting that the film thickness must be 0 at the top of the plate, i.e., that $\delta = 0$ at $x = 0$, gives:

$$h_{fg}\left[\frac{\rho_\ell(\rho_\ell - \rho_v)g\delta^4}{4\mu_\ell} + \rho_\ell \frac{\tau_v}{\mu_\ell}\frac{\delta^3}{3}\right] = k_\ell(T_s - T_w)x$$

i.e.:

$$\frac{\delta^4}{4}\frac{\rho_\ell(\rho_\ell - \rho_v)g}{\mu_\ell}h_{fg}\left[1 + \frac{\tau_v}{(\rho_\ell - \rho_v)g}\frac{4}{3\delta}\right] = k_\ell(T_s - T_w)x \quad (11.69)$$

If the shear at the surface of the liquid film can be ignored, this equation gives:

$$\frac{\delta_0^4}{4}\frac{\rho_\ell(\rho_\ell - \rho_v)g}{\mu_\ell}h_{fg} = k_\ell(T_s - T_w)x \quad (11.70)$$

Hence, Eq. (11.69) can be written as:

$$\delta^4\left[1 + \frac{\tau_v}{(\rho_\ell - \rho_v)g}\frac{4}{3\delta}\right] = \delta_0^4 \quad (11.71)$$

The effect of the shear stress on the surface of the liquid layer will be small if:

$$\frac{\tau_v}{(\rho_\ell - \rho_v)g}\frac{4}{3\delta} \ll 1 \quad (11.72)$$

This will be true in many real situations and in this case Eq. (11.71) can be approximated by:

$$\delta^4\left[1 + \frac{\tau_v}{(\rho_\ell - \rho_v)g}\frac{4}{3\delta_0}\right] = \delta_0^4$$

i.e.:

$$\frac{\delta}{\delta_0} = \left[1 + \frac{4\tau_v}{3(\rho_\ell - \rho_v)g\delta_0}\right]^{-1/4} \quad (11.73)$$

Now the heat transfer rate is given by:

$$q_x = k_\ell\frac{(T_s - T_w)}{\delta} \quad (11.74)$$

Hence, if q_{x0} is the heat transfer rate that would exist at the same position if the effects of the shear stress were negligible it follows that:

$$\frac{q_x}{q_{x0}} = \frac{\delta_0}{\delta} \quad (11.75)$$

Using Eq. (11.73) then gives:

$$\frac{q_x}{q_{x0}} = \left[1 + \frac{4\tau_v}{3(\rho_\ell - \rho_v)g\delta_0}\right]^{1/4} \tag{11.76}$$

It will be seen from Eq. (11.76) that if τ_v acts in the same direction as the gravitational force the heat transfer is increased while if τ_v acts in the opposite direction to gravity the heat transfer rate is decreased.

It might be thought that the value of the shear stress acting on the surface of the film could be obtained by assuming there is effectively a boundary layer flow in the vapor over the liquid film and that τ_v could then be obtained using standard boundary layer type relations for the wall shear stress. However, when condensation is occurring, vapor is being removed from the vapor flow over the liquid film. If the rate of momentum change in the flow direction within the vapor boundary layer on the outer surface of the liquid layer is small, the shear stress on the surface of the liquid film will be equal to the rate of decrease of vapor momentum. In this case, if the vapor has a downward velocity of U_v parallel to the surface and if the velocity of the outer surface of the liquid film is u_ℓ, there is momentum flux from the vapor to the outer surface of the liquid film that is equal to $\dot{m}_v(U_v - u_\ell)$ where $\dot{m}_v$ is the rate of vapor condensation per unit surface area at the point being considered. Momentum considerations indicate that there must be a shear stress acting on the vapor at the liquid/vapor interface that is equal to this momentum flux, i.e., that:

$$\tau_v = \dot{m}_v(U_v - u_\ell)$$

But:

$$\dot{m}_v = \frac{q_x}{h_{fg}}$$

and, in many situations:

$$U_v \gg u_\ell$$

In this case:

$$\tau_v = \frac{q_x U_v}{h_{fg}}$$

If the effects of the shear stress are small, this can be approximated by:

$$\tau_v = \frac{q_{x0} U_v}{h_{fg}}$$

Substituting this into Eq. (11.76) then gives:

$$\frac{q_x}{q_{x0}} = \left[1 + \frac{4q_{x0}U_v}{3(\rho_\ell - \rho_v)g\delta_0 h_{fg}}\right]^{1/4}$$

In general the vapor could be flowing upwards or downwards and this equation should be written as:

$$\frac{q_x}{q_{x0}} = \left[1 \pm \frac{4q_{x0}U_v}{3(\rho_\ell - \rho_v)g\delta_0 h_{fg}}\right]^{1/4}$$

CHAPTER 11: Condensation 583

where the sign of the shear stress term depends on whether the vapor flow is vertically downwards or vertically upwards.

EXAMPLE 11.5. Saturated steam at 100°C condenses on a 0.3-m high vertical plate with a surface temperature of 95°C. Assuming steady laminar flow, find the heat transfer rate at the bottom of the plate if the steam has a velocity of 10 m/s parallel to the surface of the plate.

Solution. The water properties in the film are evaluated at the mean film temperature, i.e., at:

$$T_f = \frac{(T_s + T_w)}{2} = 97.5°C$$

At this temperature for water:

$$\rho_\ell = 960 \text{ kg/m}^3, \; \mu_\ell = 2.89 \times 10^{-4} \text{ kg/m s}, \; k_\ell = 0.680 \text{ W/m-K}$$

The vapor density and latent heat are evaluated at the saturation temperature, T_s, which is 100°C and are:

$$\rho_v = 0.598 \text{ kg/m}^3, \; h_{fg} = 2257 \text{ kJ/kg}$$

In the situation being considered it will be seen that the vapor density is very small compared to the liquid density so:

$$\rho_\ell(\rho_\ell - \rho_v) = \rho_\ell^2$$

is a good approximation.

If the steam is stagnant, Eq. (11.17) gives at the bottom of the plate, i.e., at $x = 0.3$ m:

$$q_{x0} = k_\ell \left[\frac{gh_{fg}\rho_\ell^2}{4\mu_\ell k_\ell (T_s - T_w)x} \right]^{1/4} (T_s - T_w)$$

i.e.:

$$q_{x0} = 0.68 \left[\frac{9.81 \times 2257000 \times 960^2}{4 \times 0.000289 \times 0.68 \times (100 - 95) \times 0.3} \right]^{1/4} \times (100 - 95) = 38997 \text{ W/m}^2$$

When the vapor shear stress on the liquid film is accounted for, the heat transfer rate is given by:

$$\frac{q_x}{q_{x0}} = \left[1 \pm \frac{4q_{x0}U_v}{3(\rho_\ell - \rho_v)g\delta_0 h_{fg}} \right]^{1/4}$$

i.e., in the situation here being considered:

$$\frac{q_x}{q_{x0}} = \left[1 \pm \frac{4q_{x0}U_v}{3\rho_\ell g\delta_0 h_{fg}} \right]^{1/4}$$

i.e.:

$$\frac{q_x}{q_{x0}} = \left[1 \pm \frac{4 \times q_{x0} \times 10}{3 \times 960 \times 9.81 \times \delta_0 \times 2257000} \right]^{1/4} = \left[1 \pm \frac{q_{x0}}{15942000000\delta_0} \right]^{1/4} \quad (i)$$

But Eq. (11.16) gives at the bottom of the plate:

$$\delta_0 = \left[\frac{4\mu_\ell k_\ell(T_s - T_w)x}{gh_{fg}\rho_\ell(\rho_\ell - \rho_v)}\right]^{1/4} = \left[\frac{4\mu_\ell k_\ell(T_s - T_w)x}{gh_{fg}\rho_\ell^2}\right]^{1/4}$$

$$= \left[\frac{4 \times 0.000289 \times 0.68 \times (100 - 95) \times 0.3}{9.81 \times 2{,}257{,}000 \times 960^2}\right]^{1/4} = 0.000087 \text{ m}$$

Substituting this into Eq. (i) then gives:

$$\frac{q_x}{q_{x0}} = \left[1 \pm \frac{38{,}997}{1{,}594{,}200{,}000 \times 0.000087}\right]^{1/4} = [1 \pm 0.281]^{1/4}$$

Hence, if the vapor flow is downwards:

$$\frac{q_x}{q_{x0}} = [1 + 0.281]^{1/4} = 1.064$$

In this case then:

$$q_x = 1.064 \times 38997 = 41{,}490 \text{ W/m}^2$$

Similarly, if the vapor flow is upwards:

$$\frac{q_x}{q_{x0}} = [1 - 0.281]^{1/4} = 0.921$$

In this case then:

$$q_x = 0.921 \times 38{,}997 = 38210 \text{ W/m}^2$$

Therefore, the heat transfer rates with the vapor flow in the downward and upward directions are 41,490 W/m², and 38,210 W/m², respectively.

EXAMPLE 11.6. In the discussion of the effects of vapor shear stress on the heat transfer rate presented above it was assumed that the shear stress effects on the flow in the liquid film were small compared to the effects of gravity. Derive an expression for the heat transfer rate for the opposite case where the effects of gravity are negligible compared to the effects of the vapor shear stress.

Solution. If gravitational effects are negligible, Eq. (11.68) gives:

$$h_{fg}\left[\rho_\ell \frac{\tau_\ell}{\mu_\ell}\delta\right]\frac{d\delta}{dx} = k_\ell\frac{(T_s - T_w)}{\delta}$$

Integrating this equation and again noting that the film thickness must be zero at the top of the plate, i.e., that $\delta = 0$ at $x = 0$, gives:

$$h_{fg}\rho_\ell \frac{\tau_v}{\mu_\ell}\frac{\delta^3}{3} = k_\ell(T_s - T_w)x$$

i.e.:

$$\delta = \left[\frac{3\mu_\ell k_\ell(T_s - T_w)x}{h_{fg}\rho_\ell \tau_v}\right]^{1/3}$$

But:

$$q_x = k_\ell\frac{(T_s - T_w)}{\delta}$$

Hence:

$$q_x = k_\ell(T_s - T_w)\left[\frac{h_{fg}\rho_\ell \tau_v}{3\mu_\ell k_\ell(T_s - T_w)x}\right]^{1/3}$$

This can be written as:

$$\frac{q_x x}{k_\ell(T_s - T_w)} = Nu_x = \left[\frac{h_{fg}\rho_\ell \tau_v x^2}{3\mu_\ell k_\ell(T_s - T_w)}\right]^{1/3}$$

The right-hand side of this equation can be written as:

$$\left[\frac{1}{3}\frac{h_{fg}}{c_{p\ell}(T_s - T_w)}\frac{\rho_\ell \tau_v x^2}{\mu_\ell^2}\frac{\mu_\ell c_{p\ell}}{k_\ell}\right]^{1/3} = \left[\frac{1}{3}\frac{1}{Ja}Re^{*2}Pr_\ell\right]^{1/3}$$

where:

$$Re^* = \sqrt{\frac{\tau_v}{\rho_\ell}\frac{\rho_\ell x}{\mu_\ell}}$$

is the Reynolds number based on the "shear velocity" $\sqrt{\tau_v/\rho_\ell}$ and where Ja is the Jakob number and Pr_ℓ is the liquid Prandtl number.

Therefore, the heat transfer rate is given by:

$$Nu_x = \left[\frac{1}{3}\frac{1}{Ja}Re^{*2}Pr_\ell\right]^{1/3}$$

11.6
EFFECT OF NONCONDENSIBLE GASES ON FILM CONDENSATION

In practice, the vapor that is to be condensed sometimes contains noncondensible gases such as air. The presence of these noncondensible gases can significantly lower the heat transfer rate from that which would exist under the same circumstances with a pure vapor. A common example is the build-up of air in power plant condensers. These condensers usually operate at a substantial vacuum and some air entrainment is unavoidable. The continuous removal of air by specially designed ejector systems is essential to maintain the condenser vacuum and to maintain acceptable condensation rates. In some chemical plants, the separation of constituents is sometimes produced by condensing one gas from a mixture of gases and in such cases the presence of a noncondensible gas is unavoidable.

Even relatively low concentrations of noncondensible gases can substantially reduce the condensation rate. The main reason for this is that as vapor condenses, noncondensible gases get carried with the vapor toward the surface of the film. Since the film is impermeable, the concentration and partial pressure of noncondensible gases build up near the film to levels much higher than those far from the film. Since the total pressure is constant, the noncondensible gases suppress the partial pressure of the vapor at the edge of the film. This reduces the saturation temperature locally at the film/vapor interface as illustrated in Fig. 11.19. In turn, the reduction in the driving temperature difference leads to a reduction in the heat transfer and condensation

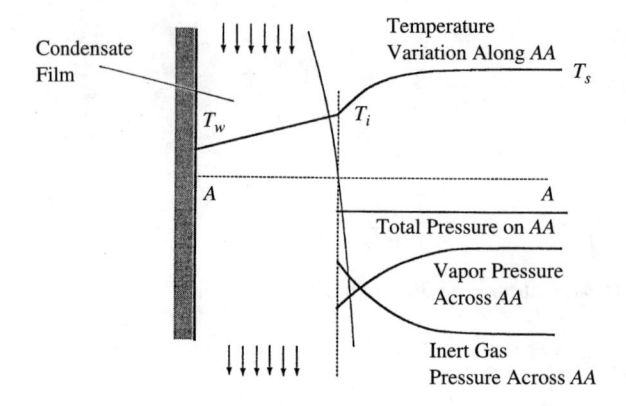

FIGURE 11.19
Changes near liquid film when a noncondensible vapor is present in vapor.

rates. A more detailed discussion and analysis of the effects of noncondensible gases is given by Rohsenow and Griffith [13].

11.7 IMPROVED ANALYSES OF LAMINAR FILM CONDENSATION

The basic Nusselt analysis ignores inertial effects in the condensed film and subcooling effects. Approximate methods of accounting for subcooling were discussed above. A method of accounting for both effects is discussed in this section. To illustrate the method, condensation on an isothermal vertical plate is again considered [52] to [54]. Interfacial shear stress effects will be neglected.

Assuming that the film is thin and that the pressure changes across the film are, therefore, negligible and that the velocity gradients in the cross-film direction (y-direction) are much greater than in the flow direction (x-direction), the flow in the film will be governed by the following set of equations:

$$\frac{\partial u}{\partial x} + \frac{\partial v}{\partial y} = 0 \tag{11.77}$$

$$u\frac{\partial u}{\partial x} + v\frac{\partial u}{\partial y} = \left(\frac{\rho_\ell - \rho_v}{\rho_\ell}\right)g + \nu_\ell \frac{\partial^2 u}{\partial y^2} \tag{11.78}$$

$$u\frac{\partial T}{\partial x} + v\frac{\partial T}{\partial y} = \frac{k_\ell}{\rho_\ell c_{p\ell}} \frac{\partial^2 T}{\partial y^2} \tag{11.79}$$

The coordinate system and velocity components are shown in Fig. 11.20.

The above set of equations are, of course, very similar to the boundary layer equations governing free convective flow over a plate.

The boundary conditions on the solution are:

$$y = 0: u = 0, v = 0, T = T_w$$

$$y = \delta: \frac{\partial u}{\partial y} = 0, T = T_{\text{sat}} \tag{11.80}$$

where T_w is the uniform plate temperature.

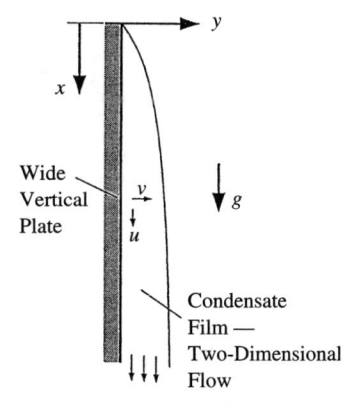

CHAPTER 11: Condensation   587

FIGURE 11.20 Coordinate system used.

The zero velocity gradient at the outer edge of the film requirement comes from the assumption that the shear stress at the outer edge of the film is zero.

The basic assumption in the solution procedure is that the velocity profiles in the film at all values of x are similar, i.e., that:

$$\frac{u}{u_\delta} = \text{function}\left(\frac{y}{\delta}\right) \tag{11.81}$$

where u_δ is the velocity at the outer edge of the film.

Now, all terms in Eq. (11.78) must have the same order of magnitude, i.e.:

$$0\left(\frac{u_\delta^2}{x}\right) = 0\left[\left(\frac{\rho_\ell - \rho_v}{\rho_\ell}\right)g\right]$$

and

$$0\left(\frac{\nu_\ell u_\delta}{\delta^2}\right) = 0\left[\left(\frac{\rho_\ell - \rho_v}{\rho_\ell}\right)g\right]$$

The first of these gives:

$$u_\delta = 0\left[\left(\frac{\rho_\ell - \rho_v}{\rho_\ell}\right)gx\right]^{1/2}$$

and the second then gives:

$$\delta = 0\left[\frac{\nu_\ell^2 x \rho_\ell}{(\rho_\ell - \rho_v)g}\right]^{1/4}$$

Hence, Eq. (11.81) can be written as:

$$\frac{u}{\left[\left(\frac{\rho_\ell - \rho_v}{\rho_\ell}\right)gx\right]^{1/2}} = \text{function}\left\{y\left[\frac{(\rho_\ell - \rho_v)g}{\nu_\ell^2 x \rho_\ell}\right]^{1/4}\right\}$$

$$= \text{function}\left\{\frac{y}{x^{1/4}}\left[\frac{(\rho_\ell - \rho_v)g}{\nu_\ell^2 \rho_\ell}\right]^{1/4}\right\} \tag{11.82}$$

i.e.:

$$\frac{u}{\left[\left(\frac{\rho_\ell - \rho_v}{\rho_\ell}\right)gx\right]^{1/2}} = \text{function}(\eta) \tag{11.83}$$

where the similarity variable, η, is here given by:

$$\eta = \frac{y}{x^{1/4}}\left[\frac{(\rho_\ell - \rho_v)g}{\nu_\ell^2 \rho_\ell}\right]^{1/4} \tag{11.84}$$

If the velocity profiles are similar, it seems logical to assume that the temperature profiles are also similar, i.e., that:

$$\frac{T - T_{\text{sat}}}{T_w - T_{\text{sat}}} = \theta(\eta) \tag{11.85}$$

the function θ thus depending only on η.

Rather than working with u and v it is, in the present situation, more convenient to express these velocity components in terms of a stream function, ψ, which is defined as before such that:

$$u = \frac{\partial \psi}{\partial y}, \quad v = -\frac{\partial \psi}{\partial x} \tag{11.86}$$

If the u velocity profiles are similar, the stream function profiles in the film must be similar, i.e.:

$$\frac{\psi}{u_\delta \delta} = \text{function}\left[\frac{y}{\delta}\right]$$

i.e.:

$$\psi = \left[\left(\frac{\rho_\ell - \rho_v}{\rho_\ell}\right)gx\right]^{1/2}\left[\frac{\nu_\ell^2 x \rho_\ell}{(\rho_\ell - \rho_v)g}\right]^{1/4} F(\eta)$$

i.e.:

$$\psi = \left[\left(\frac{\rho_\ell - \rho_v}{\rho_\ell}\right)g\nu_\ell^2 x^3\right]^{1/4} F(\eta) \tag{11.87}$$

the function F, as indicated, depending only on η.

Of course, by introducing the stream function, the continuity equation is always satisfied. The momentum equation, i.e., Eq. (11.78), can be written in terms of the stream function as:

$$\frac{\partial \psi}{\partial y}\frac{\partial^2 \psi}{\partial y \partial x} - \frac{\partial \psi}{\partial x}\frac{\partial^2 \psi}{\partial y^2} = \left(\frac{\rho_\ell - \rho_v}{\rho_\ell}\right)g + \nu_\ell \frac{\partial^3 \psi}{\partial y^3} \tag{11.88}$$

It is convenient to define:

$$A = \left[\frac{(\rho_\ell - \rho_v)g}{\rho_\ell}\right]^{1/4} \tag{11.89}$$

Eqs. (11.84) and (11.87) can then be written as:

$$\eta = \frac{A}{\nu_\ell^{1/2}} \frac{y}{x^{1/4}} \tag{11.90}$$

and

$$\psi = A\nu_\ell^{1/2} x^{3/4} F(\eta) \tag{11.91}$$

Using these, Eq. (11.88) becomes:

$$(A\nu_\ell^{1/2} x^{3/4} F') \left(\frac{A}{\nu_\ell^{1/2} x^{1/4}} \right) \left\{ \frac{\partial}{\partial y} \left[-(A\nu_\ell^{1/2} x^{3/4}) F' \left(\frac{Ay}{4\nu_\ell^{1/2} x^{5/4}} \right) + \frac{3FA\nu_\ell^{1/2}}{4x^{1/4}} \right] \right\}$$

$$- \left[-(A\nu_\ell^{1/2} x^{3/4}) F' \left(\frac{Ay}{4\nu_\ell^{1/2} x^{5/4}} \right) + \frac{3FA\nu_\ell^{1/2}}{4x^{1/4}} \right] \left[A\nu_\ell^{1/2} x^{3/4} F' \frac{A^2}{\nu_\ell x^{1/2}} \right]$$

$$= A^4 + \nu_\ell A \nu_\ell^{1/2} x^{3/4} F''' \frac{A^3}{\nu_\ell^{3/2} x^{3/4}}$$

i.e.:

$$F''' - \frac{3}{4} F F'' - \frac{(F')^2}{2} + 1 = 0 \tag{11.92}$$

In terms of the stream function, the boundary conditions on velocity are:

$$y = 0: \frac{\partial \psi}{\partial y} = 0, \frac{\partial \psi}{\partial x} = 0, \text{ i.e., } \psi = 0$$

$$y = \delta: \frac{\partial^2 \psi}{\partial y^2} = 0 \tag{11.93}$$

ψ being arbitrarily taken as 0 on the surface.

In terms of the variables introduced above, these boundary conditions become:

$$\eta = 0: F' = F = 0$$

$$\eta = \eta_\delta: F'' = 0 \tag{11.94}$$

η_δ being the value of η at the outer edge of the film.

Next, consider the energy equation, i.e., Eq. (11.79). In terms of the stream function and the temperature function, θ, this equation becomes, because T_w and T_{sat} are constants:

$$\frac{\partial \psi}{\partial y} \frac{\partial \theta}{\partial x} - \frac{\partial \psi}{\partial x} \frac{\partial \theta}{\partial y} = \frac{k_\ell}{\rho_\ell c_{p\ell}} \frac{\partial^2 \theta}{\partial y^2} \tag{11.95}$$

i.e., using Eqs. (11.89), (11.90), and (11.91):

$$\left[A\nu_\ell^{1/2} x^{3/4} F' \frac{A}{\nu_\ell^{1/2} x^{1/4}} \right] \theta' \left[-\frac{Ay}{4\nu_\ell^{1/2} x^{5/4}} \right]$$

$$- \left[\frac{3FA\nu_\ell^{1/2}}{4x^{1/4}} + A\nu_\ell^{1/2} x^{3/4} F' \left(\frac{-Ay}{4\nu_\ell^{1/2} x^{5/4}} \right) \right] \theta' \frac{A}{\nu_\ell^{1/2} x^{1/4}} = \frac{k_\ell}{\rho_\ell c_{p\ell}} \theta'' \frac{A^2}{\nu_\ell x^{1/2}}$$

590 Introduction to Convective Heat Transfer Analysis

i.e.

$$-3/4 F\theta' = \frac{\alpha_\ell}{\nu_\ell}\theta''$$

i.e.:

$$\theta'' + \frac{3Pr}{4}F\theta' = 0 \tag{11.96}$$

In terms of θ, the boundary conditions on this equation are:

$$\eta = 0: \theta = 1$$
$$\eta = \eta_\delta: \theta = 0 \tag{11.97}$$

Eqs. (11.92) and (11.96), along with the boundary conditions, constitute a pair of simultaneous ordinary differential equations in F and θ. However, the value of η_δ must be found in order to derive the solution. To do this, it is noted that if the flow up to any value of x from the top of the plate is considered, the overall energy balance requires:

$$\begin{matrix}\text{Rate of heat transfer to plate} \\ \text{up to point considered}\end{matrix} = \begin{matrix}\text{Mass flow rate} \\ \text{at section considered}\end{matrix} \times \text{Latent heat}$$

$$+ \begin{matrix}\text{Enthalpy flux relative to that} \\ \text{at the saturation temperature for} \\ \text{the film at the section considered}\end{matrix}$$

i.e.:

$$\int_0^x q_w\,dx = h_{fg}\rho_\ell \int_0^\delta u\,dy + \rho_\ell c_{p\ell}\int_0^\delta u(T_{\text{sat}} - T)\,dy \tag{11.98}$$

where q_w is the local heat transfer rate to the plate at any position on the plate. It is given by:

$$q_w = k_\ell \left.\frac{\partial T}{\partial y}\right|_{y=0} = k_\ell \theta_0' \frac{\partial \eta}{\partial y}(T_w - T_{\text{sat}}) = k_\ell \frac{A}{\nu_\ell^{1/2} x^{1/4}}\theta_0'(T_w - T_{\text{sat}}) \tag{11.99}$$

Because θ_0' is not dependent on x, this gives:

$$\int_0^x q_w\,dx = k_\ell \frac{4Ax^{3/4}}{3\nu_\ell^{1/2}}\theta_0'(T_w - T_{\text{sat}}) \tag{11.100}$$

Hence, using this and the other variables previously introduced and noting that:

$$\int_0^\delta u\,dy = \int_0^\delta \frac{\partial \psi}{\partial y}\,dy = \psi \text{ at } y = \delta$$

Eq. (11.98) becomes:

$$\frac{4k_\ell A x^{3/4}}{3\nu^{1/2}}\theta_0'(T_w - T_{\text{sat}}) = h_{fg}\rho_\ell F(\eta_\delta)A\nu_\ell^{1/2}x^{3/4}$$

$$+ \rho_\ell c_{p\ell}\left(\frac{A}{\nu^{1/2}x^{1/4}}\right)^{-1}(T_{\text{sat}} - T_w)\int_0^\delta \frac{\partial \psi}{\partial y}\theta\,d\eta$$

i.e.:

$$\frac{4}{3}\frac{k_\ell A x^{3/4}}{\nu^{1/2}}\theta_0'(T_w - T_{\text{sat}}) = h_{fg}\rho_\ell A \nu_\ell^{1/2} x^{3/4} F(\eta_\delta)$$

$$+ \rho_\ell c_{p\ell}\left(\frac{A}{\nu^{1/2}x^{1/4}}\right)^{-1}(T_{\text{sat}} - T_w)\int_0^\delta F' A \nu^{1/2} x^{3/4}\frac{A}{\nu^{1/2}x^{1/4}}\theta\, d\eta$$

i.e.:

$$\frac{4}{3}k_\ell\theta_0'(T_w - T_{\text{sat}}) = h_{fg}\rho_\ell\nu_\ell F(\eta_\delta) + \rho_\ell c_{p\ell}(T_{\text{sat}} - T_w)\nu_\ell\int_0^{\eta_\delta} F'\theta\, d\eta \quad (11.101)$$

Integrating by parts gives:

$$\int_0^{\eta_\delta} F'\theta\, d\eta = F\theta\Big|_0^{\eta_\delta} - \int_0^{\eta_\delta}\theta' F\, d\eta \quad (11.102)$$

But $F = 0$ at $\eta = 0$ and $\theta = 0$ at $\eta = \eta_\delta$ so:

$$F\theta\Big|_0^{\eta_\delta} = 0$$

Hence:

$$\int_0^{\eta_\delta} F'\theta\, d\eta = \int_0^{\eta_\delta}\theta' F\, d\eta \quad (11.103)$$

But rearranging Eq. (11.96) gives:

$$\theta' F = -\frac{4}{3Pr}\theta''$$

Hence:

$$-\int_0^{\eta_\delta}\theta' F\, d\eta = \frac{4}{3Pr}\int_0^{\eta_\delta}\theta''\, d\eta = -\frac{4}{3Pr}\left[\theta'(\eta_\delta) - \theta'(0)\right] \quad (11.104)$$

Using these results, Eq. (11.101) gives:

$$\frac{4}{3}k_\ell(T_w - T_{\text{sat}})\theta_0' = h_{fg}\rho_\ell\nu_\ell F(\eta_\delta) - \frac{4}{3}\frac{\rho_\ell c_{p\ell}(T_w - T_{\text{sat}})\nu_\ell}{Pr}\left[\theta'(\eta_\delta) - \theta_0'\right]$$

But:

$$\frac{\rho_\ell c_{p\ell}}{Pr} = \frac{\rho_\ell c_{p\ell}}{\nu_\ell}\times\frac{k_\ell}{\rho_\ell c_{p\ell}} = \frac{k_\ell}{\nu_\ell}$$

so the above equation gives:

$$0 = h_{fg}\rho_\ell\nu_\ell F(\eta_\delta) - \frac{4}{3}k_\ell(T_w - T_{\text{sat}})\theta'(\eta_\delta)$$

i.e.:

$$\frac{F(\eta_\delta)}{\theta'(\eta_\delta)} = \frac{4}{3}\frac{k_\ell}{\rho_\ell\nu_\ell c_{p\ell}}\frac{c_{p\ell}(T_{\text{sat}} - T_w)}{h_{fg}} = \frac{4}{3Pr}\frac{c_{p\ell}(T_{\text{sat}} - T_w)}{h_{fg}}$$

Hence:

$$\frac{F(\eta_\delta)}{\theta'(\eta_\delta)} = \frac{4}{3}\frac{Ja}{Pr} \tag{11.105}$$

This shows that, in a given situation, $F(\eta_\delta)/\theta'(\eta_\delta)$ is a constant. To review, the governing equations and boundary conditions are:

$$F''' + \frac{3}{4}FF'' - \frac{(F')^2}{2} + 1 = 0 \tag{11.106}$$

$$\theta'' + \frac{3Pr}{4}F\theta' = 0 \tag{11.107}$$

$$\frac{F(\eta_\delta)}{\theta'(\eta_\delta)} = \frac{4}{3}\frac{Ja}{Pr} \tag{11.108}$$

$$\eta = 0:\ F' = F = 0,\ \theta = 1 \tag{11.109}$$

$$\eta = \eta_\delta:\ F'' = 0,\ \theta = 0 \tag{11.110}$$

If the inertial terms in the momentum equation are ignored as in the Nusselt analysis, i.e., if Eq. (11.78) is approximated by:

$$\left(\frac{\rho_\ell - \rho_v}{\rho_\ell}\right)g + \nu_\ell\frac{\partial^2 u}{\partial y^2} = 0 \tag{11.111}$$

a consideration of the derivation of Eq. (11.92) shows that it becomes in this case:

$$F''' + 1 = 0 \tag{11.112}$$

Integrating this once gives:

$$F'' = -\eta + c_1$$

But $F'' = 0$ when $\eta = \eta_\delta$, so:

$$c_1 = \eta_\delta$$

Hence:

$$F'' = \eta_\delta - \eta$$

Integrating again then gives:

$$F' = \eta_\delta\eta - \frac{\eta^2}{2} + c_2$$

But $F' = 0$ when $\eta = 0$, so $c_2 = 0$, i.e.:

$$F' = \eta_\delta\eta - \frac{\eta^2}{2}$$

Integrating again then gives:

$$F = \eta_\delta\frac{\eta^2}{2} - \frac{\eta^3}{6} + c_3$$

But $F = 0$ when $\eta = 0$, so $c_3 = 0$, i.e.:

$$F = \eta_\delta \frac{\eta^2}{2} - \frac{\eta^3}{6} \tag{11.113}$$

Defining the following for convenience:

$$\zeta = \frac{\eta}{\eta_\delta} \tag{11.114}$$

Eq. (11.113) can be written as:

$$F = \eta_\delta^3 \left(\frac{\zeta^2}{2} - \frac{\zeta^3}{6} \right) \tag{11.115}$$

Substituting Eq. (11.115) into Eq. (11.107) then gives:

$$\frac{\theta''}{\theta'} = -\frac{3}{4} Pr \eta_\delta^3 \left(\frac{\zeta^2}{2} - \frac{\zeta^3}{6} \right) \tag{11.116}$$

Integrating this gives:

$$\ln\left[\frac{\theta'}{\theta'_0}\right] = -\frac{3}{4} Pr \eta_\delta^4 \left(\frac{\zeta^3}{6} - \frac{\zeta^4}{24} \right)$$

i.e.:

$$\theta' = \theta'_0 \exp\left[-\frac{3}{4} Pr \eta_\delta^4 \left(\frac{\zeta^3}{6} - \frac{\zeta^4}{24} \right) \right] \tag{11.117}$$

Integrating again gives, because $\theta = 1$ at $\eta = 0$:

$$\theta - 1 = \theta'_0 \eta_\delta \int_0^\zeta \left\{ \exp\left[-\frac{3}{4} Pr \eta_\delta^4 \left(\frac{1}{6}\zeta^3 - \frac{1}{24}\eta^4 \right) \right] \right\} d\zeta$$

But $\theta = 0$ at $\eta = \eta_\delta$ so:

$$-1 = \theta'_0 \eta_\delta \int_0^1 \left\{ \exp\left[-\frac{3}{4} Pr \eta_\delta^4 \left(\frac{1}{6}\zeta^3 - \frac{1}{24}\eta^4 \right) \right] \right\} d\zeta \tag{11.118}$$

Dividing the above two equations then gives:

$$\theta = 1 - \frac{\int_0^\zeta \left\{ \exp\left[-\frac{3}{4} Pr \eta_\delta^4 \left(\frac{1}{6}\zeta^3 - \frac{1}{24}\eta^4 \right) \right] \right\} d\zeta}{\int_0^1 \left\{ \exp\left[-\frac{3}{4} Pr \eta_\delta^4 \left(\frac{1}{6}\zeta^3 - \frac{1}{24}\eta^4 \right) \right] \right\} d\zeta} \tag{11.119}$$

Now by setting $\eta = \eta_\delta$, Eq. (11.113) gives:

$$F(\eta_\delta) = \frac{\eta_\delta^3}{3} \tag{11.120}$$

Substituting this into Eq. (11.108) then gives:

$$\frac{\eta_\delta^3}{3} = \frac{4}{3}\frac{Ja}{Pr}\theta'(\eta_\delta) = \frac{4}{3}\frac{Ja}{Pr}\frac{d\theta}{d\zeta}\bigg|_{\zeta=1}\frac{1}{\eta_\delta}$$

i.e.:

$$Pr\eta_\delta^4 = 4Ja\frac{d\theta}{d\zeta}\bigg|_{\zeta=1} \tag{11.121}$$

For any chosen value of $Pr\eta_\delta^4$, Eq. (11.119) can easily be integrated to give the variation of θ with ζ. Eq. (11.121) then allows the value of Ja corresponding to the chosen value of $Pr\eta_\delta^4$ to be found.

Eq. (11.99) gives:

$$\frac{q_w x}{k_\ell(T_\text{sat} - T_w)} = -\frac{Ax^{3/4}}{v_\ell^{1/2}}\theta_0'$$

i.e.:

$$Nu_x = -\left[\frac{(\rho_\ell - \rho_v)gx^3}{\rho_\ell v_\ell^2}\right]^{1/4}\theta_0' \tag{11.122}$$

Hence, using Eq. (11.118) the following is obtained:

$$Nu_x\left[\frac{\rho_\ell v_\ell^2}{(\rho_\ell - \rho_v)gx^3}\right]^{1/4} = 1\bigg/\left[\eta_\delta\int_0^1\left\{\exp\left[-\frac{3}{4}Pr\eta_\delta^4\left(\frac{\zeta^3}{6} - \frac{\zeta^4}{24}\right)\right]\right\}d\zeta\right]$$

i.e.:

$$Nu_x\left[\frac{\rho_\ell v_\ell^2}{(\rho_\ell - \rho_v)gx^3 Pr}\right]^{1/4}$$
$$= 1\bigg/\left[(\eta_\delta^4 Pr)^{1/4}\int_0^1\left\{\exp\left[-\frac{3}{4}(Pr\eta_\delta^4)\left(\frac{\zeta^3}{6} - \frac{\zeta^4}{24}\right)\right]\right\}d\zeta\right] \tag{11.123}$$

The left-hand side of this equation can be written as:

$$Nu_x\left[\frac{v_\ell k_\ell}{(\rho_\ell - \rho_v)gx^3 c_{p\ell}}\right]^{1/4} \tag{11.124}$$

Therefore, for any chosen value of $Pr\eta_\delta^4$, the value of Ja and then the corresponding value of the quantity given in Eq. (11.124) can be found. In this way, the variation of :

$$Nu_x\left[\frac{v_\ell k_\ell}{(\rho_\ell - \rho_v)gx^3 c_{p\ell}}\right]^{1/4}$$

with Ja can be determined. The variation so found is given in Fig. 11.21.

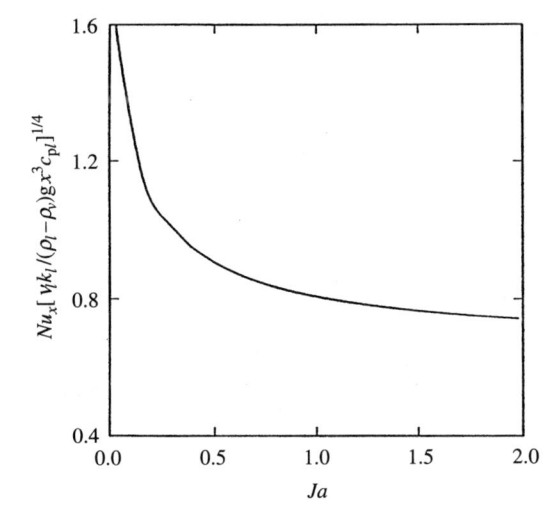

FIGURE 11.21
Variation of $Nu_x[\nu_\ell k_\ell/(\rho_\ell - \rho_v)gx^3 c_{p\ell}]^{1/4}$ with Ja when inertial effects are neglected.

Now, the result of the Nusselt analysis which ignores subcooling effects can be written as:

$$Nu_x \left[\frac{\nu_\ell k_\ell}{(\rho_\ell - \rho_v)gx^3 c_{p\ell}} \right]^{1/4} = \left[\frac{h_{fg}}{4c_{p\ell}(T_{\text{sat}} - T_w)} \right]^{1/4} = (4Ja)^{-1/4} \quad (11.125)$$

The results given by this equation agree with the results derived using the procedure described in this section for values of Ja less than approximately 0.1.

The approximate solution of Rohsenow [32] that was discussed above, see Eq. (11.35), gives:

$$Nu_x \left[\frac{\nu_\ell k_\ell}{(\rho_\ell - \rho_v)gx^3 c_{p\ell}} \right]^{1/4} = (4Ja)^{-1/4}(1 + 0.68Ja)^{1/4} \quad (11.126)$$

The results given by this equation agree with the results derived using the procedure described in this section for essentially the entire range of values of Ja considered.

The inertial terms in the momentum equation were neglected in the above analysis, i.e., the terms:

$$\frac{3}{4}FF'' - \frac{(F')^2}{2} \quad (11.127)$$

were neglected compared to the terms:

$$F''' + 1 \quad (11.128)$$

in Eq. (11.106). But the solution with these inertial terms ignored, i.e., Eq. (11.113), gives:

$$F = \eta_\delta \frac{\eta^2}{2} - \frac{\eta^3}{6} \quad (11.113)$$

596 Introduction to Convective Heat Transfer Analysis

Hence, the terms that were ignored, i.e., the terms given in Eq. (11.127), are to first order of accuracy, given by:

$$\frac{3}{4}\left[\eta_\delta \frac{\eta^2}{2} - \frac{\eta^3}{6}\right][\eta_\delta - \eta] - \left[\eta_\delta \eta - \frac{\eta^2}{3}\right]^2 \frac{1}{2}$$

i.e., by:

$$\eta_\delta^4 \left\{ \frac{3}{4}\left[\frac{\zeta^2}{2} - \frac{\zeta^3}{6}\right][1 - \zeta] - \left[\zeta - \frac{\zeta^2}{3}\right]^2 \frac{1}{2} \right\} \quad (11.129)$$

The terms that were ignored in the momentum equation thus have the order of magnitude η_δ^4. But Eq. (11.121) gives:

$$\eta_\delta^4 = 4\frac{Ja}{Pr}\frac{d\theta}{d\zeta}\bigg|_{\zeta=1} \quad (11.130)$$

This indicates that the inertial terms will only have a significant effect on the heat transfer rate if Ja is large and Pr is small. This is illustrated by the results given in Fig. 11.22 which shows the variation of:

$$Nu_x \left[\frac{\nu_\ell k_\ell}{(\rho_\ell - \rho_v)gx^3 c_{p\ell}}\right]^{1/4}$$

with Ja for various values of Pr taking account of the inertial terms together with the variation obtained by ignoring these inertial terms, this latter variation being independent of Prandtl number. It will be seen that inertial effects only have a significant influence on the heat transfer rate if $Pr < 10$.

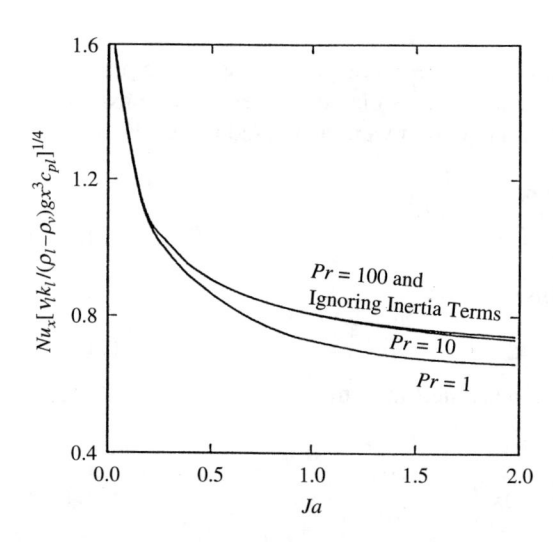

FIGURE 11.22 Variation of $Nu_x[\nu_\ell k_\ell/(\rho_\ell - \rho_v)gx^3 c_{p\ell}]^{1/4}$ with Ja for various Pr.

11.8
NONGRAVITATIONAL CONDENSATION

The rate of condensation on a vertical surface is controlled by the force of gravity acting on the condensed liquid film. A consideration of Eq. (11.20) shows for example that for a vertical plate the mean heat transfer rate from the plate with laminar flow in the film is proportional to $g^{1/4}$. Attempts have therefore been made to increase condensation rates by using centrifugal forces instead of the gravitational force to drain the condensed liquid film from the cold surface [55]. The simplest example of this would be condensation on the upper surface of a cooled circular plate rotating in a horizontal plane. This situation is shown in Fig. 11.23. A Nusselt-type analysis of this situation will be considered in the present section.

Consider the control volume shown in Fig. 11.24. The same assumptions as used in the basic Nusselt analysis are used here. With these assumptions, the forces acting on the control volume shown in Fig. 11.24 must balance, i.e.:

$$\mu_\ell \frac{\partial u}{\partial y} 2\pi r\, dr = \rho_\ell (\delta - y) 2\pi r\, dr\, \omega^2 r$$

i.e.:

$$\mu_\ell \frac{\partial u}{\partial y} = \rho_\ell (\delta - y) \omega^2 r \qquad (11.131)$$

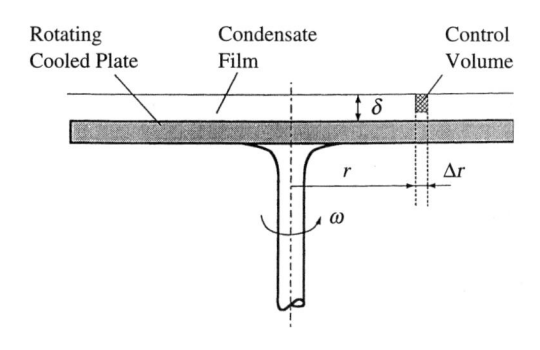

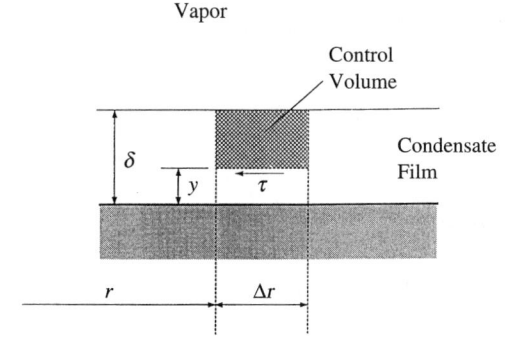

FIGURE 11.23
Laminar film condensation on a rotating horizontal circular plate.

FIGURE 11.24
Control volume used in the analysis of laminar film condensation on a rotating horizontal circular plate.

598 Introduction to Convective Heat Transfer Analysis

Integrating Eq. (11.131) with respect to y, i.e., across the liquid film and applying the no-slip condition at the surface, i.e., setting $u = 0$ at $y = 0$, gives the velocity distribution in the film as:

$$u = \frac{\rho_\ell}{\mu_\ell}\omega^2 r\left(\delta y - \frac{y^2}{2}\right) \tag{11.132}$$

The mass flow rate in the liquid film, Γ, is then given by integrating this velocity distribution. This gives:

$$\Gamma = 2\pi r \int_0^\delta \rho_\ell u\, dy = 2\pi r \frac{\rho_\ell^2 \omega^2 r \delta^3}{3\mu_\ell} \tag{11.133}$$

It is next noted that an energy balance requires that:

$$h_{fg}\frac{d\Gamma}{dr} dr = k_\ell \frac{(T_s - T_w)}{\delta} 2\pi r\, dr \tag{11.134}$$

Dividing this equation through by dr and substituting the expression for Γ given by Eq. (11.133) then gives:

$$h_{fg}\left[\frac{\rho_\ell^2 \omega^2 \delta^3}{3\mu_\ell}\right]\frac{d}{dr}(r^2 \delta^3) = k_\ell(T_s - T_w)\frac{r}{\delta} \tag{11.135}$$

i.e.:

$$\left[3r\delta^3 \frac{d\delta}{dr} + 2\delta^4\right] = \frac{3\mu_\ell k_\ell(T_s - T_w)}{h_{fg}\rho_\ell^2 \omega^2}$$

This equation indicates that δ is not dependent on r and therefore that:

$$\delta^4 = \left[\frac{3\mu_\ell k_\ell(T_s - T_w)}{2h_{fg}\rho_\ell^2 \omega^2}\right] \tag{11.136}$$

Because the local heat transfer rate per unit surface area is given by:

$$q = \frac{k_\ell(T_s - T_w)}{\delta} \tag{11.137}$$

it follows that q is the same everywhere on the plate and that the Nusselt number based on the radius of the plate R is therefore given by:

$$Nu = \frac{qR}{k_\ell(T_s - T_w)} = \frac{R}{\delta} = R\left[\frac{2h_{fg}\rho_\ell^2 \omega^2}{3\mu_\ell k_\ell(T_s - T_w)}\right]^{1/4} \tag{11.138}$$

This equation can be written in the following form:

$$Nu = \left[\frac{2}{3}\frac{h_{fg}}{c_{p\ell}(T_s - T_w)}\frac{\rho_\ell^2 \omega^2 R^2 R^2}{\mu_\ell^2}\frac{\mu_\ell c_{p\ell}}{k_\ell}\right]^{1/4} \tag{11.139}$$

i.e.:

$$Nu = \left[\frac{2}{3}\frac{Re^2 Pr_\ell}{Ja}\right]^{1/4} \quad (11.140)$$

Here Re is defined by:

$$Re = \frac{\rho_\ell(\omega R)R}{\mu_\ell} \quad (11.141)$$

ωR being, of course, the rotational velocity of the outer edge of the plate.

The condensation rate for the rotating plate is given by:

$$\dot{m}_v = \frac{q\pi R^2}{h_{fg}} = \frac{\pi R^2 k_\ell (T_s - T_w)}{h_{fg}} \left[\frac{2h_{fg}\rho_\ell^2 \omega^2}{3\mu_\ell k_\ell (T_s - T_w)}\right]^{1/4} \quad (11.142)$$

From this equation it will be seen that $\dot{m}_v$ is proportional to $\omega^{1/2}$. The condensation rate can therefore be increased by increasing the rotational speed of the plate.

EXAMPLE 11.7. Derive an expression for the ratio of the heat transfer rate to a rotating horizontal circular plate on which a vapor is condensing to the heat transfer rate that would exist if the vapor were condensing on a stationary vertical square plate. Both plates have the same surface temperature and the same surface area. Consider only one side of the square plate.

Solution. The rotating plate has a radius R and the square plate has a side length of H. Because the two plates have the same area it follows that:

$$H^2 = \pi R^2$$

If it is assumed that:

$$\rho_\ell(\rho_\ell - \rho_v) = \rho_\ell^2$$

then Eq. (11.20) gives for the vertical square plate:

$$Q_{sq} = \frac{4}{3}k_\ell \left[\frac{gh_{fg}\rho_\ell^2}{4\mu_\ell k_\ell (T_s - T_w)H}\right]^{1/4}(T_s - T_w)H^2$$

For the rotating plate, because $Q = q\pi R^2$, Eq. (11.142) gives:

$$Q_{rot} = k_\ell \left[\frac{2h_{fg}\rho_\ell^2 \omega^2}{3\mu_\ell k_\ell (T_s - T_w)}\right]^{1/4}(T_s - T_w)\pi R^2$$

Dividing the above two equations then gives:

$$\frac{Q_{rot}}{Q_{sq}} = \frac{3}{4}\left[\frac{8}{3}H\frac{\omega^2}{g}\right]^{1/4}$$

But:

$$H = \pi^{1/2}R$$

Hence:

$$\frac{Q_{rot}}{Q_{sq}} = 1.106\left[\frac{\omega^2 R}{g}\right]^{1/4}$$

11.9
CONCLUDING REMARKS

Vapor can condense on a cooled surface in two ways. Attention has mainly been given in this chapter to one of these modes of condensation, i.e., to film condensation. The classical Nusselt-type analysis for film condensation with laminar film flow has been presented and the extension of this analysis to account for effects such as subcooling in the film and vapor shear stress at the outer edge of the film has been discussed. The conditions under which the flow in the film becomes turbulent have also been discussed. More advanced analysis of laminar film condensation based on the use of the boundary layer-type equations have been reviewed.

PROBLEMS

11.1. Consider condensate forming in an insulated container with the lower surface held at constant temperature, T_w, as shown in Fig. P11.1. Since the condensate cannot drain away, the rate of condensation will decrease as the layer thickness increases. Assuming that the heat transfer is by quasi-steady conduction derive an expression for film thickness as a function of time and for the heat transfer rate as a function of time:

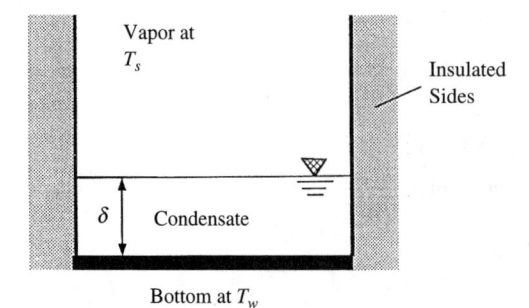

FIGURE P11.1

11.2. Consider condensate forming on a cold tube of radius, R, in a micro-gravity environment. In such an environment, the condensate will not run off the tube. Assuming that the heat transfer is by quasi-steady conduction show that the film radius as a function of time, r, is given by:

$$\frac{1}{2} + \left(\frac{r}{R}\right)^2 \left(\ln\frac{r}{R} - \frac{1}{2}\right) = \frac{2k_\ell(T_s - T_w)t}{\rho_\ell h_{fg} R^2}$$

11.3. Do film subcooling and vapor superheat cause the overall heat transfer and condensation rates to increase or decrease? Explain the answers you give.

11.4. There are several substances available commercially that can be applied to glass surfaces to prevent "fogging". The promotional literature for one of these substances for use on eye glasses states that "this long-lasting anti-fog treatment prevents surface condensation". Discuss the validity of this claim. How does the anti-fog treatment work?

11.5. Stagnant saturated steam at 100°C condenses on the outside of a horizontal copper tube with outside diameter 5.0 cm. The copper tube has a wall thickness of 1.5 mm and a thermal conductivity of 390 W/m-K. At the axial location being considered, the bulk temperature of the water flowing inside the copper tube is 80°C and the inside heat transfer coefficient is 3500 W/m². Assuming film condensation, calculate the heat transfer and condensation rate per unit length of pipe.

11.6. Stagnant saturated steam at 100°C condenses onto a vertical plate with a surface temperature of 70°C. Approximately how far down the plate does the film become wavy? At approximately what distance does the film become fully turbulent?

11.7. Stagnant saturated steam at 100°C is to be condensed on a 15-cm long vertical tube with an outside diameter of 25 mm. What tube surface temperature is needed to maintain a condensation rate of 0.6 kg per hour?

11.8. A 30-cm high by 150-cm wide plate is maintained at 5°C and is inclined at 45° from the vertical. Calculate the rate of condensation, the heat transfer rate, and the maximum film thickness when the plate is exposed to stagnant saturated water vapor at 20°C. Do the calculation both with and without the effect of film subcooling. What is the value of the Jakob number for this problem?

11.9. An 8-m high by 1.5-m wide vertical plate is maintained at 5°C. Calculate the rate of condensation and the heat transfer rate, when the plate is exposed to stagnant saturated water vapor at 20°C. Include the effect of film subcooling.

11.10. Steam is to be condensed on 25 horizontal tubes in a square array. The two configurations shown in Figure P11.10 are being considered. Explain, using only physical arguments, which configuration will give the highest condensation rate. Assume that the flow is laminar.

11.11. Assuming identical conditions, calculate the percent difference in the condensation rates for the two horizontal tube bundle configurations shown in Figure P11.11.

11.12. Steam at a saturation temperature of 50°C condenses on a wide flat vertical wall which has a height of 30 mm and which is kept at a surface temperature of 46°C. Find the heat transfer rate to the plate per unit width of the plate.

11.13. If the plate described in Problem 11.12 was 3 m high instead of 30 mm high, would the flow in the condensed film on the surface become turbulent? If it does become turbulent, find the distance from the top of the plate at which transition occurs.

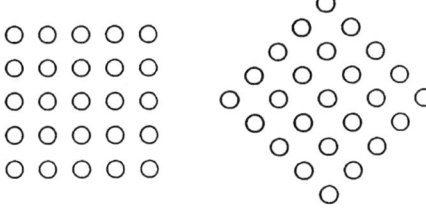

FIGURE P11.11

11.14. Steam at a saturation temperature of 60°C condenses on a wide flat vertical wall which has a height of 0.15 m and which is kept at a surface temperature of 55°C. Plot the variation of the heat transfer rate per unit wall area with distance along the plate.

11.15. Stagnant saturated steam at 80°C condenses on a 0.4-m high vertical plate with a surface temperature of 75°C. Calculate the heat transfer rate and condensation rate per meter width of the plate assuming that the flow in the liquid film is steady and laminar. Also find the maximum film thickness.

11.16. Consider the situation described in Problem 11.15 in which stagnant saturated steam at 80°C condenses on a 0.4-m high vertical plate with a surface temperature of 75°C. Calculate the heat transfer rate per meter width of plate with and without taking subcooling effects into account. Do subcooling effects have a significant influence on the condensation rate in this situation?

11.17. Stagnant saturated steam at 95°C condenses onto a vertical plate with a surface temperature of 90°C. The plate is 0.6 m high and 1.5 m wide. Determine whether the flow in the condensed liquid film remains laminar and then find the rate of condensation on the plate.

11.18. In a steam condenser there are 81 tubes arranged in a square array, i.e., there are 9 columns of tubes with 9 tubes in each column. Saturated steam at 40°C condenses on the tubes. Each tube has an outside diameter of 1 cm and has a wall temperature of 35°C. Assuming laminar flow, calculate the condensation rate per meter length of the tubes.

11.19. Consider laminar film condensation on a vertical plate when the vapor is flowing parallel to the surface in a downward direction at velocity, U. Assume that a turbulent boundary layer is formed in the vapor along the outer surface of the laminar liquid film. Determine a criterion that will indicate when the effect of the shear stress at the outer edge of the condensed liquid film on the heat transfer rate is less than 5%. Assume that $\rho_v \ll \rho_\ell$.

11.20. Steam at a saturation temperature of 50°C condenses on the upper surface of a rotating circular horizontal plate which has a diameter of 30 cm and which is kept at a surface temperature of 46°C. Plot the variation condensation rate with rotational speed for values between 50 and 1000 revolutions per minute.

REFERENCES

1. Bell, K.J. and Panchal, C.B., "Condensation," 6th. Int. Heat Transfer Conf., Toronto, Vol. 6, pp. 361-375, 1978.
2. Burmeister, L.C., *Convective Heat Transfer*, 2nd ed., Wiley-Interscience, New York, 1993.
3. Carey, V.P., *Liquid-Vapor Phase-Change Phenomena*, Hemisphere Publishing, Washington, D.C., 1992.
4. Chen, S.L., Gerner, F.M., and Tien, C.L., "General Film Condensation Correlations," *Experimental Heat Transfer*, Vol. 1, pp. 93-107, 1987.
5. Collier, J.G., *Convective Boiling and Condensation*, McGraw-Hill, New York, 1972.

6. Collier, J.G., "Heat Transfer in Condensation," in *Two-phase Flow and Heat Transfer in the Power and Process Industries,* Bergles, A.E., Collier, J.G., Delhaye, J.M., Hewitt, G.F., and Mayinger, F., Eds., pp. 330-65. Hemisphere Publishing, Washington, D.C., 1981.
7. Butterworth, D., "Condensers: Basic Heat Transfer and Fluid Flow," in *Heat Exchanger Sourcebook,* Palen, J.W., Ed., pp. 389-413, Hemisphere Publishing, Washington, D.C., 1986.
8. Eckert, E.R.G., "Pioneering Contributions to our Knowledge of Convective Heat Transfer," *ASME J. Heat Transfer,* Vol. 103, pp. 409-414, 1981.
9. Griffith, P., "Condensation," in *Handbook of Multiphase Systems,* Hestroni, G., Ed., Hemisphere Publishing, Washington, D.C., 1982.
10. Merte, H., Jr., "Condensation Heat Transfer," *Adv. Heat Transfer,* Vol. 9, pp. 181-272, 1973.
11. Silver, R.S., "An Approach to a General Theory of Surface Condensers," *Proc. Inst. Mech. Eng.,* Vol. 178, Part 1, No. 14, pp. 339-376, 1964.
12. Whalley, P.B., *Boiling, Condensation, and Gas-Liquid Flow,* Oxford University Press, New York, 1987.
13. Rohsenow, W.M. and Griffith, P., "Condensation," *Handbook of Heat Transfer Fundamentals,* Rohsenow, W.M., Hartnett, J.P., Gani, E.N., Eds., McGraw-Hill, New York, pp. 11.1-11.50, 1985.
14. Citakoglu, E. and Rose, J.W., "Dropwise Condensation: Some Factors Influencing the Validity of Heat Transfer Measurements," *Int. J. Heat Mass Transfer,* Vol. 11, p. 523, 1968.
15. Drew, T.B., Nagle, W.M., and Smith, W.Q., "The Conditions for Dropwise Condensation of Steam," *Trans. AIChE,* Vol. 31, pp. 605-621, 1935.
16. Rose, J.W., "Dropwise Condensation Theory," *Int. J. Heat Mass Transfer,* Vol. 24, p. 191, 1981.
17. Rose, J.W., "On the Mechanism of Dropwise Condensation," *Int. J. Heat Mass Transfer,* Vol. 10, pp. 755-762, 1967.
18. Tanner, D.W., Pope, D., Potter, C.J., and West, D., "Heat Tranfer in Dropwise Condensation of Low Pressures in the Absence and Presence of Non-Condensable Gas," *Int. J. Heat Mass Transfer,* Vol. 11, p. 181, 1968.
19. Tanasawa, I.T., "Dropwise Condensation – The Way to Practical Applications," Proc. 6th. Int. Heat Transfer Conf., Toronto, Vol. 6, pp. 393-405, 1978.
20. Takeyama, T. and Shimizu, S., "On the Transition of Dropwise-Film Condensation," Proc. 5th. Int. Heat Trans. Conf., Tokyo, Vol. III, 274, 1974.
21. Griffith, P., "Dropwise Condensation," in *Handbook of Heat Transfer,* Rohsenow, W.M. et al., Eds., 2nd. ed., McGraw-Hill, New York, 1985.
22. Song, Y., Xu, D., and Lin, J., "A Study of the Mechanism of Dropwise Condensation," *Int. J. Heat Mass Transfer,* Vol. 34, pp. 2827-2831, 1991.
23. Graham, C. and Griffith, P., "Drop Size Distributions and Heat Transfer in Dropwise Condensation," *Int. J. Heat Mass Transfer,* Vol. 16, p. 337, 1973.
24. Merte, H., Jr. and Son, S., "Further Consideration of Two-Dimensional Condensation Drop Profiles and Departure Sizes," *Warmeund Stoffubertragung,* Vol. 21, pp. 163-168, 1987.
25. Merte, H., Jr., Yamali, C., and Son, S., "A Simple Model for Dropwise Condensation Heat Transfer Neglecting Sweeping," Proc. Int. Heat Trans. Conf., Vol. 4, pp. 1659-1664, San Francisco Hemisphere Press, New York, 1986.
26. Griffith, P. and Lee, M.S., "The Effect of Surface Thermal Properties and Finish on Dropwise Condensation," *Int. J. Heat Mass Transfer,* Vol. 10, pp. 697-707, 1967.

27. Hannemann, R. and Mikic, B., "An Experimental Investigation into the Effect of Surface Thermal Conductivity on the Rate of Heat Transfer in Dropwise Condensation," *Int. J. Heat Mass Transfer,* Vol. 19, p. 1309, 1976.
28. Haraguchi, T., Shimada, R., Humagai, S., and Takeyama, T., "The Effect of Polyvinylidene Chloride Coating Thickness on Promotion of Dropwise Steam Condensation," *Int. J. Heat Mass Transfer,* Vol. 34, pp. 3047-3054, 1991.
29. de Gennes, P.G., Hua, X., and Levinson, P., "Dynamics of Wetting: Local Contact Angles," *J. Fluid Mech.,* Vol. 212, pp. 55-63, 1990.
30. Nusselt, W., "Die Oberflächenkondensation des Wasserdampfes," *Zeitschrift des Vereines Deutscher Ingenieure,* Vol. 60, pp. 541-569, 1916.
31. Rose, J.W., "Fundamentals of Condensation Heat Transfer: Laminar Flow Condensation," *JSME Int. J.,* Series II, Vol. 31, pp. 357-375, 1988.
32. Rohsenow, W.M., "Heat Transfer and Temperature Distribution in Laminar Film Condensation," *Trans. ASME,* Vol. 78, pp. 1645-1648, 1956.
33. Butterworth, D., "Filmwise Condensation," in *Two-Phase Flow and Heat Transfer,* Butterworth, D. and Hewitt, G.F., Eds., Oxford University Press, London, pp. 426-462, 1977.
34. Butterworth, D., "Film Condensation of Pure Vapor," in *Heat Exchanger Design Handbook,* Schlunder, E.U., Ed.-in-Chief, Vol. 2, Chapter 2.6.2, Hemisphere Publishing, New York, 1983.
35. Chen, M.M., "An Analytical Study of Laminar Film Condensation, Part 1. Flat Plates; Part 2. Single and Multiple Horizontal Tubes," *J. Heat Transfer,* Vol. 83, pp. 48-60, 1961.
36. Fujii, T., *Theory of Laminar Film Condensation,* Springer-Verlag, New York, 1991.
37. Kutateladze, S.S., "Semi-Empirical Theory of Film Condensation of Pure Vapours," *Int. J. Heat Mass Transfer,* 25, pp. 653-660, 1982.
38. Labuntsov, D.A., "Heat Transfer in Film Condensation of Pure Steam on Vertical Surfaces and Horizontal Tubes," *Teploenergetika,* Vol. 4, p. 72, 1957.
39. Rohsenow, W.M., "Film Condensation," in *Handbook of Heat Transfer,* Rohsenow, W.M. and Hartnett, J.P., Eds., Chapter 12A, McGraw-Hill, New York, 1973.
40. Shekriladze, I.G. and Gomelauri, V.I., "Theoretical Study of Laminar Film Condensation of Steam," *Int. J. Heat Mass Transfer,* Vol. 9, pp. 581-591, 1966.
41. Denny, V.E. and Mills, A.F., "Nonsimilar Solutions for Laminar Film Condensation on a Vertical Surface," *Int. J. Heat Mass Transfer,* Vol. 12, pp. 965-979, 1969.
42. Minkowycz, W.J. and Sparrow, E.M., "Condensation Heat Transfer in the Presence of Non-condensables, Interface Resistance, Superheating, Variable Properties and Diffusion," *Int. J. Heat Mass Transfer,* Vol. 9, p. 1125, 1966.
43. Poots, G. and Miles, R.G., "Effect of Variable Properties on Laminar Film Condensation of Steam," *Int. J. Heat Mass Transfer,* Vol. 10, p. 1677, 1967.
44. Kirkbridge, C.G., "Heat Transfer by Condensing Vapors on Vertical Tubes," *Trans. AIChE,* Vol. 30, p. 170, 1934.
45. Colburn, A.P., "The Calculation of Condensation Where a Portion of the Condensate Layer Is in Turbulent Flow," *Trans. AIChE,* Vol. 30, p. 187, 1933.
46. Hirshburg, R.I. and Florschuetz, L.W., "Laminar Wavy-Film Flow: Part I, Hydrodynamics Analysis, Part II, Condensation and Evaporation," *J. Heat Transfer,* Vol. 104, pp. 452-464, 1982.
47. Nakorykov, V.E., Pokusaev, B.G., and Alekseenkon, S.V., "Stationary Rolling Waves on a Vertical Film of Liquid," *J. Eng. Phys.,* Vol. 30, pp. 517-521, 1976.
48. Mills, A.F., *Heat Transfer,* Richard D. Irwin, Homewood, Il, 1992.
49. Sparrow, E.M., Minkowycz, W.J., and Saddy, M., "Forced Convection Condensation in the Presence of Non-Condensibles and Interfacial Resistance," *Int. J. Heat Mass Transfer,* Vol. 10, pp. 1829-1845, 1967.

50. Rohsenow, W.M., Weber, J.M., and Ling, A.T., "Effect of Vapor Velocity on Laminar and Turbulent Film Condensation," *Trans. ASME,* Vol. 78, pp. 1637-1644, 1956.
51. Carpenter, E.F. and Colburn, A.P., "The Effect of Vapor Velocity on Condensation Inside Tubes," in *Proceedings, General Discussion on Heat Transfer,* Inst. Mech. Eng.-ASME, New York, pp. 20-26, 1951.
52. Sparrow, E.M. and J.L. Gregg, "A Boundary-layer Treatment of Laminar Film Condensation," *J. of Heat Transfer,* Vol. 81, pp.13-18, 1959.
53. Stuhltrager, E., Naridomi, Y., Miyara, A., and Uehara, H., "Flow Dynamics and Heat Transfer of a Condensate Film on a Vertical Wall. I. Numerical Analysis and Flow Dynamics," *Int. J. Heat Mass Transfer,* Vol. 36, pp. 1677-1686, 1993.
54. Koh, J.C.Y., Sparrow, E.M., and Hartnett, J.P., "The Two-Phase Boundary Layer in Laminar Film Condensation," *Int. J. Heat Mass Transfer,* Vol. 2, p. 69, 1961.
55. Dhir, V.K. and Lienhard, J.H., "Laminar Film Condensation on Plane and Axisymmetric Bodies in Non-uniform Gravity," *J. Heat Transfer,* Vol. 93, p. 97, 1971.

APPENDIX A

Properties of Saturated Water

Properties of saturated water

T (°C)	ρ (kg/m³)	$\mu \times 10^3$ (kg/m-s)	$\nu \times 10^6$ m²/s	c_p kJ/kg-°C	k W/m-°C	$\beta \times 10^3$ 1/K	Pr
0	999.8	1.791	1.792	4.218	0.5619	-0.0853	13.45
5	1000.0	1.520	1.520	4.203	0.5723	0.0052	11.16
10	999.8	1.308	1.308	4.193	0.5820	0.0821	9.42
15	999.2	1.139	1.140	4.187	0.5911	0.148	8.07
20	998.3	1.003	1.004	4.182	0.5996	0.207	6.99
25	997.1	0.8908	0.8933	4.180	0.6076	0.259	6.13
30	995.7	0.7978	0.8012	4.180	0.6150	0.306	5.42
35	994.1	0.7196	0.7238	4.179	0.6221	0.349	4.83
40	992.3	0.6531	0.6582	4.179	0.6286	0.389	4.34
45	990.2	0.5962	0.6021	4.182	0.6347	0.427	3.93
50	998.0	0.5471	0.5537	4.182	0.6405	0.462	3.57
55	985.7	0.5043	0.5116	4.184	0.6458	0.496	3.27
60	983.1	0.4668	0.4748	4.186	0.6507	0.529	3.00
65	980.5	0.4338	0.4424	4.187	0.6553	0.560	2.77
70	977.7	0.4044	0.4137	4.191	0.6594	0.590	2.57
75	974.7	0.3783	0.3881	4.191	0.6633	0.619	2.39
80	971.6	0.3550	0.3653	4.195	0.6668	0.647	2.23
85	968.4	0.3339	0.3448	4.201	0.6699	0.675	2.09
90	965.1	0.3150	0.3284	4.203	0.6727	0.702	1.97
95	961.7	0.2978	0.3097	4.210	0.6753	0.728	1.86
100	958.1	0.2822	0.2945	4.215	0.6775	0.755	1.76
120	942.8	0.2321	0.2461	4.246	0.6833	0.859	1.44
140	925.9	0.1961	0.2118	4.282	0.6845	0.966	1.23
160	907.3	0.1695	0.1869	4.339	0.6815	1.084	1.08
180	886.9	0.1494	0.1684	4.411	0.6745	1.216	0.98
200	864.7	0.1336	0.1545	4.498	0.6634	1.372	0.91
220	840.4	0.1210	0.1439	4.608	0.6483	1.563	0.86
240	813.6	0.1105	0.1358	4.770	0.6292	1.806	0.84
260	783.9	0.1015	0.1295	4.991	0.6059	2.130	0.84
280	750.5	0.0934	0.1245	5.294	0.5780	2.589	0.86
300	712.2	0.0858	0.1205	5.758	0.5450	3.293	0.91
320	666.9	0.0783	0.1174	6.566	0.5063	4.511	1.02
340	610.2	0.0702	0.1151	8.234	0.4611	7.170	1.25
360	526.2	0.0600	0.1139	16.138	0.4115	21.28	2.35

APPENDIX B

Properties of Dry Air at Standard Atmospheric Pressure

Properties of dry air at standard atmospheric pressure

T (°C)	ρ (kg/m³)	$\mu \times 10^6$ (kg/m-s)	$\nu \times 10^6$ (m²/s)	c_p (kJ/kg-°C)	$k \times 10^3$ (W/m-°C)	Pr
−50	1.5819	14.63	9.25	1.0064	20.04	0.735
−40	1.5141	15.17	10.02	1.0060	20.86	0.731
−30	1.4518	15.59	10.81	1.0058	21.68	0.728
−20	1.3944	16.20	11.62	1.0057	22.49	0.724
−10	1.3414	16.71	12.46	1.0056	23.29	0.721
0	1.2923	17.20	13.31	1.0057	24.08	0.718
10	1.2467	17.69	14.19	1.0058	24.87	0.716
20	1.2042	18.17	15.09	1.0061	25.64	0.713
30	1.1644	18.65	16.01	1.0064	26.38	0.712
40	1.1273	19.11	16.96	1.0068	27.10	0.710
50	1.0924	19.57	17.92	1.0074	27.81	0.709
60	1.0596	20.03	18.90	1.0080	28.52	0.708
70	1.0287	20.47	19.90	1.0087	29.22	0.707
80	0.9996	20.92	20.92	1.0095	29.91	0.706
90	0.9721	21.35	21.96	1.0103	30.59	0.705
100	0.9460	21.78	23.02	1.0113	31.27	0.704
110	0.9213	22.20	24.10	1.0123	31.94	0.704
120	0.8979	22.62	25.19	1.0134	32.61	0.703
130	0.8756	23.03	26.31	1.0146	33.28	0.702
140	0.8544	23.44	27.44	1.0159	33.94	0.702
150	0.8342	23.84	28.58	1.0172	34.59	0.701
160	0.8150	24.24	29.75	1.0186	35.25	0.701
170	0.7966	24.63	30.93	1.0201	35.89	0.700
180	0.7790	25.03	32.13	1.0217	36.54	0.700
190	0.7622	25.41	33.34	1.0233	37.18	0.699
200	0.7461	25.79	34.57	1.0250	37.81	0.699
210	0.7306	26.17	35.82	1.0268	38.45	0.699
220	0.7158	26.54	37.08	1.0286	39.08	0.699
230	0.7016	26.91	38.36	1.0305	39.71	0.698
240	0.6879	27.27	39.65	1.0324	40.33	0.698
250	0.6748	27.64	40.96	1.0344	40.95	0.698
260	0.6621	27.99	42.28	1.0365	41.57	0.698

Properties of dry air at standard atmospheric pressure (*continued*)

T (°C)	ρ (kg/m³)	$\mu \times 10^6$ (kg/m-s)	$\nu \times 10^6$ (m²/s)	c_p (kJ/kg-°C)	$k \times 10^3$ (W/m-°C)	Pr
270	0.6499	28.35	43.62	1.0386	42.18	0.698
280	0.6382	28.70	44.97	1.0407	42.79	0.698
290	0.6268	29.05	46.34	1.0429	43.40	0.698
300	0.6159	29.39	47.72	1.0452	44.01	0.698
310	0.6053	29.73	49.12	1.0475	44.61	0.698
320	0.5951	30.07	50.53	1.0499	45.21	0.698
330	0.5853	30.41	51.95	1.0523	45.84	0.698
340	0.5757	30.74	53.39	1.0544	46.38	0.699
350	0.5665	31.07	54.85	1.0568	46.92	0.700
360	0.5575	31.40	56.31	1.0591	47.47	0.701
370	0.5489	31.72	57.79	1.0615	48.02	0.701
380	0.5405	32.04	59.29	1.0639	48.58	0.702
390	0.5323	32.36	60.79	1.0662	49.15	0.702
400	0.5244	32.68	62.31	1.0686	49.72	0.702
410	0.5167	32.99	63.85	1.0710	50.29	0.703
420	0.5093	33.30	65.39	1.0734	50.86	0.703
430	0.5020	33.61	66.95	1.0758	51.44	0.703
440	0.4950	33.92	68.52	1.0782	52.02	0.703
450	0.4882	34.22	70.11	1.0806	52.59	0.703
460	0.4815	34.52	71.70	1.0830	53.16	0.703
470	0.4750	34.82	73.31	1.0854	53.73	0.703
480	0.4687	35.12	74.93	1.0878	34.31	0.704
490	0.4626	35.42	76.57	1.0902	54.87	0.704
500	0.4566	35.71	78.22	1.0926	55.44	0.704
520	0.4451	36.29	81.54	1.0973	56.57	0.704
530	0.4395	36.58	83.23	1.0996	57.13	0.704
540	0.4341	36.87	84.92	1.1020	57.68	0.704
550	0.4288	37.15	86.63	1.1043	58.24	0.704
560	0.4237	37.43	88.35	1.1066	58.79	0.705
570	0.4187	37.71	90.07	1.1088	59.33	0.705
580	0.4138	37.99	91.82	1.1111	59.87	0.705
590	0.4090	38.27	93.57	1.1133	60.41	0.705
600	0.4043	38.54	95.33	1.1155	60.94	0.705
610	0.3997	38.81	97.11	1.1177	61.47	0.706
620	0.3952	39.09	98.89	1.1198	62.00	0.706
630	0.3908	39.36	100.69	1.1219	62.52	0.706
640	0.3866	39.62	102.50	1.1240	63.03	0.707
650	0.3824	39.89	104.32	1.1260	63.55	0.707

Index

Adiabatic wall temperature, 142–144
Aiding flow, 429
Air properties, 27, 607–608
Analogy Solutions, 244–245
   external flow, 254–271
   internal flow, 304–322
   Reynolds, 254–262, 304–312
   Taylor-Prandtl, 262–267
   three-layer, 267–271, 312–322
Apparent conductivity, 497
Apparent heat flux, 56–57, 229
Apparent shear stress, 54–55, 229
Apparent thermal diffusivity, 497, 546
Area averaged velocity, 488–489
Aspect ratio, 391, 537–538
Assisting flow, 429, 431
Axial heat conduction, 160–161, 180, 190

Bénard cell, 406
Bénard convection, 406
   in porous media, 544–545
Blasius solution, 83–94
Block, heated on wall, 336
Body fitted coordinates, 66
Body force, 4, 12–15
Boundary conditions, 354
Boundary layer, 7, 60–61
Boundary layer assumptions, 60–61
Boundary layer, film condensation, 586–596

Boundary layer, laminar forced
   arbitrary wall temperature, 98–105
   flat plate, 83–105
   flat plate heat transfer rate, 92–94
   flat plate temperature profile, 91
   flat plate velocity profile, 88
   integral equation solutions, 114–123
   numerical solution, 123–140
   specified heat flux, 138–140
   unheated starting length, 121–123
   uniform heat flux, 100
   uniform wall temperature, 83–94
   viscous dissipation effects, 140–150
Boundary layer, laminar, combined, 431–444
Boundary layer, laminar, natural, 349–366
Boundary layer, porous media, 498–521
   cylinder, 512, 519–521
   flat plate, forced convection, 500–503, 516–517
   integral equation solutions, 514–519
   natural convection, 526–531
   numerical solution, 507–513
   similarity solutions, 500–507, 526–531
   stagnation point, 505–507, 519
Boundary layer, turbulent forced, 255–281, 296–297
   integral equation solution, 272–281
   numerical solution, 281–299

609

Boundary layer, turbulent forced—*Cont.*
　Nusselt number, 257, 260, 265–266, 270–271, 280
　Reynolds analogy, 255–262
　Taylor-Prandtl analogy, 262–267
　three layer analogy, 267–271
　unheated leading edge section, 276–281
　viscous dissipation effects, 296–297
Boundary layer, turbulent natural, 407–414
　Nusselt number, 412–413
Boundary layer equations, 61–69
　boundary conditions, 67–69
　energy, 65–66, 70
　flow over a curved surface, 66
　integral equations, 71–80, 114–123, 408
　laminar, 61–69
　laminar, numerical solution, 123–140
　momentum, 63–65, 70
　natural convection, 349–354
　order of magnitude analysis, 62–66
　porous media, 498–499
　pressure change across, 65, 70
　turbulent, 69–71, 228–231
　turbulent kinetic energy, 71, 241–243
　two-dimensional flow equations, 66
　velocity and temperature profiles, 62
Boundary layer theory, 61–80
Boundary layer thickness
　displacement, 272, 331–332
　laminar flow, flat plate, 89
　momentum, 272, 331–333
　natural convection, 359–360, 408–412
　porous media
　　cylinder, 521
　　flat plate, 517
　　stagnation point, 519
　temperature, 61–62
　thermal, 61–62
　turbulent flow, 273, 278
　velocity, 61–62, 278
Bousinesq approximation, 342–344
Brinkman extension of Darcy model, 546
Brinkman number, 23
Buffer region (layer) 247, 268, 317–318
Bulk temperature, 7, 18
Buoyancy force, 4, 12–15
Buoyancy force parameter, 431

Centrifugal forces, 4, 342–343
　condensation, 597–599
Channel flow
　laminar, 158–219
　natural, 366–385
　porous, 524
　turbulent, 304–337
Channel flow, natural convection, 366–385
　flow rate, 379
　heat transfer, 380
　pressure variation, 368
Closure problem, 57
Coefficient of bulk expansion, 14–15
Coefficient of cubical expansion, 14–15
Combined convection, 4, 426–477
　aiding flow, 429
　assisting flow, 429
　buoyancy force parameter, 431
　correlation of heat transfer results, 449–452
　dimensional analysis, 428–429
　flow direction, effect of, 427–429
　flow separation, effect of buoyancy forces, 428–429, 449
　forced convection limit, 451
　free convection limit, 451–452
　governing equations, 430–431
　governing parameters, 426–429
　horizontal plate, 446
　indices in correlation equation, 451
　internal flow, 464–476
　　forced convection limit, 465–466
　　free convection limit, 465–466
　　full equations, solutions, 446–449
　　fully-developed flow, 466–471
　　horizontal pipe, 464, 466, 474–476
　　vertical duct, 466–471
　　vertical pipe, 465
　laminar boundary layer flow, vertical plate, 431–444
　　forced convection limit, 435–436, 444, 463
　　free convection limit, 440–441, 444, 463
　　numerical solution, 442–444
　　Nusselt number, 434–435, 440–441, 442–443
　　series solution, 433–442
　　wall shear stress, 434–435
　opposing flow, 428–429, 431

Combined convection—*Cont.*
  turbulent flow, 455–463
    forced and free convection limits, 463
Combined modes of heat transfer, 3
Compressibility effects, 20
Condensation, 555–600
  centrifugal force, 597–599
  dropwise, 555–558
  film (filmwise), 555–558
  film laminar
    boundary layer type analysis, 586–596
    effect of subcooling, 567–570
    film thickness, 562, 598
    heat transfer rate, 562, 577
    horizontal cylinders, 574–579
    horizontal rotating plate, 597–599
    improved analysis, 586–596
    inclined plate, 566–567
    mass flow rate, 561, 576
    non-gravitational, 597–599
    Nusselt analysis, 558–563
      assumptions, 558
    Nusselt number, 577, 563, 594–596, 598–599
    similarity solution, 586–596
    surface shear stress, 579–582
    temperature distribution, 558
    tubes, outside, 565, 574–579
    vapor drag, 579–582
    velocity distribution, 560, 575
    vertical cylinder, 565
    vertical plate, 558–563
  forced convection, 579–582
  heat transfer rate, 557
  inertia, effect of, 596
  non-condensable gases, 585–586
  non-gravitational, 597–599
  Nusselt analysis, 558–563
    assumptions, 558
  promoters, 557
  subcooling, 567–570
  surface shear stress, 579–582
  transition, 570–572
  tube banks, 577–578
  turbulent film:
    vertical plate, 570–572
  wavy film, 570–572
  wettability, 557
Condensation, film
  transition to turbulence, 570–572

Conduction, heat transfer by, 1, 3
Conservation principles, 31
  energy, 31
  mass, 31
  momentum, 31
Continuity equation, 31–35
  cartesian coordinates, 33
  cylindrical coordinates, 34
  numerical solution, 130–131
  porous media, 492
  turbulent flow, 52
Control volume, 32
Convection, 1
  combined, 4, 426–427
  forced, 4
  free, 4
  mixed, 4, 426
  natural, 4, 342–416
Convection, heat transfer by, 1
Convective heat transfer, 1
  forced combined with natural, 4
  interaction with radiant heat transfer, 2
  porous media, 487–545
Convective heat transfer coefficient, 5, 6
  mean, 9
Correlation of combined convection results, 449–452
  indices in correlation equation, 451
Core region, 330
Critical Rayleigh number, 404–406
Cylinder
  boundary flow over, 68
  combined convection, 428–429, 451
  film condensation, 574–579, 565
  horizontal, 574–578
  square, 151–152, 447–449, 452
  vertical, 565
Cylindrical coordinates, 34–35, 232

Damping factor, 288
Darcy model (or law) 494, 545
  deviations, 545–546
Darcy-modified Rayleigh number, 528
  critical, 545
Density, 12
  air, 27, 607–608
  water, 28, 606
Density changes, 4, 13–15
DEVDUCT, 200

DEVPIPE, 195, 197
Diffusivity, thermal, 497, 576
Dimensional analysis, 11–23
Dimensionless numbers, 16, 23
   Brinkman, 23
   Eckert, 23
   Froude, 20, 23
   Graetz, 190
   Grashof, 18, 23
   Mach, 15, 20, 23, 144–145
   Nusselt, 18, 23
   Peclet, 23
   Prandtl, 18, 23
     turbulent, 231
   Rayleigh, 23
     Darcy-modified, 528
   Reynolds, 18, 23
   Stanton, 23
   Weber, 23
Dimensionless numbers, from similarity analysis, 44–46
Dimensionless numbers, physical interpretation, 23–26
Dimensionless numbers, table listing, 23
Dimensions, 15
Dissipation function, 41
Dissipation rate, turbulent flow, 240–243
   equation for, 243
   turbulence model, 240–243
Dropwise condensation, 555–558
   promoters, 557
Duct, turn-around, 336
Duct flow, laminar, 157–220
   developing flow, 68
   fully developed flow, 59–60, 158–188
   Graetz problem, 189–197
   natural convection, 366–385
   pipe flow, developing, 189–197, 201–212
   pipe flow, fully-developed, 158–167
   pipe flow, thermally developing, 189–197
   plane duct, developing, 212–219
   plane duct, fully-developed, 169–179
   plane duct, thermally developing, 197–201
   rectangular duct, fully-developed, 179–188
   slug flow, 164–165
   transition to turbulence, 250
Duct flow, porous media, 521–525
Duct flow, turbulent, 309–327

DUCTSYM, 217
Dynamic pressure, 63
Dynamic similarity, 18

Eckert number, 23, 44–45, 66
Eddies, turbulent, 3, 49
Eddy
   conductivity, 230
   diffusivity, 230, 237
   viscosity, 230, 236
Effective leading edge, 260
Enclosures, natural convection in
   approximate solution, 401–402
   boundary layer regime, 401–402
   heating from below, 403–407
   horizontal, 403–406
   inclined, 539–540
   isotherms, 399
   porous media filled, 531–540
   streamlines, 398
   tilted (inclined) 386, 539–540
ENCLREC, 398, 404
ENCLPOR, 537
Energy equation, 31, 35–41
   boundary layer, laminar, 65–66
   boundary layer, turbulent, 229–231
   cartesian coordinates, 41
   cylindrical coordinates, 41
   integral, 278
   porous media, 495–497
Enthalpy, 36
   specific, 36
   total, 36
Entrance region
   laminar
     pipe, 189–197, 201–212
     plane duct, 197–201, 212–219
     thermal, 189–201
   turbulent, 329–336
Error function, 502, 507
Expansion coefficient, thermal, 13–14
Explicit solution, 124–125, 204–210, 373–380
   instability of, 209–210, 218–219, 379
External flow, 5
   boundary layer flow, 60–61
   combined convection, 431–464
   laminar forced convection, 83–152
   natural convection, 354–366, 526–531
   porous media, 498–521

External flow—*Cont.*
  porous media, natural convection, 526–531
  solution to full equations, 150–152, 299
  turbulent forced convection, 254–299
EXTSQCYL, 151

Falkner-Skan flows, 106–110
Film coefficient, 6
Film conductance, 6
Film (filmwise) condensation, 555–558
  boundary layer type analysis, 586–596
  horizontal cylinder, 574–579
    heat transfer rate, 577
    mass flow rate, 576
    Nusselt number, 577
    velocity distribution, 575
  improved analysis, 586–596
  inclined plate, 566–567
  non-condensible gases, effect of, 585–586
  non-gravitational, 597–599
  Nusselt analysis, 558–563
  surface shear stress, 579–582
  transition, 570–572
  vertical cylinder, 565
  vertical flat plate, 558–563
    film thickness, 562
    heat transfer rate, 562
    mass flow rate, 561
    Nusselt number, 563
    surface shear stress, 579–582
    turbulent flow, 570–572
    wavy film, 570–572
Finite differences solutions
  enclosure, 391–398
  laminar boundary layer, forced, 123–140
  laminar duct flow, forced, 179
  laminar duct flow, free, 366–380
  laminar pipe flow, forced, 167, 201–212
  turbulent boundary layer, forced, 281–296
  turbulent pipe flow, forced, 322–327
First law of thermodynamics, 31
Flat plate
  combined convection, 431–446, 451
  high speed flow, 140–150, 296–297
  laminar boundary layer flow, 83–106
  laminar natural convection, 354–365
  turbulent boundary layer flow, 254–271

Flow reversal, mixed convection in duct, 471
Flow variables, 31
Fluid Dynamics International, 336
Fluid properties, 26, 31, 606–608
  temperature for, 19, 149–150
  variation with pressure, 26
  variation with temperature, 19
Fluid temperature, 5–10
  internal flow, 7–8
  mean, 10, 19
Forced convection limit, mixed convection, 435–436, 444, 451, 465–466
Forced velocity, definition, 12
Forchheimer extension of Darcy model, 546
Form factor, 272–273, 277, 331
  equation for, 273, 331
Fourier's law, 3
Free convection, 4, 342, 366–385; *see also* Natural convection
Free convection limit, mixed convection, 440–441, 444, 451–452, 465–466
Free surface flows, 20–21
Freestream turbulence, 248–249
Friction factor, 308
  equations, 309–311
  Moody chart, 310
  relation to wall shear stress, 309
  roughness values, 310
Friction velocity, 246, 267, 288–290, 316
Froude number, 20, 23
Full equations, solutions to
  laminar flow, combined, 446–449
  laminar flow, external, 150–152
  laminar flow, internal, 219–220
  turbulent flow, external, 299, 336–337
Fully developed duct flow, 59–60
  turbulent, 304–321
  uniform heat flux, laminar, 169–174
  uniform temperature, laminar, 178–179

Governing equations, 31–80, 344–345, 349–354
Graetz number, 190
Graetz problem, 189–197
Grashof number, 18, 23, 25, 347, 470
  film condensation, 565
Gravitational forces, 4, 342

Grid lines
    laminar boundary layer solution, 124, 127, 132–133
Grid modification, 132–133

Heat conduction, 1, 3
Heat flux, 3
Heat convection, 1
Heat transfer
    combined convective, 4
    by combined modes, 1
    condensation, 557, 562
    by conduction, 1
    by convection, 1
    entrance region, 201–219
    forced convective, 4
    free convective, 4, 380
    mixed convective, 4
    modes of, 1
    natural convective, 4, 380
    porous media, 487–488
    by radiation, 1
    thermal entrance region, 193, 196–197, 200
    viscous dissipation effects, 140–150
Heat transfer coefficient, 6
    horizontal cylinder, film condensation, 574–579
        heat transfer rate, 577
        mass flow rate, 576
        Nusselt number, 577
        velocity distribution, 575
    mean, 10
    typical values, 10
    units of, 6
    variables influencing, 11–12
Heat transfer rate, 3
    condensation, 557, 562
Horizontal pipe, mixed convection, 464–466, 474–477
Horizontal porous layers, 540–545
    critical Darcy-modified Rayleigh number, 545
    experimental results, 545
    instability, 544
    small disturbance analysis, 542–545
Hydraulic diameter, 177–178, 188, 250
    film condensation, 570
Hydrostatic pressure, 13–14

Implicit solution, 124–125
Inclination, effect on natural convection:
    porous media filled enclosures, 539–540
Incompressible flow, 14
Inertia force, 23–25, 64
Inlet conditions, 202–204, 213–214, 366–367
Inner region (layer) 268, 288, 317
Instability
    Bénard flow, porous media, 545
    explicit numerical solution, 209–210, 218–219, 379
Insulating material, 487–488
Integral equation solutions
    developing turbulent duct flow, 329–335
    laminar boundary layer, forced, 114–123
    turbulent boundary layer, 272–281
    turbulent natural convection, 407–414
Integral equations for boundary layer flow, 71–80
    derivation from boundary layer equations, 78–80
    energy integral equation, 75–77
    laminar flow, 114–123
    momentum integral equation, 73–75
    natural convection, 408
    turbulent flow, 272, 278
Internal flow, 5
    combined convection, 464–476
    developing, 189–219, 329–335
    forced convection, 304–337
    fully-developed, 304–322
    laminar, 157–220
        developing, 189–219
        forced convection, 157–220
        fully-developed, 59–60
    natural convection, 366–385
    porous media, 521–525
    turbulent, 304–337
Inviscid flow, 61, 68, 492
Irrotational flow, 492
Isotherms, 538
Isotropic porous media, 492

Jakob number, 565, 569

K-E turbulence model, 242–244
Kinematic viscosity, 18, 26
Kinetic energy, turbulent, 59, 239
  equation for, 57–59, 239–242
  turbulence model, 239–244
Kinetic energy changes, 15

LAMBOUN, 135
LAMBOUQ, 140
LAMBMIX, 442
LAMBNAT, 365
Laminar boundary layer
  boundary layer thickness, 61–62, 89, 117–120
  Falkner-Skan equation, 106–110
  flat plate
    adiabatic wall temperature, 140–145
    boundary layer thickness, thermal, 92, 119–121
    flow perpendicular to, 110
    integral method, 114–123
    numerical solution, 123–140
    Nusselt number, 92–94, 100
    profiles, 88, 91
    recovery factor, 144–145
    shooting method, 88
    similarity solution, 83–104
    temperature distribution, 91
    unheated starting length, 121–123
    velocity distribution, 88
    viscous dissipation effects, 140–150
  integral method, 114–123
  numerical solutions, 123–140
  Nusselt number, 92–94, 98, 100, 104, 113, 121, 123, 134, 148–149
  recovery factor, 144–145
  Runge-Kutta procedure, 91
  similarity solution, 83–110
  unheated starting length, 121–123
  viscous dissipation, 140–150
  wedge, 106–110
Laminar duct flow
  axial conduction, 160–161
  non-circular, 179–188, 188
  Nusselt number, 164, 167, 177–179, 188, 195–197, 200, 208–210, 217, 219

pipe
  developing, 201–212,
  fully developed flow, 158–167
  slug flow, 164–165
  thermally-developing, 189–197
  wall heat flux specified, 167
  wall temperature specified, 164
plane duct
  developing, 212–219
  fully developed flow, 169–179
  natural convection, 366–385
  thermally-developing, 197–201
  wall heat flux specified, 179
  wall temperature specified, 174
  rectangular, 179–188
  simultaneous velocity and temperature development, 201–219
Laminar flow
  external, solutions to full equations, 150–152, 299
  internal, solutions to full equations, 219–220
  natural, enclosures, 385–407
Laminar flow, condensation, 558–570, 574–599
  vertical plate, 558–563
Latent heat, 562, 569
  modified, 569
Leading edge conditions, 134–135
Length scales, 19
Liquid metals, 62
Local shearing stress coefficient, 257
Logarithmic velocity profile, 247

Mach number, 15, 23
Mixed convection; see Combined convection
Mixing-cup temperature, 8
Mixing length turbulence model, 234–239, 287–289
  effect of buoyancy forces, 455–461
MIXSQCYL, 447
Models
  of porous media, 490–492, 497, 545–547
  of turbulence, 234–244, 455–461
Modes of heat transfer
  combined modes, 1
  conduction, 1
  convection, 1
  radiation, 1

Momentum equation, 31–35
Momentum equation, laminar boundary
    layer, 63–65
  derivation, 63–65
  similarity solution, 83–113, 354–361,
    498–507
Moody chart, 310

NATCHAN, 379–380
Natural convection, laminar
  Bénard cells, 406
  boundary layer equations, 349–354
  centrifugal force, 4, 342–343
  channel flow, 366–385
  combined natural and forced, 4, 426–427
  enclosures, 385–407
    boundary conditions, 387, 389–390,
      393–395
    governing equations, 388–391
    heat transfer rate, 397–399, 401–402
    horizontal, 403–407
    numerical solution procedure, 391–398
  flat plate, 354–365
    boundary layer thickness, 359–360
    combined convection, 4, 426–427
    integral method, 408–414
    Nusselt number, 359–360, 412–413
    similarity solution, 354–365
    temperature distribution, 360
    velocity distribution, 359
    vertical, 354–361
  governing equations, 344–345
  heat transfer rate, 380
  horizontal enclosure, 403–407
    Bénard cells, 406
    Bénard convection, 406
    critical Darcy-modified Rayleigh
      number, 545
    experimental results, 545
    instability, 404–406, 544
    Nusselt number variation, 405
    porous media, 540–545
    small disturbance analysis, 542–545
    stability of flow, 404–406, 544
  inlet conditions, 366–367
  integral equation solution, 408–414
  laminar-to-turbulence transition, 407
  limiting solutions, 380–384
  mean velocity, 379

Natural convection, laminar—*Cont.*
  numerical solution, 365
  Nusselt number, 18, 23, 24
    numerical, 133–134
  outlet pressure, 368
  porous medium, 526–540
  pressure variation, 368
  similarity analysis, 345–348
  temperature distribution, 360, 407
  transition to turbulence, 407
  velocity distribution, 359, 409
Natural convection, porous media
  boundary layer, 526–531
  enclosures, 531–540
  horizontal layers, 540–545
    critical Darcy-modified Rayleigh
      number, 545
    experimental results, 545
    instability, 544
    small disturbance analysis, 542–545
Natural convection, turbulent, 407–414
Navier-Stokes equations, 31–35
  cartesian coordinates, 34
  cylindrical coordinates, 35
  enclosures, natural convection, 386
  numerical solutions, 150–152, 219–220,
    299, 446–449
  turbulent flow, 52–54
Near-wall region, 244–247
Newtonian fluid, stress-strain rate relation,
  33
Nodal points, 124–127, 130, 391–392
Non-gravitational condensation, 597–599
Non-circular ducts, laminar flow, 179–188
Non-condensible gases, effect on
  condensation, 585–586
Non-Newtonian fluid, 31
Number
  Brinkman, 23
  Eckert, 23
  Froude, 20, 23
  Graetz, 190
  Grashof, 18, 23
  Nusselt, 18, 23
  Peclet, 23
  Prandtl, 18, 23
    turbulent, 231
  Rayleigh, 23
    Darcy-modified, 528
  Reynolds, 18, 23

Index   617

Number—*Cont.*
  Stanton, 23
  Weber, 23
Numerical solutions
  combined convection, boundary layer, 442–444
  full equations
    combined convection, laminar, 446–449
    laminar external flow, 150–152
    laminar internal flow, 219–220
    turbulent external flow, 299
    turbulent internal flow, 336–337
  laminar developing pipe flow, 193–197, 201–212
  laminar duct flow, 179, 182–188, 197–201, 212–219
  laminar pipe flow, 167, 193–197, 201–212
  natural convection boundary layer, 365–366
  natural convection enclosure, 391–406
  natural convection plane duct, 373–379
  porous media
    boundary layer, 508–513
    enclosure, 532–538
  turbulent boundary layer, 281–296
  turbulent developing pipe flow, 322–329
  turbulent duct flow, 322–335
  turbulent pipe flow, 322–329
Nusselt analysis, film condensation, 558–563
  assumptions, 558
Nusselt number
  combined convection
    correlation equations, 449–452
    square cylinder, 448
    turbulent flow, 461–462
    vertical plate, 434–435, 440–443
  duct flow, 164, 167, 177–179
  film condensation
    boundary layer type analysis, 594–596
    modified, 572
    non-gravitational, 598–599
    vertical plate, 563, 572

Nusselt number—*Cont.*
  laminar forced convection
    boundary layer flow, 92–94, 98, 104, 112–113
    developing duct flow, 216–217, 219
    developing pipe flow, 210
    integral equation solution, 121, 123
    pipe flow, 164, 167
    plane duct flow, 174, 177–179, 200, 219
    rectangular ducts, 187–188
    thermally developing duct flow, 197–201
    thermally developing pipe flow, 192, 196–197
    viscous dissipation effects, 148–149
  natural convection, 348
    channel flow, 382–383
    enclosures, 397–399, 401–402, 405
    flat plate, 359–361
    turbulent boundary layer, 412–413
  porous media
    cylinder, forced convection, 512–513, 521
    enclosure, 537–538
    flat plate, forced convection, 503, 517
    flat plate, natural convection, 530–531
    fully-developed duct flow, 524
    horizontal layer, 537–538
    stagnation point, forced convection, 507, 519
  turbulent forced convection
    boundary layer flow, 257, 259, 265–267, 270–271, 280, 292
    developing duct flow, 334
    developing flow, 327, 334
    developing pipe flow, 327
    integral equation solution, 280
    pipe flow, 309, 311, 320, 327

Opposing flow, 428–429, 431
Outer region, 269, 289, 318–319

Parabolic equations, 369
Parabolized equations, 369
Peclet number, 23, 501
Permeability, 491–494
  model for, 494
  typical values, 493

Pipe, laminar flow
  developing, 201–212
  fully-developed, 158–167
  thermally-developing, 189–197
Pipe, turbulent flow
  equations, 231
Pipe flow, 158–167
PIPEFLOW, 209
PIPETEM, 167
Plane duct, laminar flow
  developing, 201–212
  thermally-developing, 197–201
PORCYL, 512
Porosity, 493
  typical values, 493
Porous media, 487–488
  apparent conductivity, 497
  apparent thermal diffusivity, 497
  area averaged velocity, 488–489
  boundary layer, 498–521
    cylinder, 512–513, 519–521
    flat plate, forced convection, 500–503, 516–517
    integral equation solutions, 514–519
    natural convection, 526–531
    similarity solutions, 500–507, 526–531
    stagnation point, 505–507, 519
  Brinkman extension of Darcy model, 546
  Darcy model, 490–494, 545
  duct flow, fully-developed, 521–525
  examples, 487–488
  enclosure, 531–540
  energy equation, 495–497
  forced convection, 498–525
  Forchheimer extension of Darcy model, 546
  horizontal layers, 540–545
    critical Darcy-modified Rayleigh number, 545
    experimental results, 545
    small disturbance analysis, 542–545
  instability, 544
  isotropic, 492
  natural convection, 526–545
  non-Darcy effects, 545–547
  numerical solution, 507–513
  permeability, 491–494
    model for, 494
    typical values, 493
  pore Reynolds number, 545

Porous media—*Cont.*
  porosity, 493
    typical values, 493
  velocity at surface, 492
Potential flow, 60–61, 68, 492
Power law profiles, 320, 409
Prandtl number, 18, 23, 24–25, 264–267, 270–271, 347
  turbulent, 231, 237, 244, 269–271
Properties of fluids; *see* Fluid properties

Radiant heat transfer, 1, 2
  interaction with other modes, 2
Radiation, thermal, 1, 2
Rayleigh number, 23, 390, 399, 402, 465, 466
  Bénard flow, 406
  critical, 404–406
Rayleigh number, Darcy-modified, 528, 533
  critical, 545
RECDUCT, 187
Rectangular ducts, laminar flow, 179–188
Rectangular enclosure, 385–407
Recovery factor
  laminar flow, 144–145
  turbulent flow, 296
Relaxation factor, 395, 535–536
Reynolds analogy
  assumptions, 256–257, 306
  turbulent boundary layer flow, 255–260
  turbulent internal flow, 305–312
Reynolds number, 18, 23, 24
  condensed film, 570–572
Reynolds stresses, 54–55
Roughness values, 310
Runge-Kutta method, 91

Separation, flow, 68, 111
Separation of variables, 191–193, 197, 198
Shearing stress coefficient, 257, 273
Shearing stresses, 33
  equation for, 331, 410
  surface, effect on film condensation, 579–582
Similar flows, 41, 46
  forced convection, 41–46
  natural convection, 345–348
Similarity, 41–46

Similarity solutions
    Falkner-Skan flow, 105–113
    flat plate, 83–105
    forced convection boundary layer, 83–114, 141–143
    natural convection, 354–361
    porous media, 498–507, 526–531
    stagnation point, 112–113
SIMPLATE, 91
SIMPLCOM, 435
Simplifying assumptions used in analysis of convection, 59–61
SIMPLNAT, 359
SIMVART, 99
Skin-friction coefficient; see Shearing stress coefficient
Slug flow, 164–165
Soil, 488
Specific heat, 12
Sphere, combined convection, 451
Stability
    Bénard flow
        porous media, 545
    explicit numerical solution, 209–210, 218–219, 379
Stagnation point heat transfer, 112–113
Stanton number, 23
Stream function, 48, 388, 493, 527, 530, 532, 588
Streamlines, 538
Stress, turbulent or Reynolds, 55, 229–231
Surface temperature; see Wall temperature
Surface tension, 20

Taylor-Prandtl analogy
    assumptions, 262–264
    turbulent boundary layer flow, 262–267
Temperature
    adiabatic wall, 142–144
    bulk, 7–8
    fluid, 5–10
    for fluid properties, 19, 149–150, 297
    mean fluid, 10
    mean wall, 10
    mixing cup, 8
    wall, 5, 9

Temperature profile
    laminar, integral equation analysis, 117
    natural convection, 360, 409
    porous media, 507, 516, 523–524, 530
    power law, 276
    similarity solution, 91
Thermal conductivity, 11
    apparent, 497
Thermal diffusivity of porous media, 497, 546
Thermal dispersion, 547
Thermally developing flow, 189–201
Three-layer analogy, 267–271
Time-averaged equations, 49–59, 227–232
Time-averaging, 49–50
Transition, laminar to turbulent flow, 247–250, 258–259, 407
    boundary layer flow, 248–249
    condensed film, 570–572
    duct flow, 250
    effect of freestream turbulence, 248–249
    natural convection, 407
    pipe flow, 249
Transition region, 249, 259, 407
Triangular cross-section duct, 188
Tridiagonal matrix, 128–130, 139, 286, 295, 325, 510
Tube banks, condensation, 577–578
Tunnelling, 547
TURBINRK, 274
TURBOUND, 291
TURBOUNQ, 327
Turbulence
    kinetic energy, 59
    kinetic energy equation, 57–59
    modeling, 57
    Prandtl number, turbulent, 231
    time-averaged equations, 49–59, 227–232
    transition, laminar-to-turbulent, 247–250, 258–259, 407
Turbulence models, 57, 234–244
    dissipation of kinetic energy, 240–244
    kinetic energy based, 239–244
    mixing-length, 234–239, 246–247, 455–461
    Van Driest's relation, 288
Turbulent boundary layer
    adiabatic wall temperature, 296
    boundary layer thickness, 273, 278

Turbulent boundary layer—*Cont.*
 displacement thickness, 272
 flat plate, 254–271
 integral equation solution, 272–281
 momentum thickness, 272, 331–332
 numerical solution, 281–299
 Nusselt number, 257, 259, 265–267, 270–271, 280, 292
 Prandtl number, turbulent, 231
 recovery factor, 296
 Reynolds analogy, 255–260
 transition, 259–260
 unheated starting length, 276–281
Turbulent duct flow, 304–327
 entrance region, 329–335
 Nusselt number, 309, 311, 320, 327, 334
 relation between velocity and temperature profiles, 306–307
 Reynolds analogy, 305–312
Turbulent flow, 49; *see* Turbulence
 analogy solutions, 244–245
 closure problem, 57
 combined convection, 455–463
 condensed film, 570–572
 continuity equation, 52
 energy equation, 55–57
 equations, 49–59, 227–232
 fluctuating value of variables, 49–50, 227–228
 friction velocity, 246
 laws of averaging, 51
 mean value of variables, 49–50
 Navier-Stokes equation, 52–54, 299
 near-wall region, 245–247
 steady turbulent flow, 50
 time-averaging, 50–51
Turbulent heat transfer terms, 56, 229–231
Turbulent Prandtl number, 231, 237, 244
Turbulent shearing stress, 55, 229–231
Turbulent stresses, 54
TURDUCD, 327
TURINDEV, 335

Underrelaxation, 395–396, 535–536
Unheated starting length
 laminar flow, 121–123
 turbulent flow, 276–281
Unit thermal convective conductance, 6

Van Driest model, 288
 damping factor, 288
Velocity, area averaged, 488–489
Velocity, turbulent flow
 logarithmic, 247
 power law, 277
Velocity profile
 laminar, integral equation analysis, 115
 mixed convection, 470–471
 natural convection, 359, 409
 power law, 277
Viscosity, 11
 air, 607–608
 water, 606
Viscosity, kinematic, 18
 air, 27
 water, 28
Viscous dissipation, 38–41
 effect on heat transfer, 148–149
 laminar boundary layer, 140–150
 turbulent boundary layer, 296–297
Vorticity, 46–49, 388
Vorticity-streamfunction formulation, 388–389

Wall temperature, 5, 10
Wall layer, 268, 288
Wall region, 268, 288
Water properties, 607–608
Wavy condensed film, 570–572
Weber number, 23
Wettability, 557
Wetted perimeter, 177–178, 188, 250, 570
Work, 38